Design and Control of Concrete Mixtures

The guide to applications, methods, and materials

FIFTEENTH EDITION

by Steven H. Kosmatka and Michelle L. Wilson

Portland Cement Association
5420 Old Orchard Road
Skokie, Illinois 60077-1083
847.966.6200 Fax 847.966.9781

500 New Jersey Avenue NW, 7th Floor
Washington, DC 20001-2066
202.408.9494 Fax 202.408.0877
www.cement.org

An organization of cement companies to improve and extend the uses of portland cement and concrete through market development, engineering, research, education, and public affairs work.

KEYWORDS: admixtures, aggregates, air-entrained concrete, batching, cement, cold weather, curing, durability, fibers, finishing, high-performance concrete, hot weather, mixing, mixing water, mixture proportioning, placing, portland cement concrete, properties, special concrete, standards, supplementary cementing materials, sustainability, tests, and volume changes.

ABSTRACT: This book presents the properties of concrete as needed in concrete construction, including strength and durability. All concrete ingredients (cementing materials, water, aggregates, admixtures, and fibers) are reviewed for their optimal use in designing and proportioning concrete mixtures. Applicable ASTM, AASHTO, and ACI standards are referred to extensively. The use of concrete from design to batching, mixing, transporting, placing, consolidating, finishing, and curing is addressed. Concrete sustainability, along with special concretes, including high-performance concretes, are also reviewed.

REFERENCE: Kosmatka, Steven H. and Wilson, Michelle L., *Design and Control of Concrete Mixtures,* EB001, 15th edition, Portland Cement Association, Skokie, Illinois, USA, 2011, 460 pages.

The authors of this engineering bulletin are:
Steven H. Kosmatka, Vice President, Research and Technical Services, PCA
Michelle L. Wilson, Director of Concrete Knowledge, PCA

Cover photos show the world's tallest building, the Burj Khalifa in Dubai, U.A.E. The tower is primarily a concrete structure, with concrete construction utilized for the first 155 stories, above which exists a structural steel spire.

Fifteenth Edition Print History

First Printing 2011
Second Printing 2012
Third Printing (rev.) 2013
Fourth Printing 2014

ISBN 978-0-89312-272-0

The Library of Congress has catalaoged the previous edition as follows:

Kosmatka, Steven H.
Design and control of concrete mixtures / by Steven H. Kosmatka, Beatrix Kerkhoff, and William C. Panarese.— 14th ed.
p. cm.
ISBN 0-89312-217-3 (pbk. : alk. paper)
1. Concrete. 2. Concrete—Additives. 3. Portland cement. I. Kerkhoff, Beatrix. II. Panarese, William C. III. Title.
TA439 .K665 2002
666'.893—dc21

2001007603

Printed in the United States of America

EB001.15

WARNING: Contact with wet (unhardened) concrete, mortar, cement, or cement mixtures can cause SKIN IRRITATION, SEVERE CHEMICAL BURNS (THIRD-DEGREE), or SERIOUS EYE DAMAGE. Frequent exposure may be associated with irritant and/or allergic contact dermatitis. Wear waterproof gloves, a long-sleeved shirt, full-length trousers, and proper eye protection when working with these materials. If you have to stand in wet concrete, use waterproof boots that are high enough to keep concrete from flowing into them. Wash wet concrete, mortar, cement, or cement mixtures from your skin immediately. Flush eyes with clean water immediately after contact. Indirect contact through clothing can be as serious as direct contact, so promptly rinse out wet concrete, mortar, cement, or cement mixtures from clothing. Seek immediate medical attention if you have persistent or severe discomfort.

Table of Contents

Title Page ... i

Keywords and Abstract ... ii

Preface and Acknowledgements ... x

Foreword ... xii

Chapter 1
Introduction to Concrete ... 1
- Industry Trends ... 1
  - Cement Consumption ... 2
  - Pavements ... 3
  - Bridges ... 3
  - Buildings ... 3
- The Beginning of an Industry ... 4
- Sustainable Development ... 6
- Essentials of Quality Concrete ... 6
  - Suitable Materials ... 6
  - Water-Cementitious Materials Ratio ... 7
  - Mixing ... 8
  - Transporting ... 8
  - Placement and Consolidation ... 8
  - Finishing and Jointing ... 9
  - Hydration and Curing ... 9
  - Design-Workmanship-Environment ... 9
- References ... 10

Chapter 2
Sustainability ... 11
- Introduction to Sustainability ... 11
  - Energy Conservation and Atmosphere ... 11
  - Water Management and Resources ... 11
  - Site Selection and Development ... 12
  - Indoor Environmental Quality ... 12
  - Material Quality and Resources ... 12
  - Functional Resilience ... 12
- Rating Systems ... 12
- Concrete Sustainability ... 13
  - Durability ... 14
  - Safety ... 14
  - Disaster Resistance ... 14
  - Energy Performance ... 15
  - Indoor Air Quality ... 17
  - Acoustics ... 17
  - Stormwater Management ... 18
  - Site Remediation ... 20
  - Recycling ... 20
  - Locally Produced ... 21
  - CO_2 Sink ... 21
  - Concrete Ingredients and Sustainability ... 21
- Life-Cycle Analysis ... 22
  - Life-Cycle Cost Analysis ... 22
  - Life-Cycle Assessment and Inventory ... 22
  - LCA at the MIT Concrete Sustainability Hub ... 24
- References ... 25

Chapter 3
Portland, Blended and Other Hydraulic Cement ... 29
- Manufacture of Portland Cement ... 29
- Cement's Role in Sustainability ... 34
  - Alternative Raw Materials ... 34
  - Land Stewardship ... 34
  - Solid Waste Reduction ... 34
  - Energy and Fuel ... 35
  - Use of Tire-Derived Fuel ... 35
  - Combustion Emissions ... 35
- Types of Portland Cement ... 36
  - Type I ... 36
  - Type II ... 36
  - Type II (MH) ... 38
  - Type III ... 39
  - Type IV ... 39
  - Type V ... 39
  - Air-Entraining Portland Cements ... 39
  - White Portland Cements ... 39
- Blended Hydraulic Cements ... 40
  - Type IS ... 40
  - Type IP ... 41
  - Type IL ... 41
  - Type IT ... 41
- Performance Based Hydraulic Cements ... 41
  - Type GU ... 41
  - Type HE ... 41
  - Type MS ... 41
  - Type HS ... 41
  - Type MH ... 42
  - Type LH ... 42
- Special Cements ... 42
  - Masonry and Mortar Cements ... 42
  - Plastic Cements ... 43
  - Finely-Ground Cements (Ultrafine Cements) ... 44
  - Expansive Cements ... 44
  - Oil-Well Cements ... 45
  - Rapid Hardening Cements ... 45

Cements with Functional Additions ... 45
Water-Repellent Cements ... 45
Regulated-Set Cements ... 45
Geopolymer Cements ... 45
Ettringite Cements ... 46
Calcium Aluminate Cements ... 46
Magnesium Phosphate Cements ... 46
Sulfur Cements ... 46
Natural Cements ... 46
Other Cements ... 46
Selecting and Specifying Cements ... 46
Availability of Cements ... 46
Drinking Water Applications ... 47
Canadian and European Cement Specifications ... 47
Chemical Phases (Compounds) and Hydration of Portland Cement ... 48
Primary Phases ... 48
Water (Evaporable and Nonevaporable) ... 51
Physical Properties of Cement ... 52
Compressive Strength ... 52
Setting Time ... 54
Early Stiffening (False Set and Flash Set) ... 55
Particle Size and Fineness ... 55
Soundness ... 56
Consistency ... 57
Heat of Hydration ... 57
Loss on Ignition ... 59
Density and Relative Density (Specific Gravity) ... 59
Bulk Density ... 60
Thermal Analysis ... 60
Thermogravimetric Analysis (TGA) ... 60
Differential Thermal Analysis (DTA) ... 60
Differential Scanning Calorimetry (DSC) ... 61
Virtual Cement Testing ... 61
Transportation and Packaging ... 61
Storage of Cement ... 62
Hot Cement ... 63
References ... 63

Chapter 4
Supplementary Cementitious Materials ... 67
Fly Ash ... 68
Classification of Fly Ash ... 68
Physical Properties of Fly Ash ... 69
Chemical Properties of Fly Ash ... 69
Slag Cement ... 69
Classification of Slag Cement ... 70
Physical Properties of Slag Cement ... 70
Chemical Properties of Slag Cement ... 70
Silica Fume ... 70
Classification of Silica Fume ... 70
Physical Properties of Silica Fume ... 71
Chemical Properties of Silica Fume ... 71
Natural Pozzolans ... 71
Classification of Natural Pozzolans ... 72
Physical Properties of Natural Pozzolans ... 72
Chemical Properties of Natural Pozzolans ... 72
Reactions of SCMs ... 72
Pozzolonic Reactions ... 72
Hydraulic Reactions ... 73
Effects on Freshly Mixed Concrete ... 73
Water Demand ... 73
Workability ... 74
Bleeding and Segregation ... 74
Setting Time ... 75
Air Content ... 75
Heat of Hydration ... 76
Effects on Hardened Concrete ... 77
Strength ... 77
Impact and Abrasion Resistance ... 79
Drying Shrinkage and Creep ... 79
Permeability and Absorption ... 79
Corrosion Resistance ... 79
Carbonation ... 79
Alkali-Silica Reactivity ... 80
Sulfate Resistance ... 81
Chemical Resistance ... 81
Freeze-Thaw Resistance ... 81
Deicer-Scaling Resistance ... 81
Aesthetics ... 82
Concrete Mixture Proportions ... 84
Multi-Cementitious Systems ... 84
Availability ... 84
Storage ... 84
References ... 84

Chapter 5
Mixing Water for Concrete ... 87
Sources of Mixing Water ... 88
Municipal Water Supply ... 88
Municipal Reclaimed Water ... 89
Site-Sourced Water ... 89
Recycled Water (Water from Concrete Production) ... 89
Seawater ... 90
Effects of Impurities in Mixing Water on Concrete Properties ... 90
Alkali Carbonate and Bicarbonate ... 91
Chloride ... 91
Sulfate ... 92
Other Common Salts ... 92
Iron Salts ... 92
Miscellaneous Inorganic Salts ... 92
Acid Waters ... 92
Alkaline Waters ... 92
Industrial Wastewater ... 92
Silt or Suspended Particles ... 92
Organic Impurities ... 93
Waters Carrying Sanitary Sewage ... 93
Sugar ... 93
Oils ... 93
Algae ... 93
Interaction with Admixtures ... 93
References ... 93

Chapter 6
Aggregates for Concrete ... 95
Aggregate Geology ... 95
Aggregate Classification ... 96
Natural Aggregate ... 96

Manufactured Aggregate . . . 97
Recycled-Concrete Aggregate . . . 97
Marine-Dredged Aggregate . . . 98
Characteristics of Aggregates . . . 99
Grading . . . 100
Particle Shape and Surface Texture . . . 106
Bulk Density (Unit Weight) and Voids . . . 107
Density . . . 107
Relative Density (Specific Gravity) . . . 107
Absorption and Surface Moisture . . . 107
Resistance to Freezing and Thawing . . . 108
Wetting and Drying Properties . . . 110
Abrasion and Skid Resistance . . . 110
Strength and Shrinkage . . . 110
Resistance to Acid and Other Corrosive Substances . . . 111
Fire Resistance and Thermal Properties . . . 111
Potentially Harmful Materials . . . 111
Alkali-Aggregate Reactivity . . . 113
Alkali-Silica Reaction . . . 113
Alkali-Carbonate Reaction . . . 113
Aggregate Beneficiation . . . 113
Handling and Storing Aggregates . . . 114
References . . . 115

Chapter 7
Chemical Admixtures for Concrete . . . 117
Air-Entraining Admixtures . . . 119
Air-Entraining Materials . . . 119
Mechanism of Air Entrainment . . . 120
Control of Air Content . . . 121
Impact of Air Content on Properties of Concrete . . . 121
Water-Reducing Admixtures . . . 122
Normal (Conventional) Water Reducers . . . 123
Mid-Range Water Reducers . . . 123
High-Range Water Reducers . . . 123
Composition of Water-Reducing Admixtures . . . 124
Mechanisms of Water Reducers . . . 124
Impact of Water Reducers on Properties of Concrete . . . 125
Set Retarding Admixtures . . . 127
Types of Set Retarding Admixtures . . . 127
Mechanism of Set Retarders . . . 127
Effects of Set Retarders on Concrete Properties . . . 127
Set Accelerating Admixtures . . . 128
Types of Set Accelerating Admixtures . . . 128
Mechanism of Set Accelerating Admixtures . . . 128
Effects of Set Accelerators on Concrete Properties . . . 128
Hydration-Control Admixtures . . . 129
Workability-Retaining Admixtures . . . 129
Corrosion Inhibitors . . . 129
Shrinkage-Reducing Admixtures . . . 130
Permeability Reducing Admixtures . . . 130
Permeability Reducing Admixture – Non-Hydrostatic (PRAN) . . . 130
Permeability Reducing Admixtures – Hydrostatic (PRAH) . . . 131
Additional Considerations . . . 131
Alkali-Aggregate Reactivity Inhibitors . . . 131
Coloring Admixtures (Pigments) . . . 131
Pumping Aids . . . 132
Bonding Admixtures and Bonding Agents . . . 132
Grouting Admixtures . . . 132
Gas-Forming Admixtures . . . 132
Air Detrainers . . . 133
Fungicidal, Germicidal, and Insecticidal Admixtures . . . 133
Viscosity Modifying Admixtures . . . 133
Compatibility of Admixtures and Cementitious Materials . . . 133
Less-Than Expected Water Reduction . . . 134
Less-Than Expected Retardation . . . 134
Storing and Dispensing Chemical Admixtures . . . 134
References . . . 135

Chapter 8
Reinforcement . . . 137
Why Use Reinforcement in Concrete . . . 137
Reinforcing Bars . . . 138
Grades . . . 138
Stress-Strain Curves . . . 138
Bar Marks . . . 138
Deformed Reinforcing Bars . . . 140
Coated Reinforcing Bars and Corrosion Protection . . . 140
Welded Wire Reinforcement . . . 141
Prestressing Steel . . . 143
Fibers . . . 144
Advantages and Disadvantages of Using Fibers . . . 145
Types and Properties of Fibers and Their Effect on Concrete . . . 145
Steel Fibers . . . 145
Glass Fibers . . . 147
Synthetic Fibers . . . 148
Natural Fibers . . . 149
Multiple Fiber Systems . . . 150
References . . . 150

Chapter 9
Properties of Concrete . . . 153
Freshly Mixed Concrete . . . 153
Workability . . . 153
Bleeding and Settlement . . . 155
Air Content . . . 157
Uniformity . . . 164
Hydration, Setting, and Hardening . . . 165
Hardened Concrete . . . 166
Curing . . . 166
Drying Rate of Concrete . . . 167
Strength . . . 169
Density . . . 170
Permeability and Watertightness . . . 170
Volume Stability and Crack Control . . . 172
Durability . . . 172
Aesthetics . . . 173
References . . . 174

Chapter 10
Volume Changes of Concrete 177
Early Age Volume Changes 177
Chemical Shrinkage 177
Autogenous Shrinkage 178
Subsidence 179
Plastic Shrinkage 179
Swelling 179
Moisture Changes (Drying Shrinkage) of Hardened Concrete 180
Effect of Concrete Ingredients on Drying Shrinkage 182
Effect of Curing on Drying Shrinkage 184
Temperature Changes of Hardened Concrete 184
Low Temperatures 185
High Temperatures 185
Curling and Warping 186
Elastic and Inelastic Deformation 187
Compression Strain 187
Modulus of Elasticity 188
Deflection 188
Poisson's Ratio 188
Shear Strain 188
Torsional Strain 188
Creep 189
Chemical Changes and Effects 190
Carbonation 190
Sulfate Attack 191
Alkali-Aggregate Reactions 191
References 191

Chapter 11
Durability 195
Factors Affecting Durability 196
Permeability and Diffusion 196
Exposure Categories 199
Cracking and Durability 199
Protective Treatments 199
Deterioration Mechanisms and Mitigation 199
Abrasion and Erosion 199
Freezing and Thawing 201
Exposure to Deicers and Anti-icers 202
Alkali-Aggregate Reactivity 208
Alkali-Silica Reaction 208
Alkali-Carbonate Reaction 212
Carbonation 214
Corrosion 215
Sulfate Attack 218
Salt Crystallization or Physical Salt Attack 220
Delayed Ettringite Formation 221
Acid Attack 222
Seawater Exposure 223
References 224

Chapter 12
Designing and Proportioning Concrete Mixtures 231
Mix Design (Selecting Characteristics) 232
Strength 233
Water-Cementitious Material Ratio 234
Aggregates 235
Air Content 236
Slump 237
Water Content 238
Cementing Materials Content and Type 239
Chemical Admixtures 240
Proportioning 241
Proportioning from Field Data 241
Proportioning by Trial Mixtures 242
Measurements and Calculations 243
Examples of Mixture Proportioning 243
Example 1. Absolute Volume Method (Metric) 243
Example 2. Absolute Volume Method (Inch-Pound Units) 246
Example 3. Laboratory Trial Mixture Using the PCA Water-Cement Ratio Method (Metric) 250
Example 4. Laboratory Trial Mixture Using the PCA Water-Cement Ratio Method (Inch-Pound Units) 252
Example 5. Absolute Volume Method Using Multiple Cementing Materials and Admixtures (Metric) 256
Concrete for Small Jobs 259
Mixture Review 259
References 262

Chapter 13
Batching, Mixing, Transporting, and Handling Concrete 263
Ordering Concrete 263
Prescription Versus Performance Based Specifications 263
Batching 264
Mixing Concrete 264
Stationary Mixing 264
Ready Mixed Concrete 265
Mobile Batcher Mixed Concrete (Continuous Mixer) 265
Retempering (Remixing) Concrete 267
Transporting and Handling Concrete 268
Methods and Equipment for Transporting and Handling Concrete 268
Choosing the Best Method of Concrete Placement 272
Work At and Below Ground Level 272
Work Above Ground Level 272
References 273

Chapter 14
Placing and Finishing Concrete 275
Preparation Before Placing 275
Subgrade Preparation 276
Subbase 276
Moisture Control and Vapor Retarders 276
Formwork 278
Reinforcement 278
Depositing the Concrete 278
Placing Concrete Underwater 280
Placing on Hardened Concrete 280

Bonded Construction Joints in Structural Concrete . . . 280
Preparing Hardened Concrete . . . 281
Bonding New to Previously Hardened Concrete . . . 281
Special Placing Techniques . . . 282
Consolidation . . . 282
Vibration . . . 283
Finishing . . . 286
Screeding (Strikeoff) . . . 286
Bullfloating or Darbying . . . 287
Edging and Jointing . . . 287
Floating . . . 288
Troweling . . . 288
Brooming . . . 289
Finishing Formed Surfaces . . . 290
Special Surface Finishes . . . 290
Patterns and Textures . . . 290
Exposed-Aggregate Concrete . . . 291
Colored Finishes . . . 292
Stains, Paints, and Clear Coatings . . . 292
Curing and Protection . . . 293
Rain Protection . . . 293
Jointing Concrete . . . 293
Isolation Joints . . . 293
Contraction Joints . . . 294
Construction Joints . . . 295
Horizontal Construction Joints . . . 296
Joint Layout For Floors . . . 297
Filling Floor Joints . . . 297
Unjointed Floors . . . 298
Removing Forms . . . 298
Patching and Cleaning Concrete . . . 298
Holes, Defects, and Overlays . . . 299
Curing Patches . . . 300
Cleaning Concrete Surfaces . . . 300
Precautions . . . 301
References . . . 301

Chapter 15
Curing Concrete . . . 303
Curing Methods and Materials . . . 304
Ponding and Immersion . . . 305
Fogging and Sprinkling . . . 305
Wet Coverings . . . 305
Impervious Paper . . . 306
Plastic Sheets . . . 306
Membrane-Forming Curing Compounds . . . 307
Internal Curing . . . 308
Forms Left in Place . . . 308
Steam Curing . . . 308
Insulating Blankets or Covers . . . 309
Electrical, Oil, Microwave, and Infrared Curing . . . 309
Curing Period and Temperature . . . 310
Sealing Compounds . . . 311
References . . . 312

Chapter 16
Hot Weather Concreting . . . 315
When to Take Precautions . . . 315
Effects of High Concrete Temperatures . . . 316
Cooling Concrete Materials . . . 318
Aggregates . . . 319
Water . . . 319
Ice . . . 319
Liquid Nitrogen . . . 320
Cement . . . 321
Supplementary Cementitious Materials . . . 321
Chemical Admixtures . . . 321
Preparation Before Concreting . . . 321
Transporting, Placing, and Finishing . . . 321
Plastic Shrinkage Cracking . . . 322
Curing and Protection . . . 325
Heat of Hydration . . . 326
References . . . 326

Chapter 17
Cold Weather Concreting . . . 327
Effect of Freezing on Fresh Concrete . . . 327
Strength Gain of Concrete at Low Temperatures . . . 328
Heat of Hydration . . . 328
Special Concrete Mixtures . . . 329
Air-Entrained Concrete . . . 330
Temperature of Concrete . . . 331
Temperature of Concrete as Mixed . . . 331
Temperature Loss During Delivery . . . 333
Temperature of Concrete as Placed and Maintained . . . 333
Cooling After Protection . . . 333
Monitoring Concrete Temperature . . . 333
Concreting on Ground During Cold Weather . . . 334
Concreting Above Ground During Cold Weather . . . 335
Enclosures . . . 336
Insulating Materials . . . 338
Heaters . . . 338
Duration of Heating . . . 341
Moist Curing . . . 341
Terminating the Heating Period . . . 341
Form Removal and Reshoring . . . 341
Maturity Concept . . . 341
References . . . 343

Chapter 18
Test Methods . . . 345
Classes of Tests . . . 345
Frequency of Testing . . . 345
Testing Aggregates . . . 346
Sampling Aggregates . . . 346
Organic Impurities . . . 347
Objectionable Fine Material . . . 347
Grading . . . 347
Moisture Content of Aggregates . . . 348
Testing Freshly Mixed Concrete . . . 349
Sampling Freshly Mixed Concrete . . . 349
Consistency . . . 349
Temperature Measurement . . . 350
Density and Yield . . . 350
Air Content . . . 351
Air-Void Analysis of Fresh Concrete . . . 352

Strength Specimens 353
Time of Setting 355
Accelerated Compression Tests to Project Later-Age Strength 355
Chloride Content 355
Portland Cement Content, Water Content, and Water-Cement Ratio 356
Supplementary Cementitious Materials Content 356
Bleeding of Concrete 356
Testing Hardened Concrete 357
Strength Tests of Hardened Concrete 357
Air Content 360
Density, Absorption, and Voids 360
Portland Cement Content 361
Supplementary Cementitious Material and Organic Admixture Content 361
Chloride Content 361
Petrographic Analysis 362
Volume and Length Change 362
Durability 362
Moisture Testing 364
Carbonation 365
pH Testing Methods 365
Permeability and Diffusion 365
Nondestructive Test Methods 366
References 371

Chapter 19
High-Performance Concrete 375
High-Durability Concrete 380
Abrasion Resistance 380
Blast Resistance 380
Permeability 380
Diffusion 380
Carbonation 381
Freeze-Thaw Resistance 381
Chemical Attack 381
Alkali-Silica Reactivity 381
Resistivity 382
High-Early-Strength Concrete 382
High-Strength Concrete 383
Cement 385
Supplementary Cementing Materials 385
Aggregates 386
Admixtures 386
High-Performance Concrete Construction 387
Proportioning 387
Mixing 387
Placing, Consolidation, Finishing, and Curing 387
Temperature Control 388
Quality Control 389
Self-Consolidating Concrete 389
Ultra-High Performance Concrete 390
References 392

Chapter 20
Special Types of Concrete 395
Structural Lightweight Aggregate Concrete 395
Structural Lightweight Aggregates 396
Compressive Strength 396
Entrained Air 396
Specifications 397
Mixing 397
Workability and Finishability 397
Slump 397
Vibration 397
Placing, Finishing, and Curing 398
Insulating and Moderate-Strength Lightweight Concretes 398
Mixture Proportions 399
Workability 399
Mixing and Placing 400
Thermal Resistance 400
Strength 400
Resistance to Freezing and Thawing 400
Drying Shrinkage 401
Expansion Joints 401
Autoclaved Cellular Concrete 401
High-Density Concrete 402
High-Density Aggregates 402
Additions 403
Properties of High-Density Concrete 403
Proportioning, Mixing, and Placing 403
Mass Concrete 403
Preplaced Aggregate Concrete 406
No-Slump Concrete 406
Roller-Compacted Concrete 406
Water Control Structures 407
Pavements 407
Soil-Cement 408
Shotcrete 408
Shrinkage-Compensating Concrete 409
Pervious Concrete 409
White and Colored Concrete 410
White Concrete 410
Colored Concrete 410
Photocatalytic Concrete 411
Polymer-Portland Cement Concrete 412
Ferrocement 412
References 412

Appendix 415
Glossary 415
ASTM Standards 421
AASHTO Standards 427
Metric Conversion Factors 430
Cement and Concrete Resources 431

Index 433

Preface and Acknowledgements

Design and Control of Concrete Mixtures has been the cement and concrete industry's primary reference on concrete technology for over 85 years. Since the first edition was published in the early 1920s, the U.S. version has been updated 15 times to reflect advances in concrete technology and to meet the growing needs of architects, engineers, builders, concrete producers, concrete technologists, instructors, and students.

This fully revised 15th edition was written to provide a concise, current reference on concrete, including the many advances that occurred since the last edition was published in 2002. The text is backed by over 95 years of research by the Portland Cement Association. It reflects the latest information on standards, specifications, and test methods of ASTM International (ASTM), the American Association of State Highway and Transportation Officials (AASHTO), and the American Concrete Institute (ACI).

Besides presenting a 30% increase in new information over the prior edition within the previous chapters, this edition has added four new chapters on concrete sustainability, reinforcement, properties of concrete, and durability.

Acknowledgements. The authors wish to acknowledge contributions made by many individuals and organizations who provided valuable assistance in the writing and publishing of the 15th edition. We are particularly grateful for the assistance of the individuals listed below. This acknowledgment does not necessarily imply approval of the text by these individuals since the entire handbook was not reviewed by all those listed and since final editorial prerogatives have necessarily been exercised by the Portland Cement Association.

A special thanks to the following PCA staff for extensive technical recommendations, contributions, photography, and text edits; Wayne Adaska, Richard Bohan, Terry Collins, Jamie Farny, Connie Field, Gregory Halsted, Wes Ikezoe, Susan Lane, Brian McCarthy, Kelly McGinnis, Bruce McIntosh, John Melander, Nathaniel Mohler, Larry Novak, Basile Rabbat, David Shepherd, Stephen Szoke, Paul Tennis, Donn Thompson, and Dave Zwicke.

Additional thanks for technical assistance, references, photography, and editorial reviews goes to the following individuals and organizations: Gina Anaple, Baker Concrete Construction; Neal Anderson, Concrete Steel Reinforcing Institute (CRSI); Allen Engineering Corporation; William Baker, Skidmore, Owings & Merrill; Bob Banka, Concrete Solutions; Howard Barker; Jeffrey Beall; Bruce Blair, Lafarge North America; Mike Caldarone, Concrete Engineering Group (CEG); Nicholas Carino; Carson-Mitchell Construction; Joseph Catella; J. Crocker; Cesar Constantino, Titan America; Tim Conway, Holcim (US); James Cornell, The Beck Group; Norbert Delatte, Cleveland State University; Rex Donahey, American Concrete Institute (ACI); Thano Drimalas, University of Texas; Per Fidjestol, Elkem ASA Materials; Anthony Fiorato; Kevin Folliard, University of Texas; Sidney Freedman, Precast/Prestressed Concrete Institute (PCI); John Gajda, CTLGroup; GCC of America; Gomaco Corporation; Jack Gibbons, CRSI; Jim Grove, Iowa State University; Dale Harrington, Snyder-Associates; Todd Hawkinson, Wire Reinforcement Institute; R. Doug Hooton, University of Toronto; Kenneth Hover, Cornell University; Jason Ideker, Oregon State University; Interlocking Concrete Paving Institute (ICPI); Ara Jeknavorian, W.R. Grace; Victoria Jennings, CTLGroup; Beatrix Kerkhoff; David Lange; University of Illinois; John Kevern, University of Missouri-Kansas City; Colin Lobo, National Ready Mixed Concrete Association (NRMCA); Emily Lorenz, CTLGroup; Kevin MacDonald, Cemstone Concrete Products;

Frank Malits, Cagley and Associates; Christy Martin, Concrete Promotional Group; Ross Martin, Ross Martin Consultants; Richard McGrath, Cement Association of Canada (CAC); Michael Morrison, ACI; Michael Murray, Murray Decorative Concrete Supply; Charles Nmai, BASF Corporation; John Ochsendorf, Massachusetts Institute of Technology; Jan Olek, Purdue University; PCA Product Standards and Technology Committee; Claus Germann Petersen, Germann Instruments; Nicholas Popoff, St. Mary's Cement Company; Aleksandra Radlinska, Villanova University; Farshad Rajabipour, Penn State University; Thomas Rewerts, Thos. Rewerts & Co.; Robert Rodden, American Concrete Pavement Association (ACPA); Nicholas Santero, Massachusetts Institute of Technology; Anton Schindler, Auburn University; Michael Schneider, Baker Concrete Construction; George Seegebrecht, CEG; James Shilstone, Shilstone-Day Technologies; David Suchorski, Ash Grove Cement Company; Lawrence Sutter, Michigan Technological University; Richard S. Szecsy, Texas Aggregates and Concrete Association; Scott Tarr, Concrete Engineering Specialists; Peter Taylor, National Concrete Pavement Technology Center; Michael Thomas, University of New Brunswick; Martha VanGeem, CTLGroup; John Vaughan, Irving Materials; Leif Wathne, ACPA; W. Jason Weiss, Purdue University; John Wojakowski, Hycrete; Kari Yuers, Kryton International; Zhaozhou Zhang, Purdue University; and numerous others who provided reviews, photographs, and technical material for EB001 over the past several years.

In addition, special thanks goes to Cheryl Taylor and Arlene Zapata for their professional expertise in the design layout and artwork for this publication.

Thanks also goes to ASTM, AASHTO, and ACI for the use of their material and documents referenced throughout the book.

The authors have tried to make this edition of *Design and Control of Concrete Mixtures* a concise and current reference on concrete technology. Readers are encouraged to submit comments to improve future printings and editions of this book.

Foreword

A New Material Called Concrete

There is a new material used in construction today. It is called "concrete."

Recent advances in mix design and placement technology have led to a material that has properties we could only dream of a few years ago; the changes are so substantial that it almost should have a new name.

When we use the term concrete today, we are speaking of a material that is very different from the grey material that is in the mind of the general public. The strength, stiffness, workability and durability of modern concrete has led to expanded uses of the material and new opportunities in design.

Concrete has become the material of choice for many building and infrastructure projects, both in the United States and around the world. This trend can be particularly seen in the area of high-rise construction. The majority of tall buildings being designed and constructed today feature either primary structural systems utilizing all concrete or a composite system utilizing concrete and steel.

Of the twenty tallest buildings in the world, 13 utilize composite structures, 5 concrete, and 2 steel. Even more telling, of the 10 tallest buildings completed in 2010, there were 6 concrete systems and 4 composite systems (data as per the Council of Tall Buildings and Urban Habitat, www.ctbuh.org).

Concrete construction has always had an inherent advantage for tall building design in that it provides a significant amount of stiffness, mass, and damping for the structural system; three factors that are critical in controlling building motions and accelerations. However, even with these advantages, it has not been until the more recent advancements in concrete material technology that have allowed concrete systems to take a commanding position in highrise construction.

The high-strength and high-modulus concrete mixes required for tall buildings were previously difficult and expensive to achieve, but have now become more conventional and cost effective. Furthermore, advancements in construction technology have assisted in making concrete easier and faster to place. It is now possible to pump high-strength concrete in a single stage over 600 m (1968 ft), allowing for supertall applications. Additionally, utilizing such construction methods as self-climbing formwork systems, reusable formwork liners, panelized horizontal formwork systems, prefabricated reinforcement cages, and state-of-the-art concrete pumping systems, allows for fast construction.

Because the construction of a modern concrete building is like a vertical factory, self-climbing forms, prefabricated rebar cages and pumped high-performance concrete has greatly automated the construction process. For these reasons and others (including reduced floor-to-floor heights and inherent fire proofing), it is becoming more commonplace for designers and contractors to choose concrete for highrise building construction.

It is for these advantages that concrete construction was chosen for the world's tallest building, the Burj Khalifa in Dubai, U.A.E. The tower is primarily a reinforced concrete structure, with concrete construction utilized for the first 155 stories, above which exists a structural steel spire. The structural system can be described as a "buttressed-core", and consists of high performance concrete wall construction. The spiraling "Y" shaped plan was utilized to shape the Tower to reduce wind forces, as well as to keep the structure simple and foster constructability. High performance concrete is utilized for the tower, with wall and column concrete strengths ranging from C80 MPa to C60 MPa cube strength (9300 psi to 7000 psi cylinder strength). Additionally, the C80 MPa wall and column concrete was specified as a high-modulus concrete, in order to provide increased stiffness to the system. A system was also developed to pump concrete in a single stage to over 600 m (1968 ft), a new height record.

Utilizing concrete construction for the Burj Khalifa was a natural choice. Concrete offers higher stiffness, mass, and damping for controlling building motions and accelerations, which was critical in designing the world's tallest building. In fact, due to the stiffness of the system, SOM was able to design the tower to satisfy motion and acceleration criteria without the use of supplemental damping devices. Additionally, the tower's flat plate floor construction offers increased flexibility in shaping the building, as well as provides for the minimum possible floor thickness in order to maximize the ceiling height. Finally, the design of the structural system and the selection of its construction methods also allowed the tower to be constructed quickly and efficiently. Quite simply, the Burj Khalifa would not have been possible in its present form without the use of ever-advancing concrete.

William F. Baker, PE, SE, FIStructE
Partner
SKIDMORE, OWINGS & MERRILL LLP

CHAPTER 1

Introduction to Concrete

Concrete's versatility, durability, sustainability, and economy have made it the world's most widely used construction material. The term concrete refers to a mixture of aggregates, usually sand, and either gravel or crushed stone, held together by a binder of cementitious paste. The paste is typically made up of portland cement and water and may also contain supplementary cementing materials (SCMs), such as fly ash or slag cement, and chemical admixtures (Figure 1-1).

Figure 1-1. Concrete components: cement, water, coarse aggregate, fine aggregate, supplementary cementing materials, and chemical admixtures.

Figure 1-2. Concrete is used as a building material for many applications including high-rise (top) and pavement (bottom) construction.

Understanding the basic fundamentals of concrete is necessary to produce quality concrete. This publication covers the materials used in concrete and the essentials required to design and control concrete mixtures for a wide variety of structures.

Industry Trends

The United States uses about 180 million cubic meters (240 million cubic yards) of ready mixed concrete each year. It is used in highways, streets, parking lots, parking garages, bridges, high-rise buildings, dams, homes, floors, sidewalks, driveways, and numerous other applications (Figure 1-2).

The cement industry is the building block of the nation's construction industry (Figure 1-3). Few construction projects are viable without utilizing cement-based products geographically. U.S. cement production is widely dispersed with the operation of 97 cement plants in 36 states. The top five companies collectively operate around 57% of U.S. clinker capacity with the largest company representing around 15% of all domestic clinker capacity. An estimated 80% of U.S. clinker capacity is owned by companies headquartered outside of the U.S. (PCA 2010).

Figure 1-3. Portland cement manufacturing plant (Courtesy of GCC).

Cement Consumption

In 2009, the United States consumed 68.4 million metric tons (75.2 million tons) of portland cement (PCA 2010). Cement consumption is dependent on the time of year and prevalent weather conditions. Nearly two-thirds of U.S. cement consumption occurs in the six month period between May and October. The seasonal nature of the industry can result in large swings in cement and clinker (unfinished raw material) inventories at cement plants over the course of a year. Cement producers will typically build up inventories during the winter and then ship them during the summer.

Figure 1-4. Ready mixed concrete accounts for approximately 72% of cement consumption (PCA 2010).

The majority of all cement shipments are sent to ready mixed concrete producers (Figure 1-4) (PCA 2010). The remainder are shipped to manufacturers of concrete related products, contractors, materials dealers, oil well/mining/drilling companies, as well as government entities.

The domestic cement industry is regional in nature. The logistics of shipping cement limits distribution over long distances. As a result, customers traditionally purchase cement from local sources. About 97% of U.S. cement is shipped to customers by truck (Figure 1-5). Barge and rail account for the remaining distribution modes.

Figure 1-5. The majority of all cement is shipped to ready mixed concrete producers in cement trucks.

Concrete is used as a building material in the applications listed in Table 1-1. The apparent use of portland cement by market is provided for 2009 in Figure 1-6. The primary markets (Figure 1-7) are described further in the following sections.

Table 1-1. Markets and Applications for Concrete as a Building Material

Bridges
Buildings
Masonry
Parking Lots
Pavements
Residential
Transit and Rail
Soil Cement and Roller-Compacted Concrete
Waste Remediation
Water Resources

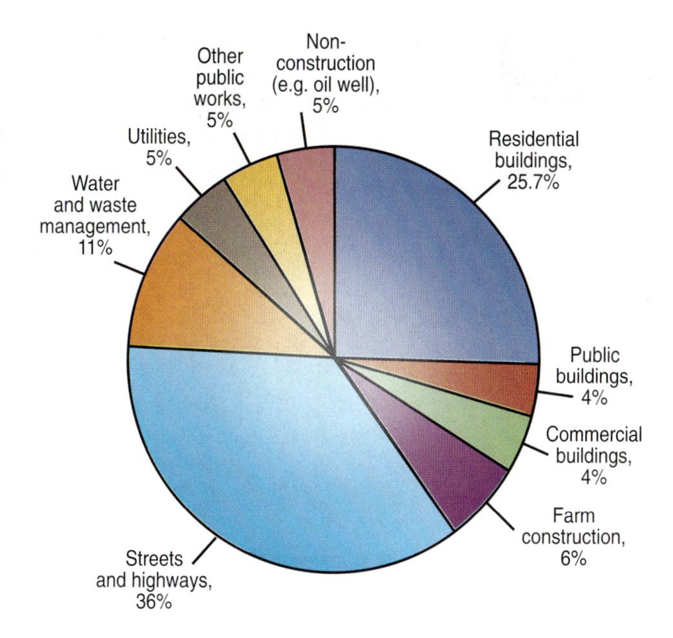

Figure 1-6. 2009 Apparent use of portland cement by market (PCA 2010).

Pavements

Concrete pavements have been a mainstay of America's infrastructure for well over 50 years. The country's first concrete street (built in Bellefontaine, Ohio, in 1891), is still in service today. Concrete can be used for new pavements, reconstruction, resurfacing, restoration, or rehabilitation. Concrete pavements generally provide the longest life, least maintenance, and lowest life-cycle cost of all alternatives.

A variety of cement-based products can be used in pavement applications including soil-cement, roller-compacted concrete, cast-in place slabs, pervious concrete, and whitetopping. They all contain the three same basic components of portland cement, soils/aggregates, and water.

While concrete pavements are best known as the riding surface for interstate highways, concrete is also a durable, economical and sustainable solution for rural roadways, residential and city streets, intersections, airstrips, intermodal facilities, military bases, parking lots and much more.

Bridges

More than 70% of the bridges throughout the U.S. are constructed of concrete. These bridges perform year-round in a wide variety of climates and geographic locations. With long life and low maintenance, concrete consistently outperforms other materials as a choice for bridge construction. A popular method to accelerate bridge construction is to use prefabricated systems and elements. These are fabricated off-site or adjacent to the actual bridge site ahead of time, and then moved into place as needed, resulting in a shorter duration for construction. Very frequently, these systems are constructed with concrete – reinforced, pretensioned, or post-tensioned (or a combination thereof).

Engineered to meet specific needs, high-performance concrete (HPC) is often used for bridge applications including: high-durability mixtures, high-strength mixtures, self-consolidating concrete, and ultra-high performance concrete.

Buildings

Reinforced concrete construction for high-rise buildings provides inherent stiffness, mass, and ductility. Occupants of concrete towers are less likely to perceive building motions than occupants of comparable tall buildings with non-concrete structural systems. A major economic consideration in high-rise construction is reducing the floor to floor height. Using a reinforced concrete flat plate system, the floor to floor height can be minimized while still providing high floor to ceiling heights. As a result, concrete has become the material of choice for many tall, slender towers.

The first reinforced concrete high-rise was the 16-story Ingalls Building, completed in Cincinnati in 1903. Greater building height became possible as concrete strength increased. In the 1950s, 34 MPa (5000 psi) was considered high strength; by 1990, two high-rise buildings were constructed in Seattle using concrete with strengths of up to 131 MPa (19,000 psi). Ultra-high-strength concrete is now manufactured with strengths in excess of 150 MPa (21,750 psi).

Slightly more than half of all low-rise buildings in the United States are constructed from concrete. Designers select concrete for one-, two-, and three-story stores, restaurants, schools, hospitals, commercial warehouses, terminals, and industrial buildings because of its durability, excellent acoustic properties, inherent fire resistance, and ease of construction. In addition, concrete is often the most economical choice: load-bearing concrete exterior walls serve not only to enclose the buildings and keep out the elements, but they also carry roof, wind, and seismic loads, eliminating the need to erect separate systems.

Four concrete construction methods are commonly used to create load-bearing walls for low-rise construction: tilt-up, precast, concrete masonry, and cast-in-place. Traditionally, precast and concrete masonry construction were the standard for low-rise construction. In recent years builders have increasingly used tilt-up construction techniques to erect low-rise commercial buildings quickly and economically.

Figure 1-7. Concrete's primary markets include: pavements, bridges, and high-rise and low-rise buildings.

The Beginning of an Industry

The oldest concrete discovered dates from around 7000 BC. It was found in 1985 when a concrete floor was uncovered during the construction of a road at Yiftah El in Galilee, Israel. It consisted of a lime concrete, made from burning limestone to produce quicklime, which when mixed with water and stone, hardened to form concrete (Brown 1996 and Auburn 2000).

A cementing material was used between the stone blocks in the construction of the Great Pyramid at Giza in ancient Egypt around 2500 BC. Some reports say it was a lime mortar while others say the cementing material was made from burnt gypsum. By 500 BC, the art of making lime-based mortar arrived in ancient Greece. The Greeks used lime-based materials as a binder between stone and brick and as a rendering material over porous limestones commonly used in the construction of their temples and palaces.

Natural pozzolans have been used for centuries. The term "pozzolan" comes from a volcanic ash mined at Pozzuoli, a village near Naples, Italy, following the 79 AD eruption of Mount Vesuvius. Sometime during the second century BC the Romans quarried a volcanic ash near Pozzuoli. Believing that the material was sand, they mixed it with lime and found the mixture to be much stronger than previously produced. This discovery was to have a significant effect on construction. The material was not sand, but a fine volcanic ash containing silica and alumina. When combined chemically with lime, this material produced what became known as pozzolanic cement. However, the use of volcanic ash and calcined clay dates back to 2000 BC and earlier in other cultures. Many of the Roman, Greek, Indian, and Egyptian pozzolan concrete structures can still be seen today. The longevity of these structures attests to the durability of these materials.

Examples of early Roman concrete have been found dating back to 300 BC. The very word concrete is derived from the Latin word "concretus" meaning grown together or compounded. The Romans perfected the use of pozzolan as a cementing material. This material was used by builders of the famous Roman walls, aqueducts, and other historic structures including the Theatre at Pompeii, Pantheon, and Colliseum in Rome (Figure 1-8). Building practices were much less refined in the Middle Ages and the quality of cementing materials deteriorated.

Figure 1-8. Coliseum in Rome, completed in 80 AD, was constructed of concrete. Much of it still stands today (Courtesy of J. Catella).

The practice of burning lime and the use of pozzolan was lost until the 1300s. In the 18th century, John Smeaton concentrated his work to determine why some limes possess hydraulic properties while others (those made from essentially pure limestones) did not. He discovered that an impure, soft limestone containing clay minerals made the best hydraulic cement. This hydraulic cement, combined with a pozzolan imported from Italy, was used in the reconstruction of the Eddystone Lighthouse in the English Channel, southwest of Plymouth, England (Figure 1-9). The project took three years to complete and began operation in 1759. It was recognized as a turning point in the development of the cement industry.

A number of discoveries followed as efforts within a growing natural cement industry were now directed to the production of a consistent quality material. Natural cement was manufactured in Rosendale, New York, in the early 1800s (White 1820). One of the first uses of natural cement was to build the Erie Canal in 1818 (Snell and Snell 2000).

The development of portland cement was the result of persistent investigation by science and industry to produce a superior quality natural cement. The invention of portland cement is generally credited to Joseph Aspdin, an English mason. In 1824, he obtained a patent for a product which he named portland cement. When set, Aspdin's product resembled the color of the natural limestone quarried on the Isle of Portland in the English Channel (Aspdin 1824). The name has endured and is now used throughout the world, with many manufacturers adding their own trade or brand names.

Figure 1-9. Eddystone lighthouse constructed of natural cement by John Smeaton.

Aspdin was the first to prescribe a formula for portland cement and the first to have his product patented (Figure 1-10). However, in 1845, I. C. Johnson, of White and Sons, Swanscombe, England, claimed to have "burned the cement raw materials with unusually strong heat until the mass was nearly vitrified," producing a portland cement as we now know it. This cement became the popular choice during the middle of the 19th century and was exported from England throughout the world. Production also began in Belgium, France, and Germany about the same time and export of these products from Europe to North America began about 1865. The first recorded shipment of portland cement to the United States was in 1868. The first portland cement manufactured in the United States was produced at a plant in Coplay, Pennsylvania, in 1871.

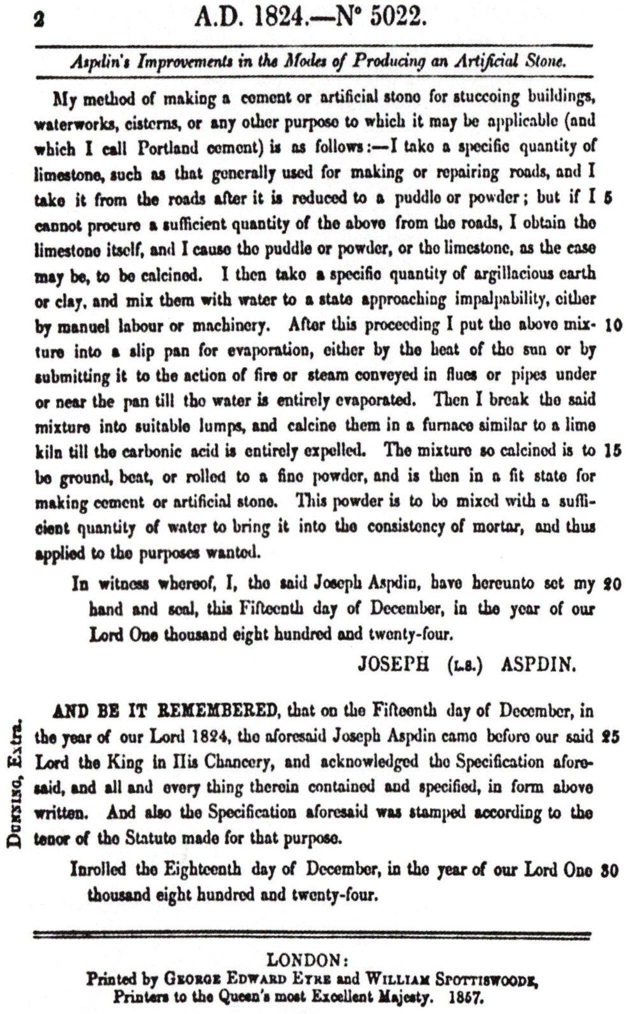

2 A.D. 1824.—N° 5022.

Aspdin's Improvements in the Modes of Producing an Artificial Stone.

My method of making a cement or artificial stone for stuccoing buildings, waterworks, cisterns, or any other purpose to which it may be applicable (and which I call Portland cement) is as follows:—I take a specific quantity of limestone, such as that generally used for making or repairing roads, and I take it from the roads after it is reduced to a puddle or powder; but if I 5 cannot procure a sufficient quantity of the above from the roads, I obtain the limestone itself, and I cause the puddle or powder, or the limestone, as the case may be, to be calcined. I then take a specific quantity of argillacious earth or clay, and mix them with water to a state approaching impalpability, either by manuel labour or machinery. After this proceeding I put the above mix- 10 ture into a slip pan for evaporation, either by the heat of the sun or by submitting it to the action of fire or steam conveyed in flues or pipes under or near the pan till the water is entirely evaporated. Then I break the said mixture into suitable lumps, and calcine them in a furnace similar to a lime kiln till the carbonic acid is entirely expelled. The mixture so calcined is to 15 be ground, beat, or rolled to a fine powder, and is then in a fit state for making cement or artificial stone. This powder is to be mixed with a sufficient quantity of water to bring it into the consistency of mortar, and thus applied to the purposes wanted.

In witness whereof, I, the said Joseph Aspdin, have hereunto set my 20 hand and seal, this Fifteenth day of December, in the year of our Lord One thousand eight hundred and twenty-four.

JOSEPH (L.S.) ASPDIN.

AND BE IT REMEMBERED, that on the Fifteenth day of December, in the year of our Lord 1824, the aforesaid Joseph Aspdin came before our said 25 Lord the King in His Chancery, and acknowledged the Specification aforesaid, and all and every thing therein contained and specified, in form above written. And also the Specification aforesaid was stamped according to the tenor of the Statute made for that purpose.

Dorsing, Extra.

Inrolled the Eighteenth day of December, in the year of our Lord One 30 thousand eight hundred and twenty-four.

LONDON:
Printed by George Edward Eyre and William Spottiswoode,
Printers to the Queen's most Excellent Majesty. 1857.

Figure 1-10. Aspdin's patent for portland cement.

Sustainable Development

Concrete is the basis of much of civilization's infrastructure and much of its physical development. Twice as much concrete is used throughout the world than all other building materials combined. It is a fundamental building material to municipal infrastructure, transportation infrastructure, office buildings, and homes. And, while cement manufacturing is resource- and energy-intensive, the characteristics of concrete make it a very low-impact construction material, from an environmental and sustainability perspective. In fact, most applications for concrete directly contribute to achieving sustainable buildings and infrastructure that are discussed in Chapter 2.

Essentials of Quality Concrete

The performance of concrete is related to workmanship, mix proportions, material characteristics, and adequacy of curing. Each of these will be discussed throughout this publication. The production of quality concrete involves a variety of materials and a number of different processes including: the production and testing of raw materials (Chapters 3-8); determining the desired properties of concrete (Chapters 9-11); proportioning of concrete constituents to meet the design requirements (Chapter 12); batching, mixing, and handling to achieve consistency (Chapter 13); proper placement, finishing, and adequate consolidation to ensure uniformity (Chapter 14); proper maintenance of moisture and temperature conditions to promote strength gain and durability (Chapters 15-17); and finally, testing for quality control and evaluation (Chapter 18).

Many people with different skills come into contact with concrete throughout its production. Ultimately, the quality of the final product depends on their workmanship. It is essential that the workforce be adequately trained for this purpose. When these factors are not carefully controlled, they may adversely affect the performance of the fresh and hardened properties.

Suitable Materials

Concrete is basically a mixture of two components: aggregates and paste. The paste, comprised of portland cement and water, binds the aggregates (usually sand and gravel or crushed stone) into a rocklike mass as the paste hardens from the chemical reaction between cement and water (Figure 1-11). Supplementary cementitious materials and chemical admixtures may also be included in the paste.

Figure 1-11. Concrete constituents include cement, water, and coarse and fine aggregates.

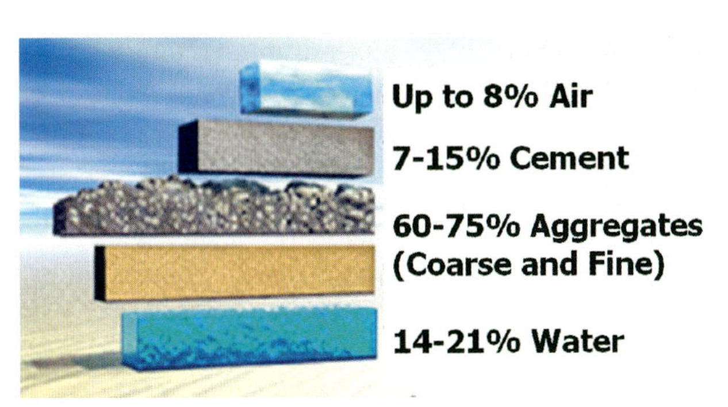

Figure 1-12. Range in proportions of materials used in concrete, by absolute volume.

The paste may also contain entrapped air or purposely entrained air. The paste constitutes about 25% to 40% of the total volume of concrete. Figure 1-12 shows that the absolute volume of cement is usually between 7% and 15% and the water between 14% and 21%. Air content in air-entrained concrete ranges from about 4% to 8% of the volume.

Aggregates are generally divided into two groups: fine and coarse. Fine aggregates consist of natural or manufactured sand with particle sizes ranging up to 9.5 mm (3⁄8 in.); coarse aggregates are particles retained on the 1.18 mm (No. 16) sieve and ranging up to 150 mm (6 in.) in size. The maximum size of coarse aggregate is typically 19 mm or 25 mm (3⁄4 in. or 1 in.). An intermediate-sized aggregate, around 9.5 mm (3⁄8 in.), is sometimes added to improve the overall aggregate gradation.

Since aggregates make up about 60% to 75% of the total volume of concrete, their selection is important. Aggregates should consist of particles with adequate strength and resistance to exposure conditions and should not contain materials that will cause deterioration of the concrete. A continuous gradation of aggregate particle sizes is desirable for efficient use of the paste.

The freshly mixed (plastic) and hardened properties of concrete may be changed by adding chemical admixtures to the concrete, usually in liquid form, during batching. Chemical admixtures are commonly used to: (1) adjust setting time or hardening, (2) reduce water demand, (3) increase workability, (4) intentionally entrain air, and (5) adjust other fresh or hardened concrete properties.

The quality of the concrete depends upon the quality of the paste and aggregate and the bond between the two. In properly made concrete, each particle of aggregate is completely coated with paste and all of the spaces between aggregate particles are completely filled with paste, as illustrated in Figure 1-13.

Specifications for concrete materials are available from ASTM International, formerly known as American Society for Testing and Materials (ASTM) and other agencies such as the American Association of State Highway and Transportation Officials (AASHTO) and the Canadian Standards Association (CSA). Other material specifications and standards for construction are available through the American Concrete Institute (ACI).

Figure 1-13. Cross section of hardened concrete made with (top) rounded siliceous gravel and (bottom) crushed limestone. Cement and water paste completely coats each aggregate particle and fills all spaces between particles.

Water-Cementitious Materials Ratio

In 1918, Duff Abrams published data that showed that for a given set of concreting materials, the strength of the concrete depends solely on the relative quantity of water compared with the cement. In other words, the strength is a function of the water to cement ratio (w/c) where *w* represents the mass of water and *c* represents the mass of cement. This became known as *Abrams law* and it remains valid today as it was in 1918. However, more often, w/cm is used and *cm* represents the mass of cementing materials, which includes the portland cement plus any supplementary cementing materials such as fly ash, slag cement, or silica fume.

Unnecessarily high water content dilutes the cement paste (the glue of concrete) and increases the volume of the concrete produced (Figure 1-14). Some advantages of reducing water content include:

- Increased compressive and flexural strength
- Lower permeability and increased watertightness
- Increased durability and resistance to weathering
- Better bond between concrete and reinforcement
- Reduced drying shrinkage and cracking
- Less volume change from wetting and drying

The less water used, the better the quality of the concrete provided the mixture can still be consolidated properly. Smaller amounts of mixing water result in stiffer mixtures; with vibration, stiffer mixtures can be easily placed. Thus, consolidation by vibration permits improvement in the quality of concrete.

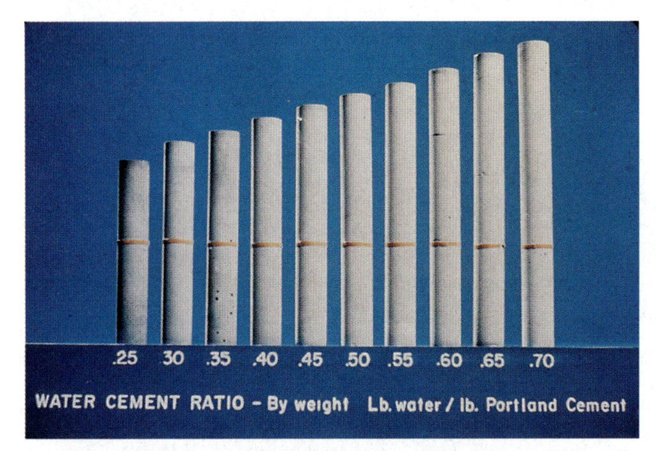

Figure 1-14. Ten cement-paste cylinders with water-cement ratios from 0.25 to 0.70. The band indicates that each cylinder contains the same amount of cement. Increased water dilutes the effect of the cement paste, increasing volume, reducing density, and lowering strength.

Reducing the water content of concrete, and thereby reducing the w/cm, leads to increased strength and stiffness, and reduced creep. The drying shrinkage and associated risk of cracking will also be reduced. The concrete will have a lower permeability or increased water tightness that will render it more resistant to weathering and the action of aggressive chemicals. The lower water to cementitious materials ratio also improves the bond between the concrete and embedded steel reinforcement.

Mixing

To ensure that the components of concrete are combined into a homogeneous mixture requires effort and care. The sequence of charging ingredients into a concrete mixer can play an important part in uniformity of the finished product. The sequence, however, can be varied and still produce a quality concrete. Different sequences require adjustments in the time of water addition, the total number of revolutions of the mixer drum, and the speed of revolution. Other important factors in mixing are the size of the batch in relation to the size of the mixer drum, the elapsed time between batching and mixing, and the design, configuration, and condition of the mixer drum and blades. Approved mixers, correctly operated and maintained, ensure an end-to-end exchange of materials by a rolling, folding, and kneading action of the batch over itself as concrete is mixed.

Concrete must be thoroughly mixed until it is uniform in appearance and all ingredients are evenly distributed through the mixture. If a concrete has been adequately mixed, samples taken from different portions of the batch will have essentially the same density, air content, slump, and coarse aggregate content.

Transporting

Concrete must be transported to the site and placed within a reasonable time frame once it leaves the batch plant. ASTM C 94, *Standard Specification for Ready-Mixed Concrete* (AASHTO M 157)*, requires that the concrete be delivered and placed within 90 minutes after the addition of water to the mixture (when hydration begins) to meet the desired setting and hardening properties, and within 300 revolutions of the mixing drum to prevent segregation and shearing of the aggregate.

Placement and Consolidation

Concrete can be placed by a variety of means such as direct chute discharge from a truck mixer or using a power buggy or wheelbarrow for easily accessible, smaller jobs. For more restrictive locations, concrete can be placed by crane and bucket or even helicopter and bucket when necessary. Also, concrete pumps and conveyors allow for placement of concrete over long distances and otherwise inaccessible heights. Specialty applications include placement by a screw spreader for pavement applications, slipforming, and underwater tremie. Selection of the appropriate placement method is dependent on the application, mix design, crew size, service environment, and economy. The contractor should determine the best placement method based on all these considerations.

Once placed, concrete must be adequately consolidated to mold it within the forms and around embedded items and reinforcement. Internal vibrators are commonly used in walls, columns, beams, and slabs. Vibrating screeds may be used on the surface of slabs or pavements. Form vibrators may be used on columns and walls where dimensions and congested reinforcement restrict the use of internal vibration.

*Many ASTM Standards include metric designations (for example ASTM C94/C94M). For brevity, these designations are omitted throughout the text. Refer to the Appendix for full ASTM Standard designations.

Finishing and Jointing

After concrete has been placed, consolidated, and screeded to remove excess concrete, the exposed surfaces require finishing. Finishing involves floating the surface to embed aggregate particles just beneath the surface, to remove slight imperfections in the surface and to compact the mortar at the surface. After floating, interior slabs may be troweled to create a hard dense surface. The surface may require texturing by brooming to produce a slip-resistant surface. Concrete pavements are frequently textured by tining the surface with stiff wires (this improves traction and reduces hydroplaning).

Slabs on grade may require jointing to provide for movement and to control cracking due to drying and thermal shrinkage. Joints are cut to induce cracks at predetermined locations. Cuts are formed in the surface by using either hand tools when the concrete is still plastic, or diamond saws after the concrete has hardened sufficiently.

Hydration and Curing

Hydration begins as soon as cement comes in contact with water. Each cement particle forms a fibrous growth on its surface that gradually spreads until it links up with the growth from other cement particles or adheres to adjacent substances. This fibrous build up results in progressive stiffening, hardening, and strength development. The stiffening of concrete can be recognized by a loss of workability that usually occurs within three hours of mixing, but is dependent upon the composition and fineness of the cement, any admixtures used, mixture proportions, and temperature conditions. Subsequently, the concrete sets and becomes hard.

Hydration continues as long as favorable moisture and temperature conditions exist and space for hydration products is available. As hydration continues, concrete becomes harder and stronger. Most of the hydration and strength development take place within the first month, but then continues slowly for a long time with adequate moisture and temperature. Continuous strength increases exceeding 30 years have been recorded (Washa and Wendt 1975 and Wood 1992).

Curing is the maintenance of a satisfactory moisture content and temperature so that the desired properties may develop. When the relative humidity within the concrete drops to about 80%, or the temperature of the concrete drops below freezing, hydration and strength gain virtually stop. If the concrete is allowed to freeze before sufficient strength is obtained it may be irreparably damaged. Concrete should not be exposed to freezing temperatures until it has achieved sufficient strength. Curing should begin immediately following finishing procedures and may also be conducted during placement operations to prevent rapid surface evaporation.

Design-Workmanship-Environment

Concrete structures are built to withstand a variety of loads and may be exposed to many different environments such as exposure to seawater, deicing salts, sulfate-bearing soils, abrasion and cyclic wetting and drying. The materials and proportions used to produce concrete will depend on the loads it is required to carry and the environment to which it will be exposed. Properly designed and built concrete structures are strong and durable throughout their service life.

After completion of proper proportioning, batching, mixing, placing, consolidating, finishing, and curing, concrete hardens into a strong, noncombustible, durable, abrasion resistant, and watertight building material that requires little or no maintenance. Furthermore, concrete is an excellent building material because it can be formed into a wide variety of shapes, colors, and textures for use in an unlimited number of applications. Some of the special types of concrete, including high-performance concrete, are covered in Chapters 19 and 20.

References

Abrams, D.A., *Design of Concrete Mixtures*, Lewis Institute, Structural Materials Research Laboratory, Bulletin No. 1, PCA LS001, http://www.cement.org/pdf_files/LS001.pdf, 1918, 20 pages.

Aspdin, Joseph, *Artificial Stone*, British Patent No. 5022, December 15, 1824, 2 pages.

Auburn, *Historical Timeline of Concrete*, AU BSC 314, Auburn University, http://www.auburn.edu/academic/architecture/bsc/classes/bsc314/timeline/timeline.htm, June 2000.

Brown, Gordon E., *Analysis and History of Cement*, Gordon E. Brown Associates, Keswick, Ontario, 1996, 259 pages.

PCA, *North American Cement Industry Annual Yearbook*, ER405, Portland Cement Association, Skokie, Illinois, 2010, 69 pages.

Snell, Luke M., and Snell, Billie G., *The Erie Canal — America's First Concrete Classroom*, http://www.sie.edu/~lsnell/erie.htm, 2000.

Washa, George W., and Wendt, Kurt F., "Fifty Year Properties of Concrete," *ACI Journal*, American Concrete Institute, Farmington Hills, Michigan, January 1975, pages 20 to 28.

White, Canvass, *Hydraulic Cement*, U. S. patent, 1820.

Wood, Sharon L., *Evaluation of the Long-Term Properties of Concrete*, RD102, Portland Cement Association, Skokie, Illinois, 1992, 99 pages.

CHAPTER 2

Sustainability

Introduction to Sustainability

Contemporary engineering, architecture, and building practices are increasingly moving toward the goal of sustainable development. In alignment with growing public awareness, manufacturers of goods and services are incorporating products and processes that reduce energy use, costs, and pollution while aiming to enhance the social value of the sector. The built environment has a significant impact on energy consumption, water and material use, and waste generation (EPA 2009). Those that create the built environment are in a unique position to take the lead and effect transformative positive changes in all areas of sustainable development.

In the pioneering report of the World Commission on Environment and Development (WCED), Our Common Future (WCED 1987), sustainable development is defined as:

> *"Meeting the needs of the present generation without compromising the ability of future generations to meet their needs."*

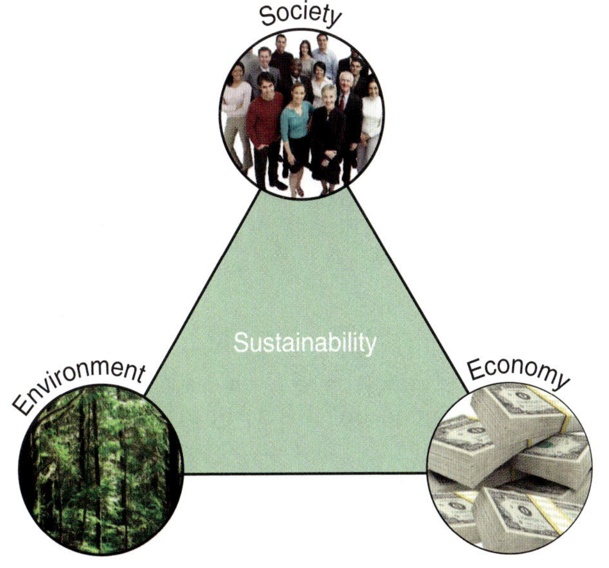

Figure 2-1. The concept of sustainability is supported by a balance of social, economic, and environmental principles.

Embedded in the concept of sustainability is a balance of social, economic, and environmental principles as illustrated in Figure 2-1. When seeking this balance, design practitioners commonly discover strong inter-relationships across factors that many previously thought to be disparate.

The leading nationally recognized sustainability programs focus primarily on site selection and development; energy conservation and atmosphere; water quality and water resources; materials quality and resources; and indoor environmental quality. It is these complex inter-relationships that present challenges to the pathway to greater sustainable achievement. The primary driving factors are generally recognized as:

Energy Conservation and Atmosphere

Energy conservation is a priority in the building sector mainly because of the amount of non-renewable fuels currently required to condition buildings and operate equipment, either directly or in the generation of electricity. There are two key sustainability aspects to reduced energy use; less greenhouse gas emissions from the consumption of fossil fuels, and less dependency on fossil fuels.

In the transportation sector, energy conservation is primarily addressed with durability, long service life, vehicle efficiency, surfaces with low rolling resistance (providing fuel efficiency), mass transit systems, and minimized traffic congestion.

Water Management and Resources

Water quality and water resources are addressed through design and construction that minimizes stormwater runoff for both the building and transportation sectors, especially in areas adjacent to wetlands and bodies of water. In buildings, water conservation is achieved with low-flow water fixtures; more efficient and effective use of rainwater; and the use of graywater where acceptable. Water resource conservation may also be addressed through the manufacturing processes (see Chapter 5 for information on water conservation in concrete production).

Site Selection and Development

The focus on site selection and site development is intended to minimize the impact of all development by encouraging the reuse of existing sites, optimal occupant density (high-rise versus low-rise construction), preservation of natural habitats, water management, and heat island mitigation. These provisions tend to be equally applicable to both the transportation and building sectors.

Indoor Environmental Quality

The intent of indoor environmental quality criteria is to improve occupant health, comfort and productivity. The focus tends to be on off-gassing from building products and finishes; particulate matter reduction; thermal comfort; lighting; and use of suitable cleaning agents. Although not presently addressed in most national certification programs and standards, consideration is also directed to exterior environmental air quality related to construction practices.

Material Quality and Resources

Material resources criteria tend to focus on landfill avoidance by encouraging pre- and post-consumer content in construction materials; and construction waste minimization. For buildings, this is accomplished by providing adequate facilities to accommodate recycling and reuse of consumer products, appliances, and equipment.

Construction systems intended to be used for sustainable construction should require minimal frequencies for routine maintenance, repair, and replacement. The impacts of maintenance repair and replacement concern more than just the construction materials. Sometimes the larger impact will be the additional energy use to maintain, repair and replace, or the interruption in service and operations.

The provisions in most currently available programs do not adequately address pollution prevention at the points of harvesting, processing, manufacturing, and assembling products. Energy use considerations are addressed by encouraging the use of locally or regionally available (indigenous) material to minimize fossil fuel consumption necessary to transport materials.

Functional Resilience

Functional resilience extends beyond material quality and resources. The key sustainability concept of functional resilience is to provide new construction that has adequate longevity for continued use and is readily adaptable to future use. Concepts of functional resilience include robustness, durability, enhanced disaster resistance, and longevity of which some components have been referred to as passive survivability. These criteria are often satisfied by exceeding the minimum requirements through design and construction that is more resistant to damage from floods, high wind events, earthquakes, fires, hail, blast, and other catastrophic events.

Rating Systems

Several organizations have developed rating systems that attempt to quantify improvements in green performance. For example, the United States Green Building Council (USGBC) developed and launched a simplified point rating system titled Leadership in Energy and Environmental Design® (LEED®) in 1998. LEED encourages designers to select from a menu of green strategies with the intent of reducing environmental impact of a project. LEED provides for four increasingly stringent levels of building certification: certified, silver, gold, and platinum. Following successful third party verification, the building becomes certified as constructed to one of these four levels of performance (www.leed.net/). In 2004, the Green Building Initiative (GBI) distributed Green Globes® in the United States. Green Globes is a building guidance and assessment program that offers a way to advance the overall environmental performance and sustainability of commercial buildings (www.thegbi.org/green-globes/). Today, work progresses to formalize sustainable construction strategies in building codes and standards.

Concrete and LEED. Concrete use can contribute credit in fifteen categories in the LEED for New Construction and Major Renovations (NC) 2009 system as summarized in Table 2-1. Although LEED-NC is the most commonly cited LEED product covering commercial structures, schools and core and shell construction of speculative buildings, LEED has similar rating systems for Neighborhood Development (ND), homes (H), retail, healthcare, and existing buildings.

It is now common for some states and municipalities to require LEED silver certification for government owned buildings. Some jurisdictions have adopted LEED or other certification programs as part of their building code. The dilemma with such regulations is that LEED is not written in, and was never intended to serve as, mandatory code language. Further, many of the provisions in LEED and other certification programs, such as using sites near mass transit facilities or redevelopment of brown fields, are outside the purview of the general building code department.

To bridge this gap, the American Society of Heating, Refrigerating and Air-Conditioning Engineers (ASHRAE), USGBC, and the Illuminating Engineering Society (IES) developed ASHRAE/USGBC/IES Standard 189.1 for the *Design of High Performance Green Buildings*. This standard provides mandatory provisions that are expressed as either prescriptive or performance requirements and may be used as a reference document in the building code.

Table 2-1. Potential Contributions by Concrete to LEED® 2009 for New Construction and Major Renovations (NC) v3

Sustainable sites (26 possible points in this credit category)		**Points**
Credit 3	Brownfield redevelopment	1
Credit 5.1	Site development, protect or restore habitat	1
Credit 5.2	Site development, maximize open space	1
Credit 6.1	Stormwater design, quantity control	1
Credit 6.2	Stormwater design, quality control	1
Credit 7.1	Heat island effect, non-roof	1
Credit 7.2	Heat island effect, roof	1
Energy and atmosphere (35 possible points in this credit category)		
Credit 1	Optimize energy performance	1 – 3
Materials and resources (14 possible points in this credit category)		
Credit 1.1	Building reuse, maintain existing walls, floors, and roof	1 – 2
Credit 1.2	Building reuse, maintain existing interior non-structural elements	1
Credit 2	Construction waste management	1 – 2
Credit 4	Recycled content	1 – 2
Credit 5	Regional materials	1 – 2
Indoor environmental quality (15 possible points in this credit category)		
Credit 4.3	Low emitting materials – flooring systems	1
Innovation in design (6 possible points in this credit category)		
Credit 1	Innovation in design	1 – 5

LEED® Certification levels: certified 40 – 49 points, silver 50 – 59 points, gold 60 – 79 points, platinum 80 – 110 points.

Concrete Sustainability

Concrete structures are the basis for much of civilization's infrastructure. Each year, approximately four metric tons of concrete are used for every one of the nearly seven billion people on our planet (USGS 2009). Concrete is a fundamental building material for municipal and transportation infrastructure, office buildings, and homes. Buildings use concrete for their foundations, walls, columns and floors. Highways and bridges are built with concrete. Airports and rail systems use concrete. Our drinking water is delivered through concrete pipes from treatment plants made with concrete.

The principles (environmental, social, and economical) of sustainable development are easily incorporated in the design and proportioning of concrete mixtures and exhibited readily through applications in service (www.sustainableconcrete.org). Particularly because of its long life, concrete is an economical, cost-effective solution. The use of concrete consumes minimal materials, energy, and other resources for construction, maintenance, and rehabilitation over its lifetime, while providing essential infrastructure to society.

Ready-mix concrete and concrete-product manufacturers also strive to improve production processes and enhance transportation efficiency and delivery methods to reduce the environmental impact of the construction process, including the use of alternate energy sources. For example, the National Ready Mixed Concrete Association (NRMCA) is committed to promoting environmental stewardship within concrete plants through the NRMCA Green-Star™ Certification Program. This program provides guidelines through the use of the environmental management systems that would ensure an environmentally friendly concrete plant (NRMCA 2010).

The Massachusetts Institute of Technology (MIT) Concrete Sustainability Hub (CSH) was formed in 2009. The mission of CSH is to advance the technology transfer from concrete science into the engineering practice, by translating the synergy of three fields of study (engineering, architecture and planning, and management) into a powerful hub for concrete sustainability studies relevant to industry and decision makers (MIT 2010). As knowledge of sustainable design and construction practices advances, so do the techniques and strategies employed to achieve even greater efficiencies or environmental impact reductions with concrete.

The following sections elaborate on the characteristics of concrete structures that contribute to innovative sustainable designs. A more in-depth look at concrete's relationship to sustainability is provided by VanGeem (2006), VanDam and Taylor (2009), Schokker (2010), reports from ACI Committee 130, and www.cement.org/SD.

Durability

Durability is the ability to resist weathering action, chemical attack, and abrasion while maintaining desired engineering properties. Concrete structures require different types of durability depending on the exposure environment and desired engineering properties, as discussed in Chapter 11. A durable material benefits the environment by conserving resources, reducing waste and the environmental impacts related to repair and replacement.

The longevity of concrete structures is readily apparent. As the most widely used building material in the world, concrete structures have withstood the test of time for many years. For example, the Hoover Dam was completed in 1936, the Glenfinnan Viaduct in Scotland was completed in 1901, and the Pantheon in Rome was completed around 125 AD. All are still in use today. Depending on the application, the design service life of building interiors is often 30 years. However, the actual average life span for a building in the U.S. is 75 years; more than double the service life. The concrete portion of structures often lasts 100 years and longer. When properly designed, concrete structures can be reused or repurposed several times in the future. Reusing concrete buildings conserves future materials and resources and reduces the construction time that new structures may require.

The oldest concrete street in America was built in 1891 in Bellefontaine, Ohio and is still in use today. Long-life concrete pavements provide 30 years or more of low maintenance service life. These pavements require less frequent repair and rehabilitation than competing materials and contribute to highway safety and congestion mitigation (Hall and others 2007). Reductions in traffic congestion reduce fuel consumption and related pollution from vehicles.

Safety

There is an inherent social value when utilizing engineered concrete structures in regard to public health and safety. People feel secure that while entering a concrete building or driving across a concrete bridge, the structure will be safe. The ACI 318 Building Code addresses the social implications of sustainable development by supplying minimum design and construction requirements necessary to provide for public health and safety. ACI 318 also permits use of environmentally responsible materials and provides durability requirements to enhance lifecycle considerations in sustainable design (ACI 318 2008).

Disaster Resistance

Properly designed reinforced concrete is resistant to storms, floods, fire, and earthquakes. These structures can also provide blast protection for occupants. The Federal Emergency Management Agency (FEMA) recognizes these attributes by promoting concrete safe rooms for providing occupant protection from natural disasters (FEMA 2008). Concrete structures provide greater functional resilience helping essential service providers housed in more robust fire and police stations, hospitals, and community shelters to continue operation after a disaster strikes. The long-term viability of private business operations can also benefit from functional resilience as well.

Tornado, Hurricane and Wind Resistance. Concrete is resistant to forces from high winds, hurricanes, and tornadoes (Figure 2-2). Compliance with the SSTD 10, *Standard for Hurricane Resistant Construction* can reduce the amount of damage observed in structures. Impact resistance of structures to tornado and hurricane debris missiles is tested using a 50 mm x 100 mm (2 in. x 4 in.) piece of wood travelling at 45 m/s (100 mph) and weighing 6.8 kg (15 pounds) (FEMA 2008). When tested against 21 MPa (3000 psi) concrete walls as thin as 50 mm (2 in.), debris missiles shattered on impact without damaging the concrete (Kiesling and Carter 2005).

Figure 2-2. The concrete home shown still standing survived a tornado in Wisconsin. This is a prime example of the resistance of concrete structures to tornado forces.

Fire Resistance. Noncombustible concrete buildings offer effective fire protection. As a separation wall, concrete helps to prevent a fire from spreading within a structure. As an exterior wall or roof, concrete helps to prevent a fire from involving other buildings. The fire endurance of concrete can be determined by its thickness and type of aggregate used by applying ACI Committee 216 procedures (ACI 216 2007).

Earthquake Resistance. In reinforced concrete construction, the combination of concrete and reinforcing steel provides the three most important properties for earthquake resistance: stiffness, strength, and ductility. Reinforced concrete walls work well because of the composite capabilities of materials within the structural system: concrete resists compression forces, and reinforcing steel resists tensile forces produced by an earthquake. See Fanella (2007) or Cleland and Ghosh (2007) for a detailed summary of seismic design and detailing requirements.

Blast Resistance. Concrete can be designed to have improved blast-resistant properties (Smith, McCann, and Kamara 2009). Blast-resistant concretes often have a compressive strength exceeding 100 MPa (14,500 psi) and typically contain steel fibers. These structures are often used in bank vaults, military applications, and buildings requiring enhanced security.

Energy Performance

Concrete can enhance overall energy performance in several applications. Structures with an exterior concrete envelope can use less energy to heat and cool than similarly insulated buildings with lighter weight wood- or steel-framed enclosures. Lower fuel consumption of vehicles traveling over rigid concrete pavement results in reductions in greenhouse gas emissions over the service life of a pavement. Moreover, use of concrete for pavements can help mitigate the urban heat-island effect, lowering ambient air temperatures. This can indirectly reduce the operational energy usage of urban buildings. In addition, indoor and outdoor lighting efficiency can be improved reducing energy costs through the use of more reflective concrete floors and hardscapes. This efficiency also leads to increased safety in buildings and on pavements.

Thermal Mass. Thermal mass is the property that enables building materials to absorb, store, and later release significant amounts of heat. Buildings constructed of cast-in-place, tilt-up, precast, autoclaved aerated concrete, insulating concrete forms (ICFs) (Figure 2-3), or masonry possess thermal mass that helps moderate indoor temperature extremes and reduces peak heating and cooling loads. These materials absorb energy slowly and hold it for much longer periods than do less thermally massive materials. This slow absorption and release of energy delays and reduces heat transfer through the material contributing to three important results:

- There are fewer spikes in the heating and cooling requirements, because the thermal mass slows the response time and moderates indoor air temperature fluctuations.
- Thermal mass can shift energy demand to off-peak time periods when utility rates are lower (Figure 2-4).
- Incorporating thermal mass can lead to a reduction in HVAC equipment capacity, resulting in upfront cost savings.

The most energy is saved when significant reversals in heat flow occur within a wall during the day, in climates with large daily temperature fluctuations above and below the balance point of the building (13°C to 18°C [55°F to 65°F]). In many climates, these buildings have lower energy consumption than non-massive buildings with walls of similar thermal resistance (Gajda 2001, Marceau and VanGeem 2005).

Figure 2-3. Insulating concrete forms (ICFs) provide thermal mass and high levels of insulation, key components for net-zero energy use buildings.

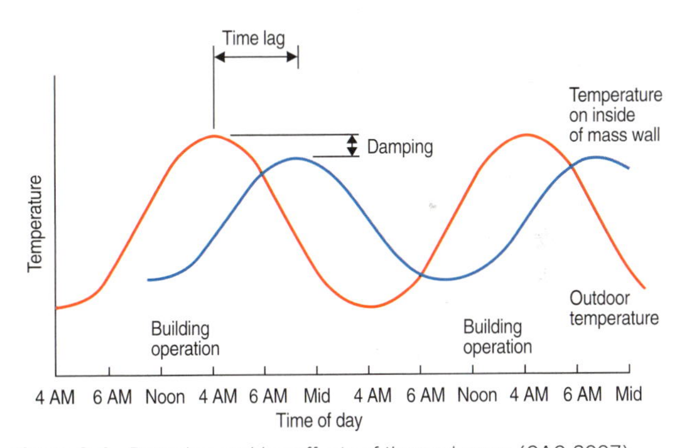

Figure 2-4. Damping and lag effects of thermal mass (CAC 2007).

ASHRAE Standard 90.1 (ASHRAE 2007), the *International Energy Conservation Code* (IECC 2009), and most other energy codes recognize the benefits of thermal mass and require less insulation for mass walls. Computer programs such as *DOE-2* and *EnergyPlus* take into account hourly heat transfer on an annual basis, which allows for more accurate determination of energy loss in buildings with mass walls and roofs. For a more detailed description of thermal mass performance and benefits, see *Thermal Mass*

Explained (The Concrete Centre 2009). For more information on thermal-mass modeling, see Marceau and VanGeem (2005).

Fuel Consumption. Studies demonstrate the lower fuel consumption for vehicles traveling on concrete pavement The improvement in fuel consumption for heavy trucks ranges from 1% to 11% (Zaniewski 1989 and Taylor and Patten 2006). Further study into the effect on smaller vehicles found that fuel consumption in city driving is reduced from 3% to 17% (Ardekani and Sumitsawan 2010). Reducing fuel consumption results in lower greenhouse gas emissions than asphalt pavements.

Heat-Island Reduction. Heat islands are areas that have higher ambient air temperatures as compared to their surrounding areas. Studies have shown that urban environments are 2°C to 4°C (3°F to 8°F) warmer than adjacent areas and this temperature difference is attributed to the replacement of natural vegetation with buildings and pavements. This additional heat causes air-conditioning systems to work harder, which uses more energy, and promotes the formation of smog. For example, in Los Angeles, the probability of smog increases 5% with each degree Celsius temperature increase (3% with every degree Fahrenheit) for temperatures that rise above 24°C (75°F) (Gajda and VanGeem 2001). Maintaining temperatures below the 24°C (75°F) threshold improves outdoor air quality, minimizes health effects such as asthma, and helps reduce the use of air-conditioning systems (Figure 2-5).

Strategies to reduce the amount of dark horizontal surfaces include white colored roofing, shade trees in parking lots, and lighter-colored surfaces for paving, parking lots, and sidewalks. Light-colored concrete materials have higher solar reflectance (also known as albedo) further reducing the heat-island effect.

Solar reflectance is the ratio of the amount of solar radiation reflected from a surface in comparison to the total amount of solar radiation reaching that surface. Ordinary portland cement concrete generally has a solar reflectance of approximately 0.35 to 0.45 (Table 2-2) although values can vary (Marceau and VanGeem 2007). These values are predominantly a result of the light color of the cement

Figure 2-5. Thermal image of a pavement in Mesa, Arizona. Note the temperature difference between the concrete pavement (foreground) and the asphalt pavement (background) (Courtesy of ACPA).

paste. This value can decrease over time as the pavement collects dirt and the cement paste at the surface is abraded. Pressure washing of the surface can restore much of the original reflectance. Surface finishing techniques and drying time also affect solar reflectance. Solar reflectance is most commonly measured using a solar reflectometer (ASTM C1549, *Standard Test Method for Determination of Solar Reflectance Near Ambient Temperature Using a Portable Solar Reflectometer*) or a pyranometer (ASTM E1918, *Standard Test Method for Measuring Solar Reflectance of Horizontal and Low-Sloped Surfaces in the Field*).

A composite index called the solar reflectance index (SRI) is used to estimate expected surface temperatures during exposure to full sun. The temperature of a surface depends on the surface's reflectance and emittance, as well as solar radiation. Emittance, also known as emissivity of a surface, is a measure of how efficiently a surface emits or releases heat. It is a value ranging from 0 to 1. Most opaque non-metallic materials encountered in the built environment (such as concrete and masonry) exhibit an emittance from 0.85 and 0.95. A value of 0.90 is commonly assumed.

The SRI is calculated using ASTM E1980, *Standard Practice for Calculating Solar Reflectance Index of Horizontal and Low-Sloped Opaque Surfaces*. The SRI is used to determine the effect of the reflectance and emittance on the surface

Table 2-2. Solar Reflectance, Emittance, and Solar Reflective Index (SRI) of Concrete Material Surfaces (Berdahl and Bretz 1994; Pomerantz, Pon, and Akbari 2000; Levinson and Akbari 2001; and Pomerantz and others 2002)

Material surface	Solar reflectance	Emittance	SRI
New concrete (ordinary)	0.35 to 0.45	0.9	38 to 52
New white portland cement concrete	0.7 to 0.8	0.9	86 to 100

temperature, and SRI values vary from 100 for a standard white surface to zero for a standard black surface. Materials with the highest SRI are the coolest and the most appropriate choice for mitigating the heat-island effect.

New concrete without added pigments has an SRI value greater than 29, the threshold value required for hardscape in most green building standards and rating systems (Marceau and VanGeem 2007).

Lighting Efficiency. Ordinary concrete and other reflective surfaces reduce energy costs associated with indoor and outdoor lighting compared to darker and less reflective materials. The more reflective surfaces have a higher brightness, which allows for a reduction in lighting power consumption or, alternatively, a reduction in the number of lighting fixtures (Adrian and Jobanputra 2005). For example, due to the differences in the light reflectance of paving materials, asphalt pavement requires 57% more energy to illuminate in comparison to portland cement concrete pavement (Novak and Bilow 2009). Even as it ages, concrete continues to reflect a significant amount of light.

The use of white cement or slag cement results in higher reflectance values (up to about 0.75). Using white cement, white fine aggregate, and white coarse aggregate for concrete provides the brightest white surface and consistent color through the entire depth of the concrete. Since natural white sands may not be available in all regions, a manufactured sand crushed from white stone provides a good alternative. In this case, both fine and coarse aggregate can be derived from a single source, resulting in uniform concrete color.

There are several options available for the construction of light reflective floors. When white concrete is placed throughout the entire slab thickness, it is referred to as full-depth construction. Alternately, a layer of white concrete can be placed over new or existing floors. In new construction, this is known as a two-course floor, while in retrofit applications, this is referred to as a topping (generally bonded to the base slab). New floor surfaces can also be made light reflective by applying proprietary shake-on materials to the fresh concrete. Each of these options; full-depth, two-course, topping, and shake-on; result in permanent color, as opposed to paints and coatings which will eventually wear off. See Farny (2001) and Tarr and Farny (2008) for information on white concrete and floor construction.

Indoor Air Quality

The need to control indoor air quality has dramatically risen as a result of more efficiently sealed or "tighter" building envelopes to accommodate increased energy performance. Air quality concerns have also increased as a result of greater sensitivity by the general public to material off-gassing from construction products and materials (similar to the odor new car purchasers experience). Concrete assemblies can provide several benefits to achieve better indoor air quality performance within these parameters.

Concrete contains low to negligible levels of volatile organic compounds (VOCs) (Budac 1998). Decorative concrete flooring (Figure 2-6), wall, and ceiling assemblies can replace carpeting, wood flooring, and painted finishes which can be a source of VOCs and other air emissions. In addition, carpeting is often linked to higher levels of allergens because it captures dust and can harbor mites and mildew. Decorative concrete finishes can be integral to the interior or exterior surface of concrete walls and to the top surface of concrete slabs. Exterior decorative finishes eliminate the need for siding materials while also providing a hard, durable exterior surface. Interior decorative finishes eliminate the need for gypsum wallboard and also provide a durable inside surface. These finishes minimize materials for construction and reduce jobsite waste (VanGeem 2008).

Figure 2-6. Decorative concrete flooring created using a sandblast stencil technique eliminates the need for carpeting or other flooring materials that can impair indoor air quality.

Uncontrolled air infiltration through a building envelope is problematic. It can allow dust, pollen, air-borne pollution, and unconditioned air into the living space or allow filtered and conditioned air to escape. Water vapor from warm, humid air can condense within wall cavities as it leaks through the wall in cold weather climates, creating moisture and mold issues. The monolithic nature of site cast and precast concrete wall systems, with fewer joints and seams, reduces uncontrolled through-wall infiltration more commonly seen in frame construction.

Acoustics

Concrete can provide excellent acoustic control. Excessive noise has an adverse effect on hearing loss, personal health and well-being, the ability to perform quiet tasks, personal comfort, and general productivity. The sound transmission class (STC) and Outside-Inside Transmission Class (OITC) are values used to rate walls, partitions, doors, and windows for their effectiveness in blocking sound. These ratings correspond to the reduction in decibels from one side of the structure to the other. As cast, unfinished concrete and masonry walls have STCs ranging from 44 to 58 (PCA 1982). With the addition of furring, insulation, and wallboard, STC values as high as 63 are obtained for 300 mm and 350 mm (6 in. and 8 in.) thick concrete walls.

High density residential and office buildings have many benefits derived from reduced energy use and increased proximity to services and public transportation. A key to desirable high density occupancy is providing adequate sound resistance between spaces. Using walls with higher resistance to sound transmission reduces undesirable noise from outdoors, between multifamily units, between work spaces and classrooms, or between hotel or patient rooms. Concrete panels also provide effective sound barriers separating buildings from highways or separating industrial areas from residential areas.

Acoustic performance also applies to concrete pavements. Noise generated by tire-pavement interaction can be a significant source of noise pollution. Many methods have been developed to produce quiet concrete pavements. These include surface grinding, grooving the surface, using enhanced porosity concrete, and use of exposed-aggregate surfaces (Hall and others 2007, Neithalath, Weiss, and Olek 2005, Rasmussen and others 2008, www.cptechcenter.org, www.surfacecharacteristics.com, and www.pavement.com).

Stormwater Management

Concrete plays a significant role in conventional surface water management by providing conveyance and treatment infrastructure that is durable and impermeable. With the move towards stormwater management on site, concrete will continue to play a major role. It is frequently the material of choice to help construct on site flow control and treatment facilities.

Since the primary cause of stormwater runoff is an increased impervious surface, sustainable site design seeks to minimize impervious surface as a first step. The second step is to manage the stormwater and minimize off-site discharge. This generally requires creating conveyance and storage features that intially hold and later allow for filtration and absorption of stormwater.

With the advent of sustainable design, there is considerable interest in replicating pre-development hydrologic site characteristics. This has resulted in strategies that emphasize conservation and use of on-site natural features integrated with small-scale hydrologic controls. These strategies in combination are known as Low Impact Development (LID) (Lehner and others 1999). Four LID strategies use concrete:

Pervious Concrete. Pervious concrete technology creates more efficient land use by eliminating the need for retention ponds, swales, and other stormwater management devices. Pervious concrete is a porous or no-fines concrete that has interconnected voids. Water percolates through these voids into the soil beneath the concrete system. If the absorptive capability of the soil is inadequate to effectively handle the anticipated runoff the pervious concrete is placed over engineered granular base material. The depth of the base is sized to provide adequate capacity to provide temporary detention of anticipated storm water to allow the surrounding soil to gradually absorb the runoff.

In pervious concrete applications, carefully controlled amounts of water and cementitious materials are used to create a paste that forms a thick coating around similarly sized aggregate particles. A pervious concrete mixture contains little or no sand, creating a substantial void content (typically between 15% and 25%). Using sufficient paste to coat and bind the aggregate particles together creates a system of highly permeable, interconnected voids that drains quickly (Figure 2-7). Water flow rates through pervious concrete are commonly measured at 0.34 cm/s (480 in./hr.), which is 200 L/m^2/min (5 gal/ft^2/min), although they can be much higher. For more information on pervious concrete, see Chapter 20 and Tennis, Leming, and Akers (2004).

Figure 2-7. a) Pervious concrete allows for infiltration of water, reducing runoff from the site (Courtesy of B. Banka).
b) Side view of water percolating through pervious concrete.

Permeable Grid Paver Systems. Permeable interlocking concrete pavers and grid paver systems are constructed with special interlocking pavers (Figure 2-8) that provide spaces between adjacent units. These spaces are typically filled with crushed granular material to allow water to infiltrate into the base and sub-base of the pavement. These base courses are designed to collect water so that it can be quickly diverted from the surface and provide a storage area that allows water to slowly percolate into the ground or to be controlled by other storm water management techniques.

Concrete pavers are made with dense concrete mixtures and can be colored or textured. As a result of their physical characteristics, pavers are durable in all climates and appropriate for a range of loading and traffic. If utility or other subsurface access is necessary, the pavers can be easily removed and then replaced without damaging the surface of the pavement.

In addition, there are proprietary cast-in-place and precast concrete grid systems that allow grass or other vegetation to grow (sometimes called cellular grassed paving systems). These systems allow a green-scape or lawn appearance for water percolation while providing a structure for vehicles to drive on (parking lots, access roads, shoulders, and driveways). They can also provide ground stabilization for embankments and level surfaces. Approximately 50% of the surface is concrete and the remaining surface contains voids for vegetation or granular fill.

Figure 2-8. Interlocking concrete pavers allow water to infiltrate between the units (Courtesy of ICPI).

Green Roofs. A green or vegetated roof is a conventional roof that also allows plants to grow on its surface. A vegetated roof includes water proofing, a drainage system, filter layer, a lightweight growing medium, and any number of species of native vegetation. An extensive green roof has a relatively shallow soil profile (25 mm to 125 mm [1 in. to 5 in.]) and is planted with ground cover adapted to the harsh conditions of the rooftop microclimate. Intensive green roofs (Figure 2-9) refer to more substantial roof gardens with deeper soil (300 mm [6 inches] or more). Intensive green roofs are often planted with shrubs, and trees as well as ground cover.

Concrete not only provides the structural strength to support the weight of the garden, but also provides a robust substrate for a water-proofing assembly to isolate the garden from the building envelope.

Figure 2-9. Chicago's Millennium Park is one of the largest green roofs and is built on top of a commuter rail station and two parking garages (Courtesy of J. Crocker).

Rainwater Catchment Systems. Concrete cisterns (Figure 2-10) collect rainfall and store it for later use in non-potable applications such as toilet flushing and irrigation. Treatment systems can purify the rainwater suitable for drinking, bathing, and other potable uses if warranted.

In addition to providing a more natural solution to stormwater management, low impact development strategies can reduce flooding and the need for large (and frequently expensive) conventional stormwater facilities such as expansive detention basins. Reducing the size of stormwater facilities can often result in more buildable land area particularly important on tighter, more densely developed sites.

Figure 2-10. This 25,450 L (7000 gallon) concrete cistern has a colored and stamped top to double as a patio (Courtesy of www.sensiblehouse.org).

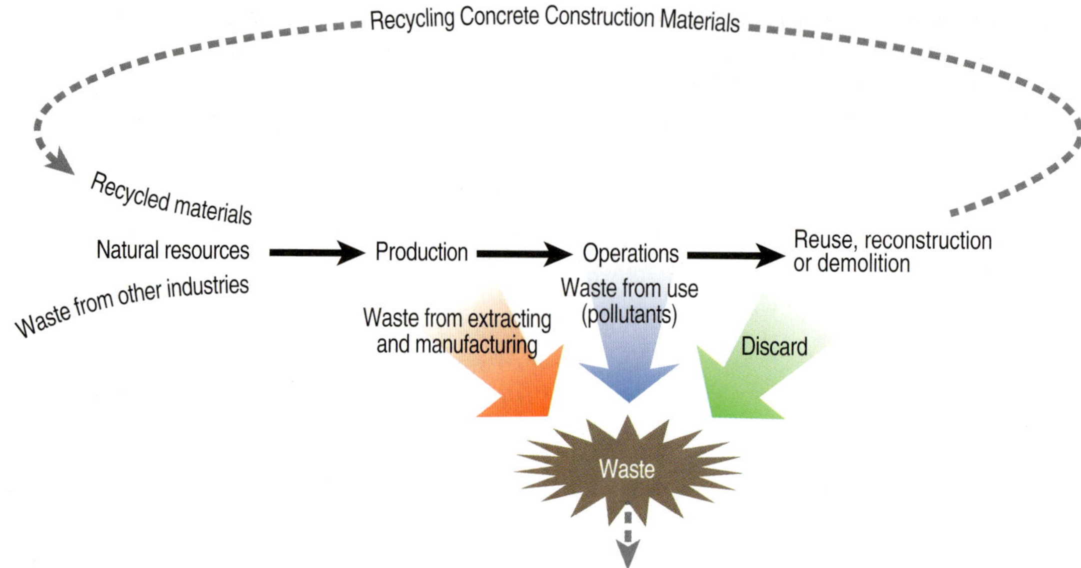

Figure 2-11. Reducing waste in concrete construction.

Site Remediation

Cement and concrete can be used to manage waste using a variety of methods. A large number of sites throughout the country are considered *brownfields*. A brownfield is a site where the expansion, redevelopment, or reuse of the site may be complicated by the presence or potential presence of a hazardous substance, pollutant, or contaminant. Cement-based solidification/stabilization (S/S) treatment is well suited for remediation of contaminated industrial properties and has already been successfully implemented throughout a number of brownfield sites (PCA 2004 and Wittenberg and Covi 2007). In S/S technology, cement is mixed into the contaminated soil or sediment to bind the organic and inorganic contaminants (Lake and others 2010). Treated materials can often be reused on site since the treatment most often improves the physical characteristics of the material. Solidification/stabilization can also be used as a viable treatment for radioactive waste (DOE 2006). For other resources, see Conner (1997) and Adaska and others (1998).

Recycling

Construction and demolition waste. In the U.S., construction and demolition waste represents approximately 123 million metric tons (135 million tons) annually, or about 1.4 kg (3 lbs) per person per day (Beachey 1998). Many concrete materials can be recycled after initial use as a building, pavement, or other structure – reducing the amount of material that is landfilled and the need for virgin materials in new construction (Figure 2-11).

Concrete pieces from demolished structures can also be reused in gabion walls or as riprap for shoreline protection. Recycled concrete in urban areas is commonly used as fill or road base (Figure 2-12). Recycled concrete can also be used as aggregate in new concrete, particularly as the coarse aggregate (Figure 2-13).

Figure 2-12. Concrete from demolished structures can be recycled and is commonly reused as fill or road base.

Figure 2-13. Paving trains can recycle existing concrete pavement. The crushed concrete can be used as aggregate in new concrete or as base material (Courtesy of NCPTC).

Recycled materials. Building products that contain constituent materials originating from recycled products reduce the need for virgin materials in new construction and reduce the environmental impacts from extracting and processing virgin materials.

Post-consumer recycled material is defined as waste generated by households or by commercial, industrial, and institutional facilities as the end-users of a product. These are products that are sold and used for a specific purpose and that require disposal. Post-consumer recycled materials include crushed concrete (Figure 2-12) and masonry from demolished buildings that are reused as aggregate for concrete in new buildings or pavements.

Pre-consumer material is defined as material collected from the waste stream of a manufacturing process. Supplementary cementitious materials (SCMs) such as fly ash, slag cement, and silica fume are examples of pre-consumer materials used in concrete.

Concrete incorporates three major types of recycled materials:

- industrial by-products, used in blended cement or as SCMs in conjunction with portland cement;
- recycled material, used as aggregates in concrete; and
- industrial by-products used as fuel or as raw materials for manufacturing portland or blended cement.

If not used in concrete, many of these materials would be treated as debris and landfilled. When used properly, SCMs may reduce the carbon and energy footprint of cement and concrete and contribute beneficially to the fresh and hardened properties of concrete (see Chapter 4).

The environmental impact of concrete can be further reduced by using aggregates derived from industrial waste or by using recycled concrete as aggregates. Blast furnace slag is a lightweight aggregate with a long history of use in the concrete industry. The FHWA reports that eleven states use recycled concrete aggregate in new concrete (FHWA 2008). These states report that concrete containing recycled aggregate performs equal to concrete containing natural aggregates. See Chapter 6 for more information on recycled-concrete aggregate.

The durability of products with recycled content materials should be carefully evaluated during the design process to ensure comparable life-cycle performance.

Locally Produced

The primary raw materials used to make cement and concrete are abundant throughout the world. The cement, aggregates, and reinforcing steel used to make concrete and the raw materials used to manufacture cement are usually manufactured or extracted from sources within 500 km (300 miles) of the concrete plant. Most ready-mixed concrete plants are within 150 km (100 miles) of the project site. Most precast concrete plants are within 300 km (200 miles) of the project site.

Reduced shipping distances associated with local building materials minimize fuel requirements and the associated energy and emissions from transportation and handling. Locally produced materials contributes to the local economy and reduces imports of materials that may have been produced in countries with much less stringent environmental regulations than in the U.S.

CO_2 Sink

During the life of a concrete structure, the concrete carbonates and absorbs much of the carbon dioxide (CO_2) initially released by calcination during the cement manufacturing process. This may be viewed simply as a loop of the complex carbon cycle. Concrete does not even necessarily have to be directly exposed to the atmosphere for this process to occur. Underground concrete piping and foundations can absorb CO_2 from air in the soil, and underground and underwater applications might absorb dissolved carbon dioxide present in groundwater, freshwaters and saltwaters (Hasselback 2009).

A recent study indicates that in countries with the most favorable recycling practices, it is appropriate to estimate that approximately 86% of the concrete is carbonated after 100 years. During this time, the concrete will absorb approximately 57% of the CO_2 emitted during the original calcination. Approximately 50% of the CO_2 is absorbed shortly after the concrete is crushed and exposed to air during recycling operations (Kjellsen and others 2005).

Concrete Ingredients and Sustainability

A key benefit of concrete is the ability to modify the concrete constituents and proportions to best meet the sustainability goals of a particular application. Concrete is composed of cementitious materials, water, aggregate, admixtures, and reinforcement. Each ingredient has sustainable attributes that contribute to the overall sustainability of concrete.

Cement, representing only 7% to 15% of the volume of concrete, provides the primary engineering and durability properties of concrete. Portland and blended cements commonly use waste fuels and by-product materials in their production to reduce energy demand, conserve natural resources, reduce emissions, and reduce the amount of material sent to landfills. The role of cement in sustainability is addressed in Chapter 3.

The use of supplementary cementitious materials (fly ash, slag, and silica fume) in the production of cement or concrete reduces the use of natural resources and energy, reduces emissions, reduces landfilled materials, and can increase the durability and strength of concrete. The role of supplementary cementing materials on sustainability is addressed in Chapter 4.

Water is essential to the hydration of cement in concrete. Municipal drinking water for use in concrete is partially replaced with water reclaimed from concrete production or municipal waste water treatment facilities, industrial waste water, and water sources not fit for human consumption. This conserves limited sources of potable water. See Chapter 5 for more information.

Aggregates, constituting 60% to 75% of concrete by volume, are traditionally sand, gravel or crushed stone. Reclaimed aggregate from concrete production and recycled hardened concrete from demolished buildings and pavements, can be used to replace a portion of new aggregate in the mix. Aggregate can also be made from industrial by-products, such as blast furnace slag aggregate. These alternative aggregate sources reduce the use of natural resources and reduce the amount of landfilled waste materials.

To conserve natural resources, the use of marginal aggregates in concrete is becoming more common. For example, a two-lift concrete pavement could effectively address the limitation of an aggregate with poor wear resistance (the marginal aggregate can still be used, just not at the surface). Similarly, reactive aggregates can be used through careful selection of cementitious materials. For more information on aggregate's contribution to sustainability, see Chapter 6.

Chemical admixtures, often made from by-products of other industries, enhance the engineering and durability properties of concrete. They also can reduce the amount of water and cementing materials in a concrete mixture resulting in a conservation of natural resources. For more information on chemical admixtures, see Chapter 7.

Reinforcement provides tensile and flexural strength to concrete elements. Reinforcing bars are primarily made from recycled steel, which conserves natural resources. Reinforcement is covered in Chapter 8.

Life-Cycle Analysis

A life-cycle analysis is a tool used to select building materials and influence design choices.

Life-Cycle Cost Analysis

A life-cycle-cost analysis (LCCA) is the practice of accounting for all expenditures incurred over the service-life of a particular structure. An LCCA is performed in units of dollars and is equal to the construction (initial or first) costs plus the present value of future utility, maintenance, insurance, and replacement costs over the service life of the building. Quite often, designs with the lowest first costs for new construction require higher maintenance costs and generate higher energy costs during the service life. Thus these structures will have a higher life-cycle cost. Conversely, durable designs using concrete, often have life-cycle costs that are less than those using other construction materials.

The service life must be accurately reflected in an LCCA study for the impact of concrete use to be correctly measured. The service life of building interiors and equipment is often considered to be 30 years, but the average life of the building shell ranges from 50 to 100 years. Studies that use too short of a service life, for example a twenty year service life, produce skewed and incorrect LCCA results. Such studies *overstate* the cost of construction materials and *understate* the cost of maintaining and operating the structure.

Life-Cycle Assessment and Inventory

A life-cycle assessment (LCA) is an environmental assessment of the life cycle of a product or process. Moving towards sustainable engineering solutions requires a better understanding of construction activities that affect the natural environment. The products and services that we consume impact the environment throughout their service life, beginning with raw materials extraction and product manufacturing, continuing on through use and operation, and finally ending with a waste management strategy (Figure 2-14). Alternately, the ultimate use may include evolution of the building or product back into its natural constituents. The LCA of a structure is a requisite measure necessary to evaluate the environmental impact of a product or structure over its useful life. Conventional assessments often overlook one or more of these phases, leading to incomplete results and indefensible conclusions.

A life cycle inventory (LCI), the first portion of an LCA, includes all of the materials and energy inputs as well as any emissions to air, water, and land (solid waste) from the stages listed in Figure 2-14. Marceau, Nisbet, and VanGeem (2007) provide a detailed life cycle inventory of portland cement concretes using a variety of mixture proportions with and without supplementary cementitious materials. Marceau, Nisbet and VanGeem (2006) also provide a life cycle inventory of portland cement manufacture. The National Renewable Energy Laboratory (NREL) *U.S. Life Cycle Inventory Database* provides an accounting of energy and material flows for materials, components, and assemblies in the United States (www.nrel.gov/lci).

An LCA provides a consistent methodology applied across all products and at all stages of their production, transport, use, and recycling at end of life or disposal, A full LCA includes the effects of operational energy such as heating and cooling (and associated emissions) and raw material use over the life of the product or structure. A full LCA categorizes these effects into impact categories such as land use, resource use, climate change, health effects,

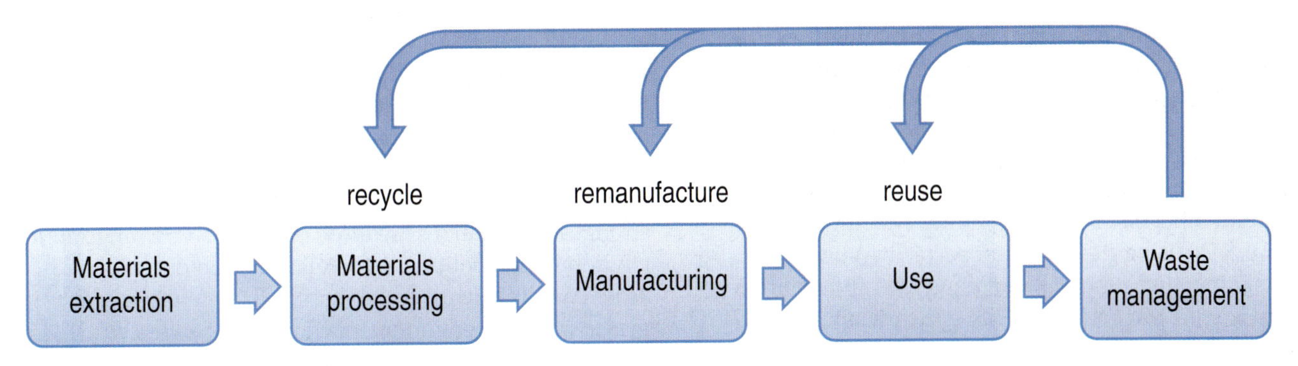

Figure 2-14. Generic product life cycle.

acidification, and toxicity. ISO Standard 14044 provides guidance on conducting a full LCA (ISO 2006).

These guidelines offer a standardized method for conducting an LCA, but do not discuss the specifics relevant to a particular product. Mapping the life cycle, developing functional units, drawing systems boundaries, and mining data are left to the discretion and challenge of individual practitioners. As more information regarding the product and its application are known, the specifics can be better described and assessed. Tables 2-3 and 2-4 demonstrate the life cycle phases and components for concrete pavements and buildings. Even though the basic material (concrete) and phases (materials, construction, use, and end of life) are the same, many of the components in the life cycle are different, especially in the use phase. Understanding and modeling the application of the product are key steps in accurately quantifying its impact.

Data for LCA come from a wide variety of sources, including government databases, industry reports, system models, and first-hand collection. Since the entire life cycle is being analyzed, the volume of necessary data is often large and overwhelming. There are a number of other LCA tools that have been developed for general building professionals. LCA software packages, such as *GaBi*, *SimaPro*, and *EIO-LCA*, can assist in the data collection process, and can provide modeling framework. The NREL U.S. LCI Database provides data for cement and concrete as well as other materials. The NIST LCA model, *Building for Environmental and Economic Sustainability (BEES)*, addresses a variety of concrete elements and structures with numerous mixture proportions (www.bfrl.nist.gov/oae/software/bees/bees.html). These types of packages are generally proficient at quantifying upstream impacts for commodities, but third-party information is often necessary to evaluate detailed processes and niche products. External models, such as those describing building energy consumption, vehicle dynamics, or electricity generation, are commonly used to complement the core LCA and provide spatial, temporal, and system-specific data. Such models are particularly useful when characterizing the operation phase of the life cycle.

It is important to use tools that consider all aspects of a comprehensive LCA, including all of the impact categories

Table 2-3. LCA Phases and Components for Concrete Pavements

Materials	Construction	Use	End of life
Aggregate quarrying	Equipment	Rolling resistance	Demolition
Cement production	Closures, traffic delay	Carbonation	Landfilling
Base, other materials	Transportation	Albedo	Recycling/reuse
Mixing		Maintenance	Carbonation
Transportation		Rehabilitation	Transportation

Table 2-4. LCA Phases and Components for Concrete Buildings

Materials	Construction	Use	End of life
Aggregate quarrying	Equipment	Plug loads	Demolition
Cement production	Temporary structures	Lighting	Landfilling
Insulation, other materials	Transportation	HVAC systems	Recycling/reuse
Mixing		Thermal mass properties	Carbonation
Transportation		Routine maintenance	Transportation

(not just primary energy and CO_2, for example), over the entire life of a building, pavement or product (including operational energy).

A unique strength of LCA is the ability to quantify environmental impact using a host of different metrics. While LEED and other environmental rating systems measure performance using a single point score, LCA breaks results into individual pollutant emissions and environmental impact categories. The life-cycle inventory (LCI) stage generates results in the form of specific inputs and outputs of the systems, such as carbon dioxide to air or copper to water. The life-cycle impact assessment step (LCIA) provides additional information to help assess the inventory results by categorizing the results into impact categories. Some widely used categories include global warming potential, eutrophication, water consumption, ecotoxicity, and human health impact. With such a large variety of ways to characterize impact, assessments can be tailored to address specific goals and objectives for an individual study.

When concrete is used, the LCA tool should allow the input of the cement content and the use of supplementary cementitious materials. This is critical as the cement content dominates the environmental impact of the concrete. Most simplified LCA tools assume a single cement content to represent all concrete and this oversimplification does not include the value derived from operational energy efficiencies. Although results for individual impacts may vary, full LCAs performed according to ISO 14044 usually show similar overall results for concrete, wood, and steel structures unless a particular system is optimized for maximum energy performance or minimal material usage. This is in contrast to simplified tools that use far too many assumptions or consider too few impact categories that one material or product will dominate the results.

LCAs can be used for any number of purposes. The flexible analysis framework and quantitative approach make LCA particularly useful for comparing alternative systems, validating and marketing green claims, establishing environmental footprints, and identifying opportunities for improvement within the life cycle. However, arguably the strongest applications are those centered on creating sound environmental policies. As a policy instrument, LCA provides comprehensive and scientifically defensible strategies to reducing environmental impact. The life-cycle approach ensures that policies are not implemented that benefit from near-term improvements at the expense of long-term deficits. Adopting such insightful policies is the key to establishing a sustainable path towards environmental goals.

LCA at the MIT Concrete Sustainability Hub

At the Massachusetts Institute of Technology Concrete Sustainability Hub, a team of professors, researchers, and students are using LCA to evaluate and improve the environmental performance of buildings and pavements. Each year, building operation and transportation activity release nearly five billion tons of carbon dioxide into the atmosphere, accounting for over two-thirds of carbon emissions in the United States. To reduce these emissions, the materials, designs, and performance of the supporting infrastructure need to be optimized. LCA provides the ideal platform for MIT to construct the life cycles of buildings and pavements in order to identify opportunities to reduce emissions.

Preliminary analyses from MIT indicate that the operation components of the infrastructure—primarily heating and cooling systems for buildings and rolling resistance for pavements—hold the most potential for large-scale improvement. While embodied emissions in the materials manufacturing and construction are important, the emissions associated with the decades of operation tend to

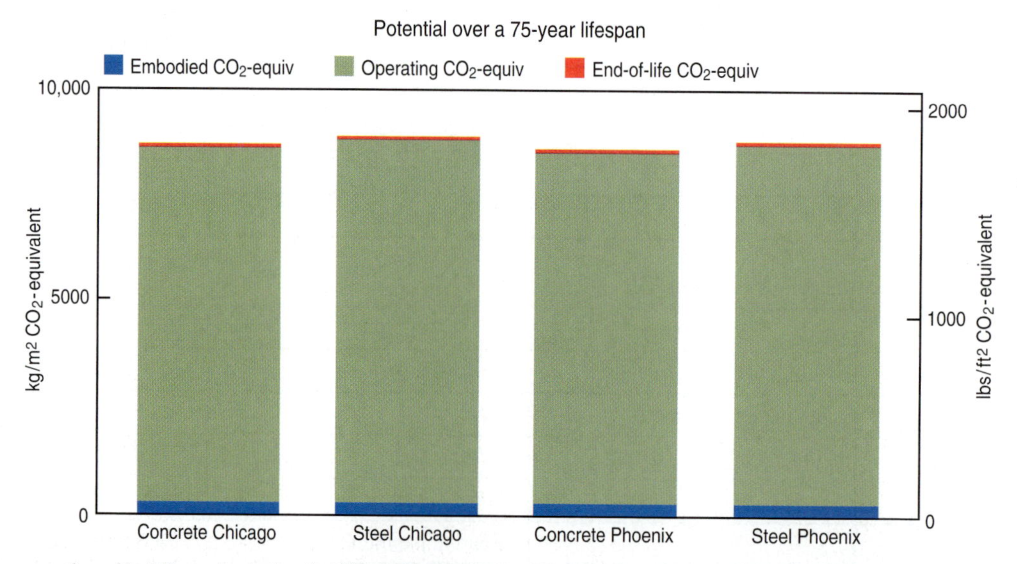

Figure 2-15. The operation of buildings dominates the life-cycle greenhouse gas emissions.

dominate the carbon footprint (Figure 2-15). However, best-practice decisions are also context-sensitive, meaning that a one-size-fits-all sustainability solution is not practical given the range of operating conditions for buildings and pavements. Determination of optimal materials, maintenance, and performance properties depend on the local climate, intended function, and performance criteria.

The case studies that MIT are evaluating for buildings and pavements are a critical step towards creating context-specific sustainability solutions. For buildings, single-family residential, multi-family residential and commercial structures are being evaluated in several locations in the United States. The differing climates and building codes affect how the buildings are designed and operated over the year, which plays a significant role in the life-cycle emission footprint. Likewise, pavement design and performance is affected by the local climate and traffic demand on a given segment, leading to dif--ferent optimal materials, service life, and maintenance schedules. Analyzing how external variables affect best-practice approaches will allow decision-makers to react with solutions that are more specific to their individual goals.

As we better understand these infrastructure systems and how they affect the environment, we achieve more insightful and effective approaches to reducing environmental impact and life cycle economic costs. Given the ubiquity of buildings and pavements, LCA conclusions have relevance in both policy discussions and design discussions, and the results are of interest to a range of audiences. MIT's LCA project strives to deliver a new level of clarity for carbon accounting, which will help the cement and concrete industry to lead the market in developing more quantitative approaches to green construction.

The MIT CS Hub Concrete Science Program is also using nano technology to explore the fundamentals of cement. Through advanced modeling of cement hydration, MIT will improve the performance of cement and enhance its sustainable properties.

References

ACI Committee 216, *Code Requirements for Determining the Fire Endurance of Concrete and Masonry Assemblies*, ACI 216.1-07, American Concrete Institute, Farmington Hills, Michigan, 2007, 28 pages.

ACI Committee 318, *Building Code Requirements for Structural Concrete (ACI 318-08) and Commentary*, American Concrete Institute, Farmington Hills, Michigan, 2008, 465 pages.

Adaska, W.S.; Tresouthick, S.W.; and West, P.B., *Solidification and Stabilization of Waste Using Portland Cement*, EB071, Portland Cement Association, Skokie, Illinois, 1998, 22 pages.

Adrian, W., and Jobanputra, R., *Influence of Pavement Reflectance on Lighting for Parking Lots*, SN2458, Portland Cement Association, Skokie, Illinois, 2005, 46 pages.

Ardekani, Siamak A., and Sumitsawan, Palinee, *Effect of Pavement Type on Fuel Consumption and Emissions in City Driving*, Ready Mixed Concrete Research and Education Foundation, Silver Spring, Maryland, USA, 2010, 81 pages.

ASHRAE, *Standard for the Design of High-Performance Green Buildings, ANSI*/ASHRAE/USGBC/IES Standard 189.1-2009, American Society of Heating, Refrigerating and Air-Conditioning Engineers, Atlanta, Georgia, 2009.

ASHRAE, *Energy Standard for Buildings Except Low-Rise Residential Buildings*, ANSI/ASHRAE/IESNA Standard 90.1-2007, American Society of Heating, Refrigerating and Air-Conditioning Engineers, Atlanta, Georgia, 2007.

Beachey, Jacob E., *Characterization of Building-Related Construction and Demolition Debris in the United States*, United States Environmental Protection Agency, Washington, D.C., USA, 1998, 94 pages.

Berdahl, P., and Bretz, S., "Spectral Solar Reflectance of Various Roof Materials", *Cool Building and Paving Materials Workshop*, Gaithersburg, Maryland, July 1994, 14 pages.

Budac, David, *Concrete's Role in the Indoor Air Environment*, SN2098, Portland Cement Association, Skokie, Illinois, 1998, 15 pages.

CAC, *Guide to Sustainable Design with Concrete*, Version 2.0, Cement Association of Canada, Ottawa, Ontario, Canada, 2007, 137 pages.

Cleland, N., and Ghosh, S.K., *Seismic Design of Precast/Prestressed Concrete Structures*, MNL 140-07, Precast/Prestressed Concrete Institute, Chicago, Illinois, 2007.

Conner, Jesse R., *Guide To Improving the Effectiveness of Cement-Based Stabilization/Solidification*, EB211, Portland Cement Association, Skokie, Illinois, 1997, 47 pages.

DOE, *Fernald Final Project Update*, U.S. Department of Energy, Washington, D.C., 2006, 11 pages.

EPA, *Buildings and their Impact on the Environment: A Statistical Summary*, Environmental Protection Agency, Washington, D.C., April 2009, 7 pages, http://www.epa.gov/greenbuilding/pubs/gbstats.pdf.

Fanella, D., *Seismic Detailing of Concrete Buildings*, SP382, Portland Cement Association, Skokie, Illinois, 2007, 80 pages.

Farny, James A., *White Cement Concrete*, EB217, Portland Cement Association, Skokie, Illinois, 2001, 32 pages.

FEMA, *Design and Construction Guidance for Community Safe Rooms*, 2nd edition, FEMA 361, Federal Emergency Management Agency, Washington, D.C., USA, 2008, 374 pages.

FHWA, *Recycled Concrete Aggregate*, Federal Highway Administration National Review, Federal Highway Administration, Washington, D.C., USA, 2008, http://www.fhwa.dot.gov/pavement/recycling/rca.cfm.

Gajda, John, *Energy Use of Single-Family Houses With Various Exterior Walls*, CD026, Portland Cement Association, Skokie, Illinois, 2001, 49 pages.

Gajda, John L., and VanGeem, Martha G., *A Comparison of Six Environmental Impacts of Portland Cement Concrete and Asphalt Cement Concrete Pavements*, SN2068, Portland Cement Association, Skokie, Illinois, 2001, 34 pages.

Hall and others, *Long-Life Concrete Pavements in Europe and Canada*, FHWA-PL-07-027, Federal Highway Administration, 2007, 80 pages.

Haselbach, L., "Potential for Carbon Dioxide Absorption in Concrete," *Journal of Environmental Engineers*, 135(6), 2009, pages 465 to 472.

ICC, *2009 International Energy Conservation Code*, International Code Council, Country Club Hills, Illinois, USA, 2009, 91 pages.

ISO, *Environmental Management – Life cycle Assessment – Requirements and Guidelines*, ISO14044:2006, International Organization for Standardization, Geneva, Switzerland, 2006, 46 pages.

Kiesling, Ernst W., and Carter, Russell (1997), Investigation of Wind Projectile Resistance of Insulating Concrete Form Homes, 2nd ed., RP122.02, Portland Cement Association, Skokie, Illinois, 2005, 17 pages.

Kjellsen, Knut O.; Guimaraes, Maria; and Nilsson, Åsa, *The CO_2 Balance of Concrete in a Life Cycle Perspective*, Nordic Innovation Centre, Oslo, Norway, December 2005, 34 pages.

Lake, Craig B.; Yuet, Pak K.; Goreham, Vincent C.; and Afshar, Ghazal A., *Validating and Quantifying Mechanisms Responsible For Successful Cement Solidification/Stabilization of Organic Contaminants*, SN3127, Portland Cement Association, Skokie, Illinois, 2010, 63 pages.

Lehner, Peter H.; Aponte, Clarke; George, P.; Cameron, Diane M.; and Frank, Andrew G., *Stormwater Strategies: Community Responses to Runoff Pollution*, Natural Resources Defense Council, New York, New York, USA, 1999.

Levinson, Ronnen, and Akbari, Hashem, *Effects of Composition and Exposure on the Solar Reflectance of Portland Cement Concrete*, Publication No. LBNL-48334, Lawrence Berkeley National Laboratory, Berkeley, California, USA, 2001, 39 pages.

Marceau, Medgar L.; Nisbet, Michael A.; and VanGeem, Martha G., *Life Cycle Inventory of Portland Cement Manufacture*, SN2095b, Portland Cement Association, Skokie, Illinois, 2006, 69 pages.

Marceau, Medgar L.; Nisbet, Michael A.; and VanGeem, Martha G., *Life Cycle Inventory of Portland Cement Concrete*, SN3011, Portland Cement Association, Skokie, Illinois, 2007, 69 pages.

Marceau, Medgar L., and VanGeem, Martha G., *Modeling Energy Performance of Concrete Buildings for LEED-NC v2.1 EA Credit 1*, R&D Serial No. 2880, Portland Cement Association, Skokie, Illinois, 2005, 54 pages.

Marceau, Medgar L., and VanGeem, Martha G., *Solar Reflectance of Concretes for LEED Sustainable Site Credit: Heat Island Effect*, SN2982, Portland Cement Association, Skokie, Illinois, 2007, 94 pages.

MIT, Concrete Sustainability Hub, http://web.mit.edu/cshub, 2010.

NREL, *U.S. Life Cycle Inventory Database*, http://www.nrel.gov/lci, 2010.

Neithalath, Narayanan; Weiss, Jason; and Olek, Jan, *Reducing the Noise Generated in Concrete Pavements through Modification of the Surface Characteristics*, SN2878, Portland Cement Association, Skokie, Illinois, 2005, 71 pages.

Novak, Lawrence C., and Bilow, David N., *A Sustainable Approach to Outdoor Lighting Utilizing Concrete Pavement*, SP393, Portland Cement Association, Skokie, Illinois, 2009, 5 pages.

NRMCA, *Green-Star Certification Program*, NRMCA Operations, Environmental and Safety Committee-Environmental Task Group, Version 1.1, National Ready Mixed Concrete Association, April 2010.

PCA, *Acoustics of Buildings*, IS159, Portland Cement Association, Skokie, Illinois, 1982, 12 pages.

PCA, *Solidification/Stabilization with Cement: Turning Liabilities into Opportunities*, PL611, Portland Cement Association, Skokie, Illinois, 2004, 4 pages.

Pomerantz, M.; Akbari, H.; Chang, S.C.; Levinson, R.; and Pon, B., *Examples of Cooler Reflective Streets for Urban Heat-Island Mitigation: Portland Cement Concrete and Chip Seals*, Lawrence Berkeley National Laboratory, Publication No. LBNL-49283, 2002, 24 pages.

Pomerantz, M.; Pon, B.; and Akbari, H., *The Effect of Pavements' Temperatures on Air Temperatures in Large Cities*, Publication No. LBNL-43442, Lawrence Berkeley National Laboratory, Berkeley, California, USA, 2000, 20 pages.

Rasmussen, Robert O.; Garber, Sabrina I.; Fick, Gary J.; Ferragut, Theordore R.; and Wiegand, Paul D., *Interim Better Practices for Constructing and Texturing Better Concrete Surfaces*, National Concrete Pavement Technology Center, Ames, Iowa, 2008, 58 pages.

Schokker, Andrea J., *The Sustainable Concrete Guide – Strategies and Examples*, U.S. Green Concrete Council, Farmington Hills, Michigan, USA, 2010, 100 pages.

Smith, S.; McCann, D.; and Kamara, M., *Blast Resistant Design Guide for Reinforced Concrete Structures*, EB090, Portland Cement Association, Skokie, Illinois, 2009, 152 pages.

SSTD 10-99, *Standard for Hurricane Resistant Residential Construction*, International Code Council, 2009.

Tarr, Scott M., and Farny, James A., *Concrete Floors on Ground*, EB075, Fourth Edition, Portland Cement Association, Skokie, Illinois, 2008, 252 pages.

Taylor, G.W., and Patten, J.D., *Effects of Pavement Structure on Vehicle Fuel Consumption – Phase III*, National Research Council of Canada, Ottawa, 2006, 98 pages [PCA SN2437c].

Tennis, Paul D.; Leming, Michael L.; and Akers, David J., *Pervious Concrete Pavements*, EB302, Portland Cement Association, Skokie, Illinois, and National Ready Mixed Concrete Association, Silver Spring, Maryland, 2004, 36 pages.

The Concrete Centre, *Thermal Mass Explained*, The Concrete Centre, Camberley, UK, 2009, 16 pages.

USGS, "Cement," *Mineral Commodity Summaries*, U.S. Geological Survey, January 2009, http://minerals.usgs.gov/minerals/pubs/commodity/cement/mcs-2009-cemen.pdf.

VanDam, T., and Taylor, P., *Building Sustainable Pavements with Concrete – Briefing Document*, National Concrete Pavement Technology Center, Iowa State University, Ames, Iowa, USA, 2009, 49 pages.

VanGeem, Martha, "Achieving Sustainability with Precast Concrete," *PCI Journal*, Vol. 51, No. 1, January-February 2006, pages 42 to 61.

VanGeem, M.G., "Integrating Insulation and Surface Finishes into Concrete Walls and Floors," F02-01, *The First International Conference on Building Energy and Environment*, Dalian, China, 2008.

WCED, *Our Common Future*, Report of the World Commission on Environment and Development, Oxford University Press, Oxford, UK, 1987, 383 pages.

Wittenberg, Roy E., and Covi, Arthur, *CQA Methodologies for In-Situ S/S at Former Manufactured Gas Plants*, SR855, Portland Cement Association, Skokie, Illinois, 2007, 4 pages.

Zaniewski, John P., *Effect of Pavement Surface Type on Fuel Consumption*, SR289, Portland Cement Association, Skokie, Illinois, 1989, 5 pages.

CHAPTER 3

Portland, Blended and Other Hydraulic Cement

Portland cements are hydraulic cements composed primarily of hydraulic calcium silicates (Figure 3-1). Hydraulic cements set and harden by reacting chemically with water and maintain their stability underwater. Hydraulic cements include portland cement and blended cements. Other types of hydraulic cements include natural cement and slag cement. They are used in all aspects of concrete construction.

Figure 3-1. Portland cement is a fine powder that when mixed with water becomes the glue that holds aggregates together in concrete.

The development of portland cement was the result of persistent investigation by science and industry to manufacture a product superior in quality to natural cement. The invention of portland cement is generally credited to Joseph Aspdin, an English mason. In 1824, he obtained a patent for his product, which he named portland cement because in its final state, it resembled the color of the natural limestone quarried on the Isle of Portland in the English Channel (Figure 3-2) (Aspdin 1824). The name has endured and is used throughout the world, with many manufacturers adding their own trade or brand names.

Aspdin was the first to prescribe a formula for portland cement and the first to have his product patented. However, in 1845, I. C. Johnson, of White and Sons, Swanscombe, England, claimed to have "burned the cement raw materials with unusually strong heat until the mass was nearly vitrified," producing a portland cement as we now know it.

Figure 3-2. Isle of Portland quarry stone (after which portland cement was named) next to a cylinder of modern concrete.

Manufacture of Portland Cement

Portland cement is manufactured by combining precise proportions of raw materials. These materials are then fired in a rotary cement kiln at high temperatures to form new chemical compounds which are hydraulic in nature. These compounds or phases are formed from oxides of calcium, silica, alumina, and iron. Table 3-1 lists the predominant sources of raw materials used in the manufacture of portland cement.

Numerous waste materials and industry byproducts are also used to supplement some or all of these components. Most of these by-products often contain some fuel component. Some examples of industry waste or by-products are tires, spent pot liners, bottom ashes, or various incinerator ashes.

Steps in the manufacture of cement are illustrated in the flow chart in Figure 3-3. While the operations of all cement plants are basically the same, no flow diagram can adequately illustrate the unique characteristics of any one specific plant. Every plant has significant differences in layout, equipment, or general appearance (Figure 3-4).

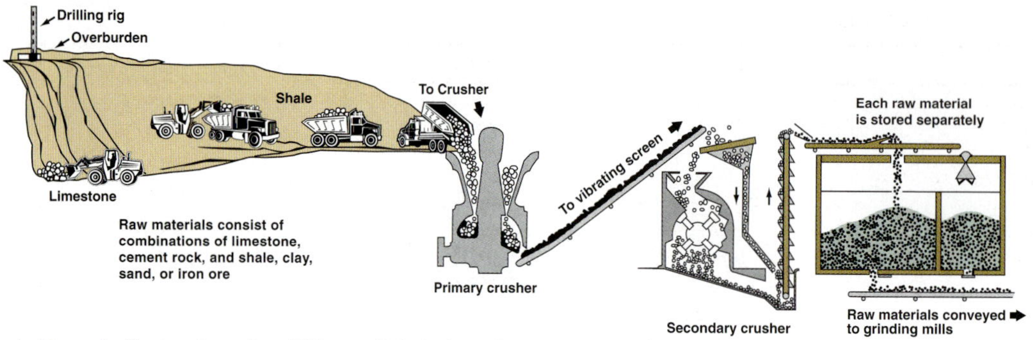

1. Stone is first reduced to 125 mm (5 in.) size, then to 20 mm (3/4 in.), and stored.

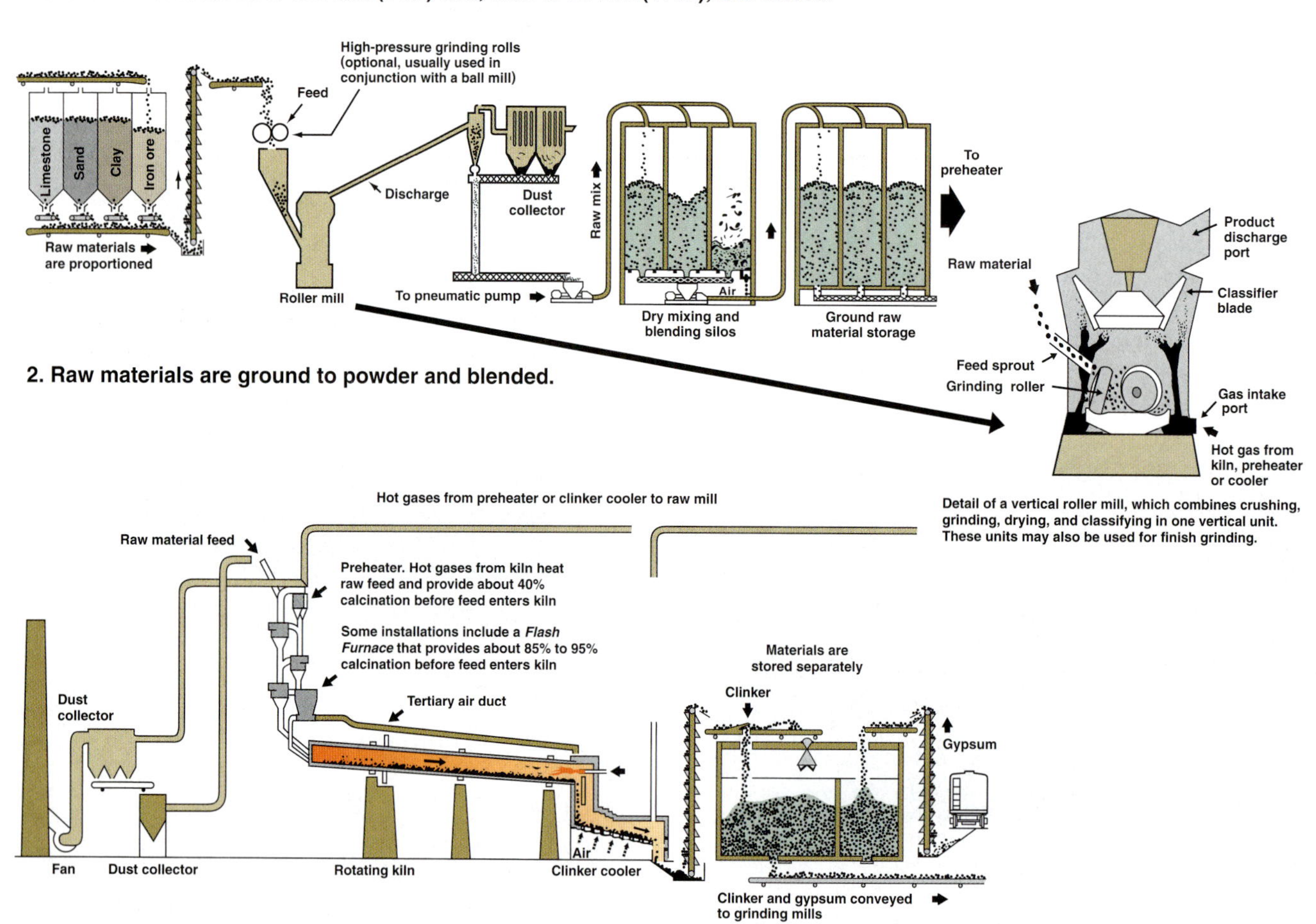

3. Burning changes raw mix chemically into cement clinker. Note four-stage preheater, flash furnaces, and shorter kiln.

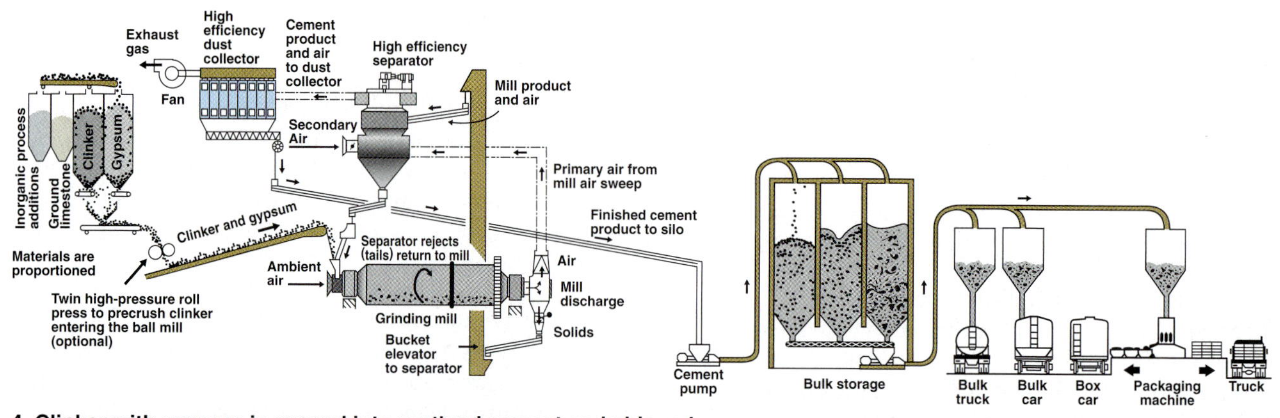

4. Clinker with gypsum is ground into portland cement and shipped.

Figure 3-3. Steps in the manufacture of portland cement.

Table 3-1. Sources of Raw Materials Used in Manufacture of Portland Cement

Calcium	Iron	Silica	Alumina	Sulfate
Alkali waste	Blast-furnace flue dust	Calcium silicate	Aluminum-ore refuse*	Anhydrite
Aragonite*	Clay*	Cement rock	Bauxite	Calcium sulfate
Calcite*	Iron ore*	Clay*	Cement rock	Gypsum*
Cement-kiln dust	Mill scale*	Fly ash	Clay*	
Cement rock	Ore washings	Fuller's earth	Copper slag	
Chalk	Pyrite cinders	Limestone	Fly ash*	
Clay	Shale	Loess	Fuller's earth	
Fuller's earth		Marl*	Granodiorite	
Limestone*		Ore washings	Limestone	
Marble		Quartzite	Loess	
Marl*		Rice-hull ash	Ore washings	
Seashells		Sand*	Shale*	
Shale*		Sandstone	Slag	
Slag		Shale*	Staurolite	
		Slag		
		Traprock		

Note: Many industrial byproducts have potential as raw materials for the manufacture of portland cement.
*Most common sources.

Figure 3-4. Aerial view of a cement plant.

Selected raw materials are transported from the quarry (Figure 3-5), crushed (Figure 3-6), milled, and proportioned so that the resulting mixture has the desired chemical composition. The raw materials are generally a mixture of calcareous (calcium carbonate bearing) material, such as limestone, and an argillaceous (silica and alumina) material such as clay, shale, fly ash, or blast-furnace slag. Most modern cement manufacturing technology uses the *dry process*, in which grinding and blending are accomplished with dry materials. In the *wet process*, the grinding and blending operations are performed with the materials mixed with water to form a slurry. In other respects, the dry and wet processes are similar.

After blending, the ground raw materials are fed into the upper or back end of a kiln (Figure 3-7). The raw mix passes through the kiln at a rate controlled by the slope and rotational speed of the kiln. Fuel (pulverized coal, petroleum coke, new or recycled petroleum oil, natural gas, tire-derived fuels, and byproduct fuel) is forced into

Figure 3-5. Limestone, a primary raw material providing calcium in making cement, is quarried near the cement plant.

Figure 3-6. Quarry rock is trucked to the primary crusher.

the lower end of the kiln where it ignites and generates material temperatures of 1400°C to 1550°C (2550°F to 2800°F). Most modern kilns (post 1980) have increased energy efficiencies by transfering a portion of the kiln into a series or string of vertical heat exchange devices known as preheater cyclones and precalciner vessels. The remaining horizontal portion of the kiln in the newest plants are only used for the final burning zone reaction.

Figure 3-7. Rotary kiln (furnace) for manufacturing portland cement clinker. Inset view inside the kiln.

The final product from the intense heating of these raw materials in the kiln is portland cement clinker. In the kiln, the materials reach approximately 1450°C (2700°F) and a number of chemical reactions occur, changing the raw material into cement clinker, grayish black pellets predominantly the size of marbles (Figure 3-8) . During manufacture, chemical analyses of all materials are made frequently to ensure a uniform, high-quality cement. When the material (clinker) emerges from the kiln it consists predominantly of four phase compounds: tricalcium silicate (C_3S), also known as alite, dicalcium silicate (C_2S), also known as belite, tricalcium aluminate (C_3A), and tetracalcium aluminoferrite (C_4AF). (These phase compounds are noted using cement chemists notation; discussed in further detail under **Chemical Phases (Compounds) and Hydration of Portland Cement**.) Figure 3-9 shows the clinker production process from raw feed to the final product.

Figure 3-8. Portland cement clinker is formed by burning calcium and siliceous raw materials in a kiln. This particular clinker is about 20 mm (3⁄4 in.) in diameter.

The clinker is rapidly cooled and then pulverized into a fine material. During this operation, small amounts of gypsum (Figure 3-10) are added to regulate the setting time of the cement and to improve shrinkage and strength development properties (Lerch 1946 and Tang 1992). Limestone and inorganic processing additions may also be added, each in amounts up to 5% by mass. Organic processing additions may be added in amounts up to 1% by mass. These additions provide significant environmental benefits as well as manufacturing improvements (Hawkins and others 2005 and Taylor 2008). In the final stage of manufacture, finish grinding, clinker and other ingredients are ground so fine that nearly all of the material passes through a 45 µm (No. 325 mesh) sieve. This extremely fine gray powder is now portland cement.

The past century has seen dramatic developments in nearly every aspect of cement manufacturing. Process improvements have greatly improved the uniformity and quality of portland cement. One particularly significant improvement is the rate of strength gain of the finished product.

Quality control has continued to improve throughout the history of portland cement production. Cement manufacturers now offer a wider range of commercial cements, including blended cements suitable for numerous applications. The cement industry has focused resources towards reducing the environmental impact of cement production, particularly carbon dioxide generation as described in the following sections.

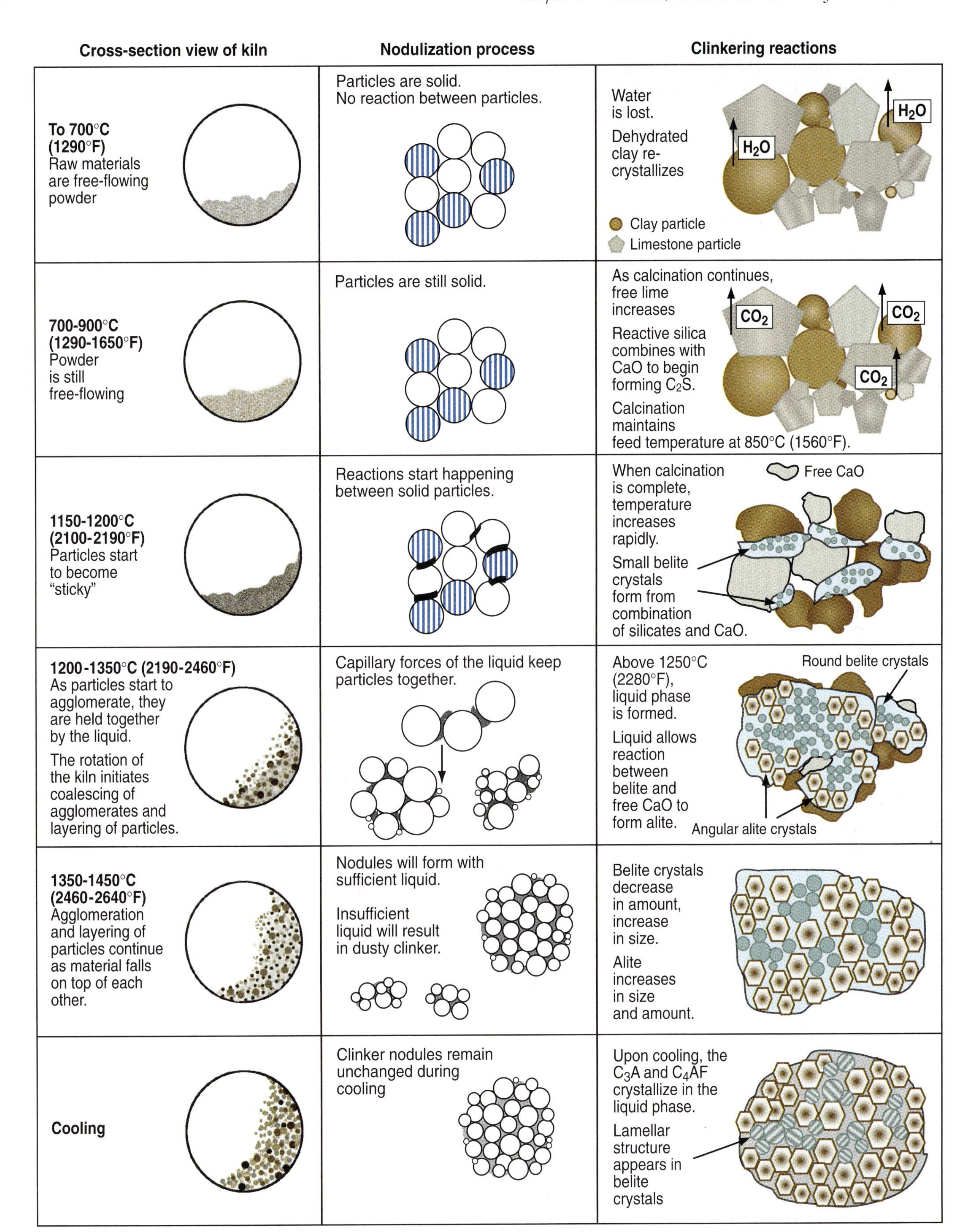

Figure 3-9. Process of clinker production from raw feed to the final product (Hills 2000).

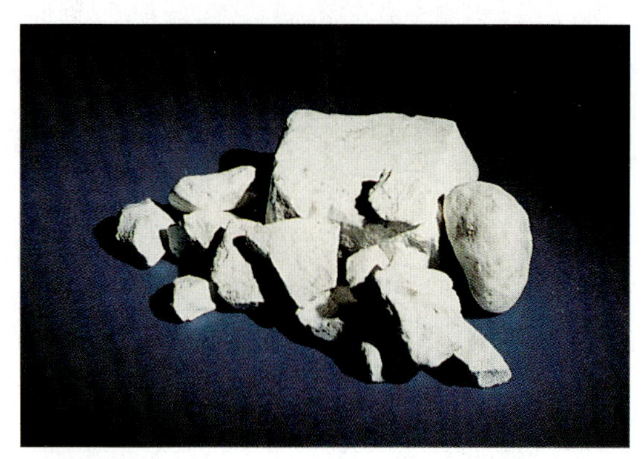

Figure 3-10. Gypsum, a source of sulfate, is interground with portland clinker to form portland cement. It helps control setting, drying shrinkage properties, and strength development.

Cement's Role in Sustainability

Cement manufacturing is energy intensive, due to the high temperature processing required. As with any industry burning fossil fuels, the cement industry also generates combustion by-products and other gaseous emissions. However, less than half of the emissions produced by the cement industry are attributable to the combustion of fuels. The calcination of limestone that occurs during a chemical reaction in the manufacturing process – fundamental to the creation of portland cement – generates approximately 60% of the CO_2 emissions from the process. The cement industry emits, on average, 0.92 tons of CO_2 for each ton of cement produced (Marceau, Nisbet, and VanGeem 2006). The emissions per ton vary because the types of equipment, process energy efficiencies, and product compositions vary from plant to plant. Blended cements and cements containing limestone typically embody less CO_2 per ton than portland cement made without mineral process additions or limestone. Marceau, Nisbet, and VanGeem (2006) provide a life cycle inventory of portland cement. For more information on cement sustainability, see the PCA Report on Sustainable Manufacturing at www.cement.org/smreport11 and the World Business Council for Sustainable Development Cement Sustainability Initiative at www.wbcsdcement.org.

Alternative Raw Materials

The U.S. cement industry is increasingly turning to the use of alternative raw materials to replace or supplement traditional materials – limestone, clay, iron ore, and sand, as sources of calcium, alumina, iron, and silica. The increased use of these alternative sources to produce cement reduce the amount of virgin materials required. Many cement plants utilize industrial by-products and wastes such as steel slag from steel plants and fly ash from coal-fired electric power plants in the manufacture of clinker.

Three by-products of the iron and steel industries can be used in the manufacture of portland cement: foundry sand, mill scale, and slag. Foundry sand can provide silica and possibly iron for the production of clinker. Mill scale contains iron oxides that can replace other iron-bearing materials in the kiln feed. Slag contains high percentages of calcium oxide and silicon dioxide along with varying amounts of aluminum oxide and iron oxides. Also, select slags can be combined with portland cement as a mineral process addition or used to produce a blended cement product.

Fly ash and bottom ash from electric power plants provide a source of silica, alumina, iron and calcium in the raw mix. Fly ash can also be used as a mineral process addition in portland cement and in the production of blended cement. Coal combustion byproducts with a high carbon content (that are not suitable for use in concrete as an SCM) also provide fuel value in the kiln.

Land Stewardship

Limestone for use in production of cement usually comes from a quarry at or near the plant while other materials such as clay, shale, iron ore, and sand are typically obtained from other nearby sources. Because these raw materials are among the most common on Earth, cement producers can mitigate environmental impact through careful site selection and operating procedures.

At the end of their useful life, cement quarries can be reclaimed as parks, recreational areas, or other developments. Many cement companies have developed closure, reclamation, and reuse plans for their quarries, which include careful soil and water contouring and landscape designs to optimize the environmental benefits of the reclaimed areas (Figure 3-11).

Figure 3-11. The Sunken Garden at the Butchart Gardens in Victoria, British Columbia was once a limestone quarry for cement production (Courtesy of Jeffrey Beall).

Solid Waste Reduction

Byproducts generated during cement making are either recycled into the process or used in other beneficial applications. Cement kiln dust (CKD) is removed from the kiln exhaust gases by pollution-control devices. Through improvements in the manufacturing process, the industry has nearly eliminated the amount of CKD that is landfilled. Recycling CKD back into the manufacturing process offsets the use of limestone and other virgin raw materials and conserves energy.

CKD that cannot be recycled back into the process can be sold for a variety of beneficial uses such as soil solidification/stabilization, waste solidification/stabilization, and agricultural soil amendment. Because of the recent manufacturing changes and advances in analytical testing equipment, the cement manufacturer is now able to control the quality of the resulting CKD from the plant. As a result, cement plants have been able to utilize previously landfilled CKD for other beneficial uses. Recently, landfilled CKD has also become a valuable commodity. The goal of the U.S. cement industry is to reduce the amount of CKD disposed per ton of clinker by 60% by 2020 (from a 1990 baseline).

Energy and Fuel

The high temperature needed for cement manufacturing makes it an energy-intensive process. The average energy input required is 4.53 million Btu per metric ton (PCA 2010). The cement industry has improved energy efficiency by 37% from 1972 (Figure 3-12). Cement production now accounts for 2.4% of U.S. energy consumption. In comparison, iron and steel mills account for 11% while paper mills account for 15% (PCA 2009).

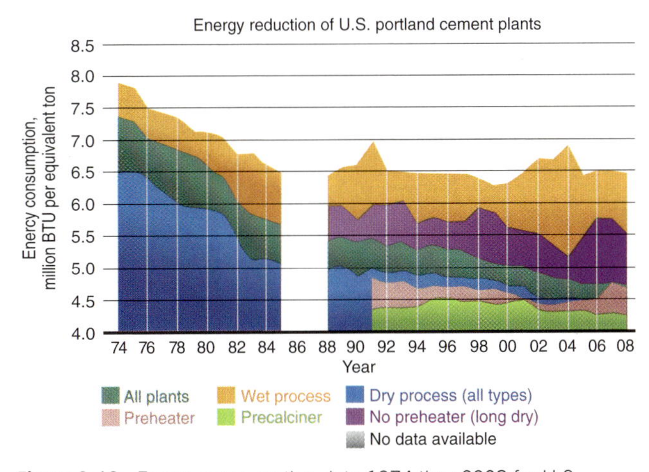

Figure 3-12. Energy consumption data 1974 thru 2008 for U.S. cement industry (PCA 2009).

Although coal, petroleum coke, and other fossil fuels have been traditionally used in cement kilns, many cement companies are turning to energy-rich alternative fuels. Many plants meet a substantial portion (20% to 70%) of their energy requirements using alternative fuels. These fuels often include consumer wastes or by-products from other industries. Cement plants carefully utilize these alternative fuels and materials so their cements conform to ASTM and AASHTO standards. Slightly more than 65% of all U.S. cement plants now incorporate alternative fuels in their energy consumption strategy (Sullivan 2009).

Alternative fuels are carefully selected for the benefit of heat value while complying with increasingly restrictive emission standards and regulations. Over the past decade, the consumption of tire-derived fuels has increased 80% with the U.S. cement industry efficiently combusting 479,000 metric tons (528,000 short tons) of tire-derived waste along with increased volumes of other waste fuels, such as plastics, diapers, roofing material, and various biosolids. In 2007, the energy obtained from waste fuels increased to 9.4% of the total energy demand at cement plants. Some plants also use a significant amount of wind-generated electricity. A goal of the U.S. cement industry is to improve the energy efficiency by 20% on a per unit basis by 2020 (from a 1990 baseline) (PCA 2009).

Use of Tire-Derived Fuel

The high temperature processing of cement manufacturing enables the industry to utilize scrap tires as an alternative fuel source. The rubber elements of tires are comprised of hydrocarbon compounds which, like coal and oil, have tremendous fuel value. Tires have 25% more fuel value than coal on an equal mass basis (PCA 2008). Approximately 300 million used tires are generated annually in the United States (RMA 2009). The EPA recognizes tire derived fuel (TDF) as an environmental best practice and encourages industries to use this resource. Using tire-derived fuel in cement manufacturing (Figure 3-13) recovers energy and conserves fossil fuel resources otherwise destined for landfills or unregulated disposal locations. The intense heat of the kiln ensures complete destruction of the tires. There are no visible emissions from the tires, rather, the use of tires as fuel can actually reduce certain emissions (PCA 2008).

Figure 3-13. Whole rubber tires introduced directly into the kiln by conveyor as an alternative fuel.

Combustion Emissions

The U.S. cement industry has a goal of reducing CO_2 emissions by 10% per ton of cement (from 1990 baseline levels) by the end of 2020. In addition, 90% of U.S. plants will adopt auditable and verifiable environmental management systems by 2020 (PCA 2009). The most recent progress in reduction of combustion emissions involves changes in cement specifications. The changes, introduced in ASTM C150, *Standard Specification for Portland Cement*, in 2004 and harmonized with AASHTO M 85 in 2007, permit up to 5% by mass of the final cement product to consist of naturally occurring, finely ground limestone. This allows for use of unburned or uncalcined ground limestone as a component in finished cement ultimately reducing CO_2 (by calcination) by approximately 2.5 million metric tons (2.8 million short tons) per year (Ashley and Lemay 2008 and Nisbet 1996). In 2009, ASTM C150 and AASHTO M 85 were again changed to also allow up to 5% inorganic processing additions, such as fly ash and slag. The use of limestone and inorganic processing additions reduces the amount of raw materials and electricity used in cement manufacture. For more on the effects of limestone and inorganic processing addition on the properties of cement and concrete, see Hawkins, Tennis, and Detwiler (2005), and Taylor 2008. Further emission reductions are achieved by using alternative waste fuels. In most cases these waste fuels would otherwise be disposed of in incinerators decreasing their value as an energy source. For more information on emission reductions, see the PCA Report on Sustainable Manufacturing at www.cement.org/smreport11 and the World Business Council for Sustainable Development Cement Sustainability Initiative at www.wbcsdcement.org.

Types of Portland Cement

Different types of portland cement are manufactured to meet various physical and chemical requirements for specific purposes. Portland cements are manufactured to meet the specifications of ASTM C150 or AASHTO M 85, (also see **Performance Based Hydraulic Cements** for portland cements meeting ASTM C1157, *Standard Performance Specification for Hydraulic Cement*). The requirements of AASHTO M 85 and ASTM C150 are equivalent. AASHTO specifications are used by some state departments of transportation in lieu of ASTM standards.

ASTM C150 (AASHTO M 85) provides for ten types of portland cement:

Type I	Normal
Type IA	Normal, air-entraining
Type II	Moderate sulfate resistance
Type IIA	Moderate sulfate resistance, air-entraining
Type II (MH)	Moderate heat of hydration and moderate sulfate resistance
Type II (MH)A	Moderate heat of hydration and moderate sulfate resistance, air-entraining
Type III	High early strength
Type IIIA	High early strength, air-entraining
Type IV	Low heat of hydration
Type V	High sulfate resistance

A detailed review of ASTM C150 (AASHTO M 85) cements follows.

Type I

Type I portland cement is a general-purpose cement suitable for all uses where the special properties of other cement types are not required. Its uses in concrete include pavements, floors, reinforced concrete buildings, bridges, tanks, reservoirs, pipe, masonry units, and precast concrete products (Figure 3-14).

Figure 3-14. Typical uses for normal or general use cements include (left to right) highway pavements, floors, bridges, and buildings.

Type II

Type II portland cement is used where protection against moderate sulfate attack is necessary. It is used in normal structures or elements exposed to soil or ground waters where sulfate concentrations are higher than normal, but not unusually severe (see Table 3-2, and Figures 3-15 to 3-17). Type II cement has moderate sulfate resistant properties because it contains no more than 8% tricalcium aluminate (C_3A).

Sulfates in moist soil or water may enter the concrete and react with the hydrated C_3A, resulting in expansion, scaling, and cracking of concrete. Some sulfate compounds, such as magnesium sulfate, directly attack calcium silicate hydrate.

The use of Type II cement in concrete must be accompanied by the use of a low water-to-cementitious materials ratio and low permeability to effectively control sulfate attack. Figure 3-15 (left) illustrates the improved sulfate resistance of Type II cement over Type I cement.

Concrete exposed to seawater is often made with Type II cement. Seawater contains significant amounts of sulfates and chlorides. Although sulfates in seawater are capable of attacking concrete, the presence of chlorides in seawater inhibits the expansive reaction that is characteristic of sulfate attack. Thus, a marine environment is an application requiring a moderate sulfate exposure class. Observations from a number of sources show that the performance

Table 3-2. Types of Cement Required for Concrete Exposed to Sulfate in Soil or Water*

Exposure class		Sulfate exposure**		Cementitious materials requirements†			Maximum w/cm
		Water-soluble sulfate in soil, % by mass	Dissolved sulfate in water, ppm	C150	C595	C1157	
S0	Negligible	$SO_4 < 0.10$	$SO_4 < 150$	NSR	NSR	NSR	None
S1	Moderate‡	$0.10 < SO_4 < 0.20$	$150 < SO_4 < 1500$	II or II(MH)	IP(MS) IS(< 70)(MS) IT(P ≥ S)(MS) IT(P < S < 70)(MS)	MS	0.50
S2	Severe	$0.20 < SO_4 < 2.00$	$1500 < SO_4 < 10,000$	V	IP(HS) IS(< 70)(HS) IT(P ≥ S)(HS) IT(P < S < 70)(HS)	HS	0.45
S3§	Very severe	$SO_4 > 2.00$	$SO_4 > 10,000$	V	IP(HS) IS(< 70)(HS) IT(P ≥ S)(HS) IT(P < S < 70)(HS)	HS	0.40

* Adapted from Bureau of Reclamation Concrete Manual, ACI 201 and ACI 318. "NSR" indicates no special requirements for sulfate resistance.

** Soil is tested per ASTM C1580 and water per ASTM D516.

† Pozzolans and slag that have been determined by testing according to ASTM C1012 or by service record to improve sulfate resistance may also be used in concrete. Maximum expansions when using ASTM C1012: Moderate exposure – 0.10% at 6 months; Severe exposure – 0.05% at 6 months or 0.10% at 12 months; Very Severe exposure – 0.10% at 18 months. Refer to ACI 201.2R for more guidance. ASTM C595 Type IT was adopted in 2009 and has not been reviewed by ACI at the time of printing.

‡ Includes seawater.

§ ACI 318 requires SCMs (tested to verify improved sulfate resistance) with Types V, IP(HS), IS(< 70)(HS) and HS cements for exposure class S3. ACI 318 requires a maximum water:cement ratio of 0.45 for exposure class S3.

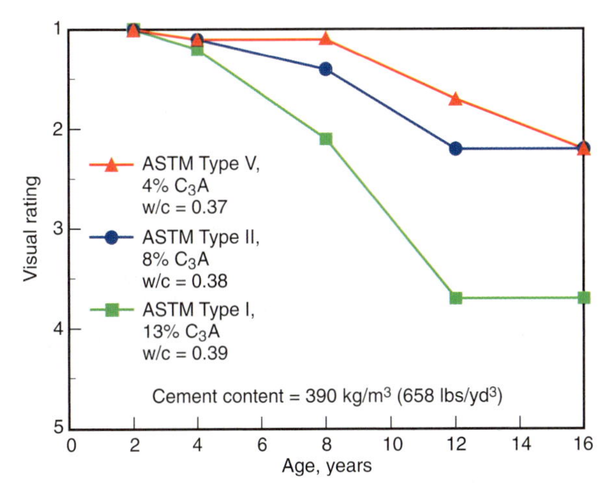

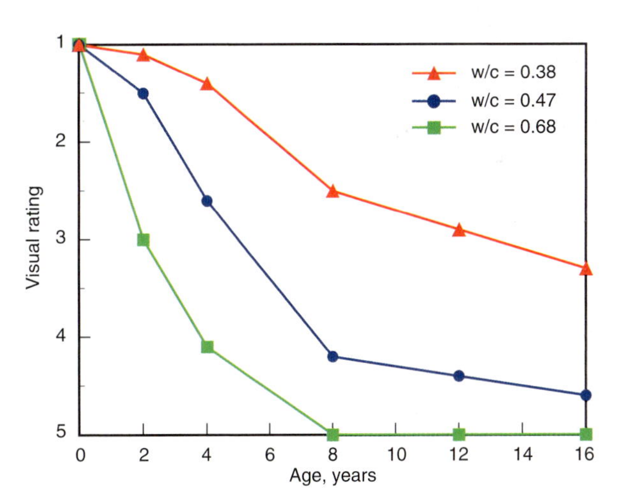

Figure 3-15. (left) Performance of concretes made with different cements in sulfate soil. Type II and Type V cements have lower C_3A contents that improve sulfate resistance. (right) Improved sulfate resistance results from low water to cementitious materials ratios as demonstrated over time for concrete beams exposed to sulfate soils in a wetting and drying environment. Shown are average values for concretes containing a wide range of cementitious materials, including cement Types I, II, V, blended cements, pozzolans, and slags. See Figure 3-17 for rating illustration and a description of the concrete beams (Stark 2002).

Figure 3-16. Moderate sulfate resistant cements and high sulfate resistant cements improve the sulfate resistance of concrete elements, such as (left to right) slabs on ground, pipe, and concrete posts exposed to high-sulfate soils.

Figure 3-17. Specimens used in the outdoor sulfate test plot in Sacramento, California, are 150 x 150 x 760-mm (6 x 6 x 30-in.) beams. A comparison of ratings is illustrated: (top) a rating of 5 for 12-year old concretes made with Type V cement and a water-to-cement ratio of 0.65; and (bottom) a rating of 2 for 16-year old concretes made with Type V cement and a water-to-cement ratio of 0.37 (Stark 2002).

of concretes in seawater with portland cements having C_3A contents as high as 10%, have demonstrated satisfactory durability, providing the permeability of the concrete is low and the reinforcing steel has adequate cover (Zhang, Bremner, and Malhotra 2003). Chapter 11 provides more information on sulfate attack.

Because of its increased availability, Type II cement is sometimes used in all aspects of construction, regardless of the need for sulfate resistance. Some cements are also labeled with more than one type designation, for example Type I/II. This simply means that such a cement meets the requirements of both cement Types I and II.

Type II (MH)

Type II (MH) cements are manufactured to generate heat at a slower rate than Type I or most Type II cements by limiting the heat index to a maximum of 100. (The heat index is the sum of $C_3S + 4.75\ C_3A$). This requirement is roughly equivalent to heat of hydration measurements of 335 kJ/kg at 7 days using ASTM C186, *Standard Test Method for Heat of Hydration of Hydraulic Cement* (Poole 2009). As an alternate to the heat index requirement, an optional requirement can be specified limiting the heat of hydration to 290 kJ/kg as determined by ASTM C186. Type II (MH) (or Type IV, if available) can often be used in structures of considerable mass, such as large piers, large foundations, and thick retaining walls (Figure 3-18). Using MH cements will reduce temperature rise, peak temperature and, minimize temperature related cracking. Thermal control is especially important when concrete is placed in warm weather (see Chapter 16).

Figure 3-18. Moderate heat and low heat cements minimize heat generation in massive elements or structures such as (left) very thick bridge supports, and (right) dams. Hoover dam, shown here, used a Type IV cement to control temperature rise.

Type II(MH) cements are also moderately sulfate resistant, as the maximum tricalcium aluminate content (C_3A) is limited to a maximum of 8% by mass.

Type III

Type III portland cement provides strength at an earlier period than normally expected. For example, strength may be achieved in a matter of days as compared to the typical expectation of 28 days. Type III is chemically similar to Type I cement, except that its particles have been ground finer. It is used when forms need to be removed as soon as possible or when the structure must be put into service quickly. In cold weather its use permits a reduction in the length of the curing period (Figure 3-19).

Type IV

Type IV portland cement is used where the rate and amount of heat generated from hydration must be minimized. It develops strength at a slower rate than other cement types. Type IV cement is intended for use in massive concrete structures, such as large gravity dams (Figure 3-18), where the temperature rise resulting from heat generated during hardening must be minimized. Type IV cement is not commonly manufactured in North America as there are other, more economical, measures for controlling heat rise in concrete available; such as the use of blended cements and supplementary cementing materials.

Type V

Type V portland cement is used in concrete exposed to severe sulfate environments – principally where soils or groundwaters have a high sulfate content. It gains strength more slowly than Type I cement. Table 3-2 lists sulfate concentrations requiring the use of Type V cement. The high sulfate resistance of Type V cement is attributed to a low tricalcium aluminate content, not more than 5%. The use of a low water to cementitious materials ratio and low permeability are critical to the performance of any concrete exposed to sulfates. Even Type V cement concrete cannot withstand a severe sulfate exposure if the concrete has a high water-cementitious materials ratio (Figure 3-17 top). Type V cement, like other portland cements, is not resistant to acids and other highly corrosive substances.

ASTM C150 (AASHTO M 85) allows both a chemical approach (for example, limiting C_3A) and a physical approach (ASTM C452 expansion test) to assure the sulfate resistance of Type V cement. Either the chemical or the physical approach can be specified, but not both.

Air-Entraining Portland Cements

Specifications for four types of air-entraining portland cement (Types IA, IIA, II (MH)A, and IIIA) are given in ASTM C150 and AASHTO M 85. They correspond in composition to ASTM Types I, II, II(MH), and III, respectively, except that small quantities of air-entraining additions are interground with the clinker during manufacture. These cements produce concrete with improved resistance to freezing and thawing. When mixed with proper intensity and for an appropriate duration, such concrete contains minute, well-distributed, and completely separated air bubbles. Air entrainment for most concretes is achieved through the use of an air-entraining admixture, rather than through the use of air-entraining cements. Air-entraining cements are available only in certain areas, so it is advisable to check on local availability.

White Portland Cements

White portland cement is a portland cement that differs from gray cement chiefly in color. It is made to conform to the specifications of ASTM C150, usually Type I or Type III. The manufacture of white cement is accomplished by limiting the amount of iron and magnesium oxides in the raw materials. These two oxides are responsible for portland cement's characteristic gray color. White portland cement is used primarily for architectural purposes in structural walls, precast and glass fiber reinforced concrete (GFRC) facing panels, terrazzo surfaces, stucco, cement paint, tile grout, and decorative concrete (Figure 3-20). Its use is recommended wherever white or colored concrete, grout, or mortar is desired and should be specified as white portland cement meeting the specifications of ASTM C150, Type [I, II, III, or V]. White cement is also used to manufacture white masonry cement meeting ASTM C91 and white plastic cement meeting ASTM C1328 (PCA 1999). White cement was first manufactured in the United States in York, Pennsylvania in 1907. See Farny (2001) for more information on white cement.

Figure 3-19. High early strength cements are used where early concrete strength is needed, such as in (left to right) cold weather concreting, fast track paving to minimize traffic congestion, and rapid form removal for precast concrete.

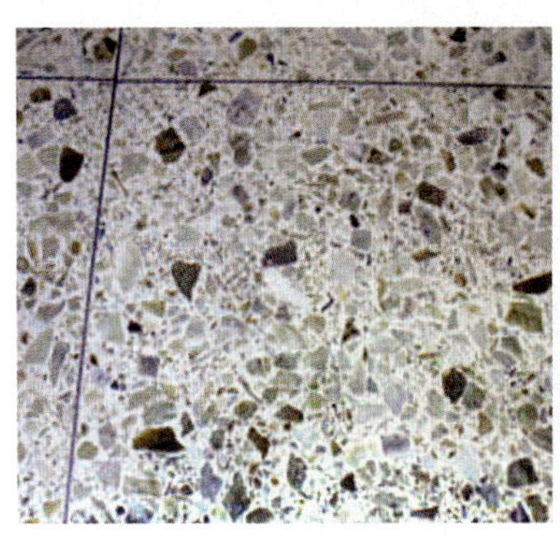

Figure 3-20. White portland cement is used in white or light-colored architectural concrete, ranging from (left to right) terrazzo for floors shown here with white cement and green granite aggregate , to decorative and structural precast and cast-in-place elements, to building exteriors. The far right photograph shows a white precast concrete building housing the ASTM Headquarters in West Conshohocken, Pennsylvania (Courtesy of ASTM).

Blended Hydraulic Cements

Blended hydraulic cements are produced by intimately and uniformly intergrinding or blending two or more types of fine materials. The primary materials are portland cement, slag cement, fly ash, silica fume, calcined clay, other pozzolans, hydrated lime, and preblended combinations of these materials (Figure 3-21). Blended hydraulic cements must conform to the requirements of ASTM C595 or AASHTO M 240, *Specification for Blended Hydraulic Cements* (also see **Performance Based Hydraulic Cements** for blended cements meeting ASTM C1157).

AASHTO M 240 also uses class designations for blended cement. The requirements of M 240 are nearly identical to those in ASTM C595.

ASTM C595 recognizes four primary classes of blended cements as follows:

Type IS (X)	Portland blast-furnace slag cement
Type IP (X)	Portland-pozzolan cement
Type IL (X)	Portland-limestone cement
Type IT(AX)(BY)	Ternary blended cement

The letters "X" and "Y" represent the nominal mass percentage of the ingredient included in the blended cement with the remaining mass percentage being portland cement. For example, a cement designated as Type IS(50) contains 50% by mass of slag cement (and 50% portland cement). For ternary blended cements, "A" is the type of ingredient present in the largest amount ("X") while "B" is the other ingredient type. "A" or "B" may be either S for slag cement, P for pozzolan, or L for limestone. "X" and "Y" are the corresponding percentages of those materials. As an example, Type IT(L10)(P10) indicates a ternary blended cement with 10% limestone and 10% pozzolan. In this example, portland cement accounts for 80% of the total cementitious content.

Types IS(<70), IP, and IT(S<70) are general purpose cements; these and their associated subcategory types are reviewed in the following discussion.

Figure 3-21. Blended cements use a combination of portland cement or clinker and gypsum blended or interground with pozzolans, slag, or fly ash. Shown is blended cement (center) surrounded by (right and clockwise) clinker, gypsum, portland cement, fly ash, slag, silica fume, and calcined clay.

Blended cements are used in all aspects of concrete construction in the same manner as portland cements. Blended cements can be used as the sole cementitious material in concrete or they can be used in combination with other supplementary cementitious materials added at the concrete plant.

A detailed review of ASTM C595 and AASHTO M 240 blended cements follows:

Type IS

Portland blast-furnace slag cement is manufactured by either: (1) intergrinding portland cement clinker and granulated blast-furnace slag (or slag cement), (2) blending portland cement and slag cement, or (3) a combination of intergrinding and blending. Portland blast-furnace slag is classified into two categories depending on its slag cement content: Type IS(< 70) has less than 70% by mass of slag cement and is suitable for general concrete construction, while Type IS(≥ 70) has 70% slag cement or more by mass, may contain hydrated lime, and is used in conjunction with portland cement in making concrete or with lime in making mortar, but is not used alone for structural concrete applications. Both types of Type IS cement include subcategories for optional special properties such as air-entrainment, moderate sulfate resistance, high sulfate resistance, or moderate heat of hydration. They may be specified by adding the suffixes A, MS, HS, or MH, respectively. For example, an air-entraining portland blast-furnace slag cement with 40% slag that has high sulfate resistance would be designated as Type IS(40)-A (HS). For more information on blended cements with slag cement, see Klieger and Isberner (1967) and PCA (1995).

Type IP

Portland-pozzolan cements are designated as Type IP. Type IP cement may be used for general construction. These cements are manufactured by intergrinding portland cement clinker with a suitable pozzolan, by blending portland cement and a pozzolan, or by a combination of intergrinding and blending. The pozzolan content of these cements is up to 40% by mass. Laboratory tests indicate that performance of concrete made with Type IP cement as a group is similar to that of Type I cement concrete. Type IP may be designated as air-entraining, moderate sulfate resistant, high sulfate resistant, low heat of hydration, or with moderate heat of hydration by adding the suffixes A, MS, HS, LH, or MH.

Type IL

Portland-limestone cements are blended cements which contain more than 5% but less than or equal to 15% by mass finely ground limestone as an ingredient. These cement generally perform similarly to Type I cements and typically offer some sustainability advantages. Type IL cements may be designated as air-entraining, moderate heat of hydration, or low heat of hydration by adding suffixes A, MH, LH respectively. Research is on-going with respect to sulfate resistance of portland-limestone cements. For more information on portland-limestone cements, see Tennis, Thomas and Weiss (2011).

Type IT

Ternary blended cements can sometimes offer advantages over binary blended cements. The combinations of different cementitious materials provided by ternary blended cements may optimize the overall performance of these cements for a specific application. For example, a rapidly reacting pozzolan like silica fume might be combined with a slower reacting material such as fly ash to enhance both early and longer term strength development. Other concrete properties might be improved as well, such as reduced permeability and resistance to ASR.

Type IT ternary blended cements meet the same chemical and physical requirements as for binary blended cements for the non-portland cement ingredient present in the largest amount. For instance, if slag cement is present in the highest amount in a Type IT cement, the provisions of Type IS apply, while if a pozzolan is present in the highest amount, then the requirements of Type IP apply. The provisions of Type IP also apply if the pozzolan and slag cement content are the same.

Performance Based Hydraulic Cements

All portland and blended cements are hydraulic cements. "Hydraulic cement" is merely a broader term. (See also ASTM C219 for terms relating to hydraulic cements.) ASTM C1157 can apply to both portland cement and blended hydraulic cements. Cements meeting the requirements of ASTM C1157 meet physical performance test requirements, as opposed to prescriptive restrictions on ingredients or cement chemistry as found in other cement specifications. ASTM C1157 provides for six types of hydraulic cement as follows:

Type GU General use
Type HE High early strength
Type MS Moderate sulfate resistance
Type HS High sulfate resistance
Type MH Moderate heat of hydration
Type LH Low heat of hydration

In addition, these cements can also include an Option R – Low Reactivity with Alkali-Reactive Aggregates – specified to help control alkali-silica reactivity. An "R" is appended to the cement type: for example, Type GU-R is a general use hydraulic cement that has low reactivity when used with alkali-reactive aggregates.

A detailed review of ASTM C1157 cements follows:

Type GU

Type GU is a general purpose cement suitable for all applications where the special properties of other types are not required. Its uses in concrete include pavements, floors, reinforced concrete buildings, bridges, pipe, precast concrete products, and other applications where Type I is used.

Type HE

Type HE cement provides higher strengths at an early age, usually a week or less. It is used in the same manner as Type III portland cement, and has the same 1- and 3-day strength requirements.

Type MS

Type MS cement is used where precaution against moderate sulfate attack is important. Applications may include drainage structures where sulfate concentrations in ground waters are higher than normal but not unusually severe (see Table 3-2). It is used in the same manner as Type II and II(MH) portland cement. Like Type II and II(MH), Type MS cement concrete must be designed with a low water-cementitious materials ratio to provide adequate sulfate resistance. ASTM C1157 also requires that mortar bar specimens demonstrate resistance to expansion while exposed to solution containing a high concentration of sulfates (ASTM C1012).

Type HS

Type HS cement is used in concrete exposed to severe sulfate action – principally where soils or ground waters have a high sulfate content (see Table 3-2). It is used in the same manner as Type V portland cement. As noted previously, low water-cementitious materials ratios are critically important to assure performance of concrete exposed to sulfate exposures. Similar to MS cements, ASTM C1157 also requires that mortar bar specimens demonstrate resistance to expansion while exposed to solution containing a high concentration of sulfates (ASTM C1012).

Type MH

Type MH cement is used in applications requiring a moderate heat of hydration and a controlled temperature rise. Type MH cement is used in the same manner as Type II(MH) portland cement.

Type LH

Type LH cement is used where the rate and amount of heat generated from hydration must be minimized. It develops strength at a slower rate than other cement types. Type LH cement is intended for use in massive concrete structures where the temperature rise resulting from heat generated during hardening must be minimized. It is used as an alternative to Type IV portland cement or in the same manner as a Type IP(LH) or IT(P≥S)(LH) blended cement.

Table 3-3 provides a matrix of commonly used cements and their typical applications in concrete construction.

Special Cements

Special cements are produced for particular applications. Table 3-4 summarizes the special cements discussed below. See Odler (2000) and Klemm (1998) for more information.

Masonry and Mortar Cements

Masonry cements and mortar cements are hydraulic cements designed for use in mortar for masonry construction (Figure 3-22). They consist of a mixture of portland cement or blended hydraulic cement and plasticizing materials (such as limestone or hydrated or hydraulic lime), together with other materials introduced to enhance one or more properties such as setting time, workability, water retention, and durability. These components are proportioned and packaged at a cement plant under controlled conditions to assure uniformity of performance.

Figure 3-22. Masonry cement and mortar cement are used to make mortar to bond masonry units together.

Masonry cements meet the requirements of ASTM C91, *Standard Specification for Masonry Cement*. ASTM C91 classifies masonry cements as Type N, Type S, and Type M. White masonry cement and colored masonry cements meeting ASTM C91 are also available in some areas. Mortar cements meet the requirements of ASTM C1329, *Standard Specification for Mortar Cement*. Mortar cements are further classified as Type N, Type S, and Type M. A brief description of each type follows:

Type N masonry cement and Type N mortar cement are used to produce ASTM C270 Type N and Type O mortars. They may also be used with portland or blended cements to produce Type S and Type M mortars.

Type S masonry cement and Type S mortar cement are used to produce ASTM C270, *Standard Specification for*

Table 3-3. Applications for Hydraulic Cements Used in Concrete Construction*

Cement specification	General purpose	Moderate heat of hydration	High early strength	Low heat of hydration	Moderate sulfate resistance	High sulfate resistance	Resistance to alkali-silica reaction (ASR)**
ASTM C150 portland cements	I	II(MH)	III	IV	II, II(MH)	V	Low-alkali option
ASTM C595 blended hydraulic cements	IP, IS(<70), IL, IT(P≥S), IT(P>L), IT(L≥S), IT(L≥P), IT(P<S<70), IT(L<S<70)	IP(MH), IS(<70)(MH), IL(MH), IT(P≥S)(MH), IT(P>L)(MH), IT(L≥S)(MH), IT(L≥P)(MH), IT(P<S<70)(MH), IT(L<S<70)(MH)	—	IP(LH), IS(<70)(LH), IL(LH), IT(P≥S)(LH), IT(P>L)(LH), IT(L≥S)(LH), IT(L≥P)(LH), IT(P<S<70)(LH), IT(L<S<70)(LH)	IP(MS) IS(<70)(MS) IT(P<S<70)(MS) IT(P≥S)(MS)	IP(HS) IS(<70)(HS) IT(P<S<70)(HS) IT(P≥S)(HS)	Low reactivity option
ASTM C1157 hydraulic cements	GU	MH	HE	LH	MS	HS	Option R

*Check the local availability of specific cements as all cements are not available everywhere.
**The option for low reactivity with ASR-susceptible aggregates can be applied to any cement type in the columns to the left.

Table 3-4. Applications for Special Cements

Special cements	Type	Application
White portland cements, ASTM C150	I, II, III, V	White or colored concrete, masonry, mortar, grout, plaster, and stucco
White masonry cements, ASTM C91	M, S, N	White mortar between masonry units
Masonry cements, ASTM C91	M, S, N	Mortar between masonry units*, plaster, and stucco**
Mortar cements, ASTM C1329	M, S, N	Mortar between masonry units*
Plastic cements, ASTM C1328	M, S	Plaster and stucco**
Expansive cements, ASTM C845	E-1(K), E-1(M), E-1(S)	Shrinkage compensating concrete
Oil-well cements, API-10A	A, B, C, D, E, F, G, H	Grouting wells
Water-repellent cements		Tile grout, paint, and stucco finish coats
Regulated-set cements		Early strength and repair***
Cements with functional additions, ASTM C595 (AASHTO M 240), ASTM C1157		General concrete construction needing special characteristics such as: water-reducing, retarding, air entraining, set control, and accelerating properties
Finely ground (ultrafine) cement		Geotechnical grouting***
Calcium aluminate cement		Repair, chemical resistance, high temperature exposures
Magnesium phosphate cement		Repair and chemical resistance
Geopolymer cement		General construction, repair, waste stabilization***
Ettringite cements		Waste stabilization***
Sulfur cements		Repair and chemical resistance
Rapid hardening hydraulic cement, ASTM C1600	URH, VRH, MRH, GRH	General paving where very rapid (about 4 hours) strength development is required
Natural cement, ASTM C10		Historic restoration of natural cement mortar, cement plaster, grout, whitewash, and concrete

* Portland cement Types I, II, and III and blended cement Types IS and IP are also used in making mortar.
** Portland cement Types I, II, and III and blended cement Types IP and IS(≤70) are also used in making plaster.
*** Portland and blended hydraulic cements are also used for these applications.

Mortar for Unit Masonry, Type S mortar. They may also be used with portland or blended cements to produce Type M mortar.

Type M masonry cement and Type M mortar cement are used to produce ASTM C270 Type M mortar without the addition of other cements or hydrated lime.

Types N, S, and M generally have increasing levels of portland cement and higher strength with Type M providing the highest strength. Type N masonry and mortar cements are most commonly used for above grade masonry.

The increased use of masonry in demanding structural applications, such as high seismic areas, resulted in the relatively recent development of mortar cement. Mortar cement is similar to masonry cement in that it is a factory-prepared cement primarily used to produce masonry mortar with higher strength and lower air content. ASTM C1329 places lower maximum air content limits on mortar cement than the limits for masonry cements, Also, ASTM C1329 is the only ASTM masonry material specification that includes bond strength performance criteria.

The workability, strength, and color of masonry cements and mortar cements remain at a high level of uniformity through the use of modern manufacturing controls. In addition to mortar for masonry construction, masonry cements and mortar cements are commonly used for parging. Masonry cements are also used in portland-cement based plaster or stucco (Figure 3-20) construction (see ASTM C926, *Standard Specification for Application of Portland Cement-Based Plaster*). Masonry cement and mortar cement are not suitable for use in concrete production.

Plastic Cements

Plastic cement is a hydraulic cement that meets the requirements of ASTM C1328, *Standard Specification for Plastic (Stucco) Cement*. It is used to produce portland cement-based plaster or stucco (ASTM C926). These cements are popular throughout the southwest and west coast of the United States (Figure 3-23). Plastic cements consist of a mixture of portland and blended hydraulic cement and plasticizing materials (such as limestone, hydrated or hydraulic lime), together with materials introduced to enhance one or more properties such as setting time, workability, water retention, and durability.

Figure 3-23. Masonry cement and plastic cement are used to make plaster or stucco for commercial, institutional, and residential buildings. Shown are a church and home with stucco exteriors. Inset shows a typical stucco texture.

ASTM C1328 defines separate requirements for Type M and Type S plastic cement with Type M having higher strength requirements. The International Building Code (IBC) does not classify plastic cement into different types, but does specify that it must meet the requirements of ASTM C1328 plastic cement. When plastic cement is used, lime or any other plasticizer may not be added to the plaster at the time of mixing.

The term "plastic" in plastic cement does not refer to the inclusion of any organic compounds in the cement. Instead, "plastic" refers to the ability of the cement to introduce a higher degree of workability ("plasticity") to the plaster. Plaster using this cement must remain workable long enough for it to be reworked to obtain the desired densification and texture. Plastic cement should not be used to make concrete. For more information on the use of plastic cement and plaster, see Melander, Farny, and Isberner (2003).

Finely-Ground Cements (Ultrafine Cements)

Finely-ground cements, also referred to as ultrafine cements, are hydraulic cements that are ground very fine for use in grouting into fine soil or thin rock fissures (Figure 3-24). The cement particles are less than 10 µm in diameter with 50% of particles less than 5 µm. Blaine surface area of these cements often exceeds 800 m^2/kg. These very fine cements consist of portland cement, slag cement, and other mineral additives.

Expansive Cements

Expansive cement is a hydraulic cement that expands slightly during the early hardening period after initial set. It must meet the requirements of ASTM C845, *Standard Specification for Expansive Hydraulic Cement*, where it is designated as Type E-1. Currently, three varieties of expansive cement are recognized. They are designated as K, M, and S. These designations are added as a suffix to the type. Type E-1(K) contains portland cement, tetracalcium trialuminosulfate, calcium sulfate, and uncombined calcium oxide (lime). Type E-1(M) contains portland cement, calcium aluminate cement, and calcium sulfate. Type E-1(S) contains portland cement with a high tricalcium aluminate content and calcium sulfate. Type E-1(K) is the most readily available expansive cement in North America.

Expansive cement may also contain formulations other than those noted. The expansive properties of each type can be varied over a considerable range.

When expansion is restrained, for example by reinforcing steel, expansive cement concrete (also called shrinkage compensating concrete) can be used to: (1) compensate for the volume decrease associated with drying shrinkage, (2) induce tensile stress in reinforcement (post-tensioning), and (3) stabilize the long-term dimensions of post-tensioned concrete structures in comparison to the original design dimensions.

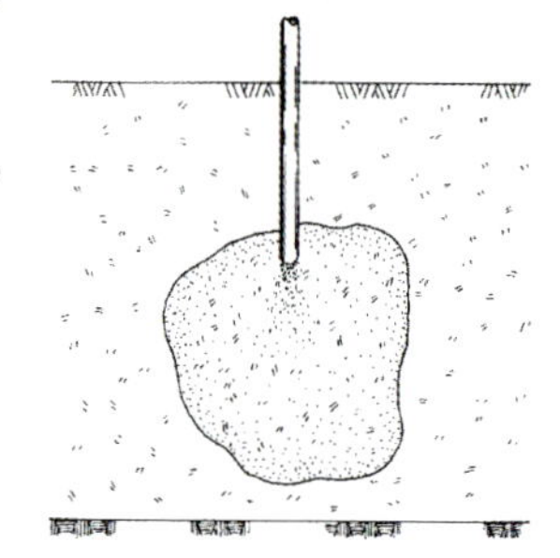

Figure 3-24. (top) A slurry of finely ground cement and water can be injected into the ground, as shown here, to stabilize in-place materials, to provide strength for foundations, or to chemically retain contaminants in soil. Illustration (right) of grout penetration in soil.

One of the major advantages of expansive cement in concrete as noted above, is the compensation for volume changes caused by drying shrinkage. Expansive cement can control and reduce drying shrinkage cracks commonly associated with the use of portland cement. Figure 3-25 illustrates the length change (early expansion and drying shrinkage) history of shrinkage-compensating concrete and conventional portland cement concrete. For more information see Pfeifer and Perenchio (1973), Russell (1978), and ACI Committee 223 (1998).

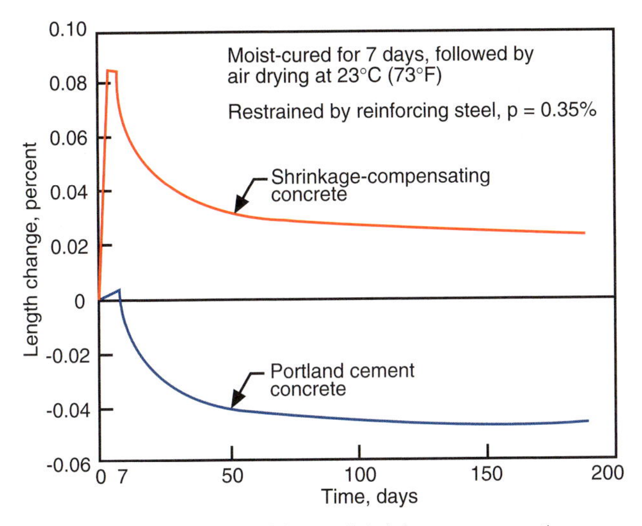

Figure 3-25. Length-change history of shrinkage compensating concrete containing Type E-1(S) cement and Type I portland cement concrete (Pfeifer and Perenchio 1973).

Oil-Well Cements

Oil-well cements are used for oil-well grouting. (This procedure is often called oil-well cementing). Oil-well cements are usually made from portland cement clinker or from blended hydraulic cements. Generally they must be slow-setting and resistant to high temperatures and pressures. *The American Petroleum Institute's Specification for Cements and Materials for Well Cementing* (API Specification 10A) includes requirements for eight classes of well cements (Classes A through H) and three grades (Grades O – ordinary, MSR – moderate sulfate resistant, and HSR – high sulfate resistant). Each class is applicable for use at a certain range of well depths, temperatures, pressures, and sulfate environments. The petroleum industry also uses conventional types of portland cement with suitable cement-modifiers. (ASTM C150 Type II or Type V cements may meet requirements for API Class G and H cements, while API Class A, Class B, and Class C cements are similar to ASTM Types I, II, and III respectively.) Expansive cements have also performed adequately as well cements.

Rapid Hardening Cements

Rapid hardening, high-early strength, hydraulic cement is used in construction applications, such as fast-track paving, where fast strength development is required (design or load-carrying strength in about four hours). They are classified according to ASTM C1600, *Standard Specification for Rapid Hardening Hydraulic Cement*, as Type URH – ultra rapid hardening, Type VRH – very rapid hardening, Type MRH – medium rapid hardening, and Type GRH – general rapid hardening.

Cements with Functional Additions

Functional additions can be interground with cement clinker to beneficially change the properties of hydraulic cement. These additions must meet the requirements of ASTM C226, *Standard Specification for Air-Entraining Additions for Use in the Manufacture of Air-Entraining Hydraulic Cement*, or ASTMC688, *Standard Specification for Functional Additions for Use in Hydraulic Cements*. ASTM C226 addresses air-entraining additions while ASTM C688 addresses the following types of additions: water-reducing, retarding, accelerating, water-reducing and retarding, water-reducing and accelerating, and set-control additions. Cement specifications ASTM C595 (AASHTO M 240) and C1157 allow functional additions. These additions can be used to enhance the performance of the cement for normal or special concrete construction, grouting, and other applications.

Water-Repellent Cements

Water-repellent cements, sometimes called waterproofed cements, are usually made by adding a small amount of water-repellent additive such as stearate (sodium, aluminum, or other) to cement clinker during final grinding (Lea 1971). Manufactured in either white or gray color, water-repellent cements reduce capillary water transmission provided there is little to no hydrostatic pressure. However, they do not stop water-vapor transmission. Water-repellent cements are used in tile grouts, paint, stucco finish coats, and in the manufacture of specialty precast units.

Regulated-Set Cements

Regulated-set cement is a calcium fluoroaluminate hydraulic cement that can be formulated and controlled to produce concrete with setting times ranging from a few minutes to one hour and with corresponding rapid early strength development (Greening and others 1971). It is a portland-based cement with functional additions that can be manufactured in the same kiln used to manufacture conventional portland cement. Regulated-set cement incorporates set control and early-strength-development components. The final physical properties of the resulting concrete are in most respects similar to comparable concretes made with portland cement.

Geopolymer Cements

Geopolymer cements are inorganic hydraulic cements that are based on the polymerization of minerals (Davidovits, Davidovits, and James 1999 and www.geopolymer.org). The term more specifically refers to alkali-activated alumino-silicate cements, also called zeolitic or polysialate cements. These cements often contain industrial by-products, such as fly ash. They have been used in general construction, high-early strength applications, and waste stabilization. These cements do not contain organic polymers or plastics.

Ettringite Cements

Ettringite cements are calcium sulfoaluminate cements that are specially formulated for particular uses, such as the stabilization of waste materials (Klemm 1998). They can be formulated to form large amounts of ettringite to stabilize particular metallic ions within the ettringite structure. Ettringite cements have also been used in rapid setting applications, including use in coal mines. Also see "Expansive Cements" above.

Calcium Aluminate Cements

Calcium aluminate cement is not portland cement based. It is used in special applications for early strength gain (for example, to achieve design strength in one day), resistance to high temperatures, and resistance to sulfates, weak acids, and seawater. Portland cement and calcium aluminate cement combinations have been used to make rapid setting concretes and mortars. Typical applications for calcium aluminate cement concrete include: chemically resistant, heat resistant, and corrosion resistant industrial floors; refractory castables; and repair applications. Standards addressing these cements include British Standard BS 915-2 and French Standard NF P15-315.

Calcium aluminate cement concrete must be used at low water-cement ratios (less than 0.40); this minimizes the potential for conversion of less stable hexagonal calcium aluminate hydrate (C_AH_{10}) to the stable cubic tricalcium aluminate hydrate (C_3AH_6), hydrous alumina (AH_3), and water. Given sufficient time and particular moisture conditions and temperatures, this conversion causes a 53% decrease in volume of hydrated material. However, this internal volume change occurs without a dramatic alteration of the overall dimensions of a concrete element, resulting in increased paste porosity and decreased compressive strength. At low water-cement ratios, there is insufficient space for all the calcium aluminate to react and form C_AH_{10}. The released water produced by the conversion reacts with additional calcium aluminate, partially compensating for the effects of conversion. Concrete design strength must therefore be based on the converted strength. Because of the unique concerns associated with the conversion phenomenon, calcium aluminate cement is often used in nonstructural applications and is rarely, if ever used within structural applications (Taylor 1997).

Magnesium Phosphate Cements

Magnesium phosphate cement is a rapid setting, early strength gain cement. It is usually used for special applications, such as repair of pavements and concrete structures, or for resistance to certain aggressive chemicals. It does not contain portland cement.

Sulfur Cements

Sulfur cement is used with conventional aggregates to make sulfur cement concrete for repairs and chemically resistant applications. Sulfur cement melts at temperatures between 113°C and 121°C (235°F and 250°F). Sulfur concrete is maintained at temperatures around 130°C (270°F) during mixing and placing. The material gains strength quickly as it cools and is resistant to acids and aggressive chemicals. Sulfur cement does not contain portland or hydraulic cement.

Natural Cements

Natural cements were used in the 1800s before portland cement became popular. It is a coarse ground hydraulic cement with slow strength gain properties and was used in mortar, plaster, whitewashing, and concrete. It was used extensively in canal, bridge, and building construction. It is still available for use in restoring historic structures and must meet ASTM C10, *Standard Specification for Natural Cement.*

Other Cements

Magnesium silicates, fly ash cement, special alkali-activated cements, and countless other cements with unique chemistry exist throughout the world. It is beyond the scope of this document to address them. For more information on special cements consult Odler (2000) and other references.

Selecting and Specifying Cements

When specifying cements for a project, the availability of cement types should be verified. Specifications should allow flexibility in cement selection. Limiting a project to only one cement type, one brand, or one standard cement specification can result in project delays and it may not allow for the best use of local materials. Cements with special properties should not be required unless special characteristics are necessary. In addition, the use of SCMs should not inhibit the use of any particular portland or blended cement. The project specifications should focus on the needs of the concrete structure and allow use of a variety of materials to accomplish those needs. A typical specification may call for portland cements meeting ASTM C150 (AASHTO M 85), blended cements meeting ASTM C595 (AASHTO M 240), or for performance based hydraulic cements meeting ASTM C1157.

If no special properties, (such as low-heat generation or sulfate resistance) are required, all general use cements should be permitted, including: Types I, GU, IS(<70), IP, IL, and IT(S<70). Also, it should be noted that some cement types meet more than one specification requirement. For example, nearly all Type II cements meet the requirements of Type I, but not all Type I cements meet the requirements of Type II. See Tables 3-3 and 3-4 for guidance on using different cements.

Availability of Cements

Some types of cement may not be readily available in all areas of the United States.

ASTM C150 (AASHTO M 85) Type I portland cement is usually carried in stock and is furnished when no other type of cement is specified. Type II cement is usually available, especially in areas of the country where moderate sulfate resistance is needed. Cement Types I and II represent about 80% of the cement shipped from plants in the United States. Type III cement and white cement are usually available in larger metropolitan areas and represent about 4% of cement produced. Type IV cement is manufactured only when specified for particular projects (massive structures like dams) and therefore is usually not readily available. Type V cement is available only in regions of the country having high sulfate environments. Air-entraining cements are sometimes difficult to obtain. Their use has decreased as the popularity of air-entraining admixtures has increased. Masonry cement is available in most areas. See PCA (2010) for statistics on cement use.

If a given cement type is not available, comparable results can frequently be obtained with another cement type that is available. For example, high-early-strength concrete can be made using concrete mixtures with higher cement contents and lower water-cement ratios. Also, the effects of heat of hydration can be minimized using leaner mixtures, smaller placing lifts, artificial cooling, or by adding an SCM to the concrete. Type V cement may at times be difficult to obtain, but because of the range of phase composition permitted in cement specifications, a Type II cement may also meet the requirements of a Type V.

Blended cements can be obtained throughout most of the United States. However, certain types of blended cement may not be available in some areas. When ASTM C595 (AASHTO M 240) blended cements are required, but are not available, similar properties may be obtained by adding pozzolans meeting requirements of ASTM C618, *Standard Specification for Coal Fly Ash and Raw or Calcined Natural Pozzolan for Use in Concrete* (AASHTO M 295) or slag cement meeting requirements of ASTM C989, *Standard Specification for Slag Cement for Use in Concrete and Mortars* (AASHTO M 302) to the concrete at a ready mix plant using normal portland cement. Like any concrete mixture, these mixes should be tested for time of set, strength gain, durability, and other properties prior to their use in construction.

Drinking Water Applications

Concrete has demonstrated decades of safe use in drinking water applications. Materials in contact with drinking water must meet special requirements to control elements entering the water supply. Some localities may require that cement and concrete meet the special requirements of the American National Standards Institute/ National Sanitation Foundation standard ANSI/NSF 61, Drinking Water System Components – Health Effects.

Canadian and European Cement Specifications

In some instances, projects in the United States designed by engineering firms from other countries refer to cement standards other than those in ASTM or AASHTO. For example, the European cement standard, EN 197-1, sometimes appears on project specifications. EN 197 cement Types CEM I, II, III, IV, and V do not correspond to the cement types in ASTM C150, nor can ASTM cements be substituted for EN specified cement without the designer's approval. CEM I is a portland cement and CEM II through V are blended cements. EN 197 also includes strength classes and ranges (32.5, 42.5, and 52.5 MPa).

Some countries outside North America use ASTM standards to specify cement and concrete. However, for other countries, there is typically no direct equivalency between ASTM and international cement standards because of differences in test methods and limits on required properties. EN 197 cements are usually not available in the United States; therefore, the best approach is to inform the designer about locally available cements and request changes in the project specifications that allow use of an ASTM or AASHTO cement.

In Canada, portland cements are manufactured to meet the specifications of the Canadian Standards Association (CSA) Standard A3001 *Cementitious Materials for Use in Concrete*. CSA A3001 includes portland cements, blended cements, portland-limestone cements, and portland-limestone blended cements.

CSA A3001 uses two-letter descriptive type designations for portland cements similar to ASTM C1157 with requirements similar to ASTM C150. Six types of portland cement are covered in CSA A3001 and are identified as follows:

GU: General use hydraulic cement
MS: Moderate sulfate-resistant hydraulic cement
MH: Moderate heat of hydration hydraulic cement
HE: High early-strength hydraulic cement
LH: Low heat of hydration hydraulic cement
HS: High sulfate-resistant hydraulic cement

The nomenclature for blended hydraulic cements is a three-letter descriptive designation (GUb, MSb, MHb, HEb, LHb, and HSb) to address its equivalent performance to portland cements with up to three supplementary cementing materials. Upon request, the designations for blended cements can also provide information on the composition of blended hydraulic cements. The designations then follow the form: BHb-Axx/Byy/Czz, where BHb is the blended hydraulic cement type, xx, yy, and zz are the supplementary materials used in the cement in proportions A, B, and C.

Supplementary cementitious materials include:

Ground granulated blast furnace slag (S),

Silica fume (SF),

Natural pozzolans (N), and

Fly ash (Classes F, CI, and CH). Class F, CI, and CH fly ashes are low (less than 15% CaO by mass), medium (between 15% and 20% CaO by mass), and high calcium oxide (more than 20% CaO by mass) contents.

Examples:

MS – portland cement (with no supplementary cementitious materials) for use when moderate sulfate resistance is required.

GUb-30F/5SF – general use blended cement containing 30% by mass Class F fly ash (F) and 5% silica fume (SF).

In addition, CSA A3001 includes provisions for portland-limestone cements (PLC) with 5% to 15% by mass of limestone. Four types of portland limestone cements are defined and are designated by an "L" suffix: Type GUL, Type HEL, Type MHL, and Type LHL. For the limestone used in these cements, requirements are that: 1) the calcium carbonate content is at least 75% by mass, 2) the methylene blue index of the limestone (an indication of clay content) in not more than 1.2 g/100 g, and 3) the Total Organic Carbon (TOC) content of the limestone is below 0.5% by mass.

CSA A3001 includes provisions for portland-limestone blended cements, which contain between 5% and 15% limestone, as well as SCMs. Type designations are GULb, HELb, MHLb, LHLb, MSLb, and HSLb.

Chemical Phases (Compounds) and Hydration of Portland Cement

During the burning operation in the manufacture of portland cement clinker, calcium combines with the other components of the raw mix to form four principal phases that make up about 85% to 90% of cement by mass. Gypsum, or other calcium sulfate forms, limestone, and grinding aids or other processing additions are also added during finish grinding.

Cement chemists use the following chemical shorthand (abbreviations):

$A = Al_2O_3$, $C = CaO$, $F = Fe_2O_3$, $H = H_2O$, $M = MgO$, $S = SiO_2$, and $\bar{S} = SO_3$.

The term "phases" is generally used to describe the components of clinker, because the word "compounds" generally refers to specific, ideal compositions, while the compositions of the phases in cement often include trace elements. Following are the four primary phases in portland cement, their approximate chemical formulas, and abbreviations:

Tricalcium silicate
$3CaO \cdot SiO_2 = C_3S$

Dicalcium silicate
$2CaO \cdot SiO_2 = C_2S$

Tricalcium aluminate
$3CaO \cdot Al_2O_3 = C_3A$

Tetracalcium aluminoferrite
$4CaO \cdot Al_2O_3 \cdot Fe_2O_3 = C_4AF$

The forms of calcium sulfate, their chemical formulas, and abbreviations are:

Anhydrous calcium sulfate (anhydrite)
$CaSO_4 = CaO \cdot SO_3 = C\bar{S}$

Calcium sulfate dihydrate (gypsum)
$CaSO_4 \cdot 2H_2O = CaO \cdot SO_3 \cdot 2H_2O = C\bar{S}H_2$

Calcium sulfate hemihydrate
$CaSO_4 \cdot ½H_2O = CaO \cdot SO_3 \cdot ½H_2O = C\bar{S}H_{½}$

Gypsum, calcium sulfate dihydrate, is the predominant source of sulfate used in cement. Finish grinding is often controlled to produce a balance of hemihydrate and gypsum. The heat generated during finish grinding will dehydrate a portion of the calcium sulfate dehydrate into calcium sulfate hemihydrate.

Primary Phases

C_3S and C_2S in clinker are also referred to as alite and belite, respectively. Alite generally constitutes 50% to 70% of the clinker, whereas belite accounts for only about 10% to 25% (Figure 3-26). Aluminate phases constitute up to about 10% of the clinker and ferrite compounds generally up to as much as 15% (Bhatty and Tennis 2008). These and other phases may be observed and analyzed through the use of microscopy (ASTM C1356, ASTM C1365, and Campbell 1999).

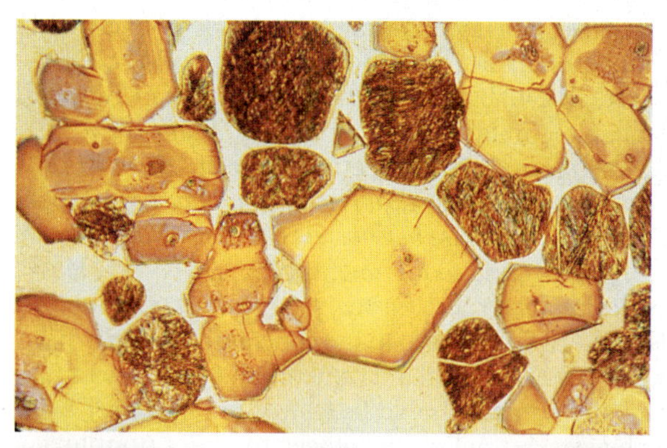

Figure 3-26. (top) Polished thin-section examination of portland clinker shows alite (C_3S) as light, angular crystals. The darker, rounded crystals are belite (C_2S). Magnification approximately 400X. (bottom) Scanning electron microscope (SEM) micrograph of alite (C_3S) crystals in portland clinker. Magnification 3000X.

In the presence of water, these phases hydrate (chemically react with water) to form new solids that are the infrastructure of hardened cement paste in concrete (Figure 3-27). The calcium silicates, C_3S and C_2S, hydrate to form calcium hydroxide and calcium silicate hydrate (archaically called tobermorite gel). Hydrated portland cement paste typically contains 15% to 25% calcium hydroxide and about 50% calcium silicate hydrate by mass. The strength and other properties of hydrated cement are due primarily to calcium silicate hydrate (Figure 3-28). C_3A, sulfates (gypsum, anhydrite, or other sulfate source), and water combine to form ettringite (calcium trisulfoaluminate hydrate), calcium monosulfoaluminate, and other related compounds. C_4AF reacts with water to form calcium aluminoferrite hydrates. These basic reactions are shown in Table 3-5. Brunauer (1957), Copeland and others (1960), Lea (1971), Powers and Brownyard (1947), Powers (1961), and Taylor (1997) addressed the pore structure and chemistry of cement paste. Figure 3-29 shows estimates of the relative volumes of the phases in hydrated portland cement pastes. The Concrete Microscopy Library (Lange and Stutzman 2009) provides a collection of micrographs of cement hydration and concrete.

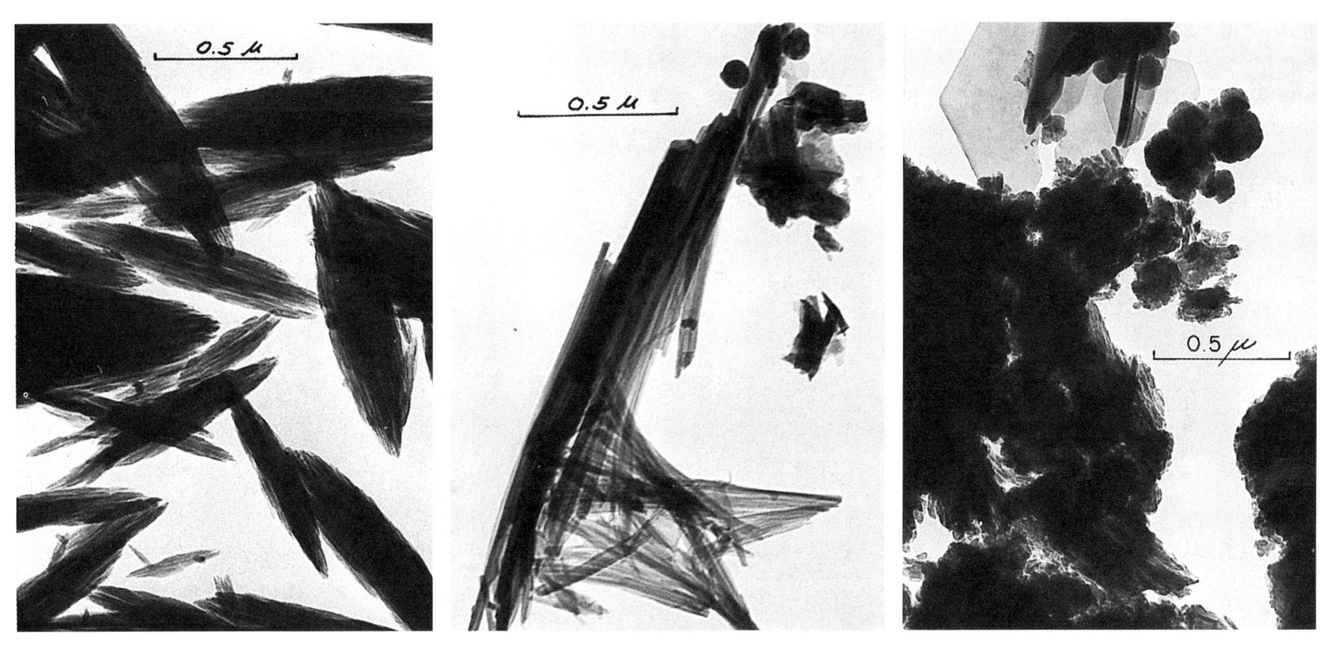

Figure 3-27. Electron micrographs of (left) dicalcium silicate hydrate, (middle) tricalcium silicate hydrate, and (right) hydrated normal portland cement. Note the fibrous nature of the calcium silicate hydrates. Broken fragments of angular calcium hydroxide crystallites are also present (right). The aggregation of fibers and adhesion of the hydration particles are responsible for the strength development of portland cement paste. Reference (left and middle) Brunauer 1962 and (right) Copeland and Schulz 1962.

Table 3-5. Portland Cement Phase Hydration Reactions (Oxide Notation)

$2\ (3CAO \cdot SiO_2)$ Tricalcium silicate	$+\ 11\ H_2O$ Water	$= 3CaO \cdot 2SiO_2 \cdot 8H_2O$ Calcium silicate hydrate (C-S-H)	$+\ 3\ (CaO \cdot H_2O)$ Calcium hydroxide
$2\ (2CaO \cdot SiO_2)$ Dicalcium silicate	$+\ 9\ H_2O$ Water	$= 3CaO \cdot 2SiO_2 \cdot 8H_2O$ Calcaium silicate hydrate (C-S-H)	$+\ CaO \cdot H_2O$ Calcium hydroxide
$3CaO \cdot Al_2O_3$ Tricalcium aluminate	$+\ 3\ (CaO \cdot SO_3 \cdot 2H_2O)$ Gypsum	$+\ 26\ H_2O$ Water	$= 6CaO \cdot Al_2O_3 \cdot 3SO_3 \cdot 32H_2O$ Ettringite
$2\ (3CaO \cdot Al_2O_3)$ Tricalcium aluminate	$+\ 6CaO \cdot Al_2O_3 \cdot 3SO_3 \cdot 32H_2O$ Ettringite	$+\ 4\ H_2O$ Water	$= 3\ (4CaO \cdot Al_2O_3 \cdot SO_3 \cdot 12H_2O)$ Calcium monosulfoaluminate)
$3CaO \cdot Al_2O_3$ Tricalcium aluminate	$+\ CaO \cdot H_2O$ Calcium hydroxide	$+\ 12\ H_2O$ Water	$= 4CaO \cdot Al_2O_3 \cdot 13H_2O$ Tetracalcium aluminate hydrate
$4CaO \cdot Al_2O_3 \cdot Fe_2O_3$ Tetracalcium aluminoferrite	$+\ 10\ H_2O$ Water	$+\ 2\ (CaO \cdot H_2O)$ Calcium hydroxide	$= 6CaO \cdot Al_2O_3 \cdot Fe_2O_3 \cdot 12H_2O$ Calcium aluminoferrite hydrate

Note: This table includes only primary reactions and not several additional minor reactions. The composition of calcium silicate hydrate (C-S-H) is not stoichiometric (Tennis and Jennings 2000).

Figure 3-28. Scanning-electron micrographs of hardened cement paste at (left) 500X, and (right) 1000X.

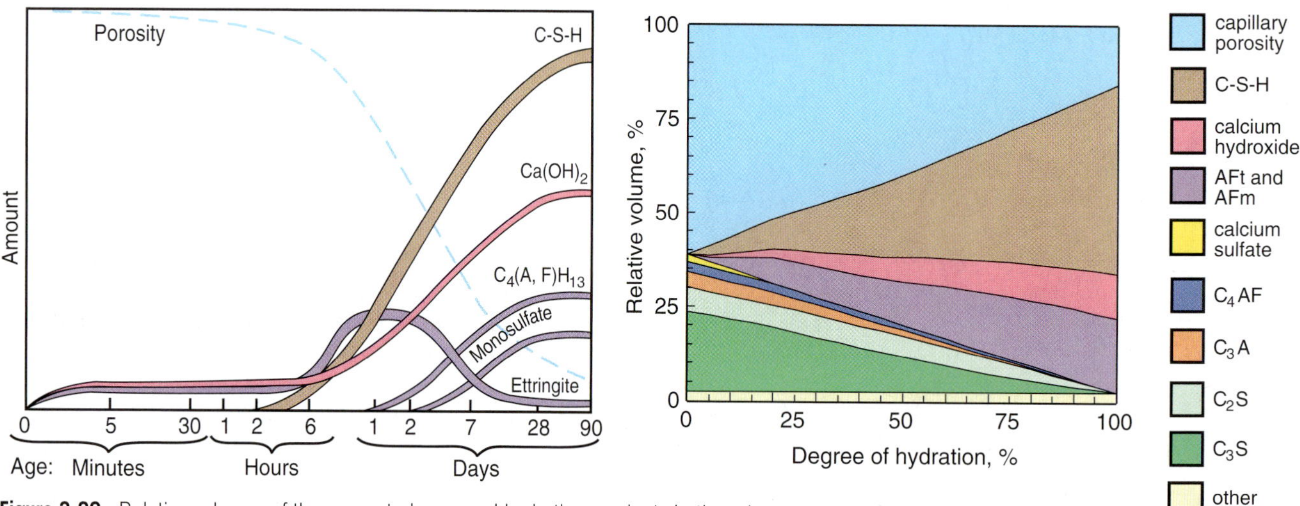

Figure 3-29. Relative volumes of the cement phases and hydration products in the microstructure of hydrating portland cement pastes (left) as a function of time (adapted from Locher, Richartz, and Sprung 1976) and (right) as a function of the degree of hydration as estimated by a computer model for a water to cement ratio of 0.50 (adapted from Tennis and Jennings 2000). Values are given for an average Type I cement composition (Gebhardt 1995): C_3S=55%, C_2S=18%, C_3A=10% and C_4AF=8%. "AFt and AFm" includes ettringite (AFt) and calcium monosulfoaluminate (AFm) and other hydrated calcium aluminate compounds. See Table 3-5 for cement phase reactions.

The percentage of each cement phase can be estimated from a chemical oxide analysis (ASTM C114, *Standard Test Methods for Chemical Analysis of Hydraulic Cement*, or AASHTO T 105) of the unhydrated cement using the Bogue equations, a form of which are provided in ASTM C150 (AASHTO M 85). X-ray diffraction (XRD) techniques (ASTM C1365, *Standard Test Method for Determination of the Proportion of Phases in Portland Cement and Portland-Cement Clinker Using X-Ray Powder Diffraction Analysis*) can generally be used to determine phase composition. Since the production of clinker depends on natural raw materials, there can be a significant level of trace elements in the major phases, which are not accounted for in the Bogue equations, Bogue calculations also assume a "perfect combination" (complete equilibrium) in the high temperature chemical reactions. XRD helps eliminate this bias. Table 3-6 shows typical oxide and phase composition and fineness for each of the principal types of portland cement.

Although elements are reported as simple oxides for consistency, they are usually not found in that oxide form in the cement. For example, sulfur from the gypsum is reported as SO_3 (sulfur trioxide), however, cement does not contain any sulfur trioxide. The amount of calcium, silica, and alumina establish the amounts of the primary phases in the cement and effectively the properties of hydrated cement. Sulfate is present to control setting, as well as drying shrinkage and strength gain (Tang 1992). Minor and trace elements and their effect on cement properties are discussed by Bhatty (1995) and PCA (1992). The primary cement phases have the following properties:

Tricalcium Silicate, C_3S, hydrates and hardens rapidly and is largely responsible for initial set and early strength (Figure 3-30). In general, the early strength of portland cement concrete is higher with increased percentages of C_3S.

Dicalcium Silicate, C_2S, hydrates and hardens slowly and contributes largely to strength increase at ages beyond one week.

Tricalcium Aluminate, C_3A, liberates a large amount of heat during the first few days of hydration and hardening. It also contributes slightly to early strength development. Cements with low percentages of C_3A are more resistant to soils and waters containing sulfates.

Tetracalcium Aluminoferrite, C_4AF, is the product resulting from the use of iron and aluminum raw materials to reduce the clinkering temperature during cement manufacture. It contributes little to strength. Most color effects that give cement its characteristic gray color are due to the presence of C_4AF and its hydrates.

Calcium Sulfate, as anhydrite (anhydrous calcium sulfate), gypsum (calcium sulfate dihydrate), or hemihydrate, (calcium sulfate hemihydrate, often called plaster of paris or bassanite) is added to cement during final grinding to provide sulfate to react with C_3A to form ettringite (calcium trisulfoaluminate). This controls the hydration of C_3A. Without sulfate, cement would set much too rapidly. In addition to controlling setting and early strength gain, the sulfate also helps control drying shrinkage and can influence strength through 28 days (Lerch 1946).

In addition to the above primary phases, portland cement may contain up to 5% limestone and numerous other lesser compounds (PCA 1997 and Taylor 1997).

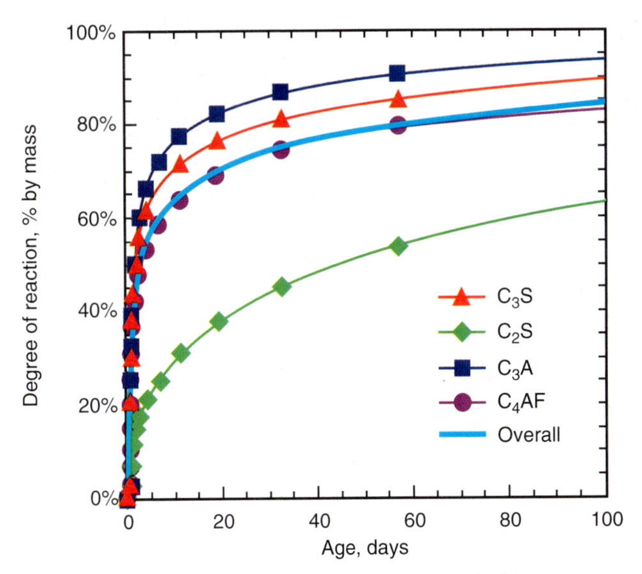

Figure 3-30. Relative reactivity of cement compounds. The curve labeled "Overall" has a composition of 55% C_3S, 18% C_2S, 10% C_3A, and 8% C_4AF, an average Type I cement composition (Tennis and Jennings 2000).

Water (Evaporable and Nonevaporable)

Water is a key ingredient of pastes, mortars, and concretes. The phases in portland cement must chemically react with water to develop strength. The amount of water added to a mixture controls the durability as well. The space initially occupied by water in a cementitious mixture is either partially or completely replaced over time as the hydration reactions proceed (Table 3-5). If more than about 35% water by mass of cement (a water to cement ratio of 0.35) is used,

Table 3-6. Composition and Fineness of Cements*

Type of portland cement	Chemical composition, %							Potential phase composition,%				Blaine fineness m²/kg
	SiO_2	Al_2O_3	Fe_2O_3	CaO	MgO	SO_3	Na_2O eq	C_3S	C_2S	C_3A	C_4AF	
I (min-max)	19.0-21.8	3.9-6.1	2.0-3.6	61.5-65.2	0.8-4.5	2.0-4.4	0.2-1.2	45-65	6-21	6-12	6-11	334-431
I (mean)	**20.2**	**5.1**	**2.7**	**63.2**	**2.5**	**3.3**	**0.7**	**57**	**15**	**9**	**8**	**384**
II (min-max)	20.0-22.5	3.8-5.5	2.6-4.4	61.3-65.6	0.6-4.5	2.1-4.0	0.2-1.2	48-68	8-25	4-8	8-13	305-461
II (mean)	**20.9**	**4.6**	**3.3**	**63.7**	**2.0**	**2.9**	**0.56**	**57**	**17**	**7**	**10**	**377**
III (min-max)	18.6-22.1	3.4-7.3	1.3-4.2	61.6-64.9	0.8-4.3	2.6-4.9	0.1-1.2	48-66	8-27	2-12	4-13	387-711
III (mean)	**20.4**	**4.8**	**2.9**	**63.3**	**2.2**	**3.6**	**0.61**	**56**	**16**	**8**	**9**	**556**
IV** (min-max)	21.5-22.8	3.5-5.3	3.7-5.9	62.0-63.4	1.0-3.8	1.7-2.5	0.29-0.42	37-49	27-36	3-4	11-18	319-362
IV (mean)**	**22.2**	**4.6**	**5.0**	**62.5**	**1.9**	**2.2**	**0.36**	**42**	**32**	**4**	**15**	**340**
V** (min-max)	20.3-22.8	3.3-4.8	3.2-5.8	62.3-65.2	0.8-4.5	2.0-2.8	0.3-0.6	47-64	12-27	0-5	10-18	312-541
V (mean)**	**21.6**	**3.8**	**3.9**	**63.9**	**2.2**	**2.3**	**0.45**	**58**	**18**	**4**	**12**	**389**
White** (min-max)	22.0-24.4	2.2-5.0	0.2-0.6	63.9-68.7	0.3-1.4	2.3-3.1	0.09-0.38	51-72	9-25	5-13	1-2	384-564
White (mean)**	**22.7**	**4.1**	**0.3**	**66.7**	**0.9**	**2.7**	**0.18**	**64**	**18**	**10**	**1**	**482**

*Values represent a summary of combined statistics. Air-entraining cements are not included. For consistency in reporting, elements are reported in a standard oxide form. This does not mean that the oxide form is present in the cement. For example, sulfur is reported as SO_3, sulfur trioxide, but, portland cement does not have sulfur trioxide present. "Potential phase composition" refers to ASTM C150 (AASHTO M 85) calculations using an oxide analysis of the cement. The actual phase composition may differ due to incomplete or alternate chemical reactions.

**Type IV and White Cement data are based on limited data in Gebhardt (1995) and PCA (1996). Type IV is not commonly available. Adapted from Bhatty and Tennis (2008).

then the porosity in the hardened material will remain, even after complete hydration. This phenomenon is called capillary porosity. Figure 3-31 shows that cement pastes with high and low water to cement ratios have equal masses after drying (thus indicating that evaporable water was removed). The cement consumed the same amount of water in both pastes resulting in more bulk volume in the higher water-cement ratio paste. As the water to cement ratio increases, the capillary porosity increases, and the strength decreases. Also, transport properties such as permeability and diffusivity are increased. This degradation in transport properties allows detrimental chemicals to more readily attack the concrete or reinforcing steel.

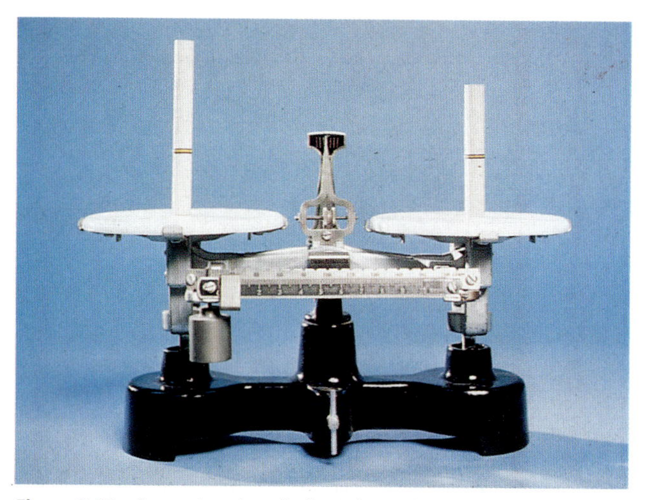

Figure 3-31. Cement paste cylinders of equal mass and equal cement content, but mixed with different water to cement ratios, after all water has evaporated from the cylinders.

Water is observed in cementitious materials in different forms. Free water includes mixing water that has not yet reacted with the cement phases. Bound water is chemically combined in the solid phases or physically bound to the solid surfaces. A reliable separation of the chemically combined water from the physically adsorbed water is not possible. Accordingly, Powers (1949) distinguished between evaporable and nonevaporable water. The nonevaporable water is the amount retained by a sample after it has been subjected to a drying procedure intended to remove all the free water (traditionally, by heating to 105°C [221°F]). Evaporable water was originally considered to be free water, but it is now recognized that some bound water is also lost as the sample is heated. All nonevaporable water is bound water, yet not all bound water is nonevaporable.

For complete hydration of portland cement to occur, only about 40% water (a water-cement ratio of 0.40) is required. If a water-cement ratio greater than about 0.40 is used, the excess water not needed for cement hydration remains in the capillary pores or evaporates. If a concrete mixture has a water-cement ratio less than about 0.40, some cement will remain unhydrated.

To estimate the degree of hydration of a hydrated material, the nonevaporable water content is often used. To convert the measured nonevaporable water into degrees of hydration, it is necessary to know the value of nonevaporable water-to-cement ratio (w_n/c) at complete hydration. This can be experimentally determined by preparing a high water-to-cement ratio cement paste (for example, 1.0) and continuously grinding the paste sample while it hydrates in a roller mill. In this procedure, complete hydration of the cement will typically be achieved after 28 days.

Alternatively, an estimate of the quantity of nonevaporable water-to-cement ratio (w_n/c) at complete hydration can be obtained from the potential Bogue composition of the cement. Nonevaporable water contents for the major phases of portland cement are provided in Table 3-7. For a typical Type I cement, these coefficients will generally result in a calculated w_n/c for completely hydrated cement somewhere between 0.22 and 0.25.

Table 3-7. Nonevaporable Water Contents for Fully Hydrated Major Cement Phases

Hydrated cement compound	Nonevaporable (combined) water content (g water/g cement compound)
C_3S hydrate	0.24
C_2S hydrate	0.21
C_3A hydrate	0.40
C_4AF hydrate	0.37
Free lime (CaO)	0.33

Physical Properties of Cement

Specifications for cement place limits on both its physical properties and often its chemical composition. An understanding of the significance of some of the physical properties is helpful in interpreting results of cement tests (Johansen, Taylor, and Tennis 2006). Tests of the physical properties of the cements should be used to evaluate the properties of the cement, rather than the concrete. Cement specifications limit the properties with respect to the type of cement. Cement should be sampled in accordance with ASTM C183 (AASHTO T 127), *Standard Practice for Sampling and the Amount of Testing of Hydraulic Cement*. During manufacture, cement is continuously monitored for its chemistry and the following properties reported.

Compressive Strength

The strength development characteristics of cement are determined by measuring the compressive strength of 50-mm (2-in.) mortar cubes tested in accordance with ASTM C109 (AASHTO T 106), *Standard Test Method for Compressive Strength of Hydraulic Cement Mortars (Using 2-in. or [50-mm] Cube Specimens* (Figure 3-32).

Compressive strength is influenced by the cement type, specifically: the phase composition, sulfate content, and fineness of the cement. Minimum strength requirements in cement specifications are typically exceeded comfortably.

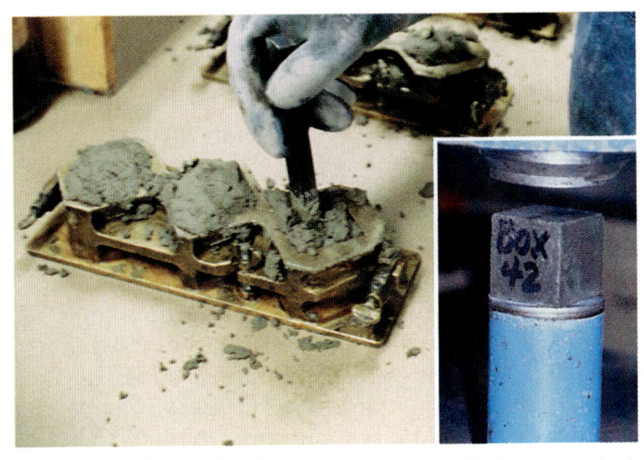

Figure 3-32. 50-mm (2-in.) mortar cubes are cast (left) and crushed (right) to determine strength characteristics of cement.

Table 3-8 shows the minimum compressive strength requirements as given in US specifications. In general, cement strengths (based on mortar-cube tests) cannot be used to predict concrete strengths with any degree of accuracy because of the many variables that influence concrete strength, including: aggregate characteristics, concrete mixture proportions, construction procedures, and environmental conditions in the field (Weaver, Isabelle and Williamson 1974 and DeHayes 1990). Therefore, it should never be assumed that two types of cement meeting the same minimum requirements will produce the same strength of mortar or concrete without modification of mix proportions.

Figures 3-33 and 3-34 illustrate the strength development for standard mortars made with various types of portland cement. Wood (1992) provides long-term strength properties of mortars and concretes made with portland and blended cements. The strength uniformity of a cement from a single source may be determined by following the procedures outlined in ASTM C917, *Standard Test Method for Evaluation of Cement Strength from a Single Source.*

Table 3-8. Minimum Compressive Strength Requirements in US Cement Specifications (MPa*)

Specification	Type	Limits			
		1 day	3 days	7 days	28 days
ASTM C150 (AASHTO M 85)	I	—**	12.0	19.0	28.0***
	II, II(MH)	—	10.0	17.0	28.0***
	III	12.0	24.0	—	—
	IV	—	—	7.0	17.0
	V	—	8.0	15.0	21.0
ASTM C595 (AASHTO M 240)	IP, IS(<70), IL, IT(P≥S), IT(P>L), IT(L≥S), IT(L≥P), IT(P<S<70), IT(L<S<70)	—	13.0	20.0	25.0
	IP(MS), IS(<70)(MS), IT(P<S<70)(MS), IT(P≥S)(MS), IP(HS), IS(<70)(HS) IT(P<S<70)(HS), IT(P≥S)(HS)	—	11.0	18.0	25.0
	IS(≥70), IT(S≥70)	—	—	5.0	11.0
	IP(LH), IL(LH), IT(P≥S)(LH)	—	—	11.0	21.0
ASTM C1157	GU	—	13.0	20.0	28.0
	HE	12.0	24.0	—	—
	MS	—	11.0	18.0	28.0***
	HS	—	11.0	18.0	25.0
	MH	—	5.00	11.0	22.0***
	LH	—	—	11.0	21.0

*For complete details, review the relevant standards. As determined by ASTM C109, which is similar to AASHTO T 106. To convert to psi, multiply by 145.0377. Air-entraining cements are not included, but have somewhat lower requirements, as do some cement types when additional optional requirements are invoked.

**Key: — = no requirement at this age.

***Optional requirement that applies only if specifically requested.

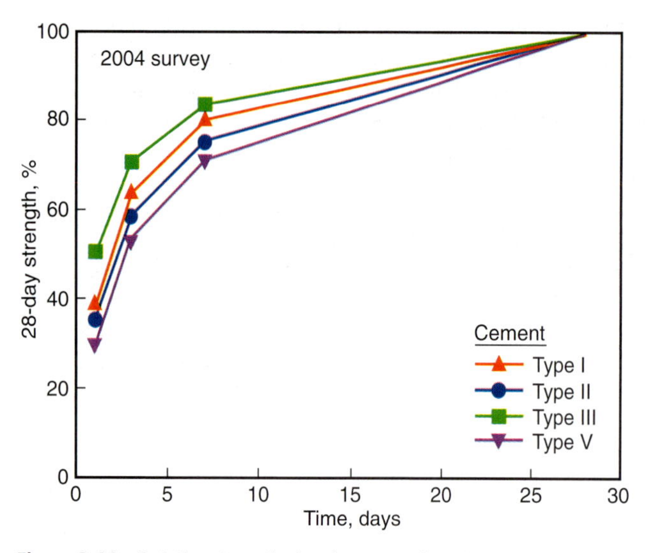

Figure 3-33. Relative strength development of portland cement mortar cubes as a percentage of 28-day strength. Mean values adapted from Bhatty and Tennis (2008).

Setting Time

The setting properties of hydraulic cement are measured to ensure that the cement is hydrating normally. Two arbitrary setting times are used: initial set – which is considered to represent the time that elapses from the moment water is added until the paste ceases to be fluid and plastic, and final set–the time required for the paste to acquire a certain degree of hardness.

To determine if a cement sets according to the time limits specified in cement specifications, tests are performed using the Vicat apparatus (ASTM C191, *Standard Test Methods for Time of Setting of Hydraulic Cement by Vicat Needle*, or AASHTO T 131), or more rarely, the Gillmore needle (ASTM C266, *Standard Test Method for Time of Setting of Hydraulic-Cement Paste by Gillmore Needles*, or AASHTO T 154).

The setting times indicate if a paste is undergoing normal hydration reactions. Sulfate (from gypsum or other sources) in the cement regulates setting time, but setting time is also affected by cement fineness, water-cement ratio, and any

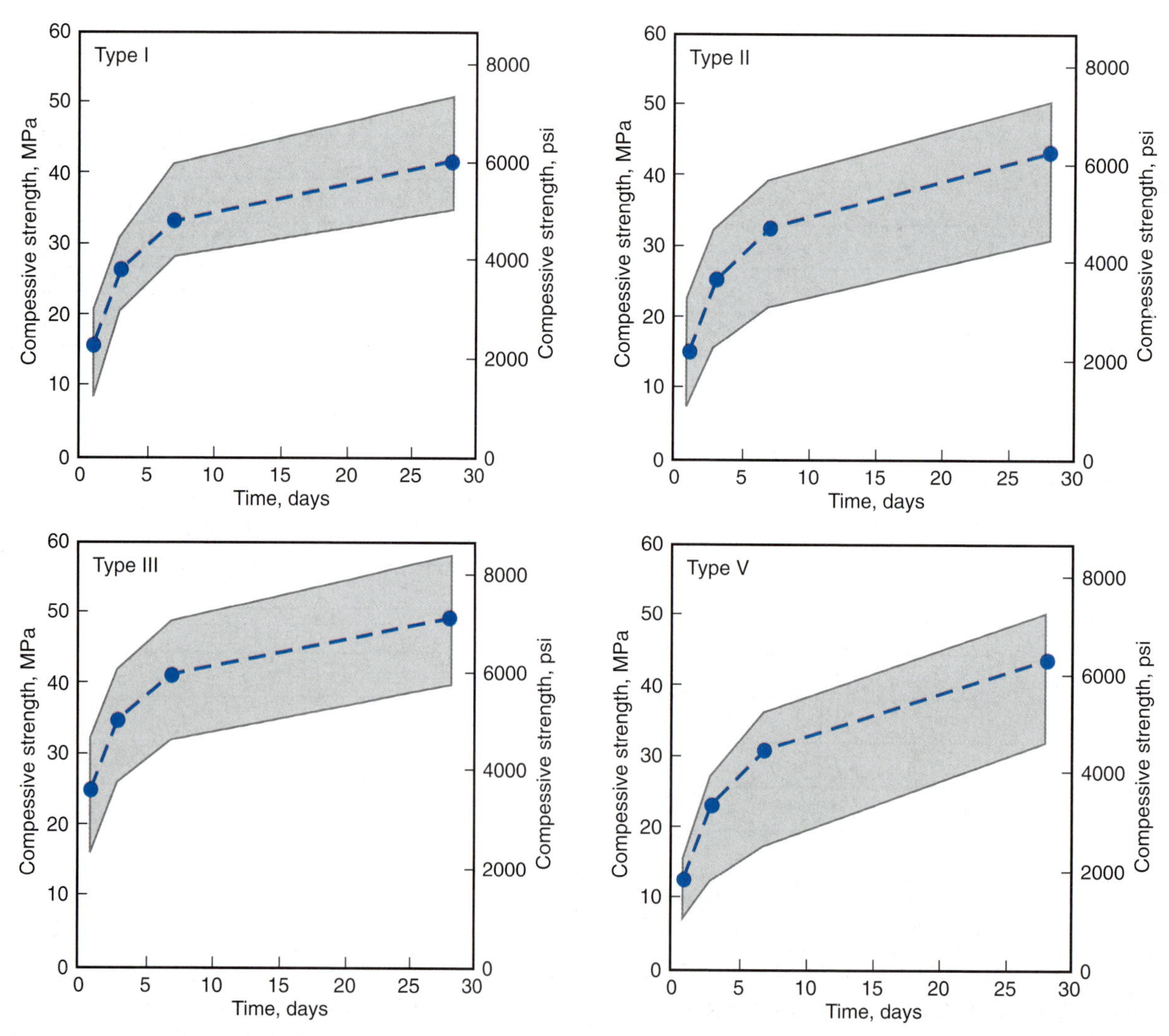

Figure 3-34. Mean values of ASTM C109 mortar cube strengths from a 2004 survey (Bhatty and Tennis 2008). Range of reported values indicated by shaded areas, and mean values by the dashed lines.

admixtures that may be used. Setting times of concretes do not correlate directly with setting times of pastes because of water loss to the air or substrate, presence of aggregate, and because of temperature differences in the field (as contrasted with the controlled temperature in a testing lab). Figure 3-35 illustrates mean set times for portland cements.

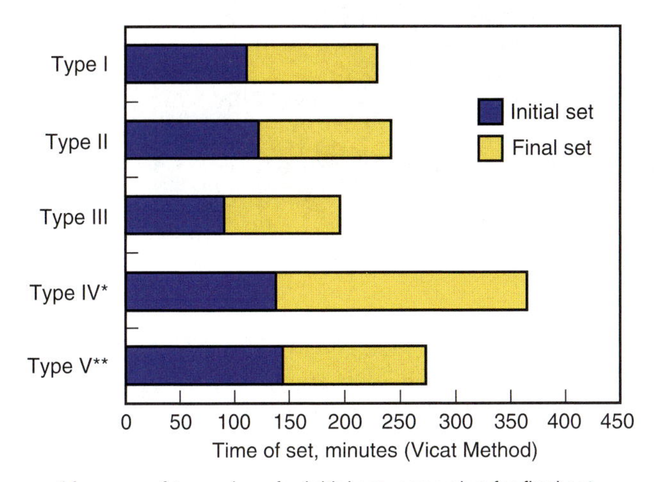

Figure 3-35. Time of set for portland cements (Bhatty and Tennis 2008). Type IV setting times are based on limited data in Gebhardt (1995).

Early Stiffening (False Set and Flash Set)

Early stiffening is the early development of stiffness in the working characteristics or plasticity of cement paste, mortar, or concrete. This includes both false set and flash set.

False set is evidenced by a significant loss of plasticity without the evolution of much heat shortly after mixing. From a placing and handling standpoint, false-set tendencies in cement will cause no difficulty provided the concrete is mixed for a longer time than usual or if it is remixed without additional water before being transported or placed. False set occurs when a significant amount of gypsum dehydrates in the cement finish mill to form plaster. Stiffening is caused by the rapid crystallization of interlocking needle-like secondary gypsum. Additional mixing without added water breaks up these crystals to restore workability. Ettringite precipitation can also contribute to false set.

Flash set or quick set is evidenced by a rapid and early loss of workability in paste, mortar, or concrete. It is usually accompanied by the evolution of considerable heat resulting primarily from the rapid reaction of aluminates. If the proper amount or form of calcium sulfate is not available to control tricalcium aluminate hydration, stiffening becomes apparent. Flash set cannot be dispelled, nor can the plasticity be regained by further mixing without the addition of water.

Proper stiffening results from the careful balance of the sulfate and aluminate compounds, as well as the temperature and fineness of the materials (which control the dissolution and precipitation rates). The amount of sulfate in the form of plaster has a significant effect. For example, with one particular cement, 2% plaster provided a 5-hour set time, while 1% plaster caused flash set to occur, and 3% demonstrated false set (Helmuth and others 1995).

Cements are tested for early stiffening using ASTM C451, *Standard Test Method for Early Stiffening of Hydraulic Cement (Paste Method),* (AASHTO T 186), and ASTM C359, *Standard Test Method for Early Stiffening of Hydraulic Cement (Mortar Method),* (AASHTO T 185), which use the penetration techniques of the Vicat apparatus. However, these tests do not address all the variables that can influence early stiffening including: mixing, placing, temperature, and field conditions. These tests also do not address early stiffening caused by interactions with other concrete ingredients. For example, concretes mixed for very short periods, less than one minute, tend to be more susceptible to early stiffening (ACI 225).

Particle Size and Fineness

Portland cement consists of individual angular particles with a range of sizes, the result of pulverizing clinker in the grinding mill (Figure 3-36 left). Approximately 95% of

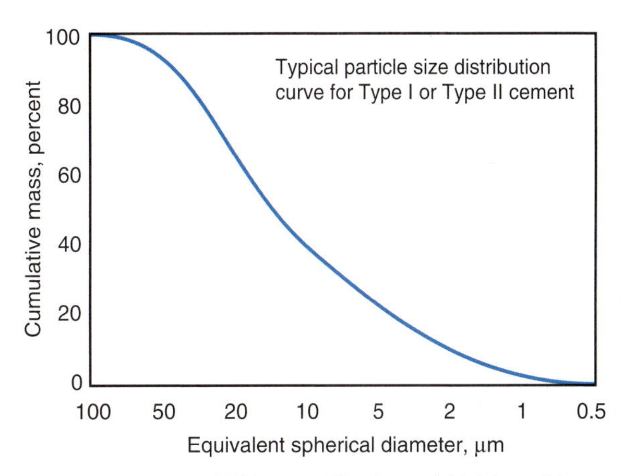

Figure 3-36. (left) Scanning-electron micrograph of powdered cement at approximately 1000x magnification and (right) particle size distribution of portland cement.

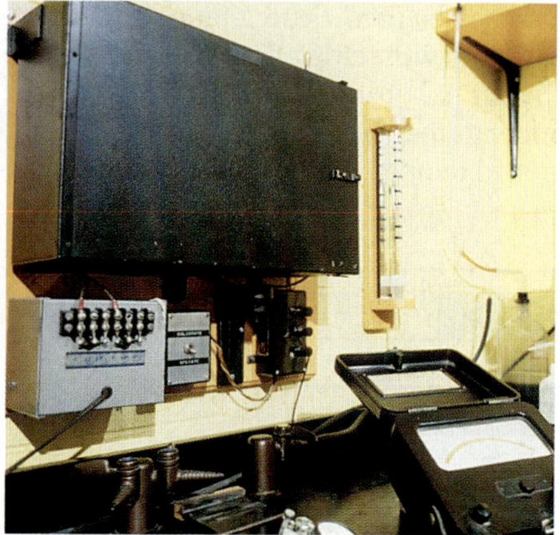

Figure 3-37. Blaine test apparatus (left) and Wagner turbidimeter (right) for determining the fineness of cement.

cement particles are smaller than 45 µm, with the average particle around 15 µm. Figure 3-36 (right) illustrates the particle size distribution for a portland cement. The fineness is, at best, a general indication of the particle size distribution of a cement. Finer cements, as indicated by higher Blaine fineness measurements, are typically indicative of finer individual cement particles. The fineness of cement affects heat released and the rate of hydration. Greater cement fineness (smaller particle size) increases the rate at which cement hydrates and thus accelerates strength development. The effects of greater fineness on paste strength are manifested principally during the first seven days.

Fineness is usually measured by the Blaine air-permeability test (ASTM C204, *Standard Test Methods for Fineness of Hydraulic Cement by Air-Permeability Apparatus*, or AASHTO T 153 – Figure 3-37) that indirectly measures the surface area of the cement particles per unit mass. Cements with finer particles have more surface area in square meters per kilogram of cement. (The use of square centimeters per gram for reporting fineness is now considered archaic.) ASTM C150 and AASHTO M 85 have minimum Blaine fineness limits of 260 m^2/kg for all cement types except Type III, and maximum limits on fineness for Type II(MH) and Type IV cements of 430 m^2/kg. The Wagner turbidimeter test (ASTM C115, *Standard Test Method for Fineness of Portland Cement by the Turbidimeter*, or AASHTO T 98 – Figure 3-37), the 45-micrometer (No. 325) sieve (ASTM C430, *Standard Test Method for Fineness of Hydraulic Cement by the 45-µm (No. 325) Sieve*, or AASHTO T 192 – Figure 3-38) and electronic (x-ray or laser) particle size analyzers (Figure 3-39) can also be used to determine values for fineness, although the results are not equivalent among different test methods. Blaine fineness data are given in Table 3-6.

Figure 3-38. Quick tests, such as washing cement over this 45-micrometer sieve, help monitor cement fineness during production. Shown is a view of the sieve holder with an inset top view of a cement sample on the sieve before washing with water.

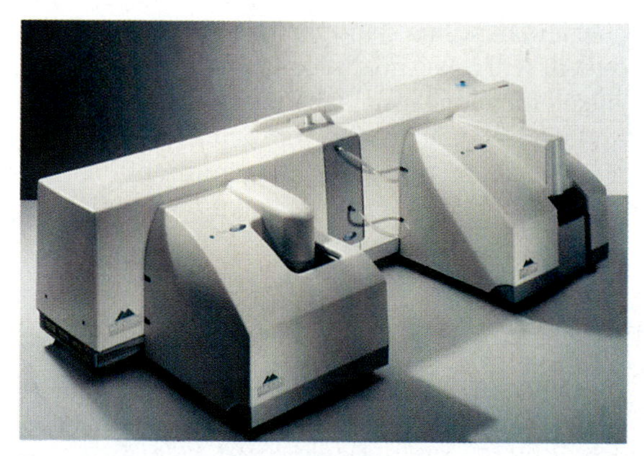

Figure 3-39. A laser particle analyzer uses laser diffraction to determine the particle size distribution of fine powders.

Soundness

Soundness refers to the ability of a hardened cement paste to retain its volume after setting. Lack of soundness or delayed destructive expansion can be caused by excessive amounts of hard-burned free lime or magnesia. Most specifications for portland cement limit the magnesia (periclase) content and require that the maximum expansion must not exceed 0.80% for portland, blended, or hydraulic cement as measured by the autoclave-expansion test. Since adoption of the autoclave-expansion test

(ASTM C151, *Standard Test Method for Autoclave Expansion of Hydraulic Cement*, or AASHTO T 107) in 1943, there have been exceedingly few cases of abnormal expansion attributed to unsound cement (Figure 3-40). See Gonnerman, Lerch, and Whiteside (1953), Helmuth and West (1998) and Klemm (2005) for extensive reviews of the autoclave expansion test.

Figure 3-40. In the soundness test, 25-mm square bars are exposed to high temperature and pressure in the autoclave to determine the volume stability of the cement paste.

Consistency

Consistency refers to the relative mobility of a freshly mixed cement paste or mortar or to its ability to flow. During cement testing, pastes are mixed to normal consistency as defined by a penetration of 10 ± 1 mm of the Vicat plunger (see ASTM C187 or AASHTO T 129 and Figure 3-41). Mortars are mixed to obtain either a fixed water-cement ratio or to yield a flow within a prescribed range. The flow is determined on a flow table as described in ASTM C230, *Standard Specification for Flow Table for Use in Tests of Hydraulic Cement*, (AASHTO M 152) and ASTM C1437, *Standard Test Method for Flow of Hydraulic Cement*, (Figure 3-42). Both the normal consistency method and the flow test are used to regulate water contents of pastes and mortars, respectively, to be used in subsequent tests; both allow comparing dissimilar ingredients with the same penetrability or flow.

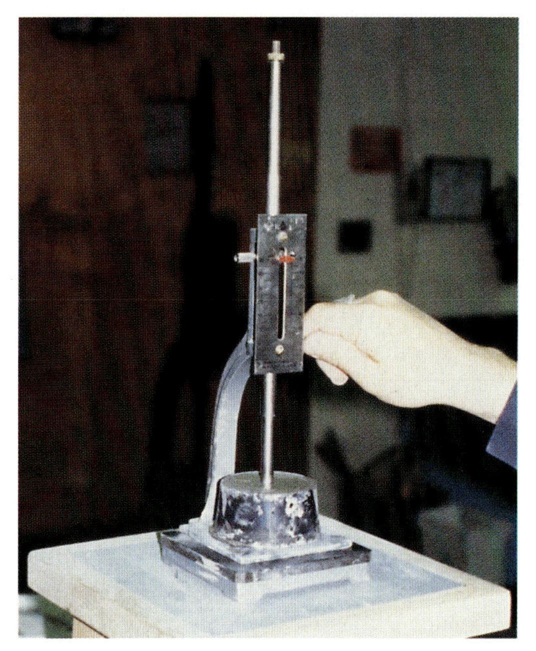

Figure 3-41. Normal consistency test for paste using the Vicat plunger.

Figure 3-42. Consistency test for mortar using the flow table. The mortar is placed in a small brass mold centered on the table (inset). For skin safety the technician wears protective gloves while handling the mortar. After the mold is removed and the table undergoes a succession of drops, the diameter of the pat is measured to determine consistency.

Heat of Hydration

Heat of hydration is the heat generated when cement and water react. The amount of heat generated is dependent chiefly upon the chemical composition of the cement, with C_3A and C_3S phases primarily responsible for high heat evolution. The water-cement ratio, fineness of the cement, and temperature of curing also are factors. An increase in the fineness, cement content, and curing temperature increases the heat of hydration. Although portland cement can evolve heat for many years, the rate of heat generation is greatest at early ages. A large amount of heat evolves within the first three days with the greatest rate of heat liberation usually occurring within the first 24 hours (Copeland and others 1960). The heat of hydration is tested in accordance with ASTM C186, *Standard Test Method for Heat of Hydration of Hydraulic Cement*, or may also be performed by conduction calorimetry according to ASTM C1702, *Standard Test Method for Measurement of Heat of Hydration of Hydraulic Cementitious Materials Using Isothermal Conduction Calorimetry* (Figure 3-43). ASTM C1679, *Standard Practice for Measuring Hydration Kinetics of Hydraulic Cementitious Mixtures Using Isothermal Calorimetry*, also provides information on isothermal conduction calorimetry testing. Often heat of hydration is also tested in both the laboratory and field using semi-adiabatic calorimetry to predict heat evolution maturity of the concrete.

For most concrete elements, such as slabs, heat generation is not a concern because the heat is quickly dissipated into the environment. However, in structures of considerable mass, greater than a meter (yard) thick, the rate and amount of heat generated are critical. If this heat is not rapidly dissipated in massive elements, a significant rise in concrete temperature can occur. This may be undesirable, since, after hardening at an elevated temperature, non-uniform cooling of the concrete to ambient temperature may create excessive tensile stresses. On the other hand, a rise in concrete temperature caused by heat of hydration is often beneficial in cold weather, as it helps maintain favorable curing temperatures.

Table 3-9 provides heat of hydration values for a variety of portland cements. These limited data show that Type III cement has a higher heat of hydration than other types.

Cements do not generate heat at a constant rate. The heat output during hydration of a typical Type I portland cement is illustrated in Figure 3-44. The first peak shown in the heat profile is caused by heat generated from the initial hydration reactions of cement compounds such as tricalcium aluminate. Sometimes called the heat of wetting, this initial heat peak is followed by a period of slow thermal activity known as the induction period. After several hours, a broad second heat peak attributed to tricalcium silicate hydration emerges, signaling the onset of the paste hardening process. Finally, a third peak due to the renewed activity of tricalcium aluminate is experienced; its intensity and location are normally dependent on the amount of tricalcium aluminate and sulfate present in the cement.

In calorimetry testing, the first heat measurements are typically obtained 5 to 7 minutes after mixing the paste; as a result, often only the downward slope of the first peak is observed (Stage 1, Figure 3-44). The second peak (C_3S peak) generally occurs between 6 and 12 hours. The third peak (renewed C_3A peak) occurs between 12 and 90 hours. This information can be helpful when trying to control temperature rise in mass concrete (Tang 1992).

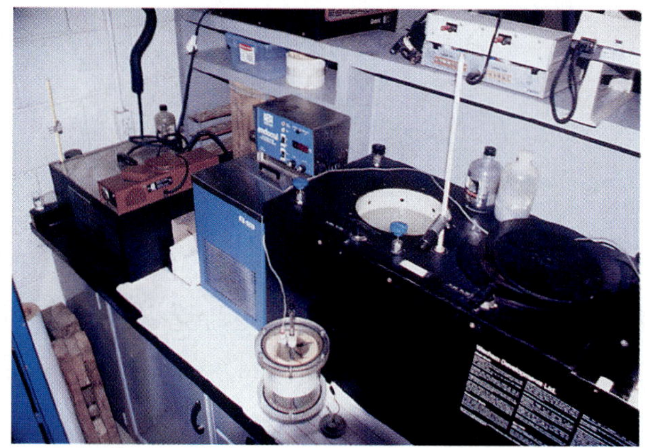

Figure 3-43. Heat of hydration can be determined by (left) ASTM C186 or by (right) a conduction calorimeter.

Table 3-9. Summary of ASTM C186 Heat of Hydration Data in 2007 Survey, kJ/kg*

	Type I		Type II**		Type III***	Type V****	
	7 days	28 days	7 days	28 days	7 days	7 days	28 days
Mean	**352**	**408**	**333**	**376**	**355**	**322**	**368**
St. Dev.	23.1	12.5	23.3	33.4	56.6	21.2	23.4
Max	380	416	393	431	399	346	398
Min	315	389	269	287	291	279	336
N	6	4	53	27	3	15	7

* Table is based on limited data for several cement types. To convert to cal/g, divide by 4.184.

** Includes 22 Type I/II cements.

*** Includes 1 Type II/III cement.

**** Includes 7 Type II/V cements.

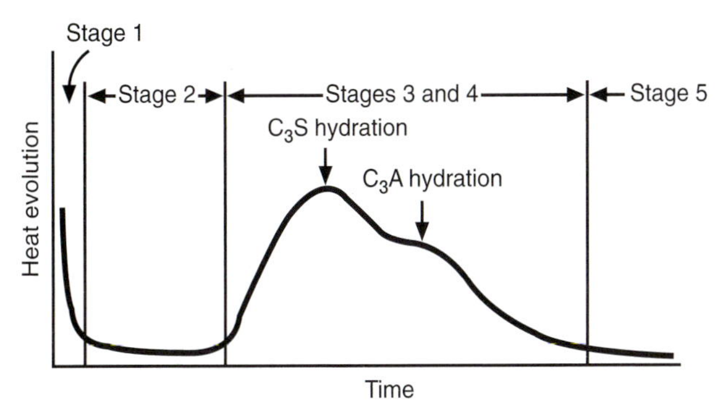

Figure 3-44. Heat evolution as a function of time for cement paste. Stage 1 is heat of wetting or initial hydrolysis (C_3A and C_3S hydration). Stage 2 is a dormant period related to initial set. Stage 3 is an accelerated reaction of the hydration products that determines rate of hardening and final set. Stage 4 decelerates formation of hydration products and determines the rate of early strength gain. Stage 5 is a slow, steady formation of hydration products establishing the rate of later strength gain.

When heat generation must be minimized in concrete, designers should choose a lower heat cement, such as an ASTM C150 (AASHTO M 85) Type II(MH) portland cement. Type IV cement can also be used to control temperature rise, but it is rarely available. Moderate-heat and low-heat cements are also available in ASTM C595 (AASHTO M 240; with (MH) or (LH) suffixes) and C1157 (Type MH or LH) specifications. Supplementary cementitious materials are also often used to reduce temperature rise.

ASTM C150 (AASHTO M 85) has both a chemical approach and a physical approach to control heat of hydration. Either approach can be specified, but not both. ASTM C595 (AASHTO M 240) and ASTM C1157 use physical limits. See Poole (2007) for more information.

Loss on Ignition

Loss on ignition (LOI) of portland cement is determined by heating a cement sample of known weight up to a temperature between 900°C and 1000°C until a constant weight is obtained. The weight loss of the sample is then determined. Loss on ignition values range from 0% to 3%. Traditionally, a high LOI has been an indication of prehydration and carbonation, which may have been caused by improper or prolonged storage, or adulteration during transport. Modern cements may have an LOI over 2% due to the use of limestone as an ingredient in portland cement. The test for loss on ignition is performed in accordance with ASTM C114 (AASHTO T 105) (Figure 3-45).

Density and Relative Density (Specific Gravity)

The density of cement is defined as the mass of a unit volume of the solids or particles, excluding air between particles. It is reported as megagrams per cubic meter or grams per cubic centimeter (the numeric value is identical for both units). The particle density of portland cement ranges from 3.10 to 3.25, averaging 3.15 Mg/m^3. Portland-blast-furnace-slag and portland-pozzolan cements have densities ranging from 2.90 to 3.15, averaging 3.05 Mg/m^3. The density of a cement, determined by ASTM C188, *Standard Test Method for Density of Hydraulic Cement*, (AASHTO T 133) (Figure 3-46), is not an indication of the cement's quality; rather, its principal use is in mixture proportioning calculations (see Chapter 12).

For mixture proportioning, it may be more useful to express the density as relative density (also called specific gravity). The relative density is a dimensionless number determined by dividing the cement density by the density of water at 4°C, which is 1.0 Mg/m^3 (1.0 g/cm^3 or 1000 kg/m^3) or 62.4 lb/ft^3.

A relative density of 3.15 is assumed for portland cement in volumetric computations of concrete mix proportioning. However, as mix proportions list quantities of concrete ingredients in kilograms or pounds, the relative density must be multiplied by the density of water at 4°C to determine the particle density in kg/m^3 or lb/ft^3. This product is then divided into the mass or weight of cement to determine the absolute volume of cement in cubic meters or cubic feet. (See Chapter 12 for more information on proportioning concrete mixtures.)

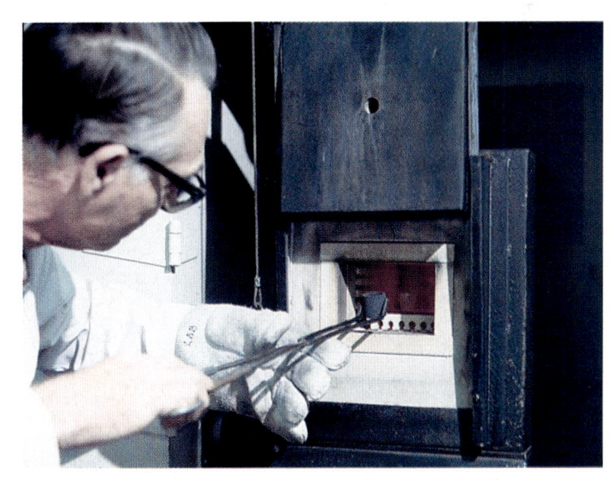

Figure 3-45. Loss on ignition test of cement.

Figure 3-46. Density of cement can be determined by (left) using a Le Chatelier flask and kerosene or by (right) using a helium pycnometer.

Bulk Density

The bulk density of cement is defined as the mass of cement particles plus air between particles per unit volume. The bulk density of cement can vary considerably depending on how it is handled and stored. Portland cement that is aerated or loosely deposited may weigh only 830 kg/m³ (52 lb/ft³), but with consolidation caused by vibration, the same cement can weigh as much as 1650 kg/m³ (103 lb/ft³) (Toler 1963). For this reason alone, good practice dictates that cement must be weighed for each batch of concrete produced, as opposed to using a volumetric measure (Figure 3-47).

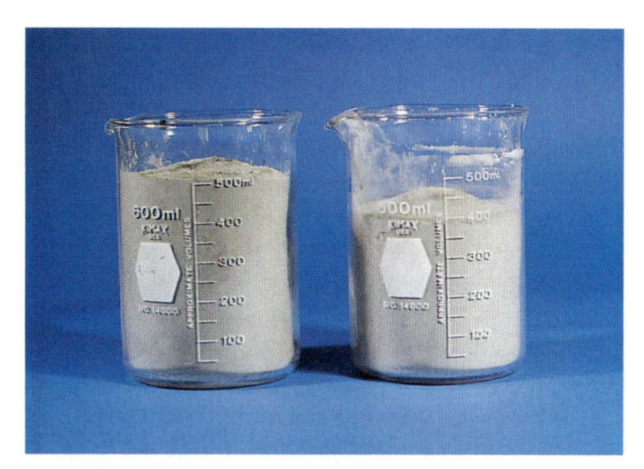

Figure 3-47. Both 500-mL beakers contain 500 grams of dry powdered cement. On the left, cement was simply poured into the beaker. On the right, cement was slightly vibrated – imitating consolidation during transport or packing while stored in a silo. The 20% difference in bulk volume demonstrates the need to measure cement by mass instead of volume for batching concrete.

Thermal Analysis

Thermal analysis techniques have been available for years to analyze hydraulic reactions and the interactions of cement with both mineral and chemical admixtures (Figure 3-48). Thermal analysis is useful for analyzing chemical and physical properties of cementitious materials and raw materials for cement manufacture (Bhatty 1993, Shkolnik and Miller 1996, and Tennis 1997).

Thermal analysis involves heating a small sample at a controlled rate to high temperature (up to 1000°C or more). As materials react or decompose, changes in the sample's weight, temperature, energy, or state (gas, solid, liquid) that occur as a function of time or temperature are recorded. Typical uses for thermal analysis include:

- identifying which hydration products have formed and in what quantities
- solving early stiffening problems
- identifying impurities in raw materials
- determining the amount of weathering in clinker or cement
- estimating the reactivity of pozzolans and slags for use in blended cements
- identifying the organic matter content and variations in a quarry
- quantifying the amount of carbonation that has taken place in an exposed sample
- analyzing durability problems with concrete.

Specific thermal analysis techniques are discussed below.

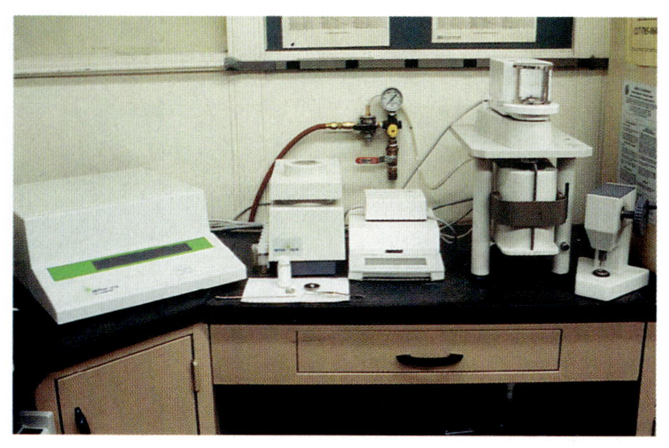

Figure 3-48. Thermal analysis equipment.

Thermogravimetric Analysis (TGA)

Thermogravimetric analysis (TGA) is a technique that measures the mass (weight) of a sample as it is being heated (or cooled) at a controlled rate. The weight change of a sample depends on the composition of the sample, the temperature, the heating rate, and the type of gas in the furnace (air, oxygen, nitrogen, argon, or other gas). A change in mass within a specific temperature range can identify the presence of a particular chemical compound. The magnitude of the weight change indicates the amount of that compound in the sample.

One example of the use of TGA is the split loss on ignition procedure, which is often used to estimate the amount of limestone in an unhdyrated portland cement (see ASTM C114). In this procedure, the CO_2 content of a cement is determined as the mass loss between 550°C and 950°C and the percentage of limestone is determined by dividing the CO_2 content of the cement by the CO_2 content of the limestone. The procedure can be done as a series of steps in furnaces at specific temperature, but TGA can provide similar data as a single procedure.

Differential Thermal Analysis (DTA)

Differential thermal analysis (DTA) is an analytical method in which the difference in temperature between a sample and a control is measured as the samples are heated. The control is usually an inert material, like powdered alumina, that does not react over the temperature range being studied. If a sample undergoes a reaction at a certain temperature, then its temperature will increase or decrease relative to the (inert) control as the reaction gives off heat (exothermic) or adsorbs heat (endothermic). A thermocouple measures the temperature of each material, allowing the difference in temperature to be recorded. DTA is ideal for monitoring the transformation of cement phases during hydration. DTA can be performed simultaneously with TGA.

Differential Scanning Calorimetry (DSC)

In differential scanning calorimetry (DSC), the heat absorbed or released is measured directly as a function of temperature or time and compared to a reference. An advantage of DTA and DSC methods is that no mass change is required; thus if a sample melts without vaporizing, measurements can still be obtained.

Like DTA, DSC can be used to determine what phases are present at different stages of hydration. Figure 3-49 shows two differential scanning calorimetry curves or thermograms. The upper curve (a) shows a portland cement paste after 15 minutes of reaction. The peaks in the curve between 100°C and 200°C are due to the endothermic (heat adsorbing) decomposition of gypsum and ettringite, while the peak around 270°C is due to syngenite (potassium calcium sulfate hydrate). Near 450°C, a smaller peak can be seen, due to calcium hydroxide.

The lower curve (b) in Figure 3-49 shows the same cement paste after 24 hours of hydration. Note the disappearance of the peak due to gypsum, the reduction in the size of the peak due to syngenite and the growth of those due to ettringite and calcium hydroxide. The size of the areas under the curves is related to the quantity of the material in the sample.

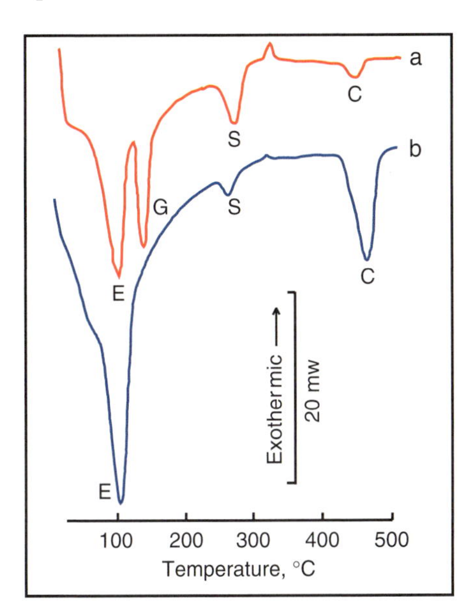

Figure 3-49. Differential scanning calorimetry thermogram of a portland cement paste after (a) 15 minutes and (b) 24 hours of hydration. C = calcium hydroxide; E = ettringite; G = gypsum; and S = syngenite.

Virtual Cement Testing

Computer technology now allows simulation of cement hydration (Figure 3-50), microstructure development, and physical properties. Combinations of materials, cement phases, or particle size distributions can be modeled to predict cement performance. Just some of the properties that can be simulated and predicted include: heat of hydration, adiabatic heat signature, compressive strength, setting time, rheology (yield stress and viscosity), percolation, porosity, diffusivity, thermal conductivity, electrical conductivity, carbonation, elastic properties, drying profiles, susceptibility to degradation mechanisms, autogenous shrinkage, and volumes of hydration reactants and products as a function of time. The effects from varying sulfate and alkali contents can also be examined, along with interaction with supplementary cementitious materials and chemical admixtures. Computer modeling predicts performance without the expense and time requirements of physical testing. Garboczi and others (2009) provide a web-based computer model for hydration and microstructure development.

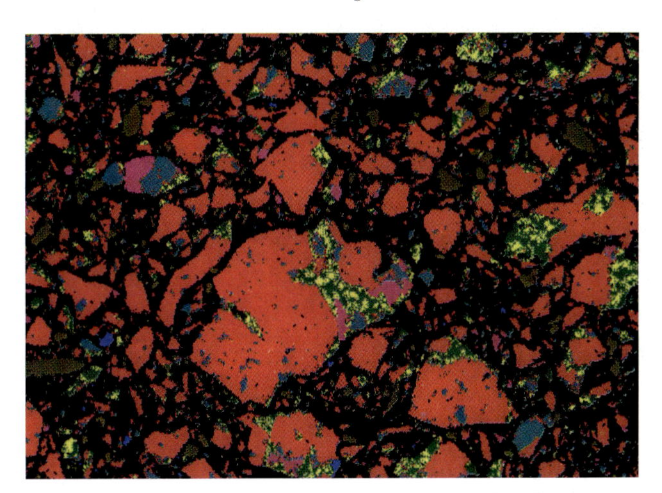

Figure 3-50. Two-dimensional image of portland cement. Colors are: red – tricalcium silicate, aqua–dicalcium silicate, green – tricalcium aluminate, yellow – tetracalcium aluminoferrite, pale green – gypsum, white – free lime, dark blue – potassium sulfate, and magenta – periclase. The image was obtained by combining a set of SEM backscattered electron and X-ray images (NIST 2007).

Transportation and Packaging

Around 100 million metric tons of cement are shipped per year throughout the United States. Most portland cements are shipped in bulk by rail, truck, barge, or ship (Figure 3-51). Pneumatic (pressurized air) loading and unloading of the transport vehicle is the most popular means of handling bulk cement. However, the recent introduction of large sacks holding one to twelve tons of cement provides a new alternative to bulk cement handling. Bulk cement is measured by the metric ton (1000 kg) or short ton (2000 lb).

Figure 3-51. Portland cements are shipped from the plant silos to the user in bulk by (left to right) rail, truck, or water.

Cement is also bagged for convenient use at construction sites (Figure 3-52) and for small jobs. In the United States a bag of portland cement has a mass of 42 kg (92.6 lb) and a volume of 28 liters. Traditionally a U.S. bag contained 94 lb (42.6 kg) and had a volume of 28 L (one cubic foot). In Canada, a bag of portland cement has a mass of 40 kg, and in other countries a bag of cement can have a mass of 25 kg or 50 kg.

Figure 3-52. A small amount of cement is shipped in bags, primarily for masonry applications and for small projects.

The mass of masonry cement and mortar cement by the bag is 36 kg for Type M, 34 kg for Type S, and 32 kg for Type N. Plastic cement has a mass of 42 kg for Type M and 35 kg for Type S. All bag weights are based on a 28 liter (1 ft^3) volume standard. Information for quantities of blended and specialty cements can be found on the bag. Partial bag sizes are also available in some areas.

Storage of Cement

Cement is a moisture-sensitive material; while dry, it will retain its quality indefinitely. Cement stored in contact with damp air or moisture sets more slowly and has less strength than cement that is kept dry. At the cement plant, and at ready mixed concrete facilities, bulk cement is stored in silos. The relative humidity in a warehouse or shed used to store bagged cement should be as low as possible. All cracks and openings in walls and roofs should be closed. Cement bags should not be stored on damp floors but should rest on pallets. Bags should be stacked closely together to reduce air circulation but should never be stacked against outside walls. Bags to be stored for long periods should be covered with tarpaulins or other waterproof covering. Bags should be stored so that the first in are the first out.

On small jobs where a shed is not available, bags should be placed on raised wooden platforms (pallets) above the ground. Waterproof coverings should fit over the pile and extend over the edges of the platform to prevent rain from reaching the cement and the platform (Figure 3-53). Rain-soaked platforms can damage the bottom bags of cement.

Figure 3-53. When stored on the job, cement should be protected from moisture.

Cement stored for long periods may develop what is called warehouse pack. This can usually be corrected by rolling the bags on the floor. At the time of use, cement should be free-flowing and free of lumps. If lumps do not break up easily, the cement should be tested before it is used in important work. Standard strength tests or loss-on-ignition tests should be made whenever the quality of cement is in question.

Ordinarily, cement does not remain in storage long, but it can be stored for long periods without deterioration. Bulk cement should be stored in weather tight concrete or steel bins or silos. Dry low-pressure aeration or vibration should be used in bins or silos to keep the cement flowable and avoid bridging. It is not uncommon for cement silos to hold only about 80% of rated capacity because of loose packing of aerated cement.

Hot Cement

When cement clinker is pulverized in the grinding mill during cement manufacture, the friction of the particle grinding mechanism generates heat. Freshly ground cement is therefore hot when placed in storage silos at the cement plant. This heat dissipates slowly. Therefore, during summer months when demand is high, cement may still be hot when delivered to a ready mixed concrete plant or jobsite. Tests have shown that the effect of hot cement on the workability and strength development of concrete is not significant (Lerch 1955). The temperatures of the mixing water and aggregates play a much greater role in establishing the final temperature of the concrete as mixed.

References

ACI Committee 223, *Standard Practice for the Use of Shrinkage Compensating Concrete*, ACI 223, American Concrete Institute, Farmington Hills, Michigan, 1998, 28 pages.

ACI Committee 225, *Guide to the Selection and Use of Hydraulic Cements*, ACI 225, ACI Committee 225 Report, American Concrete Institute, Farmington Hills, Michigan, 1999.

Ashley, Erin, and Lemay, Lionel, "Concrete's Contribution to Sustainable Development," *The Journal of Green Building*, Vol. 3, No. 4, Fall 2008, pages 37 to 49.

Aspdin, Joseph, *Artificial Stone*, British Patent No. 5022, December 15, 1824, 2 pages.

Barger, Gregory S.; Lukkarila, Mark R.; Martin, David L.; Lane, Steven B.; Hansen, Eric R.; Ross, Matt W.; and Thompson, Jimmie L., "Evaluation of a Blended Cement and a Mineral Admixture Containing Calcined Clay Natural Pozzolan for High-Performance Concrete," *Proceedings of the Sixth International Purdue Conference on Concrete Pavement Design and Materials for High Performance*, Purdue University, West Lafayette, Indiana, November 1997, 21 pages.

Bhatty, Javed I., "Application of Thermal Analysis to Problems in Cement Chemistry," Chapter 6, *Treatise on Analytical Chemistry*, Part 1, Volume 13, J. D. Winefordner editor, John Wiley & Sons, 1993, pages 355 to 395.

Bhatty, Javed I., *Role of Minor Elements in Cement Manufacture and Use*, Research and Development Bulletin RD109, Portland Cement Association, 1995, 48 pages.

Bhatty, Javed I., and Tennis, Paul D., *U.S. and Canadian Cement Characteristics: 2004*, R&D SN2879, Portland Cement Association, Skokie, Illinois, 2008, 67 pages.

Brunauer, S., *Some Aspects of the Physics and Chemistry of Cement*, Research Department Bulletin RX080, Portland Cement Association, http://www.cement.org/pdf_files/RX080.pdf, 1957.

Brunauer, Stephen, *Tobermorite Gel – The Heart of Concrete*, Research Department Bulletin RX138, Portland Cement Association, http://www.cement.org/pdf_files/RX138.pdf, 1962, 20 pages.

BS 915-2, *Specification for High Alumina Cement, British Standards Institution*, 1972.

Campbell, Donald H., *Microscopical Examination and Interpretation of Portland Cement and Clinker*, SP030, Portland Cement Association, 1999.

Copeland, L.E.; Kantro, D.L.; and Verbeck, George, *Chemistry of Hydration of Portland Cement*, Research Department Bulletin RX153, Portland Cement Association, http://www.cement.org/pdf_files/RX153.pdf, 1960.

Copeland, L.E., and Schulz, Edith G., *Electron Optical Investigation of the Hydration Products of Calcium Silicates and Portland Cement*, Research Department Bulletin RX135, Portland Cement Association, http://www.cement.org/pdf_files/RX135.pdf, 1962.

Davidovits, Joseph; Davidovits, Ralph; and James, Claude, editors, *Geopolymer '99*, The Geopolymer Institute, http://www.geopolymer.org, Insset, Université de Pic-ardie, Saint-Quentin, France, July 1999.

DeHayes, Sharon M., "C 109 vs. Concrete Strengths" *Proceedings of the Twelfth International Conference on Cement Microscopy*, International Cement Microscopy Association, Duncanville, Texas, 1990.

EN 197-1, European Standard for Cement – Part 1: *Composition, Specifications and Common Cements*, European Committee for Standardization (CEN), Brussels, 2000.

Farny, James A., *White Cement Concrete*, EB217, Portland Cement Association, 2001, 32 pages.

Garboczi, E.J.; Bentz, D.P.; Snyder, K.A.; Martys, N.S.; Stutzman, P.E.; Ferraris, C.F.; and Bullard, J.W., *An Electronic Monograph: Modeling and Measuring the Structure And Properties of Cement-Based Materials*, National Institute of Standards and Technology, Gaithersburg, Maryland, USA, 2009, http://ciks.cbt.nist.gov/~garbocz/ http://ciks.cbt.nist.gov/~garbocz/.

Gebhardt, R.F., "Survey of North American Portland Cements: 1994," *Cement, Concrete, and Aggregates*, American Society for Testing and Materials, West Conshohocken, Pennsylvania, December 1995, pages 145 to 189.

Gonnerman, H.F.; Lerch, William; and Whiteside, Thomas M., *Investigations of the Hydration Expansion Characteristics of Portland Cements*, Research Department Bulletin RX045, Portland Cement Association, 1953, 181 pages.

Greening, Nathan R.; Copeland, Liewellyn E.; and Verbeck, George J., *Modified Portland Cement and Process*, United States Patent No. 3,628,973, Issued to Portland Cement Association, December 21, 1971.

Hawkins, P.; Tennis, P.D.; and Detwiler, R.J., *The Use of Limestone in Portland Cement: A State-of-the-Art Review*, EB227, Portland Cement Association, Skokie, Illinois, USA, 2005, 44 pages.

Helmuth, R.; Hills, L.M.; Whiting, D.A.; and Bhattacharja, S., *Abnormal Concrete Performance in the Presence of Admixtures*, RP333, Portland Cement Association, Skokie, Illinois, 1995, 92 pages.

Helmuth, R., and West, P.B., "Reappraisal of the Autoclave Expansion Test," *Cement, Concrete, and Aggregates*, CCAGDP, Vol. 20, No. 1, June 1998, pages 194 to 219.

Hills, Linda M., "Clinker Formation and the Value of Microscopy," *Proceedings of the Twenty-Second International Conference on Cement Microscopy*, Montreal, 2000, pages 1 to 12.

Johansen, Vagn C.; Taylor, Peter C.; and Tennis, Paul D., *Effect of Cement Characteristics on Concrete Properties*, EB226, 2nd edition Portland Cement Association, Skokie, Illinois, 2006, 48 pages.

Klemm, Waldemar A., *Ettringite and Oxyanion-Substituted Ettringites – Their Characterization and Applications in the Fixation of Heavy Metals: A Synthesis of the Literature*, Research and Development Bulletin RD116, Portland Cement Association, 1998, 80 pages.

Klemm, W.A., *Cement Soundness and the Autoclave Expansion Test: An Update of the Literature*, SN2651, Portland Cement Association, Skokie, Illinois, 2005, 20 pages.

Klieger, Paul, and Isberner, Albert W., *Laboratory Studies of Blended Cements – Portland Blast-Furnace Slag Cements*, Research Department Bulletin RX218, Portland Cement Association, http://www.cement.org/pdf_files/RX218.pdf, 1967.

Lange, D., and Stutzman, P., *The Concrete Microscopy Library*, https://netfiles.uiuc.edu/dlange/www/CML/, 2009.

Lea, F.M., *The Chemistry of Cement and Concrete*, 3rd ed., Chemical Publishing Co., Inc., New York, 1971.

Lerch, William, *Hot Cement and Hot Weather Concrete Tests*, IS015, Portland Cement Association, http://www.cement.org/pdf_files/IS015.pdf, 1955.

Lerch, William, *The Influence of Gypsum on the Hydration and Properties of Portland Cement Pastes*, Research Department Bulletin RX012, Portland Cement Association, http:// www.cement.org/pdf_files/RX012.pdf, 1946.

Locher, F.W.; Richartz, W.; and Sprung, S., "Setting of Cement–Part I: Reaction and Development of Structure;" *ZKG INTERN.* 29, No. 10, 1976, pages 435 to 442.

Marceau, M.L.; Nisbet, M.A.; and VanGeem, M.G., *Life Cycle Inventory of Portland Cement Manufacture*, SN2095b, Portland Cement Association, Skokie, Illinois, 2006, 69 pages.

Melander, John M.; Farny, J.A.; and Isberner, Albert W. Jr., *Portland Cement Plaster/Stucco Manual*, EB049.05, Portland Cement Association, 2003, 72 pages.

NF P15-315, *Hydraulic Binders. High Alumina Melted Cement*, AFNOR Association française de normalisation, 1991.

NIST, *Virtual Cement and Concrete Testing Laboratory*, http://vcctl.cbt.nist.gov, 2007.

Nisbet, M.A., *The Reduction of Resource Input and Emissions Achieved by Addition of Limestone to Portland Cement*, SN2086, Portland Cement Association, 1996, 11 pages.

Odler, Ivan, *Special Inorganic Cements*, E&FN Spon, New York, 2000, 420 pages.

PCA, *An Analysis of Selected Trace Metals in Cement and Kiln Dust*, SP109, Portland Cement Association, 1992, 60 pages.

PCA, *Emerging Technologies Symposium on Cements for the 21st Century*, SP206, Portland Cement Association, 1995, 140 pages.

PCA, *North American Cement Industry Annual Yearbook*, ER405, Portland Cement Association, Skokie, Illinois, 2010, 69 pages.

PCA, "Portland Cement: Past and Present Characteristics," *Concrete Technology Today*, PL962, Portland Cement Association, http://www.cement.org/pdf_files/PL962.pdf, July 1996, pages 1 to 4.

PCA, "Portland Cement, Concrete, and the Heat of Hydration," *Concrete Technology Today*, PL972, Portland Cement Association, http://www.cement.org/pdf_files/PL972.pdf, July 1997, pages 1 to 4.

PCA, "What is White Cement?," *Concrete Technology Today*, PL991, Portland Cement Association, http://www.cement.org/pdf_files/PL991.pdf, April 1999, 4 pages.

PCA, *2011 Report on Sustainable Manufacturing*, Portland Cement Association, Skokie, Illinois, 2011, 26 pages, http://www.cement.org/smreport11.

PCA, *Tire-Derived Fuels*, IS325, Portland Cement Association, Skokie, Illinois, May 2008, 4 pages.

Pfeifer, Donald W., and Perenchio, W.F., *Reinforced Concrete Pipe Made with Expansive Cements*, Research and Development Bulletin RD015, Portland Cement Association, http://www.cement.org/pdf_files/RD015.pdf, 1973.

Poole, Toy, *Revision of Test Methods and Specifications for Controlling Heat of Hydration in Hydraulic Cement*, SN3007, Portland Cement Association, Skokie, Illinois, USA, 2007, 42 pages.

Poole, Toy, "Predicting Seven-Day Heat of Hydration of Hydraulic Cement from Standard Test Properties," *Journal of ASTM International*, June 2009, 10 pages.

Powers, T.C., *Some Physical Aspects of the Hydration of Portland Cement*, Research Department Bulletin RX126, Portland Cement Association, http://www.cement.org/pdf_files/RX126.pdf, 1961.

Powers, T.C., *The Nonevaporable Water Content of Hardened Portland-Cement Paste–Its Significance for Concrete Research and Its Method of Determination*, Research Department Bulletin RX029, Portland Cement Association, http://www.cement.org/pdf_files/RX029.pdf, 1949.

Powers, T.C., and Brownyard, T.L., *Studies of the Physical Properties of Hardened Portland Cement Paste*, Research Department Bulletin RX022, Portland Cement Association, http://www.cement.org/pdf_files/RX022.pdf, 1947.

RMA, *Scrap Tire Markets in the United States – 9th Biennial Report*, Rubber Manufacturers Association, Washington, D.C., USA, 2009, 105 pages.

Russell, H.G., *Performance of Shrinkage-Compensating Concretes in Slabs*, Research and Development Bulletin RD057, Portland Cement Association, http://www.cement.org/pdf_files/RD057.pdf, 1978.

Shkolnik, E., and Miller, F.M., "Differential Scanning Calorimetry for Determining the Volatility and Combustibility of Cement Raw Meal Organic Matter," *World Cement Research and Development*, May 1996, pages 81 to 87.

Stark, David, *Performance of Concrete in Sulfate Environments*, Research and Development Bulletin RD129, Portland Cement Association, http://www.cement.org/pdf_files/RD129.pdf, 2002.

Sullivan, Edward J., *U.S. and Canadian Labor-Energy Input Survey 2008*, ER394, Portland Cement Association, Skokie, Illinois, 2009, 55 pages.

Tang, Fulvio J., *Optimization of Sulfate Form and Content*, Research and Development Bulletin RD105, Portland Cement Association, 1992, 44 pages.

Taylor, P., *Specifications and Protocols for Acceptance Tests on Processing Additions in Cement Manufacturing*, NCHRP Report 607, Transportation Research Board, Washington, D.C., 2008, 96 pages. Available at www.trb.org.

Taylor, H.F.W., *Cement Chemistry*, Thomas Telford Publishing, London, 1997, 477 pages.

Tennis, Paul D., "Laboratory Notes: Thermal Analysis by TGA, DTA, and DSC," *ConcreteTechnology Today*, PL971, Portland Cement Association, http://www.cement.org/pdf_files/PL971.pdf, April 1997, pages 6 to 7.

Tennis, P.D., and Jennings, H.M., "Standard Test Method for Density of Hydraulic Cement", *Cement and Concrete Research*, Vol. 30, No. 6, 2000, pages 855 to 863.

Tennis, P.D., Thomas, M.D.A., and Weiss, W.J., *State-of-the-Art Report on Use of Limestone in Cements at Levels of up to 15%*, SN3148, Portland Cement Association, 78 pages, http://www.cement.org/pdf_files/SN3148.pdf, 2011.

Toler, H.R., *Flowability of Cement*, Research Department Report MP-106, Portland Cement Association, October 1963.

Weaver, W.S.; Isabelle, H.L.; and Williamson, F., "A Study of Cement and Concrete Correlation," *Journal of Testing and Evaluation*, American Society for Testing and Materials, West Conshohocken, Pennsylvania, January 1974, pages 260 to 280.

Wood, Sharon L., *Evaluation of the Long-Term Properties of Concrete*, Research and Development Bulletin RD102, Portland Cement Association, 1992, 99 pages.

Zhang, Min-Hong; Bremner, Theodore W.; Malhotra, V. Mohan, "The Effect of Portland Cement Types on Performance," *Concrete International*, January 2003, pages 87 to 94.

CHAPTER 4

Supplementary Cementitious Materials

Supplementary cementitious materials (SCMs), also called supplementary cementing materials, are materials that, when used in conjunction with portland or blended cement, contribute to the properties of concrete through hydraulic or pozzolanic activity or both. A hydraulic material reacts chemically with water to form cementitious compounds. A pozzolan is a siliceous or aluminosiliceous material that, in finely divided form and in the presence of moisture, chemically reacts with the calcium hydroxide to form calcium silicate hydrate and other cementitious compounds. Fly ash, slag cement, silica fume, and natural pozzolans such as calcined shale, metakaolin, or other calcined clays, are generally categorized as SCMs (Figure 4-1). Table 4-1 lists the applicable U.S. specifications these materials meet and Table 4-2 illustrates typical chemical analysis and selected properties of SCMs.

The practice of using SCMs in concrete mixtures has been growing in North America since the 1970s. Many of these materials are byproducts of other industrial processes. Their judicious use is desirable not only for their sustainability, given the environmental and energy conservation; but also for the performance benefits they provide concrete. SCMs are used in at least 60% of ready mixed concrete (PCA 2000).

Supplementary cementitious materials are added to concrete as part of the total cementitious system. Typical practice in the United States uses fly ash, slag cement, silica fume, calcined clay, or calcined shale as an addition to or as a partial replacement for hydraulic cements (including portland or blended cement) in concrete. Blended cements, which already contain pozzolans or slag, are designed to be used with or without additional SCMs. The use of these materials in blended cements is discussed in Chapter 3, and by Thomas and Wilson (2002), and Detwiler, Bhatty, and Bhattacharja (1996).

Figure 4-1. Supplementary cementitious materials. From left to right, fly ash (Class C), metakaolin (calcined clay), silica fume, fly ash (Class F), slag cement, and calcined shale.

Table 4-1. Specifications and Classes of Supplementary Cementitious Materials

- Slag Cement for Use in Concrete and Mortars — ASTM C989 (AASHTO M 302)
  - Grade 80
    - Slag cements with a low activity index
  - Grade 100
    - Slag cements with a moderate activity index
  - Grade 120
    - Slag cements with a high activity index
- Fly ash and natural pozzolans — ASTM C618 (AASHTO M 295)
  - Class N
    - Raw or calcined natural pozzolans including:
    - Diatomaceous earths
    - Opaline cherts and shales
    - Tuffs and volcanic ashes or pumicites
    - Calcined clays, including metakaolin, and shales
  - Class F
    - Fly ash with pozzolanic properties
  - Class C
    - Fly ash with pozzolanic and cementitious properties
- Silica fume — ASTM C1240 (AASHTO M 307)
- Specification for Blended Supplementary Cementitious Materials — ASTM C1697

Table 4-2. Chemical Analysis and Selected Properties of Typical Fly Ash, Slag Cement, Silica Fume, Calcined Clay, Calcined Shale, and Metakaolin

	Class F fly ash	Class C fly ash	Slag cement	Silica fume	Calcined clay	Calcined shale	Metakaolin
SiO_2, %	52	35	35	90	58	50	53
Al_2O_3, %	23	18	12	0.4	29	20	43
Fe_2O_3, %	11	6	1	0.4	4	8	0.5
CaO, %	5	21	40	1.6	1	8	0.1
SO_3, %	0.8	4.1	2	0.4	0.5	0.4	0.1
Na_2O, %	1.0	5.8	0.3	0.5	0.2	—	0.05
K_2O, %	2.0	0.7	0.4	2.2	2	—	0.4
Total Na eq. alk, %	2.2	6.3	0.6	1.9	1.5	—	0.3
Loss on ignition, %	2.8	0.5	1.0	3.0	1.5	3.0	0.7
Fineness, m^2/kg							
Blaine	420	420	400		990	730	
Nitrogen absorption				20,000			17,000
Relative density	2.38	2.65	2.94	2.40	2.50	2.63	2.50

Supplementary cementitious materials may be used to improve a particular concrete property, such as controlling heat of hydration in mass concrete or resistance to alkali-silica reactivity. The quantity used is dependent on the properties of the materials and the desired effect on concrete performance. The appropriate amount to use should be based on field performance or established by testing to determine the correct dosage rate and to verify whether the material is indeed improving the property.

Fly Ash

Fly ash, the most widely used SCM in concrete, is a byproduct of the combustion of pulverized coal in electric power generating plants. During combustion, the coal's mineral impurities (such as clay, feldspar, quartz, and shale) fuse in suspension and are carried away from the combustion chamber by the exhaust gases. In the process, the fused material cools and solidifies into glassy particles called fly ash. The fly ash is then collected from the exhaust gases by electrostatic precipitators or bag filters as a finely divided powder (Figure 4-2).

Fly ash is used in more than 50% of ready mixed concrete (PCA 2000). Class F fly ash is often used at dosages of 15% to 25% by mass of cementitious material and Class C fly ash is used at dosages of 15% to 40% by mass of cementitious material. Some concrete may contain higher volumes of fly ash. High volume fly ash (HVFA) concrete, are typically defined as > 40% by mass of cementitious material. Dosage varies with the reactivity of the ash and the desired effects on the concrete (Thomas 2007, Helmuth 1987, and ACI 232 2004).

Figure 4-2. Fly ash, a byproduct of the combustion of pulverized coal in electric power generating plants, has been used in concrete since the 1930s.

Classification of Fly Ash

Fly ashes are divided into two classes in accordance with ASTM C618 (AASHTO M 295), *Standard Specification for Coal Fly Ash and Raw or Calcined Natural Pozzolan for Use in Concrete;* Class F and Class C.

Class F fly ash has pozzolanic properties and is normally produced from burning anthracite or bituminous coal that meets the applicable requirements for this class as given herein. The silicon dioxide (SiO_2) plus aluminum oxide (Al_2O_3) plus iron oxide (Fe_2O_3) needs to be equal to or greater than 70% for Class F fly ash.

Class C fly ash in addition to having pozzolonic properties also has some cementitious properties. Class C fly ash is normally produced from lignite or subbituminous coal that meets the applicable requirements for this class as given herein. The silicon dioxide (SiO_2) plus aluminum oxide (Al_2O_3) plus iron oxide (Fe_2O_3) must be equal to or greater than 50% for Class C fly ash.

Physical Properties of Fly Ash

Most fly ash particles are fine solid spheres although some are hollow cenospheres (Figure 4-3). Also present are plerospheres, which are spherical particles containing smaller spheres. The particle sizes in fly ash vary from less than 1 µm (micrometer) to more than 100 µm with the average particle size measuring less than 20 µm (typically 10 µm). Typically only 10% to 30% of the particles by mass are larger than 45 µm. The specific surface area is typically 300 m^2/kg to 500 m^2/kg, although some fly ashes can have surface areas as low as 200 m^2/kg and some as high as 700 m^2/kg. For fly ash without close compaction, the bulk density (mass per unit volume including air between particles) can vary from 540 kg/m^3 to 860 kg/m^3 (34 lb/ft^3 to 54 lb/ft^3), whereas with close packed storage or vibration, the range can be 1120 kg/m^3 to 1500 kg/m^3 (70 lb/ft^3 to 94 lb/ft^3). The relative density (specific gravity) of fly ash generally ranges between 1.90 and 2.80 and the color is generally gray or tan.

Chemical Properties of Fly Ash

Fly ash is primarily alumino-silicate glass containing silica, alumina, iron, and calcium. Minor constituents are magnesium, sulfur, sodium, potassium, and carbon. The chemical composition of fly ash depends on the source of the coal. Because of the variety of coal types mined in North America the composition of fly ash covers a very wide range. Hard bituminous or anthracitic coals tend to produce ashes high in silica and alumina but low in calcium (Class F). Fly ashes from softer lignite of sub-bituminous coals tend to be much higher in calcium – anywhere from greater than 10% to greater than 30% CaO and are lower in silica and alumina (Class C).

Figure 4-3. Scanning electron microscope (SEM) micrograph of fly ash particles at approximately 1000x. Although most fly ash spheres are solid, some particles, called cenospheres, are hollow (as shown in the micrograph).

The performance of fly ash in concrete is strongly influenced by its chemical composition and fly ash from different sources can behave very differently. As the molten droplets of fly ash leave the burning zone in a coal plant the surface cools rapidly, preventing the formation of well-ordered crystalline phases, and the fly ash is dominated by amorphous or glassy material. The glass content can typically be as low as 50% and as high as 90%. The interior of the droplets cools more slowly, especially in the larger particles and this permits the formation of crystalline material.

The abundance and composition of the glass in fly ash is important. In low-calcium fly ash the crystalline phases that form – quartz, mullite, hematite, and magnetite – are inert in concrete, only the glass reacts and contributes to the properties of the concrete. The glass in low-calcium fly ash (typically Class F) is an aluminosilicate that will not react with water unless there is a source of calcium hydroxide or some other activator present. This material is a true pozzolan.

As the calcium content in the fly ash increases a number of things happen. Some of the calcium is incorporated into the glass. This changes the structure of the glass and makes it more reactive. It is possible for the glass to become somewhat hydraulic (it will react directly with water). In other words, Class C fly ash can be a hydraulic as well as a pozzolanic material.

ASTM C311, *Standard Test Methods for Sampling and Testing Fly Ash or Natural Pozzolans for Use in Portland-Cement Concrete*, provides test methods for fly ash (and natural pozzolans) for use as SCMs in concrete. Refer to ACI Committee 232 (ACI 232.2R-03), and Thomas (2007) for more information on using fly ash in concrete.

Slag Cement

Slag cement, previously known as ground, granulated blast-furnace slag (GGBFS), is the glassy material formed from molten slag produced in blast furnaces as a byproduct from the production of iron used in steel making. The molten slag is formed as the ingredients used to make iron melt at a temperature of about 1500°C (2730°F) and float above the denser molten iron. In order to transform the molten slag into a cementitious material, it is rapidly quenched in water to form a glassy, sand-like, granulated material, then dried and ground into a fine powder (Figure 4-4). If the slag is allowed to cool slowly in air, it will form crystalline products that have no cementitious properties. Air-cooled slag is inert, and is used for other applications, such as aggregate for structural backfill (see Chapter 6).

Slag cement, when used in general purpose concrete in North America, commonly constitutes between 30% and 50% of the cementitious material in the mixture (PCA 2000). Some slag concretes have a slag component of 70% or more of the cementitious material for applications such as mass concrete and marine environments.

Figure 4-4. Ground granulated blast-furnace slag.

Classification of Slag Cement

There are three grades of slag cement in accordance with ASTM C989, *Specification for Slag Cement for Use in Concrete and Mortars* (AASHTO M 302, *Standard Specification for Ground Granulated Blast-Furnace Slag for Use in Concrete and Mortars*). Grade 80 is slag cement with a low activity index. Grade 100 is slag cement with a moderate activity index and Grade 120 is slag cement with a high activity index. ASTM C989 defines activity based on 7- and 28-day strength results.

Physical Properties of Slag Cement

The rough and angular-shaped granulated slag is ground to produce slag cement, a fine powder material of approximately the same or greater fineness than portland cement. Most of the angular-shaped particles are ground to less than 45 µm and have a fineness of about 400 m^2/kg to 600 m^2/kg Blaine (Figure 4-5). The relative density (specific gravity) for slag cement is in the range of 2.85 to 2.95. The bulk density varies from 1050 kg/m^3 to 1375 kg/m^3 (66 lb/ft^3 to 86 lb/ft^3). The appearance of the finished product resembles white cement.

Figure 4-5. Scanning electron microscope micrograph of slag particles at approximately 2100X.

Chemical Properties of Slag Cement

The chemical composition of slag cement depends mainly on the composition of the charge to the blast furnace – oxides of silica calcium, alumina, magnesium and iron generally make up more than 95% of the slag cement. Although the composition may vary between sources, the variation from an individual plant is generally low due to the restrictive control of the iron-making process.

Slag cement in the presence of water and an activator such as NaOH or $Ca(OH)_2$ (both of which are supplied by portland cement) hydrates and sets in a manner similar to portland cement. Slag cement will hydrate and set in the absence of an activator, but the process is very slow.

ACI 233 (2003) provides an extensive review of slag cement.

Silica Fume

Silica fume is the ultrafine non-crystalline silica produced in electric-arc furnaces as an industrial byproduct of the production of silicon metals and ferrosilicon alloys. Silica fume is also known as condensed silica fume, or microsilica. In silicon metal production, a source of high purity silica (such as quartz or quartzite) together with wood chips and coal are heated in an electric arc furnace to remove the oxygen from the silica (reducing conditions). Silica fume rises as an oxidized vapor from the 2000°C (3630°F) furnaces. When it cools it condenses and is collected in bag filters. The condensed silica fume is then processed to remove impurities.

Silica fume is typically used in amounts between 5% and 10% by mass of the total cementitious material. It is used in applications where a high degree of impermeability is needed and also in high-strength concrete.

Classification of Silica Fume

Silica fume for use as a pozzolanic material in concrete may be supplied in one of three forms in accordance with ASTM C1240 (AASHTO M 307), *Standard Specification for Silica Fume Used in Cementitious Mixtures*; as-produced, as a slurry mixed with water, or as a densified or compacted product. Because of its extremely fine particle size and low bulk density, as-produced silica fume is very difficult to handle. Silica fume is most commonly used in the U.S. in densified form (Figure 4-6).

Figure 4-6. Silica fume powder.

Physical Properties of Silica Fume

Silica fume is composed of very tiny spherical amorphous (non-crystalline) particles. It is extremely fine with particles less than 1 µm in diameter and with an average diameter of about 0.1 µm, about 100 times smaller than average cement particles (Figure 4-7).

Condensed silica fume has a surface area of about 20,000 m^2/kg (nitrogen adsorption method). For comparison, tobacco smoke's surface area is about 10,000 m^2/kg. Type I and Type III cements typically have surface areas of about 300 m^2/kg to 400 m^2/kg and 500 m^2/kg to 600 m^2/kg (Blaine), respectively.

The bulk density (uncompacted unit weight) of silica fume varies from 130 kg/m^3 to 430 kg/m^3 (8 lb/ft^3 to 27 lb/ft^3). The relative density (specific gravity) is usually around 2.20, but can be as high as 2.50 if the iron content is high. The color of silica fume is gray, with the actual shade primarily dependent on the carbon content. Specially processed white silica fume is available when color is an important consideration.

Figure 4-7. Scanning electron microscope micrograph of silica fume particles at 20,000X.

Silica fume is sold more commonly in powder form but is also available internationally as a slurry. Densification may be achieved by passing compressed air through a silo containing silica fume causing the particles to tumble and collide resulting in loosely bound agglomerations. Alternatively, densification can be achieved by mechanical means – such as passing the material through a screw auger. The bulk density is increased to above 500 kg/m^3 by densification.

Chemical Properties of Silica Fume

Silica fume is predominantly composed of silicon dioxide, SiO_2, in noncrystalline, glassy (amorphous) form. When produced during the manufacture of silicon metal, the silica content is usually above 90% by mass and may even be as high as 99%. For silica fumes produced during ferrosilicon alloy production, the quantity of silica in the fume decreases as the amount of silicon in the alloy decreases. For alloys with a nominal silicon content of 75%, the resulting silica fume should still contain 85% silica or more. However, the manufacture of alloys containing only 50% silicon will generally produce silica fumes with less than 85% silica and such materials may be not be permitted by specifications for use in concrete.

ACI 234 (2006) and SFA (2008) provide an extensive review of silica fume.

Natural Pozzolans

Natural pozzolans are produced from natural mineral deposits. Some of these materials require heat treatment, known as calcining, to make them pozzolanic. Others such as volcanic ash can be used with only minimal processing (such as drying and grinding). The North American experience with natural pozzolans dates back to early 20th century public works projects, such as dams, where they were used to control temperature rise in mass concrete. In the 1930s and 1940s a blended cement containing 25% of calcined Monterey shale interground with portland cement was produced in California and was used by the State Division of Highways in several structures including the Golden Gate Bridge. In addition to controlling heat rise, natural pozzolans were used to improve resistance to sulfate attack and were among the first materials found to mitigate alkali-silica reaction.

Metakaolin. Metakaolin is a calcined clay which is produced by low-temperature calcination of kaolin clay. The clay is purified by water processing prior to very carefully controlled thermal activation at relatively low temperature (650°C to 800°C [1202°F to 1472°F]) compared to other calcined clays.

Metakaolin is used in applications where very low permeability or very high strength is required. In these applications, metakaolin is used more as an additive to the concrete rather than a replacement of cement; typical additions are around 10% of the cement mass.

Classification of Natural Pozzolans

Natural pozzolans are classified by ASTM C618 (AASHTO M 295) as Class N pozzolans. The most common are derived from shales or clays that are calcined, or heated to a sufficient temperature to transform the clay structure into a disordered amorphous aluminosilicate with pozzolanic properties. This process is often referred to as thermal activation. The optimum heat treatment varies for different materials. After heating the material is usually ground (Figures 4-8 and 4-9). Less common natural pozzolans include rice husk ash (RHA), diatomaceous earth, and volcanic ashes.

Metakaolin has been marketed under the description high-reactivity metakaolin to distinguish it from the normally less reactive calcined clays.

Figure 4-8. Scanning electron microscope micrograph of calcined shale particles at 5000X.

Figure 4-9. Scanning electron microscope micrograph of calcined clay particles at 2000X.

Figure 4-10. Metakaolin, a calcined clay.

Physical Properties of Natural Pozzolans

Calcined clays are used in general purpose concrete construction much the same as other pozzolans. Calcined clays have a relative density of between 2.40 and 2.61 with Blaine fineness ranging from 650 m^2/kg to 1350 m^2/kg.

Metakaolin is ground to a very high fineness with an average particle size of about 1 to 2 µm (Figure 4-10). It has a high Hunter L whiteness value of 90 on a scale of 0 for black to 100 for maximum whiteness.

Chemical Properties of Natural Pozzolans

Calcined shale may contain on the order of 5% to 10% calcium, which results in the material having some cementitious or hydraulic properties. Because of the amount of residual calcite that is not fully calcined, and the bound water molecules in the clay minerals, calcined shale will have a loss on ignition (LOI) of approximately 1% to 5%. The LOI value for calcined shale is not a measure or indication of carbon content as would be the case for fly ash. ACI Committee 232 (2001) provides a review of natural pozzolans.

Metakaolin is composed of kaolin clay consisting predominantly of silica (SiO_2) and alumina (Al_2O_3). The sum of the silica and alumina content is generally greater than 95% by mass. The remaining portion contains small amounts of iron, calcium, and alkalies.

Reactions of SCMs

Pozzolonic Reactions

As discussed in Chapter 3, calcium silicate hydrate (C-S-H), is the primary cementitious binder in hardened portland cement concrete and is often referred to as the "glue" that holds everything else together. Calcium hydroxide (CH) has little or no cementitious properties and contributes little to the strength of the hydrated material. It is often considered to be a weak link in concrete as it is easily leached by water and attacked by chemical agents. However, CH does help maintain the high pH necessary to stabilize C-S-H, the main binding phase.

A siliceous pozzolan reacts with CH formed by the hydration of portland cement and the reaction results in the production of more C-S-H. This C-S-H may have a slightly different composition and structure than that produced by portland cement hydration, but it contributes to the strength and reduces the permeability of the hardened material.

Many pozzolans contain reactive alumina in addition to silica, which react and produce various calcium-aluminate hydrates and calcium-alumino-silicate hydrates. The alumina may also participate in reactions with the various sulfate phases in the system.

The rate of the pozzolanic reaction is influenced by a number of parameters. The greater the surface area of the pozzolan, the faster it can react. This is what makes silica fume an extremely reactive material that contributes to concrete properties at very early ages. The composition and amount of glass in the pozzolan also affects the rate of reaction. For instance, calcium is a glass modifier and renders the glassy phases of Class C fly ash more rapidly reactive than Class F fly ash. Like most chemical reactions, the pozzolanic reaction increases with temperature. However, the pozzolanic reaction appears to be more sensitive to temperature than the normal hydration of portland cement. The solubility of glass increases with pH and this in turn increases its availability for reaction with CH. Because of this, a pozzolan will tend to react more quickly when combined with a high-alkali portland cement.

Hydraulic Reactions

Slag cement is referred to as a latent hydraulic material. This means that it will hydrate when it is mixed solely with water but the process is typically slow. Latent hydraulic materials benefit from chemical activation to promote significant hydration and the most effective activators for slag cement are calcium hydroxide or alkali compounds. This is fortunate because the hydration of portland cement releases both calcium and alkali hydroxides and serves as a very efficient activator for slag cement. There are also some high calcium fly ashes which have latent hydraulic properties.

When portland cement hydrates it produces C-S-H and CH together with other aluminate and ferrite phases. The hydration of slag cement is similar except that no calcium hydroxide is produced. A small amount of the calcium hydroxide released by the portland cement may be consumed by the hydration of the slag cement and this could be considered to be a form of pozzolanic reaction. The hydration of a portland-slag cement blend will produce more C-S-H and less CH than a straight portland cement.

As with pozzolans, the rate of reaction of slag cement in concrete is dependent on many factors. Reactivity increases with the fineness of the slag cement and the glass content of the slag, and with temperature. The composition of the slag cement is also important. Because slag cement requires activation by the portland cement, it will tend to react faster when the ratio of portland cement to slag cement is higher. Like pozzolans, the rate of dissolution of the glass in slag cement increases at higher pH. Slag cement will react faster when combined with high-alkali cement.

Slag cement hydration is significantly accelerated when the temperature is elevated and like pozzolans, the rate of slag cement reaction is generally more sensitive to temperature effects than the rate of portland cement hydration.

Effects on Freshly Mixed Concrete

This section provides a brief review of the freshly mixed concrete properties that SCMs affect and their degree of influence (assuming everything else is equal). Table 4-3 summarizes general effects that these materials have on the fresh properties of concrete mixtures. It should be noted that effects vary considerably between and often within classifications of SCMs. Specified performance should be evaluated for specific mixtures.

Water Demand

Of all the different SCMs, fly ash (Class C and F) has the most beneficial effect on water demand. Concrete mixtures containing fly ash generally require less water (about 1% to 10% less at normal dosages) for a given slump than concrete containing only portland cement. Higher dosages can result in greater water reduction (Table 4-4). However, fly ashes with a high percentage of coarse particles (larger than 45 µm) are less efficient in reducing the water demand and in extreme cases may increase the amount of water required up to 5% (Gebler and Klieger 1986). Fly ash reduces water demand in a manner similar to liquid chemical water reducers (Helmuth 1987).

The use of slag cement also typically results in a reduction in the water demand of concrete although the impact is dependent on the slag fineness and is less marked than with fly ash. In general, water demand for a given slump in concrete mixtures with slag cement will be 3% to 5% lower than ordinary portland cement concrete.

The water demand of concrete containing silica fume increases with higher amounts of silica fume. Some mixtures may not experience an increase in water demand when only a small amount (less than 5%) of silica fume is present.

Calcined clays and calcined shales generally have little effect on water demand at normal dosages; however, other natural pozzolans can significantly increase or decrease water demand. The use of finely ground calcined clay, such as metakaolin, may lead to an increased water demand, especially when used at higher dosages. The increased requirement for water can be offset by the use of water-reducing chemical admixtures.

Table 4-3. The Impact of SCM Characteristics on the Fresh Properties of Concrete*

	Fly ash		Slag cement	Silica fume	Natural pozzolans		
	Class F	Class C			Calcined shale	Calcined clay	Metakaolin
Water demand	↓	↓	↓	↑	↔	↔	↑
Workability	↑	↑	↑	↓	↑	↑	↓
Bleeding and segregation	↓	↓	↕	↓	↔	↔	↓
Setting time	↑	↕	↑	↔	↔	↔	↔
Air content	↓	↓	↔	↓	↔	↔	↓
Heat of hydration	↓	↕	↓	↔	↓	↓	↔

Key: ↓ Lowers ↑ Increases ↕ May increase or lower ↔ No impact

*The properties will change dependant on the material composition and dosage and other mixture parameters. Adapted from Thomas and Wilson (2002).

Table 4-4. Effect of Fly Ash on Mixing Water Requirements for Air-Entrained Concrete

Fly ash mix identification	Class of fly ash	Fly ash content, % by mass of cementitious material	Change in water requiremnt compared to control, %
C1A	C	25	-6
C1D	F	25	-2
C1E	F	25	-6
C1F	C	25	-8
C1G	C	25	-6
C1J	F	25	-6
C2A	C	50	-18
C2D	F	50	-6
C2E	F	50	-14
C2F	C	50	-16
C2G	C	50	-12
C2J	F	50	-10

All mixtures had cementitious materials contents of 35 kg/m^3 (564 lb/yd^3), a slump of 125 ± 25 mm (5 ± 1 in.), and an air content of 6 ± 1%. Water to cement plus fly ash ratios varied from 0.40 to 0.48 (Whiting 1989).

Workability

Generally, the use of fly ash, slag cement, and calcined clay and shale increase workability. This means that for a given slump, concrete containing these materials are generally easier to place, consolidate, and finish. The use of SCMs generally aids the pumpability of concrete.

Concrete containing SCMs will generally have equal or improved finishability compared to similar concrete mixtures without them. Mixes that contain high dosages of supplementary cementitious materials, especially silica fume, can increase the "stickiness" of a concrete mixture. Adjustments, such as the use of high-range water reducers, may be required to maintain workability and permit proper compaction and finishing.

Bleeding and Segregation

In general, the finer the supplementary cementing material, the lower the bleed rate and bleeding capacity. Increasing the SCM content will also typically lower bleeding.

Concretes using fly ash generally exhibit less bleeding and segregation than plain concretes (Table 4-5). This effect makes the use of fly ash particularly valuable in concrete mixtures made with aggregates that are deficient in fines. The reduction in bleed water is primarily due to the reduced water demand of mortars and concretes using fly ash (Gebler and Klieger 1986).

The effect of slag cement on bleeding and segregation is generally dependent on its fineness. Concretes containing ground slags of comparable fineness to that of the portland cement tend to show an increased rate and amount of bleeding than plain concretes, but this appears to have no adverse effect on segregation. Slag cements ground finer than portland cement tend to reduce bleeding.

The incorporation of silica fume in concrete has a very profound effect on bleeding. Concrete mixtures containing normal levels of silica fume (5% to 10%) and water-to-cementitious materials ratios below 0.50 may not bleed. Because of this, special attention needs to be paid to placing, finishing, and curing operations when using silica fume.

Calcined clays and calcined shales have little effect on bleeding. The effect will depend on fineness and level of replacement. For instance a very finely ground calcined clay such as metakaolin will reduce bleeding.

Table 4-5. Effect of Fly Ash on Bleeding of Concrete (ASTM C232, AASHTO T 158)*

Fly ash mixtures		Bleeding	
Identification	Class of fly ash	Percent	mL/cm²**
A	C	0.22	0.007
B	F	1.11	0.036
C	F	1.61	0.053
D	F	1.88	0.067
E	F	1.18	0.035
F	C	0.13	0.004
G	C	0.89	0.028
H	F	0.58	0.022
I	C	0.12	0.004
J	F	1.48	0.051
Average of:			
Class C		0.34	0.011
Class F		1.31	0.044
Control mixture		1.75	0.059

*All mixtures had cementitious materials contents of 307 kg/m³ (517 lb/yd³), a slump of 75 ± 25 mm (3 ± 1 in.), and an air content of 6 ± 1%. Fly ash mixtures contained 25% ash by mass of cementitious material (Gebler and Klieger 1986).

** Volume of bleed water per surface area.

Plastic Shrinkage Cracking. Concrete containing silica fume may exhibit an increase in plastic shrinkage cracking because of its low bleeding characteristics. The problem may be avoided by ensuring that such concrete is protected against drying, before, during and after finishing. Other pozzolans and slag cement generally have little effect on plastic shrinkage cracking. However, supplementary cementitious materials that significantly delay set time can increase the risk of plastic shrinkage cracking.

Setting Time

The use of SCMs will generally retard the setting time of concrete (Table 4-6). The extent of set retardation depends on many factors including the fineness and composition of the SCM and the level of replacement used, amount and composition of the portland cement or blended cement (particularly its alkali content), water-to-cementitious materials ratio (w/cm), and temperature of the concrete. Set retardation is an advantage during hot weather, allowing more time to place and finish the concrete. However, during cold weather, pronounced retardation can occur with some materials, significantly delaying finishing operations. Accelerating admixtures can be used to decrease the setting time.

Lower-calcium Class F fly ashes, slag cement, and most natural pozzolans tend to delay the time of setting of concrete. At normal laboratory temperatures the extent of the retardation is generally in the range of 15 minutes to one hour for initial set and 30 minutes to two hours for final set. Longer delays in setting will be experienced when concrete is placed at lower temperatures, while little effect may be observed at higher temperatures.

The situation for higher-calcium, Class C fly ashes is less clear. The use of some Class C ashes may lead to a slight acceleration, reducing the time to initial and final set, especially in hot weather. Other Class C ashes can cause very significant retardation of a few hours or more, especially when used at relatively high replacement levels. It is not possible to predict the effect based on the chemistry of the fly ash. Individual cement-ash-admixture combinations must be tested.

Silica fume is generally used at relatively low replacement levels and has little significant impact on the setting behavior of concrete. Likewise, metakaolin and calcined shale and clay have little effect on setting time unless used at high dosage replacement rates for portland cement.

Air Content

Supplementary cementing materials generally require an increase in the amount of air-entraining admixture necessary due to the increase in fineness of the cementitious materials content.

For fly ash concrete mixtures, the amount of air-entraining admixture required to achieve a certain air content in the concrete is a primarily a function of the carbon content, fineness, and alkali content of the fly ash. Increases in alkali contents decrease air-entraining admixture dosage requirements, while increases in the fineness and carbon content typically require an increase in dosage requirements. A Class F fly ash with high carbon content (as indicated by a high loss-on-ignition or LOI) may increase the required admixture dosage by as much as 5 times compared to a portland cement concrete without fly ash. A Class C fly ash with low carbon content typically requires 20% to 30% more admixture. The Foam Index test provides a rough indication of the required dosage of air-entraining admixture for various fly ashes relative to non-ash mixtures. It can also be used to anticipate the need to increase or decrease the dosage based on changes in the foam index (Dodson 1990).

Table 4-6. Effect of Fly Ash on Setting Time of Concrete

Fly ash test mixture		Setting time, hr:min		Retardation relative to control, hr:min	
Identification	**Classs of fly ash**	**Initial**	**Final**	**Initial**	**Final**
A	C	4:30	5:35	0:15	0:05
B	F	4:40	6:15	0:25	0:45
C	F	4:25	6:15	0:10	0:45
D	F	5:05	7:15	0:50	1:45
E	F	4:25	5:50	0:10	0:20
F	C	4:25	6:00	0:10	0:30
G	C	4:55	6:30	0:40	1:00
H	F	5:10	7:10	0:55	1:40
I	C	5:00	6:50	0:45	1:20
J	F	5:10	7:40	0:55	2:10
Average of:					
Class C		4:40	6:15	0:30	0:45
Class F		4:50	6:45	0:35	1:15
Control mixture		4:15	5:30	—	—

Concretes had a cementitious materials content of 307 kg/m^3 (517) lb/yd^3). Fly ash mixtures contained 25% ash by mass of cementitious material. Water to cement plus fly ash ratio = 0.40 to 0.45. Tested at 23°C (73°F) (Gebler and Klieger 1986).

Slag cements can have a small effect on the required dosage of air-entraining admixtures. The increase in admixture dosage required with slag cement is associated with the fine particle size of these materials as compared with portland cement.

Silica fume has a marked influence on the air-entraining admixture dosage requirements, which in most cases rapidly increases with an increase in the amount of silica fume used in the concrete. The inclusion of silica fume may result in a large increase in the air-entraining admixture requirement due to the large surface area of the silica fume – which adsorbs part of the admixture.

Heat of Hydration

The majority of supplementary cementing materials (fly ash, natural pozzolans, and slag cement) typically have a lower heat of hydration than portland cement. Consequently their use will reduce the amount of heat built up during hydration in a concrete structure. Supplementary cementitious materials are often used purposely to reduce the heat energy of mass concrete (Gajda 2007). However, the impact of SCMs on the heat of hydration must be confirmed prior to use.

When Class F fly ash is used in cementitious systems, the contribution to the heat of hydration by the fly ash is approximately 50% of that of portland cement (Gajda and VanGeem 2002). This value is an accepted approximation. Since all cements and fly ashes are different, if specific data is needed, then the heat of hydration should be measured. Class C fly ash is less commonly used in mass concrete because its heat of hydration is higher than that of Class F. The heat of hydration of Class C fly ash depends on the CaO content. For low-CaO Class C fly ashes, the heat of hydration may be similar to that of Class F fly ash. However, for higher-CaO Class C fly ashes, the heat of hydration may be similar to that of cement.

When used in sufficient quantities, slag cement reduces the temperature rise of concrete (Gajda 2007). Figure 4-11 shows the effect of 70% slag cement on the heat of hydration. The amount of reduction is based on the proportion of slag in the concrete. Different combinations of slag cement and portland and blended cements will behave differently.

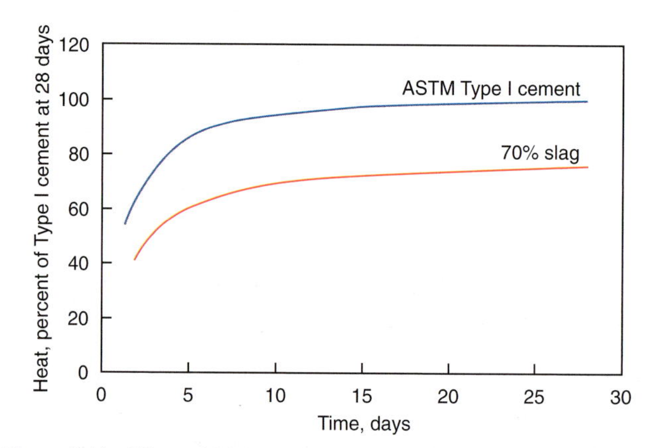

Figure 4-11. Effect of 70% slag cement on heat of hydration at 20°C (68°F) compared to a Type I cement.

Silica fume typically contributes a heat of hydration that is equal to or somewhat greater than that of portland cement depending on the dosage level. Unless specific testing is done to determine the heat of hydration of a specific mix design, an acceptable approximation is that the contribution to the heat of hydration of the cementitious system by silica fume is 100% to 120% that of portland cement (Pinto and Hover 1999). However, for a given strength, use of silica fume typically reduces total cementitious materials content which results in a lower heat of hydration.

Calcined clay imparts a heat of hydration similar to moderate heat cement (Barger and others 1997). However, metakaolin has a heat of hydration equal to or somewhat greater than that of portland cement. A good estimate is that the heat of hydration of metakaolin is 100% to 125% that of portland cement (Gajda 2007). To get a more exact determination of the heat of hydration, specific testing should be done.

Detwiler, Bhatty, and Bhattacharja (1996) provide a review of the effect of pozzolans and slag cement on heat generation.

Effects on Hardened Concrete

This section provides a summary of the hardened concrete properties that SCMs affect and their degree of influence (assuming everything else is equal). These materials vary considerably in their effect on concrete mixtures as summarized in Table 4-7. The effect of different SCMs within a classification can also vary.

Strength

The extent to which strength development of concrete is influenced by supplementary cementitious materials will depend on many factors such as: composition and amount of SCM, cement chemistry (particularly its alkali content), mixture proportions of the concrete, and temperature conditions during placement and curing.

In general, supplementary cementing materials (fly ash, slag cement, silica fume, calcined shale, and calcined clay (including metakaolin)) all contribute to the long-term strength gain of concrete. Concretes made with certain highly reactive fly ashes (especially high-calcium Class C ashes) or slag cements can equal or exceed the control strength in 1 to 28 days. Some fly ashes and natural pozzolans require 28 to 90 days to exceed control strength, depending on the mixture proportions and curing conditions. Tensile, flexural, torsion, and bond strength are affected in the same manner as compressive strength.

However, the strength of concrete containing these materials can be either higher or lower than the strength of concrete using only portland cement depending on the age and/or the ambient placement and curing temperature. Figure 4-12 illustrates this effect for various fly ashes.

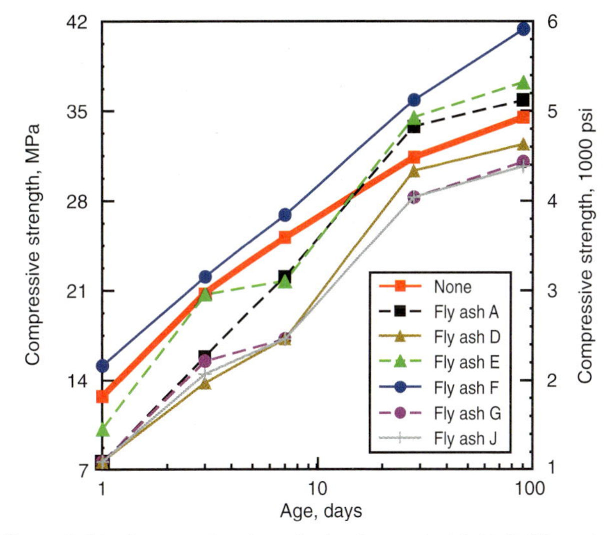

Figure 4-12. Compressive strength development at 1, 3, 7, 28, and 90 days of concrete mixtures containing 307 kg/m³ (517 lb/yd³) of cementitious materials with a fly ash dosage of 25% of the cementitious materials (Whiting 1989).

When a portion of portland cement or a blended cement is replaced by an SCM the strength-w/cm relationship changes. The direction in which it changes depends upon the nature of the SCM, its dosage level, the age of test, and other factors. Provided the relationship is known, the desired strength can be obtained with any concrete mixture by simple selection of the appropriate w/cm for that particular mixture. Therefore, the lower early-age strength observed with some SCMs can be compensated for by using a lower w/cm mixture.

Because of its increased reactivity and cementitious behavior, Class C fly ash makes a greater contribution to early age concrete strength than Class F fly ash. Class C fly ash generally achieves the same or similar strength as the control after 3 to 7 days assuming comparable ambient temperature.

Slag cement at a 35% replacement can reach the same strength as a mixture without SCM after around 7 days assuming comparable total cementitious material contents, w/cm, and ambient placement and curing temperature.

Because of its fine particle size, high surface area, and highly amorphous nature, silica fume reacts very rapidly in concrete and its contribution to strength is seen very early on. One-day strength using silica fume may be slightly increased or decreased depending on the nature of the silica fume and the alkali content of the portland cement. Between 3 and 28 days, the silica fume concrete invariably shows superior strength gain compared with the control, assuming comparable ambient placement and curing temperatures. Silica fume also aids in increasing the early strength gain of fly ash-cement concretes. Other very fine, highly reactive pozzolans, such as metakaolin, might be expected to show similar behavior.

Table 4-7. The Impact of SCM Characteristics on the Hardened Properties of Concrete*

	Fly ash		Slag cement	Silica fume	Natural pozzolans		
	Class F	Class C			Calcined shale	Calcined clay	Metakaolin
Early age strength gain	Lowers	No impact	May increase or lower	Increases	Lowers	Lowers	Increases
Long term strength gain	Increases	Increases	Increases	Increases	Increases	Increases	Increases
Abrasion resistance	No impact	No impact	No impact	No impact	No impact	No impact	No impact
Drying shrinkage and creep	No impact	No impact	No impact	No impact	No impact	No impact	No impact
Permeability and absorption	Lowers	Lowers	Lowers	Lowers	Lowers	Lowers	Lowers
Corrosion resistance	Increases	Increases	Increases	Increases	Increases	Increases	Increases
Alkali-silica reactivity	Lowers	Lowers	Lowers	Lowers	Lowers	Lowers	Lowers
Sulfate resistance	Increases	May increase or lower	Increases	Increases	Increases	Increases	Increases
Freezing and thawing	No impact	No impact	No impact	No impact	No impact	No impact	No impact
Deicer scaling resistance	May lower or have no impact	May lower or have no impact	May lower or have no impact	May lower or have no impact	May lower or have no impact	May lower or have no impact	May lower or have no impact

Key: Lowers; Increases; May increase or lower; No impact; May lower or have no impact

*The properties will change dependant on the material composition, dosage and other mixture parameters. These general trends may not apply to all materials and therefore testing should be performed to verify the impact. Adapted from Thomas and Wilson (2002).

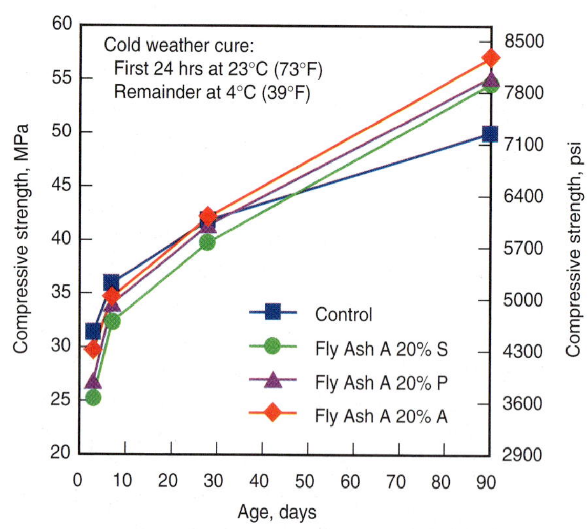

Figure 4-13. Compressive strengths for concretes cured at 23°C (73°F) for the first 24 hours and 4°C (40°F) for the remaining time. Control had a cement content of 332 kg/m³ (560 lb/yd³) and w/c of 0.45. The fly ash curves show substitution for cement (S), partial (equal) substitution for cement and sand (P), and addition of fly ash by mass of cement (A). The use of partial cement substitution or addition of fly ash increases strength development comparable to the cement-only control, even in cold weather (Detwiler 2000).

Figure 4-13 illustrates the benefit of using fly ash as an addition instead of a partial cement replacement to improve strength development in cold weather. Mass concrete design often takes advantage of the delayed strength gain of pozzolans as these structures are often not put into full service immediately. With appropriate mixture adjustments, all SCMs can be used in all seasons and still provide the necessary strength (see Chapters 16 and 17 on hot and cold weather concreting).

Supplementary cementitious materials are often used in the production of high-strength concrete. Fly ash has been used in production of concrete with strengths up to 100 MPa (15,000 psi). With silica fume, ready mix producers now have the ability to make concrete with strengths up to 140 MPa (20,000 psi), when used with high-range water reducers and appropriate aggregates (Burg and Ost 1994).

Because of the slow pozzolanic reaction of some SCMs, continuous wet curing and favorable curing temperatures may be required for periods longer than normal to ensure adequate strength gain. A seven-day moist cure or membrane cure should be adequate for concretes with normal dosages of most SCMs. As with portland cement concrete, low curing temperatures can reduce early-strength gain (Gebler and Klieger 1986).

Impact and Abrasion Resistance

The impact resistance and abrasion resistance of concrete are related to compressive strength and aggregate type. Supplementary cementing materials generally do not affect these properties beyond their influence on strength. Concretes containing fly ash are just as abrasion resistant as portland cement concretes without fly ash (Gebler and Klieger 1986). Figure 4-14 illustrates that abrasion resistance of fly ash concrete is related to strength.

Drying Shrinkage and Creep

When used in low to moderate amounts, the effect of fly ash, slag cement, calcined clay, calcined shale, and silica fume on the drying shrinkage and creep of concrete is generally small and not considered significant. Some studies indicate that silica fume may reduce specific creep (Burg and Ost 1994).

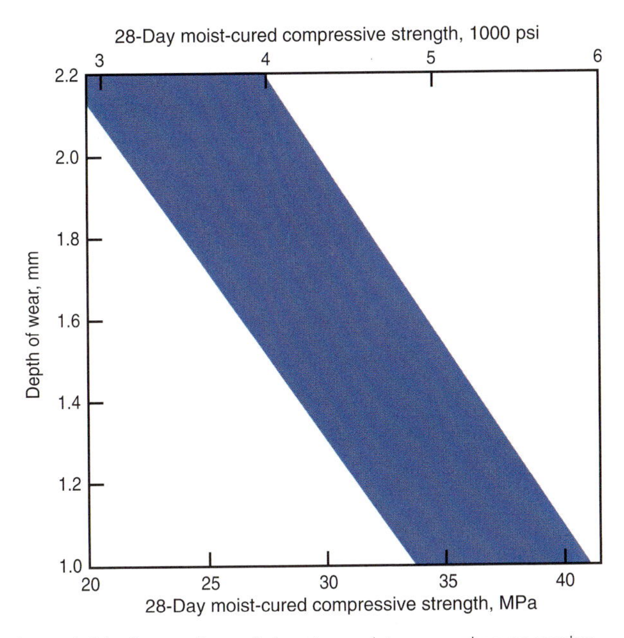

Figure 4-14. Comparison of abrasion resistance and compressive strength of various concretes with 25% fly ash. Abrasion resistance increases with strength (Gebler and Klieger 1986).

Permeability and Absorption

With appropriate design of the concrete mixture, control of w/cm, and adequate curing; fly ash, slag cement, and natural pozzolans generally reduce the permeability and absorption of concrete. Silica fume and metakaolin are especially effective in this regard. Silica fume and calcined clay can provide concrete with a chloride resistance of under 1000 coulombs using ASTM C1202, *Standard Test Method for Electrical Indication of Concrte's Ability to Resist Chloride Ion Penetration,* (Barger and others 1997). Tests show that the permeability of concrete decreases as the quantity of hydrated cementitious materials increases and is age dependent. Fly ash and slag cement plus ternary blends can provide (chloride resistance) values less than 1000 coulombs at later ages. The absorption of fly-ash concrete is about the same as concrete without ash, although some ashes can reduce absorption by 20% or more.

Corrosion Resistance

When concrete is properly cured, SCMs can help reduce reinforcing steel corrosion by reducing the permeability of concrete to water, air, and chloride ions.

Concrete with fly ash shows a slight improvement in the reduction to chloride ion ingress for concrete at an early age, but improves over time, reaching very low values at one year. Concrete with slag cement and some other pozzolans generally exhibits similar behavior.

The incorporation of silica fume can have a dramatic effect, producing concrete with very low chloride penetrability after just 28 days or so. Only very small improvements at later ages are observed. Concrete with metakaolin behaves in a similar manner to silica fume. Concrete containing silica fume or metakaolin is often used in overlays and full-depth slab placements on bridges and parking garages. These structures are particularly vulnerable to corrosion due to chloride-ion ingress.

Carbonation

The rate of carbonation is significantly increased in concretes with: a high water-cementitious materials ratio, low cement content, short curing period, low strength, and a highly permeable or porous paste. The depth of carbonation of good quality concrete is generally of little practical significance. At normal dosages, fly ash is generally reported to slightly increase the carbonation rate (Campbell and others 1991, and Stark and others 2002). However, it has been documented that concrete containing fly ash will carbonate at a similar rate compared with portland cement concrete of the same 28-day strength (Tsukayama 1980, Lewandowski 1983, Matthews 1984, Nagataki, Ohga and Kim 1986, Hobbs 1988, and Dhir 1989). This means that fly ash increases the carbonation rate provided that the basis for comparison is an equal w/cm. It has also been shown that the increase due to fly ash is more pronounced at higher levels of replacement and in poorly-cured concrete of low strength (Thomas and Matthews 1992, and Thomas and Matthews 2000). Even when concretes are compared on the basis of equal strength, concrete with fly ash (especially at high levels of replacement) may carbonate more rapidly in poorly-cured, low strength concrete (Ho and Lewis 1983, Ho and Lewis 1997, Thomas and Matthews 1992, and Thomas and Matthews 2000).

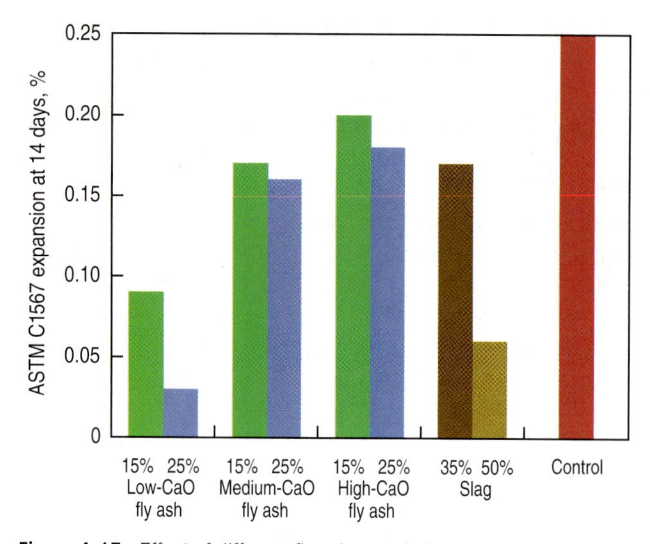

Figure 4-15. Effect of different fly ashes and slag on alkali-silica reactivity. Note that some ashes are more effective than others in controlling the reaction and that dosage of the ash or slag is critical. A highly reactive natural aggregate was used in this test. A less reactive aggregate would require less ash or slag cement to control the reaction. (Detwiler 2002).

Alkali-Silica Reactivity

Alkali-silica reactivity (ASR) can be controlled through the use of SCMs (Figures 4-15 and 4-16). Supplementary cementitious materials provide additional calcium silicate hydrate to chemically tie up the alkalies in concrete (Bhatty 1985 and Bhatty and Greening 1986) and reduce its permeability. When determining the optimum SCM content for ASR resistance, it is important to maximize the reduction in reactivity and to avoid dosages and materials that can aggravate reactivity. Dosage rates should be verified by tests, such as ASTM C1567, *Standard Test Method for Determining the Potential Alkali-Silica Reactivity of Combinations of Cementitious Materials and Aggregate (Accelerated Mortar-Bar Method)*, or ASTM C1293, *Standard Test Method for Determination of Length Change of Concrete Due to Alkali-Silica Reaction*.

The effectiveness of Class F fly ash in controlling expansion due to ASR is well established (Thomas 2007). However, the effect of fly ash varies considerably as the composition of the ash varies. The calcium content of the fly ash is an indicator of how the material behaves with regard to controlling ASR. Low-calcium fly ash (Class F) is clearly the best at controlling ASR. The efficiency of fly ash in this role decreases as the calcium oxide content increases above about 20% CaO (Figure 4-17). Low calcium Class F ashes have reduced reactivity expansion, in some cases up to 70% or more. At optimum dosage, Class C ashes can also reduce reactivity but to a lesser degree than most Class F ashes.

Fly ash can also contain significant quantities of soluble alkalis. This will tend to reduce is effectiveness in controlling ASR. Generally, the amount of fly ash required to control ASR will increase as any of the following parameters increase: calcium oxide or alkali content of the fly ash, reactivity of the aggregate, or amount of alkali available in the concrete.

ASR expansion decreases with use of slag cement. Generally, the amount of slag cement required to control expansion increases as either the reactivity of the aggregate or the amount of alkali in the mixture increases. As much as 50% slag cement may be needed to reduce the expansion to less than 0.04% in concrete with a highly-reactive aggregate. Typically, 35% slag is required with moderately reactive aggregate.

The effect of silica fume on ASR resistance appears strongly dependent on the alkali available in the concrete system. Studies have shown that lower amounts of silica fume can be used when the alkali content of the concrete is lower than typically used in the concrete prism tests (ASTM C1293, approximately 5.25 kg alkali/m^3 or 8.8 lb alkali/yd^3).

Metakaolin is also a highly reactive pozzolan and is nearly as effective on ASR resistance as silica fume – requiring a typical replacement level of somewhere between 10 to 15% to control expansion under these test conditions.

Supplementary cementitious materials that reduce alkali-silica reactions will not reduce alkali-carbonate reactions (ACR), a type of reaction involving concrete alkalies and certain dolomitic limestones.

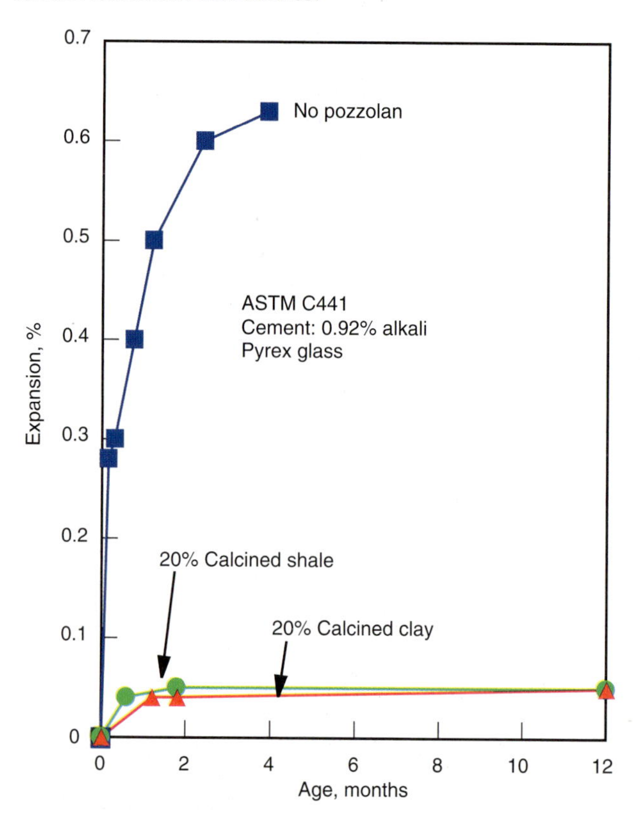

Figure 4-16. Reduction of alkali-silica reactivity by calcined clay and calcined shale (Lerch 1950).

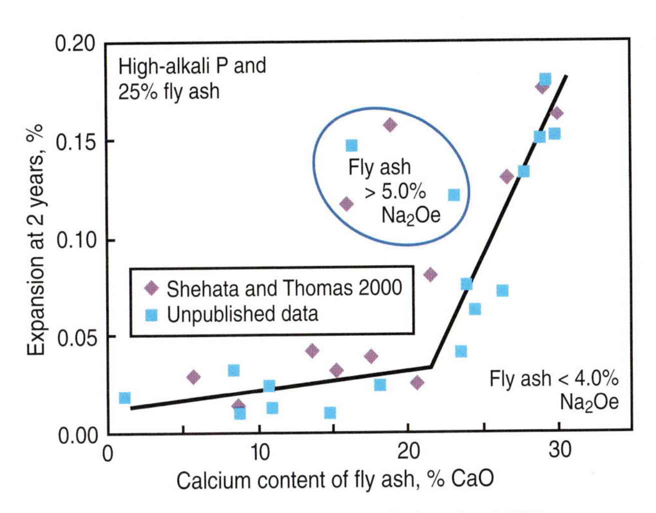

Figure 4-17. Fly ashes with CaO contents below about 20% are generally effective in reducing expansion in ASTM C1293 concrete prisms made with a reactive aggregate at 25% replacement rate. However, some fly ashes with high alkali contents are not. Without fly ash, this aggregate expanded about 0.25% at 2 years. (Thomas 2007).

Descriptions of aggregate testing and the measures needed to prevent deleterious alkali-aggregate reaction are discussed in Farny and Kerkhoff (2007), PCA (2007), AASHTO (2001) and Thomas, Fournier, and Folliard (2008). Chapter 11 provides more detailed information on ASR.

Sulfate Resistance

With proper proportioning and material selection, most supplementary cementing materials can improve the resistance of concrete to sulfate or seawater attack. This is done primarily by reducing permeability and by reducing the amount of reactive elements needed for expansive sulfate reactions. For improved sulfate resistance of lean concrete, one study (Stark 1989) showed that for a particular Class F ash, an adequate amount was approximately 20% of the cementitious materials content; this illustrates the need to determine optimum ash contents. In that case higher ash contents were found to be detrimental.

Concretes with Class F ashes are generally more sulfate resistant than those made with Class C ashes. Some Class C ashes have been shown to reduce sulfate resistance at normal dosage rates (Thomas 2007).

Slag cement is generally considered beneficial in sulfate environments. However, the Al_2O_3 content of the slag can influence its ability to mitigate sulfate attack (the higher the Al_2O_3 the less sulfate resistance). Studies indicate that properly designed concrete with slag cement can provide sulfate resistance equal to or greater than concretes made with Type V sulfate-resistant portland cement (ACI 233 and Detwiler, Bhatty, and Bhattacharja 1996).

Silica fume has been shown to provide sulfate resistance equivalent to slag and fly ash. (Fidjestøl and Frearson 1994)

Calcined clay has been demonstrated to provide sulfate resistance greater than sulfate resistant Type V cement (Barger and others 1997). Chapter 11 provides more information on concrete durability.

Chemical Resistance

Supplementary cementitious materials often reduce chemical attack by reducing the permeability of concrete. Although many of these materials may improve chemical resistance, they do not make concrete immune to chemical attack. Concrete in severe chemical exposure may need additional protection using barrier systems. Kerkhoff (2007) provides a discussion of methods and materials to protect concrete from aggressive chemicals and exposures.

Freeze-Thaw Resistance

As discussed in Chapters 6 and 11, the freeze-thaw resistance of concrete is dependent on the air void system of the paste, the strength of the concrete, the water-to-cementitious materials ratio, and the quality of aggregate relative to its freeze-thaw resistance. These properties in general are not influenced by SCMs with the exception of surface scaling.

Deicer-Scaling Resistance

Decades of field experience have demonstrated that air-entrained concretes containing normal dosages of fly ash, slag cement, silica fume, calcined clay, or calcined shale are resistant to the scaling caused by the application of deicing salts in a freeze-thaw environment (Figure 4-18). Laboratory tests also indicate that the deicer-scaling resistance of concrete made with SCMs is often equal to concrete made without SCMs.

Scaling resistance can decrease as the amount of certain SCMs increases. However, concretes that are properly designed, placed, and cured have demonstrated good scaling resistance even when made with high dosages of some SCMs.

Concrete containing supplementary cementing materials at appropriate dosages may be expected to be adequately resistant to the effects of deicing salts provided the concrete has a $w/cm \leq 0.45$, an adequate air-void system is present, and proper finishing and curing is applied. This includes insuring bleeding has ceased prior to final finishing, and the concrete is given ample opportunity to air-dry prior to first exposure to salts and freezing conditions (see Chapter 11 for more information on deicer scaling resistance).

The importance of using a low water-cement ratio for scaling resistance is demonstrated in Figure 4-19. The effect of high fly ash dosages and low cementing material contents is demonstrated in Figure 4-20. The performance of scale-resistant concretes containing fly ash at a dosage of 25% of cementing material by mass is presented in Table 4-8. The table demonstrates that well designed, placed and cured concretes with and without fly ash can be equally resistant to deicer scaling.

Figure 4-18. View of concrete slabs in PCA outdoor test plot (Skokie, Illinois) containing (A) fly ash, (B) slag, (C) calcined shale, and (D) portland cement after 30 years of deicer and frost exposure. These samples demonstrate the durability of concretes containing various cementitious materials. Source: RD124, RX157, LTS Cement No. 51, slabs with 335 kg/m³ (564 lb/yd³) of cementing material and air entrainment. (Stark and others 2002)

The ACI 318 building code states that the maximum dosage of fly ash, slag, and silica fume should be 25%, 50%, and 10% by mass of cementing materials, respectively for deicer exposures. Total SCM content should not exceed 50% of the cementitious material. Dosages less than or higher than these limits have been shown to be durable in some cases but not in others. Different materials respond differently in different environments. The selection of materials and dosages should be based on local experience and the durability should be demonstrated by field or laboratory performance.

Aesthetics

Supplementary cementitious materials may slightly alter the color of hardened concrete. Color effects are related to the color and amount of the material used in concrete. Many SCMs resemble the color of portland cement and therefore have little effect on color of the hardened concrete. Some silica fumes may give concrete a slightly bluish or dark gray tint and tan fly ash may impart a tan color to concrete when used in large quantities. Slag cement and metakaolin can make concrete whiter and may be used in place of white cement depending on dosage. While slag cement can initially impart a bluish or greenish undertone this effect will fade after time once exposed to oxygen.

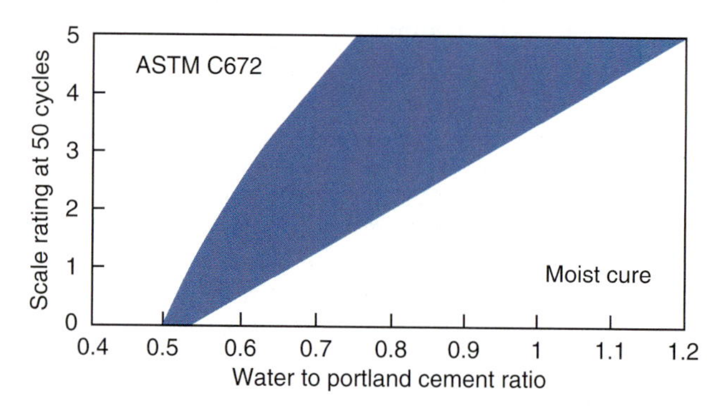

Figure 4-19. Relationship between deicer-scaling resistance and water to portland cement ratio for several air-entrained concretes with and without fly ash. A scale rating of 0 is no scaling and 5 is severe scaling (After Whiting 1989).

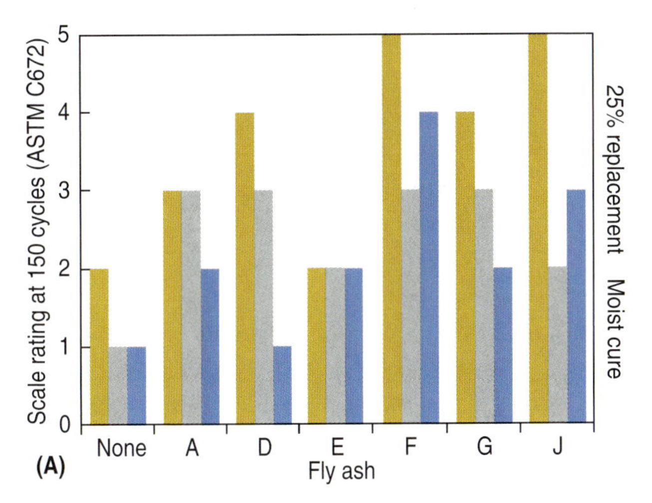

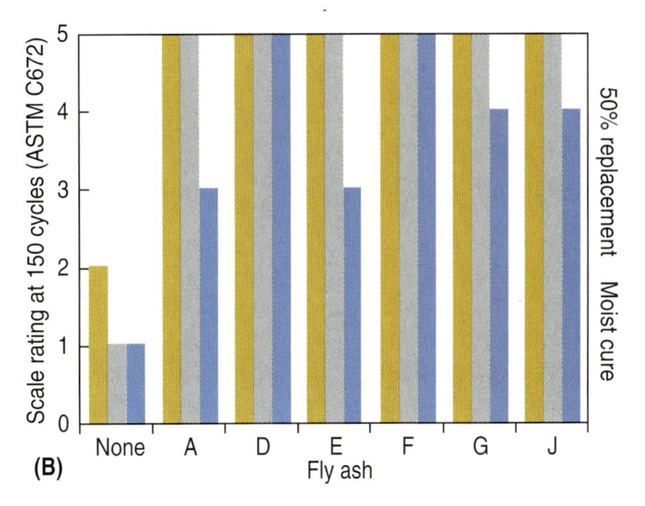

(A) and **(B)** – Cementitious material
250 kg/m³ (417 lb/yd³) 305 kg/m³ (508 lb/yd³) 335 kg/m³ (588 lb/yd³)

Figure 4-20. Relationship between deicer-scaling resistance and dosage of fly ash for air-entrained concretes made with moderate to high water-cementitious materials ratios. Replacement of portland cement with fly ash: (left) 25% and (right) 50%. A scale rating of 0 is no scaling and 5 is severe scaling (Whiting 1989).

Table 4-8. Frost and Deicer-Scaling Resistance of Fly Ash Concrete

Fly ash mixtures*		Results at 300 cycles				
		Frost resistance in water, ASTM C666 Method A (AASHTO T 161)			Deicer scaling resistance, ASTM C672**	
Identification	Class of fly ash	Expansion, %	Mass loss, %	Durability factor	Water cure	Curing compound
A	C	0.010	1.8	105	3	2
B	F	0.001	1.2	107	2	2
C	F	0.005	1.0	104	3	3
D	F	0.006	1.3	98	3	3
E	F	0.003	4.8	99	3	2
F	C	0.004	1.8	99	2	2
G	C	0.008	1.0	102	2	2
H	F	0.006	1.2	104	3	2
I	C	0.004	1.7	99	3	2
J	F	0.004	1.0	100	3	2
Average of:						
Class C		0.006	1.6	101	3	2
Class F		0.004	1.8	102	3	2
Control mixture		0.002	2.5	101	2	2

* Concrete mixtures had a cementitious materials content of 307 kg/m³ (517 lbs/yd³), a water to cementitious materials ratio of 0.40 to 0.45, a target air content of 5% to 7%, and a slump of 75 mm to 100 mm (3 in. to 4 in.). Fly ash dosage was 25% by mass of cementitious material (Gebler and Klieger 1986a).

** Scale rating (see at right)

0 = No scaling
1 = Slight scaling
2 = Slight to moderate scaling
3 = Moderate scaling
4 = Moderate to severe scaling
5 = Severe scaling

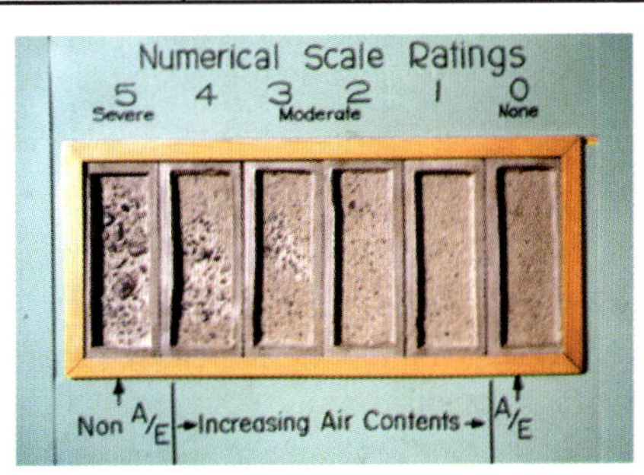

Concrete Mixture Proportions

The optimum amounts of SCMs used with portland cement or blended cement are determined by testing, taking into account the relative cost and availability of the materials, and the specified properties of the concrete.

Several test mixtures are required to determine the optimum amount of SCM to use in the concrete mixture. These test mixtures should cover a range of blends to establish the relationship between the property or properties being optimized and the SCM content. The water-to-cementitious materials ratio (w/cm) will also have a strong effect on many properties. These mixtures should be established according to ACI Standard 211.1 or 211.2, taking into account the relative densities of the supplementary cementitious materials. The results of these tests will be a series of values for each mixture tested, often as a function of age. For example, a family of compressive strength curves might be developed to evaluate the effects of SCM content and age. In other cases, the permeability at 56 days might be the property of interest. The content of a cementitious material is usually stated as a mass percentage of all the cementitious materials in a concrete mixture.

Multi-Cementitious Systems

Traditionally, fly ash, slag cement, calcined clay, calcined shale, and silica fume were used in concrete individually with portland cement. There are benefits in using more than one SCM in combination with portland cement to produce concrete mixtures. Ternary blends containing portland cement plus two SCMs are often used where special properties are required. Quaternary blends consisting of portland cement plus three SCMs have also been used.

There can be a pronounced synergy in combining either fly ash, slag cement, and/or silica fume (Thomas and Wilson 2002). Benefits include:

- fly ash (and, to a lesser extent, slag cement) offsets increased water demand of silica fume;
- Class F fly ash (and, to a lesser extent, slag cement) compensates for high heat release from blended cement with silica fume;
- silica fume compensates for low early-strength of concrete with Class F fly ash or slag cement;
- fly ash and slag cement increase long-term strength development of silica fume concrete;
- very high resistance to chloride ion penetration can be obtained with ternary blends; and
- silica fume reduces the normally high levels of Class C fly ash or slag cement required for sulfate resistance and ASR prevention.

Availability

All SCMs may not be available in all areas. Consult local material suppliers on availability. Fly ash type and availability will vary by region. Slag cements are available in most regions. Silica fume is available in most locations because only small amounts are used. Calcined clays and shales are available in select areas.

Storage

In some cases, moisture will not affect the physical performance of SCMs. In general these materials should be kept dry to avoid difficulties in handling and discharge. In particular, Class C fly ash, slag cement, and calcined shale must be kept dry as they will set and harden when exposed to moisture. Equipment for handling and storing these materials is similar to that required for portland cement. Some silica fumes are provided in liquid (slurry) form. Additional modifications may be required where using silica fume, which does not have the same flow characteristics as other SCMs.

SCMs are usually kept in bulk storage facilities or silos, although some products are available in bags. Because the materials may resemble portland cement in color and fineness, the storage facilities should be clearly marked to avoid the possibility of misuse and contamination with other materials at the batch plant. All valves and piping should also be clearly marked and properly sealed to avoid leakage and contamination. Fly ash, slag cement, and natural pozzolans should be weighed after the portland or blended cement in the batching sequence to avoid overdosing.

References

AASHTO, *Guide Specification For Highway Construction (SECTION 56X Portland Cement Concrete Resistant to Excessive Expansion Caused by Alkali-Silica Reaction* (Appendix F to ASR Transition Plan), http://leadstates.transportation.,org/ASR/library/gspec.stm, 2001.

ACI Committee 211, *Standard Practice for Selecting Proportions for Normal, Heavyweight, and Mass Concrete*, ACI 211.1-91, reapproved 2009, American Concrete Institute, Farmington Hills, Michigan, 1991, 38 pages.

ACI Committee 211, *Standard Practice for Selecting Proportions for Structural Lightweight Concrete*, ACI 211.2-98, reapproved 2004, American Concrete Institute, Farmington Hills, Michigan, 1998, 14 pages.

ACI Committee 232, *Use of Fly Ash in Concrete*, ACI 232.2R-03, American Concrete Institute, Farmington Hills, Michigan, 2004, 41 pages.

ACI Committee 232, *Use of Raw or Processed Natural Pozzolans in Concrete*, ACI 232.1R-00, reapproved 2006, American Concrete Institute, Farmington Hills, Michigan, 2001, 24 pages.

ACI Committee 233, *Slag Cement as a Cementitious Constituent in Concrete*, ACI 233R-03, American Concrete Institute, Farmington Hills, Michigan, 2003, 9 pages.

ACI Committee 234, *Guide for the Use of Silica Fume in Concrete*, ACI 234R-06, American Concrete Institute, Farmington Hill, Michigan, 2006, 63 pages.

ACI Committee 318, *Building Code Requirements for Structural Concrete (ACI 318-08) and Commentary*, ACI 318-08, American Concrete Institute, Farmington Hills, Michigan, 2008, 471 pages.

Barger, Gregory S.; Lukkarila, Mark R.; Martin, David L.; Lane, Steven B.; Hansen, Eric R.; Ross, Matt W.; and Thompson, Jimmie L., "Evaluation of a Blended Cement and a Mineral Admixture Containing Calcined Clay Natural Pozzolan for High-Performance Concrete," *Proceedings of the Sixth International Purdue Conference on Concrete Pavement Design and Materials for High Performance*, Purdue University, West Lafayette, Indiana, 1997, 21 pages.

Bhatty, M.S.Y., "Mechanism of Pozzolanic Reactions and Control of Alkali-Aggregate Expansion," *Cement, Concrete, and Aggregates*, American Society for Testing and Materials, West Conshohocken, Pennsylvania. Winter 1985, pages 69 to 77.

Bhatty, M.S.Y., and Greening, N.R., "Some Long Time Studies of Blended Cements with Emphasis on Alkali-Aggregate Reaction," *7th International Conference on Alkali-Aggregate Reaction*, 1986.

Burg, R.G., and Ost, B.W., *Engineering Properties of Commercially Available High-Strength Concretes (Including Three-Year Data)*, Research and Development Bulletin RD104, Portland Cement Association, Skokie, Illinois, 1994, 62 pages.

Campbell, D.H.; Sturm, R.D.; and Kosmatka, S.H., "Detecting Carbonation," *Concrete Technology Today*, PL911; Portland Cement Association, http://www.cement.org/pdf_files/PL911.pdf, March 1991, pages 1 to 5.

Detwiler, Rachel J., "Controlling the Strength Gain of Fly Ash Concrete at Low Temperature," *Concrete Technology Today*, CT003, Portland Cement Association, http://www.cement.org/pdf_files/CT003.pdf, 2000, pages 3 to 5.

Detwiler, Rachel J., *Documentation of Procedures for PCA's ASR Guide Specification*, SN 2407, Portland Cement Association, Skokie, Illinois, 2002.

Detwiler, Rachel J.; Bhatty, Javed I.; and Bhattacharja, Sankar, *Supplementary Cementing Materials for Use in Blended Cements*, Research and Development Bulletin RD112, Portland Cement Association, Skokie, Illinois, 1996, 108 pages.

Dhir, R.K., "Near – Surface Characteristics of Concrete: Prediction of Carbonation Resistance," *Magazine Concrete Research*, 41, No. 148, 1989, pages 137 to 143.

Dodson, V., "Air-Entraining Admixtures," Chapter 6, *Concrete Admixtures*, Van Nostrand Reinhold, New York, New York, 1990, pages 129 to 158.

Farny, J.A., and Kerkhoff, B., *Diagnosis and Control of Alkali-Aggregate Reactivity*, IS413, Portland Cement Association, Skokie, Illinois, 2007, 26 pages.

Fidjestøl, P., and Frearson, J., "High-Performance Concrete Using Blended and Triple Blended Binders," *High-Performance Concrete – Proceedings, International Conference Singapore*, 1994, SP 149, American Concrete Institute, Farmington Hills, Michigan, 1994, pages 135 to 158.

Gajda, J., and VanGeem, M., "Controlling Temperatures in Mass Concrete," *Concrete International*, Vol. 24, No. 1, American Concrete Institute, Farmington Hills, Michigan, January 2002.

Gajda, John, *Mass Concrete for Buildings and Bridges*, EB547, Portland Cement Association, Skokie, Illinois, 2007, 44 pages.

Gebler, S.H., and Klieger, P., *Effect of Fly Ash on Some of the Physical Properties of Concrete*, RD089, Portland Cement Association, http://www.cement.org/pdf_files/RD089.pdf, 1986.

Gebler, Steven H., and Klieger, Paul, *Effect of Fly Ash on the Durability of Air-Entrained Concrete*, Research and Development Bulletin RD090, Portland Cement Association, http://www.cement.org/pdf_files/RD090.pdf, 1986a, 44 pages.

Hansen, W.C., *Twenty-Year Report on the Long-Time Study of Cement Performance in Concrete*, RX175, Portland Cement Association, Skokie, Illinois, http://www.cement.org/pdf_files/RX175.pdf, 1965, 44 pages.

Ho, D.W.S., and Lewis, R.K., "Carbonation of Concrete Incorporating Fly Ash or a Chemical Admixture," *Proceedings of the First International Conference on the Use of Fly Ash, Silica Fume, Slag, and Other By-Products in Concrete*, ACI SP-79, Vol. 1, American Concrete Institute, Farmington Hills, Michigan, 1983, pages 333 to 346.

Ho, D.W.S., and Lewis, R.K., "Carbonation of Concrete and Its Prediction," *Cement and Concrete Research*, Vol. 17, No. 3, 1987, pages 489 to 504.

Hobbs, D.W., "Carbonation of Concrete Containing PFA," *Magazine Concrete Research*, 40, No. 143, 1988, pages 69 to 78.

Helmuth, Richard A., *Fly Ash in Cement and Concrete*, SP040T, Portland Cement Association, 1987, 203 pages.

Kerkhoff, Beatrix, *Effect of Substances on Concrete and Guide to Protective Treatments*, IS001, Portland Cement Association, 2007, 36 pages.

Lerch, William, *Studies of Some Methods of Avoiding the Expansion and Pattern Cracking Associated with the Alkali-Aggregate Reaction*, Research Department Bulletin RX031, Portland Cement Association, http://www.cement.org/pdf_files/RX031.pdf, 1950.

Lewandowski, R., "Effect of Different Fly-Ash Qualities and Quantities on the Properties of Concrete," *Betonwerk and Fetigteil – Technik*, Nos. 1 – 3, 1983.

Matthews, J.D., "Carbonation of Ten-Year Old Concretes With and Without PFA," *Proceedings 2nd International Conference on Fly Ash Technology and Marketing*, London, Ash Marketing, CEGB, 398A, 1984.

Nagataki, S.; Ohga, H.; and Kim, E.K., "Effect of Curing Conditions on the Carbonation of Concrete with Fly Ash and the Corrosion of Reinforcement in Long Term Tests," *Fly Ash, Silica Fume, Slag and Natural Pozzolans in Concrete*, ACI-SP91, Vol. 1, American Concrete Institute, Farmington Hills, Michigan, 1986, pages 521 to 540.

PCA, *Survey of Mineral Admixtures and Blended Cements in Ready Mixed Concrete*, Portland Cement Association, Skokie, Illinois, 2000, 16 pages.

PCA Durability Subcommittee, *Guide Specification to Control Alkali-Silica Reactions*, IS415, Portland Cement Association, Skokie, Illinois, 2007, 8 pages.

Pinto, Roberto C.A., and Hover, K.C., "Superplasticizer and Silica Fume Addition Effects on Heat of Hydration of Mortar Mixtures with Low Water-Cementitious Materials Ratio," *ACI Materials Journal*, Vol. 96, No. 5, American Concrete Institute, Farmington Hills, Michigan, September-October 1999.

SFA, *http://www.silicafume.org*, Silica Fume Association, Fairfax, Virginia, 2008.

Shehata, M.H., and Thomas, M.D.A., "The Effect of Fly Ash Composition on the Expansion of Concrete Due to Alkali Silica Reaction" *Cement and Concrete Research*, Vol. 30 (7), 2000, pages 1063 to 1072.

Stark, David, *Durability of Concrete in Sulfate-Rich Soils*, Research and Development Bulletin RD097, Portland Cement Association, http://www.cement.org/pdf_files/RD097.pdf, 1989, 14 pages.

Stark, David C.; Kosmatka, Steven H.; Farny, James A.; and Tennis, Paul D., *Performance of Concrete Specimens in the PCA Outdoor Test Facility*, RD124, Portland Cement Association, Skokie, Illinois, 2002, 36 pages.

Thomas, M.D.A., and Matthews, J.D., "Carbonation of Fly Ash Concrete," *Magazine of Concrete Research*, Vol. 44, No. 158, 1992, pages 217 to 228.

Thomas, M.D.A., and Matthews J.D., "Carbonation of Fly Ash Concrete," *Proceedings of the 4th ACI/CANMET International Conference on the Durability of Concrete*, ACI SP-192, Vol. 1, American Concrete Institute, Farmington Hills, Michigan, 2000, pages 539 to 556.

Thomas, Michael, *Optimizing the Use of Fly Ash in Concrete*, IS548, Portland Cement Association, Skokie, Illinois, 2007, 24 pages.

Thomas, M.D.A.; Fournier, B.; and Folliard, K., *Report on Determining the Reactivity of Concrete Aggregates and Selecting Appropriate Measures for Preventing Deleterious Expansion in New Concrete Construction*, HIF-09-001, Federal Highway Administration, Washington, D.C., USA, April 2008, 28 pages.

Thomas, Michael, and Wilson, Michelle L., *Supplementary Cementing Materials For Use in Concrete*, CD038, Portland Cement Association, Skokie, Illinois, 2002.

Tsukayama, R., "Long Term Experiments on the Neutralization of Concrete Mixed with Fly Ash and the Corrosion of Reinforcement," *Proceedings of the Seventh International Congressional on the Chemistry of Cement*, Paris, Vol. III, 1980, pages 30 to 35.

Whiting, D., *Strength and Durability of Residential Concretes Containing Fly Ash*, Research and Development Bulletin RD099, Portland Cement Association, http://www.cement.org/pdf_files/RD099.pdf, 1989, 42 pages.

CHAPTER 5

Mixing Water for Concrete

Water is a key ingredient in concrete, that when mixed with portland cement, forms a paste that binds the aggregates together. Water causes the hardening of concrete through hydration. Hydration is a chemical reaction between cement and water to form cementitious hydration products. Further details of hydration are discussed in Chapter 3.

Water needs to be of suitable quality for use in concrete as to not adversely impact the potential properties of concrete. Almost any water that is drinkable and has no pronounced taste or odor, also known as potable water, can be used as mixing water in concrete (Figure 5-1). However, efforts towards conservation of this important natural resource should be recognized. Many waters that are not fit for drinking are suitable for use in concrete.

Figure 5-1. Water is a key ingredient in concrete.

Acceptance criteria for water to be used in concrete are given in ASTM C1602 / C1602M, *Standard Specification for Mixing Water Used in the Production of Hydraulic Cement Concrete*.

ASTM C1602 includes provisions for:

1. Potable water – that which is fit for human consumption;
2. Non-potable water – other sources that are not potable, that might have objectionable taste or smell but not related to water generated at concrete plants. This can represent water from wells, streams, or lakes;
3. Water from concrete production operations – process (wash) water or storm water collected at concrete plants; and
4. Combined water – a combination of one or more of the above defined sources recognizing that water sources might be blended when producing concrete. All requirements in the standard apply to the combined water as batched into concrete and not to individual sources when water sources are combined.

Table 5-1. Performance Requirements for Questionable Water Sources (ASTM C1602)

	Limits	Test method
Compressive strength, minimum percentage of control at 7 days	90	ASTM C31, C39
Time of set, deviation from control, hr:min.	from 1:00 earlier to 1:30 later	ASTM C403

*Comparisons must be based on fixed proportions of a concrete mix design representative of questionable water supply and a control mix using 100% potable water.

Potable water can be used in concrete without any testing or qualification. Water of questionable suitability, including non-potable water or water from concrete production operations, can be used in concrete if it is qualified for use by requirements stated in ASTM C1602/C1602M. The primary requirements of ASTM C1602/C1602M are summarized in Table 5-1. They evaluate the impact of the questionable water on strength and setting time of concrete.

Concrete produced with the questionable water is compared to control batches produced with potable or distilled water. The 7-day strength of concrete cylinders (ASTM C39, *Standard Test Method for Compressive Strength of Cylindrical Concrete Specimens,* or AASHTO T 22) or mortar cubes (CSAA23.2-8A) must achieve at least 90% of the strength of the control batch. The setting time, as measured by ASTM C403, *Standard Test Method for Time of Setting of Concrete Mixtures by Penetration Resistance* (AASHTO T 197), of the test batch should not be accelerated by more than 60 minutes or retarded by more than 90 minutes as compared to the control batch.

The most critical water combination proposed for use in concrete by a supplier should be tested and qualified. Water combinations at levels less than that qualified can be used without testing. For example, if the concrete supplier tests water that contains 100,000 ppm (10% by mass) of solids and the concrete meets the requirements for strength and setting time, less restrictive conditions where the combined water contains less than that quantity of solids are permitted for use.

Table 5-2. Optional Chemical Limits for Combined Mixing Water (ASTM C1602)

Chemical or type of construction	Maximum concentration, ppm*	Test method
Chloride, as Cl		
Prestressed concrete or concrete in bridge decks	500**	
Other reinforced concrete in moist environments or containing aluminum embedments or dissimilar metals or with stay-in-place galvanized metal forms	1000**	ASTM C114
Sulfate, as SO_4	3000	ASTM C114
Alkalies, as (Na_2O + 0.658 K_2O)	600	ASTM C114
Total solids by mass	50,000	ASTM C1603

*ppm is the abbreviation for parts per million.

**The requirements for concrete in ACI 318 shall govern when the manufacturer can demonstrate that these limits for mixing water can be exceeded. For conditions allowing the use of calcium chloride ($CaCl_2$) accelerator as an admixture, the chloride limitation is permitted to be waived by the purchaser.

ASTM C1602 includes optional limits, as stated in Table 5-2 for limits on the chemistry and total solids content by mass in the combined mixing water. Optional limits have to be invoked in project specifications or in purchase orders. The concrete supplier is required to maintain documentation on these characteristics. The chemical composition of water is measured in accordance with methods described in ASTM C114, *Standard Test Methods for Chemical Analysis of Hydraulic Cement* and the solids content is measured in accordance with ASTM C1603, *Standard Test Method for Measurement of Solids in Water*. The density of water is measured during production of concrete to estimate the solids content using a pre-established correlation for the specific production facility as described in ASTM C1603.

ASTM C1602 also establishes minimum testing frequencies to qualify mixing water in conformance with the requirements of Table 5-1 and Table 5-2. More frequent testing is required when water has a higher concentration of solids (higher density).

AASHTO M 157, *Specification for Ready Mixed Concrete,* is referenced by some transportation agencies and includes minor differences on the requirements for water as compared to ASTM C1602. In AASHTO M 157, the chemical limits are not optional; the chloride limits stated for concrete are similar to those in ACI 318, and it refers to different test methods. (Though the same procedures would be used to analyze the chemistry of water.)

Sources of Mixing Water

When considering water quality in concrete production, it is important to account for all sources of water in the mixture. By far, the greatest volume of mixing water in concrete is from batchwater which may be from either a municipal water supply, a municipal reclaimed water supply, site-sourced water, or water from concrete production operations. Other sources of batch water include:

Free moisture on aggregates. Free moisture on aggregates (moisture adsorbed on the surface) constitutes a substantial portion of the total mixing water. It is important that the free water on the aggregate is free from harmful materials.

Ice. During hot-weather concreting, ice might be used as part of the mixing water (see Chapter 16). The ice should be completely melted by the time mixing is completed.

Jobsite addition by truck operator. Water might also be added by the truck operator at the jobsite. ASTM C94, *Standard Specification for Ready-Mixed Concrete* (AASHTO M 157), allows the addition of water on site if the slump of the concrete is less than specified, provided the maximum allowable water-cement ratio is not exceeded (see Chapter 13).

Admixtures. Water contained in admixtures must be considered part of the mixing water if the admixture's water content is sufficient to affect the water-cementitious materials ratio by 0.01 or more.

Municipal Water Supply

Municipal water supply systems get their water from a variety of locations including; aquifers, lakes and rivers, and the sea through desalination. The water is then, in most cases; purified, disinfected through chlorination, and sometimes fluoridated, prior to use as drinking water. An atomic absorption spectrophotometer can be used to detect concentration of elements in the laboratory analysis of water (Figure 5-2). Six typical analyses of city water supplies and seawater are shown in Table 5-3. These waters approximate the composition of domestic water supplies for most of the cities over 20,000 population in the United States and Canada. Water from any of these sources is suitable for use in concrete.

Table 5-3. Typical Analyses of City Water Supplies and Seawater

Chemicals	Analysis No. (Parts per million)						
	1	2	3	4	5	6	Seawater*
Silicate (SiO_2)	2.4	0.0	6.5	9.4	22.0	3.0	—
Iron (Fe)	0.1	0.0	0.0	0.2	0.1	0.0	—
Calcium (Ca)	5.8	15.3	29.5	96.0	3.0	1.3	50 to 480
Magnesium (Mg)	1.4	5.5	7.6	27.0	2.4	0.3	260 to 1410
Sodium (Na)	1.7	16.1	2.3	183.0	215.0	1.4	2190 to 12,200
Potassium (K)	0.7	0.0	1.6	18.0	9.8	0.2	70 to 550
Bicarbonate (HCO_3)	14.0	35.8	122.0	334.0	549.0	4.1	—
Sulfate (SO_4)	9.7	59.9	5.3	121.0	11.0	2.6	580 to 2810
Chloride (Cl)	2.0	3.0	1.4	280.0	22.0	1.0	3960 to 20,000
Nitrate (NO_3)	0.5	0.0	1.6	0.2	0.5	0.0	—
Total dissolved solids	31.0	250.0	125.0	983.0	564.0	19.0	35,000

* Chemical composition of seawater varies considerably depending on its source.

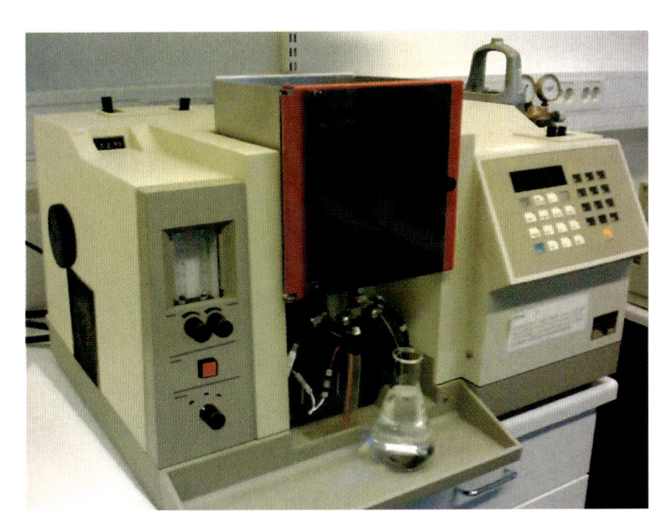

Figure 5-2. An atomic absorption spectrophotometer can be used to detect concentration of elements in the laboratory analysis of water.

Municipal Reclaimed Water

Reclaimed water is wastewater treated to remove solids and certain impurities. It is typically used for nonpotable applications uses such as irrigation, dust control, fire suppression, concrete production, and construction. Reclaimed water use supports sustainable efforts to extend our water supplies rather than discharging the treated wastewater to surface waters such as rivers and oceans (Abrams 1924).

Site-Sourced Water

Many large concrete paving projects and remote construction sites use site source water either from shallow wells, ponds or rivers. These natural sources of water are typically not a concern. When they contain significant amounts of suspended particles such as silt and contain organic impurities and algae, additional testing is warranted.

Recycled Water (Water from Concrete Production)

Recycled water from concrete production is primarily a mixture of: water, partially or completely hydrated cementitious materials, and aggregate fines resulting from processing returned concrete. Recycled water and can include truck wash water, and storm water at the concrete plant. The ready-mixed concrete industry is faced with the challenge of managing about 3% to 5% of its estimated annual production of 300 million cubic meters (400 million cubic yards) as returned concrete. In addition, about 80,000 truck mixers are washed out using about 750 to 1,500 liters (200 to 400 gallons) each of water daily (Lobo 2003). Most of this water is recirculated to keep equipment clean. However, at some point, management of process water and storm water is required for permit compliance. Given the strict regulations on discharge of water from concrete plants, the industry must look at recycling some of the process and storm water generated at ready mixed concrete plants. Environmental regulations in Europe and Japan have forced the ready mixed industry towards zero-discharge production facilities. The U.S. industry should trend in the same direction.

The U.S. Environmental Protection Agency (EPA) and state environmental agencies prohibit discharging untreated process water from concrete operations into the nation's waterways. This is water recovered from processes of concrete production that includes: wash water from mixers or that was a part of a concrete mixture, water collected in a basin as a result of storm water runoff at a concrete production facility, or water that contains quantities of concrete ingredients. In most situations, the recycled water is passed through settling ponds where the solids settle out, leaving clarified water (Figure 5-3). In some cases, the recycled water from a reclaimer unit is continually agitated to maintain the solids in suspension for reuse as a portion of the batch water in concrete. Solid content in recycled water typically varies from 2.5% to 10%. Solid contents exceeding 9 kg/m^3 (15 lb/yd^3) (represented by the 50,000 ppm limit in Table 5-2) may

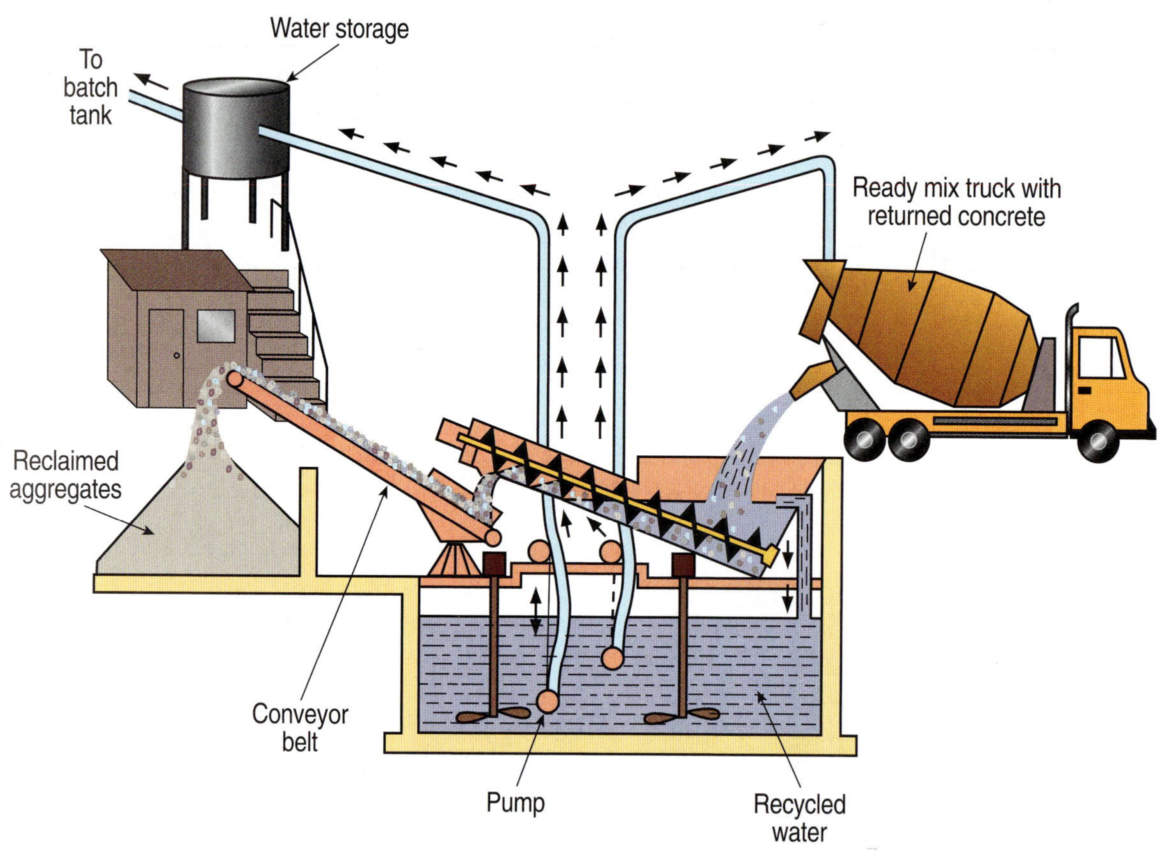

Figure 5-3. Reclaiming system allows immediate reuse of wash water in batching.

adversely impact the properties of concrete through: increased water demand, accelerated setting time, lower compressive strength, and higher permeability due to increased water demand and associated higher w/cm. (Lobo and Mullings 2003). The use of hydration control admixtures has been shown to offset the effects of higher solids contents (Table 5-4). ASTM C1602 and AASHTO M 157 permit the use of water from concrete production operations as mixing water in concrete, provided it meets the limits in Table 5-2.

Seawater

Seawater containing up to 35,000 ppm of dissolved salts is generally suitable as mixing water for concrete not containing reinforcing steel. About 78% of the salt is sodium chloride, and 15% is chloride and sulfate of magnesium. Although concrete made with seawater may have higher early strength than normal concrete, strengths at later ages (after 28 days) may be lower in comparison to normal concrete. This strength reduction can be compensated for by reducing the water-cement ratio.

Seawater is not suitable for use in production of concrete with steel reinforcement and likewise, it should not be used in prestressed concrete due to the risk of corrosion of the reinforcement. If seawater is used in plain concrete (no reinforcing steel) for marine applications moderate sulfate resistant cements, Types II or MS, should be used along with a low water-cement ratio.

Sodium or potassium in salts present in seawater used for mixing water can increase the alkali concentration in the concrete and increase the potential for deleterious expansions due to alkali-aggregate reactivity. Thus, seawater should not be used as mix water for concrete with potentially reactive aggregates.

Seawater used for mixing water also tends to cause efflorescence and dampness on concrete surfaces exposed to air and water (Steinour 1960). Marine-dredged aggregates are discussed in Chapter 6.

Effects of Impurities in Mixing Water on Concrete Properties

Excessive impurities in mixing water not only may affect setting time and concrete strength, but also may cause efflorescence, staining, corrosion of reinforcement, volume instability, and reduced durability. Therefore, certain optional limits on chlorides, sulfates, alkalis, and solids in the mixing water may be set or appropriate tests can be performed to determine the effect the impurity has on various properties (Table 5-2). Some impurities may have

Table 5-4. Effect of Recycled Water on Concrete Properties*

Recycled water with	Water demand	Setting time	Compressive strength	Permeability	Freeze-thaw resistance
Solid contents within ASTM C94 limits (≤8.9 kg/m³ [≤15 lb/yd³])	↔	↔	↔	↔	↔
High solid contents (>8.9 kg/m³ [>15 lb/yd³])	↑	↓	↓**	↑**	↔
High solid contents and treated with hydration stabilizing admixture	↔	↔	↔	no data	no data

Source: After Lobo and Mullings (2003).

* Compared to reference concrete produced with tap water.

** Strength and permeability effects were related to increased mixing water content.

Key: ↓ decreased

↑ increased

↔ no trend

little effect on strength and setting time, yet adversely affect durability and other properties.

Water containing less than 2000 parts per million (ppm) of total dissolved solids is generally satisfactory for use in concrete. Water containing more than 2000 ppm of dissolved solids should be tested for its effect on strength and time of set (Table 5-1). Additional information on the effects of mix water impurities can be found in Steinour (1960) and Abrams (1924). Over 100 different compounds and ions are discussed.

Following is a synopsis of the effects of certain impurities in mixing water on the quality of normal concrete:

Alkali Carbonate and Bicarbonate

Carbonates and bicarbonates of sodium and potassium have varying effects on the setting times of different cements. Sodium carbonate can cause very rapid setting, bicarbonates can either accelerate or retard the set depending on the chemistry of the cement used in the concrete. In large concentrations, these salts can materially reduce concrete strength. When the sum of the dissolved salts exceeds 1000 ppm, tests for their effect on setting time and 28-day strength should be made. The possibility of aggravated alkali-aggregate reactions should also be considered.

Chloride

Concern over a high chloride content in mixing water is chiefly due to the possible adverse effect of chloride ions on the corrosion of reinforcing steel or prestressing strands. Chloride ions attack the protective oxide film formed on the steel by the highly alkaline (pH greater than 13.0) chemical environment present in concrete. The acid-soluble chloride ion level at which steel reinforcement corrosion begins in concrete is about 0.2% to 0.4% by mass of cement (0.15% to 0.3% water soluble). Of the total chloride-ion content in concrete, only about 50% to 85% is water soluble; the remainder is chemically combined in hydration reactions (Whiting 1997, Whiting, Taylor, and Nagi 2002, and Taylor, Whiting, and Nagi 2000).

Chlorides can be introduced into concrete with the separate mixture ingredients – admixtures, aggregates, cementitious materials, and mixing water – or through exposure to deicing salts, seawater, or salt-laden air in coastal environments. Placing an acceptable limit on chloride content for any one ingredient, such as mixing water, is problematic considering the variety of sources of chloride ions in concrete. An acceptable limit in the concrete depends primarily upon the type of structure and the environment to which it is exposed during its service life.

A high dissolved solids content of natural water is sometimes due to a high content of sodium chloride or sodium sulfate. Both can be tolerated in rather large quantities. Concentrations of 20,000 ppm of sodium chloride are generally tolerable in concrete that will be dry in service and has low potential for corrosive reactions. Water used in prestressed concrete or in concrete designed with aluminum embedments should not contain deleterious amounts of chloride ions. The contribution of chlorides from ingredients other than water should also be considered. Calcium chloride admixtures should be avoided in steel reinforced concrete.

The ACI 318 building code and CSA Standard A23.1 limit water soluble chloride ion content in reinforced concrete to the following percentages by mass of cement (CSA's limit is based on mass of cementitious materials):

Prestressed concrete (Classes C0, C1, C2)	0.06%
Reinforced concrete exposed to chloride in service (Class C2)	0.15%

Reinforced concrete that will be dry or protected from moisture in service (Class C0)	1.00%
Other reinforced concrete construction (Class C1)	0.30%

ACI 318 and CSA Standard A23.1 do not limit the amount of chlorides in plain (unreinforced) concrete. Additional information on limits and tests can be found in ACI 222, *Corrosion of Metals in Concrete*. The acid-soluble and water-soluble chloride content of concrete can be determined using ASTM C1152, *Standard Test Method for Acid-Soluble Chloride in Mortar and Concrete*, and ASTM C1218, *Standard Test Method for Water-Soluble Chloride in Mortar and Concrete*, respectively.

Sulfate

Concern over a high sulfate content in mix water is due to possible expansive reactions and deterioration by sulfate attack (see Chapter 11). Although mixing waters containing 10,000 ppm of sodium sulfate have been used satisfactorily, the limit in Table 5-2 should be considered unless special precautions in the composition of the concrete mixture are taken.

Other Common Salts

Carbonates of calcium and magnesium are not very soluble in water and are seldom found in sufficient concentration to affect the strength of concrete. Bicarbonates of calcium and magnesium are present in some municipal waters. Concentrations up to 400 ppm of bicarbonate in these forms are not considered harmful.

Magnesium sulfate and magnesium chloride can be present in high concentrations without harmful effects on strength. Satisfactory strengths have been obtained using water with concentrations up to 40,000 ppm of magnesium chloride. Concentrations of magnesium sulfate should be less than 25,000 ppm.

Iron Salts

Natural ground waters seldom contain more than 20 ppm to 30 ppm of iron; however, acid mine waters may carry rather large quantities. Iron salts in concentrations up to 40,000 ppm do not usually affect concrete strengths adversely. The potential for staining should be evaluated.

Miscellaneous Inorganic Salts

Salts of manganese, tin, zinc, copper, and lead in mixing water can cause a significant reduction in strength and large variations in setting time. Of these, salts of zinc, copper, and lead are the most active. Salts that are especially active as retarders include sodium iodate, sodium phosphate, sodium arsenate, and sodium borate. All can greatly retard both set and strength development when present in concentrations of a few tenths percent by mass of the cement. Generally, concentrations of these salts up to 500 ppm can be tolerated in mixing water.

Another salt that may be detrimental to concrete is sodium sulfide; even the presence of 100 ppm warrants testing. Additional information on the effects of other salts can be found in the references.

Acid Waters

Acceptance of acid mixing water should be based on the concentration (in parts per million) of acids in the water. Occasionally, acceptance is based on the measured pH, a log scale measure of the hydrogen-ion concentration. The pH value is an intensity index and is not the best measure of potential acid or base reactivity. The pH of neutral water is 7.0; values below 7.0 indicate acidity and those above 7.0 alkalinity (a base).

Generally, mixing waters containing hydrochloric, sulfuric, and other common inorganic acids in concentrations as high as 10,000 ppm have no adverse effect on strength. Acid waters with pH values less than 3.0 may create handling problems and should be avoided if possible. Organic acids, such as tannic acid, can have a significant effect on strength at higher concentrations (Figure 5-4).

Alkaline Waters

Waters with sodium hydroxide concentrations of 0.5% by mass of cement do not greatly affect concrete strength provided quick set is not induced. Higher concentrations, however, may reduce concrete strength.

Potassium hydroxide in concentrations up to 1.2% by mass of cement has little effect on the concrete strength developed by some cements, but the same concentration when used with other cements may substantially reduce the 28-day strength.

The possibility for increased alkali-aggregate reactivity should be considered.

Industrial Wastewater

Most waters carrying industrial wastes have less than 4000 ppm of total solids. When such water is used as mixing water in concrete, the reduction in compressive strength is generally not greater than 10%-15%. Wastewaters such as those from tanneries, paint factories, coke plants, and chemical and galvanizing plants may contain harmful impurities. It is best to test any wastewater that contains even a few hundred parts per million of unusual solids.

Silt or Suspended Particles

About 2000 ppm of suspended clay or fine rock particles can be tolerated in mixing water. Higher amounts might not affect strength but may influence other properties of some concrete mixtures. Before use, muddy or cloudy water should be passed through settling basins or otherwise clarified to reduce the amount of silt and clay added to the mixture by way of the mix water. When cement fines are returned to the concrete in reused wash water, 50,000 ppm can be tolerated.

Organic Impurities

The effect of organic substances on the setting time of portland cement or the ultimate strength of concrete is a problem of considerable complexity. Such substances, like surface loams, can be found in natural waters. Highly colored waters, waters with a noticeable odor, or those in which green or brown algae are visible should be regarded with suspicion and tested accordingly. Organic impurities are often of a humus nature containing tannates or tannic acid (Figure 5-4).

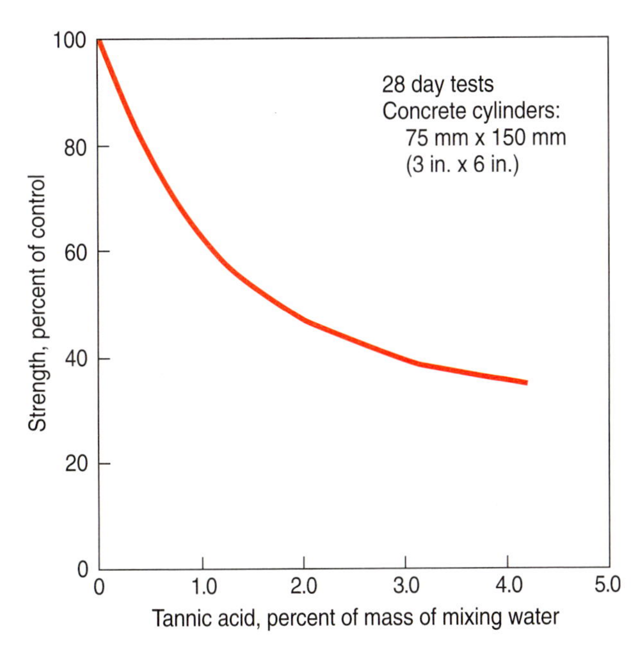

Figure 5-4. Effect of tannic acid on the strength of concrete (Abrams 1920).

Waters Carrying Sanitary Sewage

A typical sewage may contain about 400 ppm of organic matter. After the sewage is diluted in a good disposal system, the concentration is reduced to about 20 ppm or less. This amount is too low to have any significant effect on concrete strength.

Sugar

Small amounts of sucrose, as little as 0.03% to 0.15% by mass of cement, usually retard the setting of cement. The upper limit of this range varies with different cements. The 7-day strength may be reduced while the 28-day strength may be improved. Sugar in quantities of 0.25% or more by mass of cement may cause rapid setting and a substantial reduction in 28-day strength. Each type of sugar can influence setting time and strength differently.

Less than 500 ppm of sugar in mix water generally has no adverse effect on strength, but if the concentration exceeds this amount, tests for setting time and strength should be conducted.

Oils

Various kinds of oil are occasionally present in mixing water. Mineral oil (petroleum) not mixed with animal or vegetable oils may have less effect on strength development than other oils. However, mineral oil in concentrations greater than 2.5% by mass of cement may reduce strength by more than 20%. Oils may interfere with the action of air-entraining agents.

Algae

Water containing algae is unsuitable for concrete because the algae can cause an excessive reduction in strength. Algae in water leads to lower strengths either by influencing cement hydration or by causing a large amount of air to be entrained in the concrete. Algae may also be present on aggregates, in which case the bond between the aggregate and cement paste is reduced. A maximum algae content of 1000 ppm is recommended.

Interaction with Admixtures

When evaluating a water source for its effect on concrete properties, it is important to also test the water with chemical admixtures that will be used in the concrete mixture. Certain compounds in water can influence the performance and efficiency of certain admixtures. For example, the dosage of air-entraining admixture may need to be increased when used with hard waters containing high concentrations of certain compounds or minerals.

References

Abrams, Duff A., *Effect of Tannic Acid on the Strength of Concrete*, Bulletin 7, Structural Materials Research Laboratory, Lewis Institute, Chicago, http://www.cement.org/pdf_files/LS007.pdf, 1920, 34 pages (available through PCA as LS007).

Abrams, Duff A., *Tests of Impure Waters for Mixing Concrete*, Bulletin 12, Structural Materials Research Laboratory, Lewis Institute, Chicago, http://www.cement.org/pdf_files/LS012.pdf, 1924, 50 pages (available through PCA as LS012).

ACI Committee 222, *Corrosion of Metals in Concrete*, ACI 222R-01, reapproved 2010, American Concrete Institute, Farmington Hills, Michigan, 2001, 41 pages.

ACI Committee 318, *Building Code Requirements for Structural Concrete and Commentary*, ACI 318-08, American Concrete Institute, Farmington Hills, Michigan, 2008.

CSA Standard A23.1-09/A23.2-09, *Concrete Materials and Methods of Concrete Construction/Test methods and Standard Practices for Concrete*, Canadian Standards Association, Toronto, Canada, 2009.

Lobo, Colin, "Recycled Water in Concrete," *Concrete Technology Today,* Vol. 24, No. 3, CT033, Portland Cement Association, Skokie, Illinois, December 2003, pages 2 to 3.

Lobo, Colin, and Mullings, Gary M., "Recycled Water in Ready Mixed Concrete Operations," *Concrete in Focus,* http://www.nrmca.org/research_engineering/lab.htm, 2003, 10 pages.

Meininger, Richard C., Recycling Mixer Wash Water, National Ready Mixed Concrete Association, Silver Spring, Maryland, 2000.

NRMCA, *A System for 100% Recycling of Returned Concrete: Equipment, Procedures, and Effects on Product Quality*, National Ready Mixed Concrete Association, Silver Spring, Maryland, 1975.

Steinour, H.H., *Concrete Mix Water—How Impure Can It Be?*, Research Department Bulletin RX119, Portland Cement Association, http://www.cement.org/pdf_files/RX119.pdf, 1960, 20 pages.

Taylor, Peter C.; Whiting, David A.; and Nagi, Mohamad A., *Threshold Chloride Content of Steel in Concrete*, R&D Serial No. 2169, Portland Cement Association, http://www.cement.org/pdf_files/SN2169.pdf, 2000, 32 pages.

Whiting, David A., *Origins of Chloride Limits for Reinforced Concrete*, R&D Serial No. 2153, Portland Cement Association, http://www.cement.org/pdf_files/SN2153.pdf, 1997, 18 pages.

Whiting, David A.; Taylor, Peter C.; and Nagi, Mohamad A., *Chloride Limits in Reinforced Concrete*, R&D Serial No. 2438, Portland Cement Association, 2002, 96 pages.

Yelton, Rick, "Answering Five Common Questions about Reclaimers," *The Concrete Producer*, Addison, Illinois, September 1999, pages 17 to 19.

CHAPTER 6

Aggregates for Concrete

The proper selection of aggregates for use in concrete mixtures is critical to long term concrete performance. Aggregates generally occupy 60% to 75% of the concrete volume (70% to 85% by mass) and strongly influence the concrete's fresh and hardened properties, mixture proportions, and economy. Aggregates used in concrete must conform to certain standards for optimum engineering performance. Aggregates must be clean, hard, strong, and durable particles that are largely free of absorbed chemicals, coatings of clay, and other fine materials in amounts that could affect hydration and bond of the cement paste. Aggregate particles that are friable or capable of being split are undesirable. Aggregates containing appreciable amounts of soft and porous materials, including some varieties of siltstone, claystone, mudstone, shale and shaley rocks, should be avoided. Certain types of chert should be carefully avoided since they have low resistance to weathering and can cause surface defects such as popouts. Some variation in the type, quality, cleanliness, grading, moisture content, and other properties of aggregates is expected.

Fine aggregates (Figure 6-1) generally consist of natural sand or crushed stone with most particles smaller than 5 mm (0.2 in.). Coarse aggregates (Figure 6-2) typically consist of gravels, crushed stone, or a combination of both, with particles predominantly larger than 5 mm (0.2 in.) and generally between 9.5 mm and 37.5 mm (3/8 in. and 1½ in.).

Figure 6-1. Fine aggregate (sand).

Figure 6-2. Coarse aggregate. Rounded gravel (left) and crushed stone (right).

Close to half of the coarse aggregates used in concrete in North America are gravels; most of the remainder are crushed stones.

Aggregate Geology

Concrete aggregates are a mixture of rocks and minerals (see Table 6-1). A mineral is a naturally occurring solid substance with an orderly internal structure and a chemical composition that ranges within narrow limits. A rock is generally composed of several minerals. For example; granite contains quartz, feldspar, mica, and a few other minerals; and most limestones consist of calcite, dolomite, and minor amounts of quartz and clay. Rocks are classified as igneous, sedimentary, or metamorphic, depending on their geological origin.

Table 6-1. Rock and Mineral Constituents in Aggregates

Minerals	Igneous rocks	Metamorphic rocks
Silica	Granite	Marble
Quartz	Syenite	Metaquartzite
Opal	Diorite	Slate
Chalcedony	Gabbro	Phyllite
Tridymite	Peridotite	Schist
Cristobalite	Pegmatite	Amphibolite
Silicates	Volcanic glass	Hornfels
Feldspars	Obsidian	Gneiss
Ferromagnesian	Pumice	Serpentinite
Hornblende	Tuff	
Augite	Scoria	
Clay	Perlite	
Illites	Pitchstone	
Kaolins	Felsite	
Chlorites	Basalt	
Montmorillonites	**Sedimentary rocks**	
Mica		
Zeolite	Conglomerate	
Carbonate	Sandstone	
Calcite	Quartzite	
Dolomite	Graywacke	
Sulfate	Subgraywacke	
Gypsum	Arkose	
Anhydrite	Claystone, siltstone,	
Iron sulfide	argillite, and shale	
Pyrite	Carbonates	
Marcasite	Limestone	
Pyrrhotite	Dolomite	
Iron oxide	Marl	
Magnetite	Chalk	
Hematite	Chert	
Goethite		
Ilmenite		
Limonite		

For brief descriptions, see *Standard Descriptive Nomenclature for Constituents of Concrete Aggregates* (ASTM C294).

Aggregate Classification

Aggregates are classified into three categories: normalweight, lightweight, and heavyweight. The approximate bulk density of aggregate commonly used in normal weight concrete ranges from about 1200 kg/m³ to 1750 kg/m³ (75 lb/ft³ to 110 lb/ft³) while lightweight concrete aggregates range from 560 kg/m³ to 1120 kg/m³ (35 lb/ft³ to 70 lb/ft³), and heavyweight aggregates is typically over 2100 kg/m³ (130 lb/ft³).

The most commonly used normalweight aggregates-sand, gravel, and crushed stone, produce freshly mixed normalweight concrete with a density (unit weight) of 2200 kg/m³ to 2400 kg/m³ (140 lb/ft³ to 150 lb/ft³).

Lightwieght aggregates of expanded shale, clay, slate, and slag (Figure 6-3) are used to produce structural lightweight concrete with a freshly mixed density ranging from about 1350 kg/m³ to 1850 kg/m³ (90 lb/ft³ to 120 lb/ft³). Other lightweight materials such as pumice, scoria, perlite, vermiculite, and diatomite are used to produce insulating lightweight concretes ranging in density from about 250 kg/m³ to 1450 kg/m³ (15 lb/ft³ to 90 lb/ft³). See Bohan and Ries (2008) and Chapter 20 for more information on lightweight aggregates.

Heavyweight materials such as barite, limonite, magnetite, ilmenite, hematite, iron, and steel punchings or shot are used to produce heavyweight concrete and radiation-shielding concrete (ASTM C637, *Standard Specification for Aggregates for Radiation-Shielding Concrete*, and ASTM C638, *Standard Descriptive Nonmenclature of Constituents of Aggregates for Radiation-Shielding Concrete*). Heavyweight aggregates produce concretes ranging in density from 2900 kg/m³ to 6100 kg/m³ (180 lb/ft³ to 380 lb/ft³).

This chapter focuses on normalweight aggregates. Normalweight aggregates for use in concrete include natural aggregate, manufactured aggregate, recycled-concrete aggregate, and marine-dredged aggregate.

Figure 6-3. Lightweight aggregate. Expanded clay (left) and expanded shale (right).

Natural Aggregate

Gravel and sand are often a mixture of several minerals or rocks. Natural gravel and sand are usually dug or dredged from a pit, river, lake, or seabed. Weathering and erosion of rocks produces particles of stone, gravel, sand, silt, and clay. Some natural aggregate deposits of gravel and sand can be readily used in concrete with minimal processing.

The quality (or soundness) of natural aggregate depends on the bedrock from which the particles were derived and the mechanism by which they were transported. Sand and gravel derived from igneous and metamorphic rocks tend to be sound, while sand and gravel derived from rocks rich in shale and siltstone are more likely to be unsound. Natural aggregate deposited at higher elevations from glaciers may be superior to deposits in low areas. This is because rock located high in an ice sheet has been carried from higher, more mountainous areas, which tend to consist of hard, sound rocks. Sand and gravel that have been smoothed by prolonged agitation in water usually are considered higher quality because they are harder and have a more rounded shape than less abraded sand and gravel. However, the smooth surface of natural gravels can reduce the bond strength with the cement paste and reduce overall concrete strength.

Manufactured Aggregate

Manufactured aggregate (including manufactured sand) is produced by crushing sound parent rock (igneous, sedimentary, or metamorphic) at stone crushing plants. Crushed air-cooled blast-furnace slag is also used as fine or coarse aggregate.

Manufactured aggregates differ from gravel and sand in their grading, shape, and texture. As a result of the crushing operation, manufactured aggregates often have a rough surface texture, are more angular in nature, tend to be cubical or elongated in shape (depending on the method of crushing), and more uniform in grade (size) (Wigum 2004). Producers can provide high-quality material meeting specified particle shapes and gradations (Addis, Owens, and Fulton 2001). In many cases, the particle elongation and layered flakes of manufactured sands can be reduced through appropriate crushing techniques.

Manufactured aggregates are less likely than gravel and sand to be contaminated by deleterious substances such as clay minerals or organic matter (Addis, Owens, and Fulton 2001). Some specifications permit higher fines content in manufactured sands because of the expectation of less clay contamination (Addis, Owens, and Fulton 2001).

Recycled-Concrete Aggregate

The concept of recycling and reusing concrete pavements, buildings, and other structures as a source of aggregate has been demonstrated on several projects, resulting in both material and energy savings (ECCO 1999). The procedure involves demolishing and removing the existing concrete, crushing the material in primary and secondary crushers (Figure 6-4), removing reinforcing steel and other embedded items, grading and washing, and stockpiling the resulting coarse and fine aggregate (Figure 6-5) (ACI 555R-01). Dirt, gypsum board, wood, and other foreign materials should be prevented from contaminating the final product.

Figure 6-4. Heavily reinforced concrete is crushed with a beam-crusher.

Recycled concrete is simply old concrete that has been crushed to produce aggregate. Recycled-concrete aggregate (RCA) is primarily used in pavement reconstruction. It has been satisfactorily used as aggregate in granular subbases, lean-concrete subbases, soil-cement, and in new concrete as the primary source of aggregate or as a partial replacement of new aggregate.

Figure 6-5. Stockpile of recycled-concrete aggregate.

Recycled-concrete aggregate generally has a higher absorption and a lower specific gravity than conventional aggregate (ACI 555R). This results from the high absorption of the more porous hardened cement paste within the recycled concrete aggregate. Absorption values typically range from 3% to 10%, depending on the concrete being recycled; this absorption rate lies between those for natural and lightweight aggregate. Absorption rates increase as coarse particle size decreases (Figure 6-6). The high absorption of the recycled aggregate makes it necessary to add more water to achieve the same workability and slump as concrete with conventional aggregates. Dry recycled aggregate absorbs water during and after mixing. To avoid this increased absorption, RCA stockpiles should be prewetted or kept moist, as is the practice with lightweight aggregates.

The particle shape of recycled-concrete aggregate is similar to crushed rock as shown in Figure 6-7. The relative density decreases progressively as particle size decreases. The sulfate content of recycled-concrete aggregate should be determined to assess the possibility of deleterious sulfate reactivity. The chloride content should also be determined where applicable.

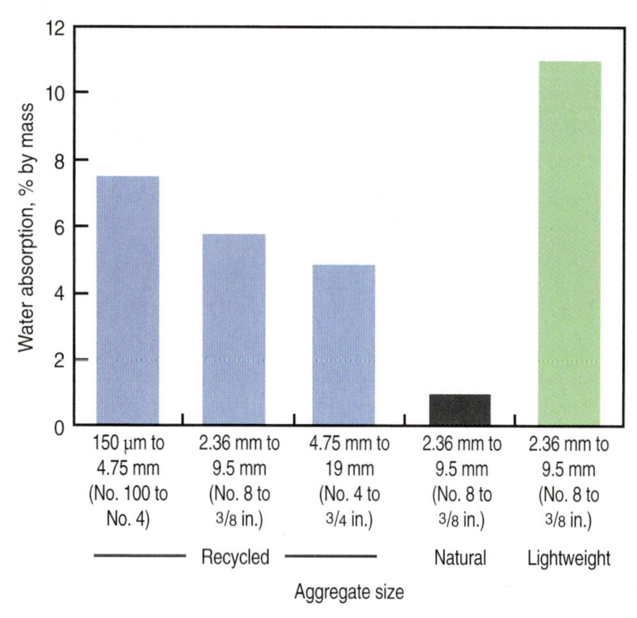

Figure 6-6. Comparison of water absorption of three different recycled aggregate particle sizes and one size of natural and lightweight coarse aggregate. (Kerkhoff and Siebel 2001)

Figure 6-7. Recycled-concrete aggregate.

New concrete made from recycled-concrete aggregate generally has good durability. As with any new aggregate source, recycled-concrete aggregate should be tested for durability, gradation, and other properties. Carbonation, permeability, and resistance to freeze-thaw action have been found to be the same or even better than concrete produced with conventional aggregates. Concrete made with recycled coarse aggregates and conventional fine aggregate can generally obtain adequate compressive strength. The use of recycled fine aggregate can result in minor compressive strength reductions, unless mixture proportions are adjusted to compensate. However, drying shrinkage and creep of concrete made with recycled aggregates is up to 100% higher than concrete made with a corresponding conventional aggregate. This is due to the large amount of existing cement paste and mortar present, especially in the fine aggregate. Therefore, considerably lower values of drying shrinkage can be achieved using recycled coarse aggregate with natural sand, as compared to fine aggregate manufactured from recycled concrete (Kerkhoff and Siebel 2001).

Additional reductions in drying shrinkage and creep can be achieved using the Equivalent Mortar Volume (EMV) method of proportioning. Using this method, the mortar fraction of the recycled concrete aggregate is treated as mortar for the new mixture, as opposed to aggregate. Values for shrinkage, creep, and other concrete properties for concrete made using this method are similar to those for concrete made with virgin aggregates (Fathifazl and others 2010). The EMV method requires additional testing of the RCA to accurately determine the amount of mortar that is present.

Concrete trial mixtures should be made with recycled-concrete aggregates to determine the proper mixture proportions. The variability in the properties of the old concrete may in turn affect the properties of the new concrete. This can partially be avoided by frequent monitoring of the properties of the old concrete that is being recycled. Adjustments in the mixture proportions may be needed.

Recycled concrete used as coarse aggregate in new concrete possesses some potential for alkali-silica-reaction (ASR) if the old concrete contained alkali-reactive aggregate. Aggregates prone to alkali-silica reactivity are listed under the section on **Alkali-Aggregate Reactivity** in this chapter. The alkali content of the cement used in the old concrete has little effect on expansion due to alkali-silica-reaction. For highly reactive aggregates made from recycled concrete, special measures discussed in Chapter 11 should be used to control ASR. Also, even if expansion from ASR did not develop in the original concrete, it cannot be assumed that it will not develop in the new concrete without appropriate control measures. Petrographic examination and expansion tests are recommended for careful evaluation (Stark 1996).

Marine-Dredged Aggregate

Marine-dredged aggregate from tidal estuaries and sand and gravel from the seashore may be used with caution in concrete applications when other aggregate sources are not available. There are two primary concerns with aggregates obtained from seabeds: (1) seashells and (2) salt. Although seashells are a hard material that can produce good quality concrete, a higher paste content may be required due to the angularity of the shells. Aggregate containing complete shells (uncrushed) should be avoided as their presence may result in voids in the concrete and may lower the compressive strength.

Marine-dredged aggregates often contain salt (chlorides) from the seawater. The highest salt content occurs in sands located just above the high-tide level, although the amount of salt on the aggregate is often not more than about 1% of the mass of the mixing water.

The presence of chlorides may affect the concrete by: (1) altering the time of set, (2) increasing drying shrinkage, (3) significantly increasing the risk of corrosion of steel reinforcement, and (4) causing efflorescence. Generally, marine aggregates containing large amounts of chloride should not be used in reinforced concrete. To reduce the chloride content, marine-dredged aggregates can be washed with fresh water.

Characteristics of Aggregates

The important characteristics of aggregates for concrete are listed in Table 6-2. Normal-weight aggregates should meet the requirements of ASTM C33, *Standard Specification for Concrete Aggregates*, or AASHTO M 6, *Standard Specification for Fine Aggregate for Hydraulic Cement Concrete* or AASHTO M 80, *Standard Specification for Coarse Aggregate*

Table 6-2. Characteristics and Tests of Aggregate

Characteristic	Significance	Test designation*	Requirement or item reported
Resistance to abrasion and degradation	Index of aggregate quality; wear resistance of floors and pavements	ASTM C131 (AASHTO T 96) ASTM C535 ASTM C779	Maximum percentage of weight loss. Depth of wear and time
Resistance to freezing and thawing	Surface scaling, roughness, loss of section, and aesthetics	ASTM C666 (AASHTO T 161) ASTM C672 (AASHTO T 103)	Maximum number of cycles or period of frost immunity; durability factor
Resistance to disintegration by sulfates	Soundness against weathering action	ASTM C88 (AASHTO T 104)	Weight loss, particles exhibiting distress
Particle shape and surface texture	Workability of fresh concrete	ASTM C295 ASTM D3398	Maximum percentage of flat and elongated particles
Grading	Workability of fresh concrete; economy	ASTM C117 (AASHTO T 11) ASTM C136 (AASHTO T 27)	Minimum and maximum percentage passing standard sieves
Fine aggregate degradation	Index of aggregate quality; Resistance to degradation during mixing	ASTM C1137	Change in grading
Uncompacted void content of fine aggregate	Workability of fresh concrete	ASTM C1252 (AASHTO T 304)	Uncompacted voids and specific gravity values
Bulk density (unit weight)	Mix design calculations; classification	ASTM C29 (AASHTO T 19)	Compact weight and loose weight
Relative density (specific gravity)	Mix design calculations	ASTM C127 (AASHTO T 85) fine aggregate ASTM C128 (AASHTO T 84) coarse aggregate	—
Absorption and surface moisture	Control of concrete quality (water-cement ratio)	ASTM C70 ASTM C127 (AASHTO T 85) ASTM C128 (AASHTO T 84) ASTM C566 (AASHTO T 255)	—
Compressive and flexural strength	Acceptability of fine aggregate failing other tests	ASTM C39 (AASHTO T 22) ASTM C78 (AASHTO T 97)	Strength to exceed 95% of strength achieved with purified sand
Definitions of constituents	Clear understanding and communication	ASTM C125 ASTM C294	—
Aggregate constituents	Determine amount of deleterious and organic materials	ASTM C40 (AASHTO T 21) ASTM C87 (AASHTO T 71) ASTM C117 (AASHTO T 11) ASTM C123 (AASHTO T 113) ASTM C142 (AASHTO T 112) ASTM C295	Maximum percentage allowed of individual constituents
Resistance to alkali reactivity and volume change	Soundness against volume change	ASTM C227 ASTM C289 ASTM C295 ASTM C342 ASTM C586 ASTM C1260 (AASHTO T 303) ASTM C1293 ASTM C1567	Maximum length change, constituents and amount of silica, and alkalinity

* The majority of the tests and characteristics listed are referenced in ASTM C33 (AASHTO M 6/M 80). ACI 221R-96 presents additional test methods and properties of concrete influenced by aggregate characteristics.

for Portland Cement Concrete. These specifications limit the permissible amounts of deleterious substances and provide requirements for aggregate characteristics. Compliance is determined using one or more of the several standard tests cited in the following sections and tables.

Identification of the constituents of an aggregate cannot alone provide a basis for predicting the behavior of aggregates in service. Visual inspection will often disclose weaknesses in coarse aggregates. Service records are invaluable in evaluating aggregates. In the absence of a performance record, the aggregates should be tested before they are used in concrete.

Grading

Grading is the particle-size distribution of an aggregate as determined by a sieve analysis (ASTM C136 or AASHTO T 27). The range of particle sizes in aggregate is illustrated in Figure 6-8. The aggregate particle size is determined using wire-mesh sieves with square openings. The seven standard ASTM C33 (AASHTO M 6/M 80) sieves for fine aggregate have openings ranging from 150 µm to 9.5 mm (No. 100 sieve to 3⁄8 in.). The 13 standard sieves for coarse aggregate have openings ranging from 1.18 mm to 100 mm (0.046 in. to 4 in.). Tolerances for the dimensions of openings in sieves are listed in ASTM E11 (AASHTO M 92).

Size numbers (grading sizes) for coarse aggregates apply to the amounts of aggregate (by mass) in percentages that pass through an assortment of sieves (Figure 6-9). For highway construction, ASTM D448 (AASHTO M 43) lists the same 13 size numbers as noted in ASTM C33 (AASHTO M 6/M 80) plus a six additional coarse aggregate size numbers. Fine aggregate or sand has the same range of particle sizes for general construction and highway work.

The grading and grading limits are usually expressed as the percentage of material passing each sieve. Figure 6-10 shows these limits for fine aggregate and for one specific size of coarse aggregate.

Figure 6-8. Range of particle sizes found in aggregate for use in concrete.

Figure 6-9. Making a sieve analysis test of coarse aggregate in a laboratory.

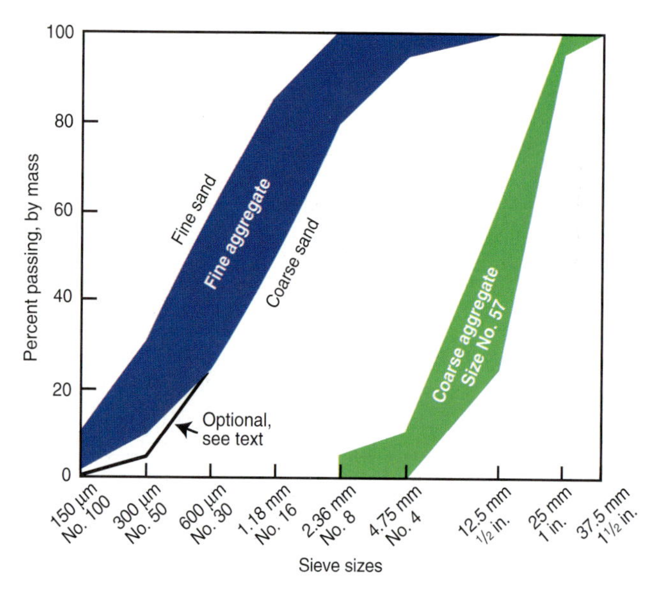

Figure 6-10. Curves indicate the limits specified in ASTM C33 for fine aggregate and for one commonly used size number (grading size) of coarse aggregate.

There are several reasons for specifying grading limits and a nominal maximum aggregate size. Grading affects relative aggregate proportions as well as cement and water requirements, workability, pumpability, economy, porosity, shrinkage, and durability of concrete. Variations in grading can seriously affect the uniformity of concrete from batch to batch. Very fine sands are often uneconomical while very coarse sands and coarse aggregate can produce harsh, unworkable mixtures.

In general, aggregates that do not have a large deficiency or excess of any size and give a smooth grading curve will produce the most satisfactory results. ASTM C33 allows for the use of aggregates with gradings falling outside the specification limits provided it can be shown that good quality concrete can be produced from those aggregates.

Fine-Aggregate Grading. Requirements of ASTM C33 or AASHTO M 6/M 43 permit a relatively wide range in fine-aggregate gradation, but specifications by other organizations are sometimes more restrictive. The most desirable fine-aggregate grading depends on the type of application, the paste content, and the maximum size of coarse aggregate. In leaner mixtures, or when small-size coarse aggregates are used, a grading that approaches the maximum recommended percentage passing each sieve is desirable for workability. In general, if the water-cement ratio is kept constant and the ratio of fine-to-coarse aggregate is suitable, a wide range in grading can be used without measurable effect on strength. However, the best economy will sometimes be achieved by adjusting the concrete mixture proportions to suit the gradation of the local aggregates.

Fine-aggregate grading within the limits of ASTM C33 (AASHTO M 6) is generally satisfactory for most concretes. The ASTM C33 (AASHTO M 6) limits with respect to sieve size are shown in Table 6-3.

Table 6-3. Fine-Aggregate Grading Limits (ASTM C33/AASHTO M 6)

Sieve size		Percent passing by mass	
9.5 mm	(3/8 in.)	100	
4.75 mm	(No. 4)	95 to 100	
2.36 mm	(No. 8)	80 to 100	
1.18 mm	(No. 16)	50 to 85	
600 μm	(No. 30)	25 to 60	
300 μm	(No. 50)	5 to 30	(AASHTO 10 to 30)
150 μm	(No. 100)	0 to 10	(AASHTO 2 to 10)

As shown in Table 6-3, AASHTO grading limits for 300 mm (No. 50) and 150 mm (No. 100) vary slightly from the ASTM C33 limits. The AASHTO specifications permit the minimum percentages (by mass) of material passing the 300 μm (No. 50) and 150 μm (No. 100) sieves to be reduced to 5% and 0% respectively (thereby matching ASTM C33 requirements), provided:

1. The aggregate is used in air-entrained concrete containing more than 237 kg/m^3 of cement (400 lb/yd^3) and having an air content of more than 3%.
2. The aggregate is used in concrete containing more than 297 kg/m^3 of cement (500 lb/yd^3) when the concrete is not air-entrained.
3. An approved supplementary cementitious material is used to supply the deficiency in material passing these two sieves.

Other requirements of ASTM C33 (AASHTO M 6) for fine aggregates include:

1. The fine aggregate must not have more than 45% retained between any two consecutive standard sieves.
2. The fineness modulus (FM) must be not less than 2.3 or more than 3.1, and not vary more than 0.2 from the average value of the aggregate source being tested. If this value is outside the required 2.3 to 3.1 range, the fine aggregate should be rejected unless suitable adjustments are made in proportions of fine and coarse aggregate. If the FM varies by more than 0.2 adjustments may need to be made with regard to coarse and fine aggregate proportions as well as the water requirements for the concrete mixture.
3. The amounts of fine aggregate passing the 300 μm (No. 50) and 150 μm (No. 100) sieves affect workability, surface texture, air content, and bleeding of concrete. Most specifications allow 5% to 30% to pass the 300 μm (No. 50) sieve. The lower limit may be sufficient for easy placing conditions or where concrete is mechanically finished, such as in pavements. However, for hand-finished concrete floors, or where a smooth surface texture is desired, fine aggregate with at least 15% passing the 300 μm (No. 50) sieve and 3% or more passing the 150 μm (No. 100) sieve should be used.

Coarse Aggregate Grading. The coarse aggregate grading requirements of ASTM C33 (AASHTO M 80) permit a wide range in grading and a variety of grading sizes (see Table 6-4). The grading for a given maximum-size coarse aggregate can be varied over a moderate range without appreciable effect on cement and water requirement of a mixture if the proportion of fine aggregate to total aggregate produces concrete of good workability. Mixture proportions should be changed to produce workable concrete if wide variations occur in the coarse-aggregate grading. Since variations are difficult to anticipate, it is often more economical to maintain uniformity in manufacturing and handling coarse aggregate than to reduce variations in gradation.

The maximum size of coarse aggregate used in concrete has a direct bearing on the economy of concrete. Usually more water and cement is required for small-size aggregates than for large sizes, due to an increase in total aggregate surface area. The water and cement required for a slump of approximately 75 mm (3 in.) is shown in Figure 6-11 for a wide range of coarse-aggregate sizes. Figure 6-11 shows that, for a given water-cement ratio, the amount of cement (or water) required decreases as the maximum size of coarse aggregate increases. Aggregates of different maximum sizes may give slightly different concrete strengths

Table 6-4. Grading Requirements for Coarse Aggregates (ASTM C33 and AASHTO M 80)

Size no.	Nominal size, sieves with square openings	Amounts finer than each laboratory sieve, mass percent passing												
		100 mm (4 in.)	90 mm (3½ in.)	75 mm (3 in.)	63 mm (2½ in.)	50 mm (2 in.)	37.5 mm (1½ in.)	25 mm (1 in.)	19 mm (¾ in.)	12.5 mm (½ in.)	9.5 mm (⅜ in.)	4.75 mm (No. 4)	2.36 mm (No. 8)	1.18 mm (No. 16)
1	90 to 37.5 mm (3½ to 1½ in.)	100	90 to 100	—	25 to 60	—	0 to 15	—	0 to 5	—	—	—	—	—
2	63 to 37.5 mm (2½ to 1½ in.)	—	—	100	90 to 100	35 to 70	0 to 15	—	0 to 5	—	—	—	—	—
3	50 to 25 mm (2 to 1 in.)	—	—	—	100	90 to 100	35 to 70	0 to 15	—	0 to 5	—	—	—	—
357	50 to 4.75 mm (2 in. to No. 4)	—	—	—	100	95 to 100	—	35 to 70	—	10 to 30	—	0 to 5	—	—
4	37.5 to 19.0 mm (1½ to ¾ in.)	—	—	—	—	100	90 to 100	20 to 55	0 to 15	—	0 to 5	—	—	—
467	37.5 to 4.75 mm (1½ in. to No. 4)	—	—	—	—	100	95 to 100	—	35 to 70	—	10 to 30	0 to 5	—	—
5	25.0 to 12.5 mm (1 to ½ in.)	—	—	—	—	—	100	90 to 100	20 to 55	0 to 10	0 to 5	—	—	—
56	25.0 to 9.5 mm (1 to ⅜ in.)	—	—	—	—	—	100	90 to 100	40 to 85	10 to 40	0 to 15	0 to 5	—	—
57	25.0 to 4.75 mm (1 in. to No. 4)	—	—	—	—	—	100	95 to 100	—	25 to 60	—	0 to 10	0 to 5	—
6	19.0 to 9.75 mm (¾ to ⅜ in.)	—	—	—	—	—	—	100	90 to 100	20 to 55	0 to 15	0 to 5	—	—
67	19.0 to 4.75 mm (¾ in. to No. 4)	—	—	—	—	—	—	100	90 to 100	—	20 to 55	0 to 10	0 to 5	—
7	12.5 to 4.75 mm (½ in. to No. 4)	—	—	—	—	—	—	—	100	90 to 100	40 to 70	0 to 15	0 to 5	—
8	9.5 to 2.36 mm (⅜ in. to No. 8)	—	—	—	—	—	—	—	—	100	85 to 100	10 to 30	0 to 10	0 to 5

for the same water-cement ratio: typically concrete with a smaller maximum-size aggregate will have a higher compressive strength. This is especially true for high-strength concrete. The optimum maximum size of coarse aggregate for higher strength depends on factors such as relative strength of the cement paste, paste-aggregate bond, and strength of the aggregate particles.

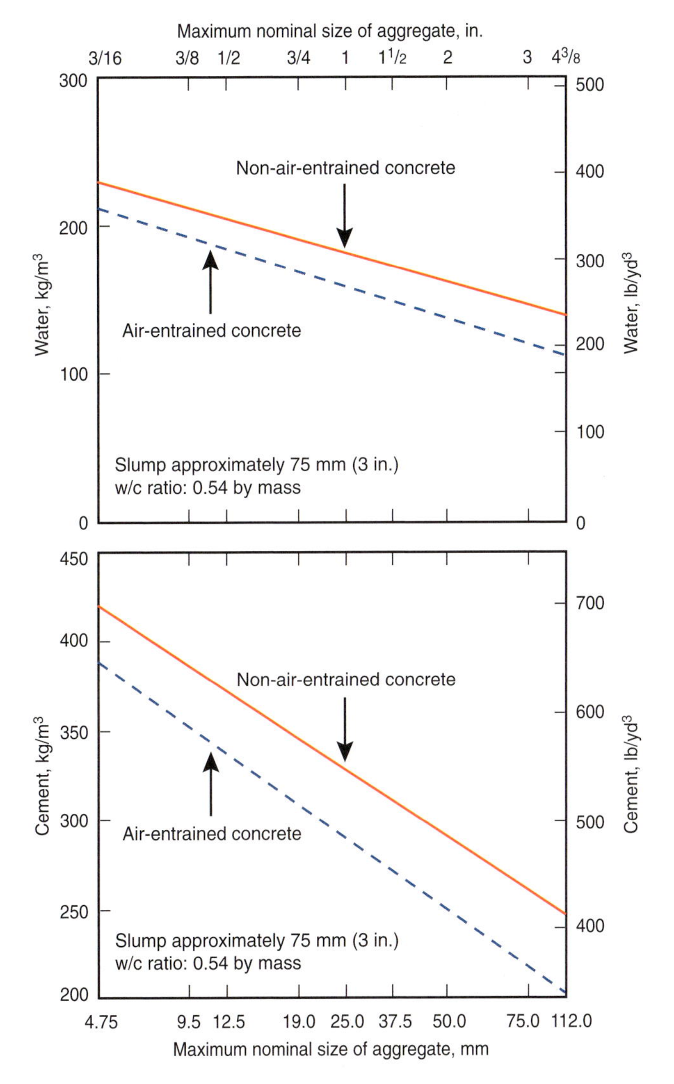

Figure 6-11. Cement and water contents in relation to maximum size of aggregate for air-entrained and non-air-entrained concrete. Less cement and water are required in mixtures having large coarse aggregate (Bureau of Reclamation 1981).

The terminology used to specify size of coarse aggregate must be chosen carefully. Particle size is determined by size of sieve and applies to the aggregate passing that sieve and retained on the next smaller sieve. When speaking of an assortment of particle sizes, the size number (or grading size) of the gradation is used. The size number applies to the collective amount of aggregate that passes through an assortment of sieves. As shown in Table 6-4, the amount of aggregate passing the respective sieves is given in percentages; it is called a sieve analysis.

Maximum Size vs. Nominal Maximum Size Aggregate. Often there is confusion surrounding the term "maximum size" of aggregate. ASTM C125, *Standard Terminology Relating to Concrete and Concrete Aggregates*, defines this term and distinguishes it from "nominal maximum size" of aggregate. The maximum size of an aggregate is the smallest sieve that ***all*** of a particular aggregate must pass through (100% passing). The nominal maximum size of an aggregate is the smallest sieve size through which the major portion of the aggregate must pass (typically 85% to 95% passing). The nominal maximum-size sieve may retain 5% to 15% of the aggregate depending on the size number. For example, aggregate size number 67 has a maximum size of 25 mm (1 in.) and a nominal maximum size of 19 mm (¾ in.) [90% to 100% of this aggregate must pass the 19-mm (¾-in.) sieve and all of the particles must pass the 25-mm (1-in.) sieve].

The maximum size of aggregate that can be used generally depends on the size and shape of the concrete member and on the amount and distribution of reinforcing steel (Figure 6-12). Requirements for limits on nominal maximum size of aggregate particles are covered by ACI 318 (ACI 318-08). The nominal maximum size of aggregate should not exceed:

1. One-fifth the narrowest dimension of a vertical concrete member: $D_{max} = 1/5\ B$
2. Three-quarters the clear spacing between reinforcing bars and between the reinforcing bars and forms: $D_{max} = 3/4\ S$, and $3/4\ C$
3. One-third the depth of slabs: $D_{max} = 1/3\ T$

These requirements may be waived if, in the judgment of the engineer, the mixture possesses sufficient workability that the concrete can be properly placed without honeycomb or voids.

Combined Aggregate Grading. Aggregate is sometimes analyzed using the combined grading of fine and coarse aggregate in the proportions they would exist in concrete. This provides a more thorough analysis of how the aggregates will perform in concrete.

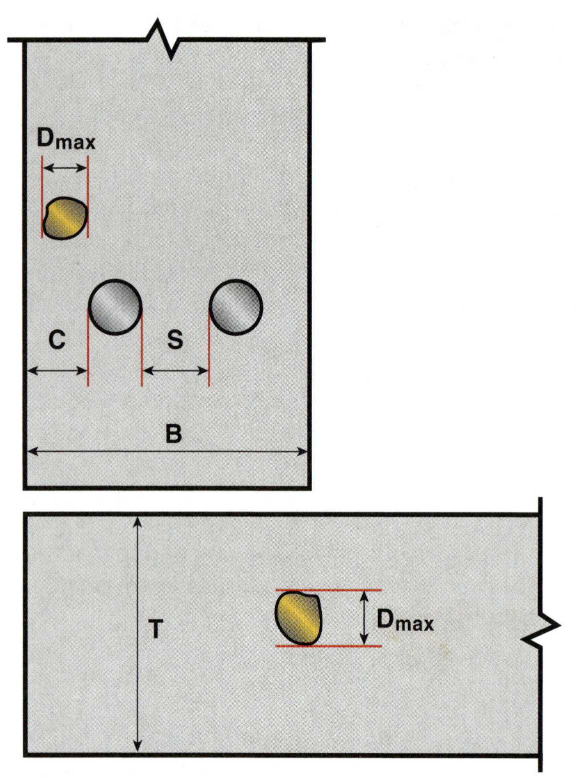

Figure 6-12. ACI 318 requirements for nominal maximum size of aggregates, D_{max}, based on concrete dimensions, B, T, and reinforcement spacing, S.

The effect of a collection of various sized aggregate particles in reducing the total volume of voids is best illustrated in Figure 6-13. A single-sized aggregate results in higher voids.* When the two aggregate sizes are combined, the void content is decreased. If several additional sizes are used, a further reduction in voids would occur. The paste requirement for concrete is dictated by the void content of the combined aggregates.

Voids in aggregates can be tested according to ASTM C29, *Standard Test Method for Bulk Density ("Unit Weight") and Voids in Aggregate,* or AASHTO T 19. In reality, the amount of cement paste required in concrete is somewhat greater than the volume of voids between the aggregates. This is illustrated in Figure 6-14. The illustration on the left represents coarse aggregates, with all particles in contact. The illustration on the right represents the dispersal of aggregates in a matrix of paste. The amount of paste necessary is greater than the void content in order to provide workability to the concrete. The actual amount is also influenced by the cohesiveness of the paste.

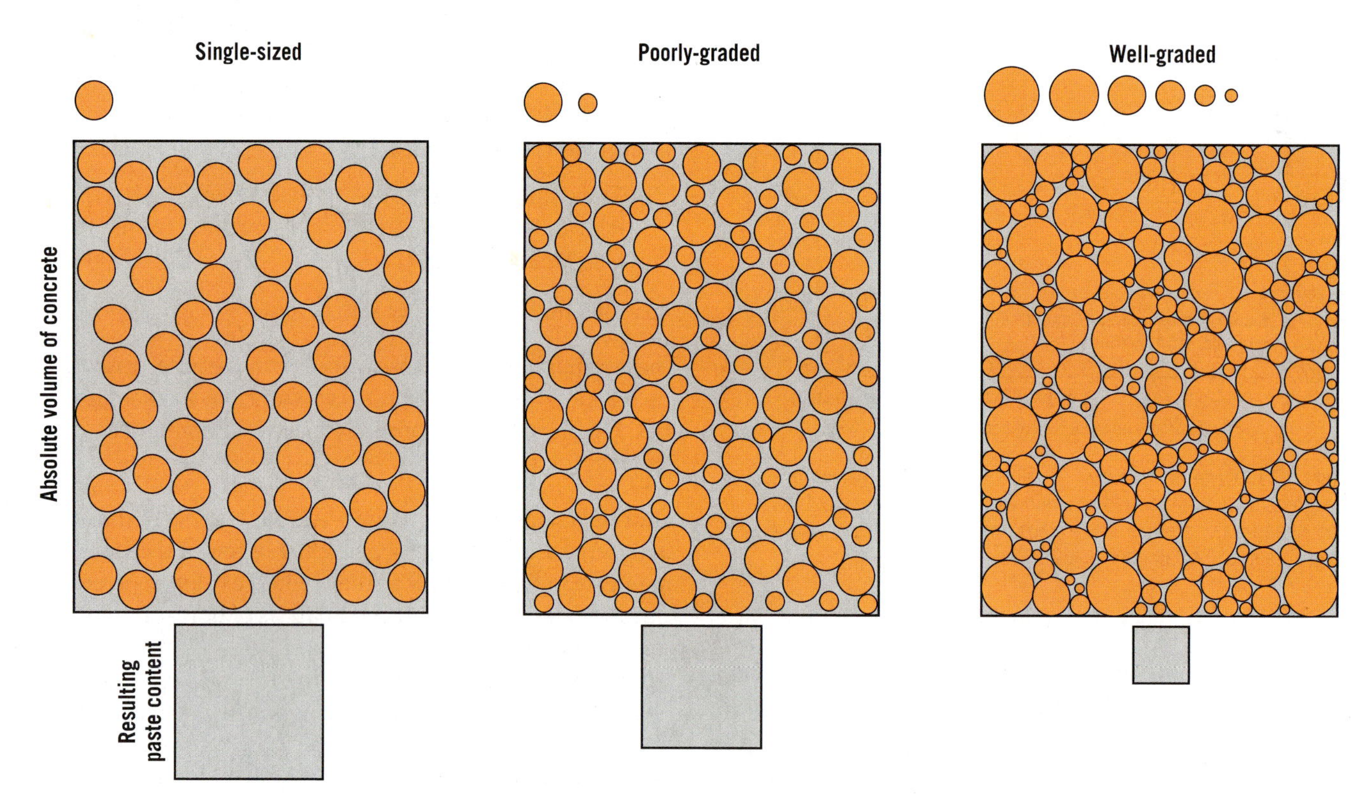

Figure 6-13. For equal absolute volumes when different sizes are combined, the void-content decreases, thus the necessary paste content decreases.

*This effect is independent of aggregate size. The voids are smaller, but the volume of voids is nearly the same (and high) when a single-size fine aggregate is used compared to a coarse aggregate. For the idealized case of spheres, the void volume is about 36% regardless of the size of particles.

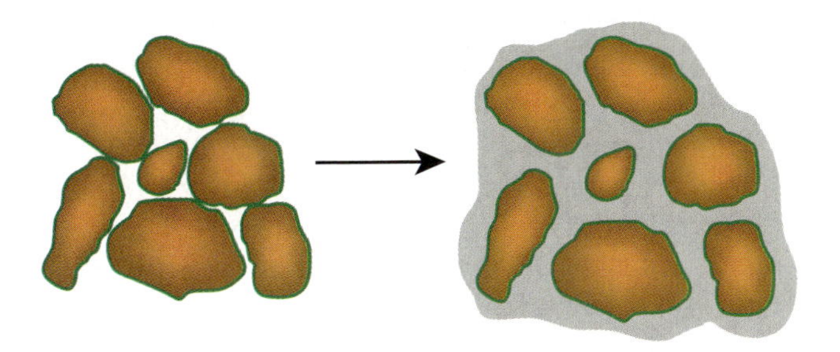

Figure 6-14. Illustration of the dispersion of aggregates in cohesive concrete mixtures.

Figure 6-15 illustrates a theoretical ideal uniform gradation (well-graded aggregate). Well-graded aggregate, having a balanced variety of sizes, maximizes the aggregate volume to the greatest extent. Sometimes mid-sized aggregate, around 9.5 mm (3⁄8 in.) in size, is lacking in an aggregate supply, resulting in a concrete with higher sand and paste requirements. The higher sand and paste requirements may cause higher water demand; resulting in poor workability, and possibly higher shrinkage. Strength and durability may also be affected. Finer aggregates require more paste because they have higher surface-to-volume ratios. Concrete mixtures that are well graded generally will have less shrinkage and permeability, and be more economical.

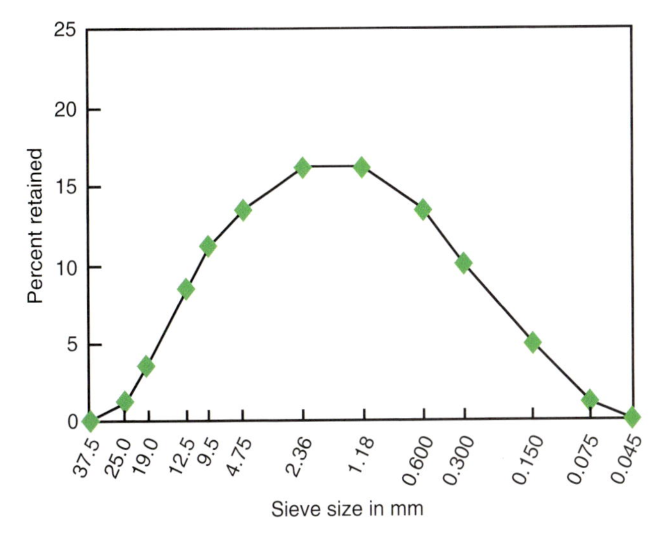

Figure 6-15. An ideal combined aggregate grading for concrete.

The combined gradation can be used to better control workability, pumpability, shrinkage, and other properties of concrete. Abrams (1918) and Shilstone (1990) demonstrate the benefits of a combined aggregate analysis:

- With constant cement content and constant consistency, there is an optimum for every combination of aggregates that will produce the most effective water to cement ratio and highest strength.
- The optimum mixture has the least particle interference and responds best to a high frequency, high amplitude vibrator.

However, this optimum mixture cannot be used for all construction due to variations in placing and finishing needs and availability. Crouch, Sauter, and Williams (2000) found in his studies on air-entrained concrete, that the water-cement ratio could be reduced by over 8% using combined aggregate gradation. Shilstone (1990) also analyzes aggregate gradation by coarseness and workability factors to improve aggregate gradation.

Studies by the NRMCA (2007) have indicated that some grading distributions that are not well graded can actually outperform well-graded aggregate blends. Particle shape and texture can impact concrete performance. If problems develop due to a poor gradation, alternative aggregates, blending, or special screening of existing aggregates should be considered.

Gap-Graded Aggregates. In gap-graded aggregates certain particle sizes are intentionally omitted. Typical gap-graded aggregates consist of only one size of coarse aggregate with all the particles of fine aggregate able to pass through the voids in the compacted coarse aggregate. Gap-graded mixtures are used in architectural concrete to obtain uniform textures in exposed-aggregate finishes. They are also used in pervious concrete mixtures to improve storm water management. They can also be used in normal structural concrete to improve other concrete properties and to permit the use of local aggregate gradations (Houston 1962 and Litvin and Pfeifer 1965).

For an aggregate of 19-mm (3⁄4-in.) maximum size, the 4.75 mm to 9.5 mm (No. 4 to 3⁄8 in.) particles can be omitted without making the concrete unduly harsh or subject to segregation. In the case of 37.5 mm (1 1⁄2 in.) aggregate, usually the 4.75 mm to 19 mm (No. 4 to 3⁄4 in.) sizes are omitted.

Care must be taken in choosing the percentage of fine aggregate in a gap-graded mixture. A poor choice can result in concrete that is likely to segregate or honeycomb because of an excess of coarse aggregate. Also, concrete with an excess of fine aggregate could have a high water demand resulting in higher shrinkage. Fine aggregate is usually 25% to 35% by volume of the total aggregate. (Lower percentages are typically used with rounded aggregates and higher with crushed material.) For a smooth off-the-form finish, a somewhat higher percentage of fine aggregate to total aggregate may be used than for an exposed-aggregate finish, but both use a lower fine aggregate content than continuously graded mixtures. Fine aggregate content also depends upon cement content, type of aggregate, and workability.

Air entrainment is usually required for workability since low-slump, gap-graded mixtures use a low fine aggregate percentage and produce harsh mixes without entrained air.

Segregation of gap-graded mixtures must be prevented by restricting the slump to the lowest value consistent with good consolidation. This may vary from zero to 75 mm (3 in.) depending on the thickness of the section, amount of reinforcement, and height of placement. However, even low slump concrete has been known to segregate when exposed to vibration, such as in a dump truck during delivery. Close control of grading and water content is also required because variations might cause segregation. If a stiff mixture is required, gap-graded aggregates may produce higher strengths than normal aggregates used with comparable cement contents. Because of their low fine-aggregate volumes and low water-cement ratios, gap-graded mixtures might be considered unworkable for some cast-in-place construction. When properly proportioned, however, these concretes are readily consolidated with vibration.

Fineness Modulus. The fineness modulus (FM) of either fine, coarse, or combined aggregate according to ASTM C125 is calculated by adding the cumulative percentages by mass retained on each of a specified series of sieves and dividing the sum by 100. The specified sieves for determining FM are: 150 µm (No. 100), 300 µm (No. 50), 600 µm (No. 30), 1.18 mm (No. 16), 2.36 mm (No. 8), 4.75 mm (No. 4), 9.5 mm (3⁄8 in.), 19.0 mm (3⁄4 in.), 37.5 mm (1½ in.), 75 mm (3 in.) and, 150 mm (6 in.).

Fineness modulus is an index of the fineness of an aggregate. In general, the higher the FM, the coarser the aggregate. However, different aggregate gradations may have the same FM. The FM of fine aggregate is useful in mixture design calculations estimating the proportions of fine and coarse aggregates. An example of how the FM of a fine aggregate is determined (with an assumed sieve analysis) is shown in Table 6-5.

Table 6-5. Determination of Fineness Modulus of Fine Aggregates

Sieve size		Percentage of individual fraction retained, by mass	Percentage passing, by mass	Cumulative percentage retained, by mass
9.5 mm	(3⁄8 in.)	0	100	0
4.75 mm	(No. 4)	2	98	2
2.36 mm	(No. 8)	13	85	15
1.18 mm	(No. 16)	20	65	35
600 µm	(No. 30)	20	45	55
300 µm	(No. 50)	24	21	79
150 µm	(No. 100)	18	3	97
	Pan	3	0	—
Total		100		283
			Fineness modulus = 283 / 100 = 2.83	

Particle Shape and Surface Texture

The particle shape and surface texture of aggregate influence the fresh concrete properties more than the properties of hardened concrete. Rough-textured, angular, elongated particles require more water to produce workable concrete than do smooth, rounded, compact aggregates. In turn, aggregate particles that are angular require more cement to maintain the same water-cementing materials ratio. Angular or poorly graded aggregates may also be more difficult to pump.

Void contents of compacted fine or coarse aggregate can be used as an index of differences in the shape and texture of aggregates of the same grading. The mixing water and cement requirement tend to increase as aggregate void content increases. Voids between aggregate particles increase with aggregate angularity.

The bond between cement paste and a given aggregate generally increases as particles change from smooth and rounded to rough and angular. This increase in bond is a consideration in selecting aggregates for concrete where flexural strength is important or where high compressive strength is needed.

Flat and elongated aggregate particles should be avoided or at least limited to about 15% by mass of the total aggregate. A particle is called flat and elongated when the ratio of length to thickness exceeds a specified value. See ASTM D4791, *Standard Test Method for Flat Particles, Elongated Particles, or Flat and Elongated Particles in Coarse Aggregate,* for determination of flat, and/or elongated particles. ASTM D3398, *Standard Test Method for Index of Aggregate Particle Shape and Texture,* provides an indirect method of establishing a particle index as an overall measure of particle shape or texture, while ASTM C295, *Standard Guide for Petrographic Examination of Aggregates for Concrete,* provides procedures for the petrographic examination of aggregate which includes evaluation of particle shape and texture.

A number of automated test machines are available for rapid determination of the particle size distribution of aggregate. These machines were designed to provide a faster alternative to the standard sieve analysis test. They can capture and analyze digital images of the aggregate particles to determine gradation. Figure 6-16 shows a videograder that measures size and shape of an aggregate by using line-scan cameras wherein two-dimensional images are constructed from a series of line images. Other machines use matrix-scan cameras to capture two-dimensional snapshots of the falling aggregate. Maerz and Lusher (2001) developed a dynamic prototype imaging system that provides particle size and shape information by using a miniconveyor system to parade individual fragments past two orthogonally oriented, synchronized cameras.

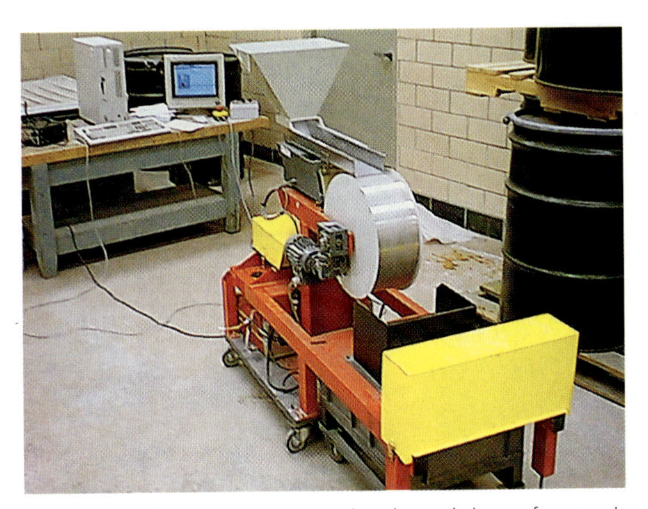

Figure 6-16. Videograder for measuring size and shape of aggregate.

Bulk Density (Unit Weight) and Voids

The bulk density or unit weight of an aggregate is the mass or weight of the aggregate required to fill a container of a specified unit volume. The volume is occupied by both aggregates and the voids between aggregate particles. The approximate bulk density of aggregate commonly used in normal-weight concrete ranges from about 1200 kg/m^3 to 1750 kg/m^3 (75 lb/ft^3 to 110 lb/ft^3). Methods of determining the bulk density of aggregates and void content are given in ASTM C29, *Standard Test Method for Bulk Density ("Unit Weight") and Voids in Aggregate,* (AASHTO T 19). In these standards, three methods are described for consolidating the aggregate in the container depending on the maximum size of the aggregate: rodding, jigging, and shoveling. The measurement of loose uncompacted void content of fine aggregate is described in ASTM C1252, *Standard Test Methods for Uncompacted Void Content of Fine Aggregate (as Influenced by Particle Shape, Surface Texture, and Grading).*

The void content between particles affects paste requirements in concrete mix design (see preceding sections, **Grading** and **Particle Shape and Surface Texture**). Void contents range from about 30% to 45% for coarse aggregates to about 40% to 50% for fine aggregate. Angularity increases void content while larger sizes of well-graded (uniform) aggregate and improved grading decreases void content.

Density

The density of aggregate particles used in mixture proportioning computations (not including voids between particles) is determined by multiplying the relative density (specific gravity) of the aggregate times the density of water. An approximate value of 1000 kg/m^3 (62.4 lb/ft^3) is often used for the density of water. The density of aggregate, along with more accurate values for water density, are provided in ASTM C127, *Standard Test Method for Density, Relative Density (Specific Gravity), and Absorption of Coarse Aggregate* (AASHTO T 85) and ASTM C128, *Standard Test Method for Density, Relative Density (Specific Gravity), and Absorption of Fine Aggregate* (AASHTO T 84).

Most natural aggregates have particle densities of between 2400 and 2900 kg/m^3 (150 and 181 lb/ft^3).

Relative Density (Specific Gravity)

The relative density (specific gravity) of an aggregate is the ratio of its mass to the mass of an equal absolute volume of water. Most natural aggregates have relative densities between 2.40 and 2.90, which means that the aggregates are 2.40 to 2.90 times as dense as water. Relative density is not generally used as a measure of aggregate quality, though some porous aggregates that exhibit accelerated freeze-thaw deterioration do have low specific gravities. It is used in certain computations for mixture proportioning and control, such as the volume occupied by the aggregate in the absolute volume method of mix design. Test methods for determining relative densities for coarse and fine aggregates are described in ASTM C127 (AASHTO T 85) and ASTM C128 (AASHTO T 84).

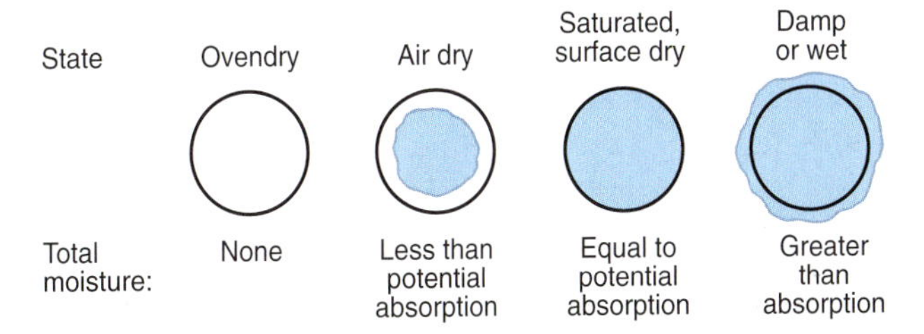

Figure 6-17. Moisture conditions of aggregate.

Absorption and Surface Moisture

The absorption and surface moisture of aggregates should be determined according to ASTM C70, *Standard Test Method for Surface Moisture in Fine Aggregate*, ASTM C127, ASTM C128, and ASTM C566, *Standard Test Method for Total Evaporable Moisture Content of Aggregate by Drying* (AASHTO T 255) so that the total water content of the concrete can be controlled and correct batch weights determined. The internal structure of an aggregate particle is made up of solid matter and voids that may or may not contain water.

The moisture conditions of aggregates are shown in Figure 6-17. They are designated as:

Ovendry — zero moisture content, fully absorbent

Air dry — dry at the particle surface but containing some interior moisture, less than potential absorption

Saturated surface dry (SSD) — neither absorbing water from nor contributing water to the concrete mixture, equal to potential absorption

Damp or wet — containing an excess of moisture on the surface (free water)

The amount of water added at the concrete batch plant must be adjusted for the moisture conditions of the aggregates in order to accurately meet the water requirement of the mix design (see Chapter 12). If the water content of the concrete mixture is not kept constant, the water-cement ratio will vary from batch to batch causing other properties, such as the compressive strength and workability to vary as well.

Coarse and fine aggregate will generally have absorption levels (moisture contents at SSD) in the range of 0.2% to 4% and 0.2% to 2%, respectively. Free-water contents will usually range from 0.5% to 2% for coarse aggregate and 2% to 6% for fine aggregate.

Bulking. Bulking is the increase in total volume of moist fine aggregate over the same mass in a dry condition. Surface tension in the moisture holds the particles apart, causing an increase in volume. Bulking of a fine aggregate can occur when it is shoveled or otherwise moved in a damp condition, even though it may have been fully consolidated previously. Figure 6-18 illustrates how the amount of bulking of fine aggregate varies with moisture content and grading; fine gradings bulk more than coarse gradings for a given amount of moisture. Figure 6-19 shows similar information in terms of weight for a particular fine aggregate. Since most fine aggregates are delivered in a damp condition, wide variations can occur in batch quantities if batching is done by volume. For this reason, good practice has long favored weighing the aggregate and adjusting for moisture content when proportioning concrete.

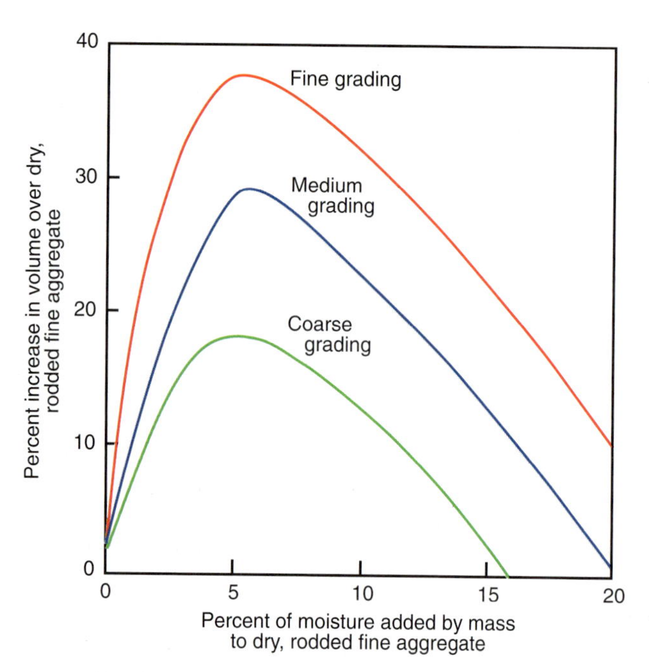

Figure 6-18. Surface moisture on fine aggregate can cause considerable bulking; the amount varies with the amount of moisture and the aggregate grading (PCA 1955 and PCA 1935).

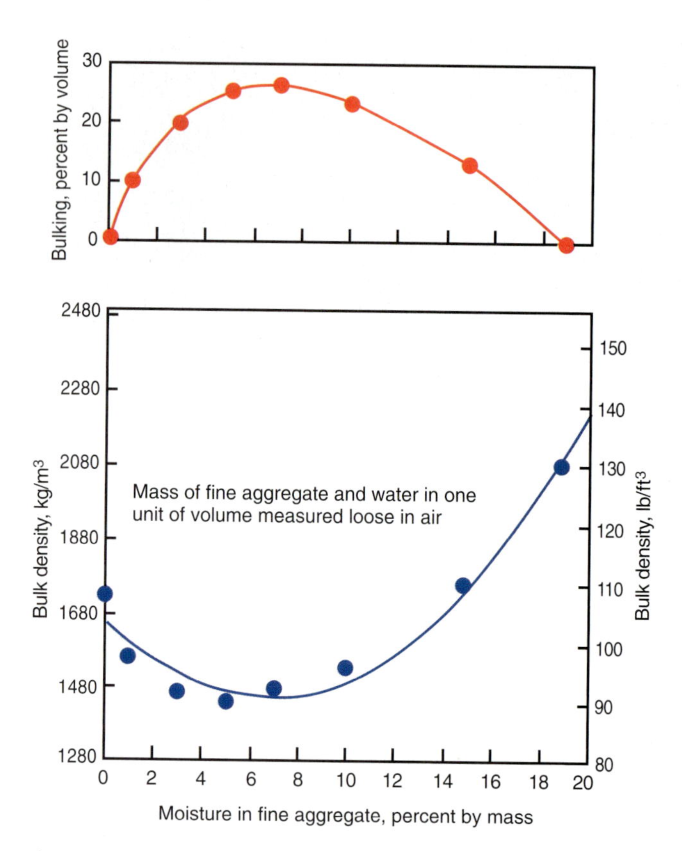

Figure 6-19. Bulk density is compared with the moisture content for a particular sand (PCA 1955).

Resistance to Freezing and Thawing

The frost resistance of an aggregate is related to its porosity, absorption, permeability, and pore structure. An aggregate particle with high absorption may not accommodate the expansion that occurs during the freezing of water if that particle becomes critically saturated. If enough offending particles are present, the result can be expansion and cracking of the aggregate and possible disintegration of the concrete. If a single problem particle is near the surface of the concrete, it can cause a popout (Figure 6-20). Popouts generally appear as conical fragments that break out of the concrete surface. The offending aggregate particle or a part of it is usually found at the bottom of the resulting void. Generally, it is coarse rather than fine aggregate particles with higher porosity values and medium-sized pores (0.1 to 5 µm) that are easily saturated and that cause concrete deterioration and popouts. Larger pores do not usually become saturated or cause concrete distress, and water in very fine pores may not freeze readily.

Figure 6-20. A popout is the breaking away of a small fragment of concrete surface due to internal pressure that leaves a shallow, typically conical depression.

The critical size at which an aggregate will fail is dependent upon the rate of freezing and the porosity, permeability, and tensile strength of the particle. For fine-grained aggregates with low permeability (cherts for example), the critical particle size may be within the range of normal aggregate sizes. It is higher for coarse-grained materials or those with capillary systems interrupted by numerous macropores (voids too large to hold moisture by capillary action). For these aggregates, the critical particle size may be sufficiently large to have no impact on freeze-thaw resistance in concrete even though the absorption may be high. Also, if potentially vulnerable aggregates are used in concrete subjected to periodic drying while in service, they may never become sufficiently saturated to cause failure.

Figure 6-21. D-cracking along a transverse joint caused by failure of carbonate coarse aggregate (Courtesy of J. Grove).

The cracking of concrete pavements caused by freeze-thaw deterioration of the aggregate is called D-cracking. D-cracks are closely spaced crack formations oriented parallel to transverse and longitudinal joints that later multiply outward from the joints toward the center of the pavement panel (Figure 6-21). This type of cracking has been observed in pavements after three to 20 years of service. D-cracked concrete resembles frost-damaged concrete caused by freeze-thaw deterioration of the hardened paste. D-cracking is a function of the pore properties of certain types of aggregate particles and the environment in which the pavement is placed. If water accumulates under pavements in the base and subbase layers, the aggregate may eventually become critically saturated. With continued freezing and thawing cycles, cracking of the concrete starts in the saturated aggregate (Figure 6-22) at the bottom of the slab and progresses upward until it reaches the wearing surface. D-cracking can be reduced either by selecting aggregates that perform better in freeze-thaw cycles or, where marginal aggregates must be used, by reducing the maximum particle size. Also, installation of permeable bases or effective drainage systems for carrying free water out from under the pavement are helpful (Harrigan 2002).

Figure 6-22. Fractured aggregate particle as a source of distress in D-cracking.

The performance of aggregates under exposure to freezing and thawing can be evaluated in two ways: (1) past performance in the field, and (2) laboratory freeze-thaw tests of concrete specimens, ASTM C1646, *Standard Practice for Making and Curing Test Specimens for Evaluating Resistance of Coarse Aggregate to Freezing and Thawing in Air-Entrained Concrete.* If similarly-sized aggregates from the same source have previously given satisfactory service when used in concrete, they may be considered suitable. Aggregates not having a service record can be considered acceptable if they perform satisfactorily in air-entrained concretes subjected to freeze-thaw tests according to ASTM C1646. In these tests concrete specimens made with the aggregate in question are subjected to alternate cycles of freezing and thawing in water according to ASTM C666, *Standard Test Method for Resistance of Concrete to Rapid Freezing and Thawing*

(AASTHO T 161). Deterioration is measured by (1) the reduction in the dynamic modulus of elasticity, (2) linear expansion, and (3) weight loss of the specimens. A failure criterion of 0.035% expansion in 350 freeze-thaw cycles or less is used by a number of state highway departments to help indicate whether an aggregate is susceptible to D-cracking. Different aggregate types may influence the criteria levels and empirical correlations between laboratory freeze-thaw tests. Field service records should be kept to select the proper criterion (Vogler and Grove 1989).

Specifications may require that the aggregates resistance to weathering be demonstrated by exposure to a sodium sulfate or magnesium sulfate solution ASTM C88, *Standard Test Method for Soundness of Aggregates by Use of Sodium Sulfate or Magnesium Sulfate* or AASHTO T 104. The test consists of a number of immersion cycles (wetting and drying) for a sample of the aggregate in a sulfate solution; this cycling creates a pressure through salt-crystal growth in the aggregate pores similar to that produced by freezing water. Upon completion of the cycling, the sample is then oven dried and the percentage of weight loss calculated. Unfortunately, this test is sometimes misleading. Aggregates behaving satisfactorily in the test might produce concrete with low freeze-thaw resistance; conversely, aggregates performing poorly might produce concrete with adequate resistance. This is attributed, at least in part, to the fact that the aggregates in the test are not confined by cement paste (as they would be in concrete) and the mechanisms of attack are not the same as in freezing and thawing. The test is most reliable for stratified rocks with porous layers or weak bedding planes.

An additional test that can be used to evaluate aggregates for potential D-cracking is the rapid pressure release method. An aggregate is placed in a pressurized chamber and the pressure is rapidly released causing the aggregate with a questionable pore system to fracture (Janssen and Snyder 1994). The amount of fracturing relates to the potential for D-cracking.

Wetting and Drying Properties

Weathering due to wetting and drying can also affect the durability of aggregates. The expansion and contraction coefficients of rocks vary with temperature and moisture content. If alternate wetting and drying occurs, severe strain can develop in some aggregates. With certain types of rock this can cause a permanent increase in the volume of the concrete and eventual concrete deterioration. Clay lumps and other friable particles can degrade rapidly with repeated wetting and drying. Popouts can also develop due to the moisture-swelling characteristics of certain aggregates, especially clay balls and shales. An experienced petrographer can assist in determining the potential for this distress.

Abrasion and Skid Resistance

The abrasion resistance of an aggregate is often used as a general index of its quality. Abrasion resistance is essential when the aggregate is to be used in concrete subject to abrasion, as in heavy-duty floors or pavements. Low abrasion resistance of an aggregate may increase the quantity of fines in the concrete during mixing. Consequently, this may increase the water requirement and require an adjustment in the water-cement ratio.

The most common test for abrasion resistance is the Los Angeles abrasion test (rattler method) performed in accordance with ASTM C131, *Standard Test Method for Resistance to Degradation of Small-Size Coarse Aggregate by Abrasion and Impact in the Los Angeles Machine* (AASHTO T 96) or ASTM C535, *Standard Test Method for Resistance to Degradation of Large-Size Coarse Aggregate by Abrasion and Impact in the Los Angeles Machine*. In this test a specified quantity of aggregate is placed in a steel drum containing steel balls, the drum is rotated, and the percentage of material worn away is measured. Specifications often set an upper limit on this mass loss. However, a comparison of the results of aggregate abrasion tests with the abrasion resistance of concrete made with the same aggregate do not generally show a clear correlation. Mass loss due to impact in the rattler is often as much as the mass loss caused by abrasion. The wear resistance of concrete is determined more accurately by abrasion tests of the concrete itself (see Chapter 9).

To provide good skid resistance on pavements, the siliceous particle content of the fine aggregate should be at least 25%. For specification purposes, the siliceous particle content is considered equal to the insoluble residue content after treatment in hydrochloric acid under standardized conditions (ASTM D3042, *Standard Test Method for Insoluble Residue in Carbonate Aggregates*). Certain manufactured sands produce slippery pavement surfaces and should be investigated for acceptance before use.

Strength and Shrinkage

The strength of an aggregate is rarely tested and generally does not influence the strength of conventional concrete as much as the strength of the paste and the paste-aggregate bond. However, aggregate strength does become important in high-strength concrete. Aggregate stress levels in concrete are often much higher than the average stress over the entire cross section of the concrete. Aggregate tensile strengths range from 2 MPa to 15 MPa (300 psi to 2300 psi) and compressive strengths from 65 MPa to 270 MPa (10,000 psi to 40,000 psi). Strength can be tested according to ASTM C170.

Different aggregate types have different compressibility, modulus of elasticity, and moisture-related shrinkage characteristics that influence the same properties in concrete. Aggregates with high absorption may have high shrinkage on drying. Quartz and feldspar aggregates, along with limestone, dolomite, and granite, are considered low shrinkage aggregates; while aggregates with sandstone, shale, slate, hornblende, and graywacke are often associated with high shrinkage in concrete (Figure 6-23).

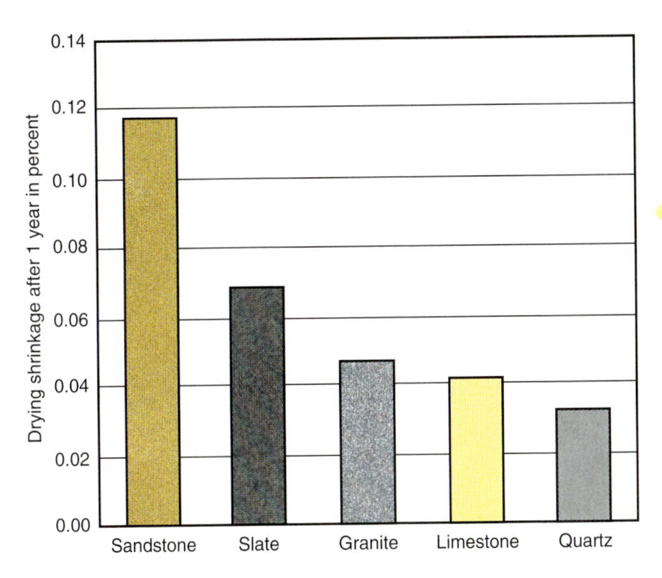

Figure 6-23. Concretes containing sandstone or slate produce a high shrinkage concrete. Granite, limestone, and quartz, are low shrinkage-producing aggregates (Carlson 1938).

Resistance to Acid and Other Corrosive Substances

Portland cement concrete is durable in most natural environments; however, concrete in service is occasionally exposed to substances that will attack it.

Although acids generally attack and leach away the hydrated cement phases of the cement paste, they may not readily attack certain aggregates, such as siliceous aggregates. Calcareous aggregates often react readily with acids. However, the sacrificial effect of calcareous aggregates may be more beneficial than siliceous aggregates in mild acid exposures or in areas where water is not flowing. With calcareous aggregate, the acid attacks the entire exposed concrete surface uniformly, reducing the rate of attack on the paste and preventing loss of aggregate particles at the surface. Calcareous aggregates also tend to neutralize the acid, especially in stagnant locations. Siliceous aggregate should be avoided when strong solutions of sodium hydroxide are present, as these solutions attack this type of aggregate.

Fire Resistance and Thermal Properties

The fire resistance and thermal properties of concrete (conductivity, diffusivity, and coefficient of thermal expansion) depend to some extent on the mineral constituents of the aggregates used. Manufactured and some naturally occurring lightweight aggregates are more fire resistant than normal-weight aggregates due to their insulating properties and high-temperature stability. In general, concrete containing a calcareous coarse aggregate performs better under fire exposure than a concrete containing quartz or siliceous aggregate such as granite or quartzite. At about 590°C (1060°F), quartz expands 0.85% causing disruption to the concrete structure (ACI Committee 216 1989 and ACI Committee 221 1996). The coefficient of thermal expansion of aggregates ranges from 0.55×10^{-6} per °C to 5×10^{-6} per °C (1×10^{-6} per °F to 9×10^{-6} per °F). For more information refer to Chapter 10 for temperature-induced volume changes and to Chapter 20 for thermal conductivity and mass concrete considerations.

Potentially Harmful Materials

Harmful substances that may be present in aggregates include organic impurities, silt, clay, shale, iron oxide, coal, lignite, mica, and certain lightweight and soft particles (Table 6-6). In addition, rocks and minerals such as some cherts, strained quartz (Buck and Mather 1984), and certain dolomitic limestones are alkali reactive (see Table 6-7). Gypsum and anhydrite may cause sulfate attack. Certain aggregates, such as some shales, will cause popouts by swelling (simply by absorbing water) or by freezing of absorbed water (Figure 6-23). Most specifications limit the permissible amounts of these substances. The performance history of an aggregate should be a determining factor in setting the limits for harmful substances. Test methods for detecting harmful substances qualitatively or quantitatively are listed in Table 6-6.

Table 6-6. Harmful Materials in Aggregates

Substances	Effect on concrete	Test designation
Organic impurities	Affects setting and hardening, may cause deterioration	ASTM C40 (AASHTO T 21) ASTM C87 (AASHTO T 71)
Materials finer than the 75-µm (No. 200) sieve	Affects bond, increases water requirement	ASTM C117 (AASHTO T 11)
Coal, lignite, or other lightweight materials	Affects durability, may cause stains and popouts	ASTM C123 (AASHTO T 113)
Soft particles	Affects durability	ASTM C235
Clay lumps and friable particles	Affects workability and durability, may cause popouts	ASTM C142 (AASHTO T 112)
Chert of less than 2.40 relative density	Affects durability, may cause popouts	ASTM C123 (AASHTO T 113) ASTM C295
Alkali-reactive aggregates	Causes abnormal expansion, map cracking, and popouts	ASTM C227 ASTM C289 ASTM C295 ASTM C342 ASTM C586 ASTM C1260 (AASHTO T303) ASTM C1293 ASTM C1567

Table 6-7. Some Potentially Harmful Reactive Minerals, Rock, and Synthetic Materials

Alkali-silica reactive substances*		Alkali-carbonate reactive substances**
Andesites Argillites Certain siliceous limestones and dolomites Chalcedonic cherts Chalcedony Cristobalite Dacites Glassy or cryptocrystalline volcanics Granite gneiss Graywackes Metagraywackes	Opal Opaline shales Phylites Quartzoses Cherts Rhyolites Schists Siliceous shales Strained quartz and certain other forms of quartz Synthetic and natural silicious glass Tridymite	Calcitic dolomites Dolomitic limestones Fine-grained dolomites

* Several of the rocks listed (granite gneiss and certain quartz formations for example) react very slowly and may not show evidence of any harmful degree of reactivity until the concrete is over 20 years old.

** Only certain sources of these materials have shown reactivity.

Aggregates are potentially harmful if they contain compounds known to react chemically with portland cement paste and produce any of the following: (1) significant volume changes of the paste, aggregates, or both; (2) interference with the normal hydration of cement; and (3) otherwise harmful byproducts.

Organic impurities may: delay setting and hardening of concrete, interfere with interaction of chemical admixtures, reduce strength gain, and in unusual cases, cause deterioration. Organic impurities such as peat, humus, and organic loam may not be as detrimental but should be avoided.

Materials finer than the 75-µm (No. 200) sieve, especially silt and clay, may be present as loose dust and may form a coating on the aggregate particles. Even thin coatings of silt or clay on gravel particles can be harmful because they may weaken the bond between the cement paste and aggregate. If certain types of silt or clay are present in excessive amounts, water requirements may increase significantly.

There is a tendency for some fine aggregates to degrade from the grinding action in a concrete mixer. This effect, which is measured using ASTM C1137, *Standard Test Method for Degradation of Fine Aggregate Due to Attrition*, may alter mixing water, air-entraining admixtures, and slump requirements.

Coal or lignite, or other low-density materials such as wood or fibrous materials, in excessive amounts will affect the durability of concrete. If these impurities occur at or near the surface, they might disintegrate, pop out, or cause stains. Potentially harmful chert in coarse aggregate can be identified using ASTM C123, *Standard Test Method for Lightweight Particles in Aggregate* (AASHTO T 113).

Soft particles in coarse aggregate are especially objectionable because they cause popouts and can affect the durability and wear resistance of concrete. If friable, they could break up during mixing and thereby increase the amount of water required. Where abrasion resistance is critical, such as in heavy-duty industrial floors, testing may indicate that further investigation or another aggregate source is warranted.

Clay lumps present in concrete may absorb some of the mixing water, cause popouts in hardened concrete, and affect durability and wear resistance. They can also break up during mixing and thereby increase the mixing-water demand.

Mica is composed of flat plates that can either act as cleavage planes in coarse aggregate or as severely flat particles in fine aggregate. Either particle size can result in a decrease in concrete strength. The test for mica content uses hazardous materials and is not commonly performed. Previous local experience with aggregate containing mica can determine if the material can be used successfully.

Aggregates can occasionally contain particles of iron oxide and iron sulfide that result in unsightly stains on exposed concrete surfaces (Figure 6-24). The aggregate should meet the staining requirements of ASTM C330, *Standard Specification for Lightweight Aggregates for Structural Concrete* (AASHTO M 195) when tested according to ASTM C641, *Standard Test Method for Iron Staining Materials in Lightweight Concrete Aggregates*; the quarry face and aggregate stockpiles should not show evidence of staining.

Figure 6-24. Iron oxide stain caused by impurities in the coarse aggregate (Courtesy of T. Rewerts).

As an additional aid in identifying staining particles, the aggregate can be immersed in a lime slurry. If staining particles are present, a blue-green gelatinous precipitate will form within 5 to 10 minutes; this will rapidly change to a brown color on exposure to air and light. The reaction should be complete within 30 minutes. If no brown gelatinous precipitate is formed when a suspect aggregate is placed in the lime slurry, there is little likelihood of any reaction taking place in concrete. These tests should be required when aggregates with no record of successful prior use are used in architectural concrete.

Alkali-Aggregate Reactivity

Alkali-Silica Reaction

Aggregates containing certain constituents can react with alkali hydroxides in concrete. The reactivity is potentially harmful only when it produces significant expansion (Mather 1975). Field performance history is the best method of evaluating the susceptibility of an aggregate to ASR. Comparisons should be made between the existing and proposed concrete mixture proportions, ingredients, and service environments. This process should tell whether special requirements are needed, are not needed, or whether testing of the aggregate or job concrete is required.

Test Methods for Identifying ASR-Aggregates. The reactivity of an aggregate is classified according to Table 6-8. The aggregate is tested using ASTM C1293, *Standard Test Method for Determination of Length Change of Concrete Due to Alkali-Silica Reaction (Concrete Prism Test)*, or CPT. Because this test takes one year to complete, a rapid assessment test, in the form of ASTM C1260 (AASHTO T 303), *Potential Alkali-Reactivity of Aggregates (Mortar-Bar Method)*, can be used to produce results in sixteen days. ASTM C1260 is less precise in determining aggregate reactivity, having been found to exhibit false positives and false negatives. These tests should not be used to disqualify use of potentially reactive aggregates, as reactive aggregates can be safely used with the careful selection of cementitious materials. See Chapter 11 for information on the mechanisms and control of ASR.

Table 6-8. Classification of Alkali-Silica Reactivity of Aggregates (Thomas, Fournier, and Folliard 2008)

Aggregate-reactivity class	Description of aggregate reactivity	One-year expansion in CPT (%)
R0	Non-reactive	< 0.040
R1	Moderately reactive	0.040 – 0.120
R2	Highly reactive	0.120 – 0.240
R3	Very highly reactive	> 0.240

Alkali-Carbonate Reaction

Reactions observed with certain dolomitic rocks are associated with alkali-carbonate reaction (ACR). Reactive rocks usually contain large crystals of dolomite scattered in and surrounded by a fine-grained matrix of calcite and clay. Calcite is one of the mineral forms of calcium carbonate; dolomite is the common name for calcium-magnesium carbonate. ACR is relatively rare because aggregates susceptible to this reaction are usually unsuitable for use in concrete for other reasons, such as strength potential. Argillaceous dolomitic limestone contains calcite and dolomite with appreciable amounts of clay and can contain small amounts of reactive silica. Alkali reactivity of carbonate rocks is not usually dependent upon its clay mineral composition (Hadley 1961). Aggregates have potential for expansive ACR if the following lithological characteristics exist (Ozol 2006 and Swenson and Gillott 1967):

- clay content, or insoluble residue content, in the range of 5% to 25%;
- calcite-to-dolomite ratio of approximately 1:1;
- increase in the dolomite volume up to a point at which interlocking texture becomes a restraining factor; and
- small size of the discrete dolomite crystals (rhombs) suspended in a clay matrix.

Test methods for identifying ACR aggregates. The three test methods commonly used to identify potentially alkali-carbonate reactive aggregate are:

- petrographic examination (ASTM C295);
- rock cylinder method (ASTM C586); and
- concrete prism test (ASTM C1105).

See Farny and Kerkhoff (2007) and Chapter 11 for more information.

Aggregate Beneficiation

Aggregate processing consists of: (1) basic processing – crushing, screening, and washing – to obtain proper gradation and cleanliness; and (2) beneficiation – improving quality by processing methods such as heavy media separation, jigging, rising-current classification, and crushing.

In heavy media separation, aggregates are passed through a heavy liquid comprised of finely ground heavy minerals and water proportioned to have a relative density (specific gravity) less than that of the desirable aggregate particles but greater than that of the deleterious particles. The heavier particles sink to the bottom while the lighter particles float to the surface. This process can be used when acceptable and harmful particles have distinguishable relative densities.

Jigging separates particles with small differences in density by pulsating water current. Upward pulsations of water through a jig (a box with a perforated bottom) move the lighter material into a layer on top of the heavier material; the top layer is then removed.

Rising-current classification separates particles with large differences in density. Light materials, such as wood and lignite, are floated away in a rapidly upward moving stream of water.

Crushing is also used to remove soft and friable particles from coarse aggregates. This process is sometimes the only means of making material suitable for use. Unfortunately, with any process some acceptable material is always lost and removal of all harmful particles may be difficult or expensive.

Handling and Storing Aggregates

Aggregates should be handled and stored in a way that minimizes segregation (separation of aggregates by size) and degradation and that prevents contamination by deleterious substances. Stockpiles should be built up in thin layers of uniform thickness to minimize segregation (Figure 6-25). The most economical and acceptable method of forming aggregate stockpiles is the truck-dump method, which discharges the loads in a way that keeps them tightly joined. The aggregate is then reclaimed with a front-end loader. The loader should remove slices from the edges of the pile from bottom to top so that every slice will contain a portion of each horizontal layer.

Figure 6-25. Stockpile of aggregate at a ready mix plant.

Crushed aggregates segregate less than rounded (gravel) aggregates and larger-size aggregates segregate more than smaller sizes. To avoid segregation of coarse aggregates, size fractions can be stockpiled and batched separately. Proper stockpiling procedures, however, should eliminate this requirement. Specifications provide a range in the amount of material permitted in any size fraction partly to accommodate some minor segregation in stockpiling and batching operations.

Washed aggregates should be stockpiled well before use so that they can drain to a uniform moisture content. Damp fine material has less tendency to segregate than dry material. When dry fine aggregate is dropped from buckets or conveyors, wind can blow away the fines; this should be avoided if possible.

Figure 6-26. Bulkheads or dividers should be used to avoid contamination of aggregate stockpiles.

Bulkheads or dividers should be used to avoid contamination of aggregate stockpiles (Figure 6-26). Partitions between stockpiles should be high enough to prevent intermingling of materials. Storage bins, or hoppers, should be circular or nearly square. Their bottoms should slope not less than 50 degrees from the horizontal on all sides to a center outlet. When loading the bin, the material should fall vertically over the outlet into the bin. Chuting the material into a bin at an angle and against the bin sides will cause segregation. Baffle plates or dividers will help minimize segregation. Bins should be kept as full as possible since this reduces breakage of aggregate particles and the tendency to segregate. Recommended methods of handling and storing aggregates are discussed at length in Matthews (1965 to 1967), NCHRP (1967), and Bureau of Reclamation (1981).

References

Abrams, D.A., *Design of Concrete Mixtures*, Lewis Institute, Structural Materials Research Laboratory, Bulletin No. 1, PCA as LS001, Chicago, http://www.cement.org/pdf_files/LS001.pdf, 1918, 20 pages.

ACI Committee 216, *Guide for Determining the Fire Endurance of Concrete Elements*, ACI 216R-89, reapproved 2001, American Concrete Institute, Farmington Hills, Michigan, 1989.

ACI Committee 221, *Guide for Use of Normal Weight Aggregates in Concrete*, ACI 221R-96, reapproved 2001, American Concrete Institute, Farmington Hills, Michigan, 1996.

ACI Committee 318, *Building Code Requirements for Structural Concrete and Commentary*, ACI 318-08, American Concrete Institute, Farmington Hills, Michigan, 2008.

ACI Committee 555, *Removal and Reuse of Hardened Concrete*, ACI 555R-01, American Concrete Institute, Farmington Hills, Michigan, 2001.

Addis, B.J.; Owens, Gill; and Fulton, F.S., *Fulton's Concrete Technology*, 8th edition, Cement and Concrete Institute, Midrand, South Africa, 2001, 330 pages.

Bohan, Richard P., and Ries, John, *Structural Lightweight Concrete*, IS032, Portland Cement Association, Skokie, Illinois, 2008, 8 pages.

Buck, Alan D., and Mather, Katharine, *Reactivity of Quartz at Normal Temperatures*, Technical Report SL-84-12, Structures Laboratory, Waterways Experiment Station, U.S. Army Corps of Engineers, Vicksburg, Mississippi, July 1984.

Bureau of Reclamation, *Concrete Manual*, 8th ed., U.S. Bureau of Reclamation, Denver, 1981.

Carlson, Roy W., "Drying Shrinkage of Concrete as Affected by Many Factors," *Proceedings*, ASTM, Vol. 38, Part II, 1938, page 419 to 437.

Crouch, L.K.; Sauter, Heather J.; and Williams, Jakob A., "92-Mpa Air-entrained HPC," *TRB-Record 1698*, Concrete 2000, page 24.

ECCO (Environmental Council of Concrete Organizations), *Recycling Concrete and Masonry*, EV 22, Skokie, Illinois, http://www.ecco.org/pdfs/ev22.pdf, 1999, 12 pages.

Farny, James A. and Kerkhoff, B., *Diagnosis and Control of Alkali-Aggregate Reactions*, IS413, Portland Cement Association, 2007, 26 pages.

Fathifazl, Gholamreza; Razaqpur, A. Ghani; Isgor, O. Burkan; Abbas, Abdelgadir; Fournier, Benoit; and Foo, Simon, "Proportioning Concrete Mixtures with Recycled Concrete Aggregate," *Concrete International*, Vol. 32, No. 3, American Concrete Institute, Farmington Hills, Michigan, March 2010, pages 37 to 43.

Hadley, D.W., *Alkali Reactivity of Carbonate Rocks—Expansion and Dedolomitization*, Research Department Bulletin RX139, Portland Cement Association, http://www.cement.org/pdf_files/RX139.pdf, 1961.

Harrigan, Edward, "Performance of Pavement Subsurface Drainage," *NCHRP Research Results Digest*, No. 268, Transportation Research Board, Washington, D.C., November 2002.

Houston, B.J., *Investigation of Gap-Grading of Concrete Aggregates; Review of Available Information*, Technical Report No. 6-593, Report 1, Waterways Experiment Station, U.S. Army Corps of Engineers, Vicksburg, Mississippi, February 1962.

Janssen, Donald J., and Snyder, Mark B., *Resistance of Concrete to Freezing and Thawing*, SHRP-C-391, Strategic Highway Research Program, Washington, D.C., 1994, 201 pages. http://gulliver.trb.org/publications/shrp/SHRP-C-391.pdf.

Kerkhoff, Beatrix, and Siebel, Eberhard, "Properties of Concrete with Recycled Aggregates (Part 2)," *Beton* 2/2001, Verlag Bau + Technik, 2001, pages 105 to 108.

Litvin, Albert, and Pfeifer, Donald W., *Gap-Graded Mixes for Cast-in-Place Exposed Aggregate Concrete*, Development Department Bulletin DX090, Portland Cement Association, http://www.cement.org/pdf_files/DX090.pdf, 1965.

Maerz, Norbert H., and Lusher, Mike, "Measurement of flat and elongation of coarse aggregate using digital image processing," *80th Annual Meeting*, Transportation Research Board, Washington D.C., 2001, pages 2 to 14.

Mather, Bryant, *New Concern over Alkali-Aggregate Reaction*, Joint Technical Paper by National Aggregates Association and National Ready Mixed Concrete Association, NAA Circular No. 122 and NRMCA Publication No. 149, Silver Spring, Maryland, 1975.

Matthews, C.W., "Stockpiling of Materials," *Rock Products*, series of 21 articles, Maclean Hunter Publishing Company, Chicago, August 1965 through August 1967.

National Cooperative Highway Research Program (NCHRP), *Effects of Different Methods of Stockpiling and Handling Aggregates*, NCHRP Report 46, Transportation Research Board, Washington, D.C., 1967.

Obla, Karthik; Kim, Haejin; and Lobo, Colin, *Effect of Continuous (Well-Graded) Combined Aggregate Grading on Concrete Performance*, Phase B: Concrete Performance, National Ready Mixed Concrete Association, Silver Spring, Maryland, May 2007, 45 pages.

Ozol, Michael A., "Alkali-Carbonate Rock Reaction," *Significance of Tests and Properties of Concrete and Concrete-Making Materials*, ASTM STP 169D, edited by Lamond, Joseph F., and Pielert, James H., American Society for Testing and Materials, Philadelphia, 2006, pages 410 to 424.

PCA, *Bulking of Sand Due to Moisture*, ST20, Portland Cement Association, Skokie, Illinois, 1935, 2 pages.

PCA, *Effect of Moisture on Volume of Sand* (1923), PCA Major Series 172, Portland Cement Association, Skokie, Illinois, 1955, 1 page.

Shilstone, James M., Sr., "Concrete Mixture Optimization," *Concrete International*, American Concrete Institute, Farmington Hills, Michigan, June 1990, pages 33 to 39.

Stark, David, *Characteristics and Utilization of Coarse Aggregates Associated with D-Cracking*, Research and Development Bulletin RD047, Portland Cement Association, http://www.cement.org/pdf_files/RD047.pdf, 1976.

Stark, David, *The Use of Recycled-Concrete Aggregate from Concrete Exhibiting Alkali-Silica Reactivity*, Research and Development Bulletin RD114, Portland Cement Association, 1996.

Swenson, E.G., and Gillott, J.E., "Alkali Reactivity of Dolomitic Limestone Aggregate," *Magazine of Concrete Research*, Vol. 19, No. 59, Cement and Concrete Association, London, June 1967, pages 95 to 104.

Thomas, Michael D.A.; Fournier, Benoit; and Folliard, Kevin J., *Report on Determining the Reactivity of Concrete Aggregates and Selecting Appropriate Measures for Preventing Deleterious Expansion in New Concrete Construction*, FHWA-HIF-09-001, Federal Highway Administration, Washington, D.C., April 2008, 28 pages.

Vogler, R.H., and Grove, G.H., "Freeze-thaw testing of coarse aggregate in concrete: Procedures used by Michigan Department of Transportation and other agencies," *Cement, Concrete, and Aggregates*, American Society for Testing and Materials, West Conshohocken, Pennsylvania, Vol. 11, No. 1, Summer 1989, pages 57 to 66.

Wigum, B.J., "Norwegian Petrographic Method – Development and Experiences During a Decade of Service," *Proceedings of the 12th International Conference on Alkali-Aggregate Reaction in Concrete*, Vol. I, International Academic Publishers – World Publishing Corporation, October 15-19, 2004, pages 444 to 452.

CHAPTER 7

Chemical Admixtures for Concrete

Chemical admixtures are those ingredients in concrete other than hydraulic cement, supplementary cementitious materials (SCMs), water, aggregates, and fiber reinforcement that are added to the mixture immediately before or during mixing (Figure 7-1). There are a variety of chemical admixtures available for use in concrete mixtures to modify fresh and hardened concrete properties. Chemical admixtures can be classified by function as follows:

1. Air-entraining
2. Normal, Mid-range, and High-range water-reducing
4. Set accelerating
5. Set retarding
6. Hydration-control
7. Rheology modifying
8. Corrosion inhibitors
9. Shrinkage reducers
10. Permeability reducing admixtures
11. Alkali-silica reactivity inhibitors
12. Coloring admixtures
13. Miscellaneous admixtures such as workability, bonding, grouting, gas-forming, anti-washout, foaming, and pumping admixtures

Figure 7-1. Liquid admixtures, from far left to right: antiwashout admixture, shrinkage reducer, water reducer, foaming agent, corrosion inhibitor, and air-entraining admixture.

The major reasons for using chemical admixtures in concrete mixtures are:

1. To achieve certain properties in concrete more effectively than by other means;
2. To maintain the quality of concrete during the stages of mixing, transporting, placing, finishing, and curing (especially in adverse weather conditions or intricate placements);
3. To overcome certain emergencies during concreting operations; and
4. Economy.

Despite these considerations, no admixture of any type or amount is a substitute for good concreting practice.

The effectiveness of an admixture depends upon factors such as its composition, addition rate, time of addition; type, brand, and amount of cementing materials; water content; aggregate shape, gradation, and proportions; mixing time; slump; and temperature of the concrete.

Chemical admixtures considered for use in concrete should meet applicable specifications as presented in Table 7-1. Trial mixtures should be made with the admixture and the other concrete ingredients at the temperature and relative humidity anticipated during placement. Then, observations can be made on the compatibility of the admixture with other ingredients, as well as its effects on the properties of the fresh and hardened concrete. The amount of admixture recommended by the manufacturer should be used, or the optimum dosage should be verified by laboratory testing. For more information on chemical admixtures for use in concrete see ACI Committee 212 (2004 and 2010), Thomas and Wilson (2002), Hewlett (1998), and Ramachandran (1995).

Table 7-1. Concrete Admixtures by Classification

Type of admixture	Desired effect	Material
Accelerators (ASTM C494 and AASHTO M 194, Type C)	Accelerate setting and early-strength development	Calcium chloride (ASTM D98 and AASHTO M 144) Triethanolamine, sodium thiocyanate, calcium formate, calcium nitrite, calcium nitrate
Air detrainers	Decrease air content	Tributyl phosphate, dibutyl phthalate, octyl alcohol, water-insoluble esters of carbonic and boric acid, silicones
Air-entraining admixtures (ASTM C260 and AASHTO M 154)	Improve durability in freeze-thaw, deicer, sulfate, and alkali-reactive environments Improve workability	Salts of wood resins (Vinsol resin), some synthetic detergents, salts of sulfonated lignin, salts of petroleum acids, salts of proteinaceous material, fatty and resinous acids and their salts, alkylbenzene sulfonates, salts of sulfonated hydrocarbons
Alkali-aggregate reactivity inhibitors	Reduce alkali-aggregate reactivity expansion	Barium salts, lithium nitrate, lithium carbonate, lithium hydroxide
Antiwashout admixtures	Cohesive concrete for underwater placements	Cellulose, acrylic polymer
Bonding admixtures	Increase bond strength	Polyvinyl chloride, polyvinyl acetate, acrylics, butadiene-styrene copolymers
Coloring admixtures (ASTM C979)	Colored concrete	Modified carbon black, iron oxide, phthalocyanine, umber, chromium oxide, titanium oxide, cobalt blue
Corrosion inhibitors (ASTM C1582)	Reduce steel corrosion activity in a chloride-laden environment	Amine carboxylates aminoester organic emulsion, calcium nitrite, organic alkyidicarboxylic, chromates, phosphates, hypophosphites, alkalis, and fluorides
Dampproofing admixtures	Retard moisture penetration into dry concrete	Soaps of calcium or ammonium stearate or oleate Butyl stearate, Petroleum products
Foaming agents	Produce lightweight, foamed concrete with low density	Cationic and anionic surfactants Hydrolized protein
Fungicides, germicides, and insecticides	Inhibit or control bacterial and fungal growth	Polyhalogenated phenols Dieldrin emulsions, Copper compounds
Gas formers	Cause expansion before setting	Aluminum powder
Grouting admixtures	Adjust grout properties for specific applications	See Air-entraining admixtures, Accelerators, Retarders, and Water reducers
Hydration control admixtures	Suspend and reactivate cement hydration with stabilizer and activator	Carboxylic acids Phosphorus-containing organic acid salts
Permeability-reducing admixture: non-hydrostatic conditions (PRAN)	Water-repellent surface, reduced water absorption	Long-chain fatty acid derivatives (stearic oleic, caprylic capric), soaps and oils, (tallows, soya-based), petroleum derivatives (mineral oil, paraffin, bitumen emulsions), and fine particle fillers (silicates, bentonite, talc)
Permeability reducing admixture: hydrostatic conditions (PRAH)	Reduced permeability, increased resistance to water penetration under pressure	Crystalline hydrophilic polymers (latex, water-soluble, or liquid polymer)
Pumping aids	Improve pumpability	Organic and synthetic polymers Organic flocculents Organic emulsions of paraffin, coal tar, asphalt, acrylics Bentonite and pyrogenic silicas Hydrated lime (ASTM C141)
Retarding admixtures (ASTM C494 and AASHTO M 194, Type B)	Retard setting time	Lignin, Borax Sugars, Tartaric acid and salts
Shrinkage reducers	Reduce drying shrinkage	Polyoxyalkylene alkyl ether Propylene glycol
Superplasticizers* (ASTM C1017, Type 1)	Increase flowability of concrete Reduce water-cement ratio	Sulfonated melamine formaldehyde condensates Sulfonated naphthalene formaldehyde condensates Lignosulfonates, Polycarboxylates

Table 7-1. Concrete Admixtures by Classification (Continued)

Type of admixture	Desired effect	Material
Superplasticizer* and retarder (ASTM C1017, Type 2)	Increase flowability with retarded set Reduce water–cement ratio	See superplasticizers and also water reducers
Water reducer (ASTM C494 and AASHTO M 194, Type A)	Reduce water content at least 5%	Lignosulfonates Hydroxylated carboxylic acids Carbohydrates (Also tend to retard set so accelerator is often added)
Water reducer and accelerator (ASTM C494 and AASHTO M 194, Type E)	Reduce water content (minimum 5%) and accelerate set	See water reducer, Type A (accelerator is added)
Water reducer and retarder (ASTM C494 and AASHTO M 194, Type D)	Reduce water content (minimum 5%) and retard set	See water reducer, Type A (retarder is added)
Water reducer—high range (ASTM C494 and AASHTO M 194, Type F)	Reduce water content (minimum 12%)	See superplasticizers
Water reducer—high range—and retarder (ASTM C494 and AASHTO M 194, Type G)	Reduce water content (minimum 12%) and retard set	See superplasticizers and also water reducers
Water reducer—mid range	Reduce water content (between 6 and 12%) without retarding	Lignosulfonates Polycarboxylates

* Superplasticizers are also referred to as high-range water reducers or plasticizers. These admixtures often meet both ASTM C494 (AASHTO M 194) and ASTM C1017 specifications.

Air-Entraining Admixtures

One of the greatest advances in concrete technology was the development of air-entrained concrete in the mid-1930s. Air-entrainment dramatically improves the durability of concrete exposed to cycles of freezing and thawing and deicer chemicals (see Chapter 11). There are also other important benefits of entrained air in both freshly mixed and hardened concrete (see Chapter 9).

Air-entraining concrete is produced by using either an air-entraining cement or by adding an air-entraining admixture during batching, or a combination of these approaches. Air-entraining cement is a portland cement with an air-entraining addition interground with the clinker during manufacture (see Chapter 3). An air-entraining admixture, on the other hand, is added directly to the concrete materials either before or during mixing. Regardless of the approach used, adequate control and monitoring is required to ensure the proper air content at all times.

Specifications and methods of testing air-entraining admixtures are given in ASTM C260, *Standard Specification for Air-Entraining Admixtures for Concrete*, and C233, *Standard Test Method for Air-Entraining Admixtures for Concrete* (AASHTO M 154 and T 157). Air-entraining additions for use in the manufacture of air-entraining cements must meet requirements of ASTM C226, *Standard Specification for Air-Entraining Additions for Use in the Manufacture of Air-Entraining Hydraulic Cement*. Applicable requirements for air-entraining cements are given in ASTM C150, *Standard Specification for Portland Cement* and AASHTO M 85. See Chapter 11, Klieger (1966), and Whiting and Nagi (1998) for more information.

Air-Entraining Materials

The primary ingredients used in air-entraining admixtures are listed in Table 7-1. Numerous commercial air-entraining admixtures, manufactured from a variety of materials, are available. Most air-entraining admixtures consist of one or more of the following materials: wood resin (Vinsol resin), sulfonated hydrocarbons, fatty and resinous acids, and synthetic materials. Chemical descriptions and performance characteristics of common air-entraining agents are shown in Table 7-2. Air-entraining admixtures are usually liquids and should not be allowed to freeze. Admixtures added at the mixer should conform to ASTM C260 (AASHTO M 154).

Air-entraining cements comply with ASTM C150 and C595 (AASHTO M 85 and M 240). To produce such cements, air-entraining additions conforming to ASTM C226 are interground with the cement clinker during manufacture. Air-entraining cements generally provide an adequate amount of entrained air to meet most job conditions; however, a specified air content may not necessarily be obtained in the concrete. If an insufficient volume of air is entrained, it may also be necessary to add an air-entraining admixture at the mixer.

Table 7-2. Classification and Performance Characteristics of Common Air-Entraining Admixtures

Classification	Chemical description	Notes and performance characteristics
Wood derived acid salts Vinsol® resin	Alkali or alkanolamine salt of: A mixture of tricyclic acids, phenolics, and terpenes.	Quick air generation. Minor air gain with initial mixing. Air loss with prolonged mixing. Mid-sized air bubbles formed. Compatible with most other admixtures.
Wood rosin	Tricyclic acids-major component. Tricyclic acids-minor component.	Same as above.
Tall oil	Fatty acids-major component. Tricyclic acids-minor component.	Slower air generation. Air may increase with prolonged mixing. Smallest air bubbles of all agents. Compatible with most other admixtures.
Vegetable oil acids	Coconut fatty acids, alkanolamine salt.	Slower air generation than wood rosins. Moderate air loss with mixing. Coarser air bubbles relative to wood rosins. Compatible with most other admixtures.
Synthetic detergents	Alkyl-aryl sulfonates and sulfates (e.g., sodium dodecylbenzenesulfonate).	Quick air generation. Minor air loss with mixing. Coarser bubbles. May be incompatible with some HRWR. Also applicable to cellular concretes.
Synthetic workability aids	Alkyl-aryl ethoxylates.	Primarily used in masonry mortars.
Miscellaneous	Alkali-alkanolamine acid salts of lignosulfonate. Oxygenated petroleum residues. Proteinaceous materials. Animal tallows.	All these are rarely used as concrete air-entraining agents in current practice.

Mechanism of Air Entrainment

Air-entraining admixtures are surfactants (surface-active agents) which concentrate at the air-water interface and reduce the surface tension encouraging the formation of microscopic bubbles during the mixing process. The air-entraining admixture stabilizes those bubbles, enhances the incorporation of bubbles of various sizes, impedes bubble coalescence, and anchors bubbles to cement and aggregate particles.

The air-entraining admixture acts at the air-water interface. Air-entraining admixtures typically have a negatively charged head which is hydrophilic and attracts water, and a hydrophobic tail which repels water. As illustrated in Figure 7-2; the hydrophobic end is attracted to the air within bubbles generated during the mixing process. The polar end, which is hydrophilic, orients itself towards water (A). The air-entraining admixture forms a tough, water-repelling film, similar to a soap film, with sufficient strength and elasticity to contain and stabilize the air bubbles. The hydrophobic film also keeps water out of the bubbles (B). The stirring and kneading action of mechanical mixing disperses the air bubbles. The charge around each bubble leads to repulsive forces, that prevent the coalescence of bubbles (C & D). The surface charge causes the air bubble to be adhered to the charged surfaces of cement and aggregate particles. The fine aggregate particles also act as a three-dimensional grid to help hold the bubbles in the mixture (E). This improves the cohesion of the mixture and further stabilizes the air bubbles (F).

Entrained air bubbles are not like entrapped air voids, which occur in all concretes as a result of mixing, handling, and placing. Entrapped air voids are largely a function of aggregate characteristics. Intentionally entrained air bubbles are extremely small in size, between 10 to 1000 µm in diameter, while entrapped voids are usually 1000 µm (1 mm) or larger. The majority of the entrained air voids in normal concrete are between 10 µm and 100 µm in diameter. As shown in Figure 7-3, the bubbles are not interconnected. They are well dispersed and randomly distributed. Non-air-entrained concrete with a 25-mm (1-in.) maximum-size aggregate has an air content of approximately 1.5%. This same mixture air entrained for severe frost exposure would require a total air content of about 6%, made up of both coarse entrapped air voids and fine entrained air voids. However, it is the finely entrained air system that is most effective at providing frost resistance.

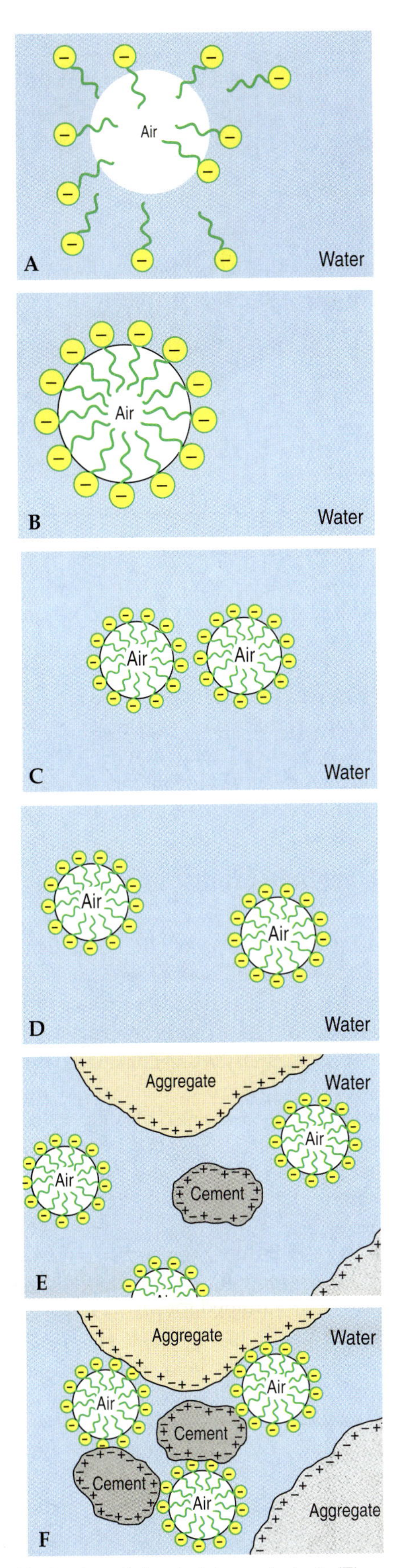

Figure 7-2. Mechanism of air-entraining surfactants (Thomas and Wilson 2002).

Figure 7-3. Polished concrete surface shows air voids clearly visible as dark circles.

Control of Air Content

The amount of air entrained in concrete for a given dose of admixture depends on materials, mixture proportions, methods of transport, placing and finishing methods, and curing. Variations in air content can be expected with variations in aggregate proportions and gradation, mixing time and intensity, temperature, and slump. The order of batching and mixing concrete ingredients when using an air-entraining admixture can also have a significant influence on the amount of air entrained. The late addition of water and extended retempering can cause clustering of air bubbles around aggregate resulting in strength reduction (Kozikowski and others 2005 and Camposagrado 2006). Therefore, consistency in batching is needed to maintain adequate control. For more information on the effects of constituent materials, mixture proportions, and placing and finishing operations on air content see Chapter 9.

When entrained air is excessive, it can be reduced by using one of the following defoaming (air-detraining) agents: tributyl phosphate, dibutyl phthalate, octyl alcohol, water-insoluble esters of carbonic acid and boric acid, and silicones. Only the smallest possible dosage of defoaming agent should be used to reduce the air content to the specified limit. An excessive amount might have adverse effects on concrete properties (Whiting and Stark 1983).

Impact of Air Content on Properties of Concrete

The presence of a finely distributed network of bubbles has a significant impact on the properties of plastic concrete. Entrained air improves the workability of concrete. It is particularly effective in lean (low cement content) mixtures that otherwise might be harsh and difficult to work. In one study, an air-entrained mixture made with natural aggregate, 3% air, and a 37-mm (1.5-in.) slump had about the same workability as a non-air-entrained concrete with 1% air and a 75-mm (3-in.) slump, even though less cement was required for the air-entrained mix (Cordon 1946). Workability of concrete mixtures with angular and poorly graded aggregates is similarly improved.

Because of improved workability with entrained air, water and sand content can be reduced significantly as shown in Figure 7-4. A volume of air-entrained concrete requires less water than an equal volume of non-air-entrained concrete of the same consistency and maximum size aggregate. Water reductions in the range of 15 L/m³ to 25 L/m³ (25 lb/yd³ to 40 lb/yd³) may be achieved with adequate air entrainment (Figure 7-4). Freshly mixed concrete containing entrained air is cohesive, looks and feels "fatty" or workable, and can usually be handled with ease. On the other hand, high air contents can make a mixture sticky and more difficult to finish. Furthermore, the air bubbles reduce the tendency for segregation and bleeding.

Improvements in the performance of hardened concrete obviously include improved resistance to freezing and thawing, and deicer-salt scaling. Furthermore, the incorporation of air also results in a reduced permeability and possibly improved resistance to sulfate attack and alkali-silica reactivity (see Chapter 11).

The single detrimental effect is that an increase in air content results in a decrease in the strength of the concrete. When the air content is maintained constant, strength varies inversely with the water-cement ratio (see Figure 7-5). A good rule of thumb is that each 1% increase in air content is accompanied by 5% to 6% reduction in strength. More information on the impact of air content on fresh and hardened properties of concrete can be found in Chapter 9.

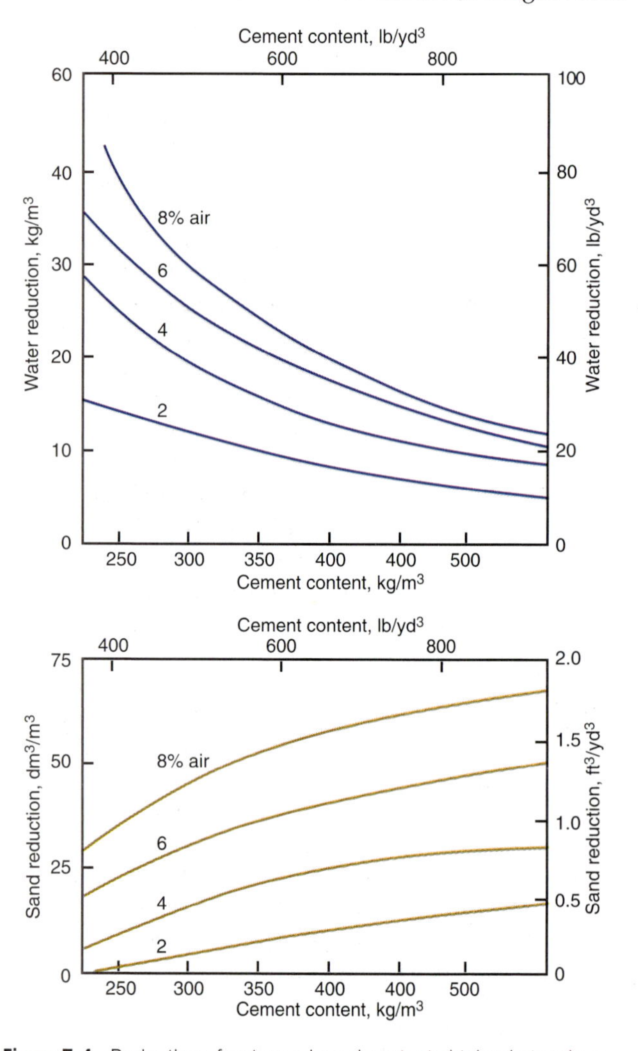

Figure 7-4. Reduction of water and sand content obtained at various levels of air and cement contents (Gilkey 1958).

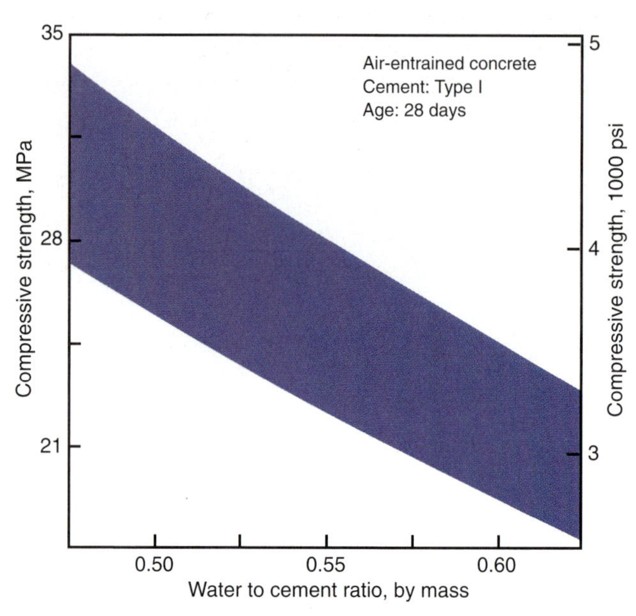

Figure 7-5. Typical relationship between 28-day compressive strength and water-cement ratio for a wide variety of air-entrained concretes using Type I cement.

Water-Reducing Admixtures

A water-reducing admixture is an admixture that increases workability without increasing the water content of concrete, or that permits a decrease in water content without decreasing the slump and conforms to ASTM C494, *Standard Specification for Chemical Admixtures for Concrete* (AASHTO M 194). Some water-reducing admixtures, typically high-range, are also plasticizers conforming to ASTM C1017, *Standard Specification for Chemical Admixtures for Use in Producing Flowing Concrete* (AASHTO M 194).

When used as a water-reducer, the water content is reduced while maintaining the slump; this reduces the water-cement ratio of the concrete and increases its strength and durability. When the same chemical is used as a plasticizer, the workability is increased while the water content is kept constant. This can improve the placing characteristics of the concrete without adversely affecting the strength and durability.

If the flow is maintained, the water content can be reduced. The conditions are ideal for the strength and durability of concrete. Here the admixture is used as a true water-reducer; allowing concrete to be produced with low water-to-cement ratios and good workability. Advantage may be taken for both purposes of a water-reducing admixture when designing concrete; producing concrete with improved fresh and hardened properties (Figure 7-6) (see Chapter 12).

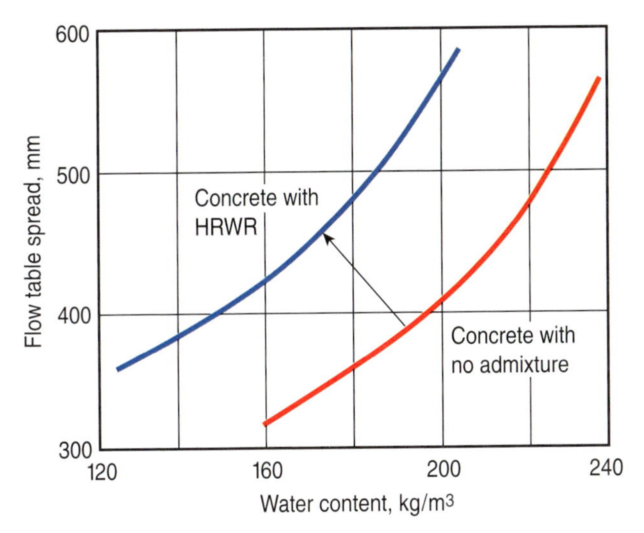

Figure 7-6. Advantage may be taken for both purposes of a water-reducing admixture (lowering w/cm, and increasing flow) when designing concrete. (Adapted from Neville 1995)

Normal (Conventional) Water Reducers

When used as a water reducer, normal range, or conventional water reducers can reduce the water content by approximately 5% to 10%. Alternatively, they may be used as plasticizers to provide a moderate increase in workability. Normal water reducers are intended for concretes with slumps up to 100 mm or 125 mm (4 in. or 5 in.), but they are also used in combination with mid-range and high-range water reducers in higher slump concrete mixtures.

Mid-Range Water Reducers

Mid-range water reducers were first introduced in 1984 to bridge the gap between normal range water reducers and high-range water reducers (superplasticizers). These admixtures provide significant water reduction (between 6% and 12%) for concretes with slumps of 125 mm to 200 mm (5 in. to 8 in.) without the retardation associated with high dosages of conventional (normal) water reducers. Mid-range water reducers can be used to reduce stickiness and improve finishability, pumpability, and placeability of concretes containing silica fume and other supplementary cementing materials. Some can also entrain air and be used in low slump concretes (Nmai, Schlagbaum, and Violetta 1998).

High-Range Water Reducers

High-range water reducers, ASTM C494 (AASHTO M 194) Types F (water reducing) and G (water reducing and retarding), can be used to impart properties induced by regular water reducers, only much more efficiently. They can greatly reduce water demand and cement contents and make low water-cement ratio, high-strength concrete with normal or enhanced workability and to generate slumps greater than 150 mm (6 in.). A water reduction of 12% to 40% can be obtained using these admixtures. The reduced water content and water-cement ratio can produce concretes with: (1) ultimate compressive strengths in excess of 70 MPa (10,000 psi), (2) increased early strength gain, (3) reduced chloride-ion penetration, and (4) other beneficial properties associated with low water-cement ratio concrete.

Figure 7-7. High-range water reducers, also known as superplasticizers, produce flowing concrete with slumps greater than 150 mm (6 in.).

When the same chemicals used for high-range water reducers are used to make flowing concrete, they are often called plasticizers or superplasticizers meeting ASTM C1017. This specification has provisions for two types of admixtures: Type 1 – plasticizing, and Type 2 – plasticizing and retarding. These admixtures are added to concrete with a low-to-normal slump and water-cement ratio to make high-slump flowing concrete (Figures 7-6 and 7-7). Flowing concrete is a highly fluid but workable concrete that can be placed with little or no vibration or compaction while still remaining essentially free of excessive bleeding or segregation. Applications where flowing concrete is used include: (1) thin-section placements (Figure 7-8), (2) areas of closely spaced and congested reinforcing steel (Figure 7-9), (3) tremie pipe (underwater) placements, (4) pumped concrete to reduce pump pressure, thereby increasing lift and distance capacity, (5) areas where conventional consolidation methods are impractical or can not be used, and (6) for reducing handling costs. The addition of a plasticizer to a 75-mm (3-in.) slump concrete can easily produce a concrete with a 230-mm (9-in.) slump. Flowing concrete is defined by ASTM C1017 as a concrete having a slump greater than 190 mm (7.5 in.), yet maintaining cohesive properties.

Figure 7-8. Plasticized, flowing concrete is easily placed in thin sections such as this bonded overlay that is not much thicker than 1.5 diameters of a quarter (shown).

Figure 7-9. Flowable concrete with a high slump (top) is easily placed (middle), even in areas of heavy reinforcing steel congestion (bottom).

Composition of Water-Reducing Admixtures

The classifications and components of water reducers are listed in Table 7-1. The chemistry of water-reducing or superplasticizing admixtures falls into broad categories: lignosulfonates, hydroxycarboxylic acid, hydroxylated polymers, salts of melamine formaldehyde sulfonates or naphthalene formaldehyde sulfonic acids, and polycarboxylates. The use of organic materials to reduce the water content or increase the fluidity of concrete dates back to the 1930s. The most recent breakthrough in this technology is the development of high-range water-reducers based on polycarboxylates – which occurred in the late 1980s.

Mechanisms of Water Reducers

Water reducers and plasticizers function as cement dispersants primarily through electrostatic and steric repulsive forces. Acidic groups within the polymer neutralize the surface charges on the cement particles (Ramachandran 1998 and Collepardi and Valente 2006). These groups bind to positive ions on the cement particle surfaces. These ions attach the polymer and give the cement a slight negative charge as well as create a layer on the surface. This negative charge and layer of adsorbed compounds create a combination of electrostatic and steric repulsion forces between individual cement particles, dispersing them, thus releasing the water tied up in agglomerations and reducing the viscosity of the paste and concrete (Figures 7-10 and 7-11). A melamine-based, naphthalene-based, or lignin-based superplasticizer uses a molecule that has a size of about 1 to 2 nm. The effect of the water reducer depends on the dosage level, sequence of addition, and molecular weight. The water reducer will also contribute to dispersion by repelling negatively charged aggregate particles and air-entrained bubbles. The electrostatic repulsion for these materials is affected far more by dissolved ions (as compared to polycarboxylates) and rapidly diminishes as the hydrating cement releases more ions into the mixture.

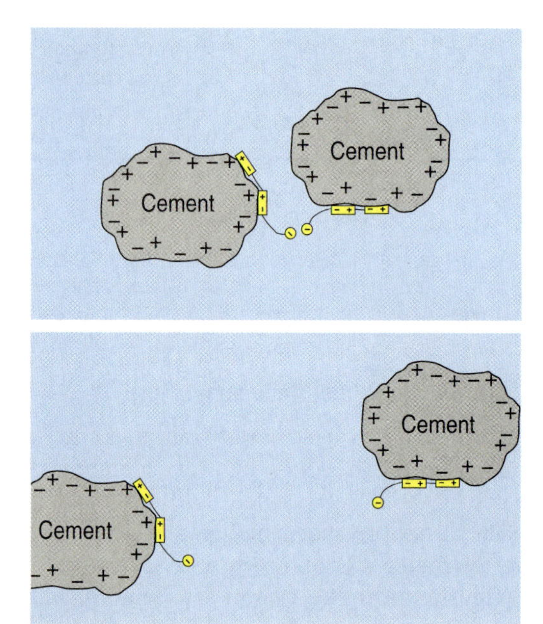

Figure 7-10. Mechanism of dispersive action of water-reducing admixtures (Thomas and Wilson 2002).

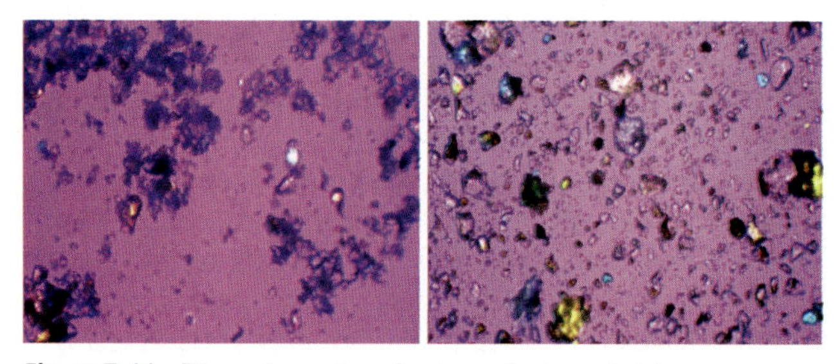

Figure 7-11. Dispersive action of water-reducing admixtures.

Polycarboxylate Technology. Polycarboxylate derivatives are the newest generation of high-range water reducers and superplasticizers. These polymers are comprised of a main carbon chain with carboxylate groups and polyethylene oxide (PEO) side chains. The number of carboxylate groups and the number and length of PEO side chains can be adjusted to change the properties of the plasticizer. The PEO side chains extend out from the cement particles and add the mechanism of steric hindrance to the typical electrostatic repulsion (Li and others 2005 and Nawa 2006).

The mechanism of steric hindrance is illustrated in Figure 7-12 (a-d). As with typical superplasticizers, the water reducer is dissolved in water, and the polar chain is absorbed at the solid-water interface (A). The long side chains physically help hold the cement grains apart allowing water to totally surround the cement grains (steric hindrance) (B). Additionally, the polar chain imparts a slight negative charge causing the cement grains to repel one another (electrostatic repulsion) (C). As the electrostatic repulsion dispersing effect wears off due to cement hydration, the long side chains still physically keep the cement dispersed (D).

The PEO chains prevent particles from agglomerating through physical separations on the order of 10 nm (Nawa 2006). This physical separation is still great enough to allow fluid to flow between the particles. This inhibition of agglomeration disperses the cement particles and allows the concrete to flow more easily. Because steric hindrance is a physical mechanism, it is not as sensitive to dissolved ions as an electrostatic repulsion mechanism. Concrete mixtures with polycarboxylate additions tend to retain fluidity for longer periods and they tend to require less water than concrete mixtures using other water reducers (Jeknavorian and others 1997). Polycarboxylate admixtures are commonly used in self-consolidating concrete (see Chapter 19 and Szecsy and Mohler 2009 for more information on self-consolidating concrete.)

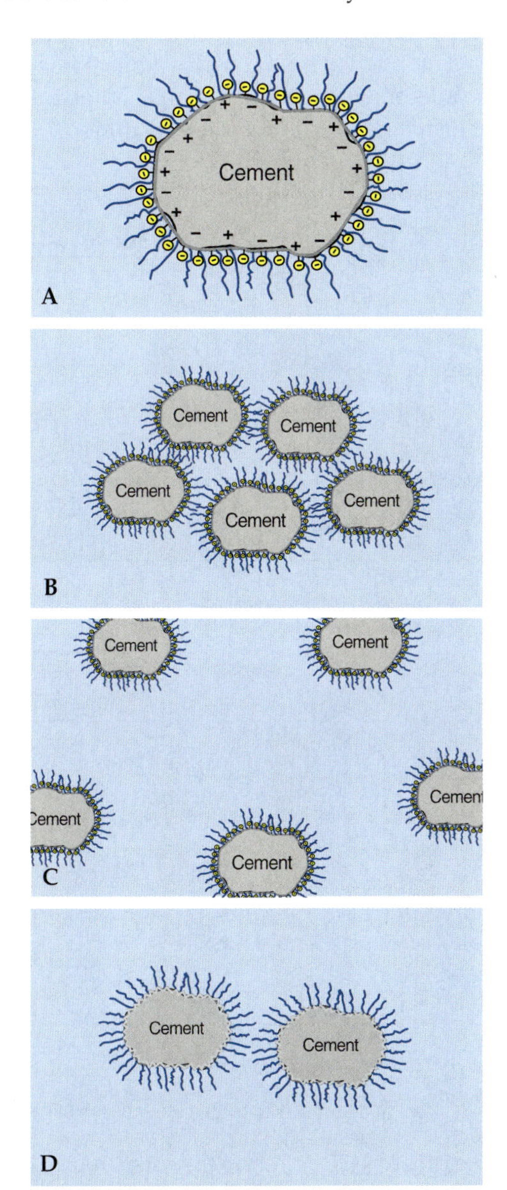

Figure 7-12. Mechanism of steric hindrance of polycarboxylate water-reducing admixtures (Thomas and Wilson 2002).

Impact of Water Reducers on Properties of Concrete

Adding a water-reducing admixture to concrete without also reducing the water content can produce a mixture with a higher slump. The rate of slump loss, however, is not reduced and in most cases is increased (Figures 7-13 and 7-14), with the exception of polycarboxylate technology. Rapid slump loss results in reduced workability and less time to place concrete.

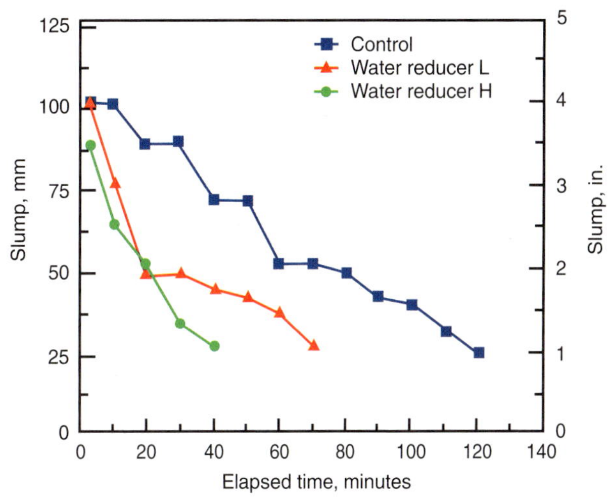

Figure 7-13. Slump loss at 23°C (73°F) in concretes containing conventional water reducers (ASTM C494 and AASHTO M 194 Type D) compared with a control mixture (Whiting and Dziedzic 1992).

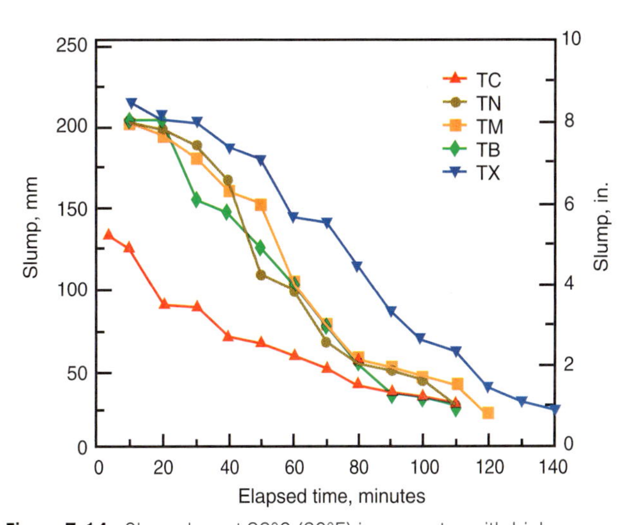

Figure 7-14. Slump loss at 32°C (90°F) in concretes with high-range water reducers (TN, TM, TB, and TX) compared with control mixture (TC) (Whiting and Dziedzic 1992).

High-range water reducers are generally more effective than regular water-reducing admixtures in producing workable concrete. The effect of certain plasticizers in increasing workability or making flowing concrete is short-lived, 30 to 60 minutes. This period is followed by a rapid loss in workability or slump loss. High temperatures can also aggravate slump loss. Due to their propensity for slump loss, these admixtures are sometimes added to the concrete mixer at the jobsite. They are available in liquid and powder form. Extended-slump-life plasticizers added at the batch plant help reduce slump-loss problems.

An increase in strength is generally obtained with water-reducing admixtures as the water-cement ratio is reduced. For concretes of equal cement content, air content, and slump, the 28-day strength of a water-reduced concrete containing a water reducer can be 10% to 25% greater than concrete without the admixture. Using a water reducer to reduce the cement and water content of a concrete mixture, while maintaining a constant water-cement ratio, can result in equal or reduced compressive strength, and can increase slump loss by a factor of two or more (Whiting and Dziedzic 1992).

Water reducers decrease, increase, or have no effect on bleeding, depending on the chemical composition of the admixture. A significant reduction of bleeding can result with large reductions of water content; this can result in finishing difficulties on flat surfaces when rapid drying conditions are present (see Chapter 16). Tests have shown that some plasticized concretes bleed more than control concretes of equal water-cement ratio (Figure 7-15); but plasticized concretes bleed significantly less than control concretes of equally high slump and higher water content.

Despite reduction in water content, water-reducing admixtures may cause increases in drying shrinkage. Usually the effect of the water reducer on drying shrinkage is small when compared to other more significant factors that affect shrinkage cracking in concrete. High-slump, low-water-content, plasticized concrete tends to develop less drying shrinkage than a high-slump, high-water-content conventional concrete. However, high slump plasticized concrete has similar or higher drying shrinkage than conventional low-slump, low-water-content concrete (Whiting 1979, Gebler 1982, and Whiting and Dziedzic 1992).

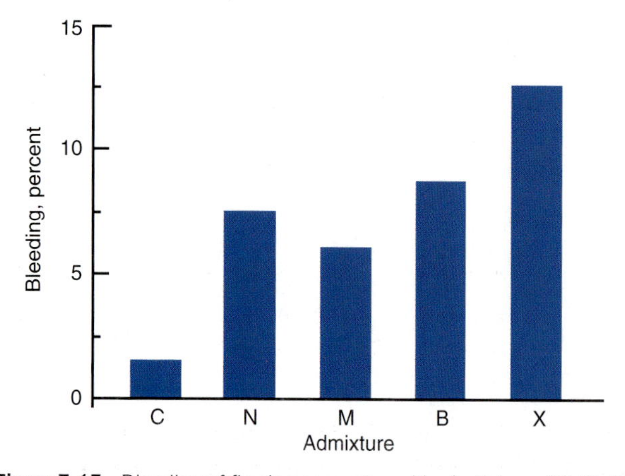

Figure 7-15. Bleeding of flowing concretes with plasticizers (N, M, B, and X) compared to control (C) (Whiting and Dziedzic 1992).

Water reducers can be modified to give varying degrees of retardation while others do not significantly affect the setting time. ASTM C494 (AASHTO M 194) Type A water reducers can have little effect on setting time at their typical dosages, while Type D admixtures provide water reduction with retardation, and Type E admixtures provide water reduction with accelerated setting. Type D water-reducing admixtures usually retard the setting time of concrete from one to four hours (Figure 7-16). Some water reducers meet the requirements of more than one category depending on the dosage rate. For example, a Type A water reducer may perform as a Type D water reducing and set retarding admixture as the dosage rate is increased. Setting time may be accelerated or retarded based on each admixture's chemistry, dosage rate, and interaction with other admixtures and cementing materials in the concrete mixture.

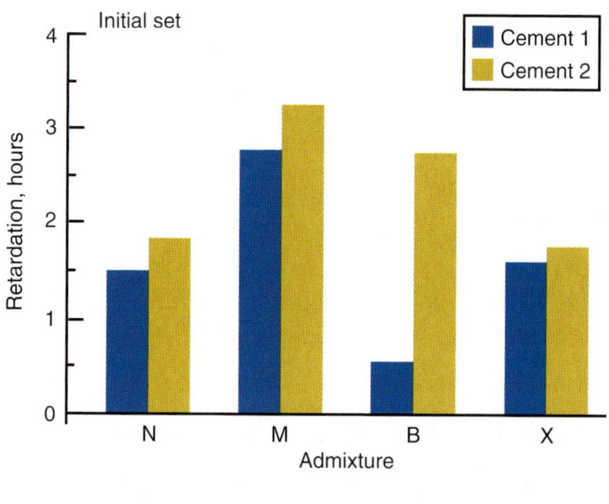

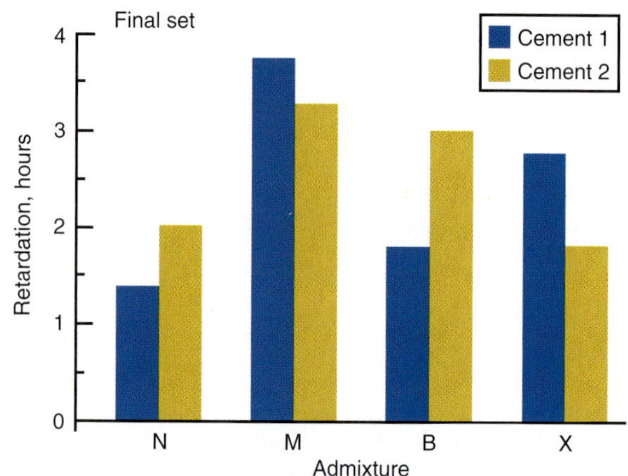

Figure 7-16. Retardation of set in cement-reduced mixtures relative to control mixture. Concretes L and H contain conventional water reducer, concretes N, M, B, and X contain high-range water reducer (Whiting and Dziedzic 1992).

Some water-reducing admixtures may also entrain some air in concrete. Lignin-based admixtures can increase air contents by 1% to 2%. Concretes with water reducers generally have good air retention. Concretes with high-range water reducers or plasticized flowing concrete can have larger entrained air voids and higher void-spacing factors than normal air-entrained concrete. Air loss can also be significant when compared to concretes without high range water reducers held at constant water-cement ratios (reduced cement and water contents) (Table 7-3). Some research has indicated poor frost- and deicer-scaling resistance for some flowing concretes when exposed to a continuously moist environment without the benefit of a drying period (Whiting and Dziedzic 1992). However, laboratory tests have shown that concretes with a moderate slump using high-range water reducers have good freeze-thaw durability, even with slightly higher void-spacing factors. This may be the result of the lower water-cement ratios often associated with these concretes.

Table 7-3. Loss of Air from Cement Reduced Concrete Mixtures

	Mixture	Initial air content, %*	Final air content, %†	Percent air retained	Rate of air loss, %/minute
C	Control	5.4	3.0	56	0.020
L	Water reducer	7.0	4.7	67	0.038
H	Water reducer	6.2	4.6	74	0.040
N	High-range water reducer	6.8	4.8	71	0.040
M	High-range water reducer	6.4	3.8	59	0.065
B	High-range water reducer	6.8	5.6	82	0.048
X	High-range water reducer	6.6	5.0	76	0.027

* Represents air content measured after addition of admixture.
† Represents air content taken at point where slump falls below 25 mm (1 in.).
(Whiting and Dziedzic 1992)

The effectiveness of water reducers on concrete is a function of their chemical composition, concrete temperature, cement composition and fineness, cement content, and the presence of other admixtures.

See Whiting and Dziedzic (1992) for more information on the effects of water reducers on concrete properties.

Set Retarding Admixtures

Set retarding admixtures are used to delay the rate of setting of concrete. Retarders are sometimes used to: (1) offset the accelerating effect of hot weather on the setting of concrete (see Chapter 16); (2) delay the initial set of concrete or grout when difficult or unusual conditions of placement occur, such as placing concrete in large piers and foundations, cementing oil wells, or pumping grout or concrete over considerable distances; (3) delay the set for special finishing techniques, such as an exposed aggregate surface, or anticipating long transport time or delays between batching and placement.

A set retarder extends the period during which concrete remains plastic. This allows a large placement to be completed before setting occurs, which helps eliminates cold joints in large or complex pours and extends the time allowed for finishing and joint preparation. The reduced hydration rate is also helpful in reducing early temperature rises, which can induce internal stresses and cracking in concrete

Types of Set Retarding Admixtures

The classifications and components of set retarders are listed in Table 7-1. Compounds used as set retarders fall into four general categories: lignosulfonates, hydroxy-carboxylic acids, sugars and their derivatives, and selected inorganic salts.

Mechanism of Set Retarders

Set retarders function by slowing the normal cement hydration through complex processes such as inhibiting the growth of certain crystalline hydration products (nucleation). Earlier research has suggested the formation of a shell of initial hydration products, which diminishes the access of water to the inner anhydrous cement surfaces. For example, lignosulfonate molecules can adsorb to the surface of calcium silicates and calcium aluminates and form a coating that slows down the dissolution of calcium from the anhydrous grains. This in turn retards the formation of the hydration products, C-S-H and ettringite, which are discussed in Chapter 3. Eventually, osmotic rupturing of the coating allows the silicates and aluminates to resume hydration.

Effects of Set Retarders on Concrete Properties

The use of a set retarding admixture delays both the initial and final set of concrete. The extent of the delay is dependent on admixture composition, dosage, time of addition, and the temperature of the concrete. Retarders do not decrease the initial temperature of concrete. The effectiveness of a set retarder can also depend when it is added to the concrete mixture.

The amount of water reduction for an ASTM C494 (AASHTO M 194) Type B retarding admixture is normally less than that obtained with a Type A water reducer. Type D admixtures provide both water reduction and retardation.

In general, some reduction in strength at early ages (one to three days) accompanies the use of retarders. However, increased long-term strength may result from retarding the initial rate of hydration. Excessive addition rates of a retarding admixture may permanently inhibit the hydration of the cement.

The effects of these materials on the other properties of concrete, such as shrinkage, may be unpredictable. The incorporation of retarders can affect some of the other properties of concrete including slump, bleeding, and early-age strength development. Therefore, acceptance tests of retarders should be made with actual job materials under anticipated job conditions.

Set Accelerating Admixtures

A set accelerating admixture is used to accelerate the rate of hydration (setting) and strength development of concrete at an early age. The strength development of concrete can also be accelerated by other methods: (1) using Type III or Type HE high-early-strength cement, (2) lowering the water-cement ratio by adding 60 kg/m^3 to 120 kg/m^3 (100 lb/yd^3 to 200 lb/yd^3) of additional cement to the concrete, (3) using a water reducer (provided it does not contribute significantly to set retardation), or (4) curing at higher temperatures. Accelerators are designated as Type C admixtures under ASTM C494 (AASHTO M 194).

Types of Set Accelerating Admixtures

Calcium chloride ($CaCl_2$) is the most common material used in set accelerating admixtures, especially for non-reinforced concrete. It should conform to the requirements of ASTM D98 (AASHTO M 144), *Standard Specification for Calcium Chloride*, and should be sampled and tested in accordance with ASTM D345, *Standard Test Method for Sampling and Testing Calcium Chloride for Roads and Structural Applications*.

Calcium chloride is highly effective in this role; however, it is limited to a dosage of 2% or less (by mass of cement) for non-reinforced concrete. Specifications may restrict its use in concrete as chlorides will accelerate corrosion of the reinforcing steel. Calcium chloride is generally available in three forms – as flake, pellets, or in solution.

The amount of calcium chloride added to concrete should be no more than is necessary to produce the desired results and in no case should be permitted to exceed 2% by mass of cementing material. When calculating the chloride content of commercially available calcium chloride, it can be assumed that regular flake contains a minimum of 77% $CaCl_2$; and concentrated flake, pellet, or granular forms contain a minimum of 94% $CaCl_2$.

A more appropriate accelerator for reinforced concrete is a non-chloride containing accelerator. These include organic compounds such as triethanolamine (TEA) and inorganic salts such as sodium and calcium salts of formate, nitrate, and nitrite (see Table 7-1). The inorganic salts tend to be less effective than calcium chloride and are used at higher dosages. TEA is not generally used on its own as an accelerating admixture, but is often used in other chemical admixtures such as normal and mid-range water reducers. There it acts as a set balancer to compensate for possible retarding effects. Certain non-chloride set accelerators are specially formulated for use in cold weather applications with ambient temperatures down to -7°C (20°F).

Mechanism of Set Accelerating Admixtures

Calcium chloride and calcium or alkali salts catalyze the hydration of the calcium silicates by weakening the barrier of initial hydration products that may form on the cement surface. In contrast to those inorganic salts, TEA, acts on the C_3A component of the cement accelerating its reaction with gypsum and the production of ettringite and also promoting the subsequent conversion to monosulfate (see Chapter 3). At high dosage levels, TEA may cause flash set to occur. It can also retard, even permanently, the hydration of C_3S leading to reduced long-term strength. Calcium formate, another organic compound, works by accelerating the hydration of C_3S.

Effects of Set Accelerators on Concrete Properties

Set accelerators cause a reduction in the time to both the initial and final set, the effect generally increasing with admixture dosage. There will also typically be an increase in the early-age strength development dependant on the admixture dosage, composition, and time of addition. The increase in strength due to the addition of calcium chloride is particularly noticeable at low temperatures and early ages.

The widespread use of calcium chloride as an accelerating admixture has provided much data and experience on the effect of this chemical on the properties of concrete. Besides accelerating strength gain, calcium chloride causes an increase in potential reinforcement corrosion and may lead to discoloration (a darkening of concrete). The incorporation of calcium chloride can also affect other properties of concrete including shrinkage, long-term strength and resistance to freezing and thawing, sulfates, and ASR.

An overdose of calcium chloride can result in placement problems and can be detrimental to concrete. It may cause: rapid stiffening, a large increase in drying shrinkage, corrosion of reinforcement, and loss of strength at later ages (Abrams 1924 and Lackey 1992).

Applications where calcium chloride should be used with caution:

1. Concrete reinforced with steel
2. Concrete subjected to steam curing
3. Concrete containing embedded dissimilar metals, especially if electrically connected to steel reinforcement
4. Concrete slabs supported on permanent galvanized-steel forms
5. Colored concrete

Calcium chloride or admixtures containing soluble chlorides should not be used in the following:

1. Construction of parking garages

2. Prestressed concrete because of possible steel corrosion hazards
3. Concrete containing embedded aluminum (for example, conduit) since serious corrosion of the aluminum can result, especially if the aluminum is in contact with embedded steel and the concrete is in a humid environment
4. Concrete containing aggregates that, under standard test conditions, have been shown to be potentially deleteriously reactive
5. Concrete exposed to soil or water containing sulfates
6. Floor slabs intended to receive dry-shake metallic finishes
7. Hot weather
8. Massive concrete placements

The maximum chloride-ion content for corrosion protection of prestressed and reinforced concrete as recommended by the ACI 318 building code is presented in Chapter 11. Resistance to the corrosion of embedded steel is further improved with an increase in the depth of concrete cover over reinforcing steel, and a lower water-cement ratio. Stark (1989) demonstrated that concretes made with 1% $CaCl_2 \cdot 2H_2O$ by mass of cement developed active steel corrosion when stored continuously in fog. When 2% $CaCl_2 \cdot 2H_2O$ was used, active corrosion was detected in concrete stored in a fog room at 100% relative humidity. Risk of corrosion was greatly reduced at a lower relative humidity (50%). Gaynor (1998) demonstrates how to calculate the chloride content of fresh concrete and compare it with recommended limits.

Calcium chloride is not an antifreeze agent. When used in allowable amounts, it will not reduce the freezing point of concrete by more than a few degrees. Attempts to protect concrete from freezing by this method are not effective. Instead, proven reliable precautions should be taken during cold weather (see Chapter 14).

Calcium chloride should be added to the concrete mixture in solution form as part of the mixing water. If added to the concrete in dry flake form, all of the dry particles may not completely dissolve during mixing. Undissolved lumps in the mixture can cause popouts or dark spots in the hardened concrete.

Hydration-Control Admixtures

Hydration controlling admixtures became available in the late 1980s. They consist of a two-part chemical system: (1) a stabilizer or retarder that essentially stops the hydration of cementing materials, and (2) an activator that reestablishes normal hydration and setting once added to the stabilized concrete. The stabilizer can suspend hydration for extended periods (for example, 72 hours). A set activator is added to the mixture just before the concrete is ready for use. These admixtures make it possible to reuse concrete returned in a ready-mix truck by suspending setting overnight. The admixture is also useful in maintaining concrete in a stabilized non-hardened state during long hauls. The concrete is reactivated when it arrives at the project. This admixture presently does not have a standard specification (Kinney 1989).

Workability-Retaining Admixtures

Workability retaining admixtures provide varying degrees of workability retention without affecting the initial set of concrete or early-age strength development, as is the case with retarding admixtures. These admixtures can be used with mid-range or high-range water reducers to provide desired levels of workability retention in moderate to high slump concrete mixtures, including self-consolidating concrete (SCC). Their main benefit is reducing the need for slump adjustments prior to concrete placement, thus helping to maintain consistency in concrete performance throughout a project. Workability-retaining admixtures should meet the requirements of ASTM C494/C494M, Type S.

Corrosion Inhibitors

Corrosion inhibitors are chemical admixtures that are added to concrete to limit the corrosion of steel reinforcement. Corrosion inhibitors are used in concrete for parking structures, marine structures, and bridges where chloride salts are present. Chlorides can cause corrosion of steel reinforcement in concrete (Figure 7-17).

Figure 7-17. The damage to this concrete parking structure resulted from chloride-induced corrosion of steel reinforcement.

Commercially available corrosion inhibitors include: calcium nitrite, sodium nitrite, dimethyl ethanolamine, amines, phosphates, and ester amines as listed in Table 7-1.

Corrosion-inhibiting admixtures chemically arrest the corrosion reaction and reinstate the passive layer, which provides protection to the steel. When calcium nitrite is used as an admixture, nitrite anions are in solution with hydroxyl and chloride ions. The nitrite-ions cause the ferric oxide of the passivation layer around the steel reinforcement to become more stable. This ferric oxide film is created by the high pH environment in concrete. In effect, the chloride-ions are prevented from penetrating the passive film and making contact with the steel. A certain amount of nitrite can stop corrosion up to some level of chloride-ion. Therefore, increased chloride levels require increased levels of nitrite to stop corrosion.

Organic inhibitors, based on a combination of amines and esters in a water medium, act in two ways. First, the esters provide some water repellency (see section on PRAN-type permeability reducing admixtures), thereby restricting the ingress of water soluble chlorides. Second, the amines adsorb onto the steel surface forming a tight film which repels moisture and acts as a barrier to chemical attack. The result of this dual action is an increase in the time to corrosion initiation and a decrease in the rate of corrosion once it has started.

Cathodic inhibitors react with the steel surface to interfere with the reduction of oxygen. The reduction of oxygen is the principal cathodic reaction in alkaline environments (Berke and Weil 1994). Corrosion inhibitors should conform to ASTM C1582, *Standard Specification for Admixtures to Inhibit Chloride-Induced Corrosion of Reinforcing Steel in Concrete.*

Shrinkage-Reducing Admixtures

Shrinkage-reducing admixtures (SRAs), introduced in the 1980s, have potential uses in bridge decks, critical floor slabs, and buildings where cracks, curling, and warping must be minimized for durability or aesthetic reasons (Figure 7-18). As concrete dries, water is removed from the capillary pores and a meniscus is formed at the air interface due to surface tension. Surface tension forces also act on the solid phases and tend to draw the walls of the pore together. As the water meniscus recedes into smaller and smaller pores the surface tension forces increase, causing the concrete to shrink more. Shrinkage reducing admixtures reduce the surface tension of the liquid phase, which reduces the forces exerted on the pore walls thereby producing less drying shrinkage.

Propylene glycol and polyoxyalkylene alkyl ether have been used as shrinkage reducers. Drying shrinkage reductions between 25% and 50% have been demonstrated in laboratory tests. These admixtures have negligible effects on slump, but can impact air content and may possibly require an increase in the dose of air-entraining admixture to achieve a target air content. A delay in time of set and slower bleed rate may also result from the use of SRAs. They are generally compatible with other admixtures (Nmai and others 1998 and Shah, Weiss, and Yang 1998). The manufacturer's recommendations should be followed, particularly when used in air-entrained concrete. Due to their potential effects on bleeding and setting time, caution and proper planning are required when SRAs are used in slabs that receive a hard-trowelled finish. Premature finishing of a SRA-treated concrete slab can trap bleed-water and, subsequently, lead to delamination of the concrete surface. Because of their effectiveness in reducing drying shrinkage, SRAs are also generally effective in reducing curling and cracking in slabs. SRAs should meet the requirements of ASTM C 494/C 494M, Type S.

Figure 7-18. Shrinkage cracks, such as shown on this bridge deck, can be reduced with the use of good concreting practices and shrinkage reducing admixtures.

Permeability Reducing Admixtures

The ingress of water and water borne chemicals into concrete can have many undesirable effects; including damp, leaking structures, corrosion of steel reinforcement, and concrete deterioration. Water can enter concrete through two primary mechanisms; capillary absorption under non-hydrostatic conditions (often referred to as wicking) and the direct ingress of water under pressure. Technically, the term permeability only refers to concrete exposed to water under pressure. However, permeability is often used informally to describe any passage of water through concrete, whether by pressure driven ingress or by wicking.

Considering these two mechanisms of water ingress, permeability reducing admixtures (PRAs) can be divided into two categories; Permeability Reducing Admixture – Non-Hydrostatic (PRAN) and Permeability Reducing Admixtures – Hydrostatic (PRAH) (ACI 212.3R-10).

Permeability Reducing Admixture – Non-Hydrostatic (PRAN)

Permeability reducing admixtures that are non-hydrostatic (PRAN) have traditionally been referred to as "damp-proofers". Most PRANs are hydrophobic in nature. PRANs give concrete a water repellent property and they provide reduced absorption (wicking). Common materials include soaps such as stearates and other long chain fatty acids, or their derivatives, as well as petroleum products. These PRANs are sometimes used to reduce the transmission of moisture through concrete in contact with water or damp soil. However, hydrophobic admixtures are usually not effective when the concrete is in contact with water under pressure. Some PRANs contain finely divided solids such as bentonite or siliceous powders that restrict water absorption. Often referred to as "densifiers", finely divided solids may reduce permeability slightly, although the effect is relatively small. In practice, fine solid fillers are usually used for non-hydrostatic applications similar to hydrophobic admixtures.

Permeability Reducing Admixtures – Hydrostatic (PRAH)

Hydrostatic permeability reducing admixtures (PRAH) have often been referred to as "waterproofers", although permeability reducing admixture for hydrostatic conditions is a more technically correct term. PRAHs contain materials that act to block the pores and capillaries in concrete. These materials have been shown to be effective in reducing permeability to water under pressure. They have also been shown to reduce concrete corrosion in chemically aggressive environments. Products usually consist of hydrophilic crystalline materials that react in concrete to produce pore blocking deposits, or polymeric materials that coalesce in the concrete's pores. Reactive PRAHs have also been shown to increase the autogenous sealing of leaking, hairline cracks.

Additional Considerations

Permeability reducing admixtures will not correct for a poorly designed concrete mixture. Proper proportioning, placement, and curing are needed for effective performance from a PRA. Concrete joints should be treated with a suitable waterstop. Also, the selection of a PRA must take into account the expected service conditions as well as the features of the admixture (PRAN or PRAH). PRANs are usually evaluated using an absorption based test method such as ASTM C1585, *Standard Test Method for Measurement of Rate of Absorption of Water by Hydraulic-Cement Concretes*. PRAHs are best evaluated using a pressure driven penetration test such as the U.S. Army Corp of Engineers CRC C48-92, *Standard Test Method for Water Permeability of Concrete*, or the European standard BS EN 12390-8, *Testing Hardened Concrete – Depth of Penetration of Water under Pressure*. PRAs are discussed in detail in ACI 212.3R-10, *Report on Chemical Admixtures for Concrete*.

Alkali-Aggregate Reactivity Inhibitors

In the 1950s, McCoy and Caldwell discovered that lithium based compounds when used in sufficient quantity were capable of inhibiting damage due to alkali-silica reactivity (ASR).

The use of lithium nitrate, lithium carbonate, lithium hydroxide, lithium aluminum silicate (decrepitated spodumene), and barium salts have shown reductions of alkali-silica reaction (ASR) in laboratory tests (Figure 7-19) (Thomas and Stokes 1999 and AASHTO 2001). Some of these materials may have potential for use as an additive to cement (Gajda 1996).

As discussed in Chapter 11, ASR gel has a great capacity to absorb moisture from within the concrete pores. This can cause a volumetric expansion of the gel, which in turn leads to the build up of internal stresses and the eventual disruption of the cement paste surrounding the aggregate particle. If lithium nitrate is used as an admixture, the pore solution will contain lithium and nitrate ions in addition to sodium, potassium, and hydroxyl ions. Likewise the reaction product (ASR gel) that forms will also contain appreciable quantities of lithium. The reduction in ASR by lithium salts appears to result from an exchange of the lithium ion with sodium and potassium. The resulting lithium-bearing reaction product does not have the same propensity to absorb water and expand.

There has been considerable research in the last 20 years in the use of lithium to combat ASR in concrete – including a major study conducted under the support of the Federal Highway Administration (Thomas and others 2007). Alkali-aggregate reactivity inhibitors should meet the requirements of ASTM C494/C494M, Type S.

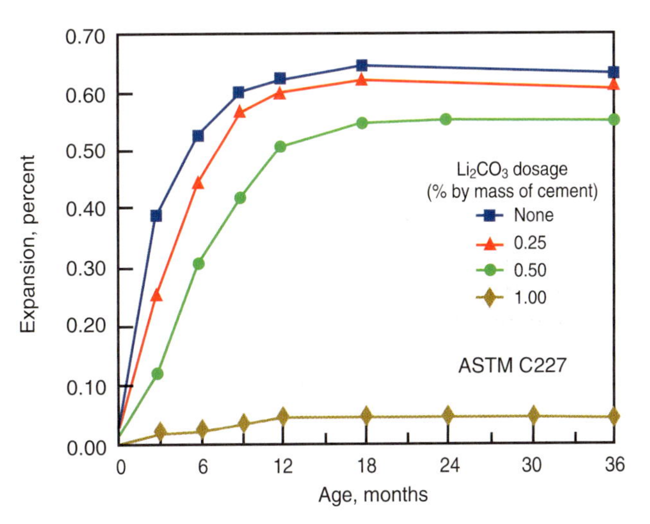

Figure 7-19. Expansion of specimens made with lithium carbonate admixture (Stark 1992).

Coloring Admixtures (Pigments)

Natural and synthetic materials are used to color concrete for aesthetic and safety reasons (Figure 7-20). Pigments used in amounts less than 6% generally do not affect concrete properties. Unmodified carbon black substantially reduces air content. Most carbon black for coloring

Figure 7-20. Red and blue pigments were used to color this terrazzo floor.

concrete contains an admixture to offset this effect on air. Generally, the amount of pigments used in concrete should not exceed 10% by weight of the cement.

Before a coloring admixture is used on a project, it should be tested for color fastness in direct sunlight and autoclaving, chemical stability in cement, and effects on concrete properties. To avoid color distortions, calcium chloride should not be used with pigments. Pigments should conform to ASTM C979, *Standard Specification for Pigments for Integrally Colored Concrete*.

Pumping Aids

Pumping aids are added to concrete mixtures to improve pumpability. Pumping aids are not a cure-all; they are best used to make marginally pumpable concrete more pumpable. These admixtures increase viscosity or cohesion in concrete to reduce dewatering of the paste while under pressure from the pump.

Some pumping aids may increase water demand, reduce compressive strength, cause air entrainment, or retard setting time. These side effects can be corrected by adjusting the mixture proportions or by adding another admixture to offset the side effect.

A partial list of materials used in pumping aids is given in Table 7-1. Some admixtures that serve other primary purposes but also improve pumpability are air-entraining admixtures and some water-reducing and retarding admixtures.

Bonding Admixtures and Bonding Agents

Bonding admixtures are usually water emulsions of organic materials including rubber, polyvinyl chloride, polyvinyl acetate, acrylics, styrene butadiene copolymers, and other polymers. They are added to portland cement mixtures to increase the bond strength between old and new concrete. Flexural strength and resistance to chloride-ion ingress are also improved. They are added in proportions equivalent to 5% to 20% by mass of the cementing materials; the actual quantity depending on job conditions and type of admixture used. Some bonding admixtures may increase the air content of mixtures. Non-reemulsifiable types are resistant to water and are better suited to exterior applications and in applications where moisture is present.

A bonding admixture is only as good as the surface to which the concrete is applied. The surface must be dry, clean, sound, free of dirt, dust, paint, and grease, and at the proper temperature. Organic or polymer-modified concretes are acceptable for patching and thin-bonded overlayment, particularly where feather-edged patches are desired.

Bonding agents should not be confused with bonding admixtures. Admixtures are an ingredient in the concrete; bonding agents are applied to existing concrete surfaces immediately before the new concrete is placed. Bonding agents help "glue" the existing and the new materials together. Bonding agents are often used in restoration and repair work; they consist of portland cement or latex-modified portland cement grout or polymers such as epoxy resins (ASTM C881, *Standard Specification for Epoxy-Resin-Base Bonding Systems for Concrete*, or AASHTO M 235, *Standard Specification for Epoxy Resin Adhesives*) or latex (ASTM C1059, *Standard Specification for Latex Agents for Bonding Fresh to Hardened Concrete*).

Grouting Admixtures

Portland cement grouts are used for a variety of purposes: to stabilize foundations, set machine bases, fill cracks and joints in concrete work, cement oil wells, fill cores of masonry walls, grout prestressing tendons and anchor bolts, and fill the voids in preplaced aggregate concrete. To alter the properties of grout for specific applications, various air-entraining admixtures, accelerators, retarders, and nonshrink admixtures are often used.

Gas-Forming Admixtures

Aluminum powder and other gas-forming materials are sometimes added to concrete and grout in very small quantities. These materials cause a slight expansion of the mixture prior to hardening. This may be of benefit where the complete grouting of a confined space is essential, such as under machine bases or in post-tensioning ducts of prestressed concrete. These materials are also used in larger quantities to produce autoclaved cellular concretes. The amount of expansion that occurs is dependent upon the amount of gas-forming material used, the temperature of the fresh mixture, the alkali content of the cement, and other variables. Where the amount of expansion is critical, careful testing through trial batching and jobsite control of mixtures and temperatures must be exercised. Gas-forming agents will not overcome shrinkage once hardening caused by drying or carbonation occurs.

Air Detrainers

Air-detraining admixtures reduce the air content in concrete. They are used when the air content cannot be reduced by adjusting the mixture proportions or by changing the dosage of the air-entraining admixture and other admixtures. Air-detrainers are rarely used and their effectiveness and dosage rate should be established on trial mixtures prior to use in actual job mixtures. Materials used in air-detraining agents are listed in Table 7-1.

Fungicidal, Germicidal, and Insecticidal Admixtures

Bacteria and fungal growth on concrete surfaces or in hardened concrete may be partially controlled through the use of fungicidal, germicidal, and insecticidal admixtures. The most effective materials are polyhalogenated phenols, dieldrin emulsions, and copper compounds. The effectiveness of these materials is generally temporary. In high dosages they may reduce the compressive strength of concrete.

Viscosity Modifying Admixtures

Another group of chemical admixtures important to self-consolidating concrete production is viscosity modifying admixtures (VMA). These are also used as antiwashout admixtures for concrete placed underwater. Antiwashout admixtures increase the cohesiveness of concrete to a level that allows limited exposure to water with little loss of cement paste. This cohesiveness allows placement of concrete in water and under water without the use of tremies. These admixtures increase the viscosity of water resulting in a mixture with increased thixotropy and resistance to segregation. VMAs usually consist of water soluble cellulose ethers, acrylic polymers, or high molecular weight biogums. Viscosity modifying admixtures should meet the requirements of ASTM C494/C494M, Type S.

The two basic types of VMAs include thickening type and binding type VMAs. The thickening type increases the viscosity through molecular obstruction. The technology is based on an addition of a large polymer molecule into the paste. Thickened paste then translates to increased cohesion of the mortar system and the concrete as a whole. The binding type of VMA is much more effective than the thickening type (Bury and Buehler 2002a). A binding type VMA chemically combines with water molecules, as opposed to just obstructing them. Binder VMAs are typically inorganic materials that produce a gel. The gel promotes thixotropic behavior. In both types, the increase in viscosity may also be accompanied by an increase in yield stress. VMAs dampen the changes in viscosity potentially caused by material and process variations. However, VMAs should not be considered a substitute for good concrete quality control (EFNARC 2006).

VMAs can be used to replace fines or to supplement them. Trial batch evaluation, using the recommended dosage from the manufacturer as a starting point, is the best method to determine the appropriate use for each mixture. An increase in high-range water reducer dosage may be necessary when a VMA is used to counteract an increase in yield stress.

Compatibility of Admixtures and Cementitious Materials

Fresh concrete problems of varying degrees of severity are encountered as a result of cement-admixture incompatibility and incompatibility between admixtures. Incompatibility between supplementary cementing materials and admixtures or cements can also occur. Slump loss, air loss, early stiffening, and other factors affecting fresh concrete properties can result from incompatibilities. While these problems primarily affect the plastic-state performance of concrete, long-term hardened concrete performance may also be adversely affected. For example, early stiffening can cause difficulties with consolidation of concrete which may also compromise strength.

When incompatibility is encountered, it can often be solved by changing the admixture dosage rate or the sequence of addition to the mixture. However, some incompatibility issues may be solved by modifying the composition of the cement, particularly the C_3A, alkali or sulfate content; or modifying the composition of the admixture. In practice, changing the source of the cement or the admixture may be the most direct solution to achieving the desired performance.

Taylor and others (2006 and 2006a) have developed protocols for testing the compatability of various combinations of materials. This preconstruction testing can reduce the likelihood of performance problems in the field. However, reliable test methods are not available to adequately address all incompatibility issues due to variations in materials, mixing equipment, mixing time, and environmental factors. When incompatibility is discovered in the field, a common solution is to simply change admixtures or cementing materials. For more information on incompatibility of cement and chemical admixtures refer to Taylor, Kosmatka, and Voight 2008, Helmuth and others 1995, Tagni-Hamou and Aïtcin 1993, and Tang and Bhattacharja 1997.

Less-Than Expected Water Reduction

If the water reduction achieved using an admixture is less than expected based on previous experience with the same admixture, this may be caused by: the composition of the cementitious materials, the presence of other set-control admixtures, the temperature of the concrete, clay minerals in the aggregates, and the dose of the admixture itself.

Slump loss. High-range water reducers are only effective for a limited period before they are overwhelmed by the hydration products (particularly ettringite) in the very early stages of hydration. The rate of loss depends on: C_3A, SO_3 and alkalis, temperature, and fineness of cement and pozzolans.

Slump loss can often be offset by delaying the time of addition of the admixture. For example, a high-range water reducer is often added on site rather than during batching. Concrete batched from a remote location, such as in municipal paving projects that employ a stationary ready-mix source are a prime example.

Incompatibility between some high-range water-reducers and cementing materials can result in very rapid losses in workability, shortly after mixing. While this can often be attributed to the temperature of the concrete, the reactivity of the cement and the continuous availability of admixture to disperse the hydrating cement grains is a key factor.

Certain minerals found in various aggregate sources such as expansive clays, have been found to rapidly adsorb polycarboxylate-type superplasticizers, thus significantly reducing their effectiveness (Jeknavorian and others 2003).

Cement admixture compatibility, with regards to slump loss, can be examined in the laboratory by testing the flow properties of pastes. A suitable test (Tang and Bhattacharja 1997) is the mini-slump test. This test gives an indication of how long the plasticizing action of a high-range water reducer can be maintained.

Another test used for this purpose is the Marsh cone method(ASTM D6910 and API 13B-1). This test can also be used to gauge the saturation point for a particular cement/admixture combination. The saturation point is the level at which further admixture addition will no longer produce any benefit.

ASTM C1679, *Standard Practice for Measuring Hydration Kinetics of Hydraulic Cementitious Mixtures Using Isothermal Calorimetry*, may be another useful approach to evaluate the compatibility of cementitious mixtures containing chemical admixtures. Changes in the thermal power curve obtained from this practice may indicate changes in a material property.

Less-Than Expected Retardation

If the length of retardation is less than expected, this may be due to an increase in the C_3A content of the cement. Abnormally retarded set may be caused by: a low C_3A content or low cement reactivity, excessive admixture with retarding properties, high levels of SCMs, or low temperature. Testing an admixture over a range of addition rates can often identify a critical dosage above or below which unacceptable set performance can result.

Problems on site can be avoided by trial batching with the actual materials in environments that are as close as possible to field conditions.

The performance of concrete produced with cements of abnormally low C_3A and SO_3 content should be carefully observed when water-reducing or set retarding admixtures are used. Any changes in the alkali content of the cement should alert the concrete producer to potential changes in admixture performance.

Storing and Dispensing Chemical Admixtures

Liquid admixtures can be stored in barrels or bulk tankers. Powdered admixtures can be placed in special storage bins and some are available in premeasured plastic bags. Admixtures added to a truck mixer at the jobsite are often in plastic jugs or bags. Powdered admixtures, such as certain plasticizers, or an admixture drum or barrel may be stored at the project site.

Dispenser tanks at concrete plants should be properly labeled for specific admixtures to avoid contamination and avoid dosing the wrong admixture. Most liquid chemical admixtures should not be allowed to freeze; therefore, they should be stored in heated environments. Consult the admixture manufacturer for proper storage temperatures. Powdered admixtures are usually less sensitive to temperature restrictions, but may be sensitive to moisture.

Figure 7-21. Liquid admixture dispenser at a ready mix plant provides accurate volumetric measurement of admixtures.

Liquid chemical admixtures are usually dispensed individually in the batch water by volumetric means (Figure 7-21). Liquid and powdered admixtures can be measured by mass, but powdered admixtures should not be measured by volume. Care should be taken to keep certain admixtures separate before they are dispensed into the batch. Some combinations may neutralize the effects desired. Consult the admixture manufacturer concerning compatible admixture combinations or perform laboratory tests to document performance.

References

AASHTO, "Portland Cement Concrete Resistant to Excessive Expansion Caused by Alkali-Silica Reaction," Section 56X, *Guide Specification For Highway Construction*, http://leadstates.tamu.edu/ASR/library/gspec.stm, American Association of State Highway and Transportation Officials, Washington, D.C., 2001.

Abrams, Duff A., *Calcium Chloride as an Admixture in Concrete*, Bulletin 13 (PCA LS013), Structural Materials Research Laboratory, Lewis Institute, Chicago, http://www.cement.org/pdf_files/LS013.pdf, 1924.

ACI Committee 212, *Chemical Admixtures for Concrete*, ACI 212.3R-10, American Concrete Institute, Farmington Hills, Michigan, 2010, 30 pages.

ACI Committee 212, *Guide for the Use of High-Range Water-Reducing Admixtures (Superplasticizers) in Concrete*, ACI 212.4R-04, American Concrete Institute, Farmington Hills, Michigan, 2004, 13 pages.

ACI Committee 318, *Building Code Requirements for Structural Concrete and Commentary*, ACI 318-08, American Concrete Institute, Farmington Hills, Michigan, 2008

Berke, N.S., and Weil, T.G., "World Wide Review of Corrosion Inhibitors in Concrete," *Advances in Concrete Technology*, CANMET, Ottawa, 1994, pages 891 to 914.

Bury, M.A., and Buehler, E., "Methods and Techniques for Placing Self-Consolidating Concrete – An Overview of Field Experiences in North American Applications," *Conference Proceedings: First North American Conference on the Design and Use of Self-Consolidating Concrete*, Advanced Cement-Based Materials Center, Evanston, Illinois, 2002, pages 281 to 286.

Camposagrado, G., *Investigation of the Cause and Effect of Air Void Coalescence in Air Entrained Concrete Mixes*, SN2624, Portland Cement Association, 2006.

Collepardi, M., and Valente, M., "Recent Developments in Superplasticizers," *Eighth CANMET/ACI International Conference on Superplasticizers and Other Chemical Admixtures in Concrete*, SP239, American Concrete Institute, Farmington Hills, Michigan, 2006, pages 1 to 14.

Cordon, W.A., *Entrained Air—A Factor in the Design of Concrete Mixes*, Materials Laboratories Report No. C-310, Research and Geology Division, Bureau of Reclamation, Denver, March 15, 1946.

European Federation of Producers and Contractors of Specialist Products for Structures (EFNARC), *Guidelines for Viscosity Modifying Admixtures For Concrete*, September 2006, http://www.efnarc.org/pdf/Guidelines%20for%20VMA%20 (document%20180).pdf, (Accessed November 2008).

Gajda, John, *Development of a Cement to Inhibit Alkali-Silica Reactivity*, Research and Development Bulletin RD115, Portland Cement Association, 1996, 58 pages.

Gaynor, Richard D., "Calculating Chloride Percentages," *Concrete Technology Today*, PL983, Portland Cement Association, http://www.cement.org/pdf_files/PL983.pdf, 1998, pages 4 to 5.

Gebler, S.H., *The Effects of High-Range Water Reducers on the Properties of Freshly Mixed and Hardened Flowing Concrete*, Research and Development Bulletin RD081, Portland Cement Association, http://www.cement.org/pdf_files/RD081.pdf, 1982.

Gilkey, H.J., "Re-Proportioning of Concrete Mixtures for Air Entrainment," *Journal of the American Concrete Institute, Proceedings*, vol. 29, no. 8, Farmington Hills, Michigan, February 1958, pages 633 to 645.

Helmuth, Richard; Hills, Linda M.; Whiting, David A.; and Bhattacharja, Sankar, *Abnormal Concrete Performance in the Presence of Admixtures*, RP333, Portland Cement Association, http://www.cement.org/pdf_files/RP333.pdf, 1995.

Hewlett, P.C., *Lea's Chemistry of Cement and Concrete*, 1998, 4th Edition, Arnold, London, 1998.

Jeknavorian, Ara A.; Jardine, Leslie; Ou, Chia-Chih; Koyata, Hideo; and Folliard, Kevin, W.R. Grace and Co. Conn., Cambridge, Massachusetts, USA. American Concrete Institute, SP (2003), SP-217 *Seventh CANMET/ACI International Conference on Superplasticizers and Other Chemical Admixtures in Concrete*, 2003, pages 143 to 159.

Jeknavorian, A.A.; Roberts, L.R.; Jardine, L.; Koyata, H.; Darwin, D.C., *Condensed polyacrylic acid-aminated polyether polymers as superplasticizers for concrete. Superplasticizers and Other Chemical Admixtures in Concrete*, SP-173, American Concrete Institute, Farmington Hills, Michigan, pages 55 to 81.

Kinney, F.D., "Reuse of Returned Concrete by Hydration Control: Characterization of a New Concept," Superplasticizers and Other Chemical Admixtures in Concrete, SP119, American Concrete Institute, Farmington Hills, Michigan, 1989, pages 19 to 40.

Klieger, Paul, *Air-Entraining Admixtures*, Research Department Bulletin RX199, Portland Cement Association, http://www.cement.org/pdf_files/RX199.pdf, 1966, 12 pages.

Kozikowski, Jr., R.L.; Vollmer, D.B.; Taylor, P.C.; and Gebler, S.H., *Factor(s) Affecting the Origin of Air-Void Clustering*, SN2789, Portland Cement Association, 2005.

Lackey, Homer B., "Factors Affecting Use of Calcium Chloride in Concrete," *Cement, Concrete, and Aggregates, American Society for Testing and Materials*, West Conshohocken, Pennsylvania, Winter 1992, pages 97 to 100.

Li, C.Z.; Feng, N.Q.; Li, Y.D.; and Chen, R.J., "Effects of Polyethlene Oxide Chains on the Performance of Polycarboxylate-Type Water-Reducers," *Cement and Concrete Research*, Vol. 35, No. 5, May 2005, pages 867 to 873.

McCoy, W.J., and Caldwell, A.G., "New Approach to inhibiting Alkali-Aggregate Expansion", *Journal of the American Concrete Institute*, Vol. 22, pages 693 to 706.

Nawa, T., "Effect of Chemical Structure on Steric Stabilization of Polycarboxylate-based Superplasticizer," *Journal of Advanced Concrete Technology*, Vol. 4, No. 2, June 2006, pages 225 to 232.

Neville, A.M., *Properties of Concrete*, Pearson Education Limited, Essex, England, 1995.

Nmai, Charles K.; Schlagbaum, Tony; and Violetta, Brad, "A History of Mid-Range Water-Reducing Admixtures," *Concrete International*, American Concrete Institute, Farmington Hills, Michigan, April 1998, pages 45 to 50.

Nmai, Charles K.; Tomita, Rokuro; Hondo, Fumiaki; and Buffenbarger, Julie, "Shrinkage-Reducing Admixtures," *Concrete International*, American Concrete Institute, Farmington Hills, Michigan, April 1998, pages 31 to 37.

Ramachandran, V.S, *Superplasticizers: Properties and Applications in Concrete*, Canada Centre for Mineral and Energy Technology, Ottawa, Ontario,1998, 404 pages.

Ramachandran, V.S., *Concrete Admixtures Handbook*, Noyes Publications, Park Ridge, New Jersey, 1995.

Shah, Surendra P.; Weiss, W. Jason; and Yang, Wei, "Shrinkage Cracking—Can it be Prevented?," *Concrete International*, American Concrete Institute, Farmington Hills, Michigan, April 1998, pages 51 to 55.

Stark, David, *Influence of Design and Materials on Corrosion Resistance of Steel in Concrete*, Research and Development Bulletin RD098, Portland Cement Association, http://www.cement.org/pdf_files/RD098.pdf, 1989, 44 pages.

Stark, David C., *Lithium Salt Admixtures—An Alternative Method to Prevent Expansive Alkali-Silica Reactivity*, RP307, Portland Cement Association, 1992, 10 pages.

Szecsy, R., and Mohler, N., *Self-Consolidating Concrete*, IS546, Portland Cement Association, 2009, 24 pages

Tagnit-Hamou, Arezki, and Aïtcin, Pierre-Claude, "Cement and Superplasticizer Compatibility," *World Cement*, Palladian Publications Limited, Farnham, Surrey, England, August 1993, pages 38 to 42.

Tang, Fulvio J., and Bhattacharja, Sankar, *Development of an Early Stiffening Test*, RP346, Portland Cement Association, 1997, 36 pages.

Taylor, Peter C.; Johansen, Vagn C.; Graf, Luis A.; Kozikowski, Ronald L.; Zemajtis, Jerzy Z.; and Ferraris, Chiara F., *Identifying Incompatible Combinations of Concrete Materials: Volume I-Final Report*, FHWA HRT-06-079, Federal Highway Administration, Washington D.C., 2006, 157 pages. http://www.fhwa.dot.gov/pavement/concrete/pubs/06079/index.cfm.

Taylor, Peter C.; Johansen, Vagn C.; Graf, Luis A.; Kozikowski, Ronald L.; Zemajtis, Jerzy Z.; and Ferraris, Chiara F., *Identifying Incompatible Combinations of Concrete Materials: Volume II-Test Protocol*, FHWA HRT-06-080, Federal Highway Administration and Portland Cement, Washington D.C., 2006a, 83 pages. http://www.fhwa.dot.gov/pavement/concrete/pubs/06080/index.cfm.

Taylor, P.; Kosmatka, S.; and Voigt, G., *Integrated Materials and Construction Practices for Concrete Pavement: A State-of-the-Practice Manual*, FHWA Publication No. HIF-07-004; SN3005, PCA, 2006, 350 pages.

Thomas, Michael D.A.; Fournier, Benoit; Folliard, Kevin J.; Ideker, Jason H.; and Resendez, Yadhira, *The Use of Lithium To Prevent or Mitigate Alkali-Silica Reaction in Concrete Pavements and Structures*, FHWA-HRT-06-133, Federal Highway Administration, Turner-Fairbank Highway Research Center, Maclean, Virginia, USA, March 2007, 50 pages.

Thomas, M.D.A., and Wilson, M.L., *Admixtures Use in Concrete*, CD039, PCA, Skokie, IL, 2002.

Thomas, Michael D.A., and Stokes, David B., "Use of a Lithium-Bearing Admixture to Suppress Expansion in Concrete Due to Alkali-Silica Reaction," *Transportation Research Record No. 1668*, Transportation Research Board, Washington, D.C., 1999, pages 54 to 59.

Whiting, David, *Effects of High-Range Water Reducers on Some Properties of Fresh and Hardened Concretes*, Research and Development Bulletin RD061, Portland Cement Association, http://www.cement.org/pdf_files/RD061.pdf, 1979.

Whiting, D., and Dziedzic, W., *Effects of Conventional and High-Range Water Reducers on Concrete Properties*, Research and Development Bulletin RD107, Portland Cement Association, 1992, 25 pages.

Whiting, David A., and Nagi, Mohamad A., *Manual on the Control of Air Content in Concrete*, EB116, National Ready Mixed Concrete Association and Portland Cement Association, 1998, 42 pages.

Whiting, D., and Stark, D., *Control of Air Content in Concrete*, National Cooperative Highway Research Program Report No. 258 and Addendum, Transportation Research Board and National Research Council, Washington, D.C., May 1983.

CHAPTER 8

Reinforcement

Why Use Reinforcement in Concrete

Concrete is very strong in compression, but relatively weak in tension. The tensile capacity of concrete is only about one tenth of its compressive strength. Consider the beam shown in Figure 8-1. Load applied at the center of the beam produces compression in the top of the beam, and tension in the bottom at midspan. Unreinforced, plain concrete, beams can fail suddenly upon cracking, with little warning. The addition of the steel reinforcement significantly increases the load capacity of that same beam as shown in Figure 8-2.

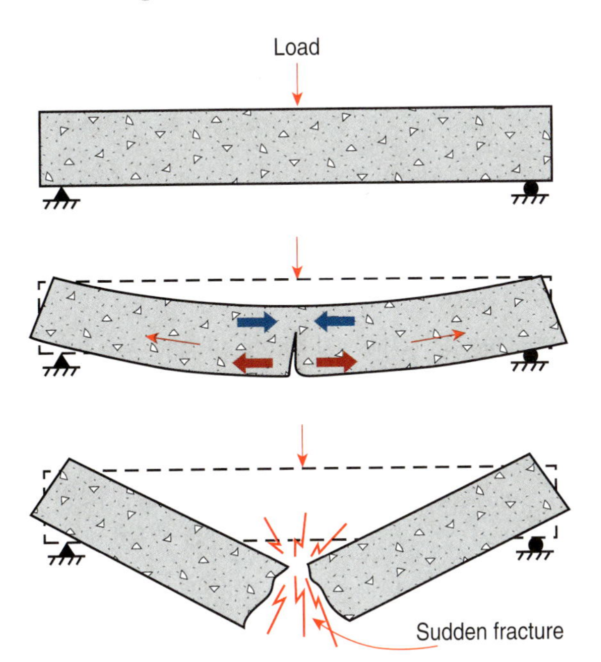

Figure 8-1. Unreinforced concrete sudden failure under loading.

Reinforcing steel is strong in tension and can be utilized in reinforced concrete to offset the relatively low tensile strength of concrete. Placing reinforcement in the tension zone of concrete also provides crack control in addition to strength and ductility to concrete structures (Figure 8-2).

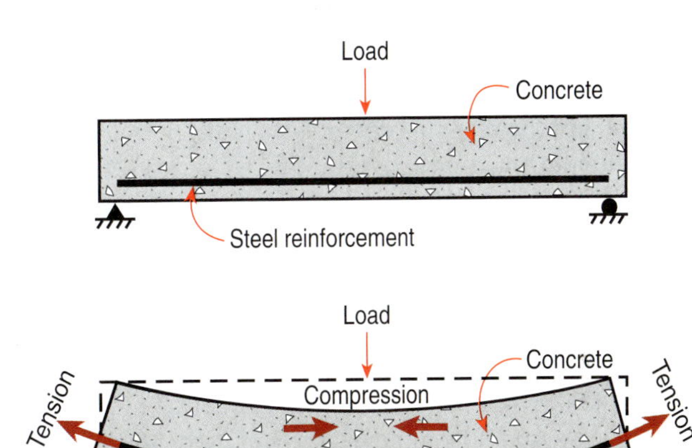

Figure 8-2. Reinforcement provides ductility and crack control in concrete.

Without reinforcement, most modern concrete structures would not be possible.

In addition to resisting tensile forces in structural members, reinforcement is also used in concrete construction for the following reasons:

- To resist a portion of compression loading. For example in columns, thereby reducing the size of the structure or increasing the structure's load carrying capacity. Compression in steel also reduces long term creep deflections.
- To resist diagonal tension or shear in beams, walls, and columns. This type of reinforcement is commonly in the form of stirrups, ties, hoops, or spirals.
- To resist internal pressures in circular structures such as tanks, pipes, bins, and silos.
- To reduce the size of cracks in concrete by distributing stresses resulting in numerous small cracks in place of a few large cracks.
- To limit crack widths and control spacing of cracks due to stresses induced by temperature changes and shrinkage.

Reinforced concrete includes many types of structures and components: elevated floor slabs, pavements, walls, beams, columns, mats, frames, and more. Reinforcement typically comes in four distinctly separate types: reinforcing bars, welded wire reinforcement, prestressing steel, and fibers. See Corrosion in Chapter 11 for fiber reinforced plastic and special steels.

Reinforcing Bars

Reinforcing steel bars (also referred to as rebar) are remarkably well-suited for concrete reinforcement because they have high tensile strength and strain compatibility. Because of reinforcement's high tensile strength, relatively small amounts are required. For example, steel reinforcing bars are generally about 15 times stronger than conventional concrete (420 MPa [60,000 psi] versus 28 MPa [4,000 psi]). A flexural beam typically requires only a small amount of reinforcement to provide tensile strength to the concrete. Reinforcement ratios (area of steel divided by the gross area of concrete) typically vary from 0.0018 to 0.08 in most structures.

The bond between concrete and steel allows for an effective transfer of stress or load between the steel and concrete so that both materials act together in a composite action. As concrete bonds to steel, both materials expand and contract to about the same degree with temperature changes. Because of this unique compatibility, steel is the most common material used to reinforce concrete. Reinforcing bars are available in different grades and sizes for use in concrete.

Grades

Reinforcing bar is specified by ASTM A615, *Standard Specification for Deformed and Plain Carbon-Steel Bars for Concrete Reinforcement*. The most common form is grade 60, which has a minimum yield strength of 420 MPa (60,000 psi) and a minimum tensile strength of 620 MPa (90,000 psi). Grades up to 80 are often utilized for columns, and Grade 100 rebar can be utilized for column confinement reinforcement in high seismic regions. Table 8-1 lists the minimum yield and tensile strengths for the most common ASTM types of steel reinforcement.

ASTM reinforcement specifications define the material properties as well as ductility requirements. Non-prestressed reinforcing steel with a minimum yield of 550 MPa (80,000 psi) is permitted for reinforced concrete designed by ACI 318. Yields up to 700 MPa (100,000 psi) are permitted for the confinement reinforcement and for columns spirals in seismic regions.

Welding of reinforcing bars must be performed with caution, due to possible changes in metallurgy that could be detrimental to the strength and ductility of the bar. If weldablity is important for construction purposes, ASTM A706, *Standard Specification for Low-Alloy Steel Deformed and Plain Bars for Concrete Reinforcement*, Grade 60, provides provisions for rebar that allows for reliable welding. All welding of reinforcement bars must be in accordance with AWS D1.4.

Stress-Strain Curves

Reinforcing bars have distinct stress-strain curves with a definable yield plateau for the lower strength low-carbon steels (Figure 8-3) (Helgason and others 1976). The observed strain hardening of reinforcing bars is not considered in the design of structural reinforced concrete. The slope of the initial elastic portion of the steel stress-strain curve is the modulus of elasticity (E_s) and is taken as 200 GPa (29,000,000 psi) (ACI 318 2008).

Some reinforcing bars, particularly those steels with a grade of 75 and higher, do not always exhibit a well-defined yield point (Fintel 1985). Even Grade 40 and Grade 60 steels exhibit strain hardening and the length of the yield plateau can vary.

Bar Marks

To uniquely identify the reinforcement, reinforcing bars are marked during production with a letter and number signifying the type and grade of steel (Figure 8-4). The bar mark also contains a letter or symbol identifying the mill in which the bar was produced. Additionally, the bars are marked with their bar size, such as the number '4' indicating a No. 4 bar. The bar size number is approximately the bar nominal diameter in 8ths of an inch (that is, a No. 4 bar has approximately a 4/8-in., or ½-in., diameter). Bars can also be marked with an approximately equivalent metric size called "soft metric bar" sizes, directly converted from inch-pound bar sizes (CRSI 2008). Alternately, rebar may be produced and marked to true metric sizes. Table 8-2 lists typical bar sizes for standard ASTM reinforcing bars.

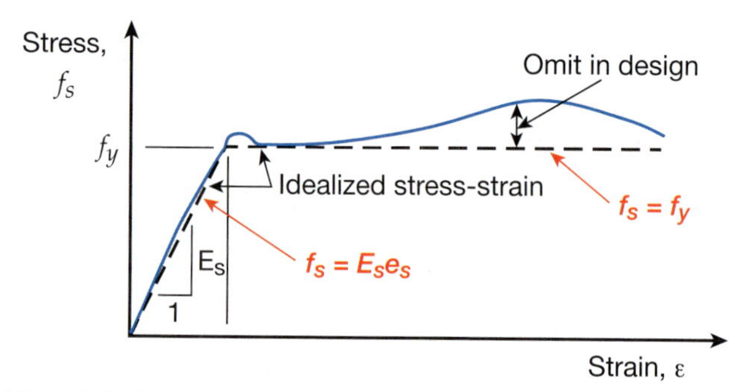

Figure 8-3. Stress–strain curve for reinforcing bars.

Table 8-1. Summary of ASTM Strength Requirements for Reinforcements

Reinforcement	ASTM	Designation	Minimum	Minimum
Type	Specification		Yield (MPa [ksi])	Strength (MPa [ksi])
Reinforcing bars	A615	Grade 40	280 (40)	420 (60)
		Grade 60	420 (60)	620 (90)
		Grade 75	520 (75)	690 (100)
		Grade 80	550 (80)	725 (105)
	A706	Grade 60	420 (60), 540 (78) max	550 (80), not less than 1.25 actual yield
		Grade 80	550 (80), 675 (98) max	690 (100), not less than 1.25 actual yield
	A1035	Grade 100	690 (100)	1030 (150)
		Grade 120	830 (120)	1030 (150)
Wire – plain	A1064	Grade 70	485 (70)	550 (80)
Wire – deformed	A1064	Grade 75	515 (75)	585 (85)
Welded wire reinforcement				
Plain W1.2 and larger	A1064	Grade 65	450 (65)	515 (75)
Deformed	A1064	Grade 70	485 (70)	550 (80)
Prestressing tendons				
Seven-wire strand	A416	Grade 250 (stress-relieved)	85% of tensile	1725 (250)
		Grade 250 (low-relaxation)	90% of tensile	1725 (250)
		Grade 270 (stress-relieved)	85% of tensile	1860 (270)
		Grade 270 (low-relaxation)	90% of tensile	1860 (270)
Wire	A421	Type BA (stress-relieved)	85% of tensile	Varies by size from 1655 (240) to 1620 (235)
		Type BA (low-relaxation)	90% of tensile	Varies by size from 1655 (240) to 1620 (235)
		Type WA (stress-relieved)	85% of tensile	Varies by size from 1725 (250) to 1620 (235)
		TypeWA (low-relaxation)	90% of tensile	Varies by size from 1725 (250) to 1620 (235)
Bars	A722	Type I (plain)	85% of tensile	1035 (150)
		Type II (deformed	80% of tensile	1035 (150)

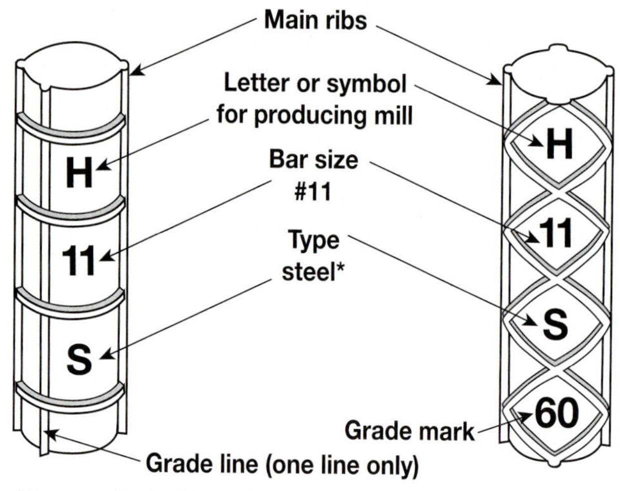

Figure 8-4. Bar marks for a Grade 60 reinforcing bars (Courtesy of CRSI).

Deformed Reinforcing Bars

Reinforced concrete design is predicated on the fact that once the plastic concrete has hardened the concrete and reinforcing bars undergo the same strains (that is, they behave as a composite material). Between concrete and steel, a chemical bond also develops while the concrete hardens. In order to enhance bond, reinforcing bars are produced with deformations, as shown in Figure 8-4, to provide mechanical bond between the bars and the concrete (Zuo and Darwin 2000).

Non-deformed bars, referred to as plain bars, are available. However, the current ACI 318 Building Code permits their usage only as spiral reinforcement.

Coated Reinforcing Bars and Corrosion Protection

In exterior environments exposed steel corrodes. Fortunately, when concrete surrounds the steel, it can provide some level of corrosion resistance as cover. Corrosion causes the steel reinforcement to rust and expand in volume. The expansion generates great internal stresses that may result in spalling of the concrete cover over the steel reinforcement (Figure 8-5) (see Chapter 11). In extreme cases, the corrosion can result in a reduction in the strength of the element. Stainless steel reinforcing bars (ASTM A955, *Standard Specification for Deformed and Plain Stainless-Steel Bars for Concrete Reinforcement*) may be specified to limit corrosion in extreme environments.

Figure 8-5. Reinforcing steel corrosion and concrete spalling due to reduced cover.

Table 8-2. ASTM Standard Reinforcing Bars

ASTM standard metric reinforcing bars			
Bar size designation	Nominal dimensions		
	Area (mm²)	Weight (kg/m)	Diameter (mm)
#10	71	0.560	9.5
#13	129	0.994	12.7
#16	199	1.552	15.9
#19	284	2.235	19.1
#22	387	3.042	22.2
#25	510	3.973	25.4
#29	645	5.060	28.7
#32	819	6.404	32.3
#36	1006	7.907	35.8
#43	1452	11.38	13.0
#57	2581	20.24	57.3

The current A615M specification covers bar sizes #43 and #57 in Grade 420, and bar sizes #36, #43, and #57 in Grade 520. The current A706 specification also covers bar sizes #43 and #57. Bar sizes #29 through #57 are not included in the A996M specification.

ASTM standard inch-pound reinforcing bars			
Bar size designation	Nominal dimensions		
	Area (in.²)	Weight (lb/ft)	Diameter (in.)
#3	0.11	0.376	0.375
#4	0.20	0.668	0.500
#5	0.31	1.043	0.625
#6	0.44	1.502	0.750
#7	0.60	2.044	0.875
#8	0.79	2.670	1.000
#9	1.00	3.400	1.128
#10	1.27	4.303	1.270
#11	1.56	5.313	1.410
#14	2.25	7.65	1.693
#18	4.00	13.60	2.257

The current A615 specification covers bar sizes #14 and #18 in Grade 60, and bar sizes #11, #14, and #18 in Grade 75. The current A706 specification also covers bar sizes #14 and #18. Bar sizes #9 through #18 are not included in the A996 specification.

To improve the corrosion resistance of reinforcing bars under severe conditions, the bars may be epoxy coated or zinc coated (galvanized).

Epoxy coating provides a waterproof surface to enhance corrosion resistance. As shown in Figures 8-6 and 8-7 epoxy-coated reinforcement is generally recognized in the field by a green coating. ASTM A775, *Standard Specification for Epoxy-Coated Reinforcing Steel Bars*, and ASTM A934, *Standard Specification for Epoxy-Coated Prefabricated Steel Reinforcing Bars*, cover the requirements for bars coated before or after fabrication (respectively) including surface preparation. Zinc and epoxy dual coated bar requirements are covered in ASTM A1055, *Standard Specification for Zinc and Epoxy Dual-Coated Steel Reinforcement Bars*. Care must be taken to minimize damage to the epoxy coating during shipping and placement (Figure 8-6). Any damage to in the epoxy coating can result in localized corrosion. Repairs to the coating should be made prior to concrete placement.

Special design requirements are necessary when using epoxy-coated reinforcing bars because the epoxy coating interferes with the chemical bond between concrete and rebar. Additional design requirements include longer lap splices between individual reinforcing bars (ACI 318 2008).

Figure 8-6. Epoxy-coated reinforcing bars.

Figure 8-7. Shipping of epoxy-coated reinforcing bars.

Zinc coating provides a sacrificial physical barrier and cathodic protection which can also protect across scratches in the coating. Galvanized bars are covered by ASTM A767, *Standard Specification for Zinc-Coated (Galvanized) Steel Bars for Concrete Reinforcement*. Bars are typically galvanized after cutting and bending to the project requirements. The zinc coating does not interfere with the chemical bond between concrete and reinforcement (ACI 318 2008).

Figure 8-8. Welded wire reinforcement (Courtesy of BASF).

Welded Wire Reinforcement

For reinforcing thin slabs and other structural elements, an alternative to reinforcing bars is welded wire reinforcement (WWR), formerly known as welded wire mesh or fabric. Welded steel wire reinforcement is the predominant form of reinforcement used for the topping slab of composite steel deck slab systems. A grid of orthogonal longitudinal and transverse cold-drawn steel wires is welded together at every wire intersection (Figure 8-8). The size and spacing of the wires can vary in each direction, based on the requirements of the project. Welded wire reinforcement can be epoxy coated or zinc coated (galvanized).

Welded wire reinforcement is typically designated by the spacing of the wires in each direction along with the wire diameter. For example, 152 x 152-MW 26 x MW 26 (6 x 6-W4 x W4), where the spacing is in millimeters (inches) and the wire size is the cross-sectional area in square millimeters (hundredths of a square inch). Plain (smooth) wires are designated with an "MW" ("W") and deformed wires are designated with an "MD" ("D"). As per Table 8-3, a 152 x 152-MW 26 x MW 26 (6 x 6-W4 x W4) would provide 170 mm^2 per meter (0.08 $in.^2$ per foot) of width of welded wire reinforcement (WRI 2010). Typically the yield strength of welded wire reinforcement is taken as 420 MPa (60 ksi) but is available with higher yield strengths, including 551 MPa to 689 MPa (80 ksi to 100 ksi).

Plain and deformed welded wire reinforcement is covered in a combined standard, ASTM A1064, *Standard Specification for Steel Wire and Welded Wire Reinforcement, Plain and Deformed, for Concrete.* Plain steel wires are available in sizes ranging from MW9 to MW290 (W1.4 to W45) and deformed steel wires are available in sizes ranging from MD26 to MD290 (D4 to D45). Stainless steel wires are specified according to ASTM A1022, *Standard Specification for Deformed and Plain Stainless Steel Wire and Welded Wire for Concrete*. Epoxy-coated WWR is available in accordance with ASTM A884, *Standard Specification for Epoxy-Coated Steel Wire and Welded Reinforcement*. Galvanized WWR is available in accordance with ASTM A1060, *Standard Specification for Zinc-Coated (Galvanized) Steel Welded Wire Reinforced, Plain and Deformed for Concrete.*

Even plain wire used in welded wire reinforcement has both chemical bond and mechanical bond to the concrete. The mechanical bond results from bearing of the welded cross wires against the concrete in the grid of reinforcement.

Table 8-3. Welded Wire Reinforcement

W & D wire size* plain*	W & D metric wire size (conversion) plain**	Customary units			Metric units (conversion)		
		Area (sq in.)	Diameter (in.)	Nominal weight (lb/ft)	Nominal area (mm^2)	Nominal diameter (mm)	Nominal mass (kg/m)
W45	MW 290	.45	.757	1.530	290	19.23	2.28
W31	MW 200	.31	.628	1.054	200	15.96	1.57
W20	MW 130 MW 122	.200 .189	.505 .490	.680 .643	129 122	12.8 12.4	1.01 0.96
W18	MW 116 MW 108	.180 .168	.479 .462	.612 .571	116 108	12.2 11.7	0.91 0.85
W16	MW 103 MW 94	.160 .146	.451 .431	.544 .495	103 94	11.5 10.9	0.81 0.74
W14	MW 90 MW 79	.140 .122	.422 .394	.476 .414	90 79	10.7 10.0	0.71 0.62
W12	MW 77	.120	.391	.408	77	9.9	0.61
W11	MW 71	.110	.374	.374	71	9.5	0.56
W10.5	MW 68 MW 67	.105 .103	.366 .363	.357 .351	68 67	9.3 9.2	0.53 0.52
W10	MW 65	.100	.357	.340	65	9.1	0.51
W9.5	MW 61	.095	.348	.323	61	8.8	0.48
W9	MW 58 MW 56	.090 .086	.338 .331	.306 .292	58 55.5	8.6 8.4	0.45 0.43
W8.5	MW 55	.085	.329	.289	54.9	8.4	0.43
W8	MW 52	.080	.319	.272	52	8.1	0.40
W7.5	MW 48	.075	.309	.255	48.4	7.8	0.38
W7	MW 45	.070	.299	.238	45	7.6	0.35
W6.5	MW 42 MW 41	.065 .063	.288 .283	.221 .214	42 41	7.3 7.2	0.33 0.32
W6	MW 39	.060	.276	.204	39	7.0	0.30
W5.5	MW 36 MW 35	.055 .054	.265 .263	.187 .184	35.5 34.8	6.7 6.7	0.28 0.27
W5	MW 32 MW 30 MW 29	.050 .047 .045	.252 .244 .239	.170 .158 .153	32 30 29	6.4 6.2 6.1	0.25 0.24 0.23
W4	MW 26	.040	.226	.136	26	5.7	0.20
W3.5	MW 23	.035	.211	.119	23	5.4	0.18
W2.9	MW 19	.029	.192	.098	19	4.9	0.15
W2.0	MW 13	.020	.160	.068	13	4.1	0.10
W1.4	MW 9	.014	.135	.048	9	3.4	0.07

* U.S. customary sizes can be specified in 0.001 sq in increments.
** Metric wire sizes can be specified in 1 sq mm increments.
(Courtesy of Wire Reinforcement Institute)

Prestressing Steel

Prestressing a concrete structure takes maximum advantage of the beneficial aspects of both the concrete and steel. In prestressed members, compressive stresses are purposefully introduced into the concrete to reduce tensile stresses resulting from applied loads, including the self weight of the member (dead load). Prestressing steel, such as strands, bars, and wires, transfer the compressive stresses to the concrete.

Pretensioning is a method of prestressing in which the tendons are tensioned before the concrete is placed. The prestressing force is primarily transferred to the concrete through bond. Post-tensioning is a method of prestressing in which the tendons are tensioned after the concrete has hardened. The prestressing force is primarily transferred to the concrete through the end anchorages. In both methods, prestressing is transferred to the concrete at an early age. In pretensioned applications, steam curing helps achieve 28 MPa to 55 MPa (4000 psi to 8000 psi) in about 16 hours. In post-tensioned construction, concrete strengths of 21 MPa to 28 MPa (3000 to 4000 psi) are reached in two to three days, depending on the application.

Pretensioning consists of tensioning a steel tendon in the forms prior to placing the concrete and cutting the stressed wires after the concrete hardens. Pretensioning then adds a precompression to the concrete to offset tension stresses induced later during loading (Figure 8-9) (PCI 2010).

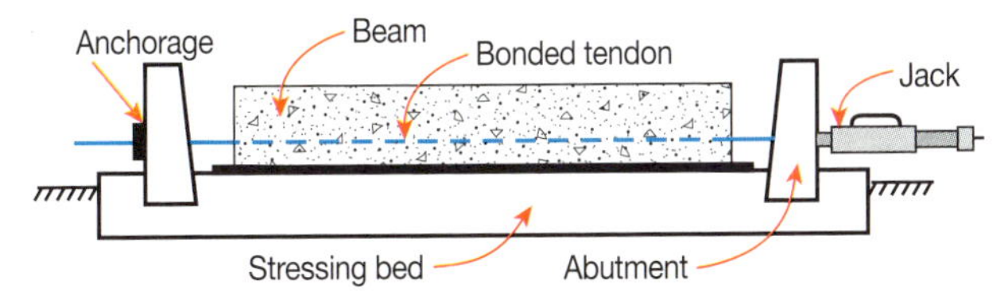

Figure 8-9. Pretensioning prestressed concrete.

Post-tensioning consists of casting the concrete around tendons placed are in tubes or ducts. After the concrete hardens, the tendons are stressed (Figures 8-10 and 8-11). After stressing, the tubes or ducts maybe grouted (bonded tendons) or left ungrouted (unbounded tendons) (PTI 2006).

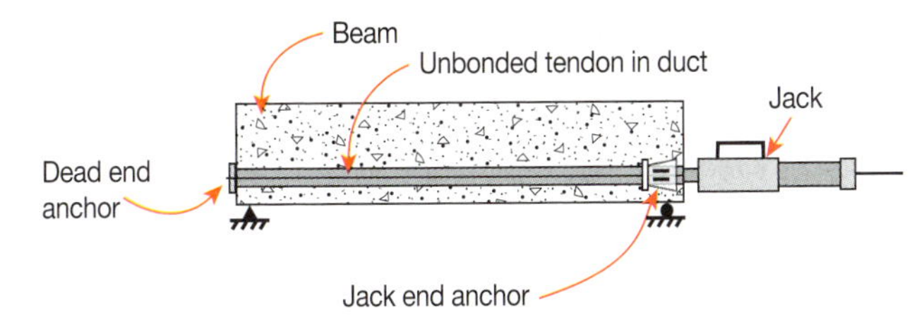

Figure 8-10. Post-tensioning prestressed concrete.

Figure 8-11. Post-tensioning tendon layout.

Figure 8-12. 'Blow-out' of post-tensioning tendon during stressing.

Tendons must be properly placed and stressed only after concrete has developed enough strength to withstand the load. If the tendon is improperly placed, there is inadequate concrete cover, the concrete is not strong enough, or the tendon is over stressed, a blow-out can occur (Figure 8-12). Tensioning operations can be dangerous unless appropriate safety precautions are taken.

Prestressing steel comes in three standard types: wires, tendons composed of several stands of wires, and high strength alloy steel bars. The prestressing wires are cold-drawn, high-strength, high-carbon steel with diameters ranging from 4.9 to 7.0 mm (0.192 to 0.276 in.). Alloy plain (smooth) steel bars range from 19 to 35 mm (0.750 to 1.375 in.) in diameter. The requirements for prestressing steels are covered by ASTM A421, *Standard Specification for Uncoated Stress-Relieved Steel Wire for Prestressed Concrete*, ASTM A416, *Standard Specification for Steel Strand, Uncoated Seven-Wire Stress Relieved for Prestressed Concrete*, and ASTM A722, *Standard Specification for Uncoated High-Strength Steel Bar for Prestressed Concrete*. The most common form of tendon consists of low-relaxation seven-wire strand ASTM A421 Grade 270 (f_{pu} = 1,750 MPa [250,000 psi]) (Table 8-4).

Table 8-4A. Standard Prestressing Tendons

Type*	Nominal diameter, mm	Nominal area, mm²	Nominal mass, kg/m
Seven-wire strand (Grade 1725)	6.4	23.2	0.182
	7.9	37.4	0.294
	9.5	51.6	0.405
	11.1	69.7	0.548
	12.7	92.9	0.730
	15.2	139.4	1.094
Seven-wire strand (Grade 1860)	9.53	54.8	0.432
	11.1	74.2	0.582
	12.70	98.7	0.775
	15.24	140.0	1.102
Prestressing wire	4.88	18.7	0.146
	4.98	19.5	0.149
	6.35	31.7	0.253
	7.01	38.6	0.298
Prestressing bars (plain)	19	284	2.23
	22	387	3.04
	25	503	3.97
	29	639	5.03
	32	794	6.21
	35	955	7.52
Prestressing bars (deformed)	15	181	1.46
	20	271	2.22
	26	548	4.48
	32	806	6.54
	36	1019	8.28

*Availability of some tendon sizes should be investigated in advance.

Table 8-4B. Standard Prestressing Tendons

Type*	Nominal diameter, in.	Nominal area, in.²	Nominal weight, lb/ft
Seven-wire strand (Grade 250)	1/4 (250)	0.036	0.122
	5/6 (0.313)	0.058	0.197
	3/8 (0.375)	0.080	0.272
	7/16 (0.438)	0.108	0.367
	1/2 (0.500)	0.144	0.490
	(0.600)	0.216	0.737
Seven-wire strand (Grade 270)	3/8 (0.375)	0.085	0.290
	7/16 (0.438)	0.115	0.390
	1/2 (0.500)	0.153	0.520
	(0.600)	0.217	0.740
Prestressing wire	0.192	0.029	0.098
	0.196	0.030	0.100
	0.250	0.049	0.170
	0.276	0.060	0.200
Prestressing bars (plain)	3/4	0.44	1.50
	7/8	0.60	2.04
	1	0.78	2.67
	1 1/8	0.99	3.38
	1 1/4	1.23	4.17
	1 3/8	1.48	5.05
Prestressing bars (deformed)	5/8	0.28	0.98
	3/4	0.42	1.49
	1	0.85	3.01
	1 1/4	1.25	4.39
	1 3/8	1.58	5.56

Prestressing wires, tendons, and alloy bars do not exhibit the same elastic plastic stress-strain curves of conventional reinforcing steel. The yielding of the prestressing steel occurs gradually. The ratio of the fracture strain to yield strain for a prestressing wire is significantly less than that for Grade 60 reinforcing bar. Hence, prestressing tendons tend to be less ductile than conventional reinforcement. Prestressing tendons also exhibit some relaxation under service loads which needs to be accounted for in design.

Since a prestressed structure counteracts the loading induced tensions with a compressive force, prestressed structures tend to have reduced cracking compared to non-prestressed structures. The induced tension also results in less deflection under load. For extremely aggressive environments, epoxy-coated strand is available and is covered under ASTM A882, *Standard Specification for Filled Epoxy-Coated Seven-Wire Prestressing Steel Strand.*

Fibers

Fibers have been used in construction materials for centuries. The past three decades have seen a growing interest in the use of fibers in ready mixed concrete, precast concrete, and shotcrete. Fibers made from steel, plastic, glass, and natural materials (such as wood cellulose) are available in a variety of shapes, sizes, and thicknesses. They may be round, flat, crimped, and deformed with typical lengths of 6 mm to 150 mm (0.25 in. to 6 in.) and thicknesses ranging from 0.005 mm to 0.75 mm (0.0002 in. to 0.03 in.) (Figure 8-13). They are added to concrete during mixing. The main factors that control the performance of the composite material are:

1. Physical properties of fibers and matrix
2. Strength of bond between fibers and matrix

Figure 8-13. Steel, glass, synthetic and natural fibers with different lengths and shapes can be used in concrete.

Although the basic governing principles are the same, there are several characteristic differences between conventional reinforcement and fiber systems:

1. Fibers are generally randomly distributed throughout a given cross section whereas reinforcing bars or wires are placed only where required
2. Most fibers are relatively short and closely spaced as compared with continuous reinforcing bars or wires
3. It is generally not possible to achieve the same reinforcement ratio of area of reinforcement to area of concrete using fibers as compared to using a network of reinforcing bars or wires

Fibers are typically added to concrete in low volume dosages (often less than 1%), and have been effective in reducing plastic shrinkage cracking.

Fibers typically do not significantly alter free shrinkage of concrete. However at high enough dosages fibers can increase the resistance to cracking and decrease crack width (Shah, Weiss, and Yang 1998).

Advantages and Disadvantages of Using Fibers

Fibers are generally distributed throughout the concrete cross section. Therefore, many fibers are inefficiently located for resisting tensile stresses resulting from applied loads. Depending on fabrication method, random orientation of fibers may be either two-dimensional (2-D) or three-dimensional (3-D). Typically, the spray-up fabrication method has a 2-D random fiber orientation while the premix (or batch) fabrication method typically has a 3-D random fiber orientation. Also, many fibers extend across cracks at angles other than 90° or they may have less than the required embedment length for development of adequate bond. Therefore, only a small percentage of the fiber content may be efficient in resisting tensile or flexural stresses. Efficiency factors can be as low as 0.4 for 2-D random orientation and 0.25 for 3-D random orientation. The efficiency factor depends on fiber length and critical embedment length. Conceptually, reinforcing with fibers is an inefficient means of obtaining composite strength.

Fiber concretes are best suited for thin section shapes where correct placement of conventional reinforcement would be extremely difficult. In addition, spraying of fiber concrete accommodates the fabrication of irregularly shaped products. A substantial weight savings can be realized using relatively thin fiber concrete sections with the equivalent strength of thicker conventionally reinforced concrete sections.

Types and Properties of Fibers and Their Effect on Concrete

Steel Fibers

Steel fibers are short, discrete lengths of steel with an aspect ratio (ratio of length to diameter) from about 20 to 100, and with a variety of cross sections and profiles. Some steel fibers have hooked ends to improve resistance to pullout from a cement-based matrix (Figure 8-14).

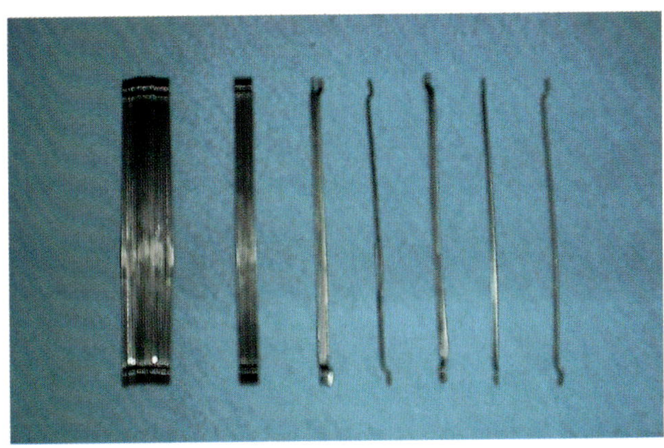

Figure 8-14. Steel fibers with hooked ends are collated into bundles to facilitate handling and mixing. During mixing the bundles separate into individual fibers.

ASTM A820, *Standard Specification for Steel Fibers for Fiber-Reinforced Concrete*, classifies four different types based on their manufacture. Type I – Cold-drawn wire fibers are the most commercially available, manufactured from drawn steel wire. Type II – Cut sheet fibers are manufactured as the name implies: steel fibers are laterally sheared off steel sheets. Type III – Melt-extracted fibers are manufactured with a relatively complicated technique where a rotating wheel is used to lift liquid metal from a molten metal surface by capillary action. The extracted molten metal is then rapidly frozen into fibers and spun off the wheel. The resulting fibers have a crescent-shaped cross section. Type IV – Other fibers. For tolerances for length, diameter, and aspect ratio, as well as minimum tensile strength and bending requirement, see ASTM A820.

Steel-fiber volumes used in concrete typically range from 0.25% to 2%. Volumes of more than 2% generally reduce workability and fiber dispersion and require special mix design or concrete placement techniques.

The compressive strength of concrete is only slightly affected by the presence of fibers. The addition of 1.5% by volume of steel fibers can increase the direct tensile strength by up to 40% and the flexural strength up to 150% in comparison to concrete without steel fibers.

Steel fibers do not affect free shrinkage. Steel fibers delay the fracture of restrained concrete during shrinkage and they improve stress relaxation induced by creep mechanisms (Altoubat and Lange 2001).

The durability of steel-fiber concrete is contingent on the same factors as conventional concrete. Freeze-thaw durability is not diminished by the addition of steel fibers provided the mixture is adjusted to accommodate the fibers. The concrete must also be air-entrained and properly consolidated during placement. Steel fibers have a relatively high modulus of elasticity (Table 8-5). Their bond to the cement matrix can be enhanced by mechanical anchorage or surface roughness. They are protected from corrosion by the alkaline environment in the cement matrix (ACI 544.1R).

Steel fibers are most commonly used in airport pavements and runway/taxi overlays. They are also used in bridge decks (Figure 8-15), industrial floors, and highway pavements. Structures exposed to high-velocity water flow

Table 8-5. Properties of Selected Fiber Types

Fiber type	Relative density (specific gravity)	Diameter, μm (0.001 in.)	Tensile strength, MPa (ksi)	Modulus elasticity, MPa (ksi)	Strain at failure, %
Steel	7.80	100–1000 (4–40)	500–2600 (70–380)	210,000 (30,000)	0.50–3.5
Glass					
E	2.54	8–15 (0.3–0.6)	2000–4000 (290–580)	72,000 (10,400)	3.0–4.8
AR	2.70	12–20 (0.5–0.8)	1500–3700 (220–540)	80,000 (11,600)	2.5–3.6
Synthetic					
Acrylic	1.18	5–17 (0.2–0.7)	200–1000 (30–145)	17,000–19,000 (2500–2800)	28–50
Aramid	1.44	10–12 (0.4–0.47)	2000–3100 (300–450)	62,000–120,000 (9000–17,000)	2–3.5
Carbon	1.90	8–9 (0.3–0.35)	1800–2600 (260–380)	230,000–380,000 (33,400–55,100)	0.5–1.5
Nylon	1.14	23 (0.9)	1000 (140)	5200 (750)	20
Polyester	1.38	10–80 (0.4–3.0)	280–1200 (40–170)	10,000–18,000 (1500–2500)	10–50
Polyethylene	0.96	25–1000 (1–40)	80–600 (11–85)	5000 (725)	12–100
Polypropylene	0.90	20–200 (0.8–8)	450–700 (65–100)	3500–5200 (500–750)	6–15
Natural					
Wood cellulose	150	25-125 (1–5)	350–2000 (51–290)	10,000–40,000 (1500–5800)	
Sisal			280–600 (40–85)	13,000–25,000 (1900–3800)	3.5
Coconut	1.12–1.15	100–400 (4–16)	120–200 (17–29)	19,000–25,000 (2800–3800)	10–25
Bamboo	1.50	50–400 (2–16)	350–500 (51–73)	33,000–40,000 (4800–5800)	
Jute	1.02–1.04	100–200 (4–8)	250–350 (36–51)	25,000–32,000 (3800–4600)	1.5–1.9
Elephant grass		425 (17)	180 (26)	4900 (710)	3.6

Adapted from PCA (1991) and ACI 544.1R-96.

have lasted about three times longer than conventional concrete alternatives. Steel fiber concrete is also used for many precast concrete applications for the improved impact resistance or toughness imparted by the fibers. In utility boxes and septic tanks, steel fibers have been used to replace conventional reinforcement. Use of steel fibers for shear reinforcement in beams is addressed in ACI 318.

Figure 8-15. Bridge deck with steel fibers.

Steel fibers are also widely used with shotcrete in thin-layer applications, especially rock-slope stabilization and tunnel linings. Silica fume and accelerators have enabled shotcrete to be placed in thicker layers. Silica fume also reduces the permeability of the shotcrete material (Morgan 1987). Steel-fiber shotcrete has been successfully applied with fiber volumes up to 2%.

Figure 8-16. Tightly bunched steel fibers are placed in a form before cement slurry is poured into this application of slurry-infiltrated steel-fiber concrete (SIFCON).

Slurry-infiltrated concrete (SIFCON) with fiber volumes up to 20% has been used since the late 1970s. Slurry-infiltrated concrete can be used to produce a component or structure with strength and ductility that far exceeds that of conventionally mixed or sprayed fiber concrete. SIFCON can be expensive and requires refinement, but it holds great potential for applications exposed to severe conditions and requiring very high strength and toughness. These applications include impact and blast-resistant structures, refractories, protective revetment, and taxiway and pavement repairs (Figure 8-16). Table 8-6 shows a SIFCON mix design.

Table 8.6. SIFCON Mix Design

Cement	1000 kg/m³ (1686 lb/yd³)
Water	330 kg/m³ (556 lb/yd³)
Siliceous sand ≤ 0.7 mm (≤0.028 in.)	860 kg/m³ (1450 lb/yd³)
Silica slurry	13 kg/m³ (1.3 lb/yd³)
High-range water reducer	35 kg/m³ (3.7 lb/yd³)
Steel fibers (about 10 Vol.-%)	800 kg/m³ (84 lb/yd³)

Glass Fibers

Initial research on glass fibers in the early 1960s used conventional borosilicate glass (E-glass) (Table 8-5) and soda-lime-silica glass fibers (A-glass). The test results showed that alkali reactivity between the E-glass fibers and the cement-paste reduced the strength of the concrete. Continued research resulted in alkali-resistant glass fibers (AR-glass) (Table 8-5), that improved long-term durability, but sources of other strength-loss trends were observed. The standard for AR-glass fibers is ASTM C1666, *Standard Specification for Alkali Resistant (AR) Glass Fiber for GFRC and Fiber-Reinforced Concrete and Cement*. One acknowledged source of strength loss was fiber embrittlement stemming from infiltration of calcium hydroxide particles and by-products of cement hydration into fiber bundles. Alkali reactivity and cement hydration are the basis for the following two widely held theories explaining strength and ductility loss, particularly in exterior glass fiber concrete:

- Alkali attack on glass-fiber surfaces reduces fiber tensile strength and, subsequently, lowers compressive strength.
- Ongoing cement hydration causes calcium hydroxide penetration of fiber bundles, thereby increasing fiber-to-matrix bond strength and embrittlement; the latter lowers tensile strength by inhibiting fiber pullout.

Fiber modifications to improve long-term durability involve (1) specially formulated chemical coatings to help combat hydration-induced embrittlement, and (2) employment of a dispersed microsilica slurry to adequately fill fiber voids, thereby reducing potential for calcium hydroxide infiltration.

A low-alkaline cement has been developed in Japan that produces no calcium hydroxide during hydration. Accelerated tests with the cement in alkali-resistant-glass fiber-reinforced concrete samples have shown greater long-term durability than previously achieved.

Metakaolin can be used in glass-fiber-reinforced concrete without significantly affecting flexural strength, strain, modulus of elasticity, and toughness. (Marikunte, Aldea, and Shah 1997).

The single largest application of glass-fiber concrete has been the manufacture of exterior building façade panels (Figure 8-17). Other applications are listed in PCA(1991).

Figure 8-17. (top) Glass-fiber-reinforced concrete panels are light and strong enough to reduce this building's structural requirements. (bottom) Spray-up fabrication made it easy to create their contoured profiles.

Synthetic Fibers

Synthetic fibers are man-made fibers developed during research and development in the petrochemical and textile industries. Fiber types that are used in portland cement concrete are: acrylic, aramid, carbon, nylon, polyester, polyethylene, and polypropylene. Table 8-5 summarizes the range of physical properties of these fibers.

Synthetic fibers can reduce plastic shrinkage and subsidence cracking and may help strengthen concrete after it cracks. Ultra-thin white topping often uses synthetic fibers for potential containment to delay pothole development. Problems associated with synthetic fibers include: (1) low fiber-to-matrix bonding; (2) inconclusive performance testing for low fiber-volume usage with polypropylene, polyethylene, polyester and nylon; (3) a low modulus of elasticity for polypropylene and polyethylene; and (4) the high cost of carbon and aramid fibers.

Polypropylene fibers (Figure 8-18), the most popular of the synthetics, are chemically inert, hydrophobic, and lightweight. They are produced as continuous cylindrical monofilaments that can be cut to specified lengths or cut as films and tapes and formed into fine fibrils of rectangular cross section (Figure 8-19).

Figure 8-18. Polypropylene fibers.

Figure 8-19. Polypropylene fibers are produced either as (left) fine fibrils with rectangular cross section or (right) cylindrical monofilament.

Used at a rate of at least 0.1 percent by volume of concrete, polypropylene fibers reduce plastic shrinkage cracking and subsidence cracking over steel reinforcement (Suprenant and Malisch 1999). The presence of polypropylene fibers in concrete may reduce settlement of aggregate particles, thus reducing capillary bleed channels. Polypropylene fibers can help reduce spalling of high-strength, low-permeability concrete exposed to fire in a moist condition.

New developments show that monofilament fibers are able to fibrillate during mixing if produced using both polypropylene and polyethylene resins. The two polymers are incompatible and tend to separate when manipulated. During the mixing process each fiber turns into a unit with several fibrils at its end. The fibrils provide better mechanical bonding than conventional monofilaments. The high number of fine fibrils also reduces plastic shrinkage cracking and may increase the ductility and toughness of the concrete (Trottier and Mahoney 2001).

Acrylic fibers are generally considered the most promising replacement for asbestos fibers. They are used in cement board and roof-shingle production, where fiber volumes of up to 3% can produce a composite with mechanical properties similar to that of an asbestos-cement composite. Acrylic-fiber concrete composites exhibit high post-cracking toughness and ductility. Although lower than that of asbestos-cement composites, acrylic-fiber-reinforced concrete's flexural strength is ample for many building applications.

Aramid fibers have high tensile strength and a high tensile modulus. Aramid fibers are two and a half times as strong as E-glass fibers and five times as strong as steel fibers. In addition to excellent strength characteristics, aramid fibers have excellent strength retention up to 160°C (320°F), dimensional stability up to 200°C (392°F), static and dynamic fatigue resistance, and creep resistance. Aramid strand is available in a wide range of diameters.

Carbon fibers were developed primarily for their high strength and elastic modulus and stiffness properties for applications within the aerospace industry. Compared with most other synthetic fibers, the manufacture of carbon fibers is expensive and this has limited their commercial development. Carbon fibers have high tensile strength and modulus of elasticity (Table 8-5). They are also inert to most chemicals. Carbon fiber is typically produced in strands that may contain up to 12,000 individual filaments. The strands are commonly spread prior to incorporation in concrete to facilitate cement matrix penetration and to maximize fiber effectiveness.

Nylon fibers are produced for use in apparel, home furnishing, industrial, and textile applications. Only two types of nylon fiber are currently marketed for use in concrete, nylon 6 and nylon 66. Nylon fibers are spun from nylon polymer and transformed through extrusion, stretching, and heating to form an oriented, crystalline, fiber structure. For concrete applications, high tenacity (high tensile strength), heat and light stable yarn is spun and subsequently cut into shorter lengths. Nylon fibers exhibit good tenacity, toughness, and elastic recovery. Nylon is hydrophilic, with moisture retention of 4.5 percent. Their use increases the water demand of concrete. While this does not affect concrete hydration or workability at low prescribed contents ranging from 0.1 to 0.2 percent by volume, increased water demand should be considered at higher fiber volume contents. This comparatively small dosage has potentially greater reinforcing value than low volumes of polypropylene or polyester fiber. Nylon is relatively inert and resistant to a wide variety of organic and inorganic materials including strong alkalis.

Synthetic fibers are also used in stucco and mortar. For this use the fibers are shorter than synthetic fibers used in concrete. Usually small amounts of 13-mm (½-in.) long alkali-resistant fibers are added to base coat plaster mixtures. They can be used in small diameter stucco and mortar pumps and spray guns. They should be added to the mixture in accordance with manufacturer's recommendation.

For further details about chemical and physical properties of synthetic fibers and properties of synthetic fiber concrete, see ACI 544.1R. ASTM C1116, *Standard Specification for Fiber-Reinforced Concrete*, classifies steel, glass, and synthetic fiber concrete and shotcrete.

The technology of interground fiber cement takes advantage of the fact that some synthetic fibers are not destroyed or pulverized in the cement finishing mill. The fibers are mixed with dry cement during grinding where they are uniformly distributed. The surface of the fibers is roughened during grinding, which offers a better mechanical bond to the cement paste (Vondran 1995).

Natural Fibers

Natural fibers were used as reinforcement long before the advent of conventional reinforced concrete. Mud bricks reinforced with straw and mortars reinforced with horsehair are just a few examples of how natural fibers were used historically as a form of reinforcement. Many natural reinforcing materials can be obtained at low cost and energy using local manpower and technical know-how. Such fibers are used in the manufacture of low-fiber-content concrete and occasionally have been used in thin-sheet concrete with high-fiber content. For typical properties of natural fibers see Table 8-5. Relevant ASTM standards include ASTM D7357, *Standard Specification for Cellulose Fibers for Fiber-Reinforced Concrete*, and D6942, *Standard Test Method for Stability of Cellulose Fibers in Alkaline Environments*.

Unprocessed Natural Fibers In the late 1960s, research on the engineering properties of natural fibers, and concrete made with these fibers was undertaken. The results indicated that these fibers can be used successfully to make thin sheets for walls and roofs. Products were made using portland cement and unprocessed natural fibers such as coconut coir, sisal, bamboo, jute, wood, and vegetable fibers. Although the concretes made with unprocessed natural fibers show good mechanical properties, they have some deficiencies in durability. Many of the natural fibers are highly susceptible to volume changes due to variations in fiber moisture content. Fiber volumetric changes that accompany variations in fiber moisture content can drastically affect the bond strength between the fiber and cement matrix.

Wood Fibers (Processed Natural Fibers). The properties of wood cellulose fibers are greatly influenced by the method by which the fibers are extracted and the refining processes involved. The process by which wood is reduced to a fibrous mass is called pulping. The kraft process is most commonly used for producing wood cellulose fibers. This process involves cooking wood chips in a solution of sodium hydroxide, sodium carbonate, and sodium sulfide. Wood cellulose fibers have relatively good mechanical properties compared to many man-made fibers such as polypropylene, polyethylene, polyester, and acrylic. Delignified cellulose fibers (lignin removed) can be produced with a tensile strength of up to approximately 2000 MPa (290 ksi) for selected grades of wood and pulping processes. Fiber tensile strength of approximately 500 MPa (73 ksi) can be routinely achieved using a chemical pulping process with the more common, less expensive grades of wood.

Multiple Fiber Systems

For a multiple fiber system, two or more fibers are blended into one system. The hybrid-fiber concrete combines macro- and microsteel fibers. A common macrofiber blended with a newly developed microfiber, which is less than 10 mm (0.4 in.) long and less than 100 micrometers (0.004 in.) in diameter, leads to a closer fiber-to-fiber spacing, which reduces microcracking and increases tensile strength. The intended applications include thin repairs and patching (Banthia and Bindiganavile 2001). A blend of steel and polypropylene fibers has also been used for some applications. This system is purported to combine the toughness and impact-resistance of steel fiber concrete with the reduced plastic cracking of polypropylene fiber concrete. For a project in the Chicago area (Wojtysiak, Borden, and Harrison 2001), a blend of 30 kg/m^3 (50 lb/yd^3) of steel fibers and 0.9 kg/m^3 (1½ lb/yd^3) of fibrillated polypropylene fibers were used for slabs on grade. The concrete with blended fibers had a lower slump compared to plain concrete but seemed to have enhanced elastic and post-elastic strength.

References

ACI Committee 318, *Building Code Requirements for Structural Concrete and Commentary*, ACI 318R-08, American Concrete Institute, Farmington Hills, Michigan, 2008, 465 pages.

ACI Committee 544, *State-of-the-Art Report on Fiber Reinforced Concrete*, ACI 544.1R-96, reapproved 2009, American Concrete Institute, Farmington Hills, Michigan, 1997, 66 pages.

Altoubat, Salah A., and Lange, David A., "Creep, Shrinkage, and Cracking of Restrained Concrete at Early Age," *ACI Materials Journal*, American Concrete Institute, Farmington Hills, Michigan, July-August 2001, pages 323 to 331.

AWS, *Structural Welding Code – Reinforcing Steel*, AWS D1.4/D1.4M:2005, American Welding Society, Miami, Florida, 2005.

Banthia, Nemkumar, and Bindiganavile, Vivek, "Repairing with Hybrid-Fiber-Reinforced Concrete," *Concrete International*, American Concrete Institute, Farmington Hills, Michigan, June 2001, pages 29 to 32.

CRSI, *CRSI Design Handbook*, Concrete Reinforcing Steel Institute, Schaumburg, Illinois, 2008.

CRSI, *Placing Reinforcing Bars*, 8th Edition, Concrete Reinforcing Steel Institute, Schaumburg, Illinois, 2009, 254 pages.

Fintel, *Handbook of Concrete Engineering*, 2nd edition, Van Nostrand Reinhold, New York, 1985, 892 pages.

Helgason, Thorsteinn; Hanson, J.M.; Somes, N.F.; Corley, W.G.; and Hognestad, Eivind, *Fatigue Strength of High-Yield Reinforcing Bars*, Research and Development Bulletin RD045, Portland Cement Association, 1976, 34 pages.

Marikunte, S.; Aldea, C.; and Shah, S., "Durability of Glass Fiber Reinforced Cement Composites: Effect of Silica Fume and Metakaolin," *Advanced Cement Based Materials*, Volume 5, Numbers 3/4, April/May 1997, pages 100 to 108.

Morgan, D.R., "Evaluation of Silica Fume Shotcrete," *Proceedings, CANMET/CSCE International Workshop on Silica Fume in Concrete*, Montreal, May 1987.

PCA, *Fiber Reinforced Concrete*, SP039, Portland Cement Association, 1991, 54 pages.

PCI, *PCI Design Handbook, Precast and Prestressed Concrete*, 7th Edition, Precast/Prestressed Concrete Institute, 2010.

PTI, *Post-Tensioning Manual*, Sixth Edition, Post-Tensioning Institute, 2006, 354 pages.

Shah, S.P.; Weiss, W.J.; and Yang, W., "Shrinkage Cracking – Can it be prevented?," *Concrete International*, American Concrete Institute, Farmington Hills, Michigan, April 1998, pages 51 to 55.

Suprenant, Bruce A., and Malisch, Ward R., "The fiber factor," *Concrete Construction*, Addison, Illinois, October 1999, pages 43 to 46.

Trottier, Jean-Francois, and Mahoney, Michael, "Innovative Synthetic Fibers," *Concrete International*, American Concrete Institute, Farmington Hills, Michigan, June 2001, pages 23 to 28.

Vondran, Gary L., "Interground Fiber Cement in the Year 2000," *Emerging Technologies Symposium on Cements for the 21st Century*, SP206, Portland Cement Association, March 1995, pages 116 to 134.

Wojtysiak, R.; Borden, K.K.; and Harrison P., *Evaluation of Fiber Reinforced Concrete for the Chicago Area – A Case Study*, 2001.

WRI, *Structural Welded Wire Reinforcement Manual of Standard Practice*, Wire Reinforcement Institute, Hartford, Connecticut, 8th Edition, July 2010, 38 pages.

Zuo, Jun, and Darwin, David, "Splice Strength of Conventional and High Relative Rib Area Bars in Normal and High Strength Concrete", *ACI Structural Journal*, Vol. 97, no. 4, July-August 2000, 12 pages.

CHAPTER 9

Properties of Concrete

Quality concrete possesses well defined and accepted principal requirements. For freshly mixed concrete, those requirements include:

Consistency – The ability to flow.

Stability – The resistance to segregation.

Uniformity – Homogeneous mixture, with evenly dispersed constituents.

Workability – Ease of placing, consolidating, and finishing.

Finishability – Ease of performing finishing operations to achieve specified surface characteristics.

For hardened concrete, they include:

Strength – Resists strain or rupture induced by external forces (compressive, flexural, tensile, torsion, and shear).

Durability – Resists weathering, chemical attack, abrasion, and other service conditions.

Appearance – Meets the desired aesthetic characteristics.

Economy – Performs as intended within a given budget.

For different applications, concrete properties need to be controlled within certain ranges. By understanding the nature and basic characteristics of concrete, fresh and hardened properties can be more readily met. The following sections discuss the properties of freshly mixed and hardened concrete.

Freshly Mixed Concrete

Freshly mixed concrete should be plastic or semifluid and generally capable of being molded by hand. A very wet concrete mixture can be molded in the sense that it can be cast in a mold, but this is not within the definition of *plastic* – that which is pliable and capable of being molded or shaped like a lump of modeling clay.

In a plastic concrete mixture, all grains of sand and particles of gravel or stone are encased and held in suspension. The ingredients are not apt to segregate during transport. When the concrete hardens, it becomes a homogeneous mixture of all the components. During placing, concrete of plastic consistency does not crumble but flows cohesively without segregation maintaining stability.

In construction practice, thin concrete members and heavily reinforced concrete members require more workable mixtures for ease of placement. A plastic mixture is required for strength and homogeneity during handling and placement. While a plastic mixture with a slump ranging from 75 mm to 150 mm (3 in. to 6 in.) is suitable for most concrete work, plasticizing admixtures may be used to make concrete more flowable in thin or heavily reinforced concrete members.

In general, fresh concrete must be capable of satisfying the following requirements:

- it must be easily mixed and transported
- it must be uniform throughout a given batch (and consistent between batches)
- it should have flow properties such that it is capable of completely filling the forms for which it was designed
- it must have the ability to be compacted without requiring an excessive amount of energy
- it must **not** segregate during transportation, placing and consolidation
- it must be capable of being finished properly (either against the forms by means of trowelling or other surface treatment)

Workability

The ease of placing, consolidating, and finishing freshly mixed concrete and the degree to which it resists segregation is called workability. Concrete should be workable, but the ingredients should not separate during transport and handling (Figure 9-1). Concrete properties related to workability include the *consistency* (flow) and *stability* (segregation resistance). Further defined flow characteristics (used in high-performance concrete mixtures) include: unconfined flowability, called the *filling ability*; and confined flowability, called the *passing ability*. *Dynamic stability* is the ability to resist separation during transport and placement. *Static stability* is the ability to maintain a uniform distribution of all mixture components after the fluid concrete has stopped moving.

Figure 9-1. Workable concrete should flow sluggishly into place without segregation.

The degree of workability required for proper placement of concrete is controlled by the placement method, type of consolidation, and type of concrete. For example, self-consolidating concrete has the unique properties of high workability without loss of stability, and allows for complex shapes and rigorous construction schedules (Sceszy and Mohler 2008). Different types of placements require different levels of workability (see Chapter 14).

Factors that influence the workability of concrete are: (1) the method and duration of transportation; (2) quantity and characteristics of cementitious materials; (3) concrete consistency (slump); (4) grading, shape, and surface texture of fine and coarse aggregates; (5) entrained air; (6) water content; (7) concrete and ambient air temperatures; and (8) admixtures. A uniform distribution of aggregate particles and the presence of entrained air significantly help control segregation and improve workability. Figure 9-2 illustrates the effect of casting temperature on the consistency, or slump, and potential workability of concrete mixtures.

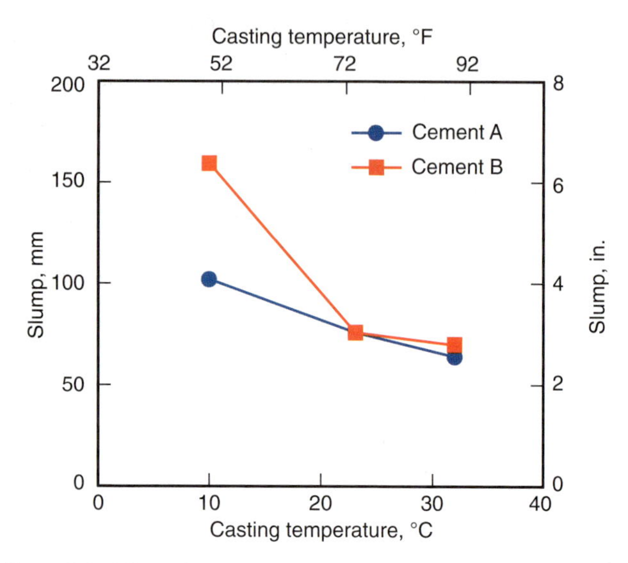

Figure 9-2. Effect of casting temperature on the slump (and relative workability) of two concretes made with different cements (Burg 1996).

Consistency. Consistency is considered a close indication of workability. The slump test, ASTM C143, *Standard Test Method for Slump of Hydraulic-Cement Concrete* (AASHTO T 119), is the most generally accepted method used to measure the consistency of concrete. The primary benefit of the slump test is that it measures the consistency from one batch of concrete to the next. However, it does not characterize the rheology or workability of a concrete mixture quantitatively. Other test methods are available for measuring consistency of concrete (see Chapter 18). A low-slump concrete has a stiff consistency (Figure 9-3). If the consistency is too dry and harsh, the concrete will be difficult to place and compact and larger aggregate particles may separate from the mixture. However, it should not be assumed that a wetter, more fluid mixture is necessarily

Figure 9-3. Concrete of a stiff consistency (low slump) is suitable for a paving or slip-form applications.

more workable. Segregation, honeycombing, and reduced hardened properties can occur if the mixture is too wet. Excessive water used to produce high slump is a primary cause of poor concrete performance, as it leads to bleeding, segregation, and increased drying shrinkage. If a finished concrete surface is to be level, uniform in appearance, and wear resistant, all batches placed in the floor must have nearly the same slump and must meet specification criteria. The consistency should be the lowest water content practical for placement using the available consolidation equipment. See Powers (1932) and Daniel (2006).

Rheology. To understand fully the concepts of flowability and stability of concrete mixtures, the science of rheology has been widely used. Rheology is the study of material deformation and flow. Rheology allows researchers, practitioners, mixture developers and others, a more scientific approach to determine the flow and workability of concrete was developed using the Bingham model (Bingham 1916).

The Bingham model describes two properties of the material, the yield stress and the viscosity (Figure 9-4). In fresh concrete, *yield stress* defines the threshold between static and fluid behavior. Consider a laborer trying to pull a come-along through a pile of freshly mixed concrete. As soon as the laborer exerts enough force on the pile of concrete, it will begin to move, indicating that the laborer has overcome the yield stress of the concrete. As soon as the stress applied to the concrete no longer exceeds the yield stress, the concrete stops moving. The *viscosity* of the concrete determines how fast it moves (rate of deformation). In order for concrete to flow without assistance, the yield stress of the freshly mixed concrete must be low enough that it can move by the effects of gravity. While concrete does not completely follow Bingham model behavior, it is sufficiently close at low shear rates to be useful in understanding the rheological behavior of concrete.

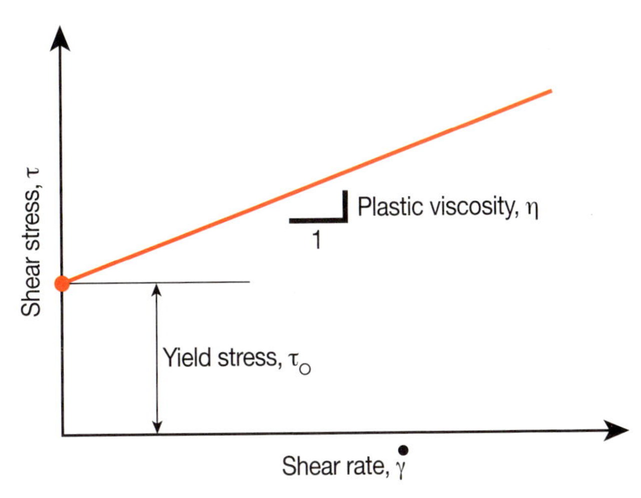

Figure 9-4. Basic Bingham behavior curve.

The Bingham model is a means to characterize concrete by two measurable parameters determined with established engineering principles. From a practical standpoint, those two parameters give contractors and producers a means to explicitly specify the fresh property performance parameters best suited for the placement application.

Bleeding and Settlement

Bleeding is the development of a layer of water at the top or surface of freshly placed concrete. It is caused by sedimentation (settlement) of solid particles (cement and aggregate) and the simultaneous upward migration of water (Figure 9-5). Some bleeding is normal and it should not diminish the quality of properly placed, finished, and cured concrete.

Figure 9-5. Bleed water on the surface of a freshly placed concrete slab.

Excessive bleeding increases the water-cement ratio near the top surface which creates a weak top layer with poor durability, particularly if finishing operations take place while bleed water is present. A water pocket or void can develop under a prematurely finished surface (which can cause a future surface delamination). Bleed water can accumulate under and alongside coarse aggregate particles (Figure 9-6). This is especially likely when differential settlement occurs between the aggregate and paste, or between the paste and reinforcement. Once the aggregate can no longer settle, the paste continues to settle allowing bleed water to rise and collect under the aggregate. Bleed-water channels also tend to migrate along the sides of coarse aggregate. This reduction of paste-aggregate bond reduces concrete strengths.

The bleeding properties of fresh concrete can be determined by two methods described in ASTM C232, *Standard Test Methods for Bleeding of Concrete* (AASHTO T 158). Because most concrete ingredients today provide concrete with a normal and acceptable level of bleeding. Bleeding is usually not a concern and bleeding tests are rarely performed (see Chapter 18). However, there are situations in which bleeding properties of concrete should be reviewed prior to construction. In some instances, lean concretes placed in very deep forms have accumulated large amounts of bleed water at the surface.

On the other hand, lack of bleed water on concrete flat work can sometimes lead to plastic shrinkage cracking or a dry surface that is difficult to finish. Some bleeding may be helpful to control plastic shrinkage cracking.

Figure 9-6. Trapped bleed water pocket left void under a coarse aggregate particle.

The bleeding rate and bleeding capacity (total settlement per unit of original concrete height) increases with initial water content, concrete height, and pressure (Figure 9-7). The accumulation of water at the surface of a concrete mixture can occur slowly by uniform seepage over the entire surface or at localized channels carrying water to

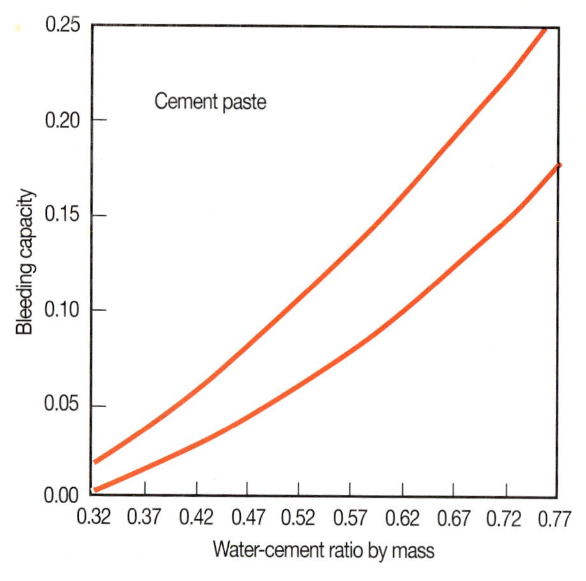

Figure 9-7. Range in relationship between bleeding rate and water-cement ratio of pastes made with normal portland cement and water. The range is attributed to different cements having different chemical composition and fineness (Steinor 1945).

the surface. Uniform seepage is referred to as normal bleeding. Water rising through the concrete in discrete paths, sometimes carrying fine particles with it, is termed *channel bleeding*. This usually occurs only in concrete mixtures with very low cement contents, high water contents, or concretes with very high bleeding properties (Figure 9-8).

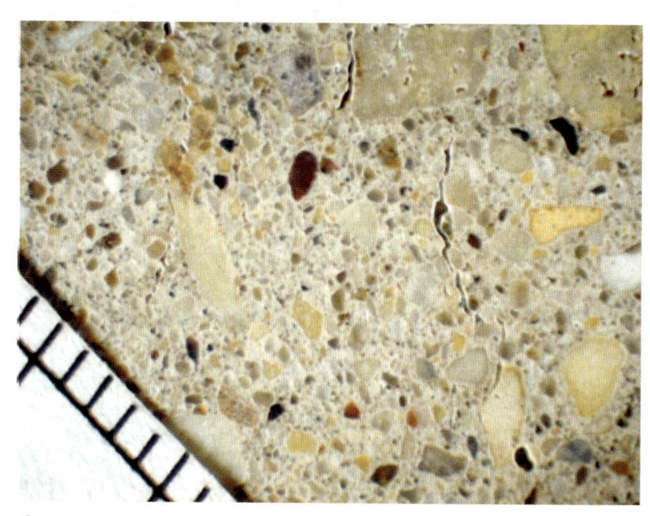

Figure 9-8. Large bleed channels may form from concrete mixtures with very low cement contents, high water contents, or other mixtures with high bleeding properties.

As bleeding proceeds, the water layer at the surface maintains the original height of the concrete sample in a vessel, assuming that there is no pronounced temperature change or evaporation. The surface subsides as the solids settle through the liquid. After evaporation of all bleed water, the hardened surface will be slightly lower than the freshly placed surface. This decrease in height from time of placement to initial set is called settlement shrinkage.

Reduced bleeding may be required for a variety of reasons including facilitating finishing operations, minimizing the formation of weak concrete at the top of lifts, reducing sand streaking in wall forms, or to stabilize the hardened volume with respect to the plastic volume of the concrete.

The most effective means of reducing bleeding in concrete include:

1. Reduce the water content, water-cementitious material ratio, and slump.
2. Increase the amount of cement resulting in a reduced water-cement ratio.
3. Use finer cementitious materials (Figure 9-9).
4. Increase the amount of fines in the sand.
5. Use or increase the amount of supplementary cementing materials such as fly ash, slag cement, or silica fume.
6. Use blended hydraulic cements.

7. Use chemical admixtures that permit reduced water-to-cementititous materials ratios or provide other means capable of reducing the bleeding of concrete.
8. Use air-entrained concrete.

For more information on bleeding see Powers (1939), Steinour (1945), and Kosmatka (2006).

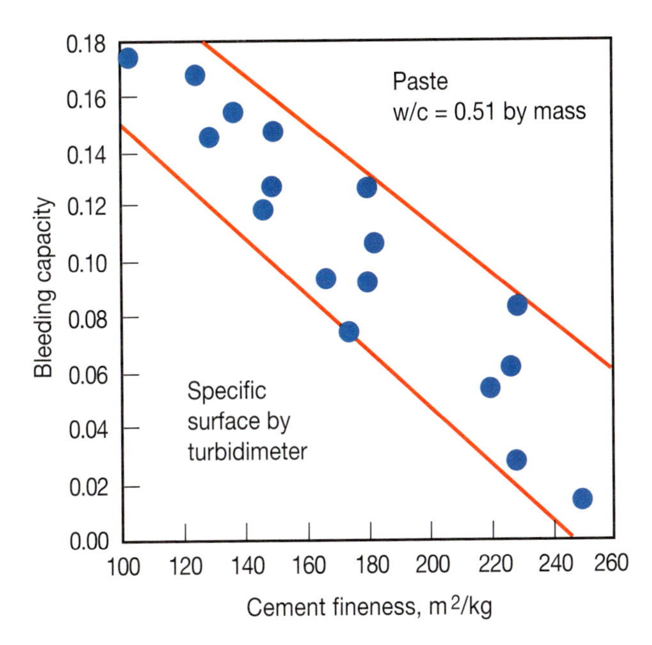

Figure 9-9. Effect of cement fineness by Wagner turbidimeter on bleeding capacity of paste. Note that Wagner values are slightly more than half of Blaine values (Steinor 1945).

Air Content

Air entrainment is recommended for nearly all exterior concretes, principally to improve freeze-thaw resistance when exposed to freezing water and deicing chemicals (see Chapter 11). A small amount of entrained air is sometimes useful for concrete that does not require freeze-thaw protection because it reduces bleeding and increases plasticity. There are also other important benefits of entrained air in both freshly mixed and hardened concrete.

Air-entrained concrete is produced using either an air-entraining cement or adding an air-entraining admixture during batching. The air-entraining admixture stabilizes bubbles formed during the mixing process, enhances the incorporation of bubbles of various sizes by lowering the surface tension of the mixing water, impedes bubble coalescence, and adheres bubbles to cement and aggregate particles (see Chapter 7).

While minimum air contents are well established for durability, there is also a reason to consider setting a maximum air content to control strength and potential surface delaminations. Entrained air lowers the compressive strength of concrete (a general rule is 5%-6% strength reduction for every percent of entrained air). When floor finishing operations include steel troweling, a maximum total air content of 3% has been established to reduce the possibility of blistering (ACI 302). This occurs because steel trowels can seal the surface and trap air pockets beneath it, especially when monolithic surface treatments are used.

The total air content developed in hardened concrete is impacted by constituent materials, mixture proportions, production and handling, delivery, placing and finishing methods, and the environment as discussed in the following sections.

Concrete Constituents. As summarized in Table 9-1, the constituent materials may have a significant effect on air content.

As cement content increases, the air content decreases for a fixed dosage of air-entraining admixture per unit of cement within the normal range of cement contents (Figure 9-10). In going from 240 kg/m³ to 360 kg/m³ of cement (400 lb/yd³ to 600 lb/yd³), the dosage rate may have to be doubled to maintain a constant air content. However, studies indicate that when dosage is increased the air-void spacing factor generally decreases. For a given air content the specific surface increases, thus improving durability.

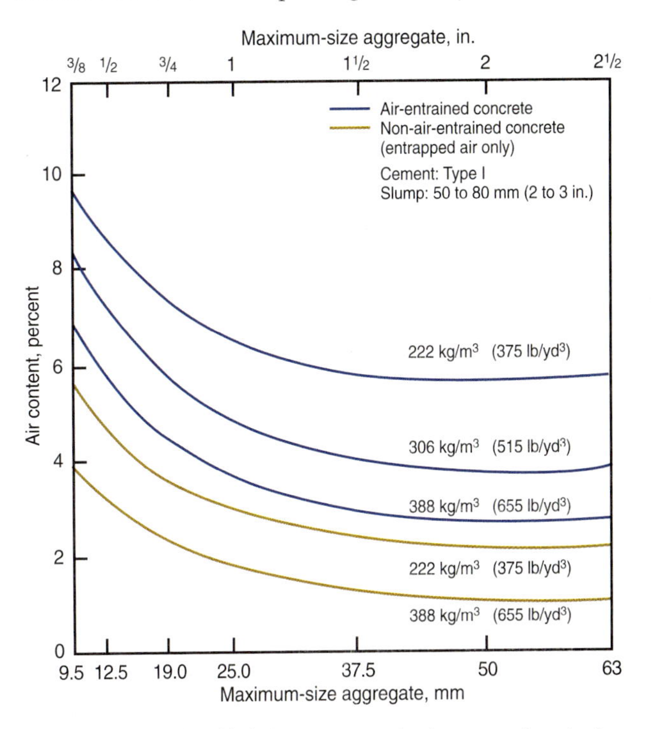

Figure 9-10. Relationship between aggregate size, cement content, and air content of concrete. The air-entraining admixture dosage per unit of cement was constant for air-entrained concrete (PCA Major Series 336).

An increase in cement fineness will result in a decrease in the amount of air entrained. Type III cement, a very finely ground material, may require twice as much air-entraining admixture as a Type I cement of normal fineness. High-alkali cements may entrain more air than low alkali

Table 9-1. Effect of Concrete Constituents on Control of Air Content in Concrete

	Characteristic/Material	Effects	Guidance
Portland cement	Alkali content	Air content increases with increase in cement alkali level. Less air-entraining admixture dosage needed for high-alkali cements. Air-void system may be more unstable with some combinations of alkali level and air-entraining admixture used.	Changes in alkali content or cement source require that air-entraining admixture dosage be adjusted. Decrease dosage as much as 40% for high-alkali cements.
Portland cement	Fineness	Decrease in air content with increased fineness of cement.	Use up to 100% more air-entraining admixture for very fine (Type III) cements. Adjust admixture if cement source or fineness changes.
Portland cement	Cement content in mixture	Decrease in air content with increase in cement content. Smaller and greater number of voids with increased cement content.	Increase air-entraining admixture dosage rate as cement content increases.
Portland cement	Contaminants	Air content may be altered by contamination of cement with finish mill oil.	Verify that cement meets ASTM C150 (AASHTO M 85) requirements on air content of test mortar.
Supplementary cementitious materials	Fly ash	Air content decreases with increase in loss on ignition (carbon content). Air-void system may be more unstable with some combinations of fly ash/cement/air-entraining admixture.	Changes in LOI or fly ash source require that air-entraining admixture dosage be adjusted. Perform "foam index" test to estimate increase in dosage. Prepare trial mixes and evaluate air-void systems.
Supplementary cementitious materials	Slag cement	Decrease in air content with increased fineness of slag cement.	Use up to 100% more air-entraining admixture for finely ground slags.
Supplementary cementitious materials	Silica fume	Decrease in air content with increase in silica fume content.	Increase air-entraining admixture dosage up to 100% for fume contents up to 10%.
Supplementary cementitious materials	Metakaolin	No apparent effect.	Adjust air-entraining admixture dosage if needed.
Chemical admixtures	Water reducers	Air content increases with increases in dosage of lignin-based materials. Spacing factors may increase when water-reducers used.	Reduce dosage of air-entraining admixture. Select formulations containing air-detraining admixtures. Prepare trial mixes and evaluate air-void systems.
Chemical admixtures	Retarders	Effects similar to water-reducers.	Adjust air-entraining admixture dosage.
Chemical admixtures	Accelerators	Minor effects on air content.	No adjustments normally needed.
Chemical admixtures	High-range water reducers (Plasticizers)	Moderate increase in air content when formulated with lignosulfonate. Spacing factors increase.	Only slight adjustments needed. No significant effect on durability.
Aggregate	Maximum size	Air content requirement decreases with increase in maximum size. Little increase over 37.5 mm (1½ in.) maximum size aggregate.	Decrease air content.
Aggregate	Sand-to-total aggregate ratio	Air content increases with increased sand content.	Decrease air-entraining admixture dosage for mixtures having higher sand contents.
Aggregate	Sand grading	Middle fractions of sand promote air-entrainment.	Monitor gradation and adjust air-entraining admixture dosage accordingly.

cements with the same amount of air-entraining material. A low-alkali cement may require 20% to 40% (occasionally up to 70%) more air-entraining admixture than a high-alkali cement to achieve an equivalent air content. Precautions are necessary when using more than one cement source in a batch plant. (Greening 1967).

The effect of fly ash on the required dosage of air-entraining admixtures can range from no effect to an increase in required dosage of up to five times the normal amount (Gebler and Klieger 1986). Class C ash typically requires less air-entraining admixture than Class F ash and tends to lose less air during mixing (Thomas 2007). Ground slags have variable effects on the required dosage rate of air-entraining admixtures. Silica fume has a marked influence on the air-entraining admixture requirement. In most cases, dosage rapidly increases with an increase in the amount of silica fume used in the concrete. Large quantities of slag and silica fume can double the dosage of air-entraining admixtures required (Whiting and Nagi 1998). The inclusion of both fly ash and silica fume in non-air-entrained concrete will generally reduce the amount of entrapped air.

Water-reducing and set-retarding admixtures generally increase the efficiency of air-entraining admixtures by 50% to 100%. Therefore, less air-entraining admixture will usually give the desired air content. Also, the time of addition of these admixtures into the mixture affects the amount of entrained air. Delayed additions generally increase air content.

Set retarders may increase the air-void spacing in concrete. Some water-reducing or set-retarding admixtures are not compatible with some air-entraining admixtures. If they are added together to the mixing water before being dispensed into the mixer, a precipitate may form. This will settle out and result in large reductions in entrained air. The fact that some individual admixtures interact in this manner does not mean that they will not be fully effective if dispensed separately into a batch of concrete.

Superplasticizers (high-range water reducers) may increase or decrease the air content of a concrete mixture. The effect is based on the admixture's chemical formulation and the slump of the concrete. Naphthalene-based superplasticizers tend to increase the air content while melamine based materials may decrease or have little effect on air content. The normal air loss in flowing concrete during mixing and transport is about 2% to 4% (Whiting and Dziedzic 1992).

Superplasticizers also affect the air-void system of hardened concrete by increasing the general size of the entrained air voids. This results in a higher-than-normal spacing factor, occasionally higher than what may be considered desirable for freeze-thaw durability. However, tests on superplasticized concrete with slightly higher spacing factors have indicated that superplasticized concretes can demonstrate good freeze-thaw durability. This may be caused by the reduced water-cement ratio often associated with superplasticized concretes.

A small quantity of calcium chloride is sometimes used in cold weather to accelerate the hardening of concrete. It can be used successfully with air-entraining admixtures if it is added separately in solution form to the mix water. However, calcium chloride is not intended for use with reinforced concrete as it will corrode the reinforcing steel. Calcium chloride will slightly increase air content. However, if calcium chloride comes in direct contact with some air-entraining admixtures, a chemical reaction can take place that makes the admixture less effective. Non-chloride accelerators may increase or decrease air content, depending upon the chemistry of the specific admixture. Generally they have little effect on air content.

Coloring agents such as carbon black usually decrease the amount of air entrained for a given amount of admixture. This is especially true for coloring materials with increasing percentages of carbon (Taylor 1948).

The size of coarse aggregate has a pronounced effect on the air content of both air-entrained and non-air-entrained concrete, as shown in Figure 9-10. There is little change in air content when the size of aggregate is increased above 37.5 mm (1.5 in.).

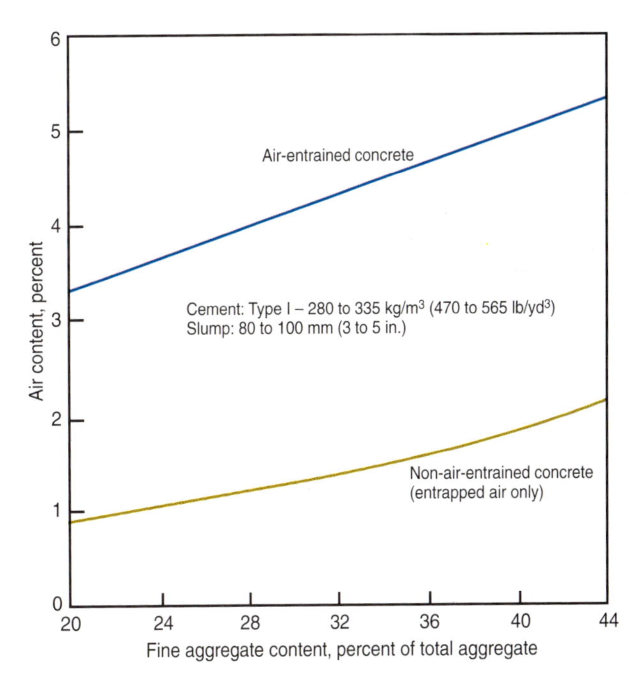

Figure 9-11. Relationship between percentage of fine aggregate and air content of concrete (PCA Major Series 336).

The fine-aggregate content of a mixture affects the percentage of entrained air. As shown in Figure 9-11, increasing the amount of fine aggregate causes more air to be entrained for a given amount of air-entraining cement or admixture (more air is also entrapped in non-air-entrained concrete).

Fine-aggregate particles passing the 600 µm to 150 µm (No. 30 to No. 100) sieves entrap more air than either very fine or coarser particles. Appreciable amounts of material passing the 150 µm (No. 100) sieve will result in a significant reduction of entrained air.

Fine aggregates from different sources may entrap different amounts of air even though they have identical gradations. This may be due to differences in shape and surface texture or as a result of contamination by organic materials.

The mixing water used may also affect air content. Algae-contaminated water increases air content. Highly alkaline wash water from truck mixers can affect air contents. The effect of water hardness in most municipal water supplies is generally insignificant. Very hard water from wells used in many communities, may decrease the air content in concrete.

Mixture Design. The concrete mixture design's effect on the air content is summarized in Table 9-2.

An increase in the mixing water makes more water available for the generation of air bubbles, thereby increasing the air content as slumps increase up to about 150 mm or 175 mm (6 in. or 7 in.). An increase in the water-cement ratio from 0.4 to 1.0 can increase the air content by 4%. A portion of the air increase is due to the relationship between slump and air content. Air content increases with slump even when the water-cement ratio is held constant. The spacing factor, $\bar{L}$, of the air-void system also increases. That is, the voids become coarser at higher water-cement ratios, thereby reducing concrete freeze-thaw durability (Stark 1986).

The addition of 5 kg of water per cubic meter of concrete (8.4 lb/yd^3) can increase the slump by 25mm (1 in.). A 25-mm (1-in.) increase in slump increases the air content by approximately 0.5% to 1% for concretes with a low-to-moderate slump and constant air-entraining admixture dosage.

This approximation is greatly affected by concrete temperature, slump, and the type and amount of cement and admixtures present in the concrete. A low slump concrete with a high dosage of water-reducing and air-entraining admixtures can undergo large increases in slump and air content with a small addition of water. Alternatively, a very fluid concrete mixture with a 200-mm to 250-mm (8-in. to 10-in.) slump may lose air with the addition of water.

Production Procedures. The effect of the production procedures on the air content is summarized in Table 9-3.

Mixing action is one of the most important factors in the production of entrained air in concrete. Uniform distribution of entrained air voids is essential to produce scale resistant concrete. Non-uniformity might result from inadequate dispersion of the entrained air during short mixing periods. In production of ready mixed concrete, it is especially important that adequate and consistent mixing be maintained at all times.

The amount of entrained air varies with the type and condition of the mixer, the amount of concrete being mixed, and the rate and duration of mixing. The amount of air entrained in a given mixture will decrease appreciably as the mixer blades become worn, or if hardened concrete is allowed to accumulate in the drum or on the blades. Because of differences in mixing action and time, concretes made in a stationary mixer and those made in a transit mixer may differ significantly in amounts of air entrained. The air content may increase or decrease when the size of the batch departs significantly from the rated capacity of the mixer. Little air is entrained in very small batches in a large mixer. However, the air content increases as the mixer capacity is approached.

Figure 9-12 shows the effect of mixing speed and duration of mixing on the air content of freshly mixed concretes made in a transit mixer. Generally, more air is entrained as the speed of mixing is increased up to about 20 rpm, beyond which air entrainment decreases. In the tests from which the data in Figure 9-12 were derived, the air content reached an upper limit during mixing and a gradual decrease in air content occurred with prolonged mixing. Mixing time and speed will have different effects on the air content of different mixtures. Significant amounts of air can be lost during mixing with certain types of mixing equipment and mixture proportions.

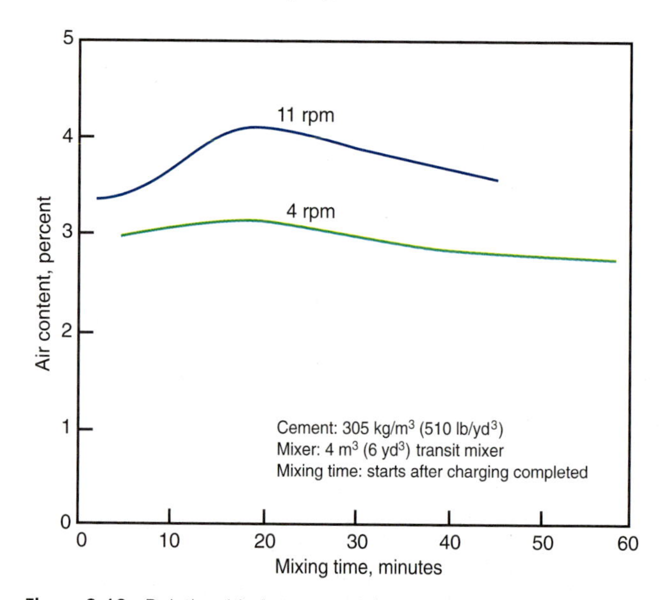

Figure 9-12. Relationship between mixing time and air content of concrete (PCA Major Series 336).

Table 9-2. Effect of Concrete Mixture Design on Control of Air Content in Concrete

	Characteristic/Material	Effects	Guidance
Mix water and slump	Water chemistry	Very hard water reduces air content. Batching of admixture into concrete wash water decreases air. Algae growth may increase air.	Increase air entrainer dosage. Avoid batching into wash water.
	Water-to-cement ratio	Air content increases with increased water to cement ratio.	Decrease air-entraining admixture dosage as water to cement ratio increases.
	Slump	Air increases with slumps up to about 150 mm (6 in.). Air decreases with very high slumps. Difficult to entrain air in low-slump concretes.	Adjust air-entraining admixture dosages for slump. Avoid addition of water to achieve high-slump concrete. Use additional air-entraining admixture; up to ten times normal dosage.

Table 9-3. Effect of Production Procedures on Control of Air Content in Concrete

	Procedure/Variable	Effects	Guidance
Production procedures	Batching sequence	Simultaneous batching lowers air content. Cement-first raises air content.	Add air-entraining admixture with initial water or on sand.
	Mixer capacity	Air increases as capacity is approached.	Run mixer close to full capacity. Avoid overloading.
	Mixing time	Central mixers: air content increases up to 90 seconds of mixing. Truck mixers: air content increases with mixing. Short mixing periods (30 seconds) reduce air content and adversely affect air-void system.	Establish optimum mixing time for particular mixer. Avoid overmixing. Establish optimum mixing time (about 60 seconds).
	Mixing speed	Air content gradually increases up to approximately 20 rpm. Air may decrease at higher mixing speeds.	Follow truck mixer manufacturer recommendations. Maintain blades and clean truck mixer.
	Admixture metering	Accuracy and reliability of metering system will affect uniformity of air content.	Avoid manual-dispensing or gravity-feed systems and timers. Positive-displacement pumps interlocked with batching system are preferred.

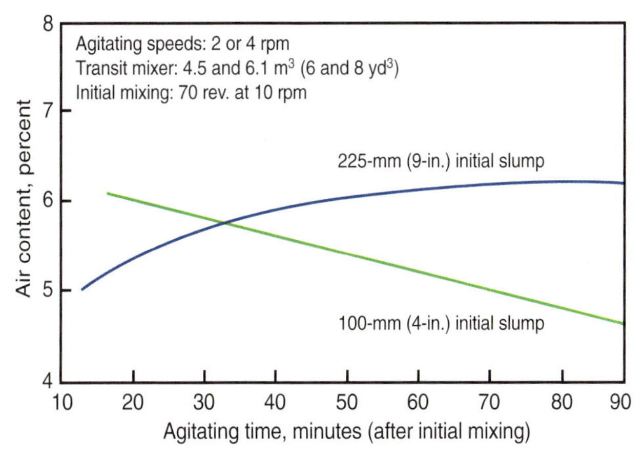

Figure 9-13. Relationship between agitating time, air content, and slump of concrete (PCA Major Series 336).

Figure 9-13 shows the effect of continued mixer agitation on air content. The changes in air content with prolonged agitation can be explained by the relationship between slump and air content. For high-slump concretes, the air content increases with continued agitation as the slump decreases to about 150 mm or 175 mm (6 in. or 7 in.). Prolonged agitation will decrease slump further and will also decrease air content. For initial slumps lower than 150 mm (6 in.), both the air content and slump decrease with continued agitation. When concrete is retempered (the addition of water and remixing to restore original slump), the air content is increased. However, after 4 hours, retempering is ineffective in increasing air content and may cause clustering of air bubbles (Kozikowski and others 2005). Prolonged mixing or agitation of concrete is accompanied by a progressive reduction in slump.

Table 9-4. Effect of Transportation and Delivery on Control of Air Content in Concrete

Transport and delivery	Transport and delivery	Some air (1% to 2%) normally lost during transport. Loss of air in nonagitating equipment is slightly higher.	Normal retempering with water to restore slump will restore air. If necessary, retemper with air-entraining admixture to restore air. Dramatic loss in air may be due to factors other than transport.
	Haul time and agitation	Long hauls, even without agitation, reduce air, especially in hot weather.	Optimize delivery schedules. Maintain concrete temperature in recommended range.
	Retempering	Regains some of the lost air. Does not usually affect the air-void system. Retempering with air-entraining admixtures restores the air-void system. May cause clustering of air bubbles.	Retemper only enough to restore workability. Avoid addition of excess water. Higher admixture dosage is needed for jobsite admixture additions.

Transportation and Delivery. The effect of transportation and delivery of concrete on the air content is summarized in Table 9-4.

Generally, some air, approximately 1% to 2%, is lost during transportation of concrete from the mixer to the jobsite. The stability of the air content during transport is influenced by several variables including concrete ingredients, haul time, amount of agitation or vibration during transport, temperature, slump, and amount of retempering.

Placement and Consolidation. The effect of placing techniques and internal vibration on air content is summarized in Table 9-5.

Once at the jobsite, the concrete air content remains essentially constant during handling by chute discharge, wheelbarrow, power buggy, and shovel. However, concrete pumping, crane and bucket, and conveyor-belt handling can cause some loss of air, especially with high-air-content mixtures. Pumping concrete can cause a loss of up to 3% of air (Whiting and Nagi 1998).

The effect of slump and vibration on the air content of concrete is shown in Figure 9-14. For a constant amount of air-entraining admixture, air content increases as slump increases up to about 150 mm or 175 mm (6 in. or 7 in.). Beyond that, air content begins to decrease with further increases in slump. At all slumps, however, even 15 seconds of vibration (ACI 309) will cause a considerable reduction in air content. Prolonged vibration of concrete should be avoided.

The greater the slump, air content, and vibration time, the larger the percentage of reduction in air content during vibration (Figure 9-14). However, if vibration is properly applied, little of the intentionally entrained air is lost. The air lost during handling and moderate vibration consists mostly of the larger bubbles. These are usually undesirable from the standpoint of strength. While the average size of the air voids is reduced, the air-void spacing factor remains relatively constant.

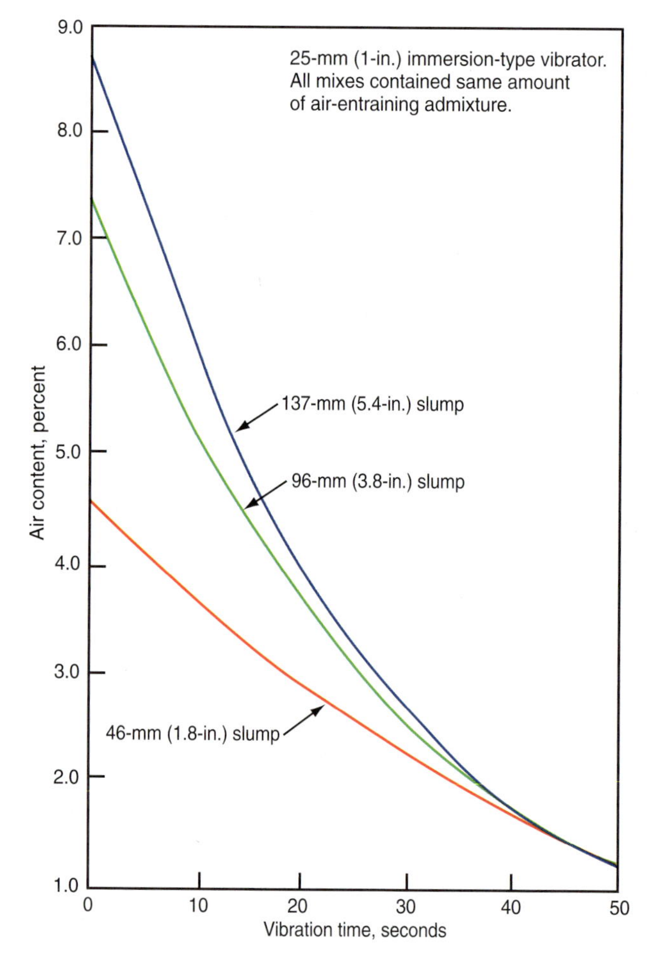

Figure 9-14. Relationship between slump, duration of vibration, and air content of concrete (Brewster 1949).

Internal vibrators reduce air content more than external vibrators. The air loss due to vibration increases as the volume of concrete is reduced or as the vibration frequency is significantly increased. Lower vibration frequencies (8000 vpm) have less effect on spacing factors and air contents than high vibration frequencies (14,000 vpm). High frequencies can significantly increase spacing factors and decrease air contents after 20 seconds of vibration (Brewster 1949 and Stark 1986).

Table 9-5. Effect of Placement Techniques and Internal Vibration on Control of Air Content in Concrete

	Procedure/Variable	Effects	Guidance
Placement techniques	Belt conveyors	Reduces air content by an average of 1%.	Avoid long conveyed distance if possible. Reduce the free-falling effect at the end of conveyor.
	Pumping	Reduction in air content ranges from 2% to 3%. Does not significantly affect air-void system. Minimum effect on freeze-thaw resistance.	Use of proper mix design provides a stable air-void system. Avoid high-slump, high-air-content concrete. Keep pumping pressure as low as possible. Use loop in descending pump line.
	Shotcrete	Generally reduces air content in wet-process shotcrete.	Air content of mixture should be at high end of target zone.

Specified air contents and uniform air void distributions can be achieved in pavement construction by operating paving machine speeds at 1.22 meters/min. to 1.88 meters/min. (4 ft/min to 6 ft/min) and by using vibrator frequencies of 5,000 vibrations/min to 8,000 vibrations/min. The most uniform distribution of air voids throughout the depth of concrete, in and out of the vibrator trails, is obtained with the combination of a vibrator frequency of approximately 5,000 vibrations per minute and a slipform paving machine forward track speeds of 1.22 meters per minute (4 feet per minute). Higher frequency speeds, singularly or in combination can result in discontinuities and lack of required air content in the upper portion of the concrete pavement. This in turn provides a greater opportunity for water and salt to enter the pavement and reduce the durability and life of the pavement (Cable and others 2000).

Finishing and Environment. The effect of finishing and environment on air content is summarized in Table 9-6.

Proper screeding, floating, and general finishing practices should not affect the air content. McNeal and Gay (1996) and Falconi (1996) demonstrated that the sequence and timing of finishing and curing operations are critical to surface durability. Overfinishing (excessive finishing) may reduce the amount of entrained air in the surface region of slabs—thus making the concrete surface vulnerable to scaling. However, as shown in Figure 9-15, early finishing does not necessarily affect scale resistance unless bleed water is present (Pinto and Hover 2001). Concrete to be exposed to deicers should never be steel troweled.

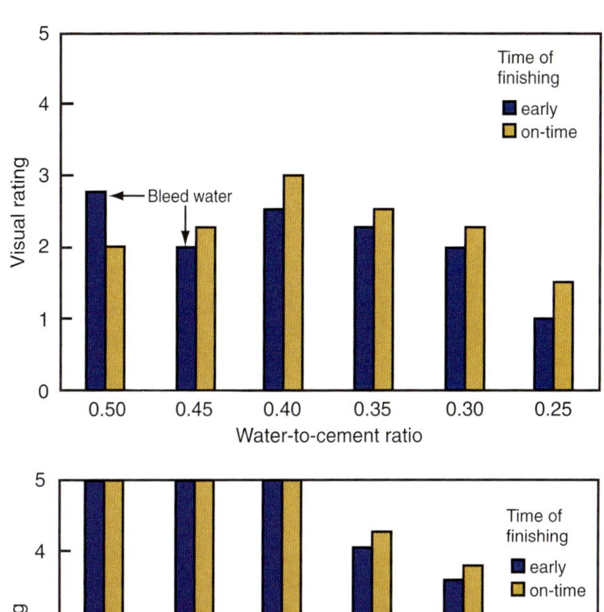

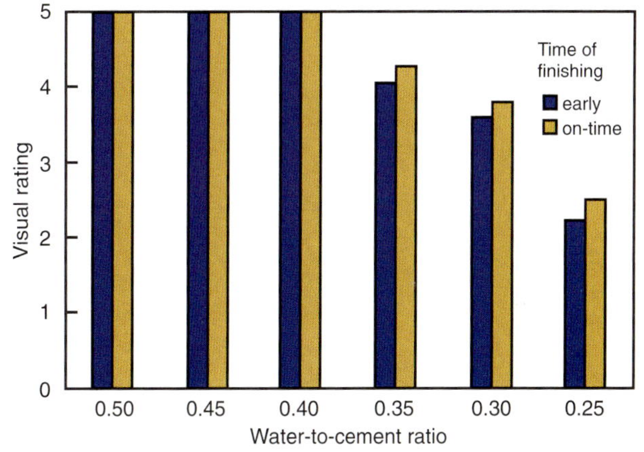

Figure 9-15. Effect of early finishing – magnesium floating 20 minutes after casting – on scale resistance for: (top) 6% air entrained concrete; (bottom) non-air-entrained concrete.

Table 9-6. Effect of Finishing and Environment on Control of Air Content in Concrete

Finishing and environment	Internal vibration	Air content decreases under prolonged vibration or at high frequencies. Proper vibration does not influence the air-void system.	Do not overvibrate. Avoid high-frequency vibrators (greater than 10,000 vpm). Avoid multiple passes of vibratory screeds. Closely spaced vibrator insertion is recommended for better consolidation.
	Finishing	Air content reduced in surface layer by excessive finishing.	Avoid finishing with bleed water still on surface. Avoid overfinishing. Do not sprinkle water on surface prior to finishing. Do not steel trowel exterior slabs.
	Temperature	Air content decreases with increase in temperature. Changes in temperature do not significantly affect spacing factors.	Increase air-entraining admixture dosage as temperature increases.

Temperature of the concrete affects air content, as shown in Figure 9-16. Less air is entrained as the temperature of the concrete increases, particularly as slump is increased. This effect is especially important during hot-weather concreting when the concrete might be quite warm. A decrease in air content can be offset, when necessary, by increasing the quantity of air-entraining admixture.

During cold-weather concreting, the air-entraining admixture may lose some of its effectiveness if hot mix water is used during batching. To offset this loss, these admixtures should be added to the batch after the temperature of the concrete ingredients have equalized.

Although increased concrete temperature during mixing generally reduces air volume, the spacing factor and specific surface are only slightly affected.

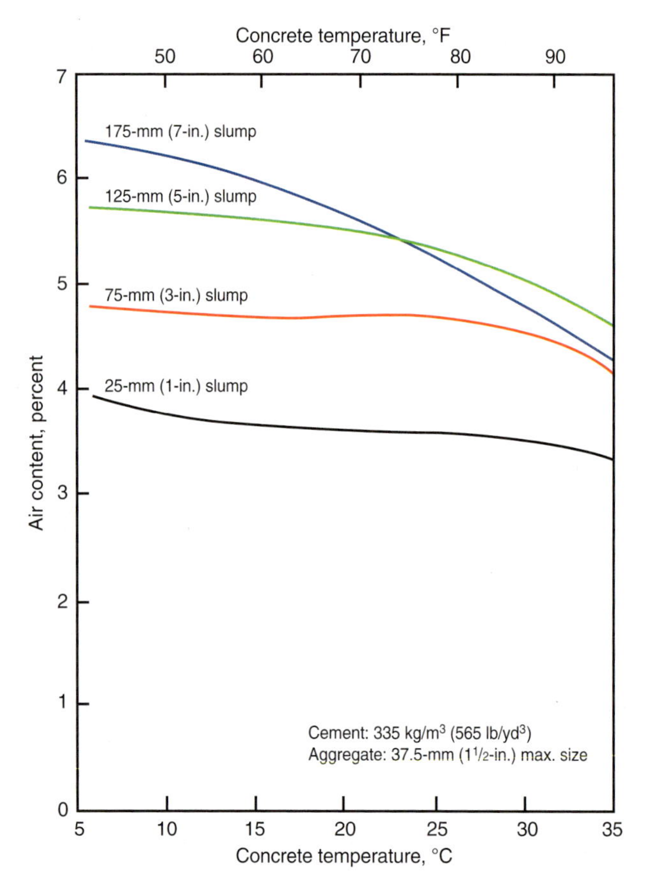

Fig. 9-16. Relationship between temperature, slump, and air content of concrete (PCA Major Series 336 and Lerch 1960).

Uniformity

Uniformity is a measure of the homogeneity of the concrete. This measurement includes within-batch uniformity and between batching of concrete mixtures. Samples of fresh concrete collected at the point of discharge are tested on site to determine properties such as: slump, temperature, air content, unit weight, and yield. Sampling of fresh concrete is performed in accordance with ASTM C172, *Standard Practice for Sampling Freshly Mixed Concrete.*

Test specimens may be cast for later testing of strength and other hardened concrete properties. The results of these tests are used to determine that the concrete meets specification requirements. These results also provide a measure of the uniformity of the concrete both between batches and, when necessary, within a batch.

ASTM C94, *Standard Specification for Ready-Mixed Concrete*, covers criteria for determining the within-batch uniformity of concrete. If the within-batch uniformity is low, this is indicative of inadequate or inefficient mixing. Samples of concrete are taken at two locations within the batch to represent the first and last portions on discharge. The samples are tested separately for density, air, slump, and strength and the difference between the test results must be less than the requirements of the specification. For example, if the average of the slump for the two samples is 90 mm (3.5 in.), the difference between the measured individual values cannot vary more than 25 mm (1 in.).

Consolidation. Uniformity of concrete is typically achieved by consolidation. Vibration sets into motion the particles in freshly mixed concrete, reducing friction between them and giving the mixture the mobile qualities of a thick fluid. The vibratory action permits use of a stiffer mixture containing a larger proportion of coarse aggregate and a smaller proportion of fine aggregate. The larger the maximum size aggregate in concrete with a well-graded aggregate, the less volume there is to fill with paste and the less aggregate surface area there is to coat with paste. Thus less water and cement are needed. Concrete with an optimally graded aggregate will be easier to consolidate and place. Consolidation of coarser as well as stiffer mixtures results in improved quality and economy. On the other hand, poor consolidation results in porous, weak concrete (Figure 9-17) with poor durability (Figure 9-18).

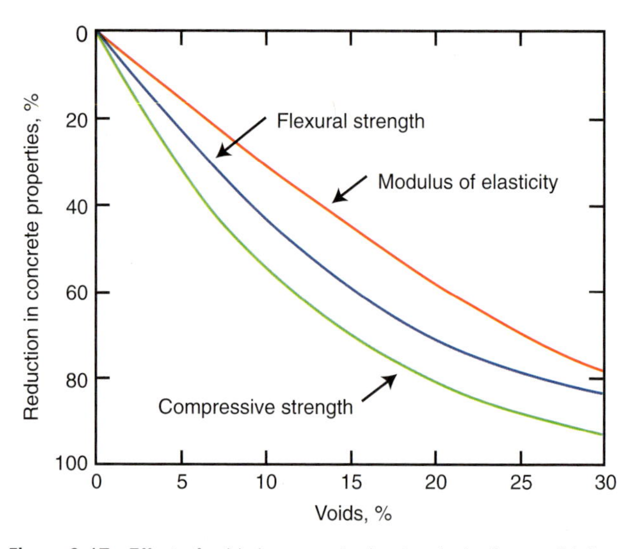

Figure 9-17. Effect of voids in concrete due to a lack of consolidation on modulus of elasticity, compressive strength, and flexural strength.

Figure 9-18. Poor consolidation can result in early corrosion of reinforcing steel and low compressive strength.

Mechanical vibration has many advantages. Vibrators make it possible to economically place mixtures that are impractical to consolidate by hand. For more information on consolidation see Chapter 14.

Hydration, Setting, and Hardening

The binding quality of portland cement paste is due to the chemical reaction between the cement and water, called hydration. As discussed in Chapter 3, portland cement is not a simple chemical compound, it is a mixture of many compounds. The two calcium silicates, which constitute about 75% of the weight of portland cement, react with water to form new compounds: calcium hydroxide and calcium silicate hydrate. The latter is by far the most important cementing component in concrete. The engineering properties of concrete – setting and hardening, strength, and dimensional stability – depend primarily on calcium silicate hydrate. It is the heart of concrete.

The chemical composition of calcium silicate hydrate is somewhat variable, but it contains lime (CaO) and silicate (SiO_2) in a ratio on the order of 3 to 2. The surface area of calcium silicate hydrate is around 300 m^2/g. In hardened cement paste, the calcium silicate hydrate forms dense, bonded aggregations between the other crystalline phases and the remaining unhydrated cement grains. These aggregates also adhere to grains of sand and to pieces of coarse aggregate, cementing all materials together (Copeland and Schulz 1962).

As concrete hardens, its gross volume remains almost unchanged, but hardened concrete contains pores filled with water and air that have no strength. The strength resides in the solid part of the paste, mostly in the calcium silicate hydrate and crystalline compounds.

The less porous the cement paste, the stronger the concrete. When mixing concrete, therefore, no more water than is absolutely necessary to make the concrete plastic and workable should be used. Even then, the water used is usually more than is required for complete hydration of the cement. About 0.4 times as much water (by mass) as cement is needed to completely hydrate cement (Powers 1948 and 1949). However, complete hydration is rare in field concrete placements due to a lack of moisture and the long time (decades) required to achieve complete hydration.

Knowledge of the amount of heat released as cement hydrates can be useful in planning construction. In winter, the heat of hydration will help protect the concrete against damage from freezing temperatures. The heat may be harmful, however, in massive structures such as dams because it may produce undesirable temperature differentials.

Knowledge of the rate of reaction between cement and water is important because it determines the rate of hardening. The initial reaction must be slow enough to allow time for the concrete to be transported and placed. However, rapid hardening is typically desirable once the concrete has been placed and finished. Gypsum, added at the cement mill when clinker is ground, acts as a regulator of the initial rate of setting of portland cement. Other factors that influence the rate of hydration include cement fineness, admixtures, amount of water added, and temperature of the materials at the time of mixing. Figure 9-19 illustrates the setting properties of a concrete mixture at different temperatures.

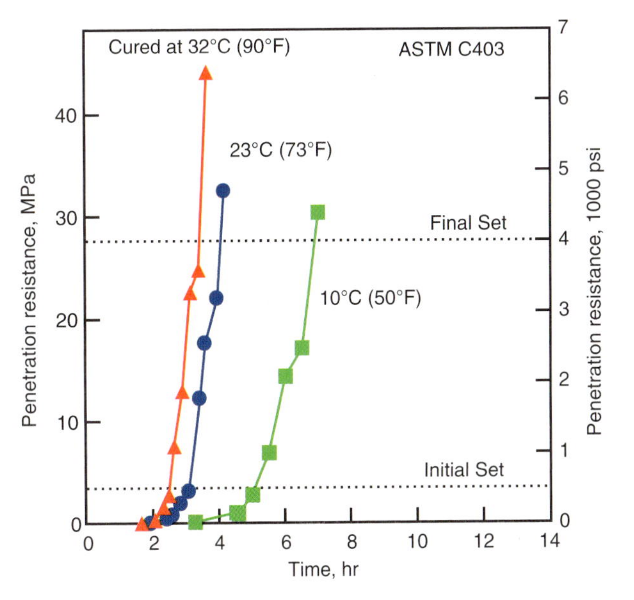

Figure 9-19. Initial and final set times for a concrete mixture at different temperatures (Burg 1996).

The setting and hardening of portland cement can be explained using a simple model showing unhydrated cement grains dispersed in water (Figure 9-20). In this example, time starts when the water is first added to the cement. Upon the addition of water, a chemical reaction occurs between the water and cement – the reaction is called hydration. The solid products resulting from hydration occupy a greater volume than the original cementitious materials and consequently some of the space between the cement grains is filled in. Eventually the hydration products will connect adjacent grains and a continuous solid network is formed. This is referred to as initial set.

Between the addition of water and just before initial set occurs the paste has little rigidity. If the rigidity of the paste was plotted against time – there would be only a small increase in stiffness during this period. This is referred to as the dormant period. During the dormant period the paste is still plastic and the concrete can still be handled and placed. As the cement continues to hydrate, more hydration products are formed and the solid matrix becomes more dense and rigid. This period is called the setting or transition period as it represents the period during which the paste transforms from a fluid to a solid. Eventually the paste can be considered a rigid and solid material with mechanical properties such as strength and stiffness.

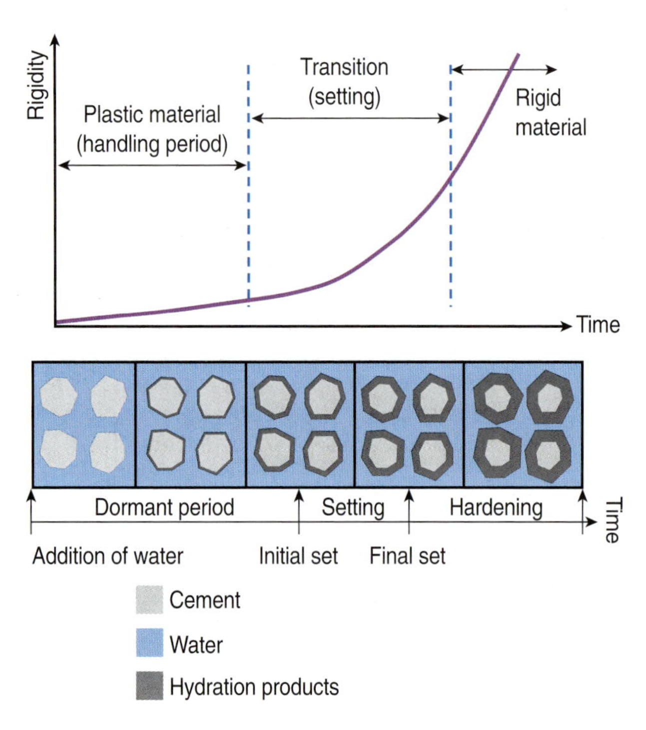

Figure 9-20. Setting of concrete (Adapted from Young and others 1998).

Setting times are determined in accordance with ASTM C403, *Standard Test Method for Time of Setting of Concrete Mixtures by Penetration Resistance*. The setting time is determined by measuring the resistance of a sieved mortar sample to the penetration of a standard needle. Initial and final set are defined as a penetration resistance of 3.5 MPa and 27.5 MPa (500 psi and 4000 psi), respectively.

The setting time of concrete is affected by the type and amount of portland cement used, the type and level of supplementary cementing materials, the presence of set modifying admixtures, the water-to-cementitious materials ratio and the temperature of the concrete.

Hardened Concrete

The following sections will discuss the properties of hardened concrete including: curing, drying rate, strength, density, permeability and watertightness, volume stability and crack control, durability, and aesthetics in more detail.

Curing

Increase in strength with age continues provided (1) unhydrated cement is still present, (2) the concrete remains moist or has a relative humidity above approximately 80% (Powers 1948), (3) the concrete temperature remains favorable, and (4) sufficient space is available for hydration products to form. When the relative humidity within the concrete drops to about 80% or the temperature of the concrete drops below 10°C (14°F), hydration and strength gain virtually stop. Figure 9-21 illustrates the relationship between strength gain and curing temperature.

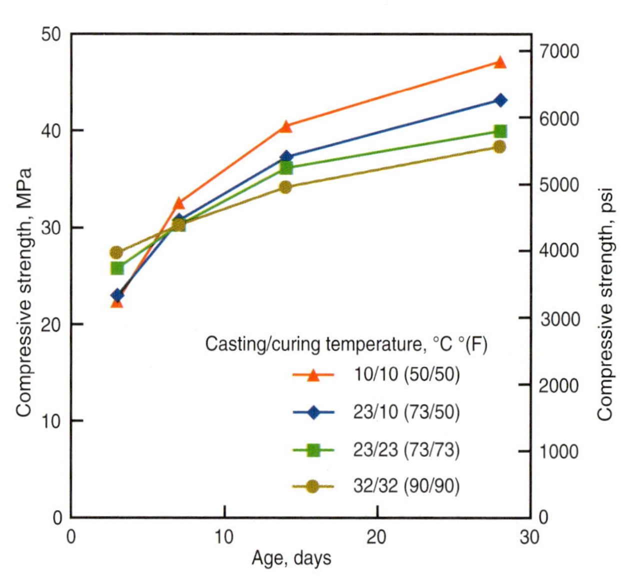

Figure 9-21. Effect of casting and curing temperature on strength development. Note that cooler temperatures result in lower early strength and higher later strength (Burg 1996).

If concrete is resaturated after a drying period, hydration is resumed and strength will again increase. However, it is best to moist-cure concrete continuously from the time it is placed until it has attained the desired quality. Once concrete has dried it is difficult to resaturate. Figure 9-22 illustrates the long-term strength gain of concrete in an outdoor exposure. Outdoor exposures often continue to provide moisture through ground contact and rainfall. Indoor concretes often dry out after curing and do not continue to gain strength. See Chapter 15 for more information on curing concrete.

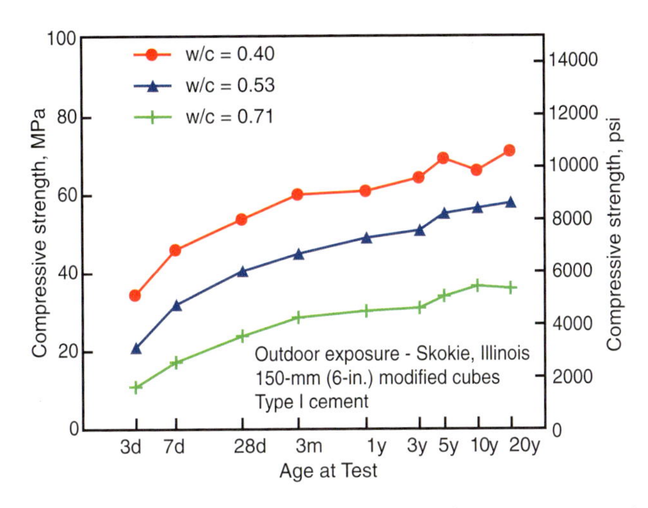

Figure 9-22. Concrete strength gain versus time for concrete exposed to outdoor conditions. Concrete continues to gain strength for many years when moisture is provided by rainfall and other environmental sources (Wood 1992).

Drying Rate of Concrete

Knowledge of the rate of drying is helpful in understanding the properties or physical condition of concrete. Concrete must continue to hold enough moisture throughout the curing period for the cement to hydrate to the extent that desired properties are achieved. Freshly cast concrete usually has an abundance of water, but as drying progresses from the surface inward, strength gain will continue at each depth only as long as the relative humidity at that point remains above about 80%.

During the first stage of drying, liquid water is present at the surface and evaporates into the air over the concrete (Figure 9-23). The rate of evaporation at the surface depends upon temperature, relative humidity, and air flow over the surface. Warm, dry, rapidly moving air will cause faster evaporation than cool, stagnant air. As the liquid water evaporates, it is replenished with water from within the body of the concrete. As liquid water moves from within the body of the concrete to replace water that has evaporated at the surface, the concrete must shrink to make up the volume of water that has left. If the rate of evaporation is very high, the concrete may shrink excessively before the cement paste has developed much strength. This is the cause of plastic shrinkage cracking that may occur within the first few hours after the concrete is placed.

A common example is the surface of a concrete floor that has not had sufficient moist curing. Because it has dried quickly, concrete at the surface is weak and traffic on it creates dusting. Also, when concrete dries, it shrinks as it loses water. Drying shrinkage is a primary cause of cracking, and the width of cracks is a function of the degree of drying, spacing or frequency of cracks, and the

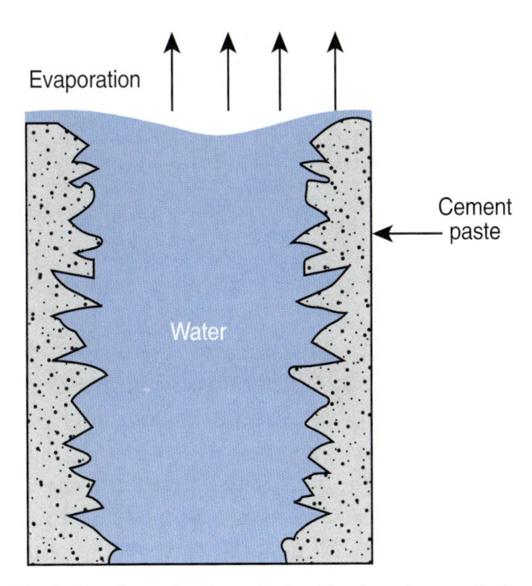

Figure 9-23. Drying Stage 1 – Pores in freshly placed concrete (or concrete that has been re-wetted) are saturated with liquid water and drying begins by evaporation from the exposed surface (adapted from Scherer 1990).

age at which the cracks occur. While the surface of a concrete element will dry quite rapidly, it takes a much longer time for concrete in the interior to dry.

When the concrete can no longer shrink to accommodate the volume lost due to water evaporation, the second stage of drying begins (Figure 9-24). Liquid water recedes from the exposed surface of the concrete into the pores. Within each pore, water clings to the sidewalls and forms a curved surface called the meniscus. At the surface of the concrete, water evaporates from the meniscus in each pore into the air over the concrete. Therefore, the rate of evaporation still depends mostly on the temperature, relative humidity, and air flow over the concrete surface. At this point, water still fills the pore structure of the concrete.

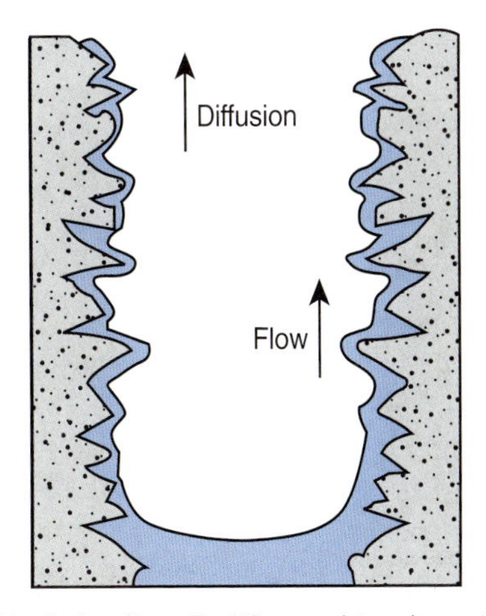

Figure 9-24. Drying Stage 2 – When moisture has retreated below the surface, movement depends on fluid flow along the surface of pores and evaporation into the pores (adapted from Scherer 1990).

There are continuous paths for liquid water to flow from within the body of concrete to the partially filled pores at the surface where the water can evaporate. The surface may appear to be dry, but the concrete is just beginning to dry in a very thin layer. The rate of drying during this period steadily decreases.

The third stage of drying begins when enough water has evaporated from just below the surface that the pores are no longer continuously filled with liquid (Figure 9-25). Pockets of liquid water exist but moisture must now move by vapor diffusion within the body of the concrete before arriving at the surface where it can evaporate. This stage is called the second falling rate period because the rate of drying continuously decreases over time and is slower than the previous stage of drying (Figure 9-26). The rate of drying depends less on temperature, relative humidity, and air flow above the concrete surface because moisture must evaporate and diffuse within the body of the concrete before arriving at the surface. The rate of drying during this stage is determined by the quality of the cement paste: low water-cement ratio cement paste offers more resistance to vapor diffusion than high water-cement ratio paste. Concrete made with a water-cement ratio, greater than approximately 0.65, will have a continuously connected capillary pore system. In these concretes moisture vapor moves with much less resistance than concretes made with lower water cement ratios.

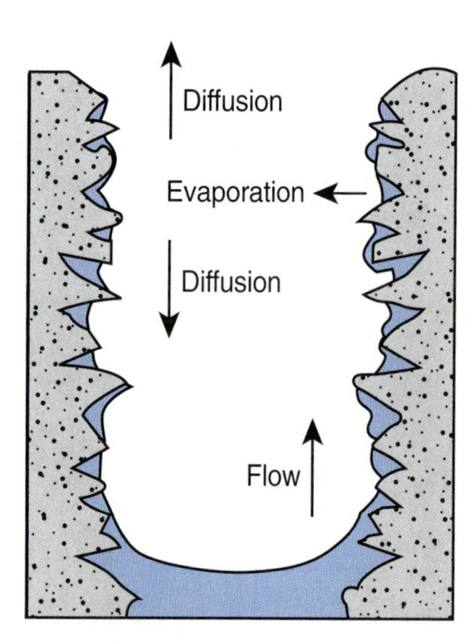

Figure 9-25. Drying Stage 3 – When moisture is no longer continuously wetting the surface of pores, moisture must evaporate within the body of the paste and diffuse toward the surface (adapted from Scherer 1990).

The moisture content of concrete depends on the concrete's constituents, original water content, drying conditions, and the size of the concrete element (Hedenblad 1996 and 1997). Size and shape of a concrete member

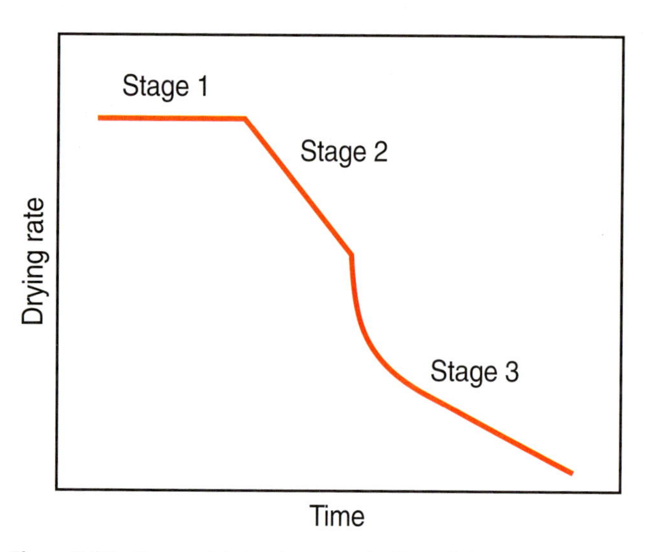

Figure 9-26. Stages of drying for concrete. Stage 1 has a constant rate and depends on air movement and relative humidity over the slab while Stages 2 and 3 depend more on the properties of the cement paste (adapted from Hughes 1966).

have an important bearing on the rate of drying. Concrete elements with large surface area in relation to volume (such as floor slabs) dry faster than large volume concrete members with relatively small surface areas (such as bridge piers). After several months of drying in air with a relative humidity of 50% to 90%, moisture content is about 1% to 2% by mass of the concrete.

Concrete to receive flooring material must be dry enough to permit the adhesive to bond properly and to prevent damage to the flooring. For coatings, the concrete must be sufficiently dry to develop adequate bond and to allow the coating to chemically cure. A concrete surface may look dry, but the slab can still contain sufficient moisture to cause problems after it is covered (Kanare 2008).

Theoretically, it is possible to calculate the drying time for a given concrete (Hall 1997). This calculation requires the absorption characteristics, diffusion coefficients for water and water vapor, porosity and pore size distribution, and degree of hydration. Since this information usually is not available, current practice relies on experimental data combined with measurements of the actual moisture condition of the concrete slab in the field (Kanare 2008).

Many other properties of hardened concrete also are affected by its moisture content: elasticity, creep, insulating value, fire resistance, abrasion resistance, electrical conductivity, frost resistance, scaling resistance, and resistance to alkali-aggregate reactivity.

Also see the section on **Density** in this chapter and see Chapter 10 on **Volume Change of Concrete** (Figure 10-11) for additional information on drying effects, mass loss, and shrinkage.

Strength

Compressive strength is the measured maximum resistance of a concrete specimen to axial loading. It is generally expressed in megapascals (MPa) or pounds per square inch (psi) at an age of 28 days. Other test ages are also used. However, it is important to realize the relationship between the 28-day strength and other test ages. Seven-day strengths are often estimated to be about 75% of the 28-day strength while 56-day and 90-day strengths are about 10% to 15% greater than 28-day strengths, as shown in Figure 9-27. The specified compressive strength is designated by the symbol f'_c, and ideally is exceeded by the actual compressive strength, designated by f'_{cr}.

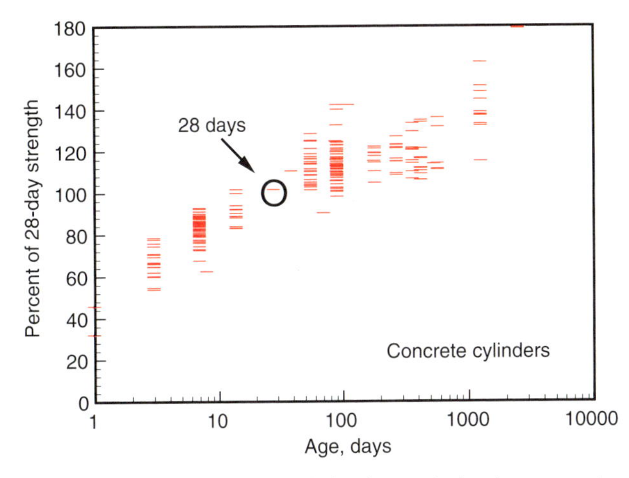

Figure 9-27. Compressive strength development of various concretes illustrated as a percentage of the 28-day strength (Lange 1994).

The compressive strength that a concrete achieves is influenced by the water-cementitious materials ratio, the extent to which hydration has progressed, the curing and environmental conditions, and the age of the concrete. The relationship between strength and water-cement ratio has been studied since the late 1800s and early 1900s (Feret 1897 and Abrams 1918). Figure 9-28 shows compressive strengths for a wide range of concrete mixtures and water-cement ratios at an age of 28 days. Note that strengths increase as the water-cement ratios decrease. These factors also affect flexural and tensile strengths and bond of concrete to steel.

The water-cement ratio compressive strength relationships in Figure 9-28 are for typical non-air-entrained concretes. When more precise values for concrete are required, graphs should be developed for the specific materials and mix proportions to be used on the job.

For a given workability and a given amount of cement, air-entrained concrete requires less mixing water than non-air-entrained concrete. The lower water-cement ratio possible for air-entrained concrete tends to offset the somewhat lower strengths of air-entrained concrete, particularly in lean to medium cement content mixtures.

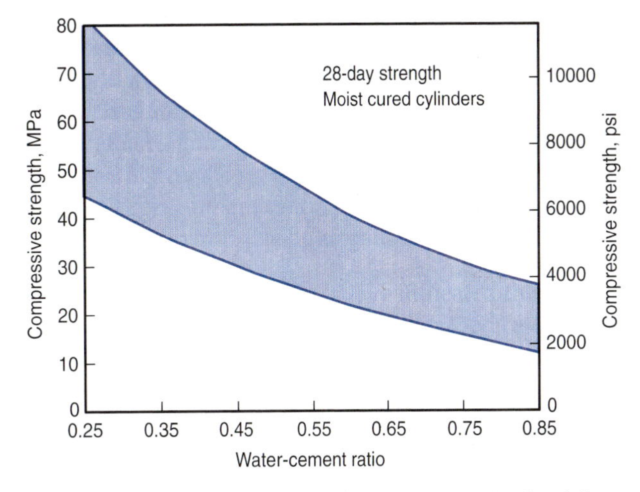

Figure 9-28. Range of typical strength to water-cement ratio relationships of portland cement concrete based on over 100 different concrete mixtures cast between 1985 and 1999.

To determine compressive strength, tests are made on specimens of mortar or concrete. The United States, unless otherwise specified, compression tests of mortar are made on 50-mm (2-in.) cubes, while compression tests of concrete are made on cylinders 150 mm (6 in.) in diameter and 300 mm (12 in.) high (Figure 9-29) or smaller cylinders sized at 100 mm x 200 mm (4 in. x 8 in.).

The compressive strength of concrete is a fundamental physical property frequently used in design calculations for bridges, buildings, and other structures. Most general-use concrete has a compressive strength between 20 MPa and 40 MPa (3000 psi and 6000 psi). Compressive strengths of 70 MPa to 140 MPa (10,000 psi to 20,000 psi) have been used in special bridge and high-rise building applications.

Figure 9-29. Testing a 150-mm x 300-mm (6-in. x 12-in.) concrete cylinder in compression. The load on the test cylinder is registered on the display.

The flexural strength or modulus of rupture of concrete is used to design pavements and other slabs on ground. Compressive strength, which is easier to measure than flexural strength, can be used as an index of flexural strength, once the empirical relationship between them has been established for the materials and the size of the element involved. The flexural strength of normal-weight concrete is often approximated as 0.7 to 0.8 times the square root of the compressive strength in megapascals (7.5 to 10 times the square root of the compressive strength in pounds per square inch). Wood (1992) illustrates the relationship between flexural strength and compressive strength for concretes exposed to moist curing, air curing, and outdoor exposure.

The direct tensile strength of concrete is about 8% to 12% of the compressive strength and is often estimated as 0.4 to 0.7 times the square root of the compressive strength in megapascals (5 to 7.5 times the square root of the compressive strength in pounds per square inch). Splitting tensile strength is 8% to 14% of the compressive strength (Hanson 1968). Splitting tensile strength versus time is presented by Lange (1994).

The torsional strength for concrete is related to the modulus of rupture and the dimensions of the concrete element. Hsu (1968) presents torsional strength correlations.

Shear strength to compressive strength relationships are discussed in the ACI 318 building code. The correlation between compressive strength and flexural, tensile, torsional, and shear strength varies with concrete ingredients and environment.

Modulus of elasticity, denoted by the symbol *E*, may be defined as the ratio of normal stress to corresponding strain for tensile or compressive stresses below the proportional limit of a material. For normal-weight concrete, *E* ranges from 1 GPa to 4 GPa (2 million psi to 6 million psi) and can be approximated as 5,000 times the square root of the compressive strength in megapascals (57,000 times the square root of the compressive strength in pounds per square inch). Like other strength relationships, the modulus of elasticity to compressive strength relationship is mixture specific and should be verified in a laboratory (Wood 1992).

Density

Conventional concrete, normally used in pavements, buildings, and other structures, has a density (unit weight) in the range of 2200 kg/m^3 to 2400 kg/m^3 (137 lb/ft^3 to 150 lb/ft^3). The density of concrete varies depending on the amount and density of the aggregate, the amount of air that is entrapped or purposely entrained, and the water and cement contents, which in turn are influenced by the maximum size of the aggregate. Reducing the cement paste content (increasing aggregate volume) increases density. Values of the density of fresh concrete are given in Table 9-7. For the design of reinforced concrete structures, the combination of conventional concrete and reinforcing steel is commonly assumed to weigh 2400 kg/m^3 (150 lb/ft^3).

The weight of dry concrete equals the weight of the freshly mixed concrete ingredients less the weight of mix water that evaporates during drying. Some of the mix water combines chemically with the cement during the hydration process, converting the cement phases into hydrates. Also, some of the water remains tightly held in pores and capillaries and does not evaporate under normal conditions. The amount of mix water that will evaporate from concrete exposed to ambient air at 50% relative humidity is about 0.5% to 3% of the concrete weight. The actual amount depends on initial water content of the concrete, absorption characteristics of the aggregates, and size and shape of the concrete element.

There is a wide spectrum of special concretes to meet various needs. Their densities range from lightweight insulating concretes with a density of as little as 240 kg/m^3 (15 lb/ft^3) to heavyweight concrete with a density of up to 6000 kg/m^3 (375 lb/ft^3) used for counterweights or radiation shielding.

Permeability and Watertightness

Concrete used in water-retaining structures or exposed to weather or other severe exposure conditions must be of low permeability or watertight. Watertightness is the ability of concrete to hold back or retain water without visible leakage. Permeability refers to the amount of water migration through concrete when the water is under pressure or to the ability of concrete to resist penetration by water or other substances (liquid, gas, or ions). Generally, the same properties of concrete that make it less permeable also make it more watertight.

The overall permeability of concrete to water migration is a function of: (1) the permeability of the paste; (2) the permeability and gradation of the aggregate; (3) the quality of the paste and aggregate transition zone; and (4) the relative proportion of paste to aggregate. Decreased permeability improves concrete's resistance to freezing and thawing, resaturation, sulfate attack, chloride-ion penetration, and other chemical attack (see Chapter 11).

The permeability of the paste is particularly important because the paste envelops all constituents in the concrete.

Table 9-7A. Observed Average Density of Fresh Concrete (SI Units)*

Maximum size of aggregate, mm	Air content, percent	Water, kg/m3	Cement, kg/m3	Density, kg/m3** Relative density of aggregate†				
				2.55	2.60	2.65	2.70	2.75
19	6.0	168	336	2194	2227	2259	2291	2323
37.5	4.5	145	291	2259	2291	2339	2371	2403
75	3.5	121	242	2307	2355	2387	2435	2467

* Source: Bureau of Reclamation 1981, Table 4.
** Air-entrained concrete with indicated air content.
† On saturated surface-dry basis. Multiply relative density by 1000 to obtain density of aggregate particles in kg/m3.

Table 9-7B. Observed Average Density of Fresh Concrete (Inch-Pound Units)*

Maximum size of aggregate, in.	Air content, percent	Water, lb/yd3	Cement, lb/yd3	Density, lb/ft3** Specific gravity of aggregate†				
				2.55	2.60	2.65	2.70	2.75
¾	6.0	283	566	137	139	141	143	145
1½	4.5	245	490	141	143	146	148	150
3	3.5	204	408	144	147	149	152	154

* Source: Bureau of Reclamation 1981, Table 4.
** Air-entrained concrete with indicated air content.
† On saturated surface-dry basis. Multiply specific gravity by 62.4 to obtain density of aggregate particles in lb/ft3.

Paste permeability is related to water-cement ratio, degree of cement hydration, and length of moist curing. A low-permeability concrete requires a low water-cement ratio and an adequate moist-curing period. Air entrainment aids watertightness but has little effect on permeability. Permeability increases with drying.

The permeability of mature hardened cement paste kept continuously moist ranges from 0.1×10^{-12} to 120×10^{-12} cm/s for water-cement ratios ranging from 0.3 to 0.7 (Powers and others 1954).The permeability of rock commonly used as concrete aggregate varies from approximately 1.7×10^{-9} to 3.5×10^{-13} cm/s. The permeability of mature, good-quality concrete is approximately 1×10^{-10} cm/s.

Test results obtained by subjecting 25-mm (1-in.) thick non-air-entrained mortar disks to 140-kPa (20-psi) water pressure are given in Figure 9-30. Mortar disks that had a water-cement ratio of 0.50 by weight or less and were moist-cured for seven days showed no water leakage. Where leakage occurred, it was greater in mortar disks made with high water-cement ratios. Also, for each water-cement ratio, leakage was less as the length of the moist-curing period increased. In disks with a water-cement ratio of 0.80, the mortar still permitted leakage after being moist-cured for one month. These results clearly show that a low water-cement ratio and a reasonable period of moist curing significantly reduce permeability.

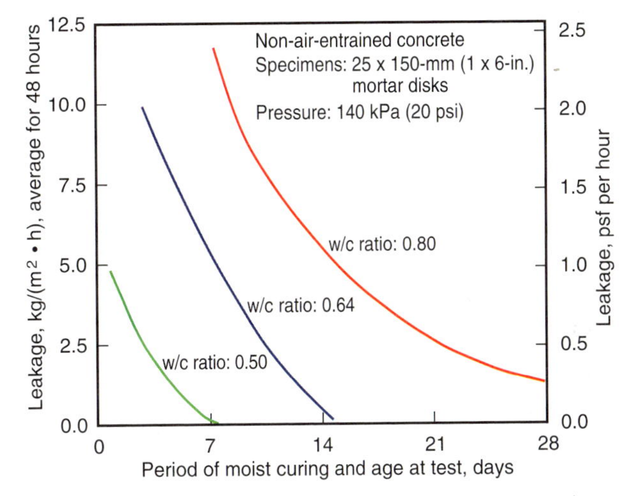

Figure 9-30. Effect of water-cement ratio (w/c) and curing duration on permeability of mortar. Note that leakage is reduced as the water-cement ratio is decreased and the curing period increased (McMillan and Lyse 1929 and PCA Major Series 227).

Figure 9-31 illustrates the effect of different water cement ratios on concrete's resistance to chloride ion penetration as indicated by electrical conductance. The total charge in coulombs was significantly reduced with a low water-cement ratio. Also, the results showed that a lower charge passed when the concrete contained a higher air content.

A low water-cement ratio also reduces segregation and bleeding, further contributing to watertightness. Of course watertight concrete must also be free from cracks, honeycomb, or other large visible voids.

Occasionally, pervious concrete – no-fines concrete that readily allows passage of water – is designed for special applications. In these concretes, the fine aggregate is either greatly reduced or completely removed. This condition causes a high volume of interconnected air voids. Flow rates in pervious concrete are orders of magnitude higher – often 0.2 cm/s or higher. Pervious concrete has been used in tennis courts, pavements, parking lots, greenhouses, and drainage structures. Pervious concrete has also been used in buildings because of its thermal insulation properties.

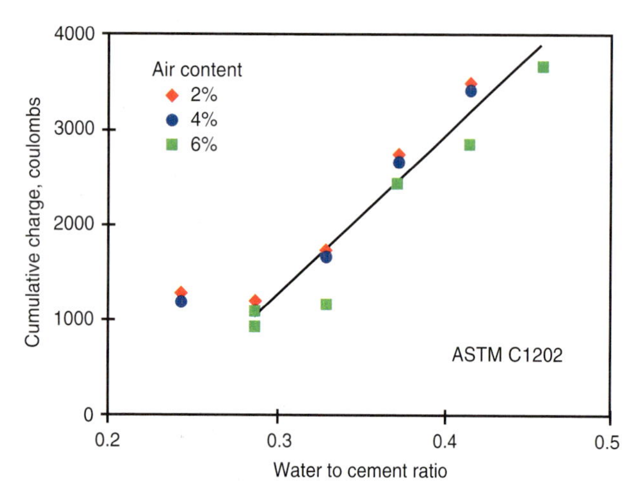

Figure 9-31. Total charge at the end of the rapid chloride permeability test as a function of water to cement ratio (Pinto and Hover 2001).

Volume Stability and Crack Control

Hardened concrete changes volume due to changes in temperature, moisture, and stress. These volume or length changes may range from about 0.01% to 0.08%. Thermal volume changes of hardened concrete are about the same as those for steel.

Concrete under stress deforms elastically. Sustained stress results in an additional deformation called creep. The rate of creep (deformation per unit of time) decreases with time.

Concrete kept continually moist will expand slightly. When permitted to dry, concrete will shrink. The primary factor influencing the amount of drying shrinkage is the water content of the freshly mixed concrete. Drying shrinkage increases directly with increases in this water content. The amount of shrinkage also depends upon several other factors, such as: (1) the amount of aggregate used; (2) properties of the aggregate; (3) size and shape of the concrete element; (4) relative humidity and temperature of the ambient air; (5) method of curing; (6) degree of hydration; and (7) time.

Two basic causes of cracks in concrete are: (1) stress due to applied loads and (2) stress due to drying shrinkage, temperature changes, durability related distress, and restraint.

Drying shrinkage is an inherent, unavoidable property of concrete. However, properly positioned reinforcing steel is used to reduce crack widths, or joints are used to predetermine and control the location of cracks. Thermal stress due to fluctuations in ambient temperature also can cause cracking, particularly at an early age.

Concrete shrinkage cracks are often the result of restraint. As drying shrinkage occurs, if there is no restraint, the concrete will not crack. Restraint comes from several sources. Drying shrinkage is always greater near the surface of concrete; the moist inner portions restrain the concrete near the surface, which can cause cracking. Other sources of restraint are reinforcing steel embedded in concrete, the interconnected parts of a concrete structure, and the friction of the subgrade on which concrete is placed.

Joints. Joints are the most effective method of controlling unsightly cracking. If a sizable expanse of concrete (a wall, slab, or pavement) is not provided with properly spaced joints to accommodate drying shrinkage and temperature contraction, the concrete will crack in a random manner.

Contraction (shrinkage control) joints are grooved, formed, or sawed into sidewalks, driveways, pavements, floors, and walls so that cracking will occur in these joints rather than in a random manner. Contraction joints permit movement in the plane of a slab or wall. They extend to a depth of approximately one-quarter the concrete thickness.

Isolation joints separate a concrete placement from other parts of a structure and permit horizontal and vertical movements. They should be used at the junction of floors with walls, columns, footings, and other points where restraint can occur. They extend the full depth of slabs and include a premolded joint filler.

Construction joints occur where concrete work is concluded for the day. They separate areas of concrete placed at different times. In slabs-on-ground, construction joints usually align with, and function as, control or isolation joints. They may also require dowels for load transfer.

For more information on volume changes in concrete, see Chapter 10.

Durability

The durability of concrete may be defined as the ability of concrete to resist weathering action, chemical attack, and abrasion while maintaining its desired engineering properties. Concrete is exposed to a greater variety of potentially harmful exposure conditions than any other construction material.

There are many causes of concrete deterioration and most of these involve either the movement of moisture or the movement of species, such as chlorides and sulfates, dissolved in the water. Generally, the greater the resistance of the concrete to the movement of water, the lower its permeability and the greater its resistance to deterioration. The following sections discuss several deterioration mechanisms. For more information on each of these topics and preventive measures see Chapter 11.

Freeze thaw and deicer salts. Deterioration due to freezing and thawing is a result of the expansive forces that are generated when the water in saturated concrete freezes. If the concrete is not designed to resist freeze-thaw cycles, cracking will occur. As the number of freeze-thaw cycles increase, the cracking will become more advanced and eventually severe deterioration may occur. Damage due to freezing and thawing is exacerbated in the presence of deicing salts. When concrete is exposed to these conditions particular attention must be paid to ensure that the mixture proportions are appropriate and that the concrete is finished and cured properly. D-cracking or durability cracking occurs when frost-susceptible aggregates are used in concrete exposed to freezing and thawing. In such cases, it is the expansion of water in the aggregate that leads to cracking of the concrete.

Corrosion. Concrete may be exposed to chloride ions during service. Common sources of chlorides include: deicing salt, seawater or chloride-contaminated groundwater. Over time, the chlorides will penetrate through the concrete cover and eventually reach the embedded steel reinforcement. The chlorides breakdown the passive layer, allowing corrosion of the steel to occur. Because the products of corrosion, rust, occupy more volume than the metallic steel, expansive forces develop which can lead to cracking in the concrete. Eventually corrosion can lead to spalling and delamination of the concrete cover. Corrosion of embedded steel reinforcement is the most prevalent form of deterioration of reinforced concrete structures.

Carbonation. Another form of deterioration of concrete due to corrosion is caused by carbonation. Carbonation of concrete occurs when carbon dioxide from the atmosphere penetrates concrete and reacts with the products of cement hydration and reduces the alkalinity of the concrete. When the carbonation depth reaches the steel the protective layer becomes unstable and the steel starts to corrode. Carbonation is typically a very slow process.

Alkali-silica reactivity. Alkali-silica reaction (ASR) is the reaction between the alkalies (sodium and potassium) in portland cement and certain siliceous rocks or minerals, such as opaline chert, strained quartz, and acidic volcanic glass, present in some aggregates. The reaction product is an alkali-silica gel which has the capacity to adsorb water and swell. Under certain conditions the products of the reaction may cause abnormal expansion and cracking of concrete in service.

Abrasion. Floors, pavements, and hydraulic structures are subjected to abrasion. In these applications concrete must have a high abrasion resistance. Test results indicate that abrasion resistance is closely related to the compressive strength of concrete. Strong concrete has more resistance to abrasion than weak concrete. The type of aggregate and surface finish or treatment used also have a strong influence on abrasion resistance. Hard aggregate is more wear resistant than soft aggregate and a steel-troweled surface resists abrasion better than an untroweled surface.

Sulfate attack. Excessive amounts of sulfates in soil or water can attack and destroy a concrete that is not properly designed. Sulfates (for example, calcium sulfate, sodium sulfate, and magnesium sulfate) can attack concrete by reacting with hydrated compounds in the hardened cement paste. These reactions can induce sufficient pressure to disrupt the cement paste, resulting in disintegration of the concrete (loss of paste cohesion and strength).

Other forms of concrete deterioration. There are other forms of concrete deterioration less common or which occur only in special conditions:

1. Thaumasite form of sulfate attack – which differs from classical sulfate salt attack
2. Delayed ettringite formation (DEF) – which only occurs in concrete exposed to excessive temperatures at early ages
3. Alkali-carbonate reaction (ACR) – which involves the attack by alkalis on carbonate phases of the rock and is much less widespread than alkali-silica reaction
4. Salt crystallization and attack by chemicals other than sulfates

Different concretes require different degrees of durability depending on the exposure environment and the properties desired. The concrete ingredients, proportioning of those ingredients, interactions between the ingredients, and placing and curing practices determine the ultimate durability and service life of the concrete. For more information on the durability of concrete, see Chapter 11.

Aesthetics

Pleasing decorative finishes can be built into concrete during construction. Variations in the color and texture of concrete surfaces are limited only by the imagination of the designer and the skill of the concrete craftsman.

Color may be added to the concrete through the use of white cement and pigments, exposure of colorful aggregates, or addition of score lines to create borders for the application of penetrating or chemically reactive stains. Desired textured finishes can be varied, from a smooth polish to the roughness of gravel. Geometric patterns can be scored, stamped, rolled, or inlaid into the concrete to resemble stone, brick, or tile paving (Figure 9-32). Other interesting patterns are obtained using divider strips (commonly redwood) to form panels of various sizes and shapes – rectangular, square, circular, or diamond. Special techniques are also available to make concrete slip-resistant and sparkling.

Figure 9-32. Pattern-stamped finish and colored surfaces are popular for decorative concretes.

These surface treatments are just as pleasing in the interior as they are on the exterior of a home or commercial building. Colored and imprinted concrete is an excellent flooring material combining the economy, durability, decorative qualities, and strength of concrete and the thermal mass needed for passive solar buildings. Special concrete finishes (interior or exterior) enhance the aesthetic appeal and value of any property. For more information on decorative concrete see ACI Committees 303 and 310, and Kosmatka and Collins 2004.

References

Abrams, D. A., *Design of Concrete Mixtures*, Lewis Institute, Structural Materials Research Laboratory, Bulletin No. 1, PCA LS001, http://www.cement.org/pdf_files/LS001.pdf, 1918, 20 pages.

ACI Committee 302, *Guide for Concrete Floor and Slab Construction*, 302.1R-04, ACI Committee 302 Report, American Concrete Institute, Farmington Hills, Michigan, 2004.

ACI Committee 303, *Guide to Cast-in-Place Architectural Concrete Practice*, ACI 303R-04, ACI Committee 303 Report, American Concrete Institute, Farmington Hills, Michigan, 2004.

ACI Committee 309, *Guide for Consolidation of Concrete*, ACI 309R-05, ACI Committee 309 Report, American Concrete Institute, Farmington Hills, Michigan, 2006.

ACI Committee 318, *Building Code Requirements for Structural Concrete and Commentary*, ACI 318-08, American Concrete Institute, Farmington Hills, Michigan, 2008, 465 pages.

Bingham, E.C., "An Investigation of the Laws of Plastic Flow," *Bulletin of the Bureau of Standards*, Vol. 13, No. 2, August 1916, pages 309-353.

Brewster, R.S., *Effect of Vibration Time upon Loss of Entrained Air from Concrete Mixes*, Materials Laboratories Report No. C-461, Research and Geology Division, Bureau of Reclamation, Denver, November 25, 1949.

Bureau of Reclamation, *Concrete Manual*, 8th Edition, Bureau of Reclamation, Denver, Colorado, 1981, page 33.

Burg, Ronald G., *The Influence of Casting and Curing Temperature on the Properties of Fresh and Hardened Concrete*, Research and Development Bulletin RD113, Portland Cement Association, 1996, 20 pages.

Cable, J.K.; McDaniel, L.; Schlorholtz, S.; Redmond, D.; and Rabe, K., *Evaluation of Vibrator Performance vs. Concrete Consolidation and Air Void System*, Serial No. 2398, Portland Cement Association, http://www.cement.org/pdf_files/ SN2398.pdf, 2000, 60 pages.

Copeland, L.E., and Schulz, Edith G., *Electron Optical Investigation of the Hydration Products of Calcium Silicates and Portland Cement*, Research Department Bulletin RX135, Portland Cement Association, http://www.cement.org/pdf_files/RX135.pdf, 1962, 12 pages.

Daniel, D. Gene, "Factors Influencing Concrete Workability," *Significance of Tests and Properties of Concrete and Concrete-Making Materials*, STP 169D, American Society for Testing and Materials, West Conshohocken, Pennsylvania, 2006, pages 59 to 72.

Falconi, M I., *Durability of Slag Cement Concretes Exposed to Freezing and Thawing in the Presence of Deicers*, Masters of Science Thesis, Cornell University, Ithaca, NewYork, 1996, 306 pages.

Feret, R., "Etudes Sur la Constitution Intime Des Mortiers Hydrauliques" [Studies on the Intimate Constitution of Hydraulic Mortars], *Bulletin de la Societe d'Encouragement Pour Industrie Nationale*, 5th Series, Vol. 2, Paris, 1897, pages 1591 to 1625.

Gebler, Steven H., and Klieger, Paul, *Effect of Fly Ash on Durability of Air-Entrained Concrete*, Research and Development Bulletin RD090, Portland Cement Association, http://www.cement.org/pdf_files/RD090.pdf, 1986.

Greening, Nathan R., *Some Causes for Variation in Required Amount of Air-Entraining Agent in Portland Cement Mortars*, Research Department Bulletin RX213, Portland Cement Association, http://www.cement.org/pdf_files/RX213.pdf, 1967.

Hall, C., "Barrier Performance of Concrete: A Review of Fluid Transport Theory," *Penetration and Permeability of Concrete: Barriers to Organic and Contaminating Liquids: State-of-the-Art Report Prepared by Members of the RILEM Technical Committee 146-TCF*, RILEM Report 16, E&FN Spon, London, 1997, pages 7 to 40.

Hanson, J.A., *Effects of Curing and Drying Environments on Splitting Tensile Strength of Concrete*, Development Department Bulletin DX141, Portland Cement Association, http://www.cement.org/pdf_files/DX141.pdf, 1968, page 11.

Hedenblad, G., *Drying of Construction Water in Concrete-Drying Times and Moisture Measurement*, Stockholm, Byggforskningsrådet, 1997.

Hedenblad, Goren, *Data for Calculation of Moisture Migration*, Swedish Council for Building Research, S- 11387, Stockholm, Sweden, 1996.

Hsu, Thomas T.C., *Torsion of Structural Concrete—Plain Concrete Rectangular Sections*, Development Department Bulletin DX134, Portland Cement Association, http://www.cement.org/pdf_files/DX134.pdf, 1968.

Hughes B.P., et al, "The Diffusion of Water in Concrete at Temperatures Between 50 and 95°C," *British Journal of Applied Physics*, Vol. 17, No. 12, December 1966, pages 1545 to 1552.

Kanare, Howard M., *Concrete Floors and Moisture*, EB119, Portland Cement Association, Skokie, Illinois, and National Ready Mixed Concrete Association, Silver Spring, Maryland, 2008, 176 pages.

Kosmatka, Steven H., "Bleed Water," *Significance of Tests and Properties of Concrete and Concrete-Making Materials*, STP 169D, American Society for Testing and Materials, West Conshohocken, Pennsylvania, 2006, pages 99 to 122. [PCA RP328]

Kosmatka, Steven H., and Collins, Terry C., *Finishing Concrete with Color and Texture*, PA124, Portland Cement Association, Serial No. 2416a, 2004, 76 pages.

Kozikowski, Jr., R.L.; Vollmer, D.B.; Taylor, P.C.; and Gebler, S.H., *Factor(s) Affecting the Origin of Air-Void Clustering*, SN2789, Portland Cement Association, 2005.

Lange, David A., *Long-Term Strength Development of Concrete*, RP326, Portland Cement Association, 1994.

Lerch, William, *Basic Principles of Air-Entrained Concrete*, T-101, Portland Cement Association, 1960.

McMillan, F.R., and Lyse, Inge, "Some Permeability Studies of Concrete," *Journal of the American Concrete Institute, Proceedings*, Vol. 26, Farmington Hills, Michigan, December 1929, pages 101 to 142.

McNeal, F., and Gay, F., "Solutions to Scaling Concrete," *Concrete Construction*, Addison, Illinois, March 1996, pages 250 to 255.

Pinto, Roberto C.A., and Hover, Kenneth C., *Frost and Scaling Resistance of High-Strength Concrete*, Research and Development Bulletin RD122, Portland Cement Association, 2001, 70 pages.

Powers, T.C., "Studies of Workability of Concrete," *Journal of the American Concrete Institute*, Vol. 28, Farmington Hills, Michigan, February 1932, page 419.

Powers, T.C., *A Discussion of Cement Hydration in Relation to the Curing of Concrete*, Research Department Bulletin RX025, Portland Cement Association, http://www.cement.org/pdf_files/RX025.pdf, 1948, 14 pages.

Powers, T.C., *The Bleeding of Portland Cement Paste, Mortar, and Concrete*, Research Department Bulletin RX002, Portland Cement Association, http://www.cement.org/pdf_files/RX002.pdf, 1939.

Powers, T.C., *The Nonevaporable Water Content of Hardened Portland Cement Paste—Its Significance for Concrete Research and Its Method of Determination*, Research Department Bulletin RX029, Portland Cement Association, http://www.cement.org/pdf_files/RX029.pdf, 1949, 20 pages.

Powers, T.C.; Copeland, L.E.; Hayes, J.C.; and Mann, H.M., *Permeability of Portland Cement Pastes*, Research Department Bulletin RX053, Portland Cement Association, http://www.cement.org/pdf_files/RX053.pdf, 1954.

Scherer, George W., "Theory of Drying," *Journal of the American Ceramic Society*, Vol 73, No. 1, 1990, pages 3 to 14.

Stark, David C., *Effect of Vibration on the Air-System and Freeze-Thaw Durability of Concrete*, Research and Development Bulletin RD092, Portland Cement Association, http://www.cement.org/pdf_files/RD092.pdf, 1986.

Steinour, H.H., *Further Studies of the Bleeding of Portland Cement Paste*, Research Department Bulletin RX004, Portland Cement Association, http://www.cement.org/pdf_files/RX004.pdf, 1945.

Szecsy, Richard, and Mohler, Nathaniel, *Self-Consolidating Concrete*, IS546D, Portland Cement Association, Skokie, Illinois, 2009, 24 pages.

Taylor, Thomas G., *Effect of Carbon Black and Black Iron Oxide on Air Content and Durability of Concrete*, Research Department Bulletin RX023, Portland Cement Association, http://www.cement.org/pdf_files/RX023.pdf, 1948.

Thomas, Michael, *Optimizing the Use of Fly Ash in Concrete*, IS548, Portland Cement Association, Skokie, Illinois, 2007, 24 pages.

Whiting, David A., and Nagi, Mohamad A., *Manual on Control of Air Content in Concrete*, EB116, National Ready Mixed Concrete Association and Portland Cement Association, 1998, 42 pages.

Whiting, D., and Dziedzic, D., *Effects of Conventional and High-Range Water Reducers on Concrete Properties*, Research and Development Bulletin RD107, Portland Cement Association, 1992, 25 pages.

Wood, Sharon L., *Evaluation of the Long-Term Properties of Concrete*, Research and Development Bulletin RD102, Portland Cement Association, 1992, 99 pages.

Young, J. Francis; Mindess, Sidney; Gray, Robert J.; and Bentur, Arnon, *The Science and Technology of Civil Engineering Materials*, Prentice Hall, Upper Saddle River, New Jersey, 1998, 398 pages.

CHAPTER 10

Volume Changes of Concrete

Concrete undergoes slight changes in volume for various reasons, and understanding the nature of these changes is useful in planning or analyzing concrete work. If concrete is free of any restraints to deform, normal volume changes would be of little consequence. But since concrete in service is usually restrained by foundations, subgrades, reinforcement, or connecting members and significant stresses can develop. This is particularly true of tensile stresses.

While quite strong in compression, cracks develop because concrete is relatively weak in tension. Controlling the variables that affect volume changes can minimize high stresses and cracking. Tolerable crack widths should be considered in the structural design.

The subject of concrete volume changes deals with linear expansion and contraction due to the hydration reaction, temperature change, and moisture change. But chemical effects such as carbonation shrinkage, sulfate attack, and the disruptive expansion of alkali-aggregate reactions also cause volume changes. In addition, creep is a volume change or deformation caused by sustained stress or load. Equally important is the elastic or inelastic change in dimensions or shape that occurs instantaneously under applied load.

For convenience, the magnitude of volume changes is generally stated in linear rather than volumetric units (Pickett 1947). Changes in length are often expressed as a coefficient of length in parts per million, or simply as millionths. It is applicable to any length unit (for example, m/m or ft/ft); one millionth is 0.000001 m/m (0.000001 in./in.) and 600 millionths is 0.000600 m/m (0.000600 in./in.). Change of length can also be expressed as a percentage; thus 0.06% is the same as 0.000600, which is also the same as 6 mm per 10 m (¾ in. per 100 ft). The volume changes that ordinarily occur in concrete are small, ranging in length change from perhaps 10 millionths up to about 1000 millionths.

Early Age Volume Changes

The volume of concrete begins to change immediately after the water comes into contact with the cement. Early volume changes, within 24 hours, can influence the long-term volume changes and potential crack formation in hardened concrete. Following are discussions on various forms of early volume change:

Chemical Shrinkage

Chemical shrinkage refers to the reduction in absolute volume of solids and liquids in paste resulting from cement hydration. The absolute volume of hydrated cement products is less than the absolute volume of cement and water before hydration. This change in volume of cement paste during the plastic state is illustrated by the first two bars in Figure 10-1. This does not include air formed during mixing. Chemical shrinkage continues to occur at a microscopic scale as long as cement hydrates. After initial set, the paste cannot deform as much as when it was in a plastic state. Therefore, further hydration and chemical shrinkage is compensated by the formation of voids in the microstructure (Figure 10-1). Most of this volume change is internal and does not significantly change the visible external dimensions of a concrete element.

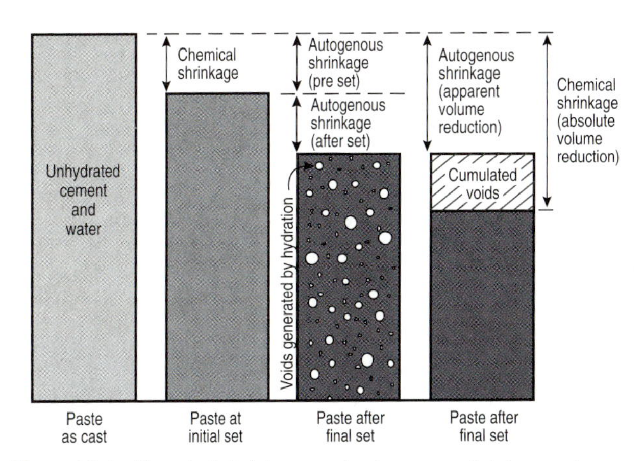

Figure 10-1. Chemical shrinkage and autogenous shrinkage volume changes of fresh and hardened paste. Not to scale.

The amount of volume change due to chemical shrinkage can be estimated from the hydrated cement phases and their densities or it can be determined by physical test as illustrated in Figure 10-2 in accordance with ASTM C1608, *Standard Test Method for Chemical Shrinkage of Hydraulic Cement Paste*. An example of long-term chemical shrinkage for portland cement paste is illustrated in Figure 10-3. Early researchers sometimes referred to chemical shrinkage as the absorption of water during hydration (Powers 1935). Le Chatelier (1900) was the first to study chemical shrinkage of cement pastes. It is important to recognize that the shrinkage potential of a concrete mixture is affected by the water content and the quantity, gradation, and volume stability of the aggregates.

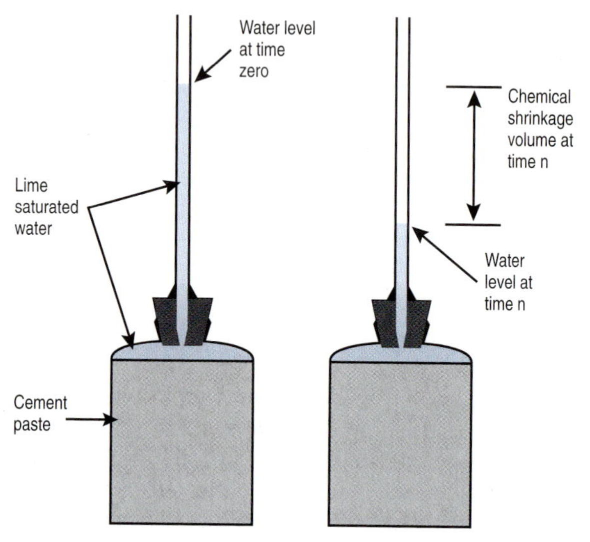

Figure 10-2. Test for chemical shrinkage of cement paste showing flask for cement paste and pipet for absorbed water measurement (ASTM C1608).

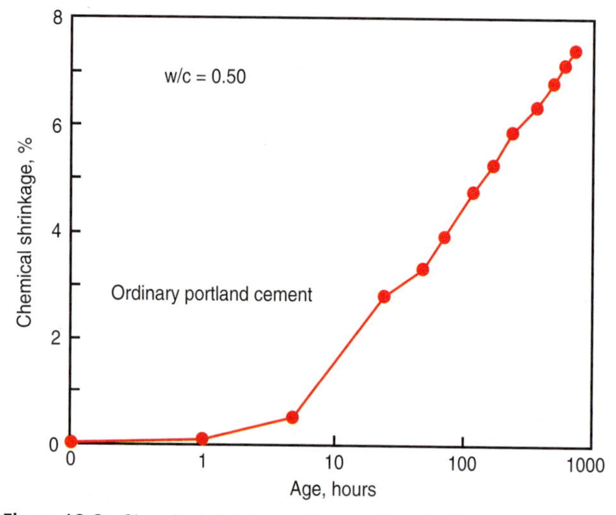

Figure 10-3. Chemical shrinkage of cement paste (Tazawa 1999).

Autogenous Shrinkage

Autogenous shrinkage is the macroscopic volume reduction (visible dimensional change) of cement paste, mortar, or concrete caused by cement hydration and is measured in accordance with ASTM C1698, *Standard Test Method for Autogenous Strain of Cement Paste and Mortar*. The macroscopic volume reduction of autogenous shrinkage is much less than the absolute volume reduction of chemical shrinkage because of the rigidity of the hardened paste structure. Chemical shrinkage is the driving force behind autogenous shrinkage. The relationship between autogenous shrinkage and chemical shrinkage is illustrated in Figures 10-1, 10-4, and 10-5. Some researchers and organizations consider that autogenous shrinkage starts at initial set while others evaluate autogenous shrinkage from time of placement (Bentur 2003).

When external water is available, autogenous shrinkage cannot occur. If external water is not available (due to a lack of moist curing or thick sections that limit the transport of curing water) cement hydration consumes pore water resulting in self desiccation of the paste and a uniform reduction of volume (Copeland and Bragg 1955 and Radlinska and others 2008). Autogenous shrinkage

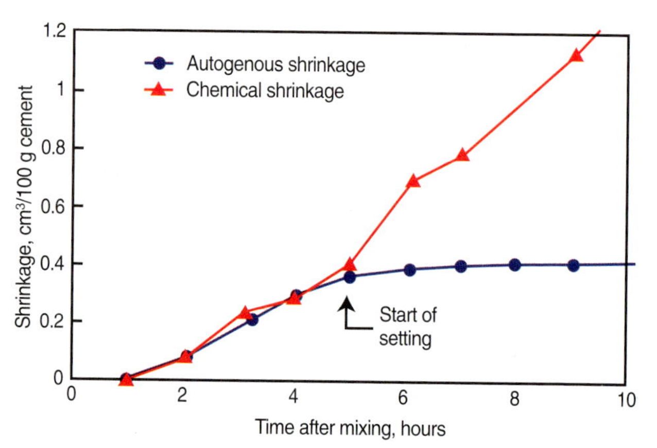

Figure 10-4. Relationship between autogenous shrinkage and chemical shrinkage of cement paste at early ages (Hammer 1999).

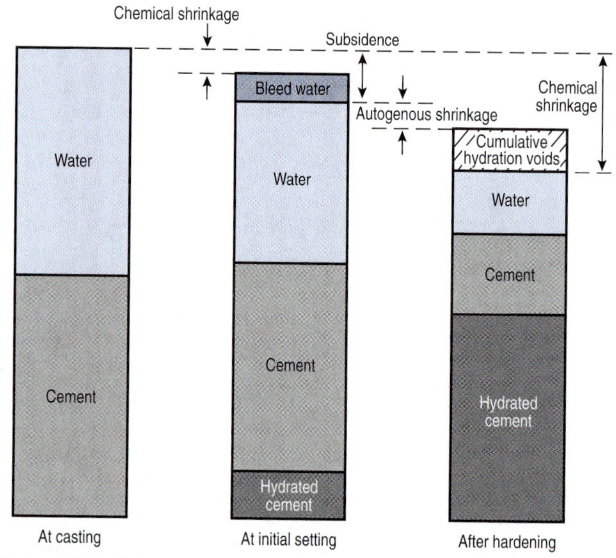

Figure 10-5. Volumetric relationship between subsidence, bleed water, chemical shrinkage, and autogenous shrinkage. Only autogenous shrinkage after initial set is shown. Not to scale.

increases with a decrease in water to cement ratio and with an increase in the fineness and amount of cementitious material. Autogenous shrinkage is most prominent in concrete with a water to cement ratio under 0.42 (Holt 2001). High-strength, low water to cement ratio (0.30) concrete can experience 200 millionths to 400 millionths of autogenous shrinkage. Autogenous shrinkage can be half that of drying shrinkage for concretes with a water to cement ratio of 0.30 (Holt 2001 and Aïtcin 1999).

Recent use of high performance, low water to cement ratio concrete in bridges and other structures has renewed interest in designing for autogenous shrinkage to control crack development. Concretes susceptible to large amounts of autogenous shrinkage should be cured with external water for at least 7 days to minimize crack development, and fogging should be provided as soon as the concrete is cast. The hydration of supplementary cementing materials also contributes to autogenous shrinkage, although at different levels than with portland cement. In addition to adjusting paste content and water to cement ratios, autogenous shrinkage can be reduced by using shrinkage reducing admixtures or internal curing techniques. Some cementitious systems may experience autogenous expansion. Tazawa (1999) and Holt (2001) review techniques to control autogenous shrinkage (see ACI 231).

Test methods for autogenous shrinkage and expansion of cement paste, mortar, and concrete and tests for autogenous shrinkage stress of concrete are presented by Tazawa (1999).

Subsidence

Subsidence refers to the vertical shrinkage of fresh cementitious materials before initial set. It is caused by bleeding (settlement of solids relative to liquids), air voids rising to the surface, and chemical shrinkage. Subsidence is also called settlement shrinkage. Subsidence of well-consolidated concrete with minimal bleed water is insignificant. The relationship between subsidence and other shrinkage mechanisms is illustrated in Figure 10-5. Excessive subsidence is often caused by a lack of consolidation of fresh concrete. Excessive subsidence over embedded items, such as supported steel reinforcement, can result in cracking over embedments. Concretes made with air entrainment, sufficient fine materials, and low water contents can help minimize subsidence cracking. Also, fibers have been reported to reduce subsidence cracking (Suprenant and Malisch 1999).

Plastic Shrinkage

Plastic shrinkage refers to volume change occurring while the concrete is still fresh, before hardening. It is usually observed in the form of plastic shrinkage cracks occurring before or during finishing (Figure 10-6). The cracks often resemble tears in the surface. Plastic shrinkage results from a combination of chemical and autogenous shrinkage and

Figure 10-6. Plastic shrinkage cracks resemble tears in fresh concrete.

rapid evaporation of moisture from the surface that exceeds the bleeding rate. Plastic shrinkage cracking can be controlled by minimizing surface evaporation through use of fogging, wind breaks, shading, plastic sheet covers, wet burlap, spray-on finishing aids (evaporation retarders), and fibers (see Chapter 16).

Swelling

Concrete, mortar, and cement paste swell in the presence of external water. When water drained from capillaries by chemical shrinkage is replaced by external water, the volume of the concrete mass initially increases. As there is no self desiccation, there is no autogenous shrinkage. External water can come from wet curing or submersion. Swelling occurs due to a combination of crystal growth, absorption of water, and osmotic pressure. The swelling is not large, only about 50 millionths at early ages (Figure 10-7). When the external water source is removed, autogenous shrinkage and drying shrinkage reverse the volume change.

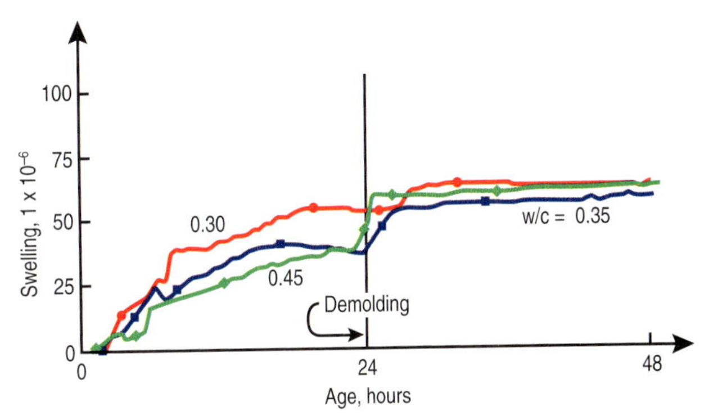

Figure 10-7. Early age swelling of 100 x 100 x 375-mm (4 x 4 x 15-in.) concrete specimens cured under water (Aïtcin 1999).

As cement hydrates, the exothermic reaction provides a significant amount of heat. In large elements the heat is retained, rather than dissipated as happens with thin elements. This temperature rise, occurring over the first few hours and days, can induce a small amount of expansion that counteracts autogenous and chemical shrinkage (Holt 2001).

Moisture Changes (Drying Shrinkage) of Hardened Concrete

Hardened concrete expands slightly with a gain in moisture and contracts with a loss in moisture. The effects of these moisture cycles are illustrated schematically in Figure 10-8. Specimen A represents concrete stored continuously in water from time of casting. Specimen B represents the same concrete exposed first to drying in air and then to alternate cycles of wetting and drying. For comparative purposes, it should be noted that the swelling that occurs during continuous wet storage over a period of several years is usually less than 150 millionths; this is about one-fourth of the shrinkage of air-dried concrete for the same period. Figure 10-9 illustrates swelling of concretes wet cured for 7 days followed by shrinkage when sealed or exposed to air drying. Autogenous shrinkage reduces the volume of the sealed concretes to a level about equal to the amount of swelling at 7 days. Note that the concretes wet cured for 7 days had less shrinkage due to drying and autogenous effects than the concrete that had no water curing. This illustrates the importance of early, wet curing to minimize shrinkage (Aïtcin 1999).

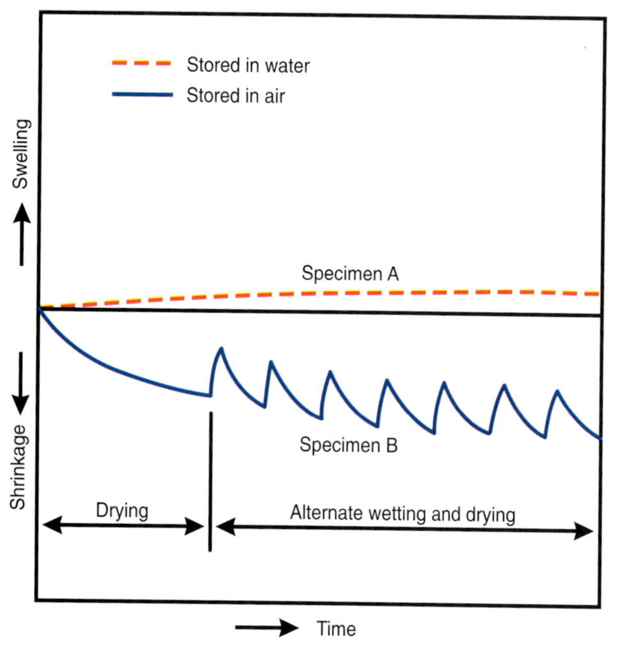

Figure 10-8. Schematic illustration of moisture movements in concrete. If concrete is kept continuously wet, a slight expansion occurs. However, drying usually takes place, causing shrinkage. Further wetting and drying causes alternate cycles of swelling and shrinkage (Roper 1960).

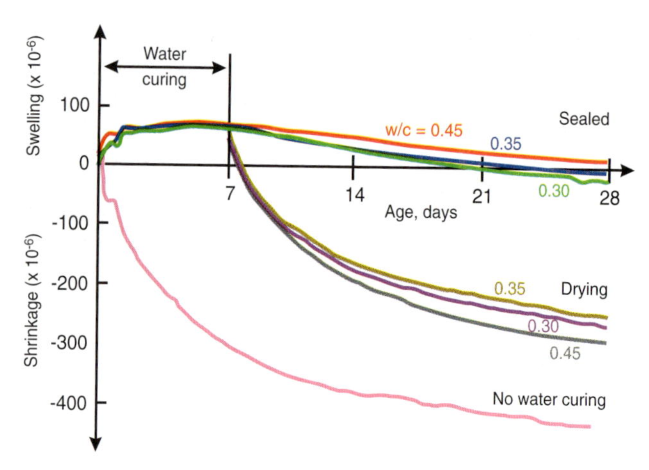

Figure 10-9. Length change of concrete samples exposed to different curing regimes (Aïtcin 1999).

Tests indicate that the drying shrinkage of small, plain concrete specimens (without reinforcement) ranges from about 400 millionths to 800 millionths when exposed to air at 50% relative humidity. Concrete with a unit drying shrinkage of 550 millionths shortens about the same amount as the thermal contraction caused by a decrease in temperature of 55°C (100°F). Preplaced aggregate concrete has a drying shrinkage of 200 millionths to 400 millionths; this is considerably less than normal concrete due to point-to-point contact of aggregate particles in preplaced aggregate concrete. The drying shrinkage of structural lightweight concrete ranges from slightly less than that of normal-density concrete, to 30 percent more depending on the type of aggregate used.

The drying shrinkage of reinforced concrete is less than that for plain concrete, the difference depending on the amount of reinforcement. Steel reinforcement restricts but does not prevent drying shrinkage. In reinforced concrete structures with normal amounts of reinforcement, drying shrinkage is assumed to be 200 to 300 millionths. Similar values are found for slabs on ground restrained by subgrade. ACI Committee 209 (1992 and 2008) provides approaches to predicting and modeling shrinkage in reinforced structures.

For many outdoor applications, concrete reaches its maximum moisture content in winter; so in winter the volume changes due to increase in moisture content and the decrease in average temperature tend to offset each other.

The amount of moisture in concrete is affected by the relative humidity of the ambient air. The free moisture content of concrete elements after drying in air at relative humidities of 50% to 90% for several months is about 1% to 2% by weight of the concrete. The actual amount depends on the concrete's constituents, original water content, drying conditions, and the size and shape of the concrete element.

After concrete has dried to a constant moisture content at one relative humidity condition, a decrease in humidity causes it to lose moisture while an increase causes it to gain moisture. The concrete shrinks or swells with each such change in moisture content due primarily to responses of the cement paste. Most aggregates show little response to changes in moisture content, although there are a few aggregates that swell or shrink in response to such changes.

As drying takes place, concrete shrinks. Where there is no restraint, movement occurs freely and no stresses or cracks develop (Figure 10-10a top). If the tensile stress that results from restrained drying shrinkage exceeds the tensile strength of the concrete, cracks can develop (Figure 10-10a bottom). The magnitude of the tensile stress increases with an increased drying rate (Grasley 2003). Random cracks may develop if joints are not properly provided and the concrete element is restrained from shortening (Figure 10-10b). Contraction joints for slabs on ground (Figure 10-10c) should be spaced at distances of 24 to 30 times the slab thickness to control random cracks (Tarr and Farny 2008). Joints in walls are equally important for crack control (Figure 10-10d). Figure 10-11 illustrates the relationship between drying rate at different depths, drying shrinkage, and mass loss for normal-density concrete (Hanson 1968).

Shrinkage may continue for a number of years, depending on the size and shape of the concrete. The rate and ultimate amount of shrinkage are usually smaller for large masses of concrete than for small masses; on the other hand, shrinkage continues longer for large masses. Higher volume-to-surface ratios (larger elements) experience lower shrinkage as shown in Figure 10-12.

The rate and amount of drying shrinkage for small concrete specimens made with various cements are shown in Figure 10-13. Specimens were initially moist-cured for 14 days at 21°C (70°F), then stored for 38 months in air at the same temperature and 50% relative humidity. Shrinkage recorded at the age of 38 months ranged from 600 to 790 millionths. An average of 34% of this shrinkage occurred within the first month. At the end of 11 months an average of 90% of the 38-month shrinkage had taken place.

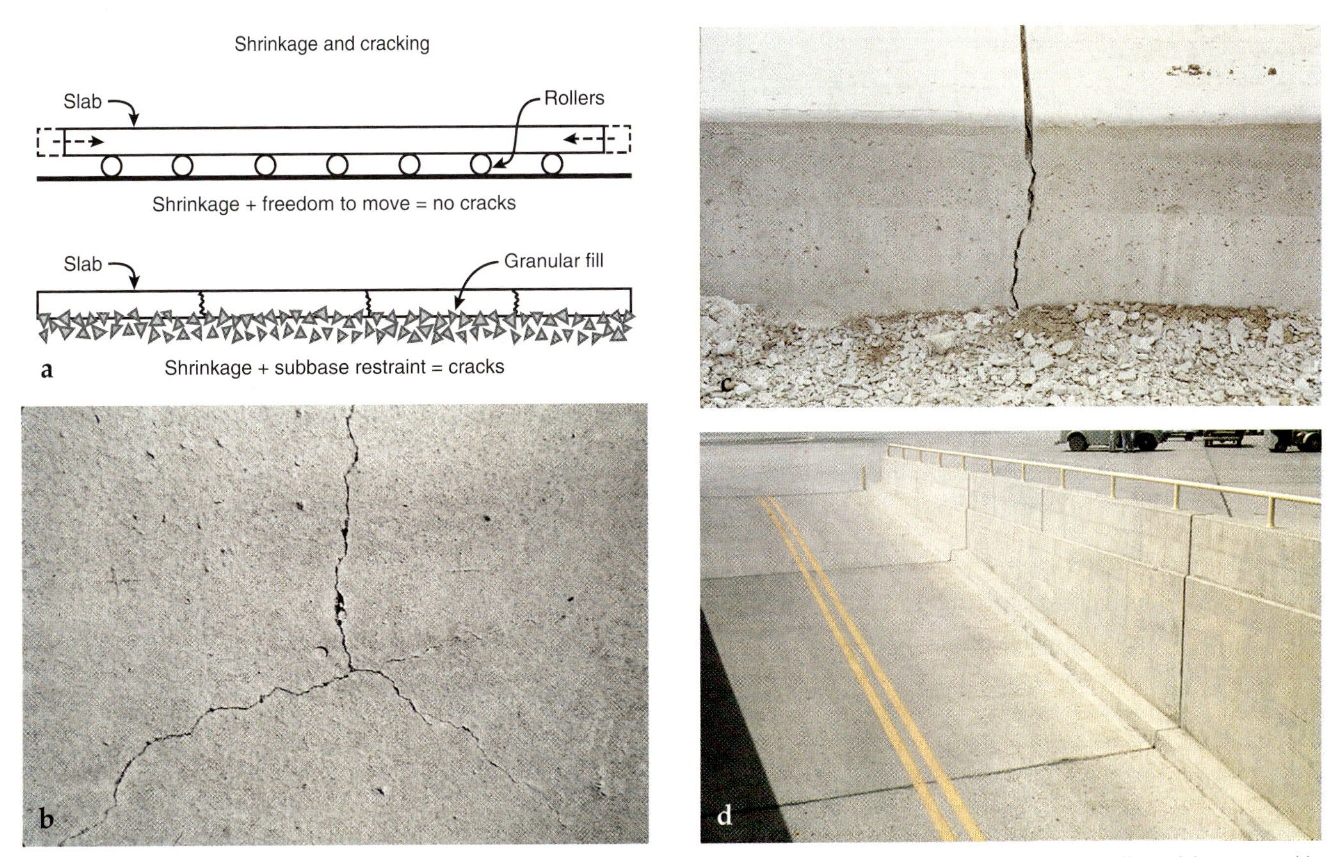

Figure 10-10. (a) Illustration showing no crack development in concrete that is free to shrink (slab on rollers); however, in reality a slab on ground is restrained by the subbase (or other elements) creating tensile stresses and cracks. (b) Typical shrinkage cracks in a slab on ground. (c) A properly functioning contraction joint controls the location of shrinkage cracking. (d) Contraction joints in the slabs and walls shown will minimize the formation of cracks.

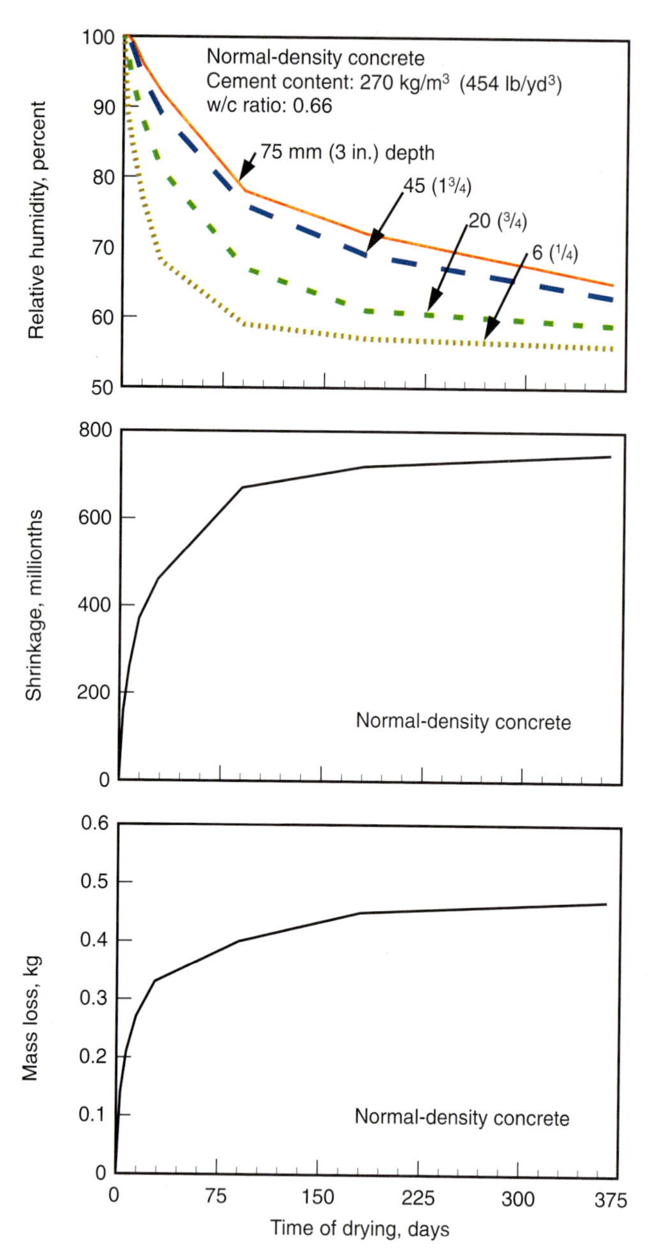

Figure 10-11. Relative humidity distribution at various depths, drying shrinkage, and mass loss of 150 x 300-mm (6 x 12-in.) cylinders moist-cured for 7 days followed by drying in laboratory air at 23°C (73°F) and 50% RH (Hanson 1968).

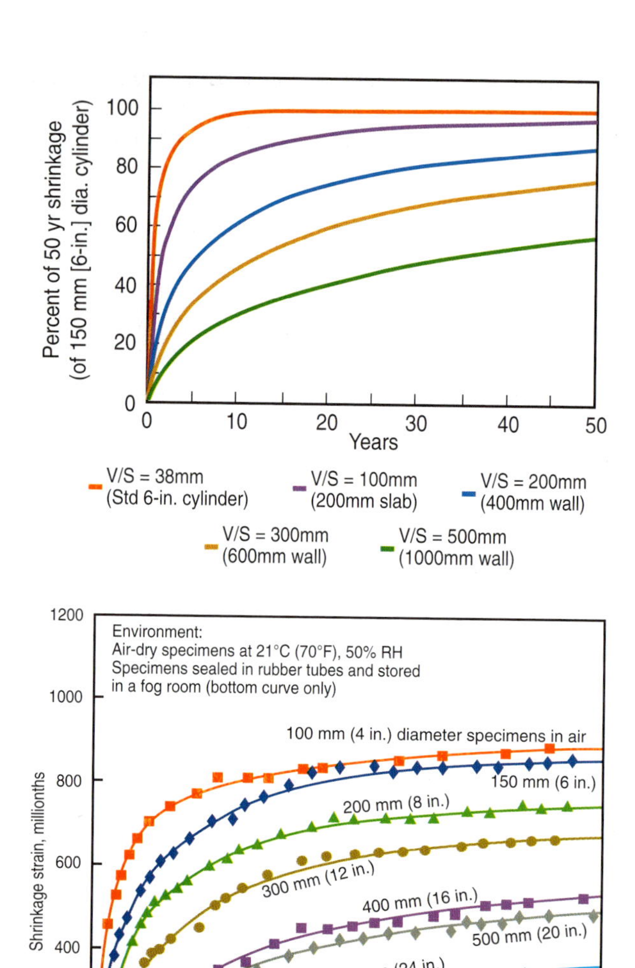

Figure 10-12. (top) Drying shrinkage of various volume/surface ratios based on the 50-year shrinkage of a standard 150-mm x 300-mm (6-in. x 12-in.) cylinder (Baker and others 2007). (bottom) Drying shrinkage of various sizes of cylindrical specimens made of Elgin, Illinois gravel concrete (Hansen and Mattock 1966).

Effect of Concrete Ingredients on Drying Shrinkage

The most important controllable factor affecting drying shrinkage is the amount of water per unit volume of concrete. Typically, about half the water in concrete mixtures is consumed by the chemical reaction of hydration and the other half is necessary to achieve workability and finishability during placement. Some of this excess water is expelled from the concrete during the bleeding phase. However, water that is not bled during placement, remains in the hardened concrete to eventually evaporate and contribute to drying shrinkage. The results of tests illustrating the water content to shrinkage relationship are shown in Figure 10-14. Shrinkage can be minimized by keeping the water content of concrete as low as possible. This is achieved by keeping the total coarse aggregate content of the concrete as high as possible (minimizing paste content). Use of low slumps and placing methods that minimize water requirements are thus major factors in controlling concrete shrinkage. Any practice that increases the water requirement of the cement paste, such as the use of high slumps (without superplasticizers), excessively high freshly mixed concrete temperatures, high fine-aggregate contents, or use of small-size coarse aggregate, will increase shrinkage.

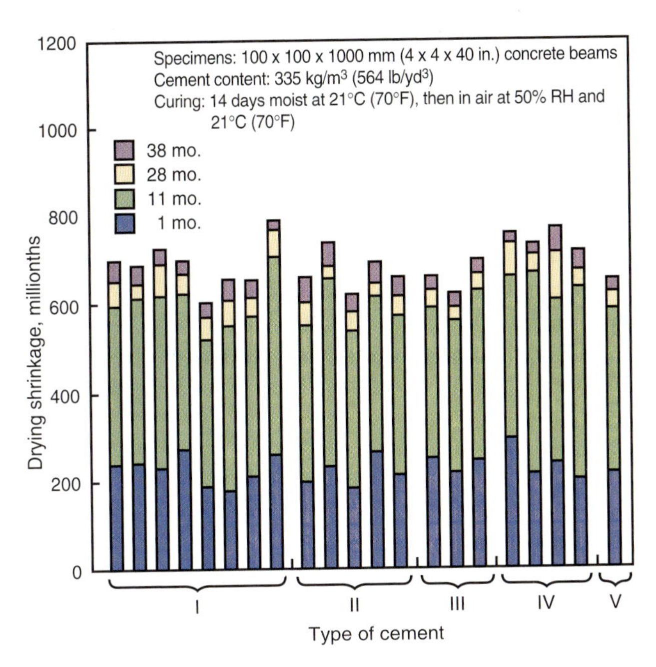

Figure 10-13. Results of long-term drying shrinkage tests by the U.S. Bureau of Reclamation. Shrinkage ranged from 600 to 790 millionths after 38 months of drying. The shrinkage of concretes made with air-entraining cements was similar to that for non-air-entrained concretes in this study (Bureau of Reclamation 1947 and Jackson 1955).

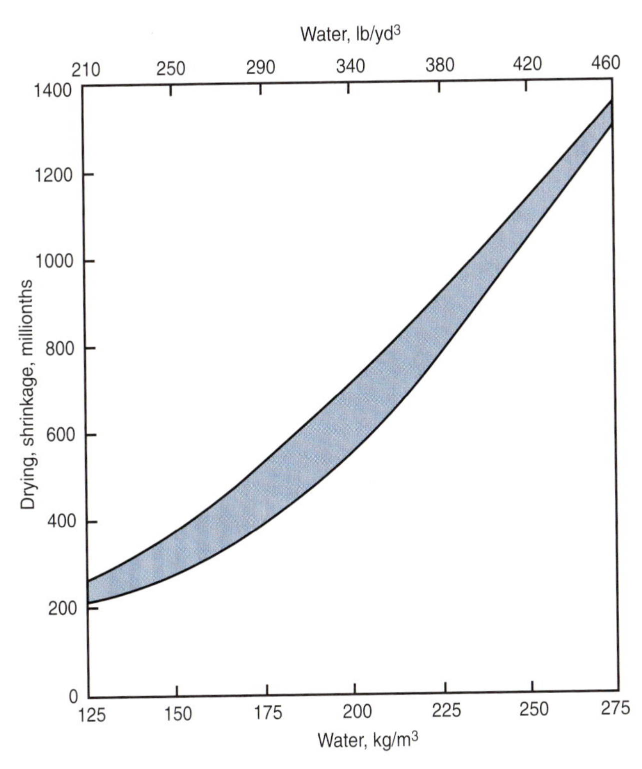

Figure 10-14. Relationship between total water content and drying shrinkage. A large number of mixtures with various proportions is represented within the shaded area of the curves. Drying shrinkage increases with increasing water contents.

Also, anything that decreases the bleed such as accelerators, air entrainment, and fibers can result in greater drying shrinkage. The effect of accelerators is generally offset by low temperatures which increase the bleed period. It has been found that a small amount of water can be added to ready mixed concrete at the jobsite without affecting drying shrinkage properties as long as the additions are within mix specifications (Suprenant and Malisch 2000).

The general uniformity of shrinkage of concretes with different types of cement at different ages is illustrated in Figure 10-13. However, this does not mean that all cements or cementing materials have similar shrinkage. Cements containing limestone may exhibit similar or less drying shrinkage depending on the fineness (Bucher, Radlinska, and Weiss 2008 and Bentz and others 2009). Fly ash and slag cement have little effect on shrinkage at normal dosages. Figure 10-15 shows that concretes with normal dosages of selected fly ashes performed similar to the control concrete made with only portland cement as the cementing material. However, very fine SCMs, such as silica fume that increase water demand can also increase shrinkage properties in concrete mixtures if not accounted for in the design.

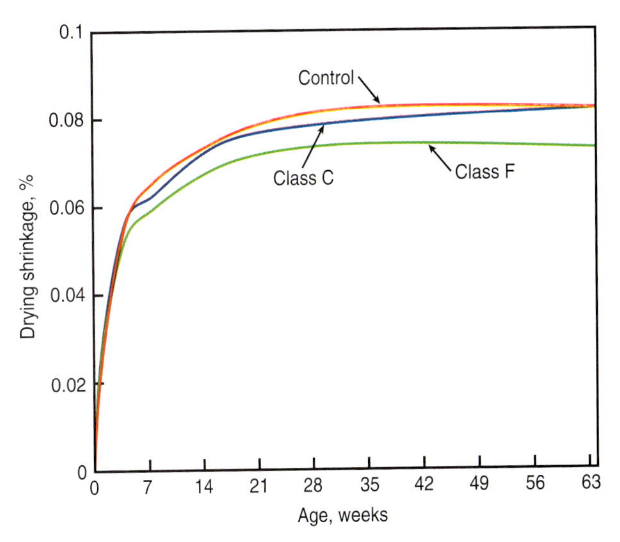

Figure 10-15. Drying shrinkage of fly ash concretes compared to a control mixture. The graphs represent the average of four Class C ashes and six Class F ashes, with the range in drying shrinkage rarely exceeding 0.01 percentage points. Fly ash dosage was 25% of the cementing material (Gebler and Klieger 1986).

Aggregates in concrete, especially coarse aggregate, physically restrain the shrinkage of hydrating and drying cement paste. Paste content affects the drying shrinkage of mortar more than that of concrete. Drying shrinkage is also highly dependent on the type of aggregate. Except for total water content, the type of aggregate has the greatest influence on the drying shrinkage potential. Studies have shown that the type of aggregate can increase the drying shrinkage by as much as 120% to 150% (Powers 1959, Meininger 1966, and Tremper and Spellmen 1963). Hard, rigid aggregates are difficult to compress and provide more restraint to shrinkage than softer, less rigid aggregates. As an extreme example, if steel balls were substituted for ordinary coarse aggregate, shrinkage would be reduced 30% or more. Drying shrinkage can also be reduced by avoiding aggregates that have high drying shrinkage properties such as some sandstone and graywacke, and aggregates containing excessive amounts of clay. Quartz, granite, feldspar, limestone, and dolomite aggregates generally produce concretes with lower drying shrinkages (ACI Committee 224 2001). Steam curing will also reduce drying shrinkage.

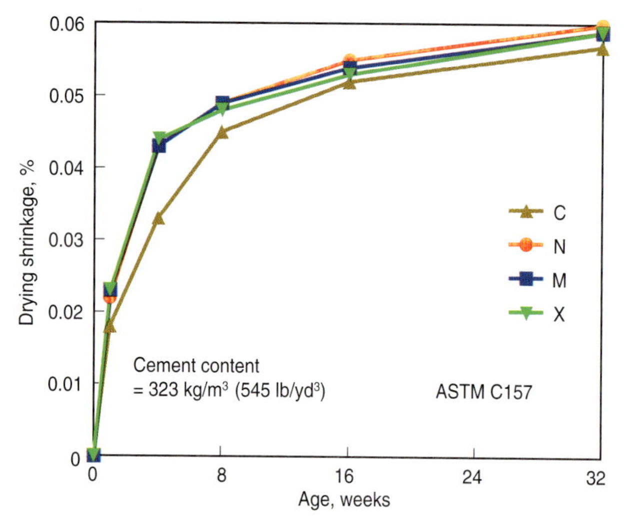

Figure 10-16. Drying shrinkage of concretes made with selected high-range water reducers (N,M, and X) compared to a control mixture (C) (Whiting and Dziedzic 1992).

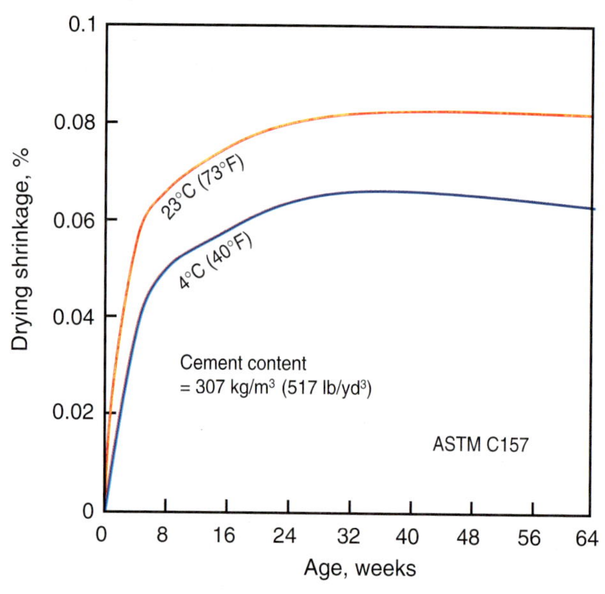

Figure 10-17. Effect of initial curing on drying shrinkage of portland cement concrete prisms. Concrete with an initial 7-day moist cure at 4°C (40°F) had less shrinkage than concrete with an initial 7-day moist cure at 23°C (73°F). Similar results were found with concretes containing 25% fly ash as part of the cementing material (Gebler and Klieger 1986).

Most chemical admixtures have little effect on shrinkage. However, the use of accelerators such as calcium chloride will increase drying shrinkage of concrete. Despite reductions in water content, some water-reducing admixtures can increase drying shrinkage, particularly those that contain an accelerator to counteract the retarding effect of the admixture. Air entrainment has little or no effect on drying shrinkage. High-range water reducers usually have little effect on drying shrinkage (Figure 10-16). Drying shrinkage can be evaluated in accordance with ASTM C157, *Standard Test Method for Length Change of Hardened Hydraulic-Cement Mortar and Concrete* (AASHTO T 160).

Effect of Curing on Drying Shrinkage

The amount and type of curing can affect the rate and ultimate amount of drying shrinkage. Curing compounds, sealers, and coatings can trap free moisture in the concrete for long periods, resulting in delayed shrinkage. Wet curing methods, such as fogging or wet burlap, hold off shrinkage until curing is terminated, after which the concrete dries and shrinks at a normal rate. Cooler initial curing temperatures can reduce shrinkage (Figure 10-17). Steam curing will also reduce drying shrinkage. Computer software is available to predict the effect of curing and environmental conditions on shrinkage and cracking (FHWA and Transtec 2001). Hedenblad (1997) provides tools to predict the drying of concrete as effected by different curing methods and type of construction.

Temperature Changes of Hardened Concrete

Concrete expands slightly as temperature rises and contracts as temperature falls, although it can expand slightly as free water in the concrete freezes. Temperature changes may be caused by environmental conditions or by cement hydration. An average value for the coefficient of thermal expansion of concrete is about 10 millionths per degree Celsius (5.5 millionths per degree Fahrenheit), although values ranging from 6 to 13 millionths per degree Celsius (3.2 to 7.0 millionths per degree Fahrenheit) have been observed. This amounts to a length change of 5 mm for 10 m of concrete (2⁄3 in. for 100 ft of concrete) subjected to a rise or fall of 50°C (100°F). The coefficient of thermal expansion for structural low-density (lightweight) concrete varies from 7 to 11 millionths per degree Celsius (3.6 to 6.1 millionths per degree Fahrenheit). The coefficient of thermal expansion of concrete can be determined by AASHTO T336, *Standard Method of Test for Coefficient of Thermal Expansion of Hydraulic Cement Concrete.*

Thermal expansion and contraction of concrete varies with factors such as aggregate type, cement content, water-cement ratio, temperature range, concrete age, and relative humidity. Of these, aggregate type has the greatest influence.

Table 10-1. Effect of Aggregate Type on Thermal Coefficient of Expansion of Concrete

Aggregate type (from one source)	Coefficient of expansion, mllionths per °C	Coefficient of expansion, mllionths per °F
Quartz	11.9	6.6
Sandstone	11.7	6.5
Gravel	10.8	6.0
Granite	9.5	5.3
Basalt	8.6	4.8
Limestone	6.8	3.8

Coefficients of concretes made with aggregates from different sources may vary widely from these values, especially those for gravels, granites, and limestones (Davis 1930).

Table 10-1 shows some experimental values of the thermal coefficient of expansion of concretes made with aggregates of various types. These data were obtained from tests on small concrete specimens in which all factors were the same except aggregate type. In each case, the fine aggregate was of the same material as the coarse aggregate.

The thermal coefficient of expansion for steel is about 12 millionths per degree Celsius (6.5 millionths per degree Fahrenheit), which is comparable to that for concrete. The coefficient for reinforced concrete can be assumed as 11 millionths per degree Celsius (6 millionths per degree Fahrenheit), the average for concrete and steel.

Temperature changes that result in shortening can crack concrete members that are highly restrained by another part of the structure or by ground friction. Consider a long restrained concrete member cast without joints that, after moist curing, is allowed to drop in temperature. As the temperature drops, the concrete wants to shorten, but cannot because it is restrained longitudinally. The resulting tensile stresses cause the concrete to crack. Tensile strength and modulus of elasticity of concrete both may be assumed proportional to the square root of concrete compressive strength. Calculations show that a large enough temperature drop will crack concrete regardless of its age or strength, provided the coefficient of expansion does not vary with temperature and the concrete is fully restrained (FHWA and Transtec 2001 and PCA 1982).

Precast wall panels and slabs on ground are susceptible to bending and curling caused by temperature gradients that develop when concrete is cool on one side and warm on the other. The calculated amount of curling in a wall panel is illustrated in Figure 10-18.

For the effect of temperature changes in mass concrete due to heat of hydration, see Chapter 20.

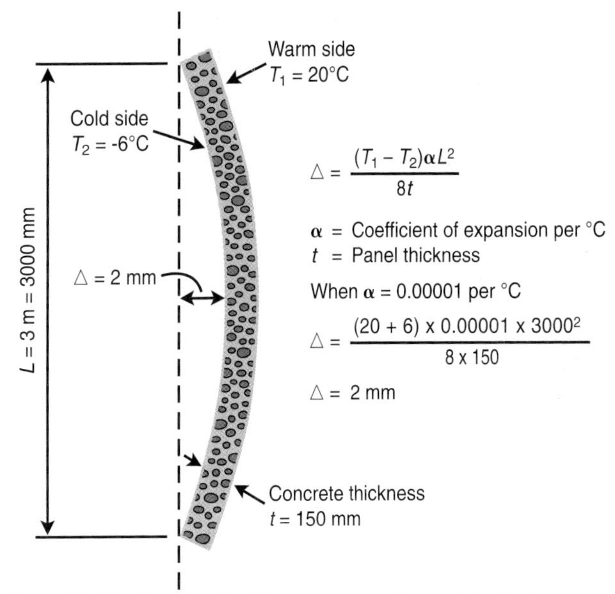

Figure 10-18. Curling of a plain concrete wall panel due to temperature that varies uniformly from inside to outside.

Low Temperatures

Concrete continues to contract as the temperature is reduced below freezing. The amount of volume change at subfreezing temperatures is greatly influenced by the moisture content, behavior of the water (physical state – ice or liquid), and type of aggregate in the concrete. In one study, the coefficient of thermal expansion for a temperature range of 24°C to -157°C (75°F to –250°F) varied from 6×10^{-6} per °C (3.3×10^{-6} per °F) for a low density (lightweight) aggregate concrete to 8.2×10^{-6} per °C (4.5×10^{-6} per °F) for a sand and gravel mixture. Subfreezing temperatures can significantly increase the compressive and tensile strength and modulus of elasticity of moist concrete. Dry concrete properties are not as affected by low temperatures. In the same study, moist concrete with an original compressive strength of 35 MPa at 24°C (5000 psi at 75°F) achieved over 117 MPa (17,000 psi) at –101°C (–150°F). The same concrete tested ovendry or at a 50% internal relative humidity had strength increases of only about 20%. The modulus of elasticity for sand and gravel concrete with 50% relative humidity was only 8% higher at –157°C than at 24°C (–250°F than at 75°F), whereas the moist concrete had a 50% increase in modulus of elasticity. Cooling from 24°C to –157°C (75°F to –250°F), the thermal conductivity of normal-weight concrete also increased, especially for moist concrete. The thermal conductivity of lightweight aggregate concrete is little affected (Monfore and Lentz 1962 and Lentz and Monfore 1966).

High Temperatures

Temperatures greater than 95°C (200°F) that are sustained for several months or even several hours can have significant effects on concrete. The total amount of volume change of concrete is the sum of volume changes of the cement paste and aggregate. At high temperatures, the paste shrinks due to dehydration while the aggregate expands. For normal-aggregate concrete, the expansion of the aggregate exceeds the paste shrinkage resulting in an overall expansion of the concrete. Some aggregates such as expanded shale, andesite, or pumice with low coefficients of expansion can produce a very volume-stable concrete in high-temperature environments (Figure 10-19). On the other hand, some aggregates undergo extensive and abrupt volume changes at a particular temperature, causing disruption in the concrete. For example, in one study a dolomitic limestone aggregate containing an iron sulfide impurity caused severe expansion, cracking, and disintegration in concrete exposed to a temperature of 150°C (302°F) for four months; yet at temperatures above and below 150°C (302°F) there was no detrimental expansion (Carette, Painter, and Malhotra 1982). The coefficient of thermal expansion tends to increase with temperature rise.

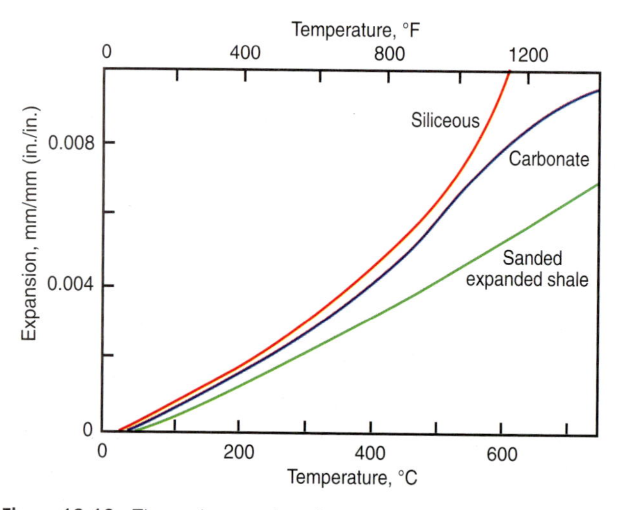

Figure 10-19. Thermal expansion of concretes containing various types of aggregate (Abrams 1977).

Besides volume change, sustained high temperatures can also have other, usually irreversible, effects such as a reduction in strength, modulus of elasticity, and thermal conductivity. Creep increases with temperature. Above 100°C (212°F), the paste begins to dehydrate (lose chemically combined water of hydration) resulting in significant strength losses. Strength decreases with increases in temperature until the concrete loses essentially all its strength. The effect of high-temperature exposure on compressive strength of concretes made with various types of aggregate is illustrated in Figure 10-20. Several factors including concrete moisture content, aggregate type and stability, cement content, exposure time, rate of temperature rise, age of concrete, restraint, and existing stress all influence the behavior of concrete at high temperatures. Figure 10-21 shows the difference in response between normal-strength concrete and high-strength concrete (HSC) (see Chapter 19).

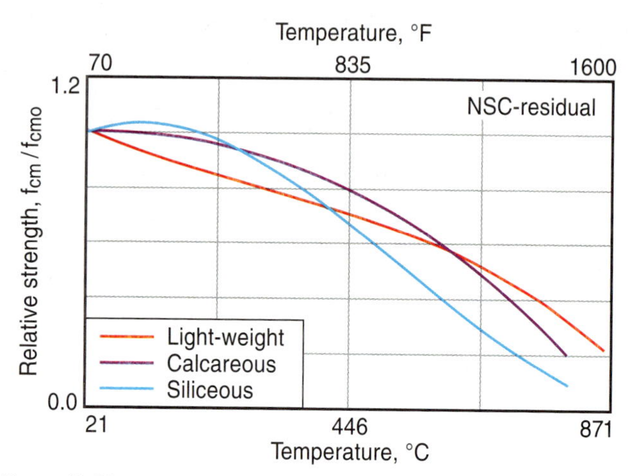

Figure 10-20. Relationships between high temperatures and the residual compressive strength of normal-strength concretes containing various types of aggregate (Knaack, Kurama, and Kirkner 2009).

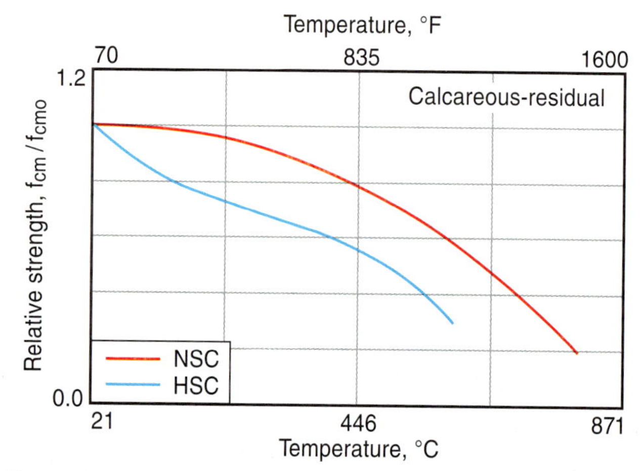

Figure 10-21. Relationships between high temperatures and the residual compressive strength of normal-strength and high-strength concretes containing calcareous aggregate (Knaack, Kurama, and Kirkner 2009).

If stable aggregates are used and strength reduction and the effects on other properties are accounted for in the mix design, high quality concrete can be exposed to temperatures of 90°C to 200°C (200°F to 400°F) for long periods. Some concrete elements have been exposed to temperatures up to 250°C (500°F) for long periods; however, special steps should be taken or special materials (such as heat-resistant calcium aluminate cement) should be considered for exposure temperatures greater than 200°C (400°F). Before any structural concrete is exposed to high temperatures (greater than 90°C or 200°F), laboratory testing should be performed to determine the thermal properties. This will avoid any unexpected distress.

Curling and Warping

In addition to horizontal movement caused by changes in moisture and temperature, curling and warping of slabs on ground can be a problem. This is caused by differences in moisture content and temperature between the top and bottom of slabs (Figure 10-22).

Curling and warping are intimately related to the shrinkage potential of the concrete mixture. While the terms have been used interchangeably for the past century, there is a difference between "curling" and "warping" due to the difference in the cause of the distortion (Tarr 2004). Curling is the deformation of the slab due to a difference in temperature between the surface and the bottom of the slab. Like most materials, concrete expands and contracts with change in temperature. If the slab surface is cooler than the slab bottom, the surface contracts causing the slab edges to curl upward. Slab "warping" is the deformation of the slab surface profile due to a difference in moisture between the surface and bottom of the slab. As with a sponge, if the slab surface is allowed to dry and the bottom is kept moist, the edges will tend to warp upward. Exterior pavement slabs typically have a permanent upward edge warp and experience curling on a daily basis due to surface warming and cooling cycles related to exposure to the sun. In general, the edges of interior concrete floor slab panels warp upward due to a moisture difference between the top and bottom of the slab (Tarr and Farny 2008).

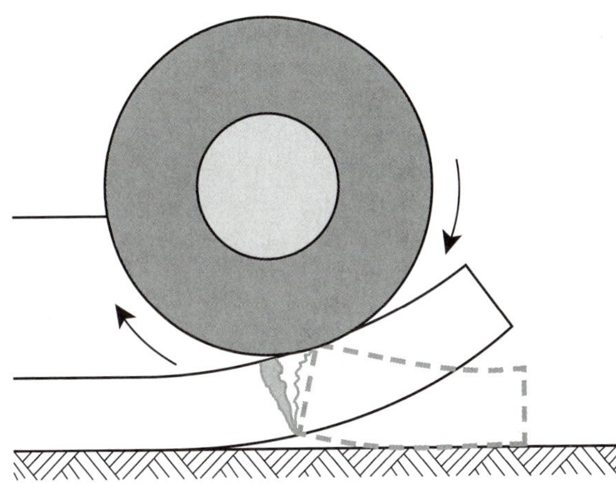

Figure 10-22. Illustration of curling of a concrete slab on ground. The edge of the slab at a joint or free end lifts off the subbase creating a cantilevered section of concrete that can break off under heavy wheel loading.

A slab will assume a reverse curl when the surface is wetter or warmer than the bottom. However, enclosed slabs, such as floors on ground, only curl upward. When the edges of an industrial floor slab are curled upward they lose support from the subbase and become a cantilever. Lift-truck traffic passing over joints causes a repetitive vertical deflection that creates a great potential for fatigue cracking in the slab. The amount of vertical upward curl (curling) is small for a short, thick slab.

Curling can be reduced or eliminated by using design and construction techniques that minimize shrinkage potential and by using techniques described earlier to reduce temperature and moisture-related volume changes. Thickened edges, shorter joint spacings, permanent vapor-impermeable sealers, and large amounts of reinforcing steel placed 50 mm (2 in.) below the surface all help reduce curling (Ytterberg 1987). Design options such as post-tensioning and shrinkage-compensating concrete can also be used to control curling/warping (Tarr and Farny 2008).

Elastic and Inelastic Deformation

Compression Strain

The series of curves in Figure 10-23 illustrate the amount of compressive stress and strain that results instantaneously due to loading of unreinforced concrete. With water-cement ratios of 0.50 or less and strains up to 1500 millionths, the upper three curves show that strain is closely proportional to stress; in other words, the concrete is almost elastic. The upper portions of the curves and beyond show that the concrete is inelastic. The curves for high-strength concrete have sharp peaks, whereas those for lower-strength concretes have long and relatively flat peaks. Figure 10-23 also shows the sudden failure characteristics of higher strength, low water to cement ratio, concrete cylinders.

When load is removed from concrete in the inelastic zone, the recovery line usually is not parallel to the original line for the first load application. Therefore, the amount of permanent set may differ from the amount of inelastic deformation (Figure 10-24).

The term "elastic" is not favored for general discussion of concrete behavior because frequently the strain may be in the inelastic range. For this reason, the term "instantaneous strain" is often used.

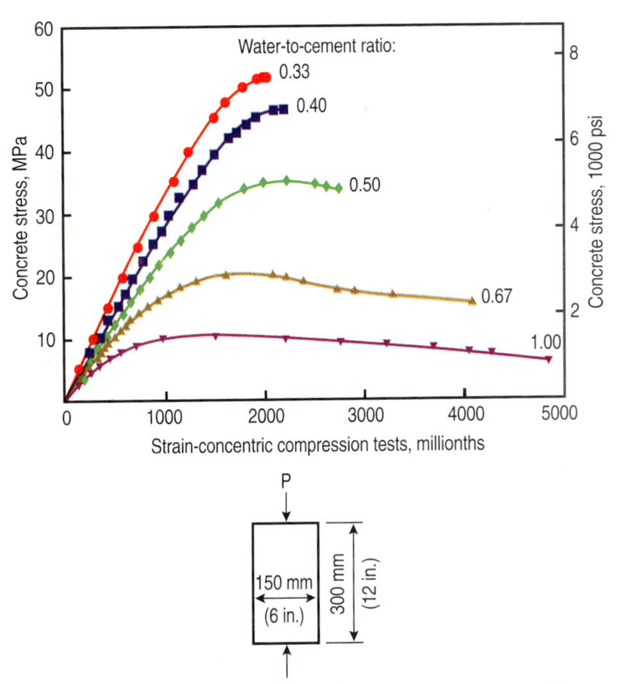

Figure 10-23. Stress-strain curves for compression tests on 150 x 300-mm (6 x 12-in.) concrete cylinders at an age of 28 days (Hognestad, Hanson, and McHenry 1955).

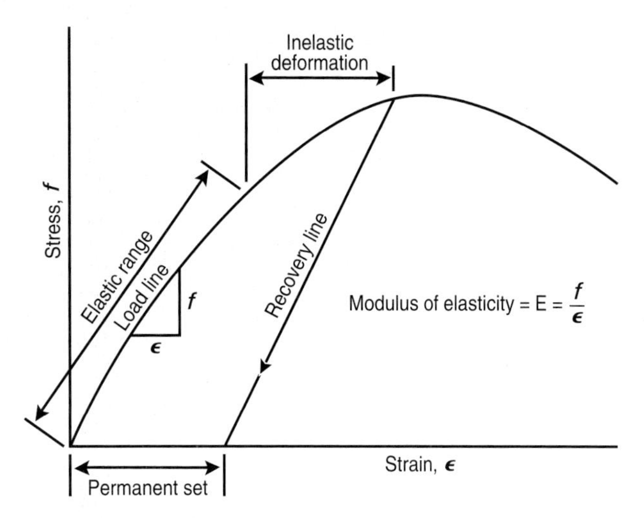

Figure 10-24. Generalized stress-strain curve for concrete.

Modulus of Elasticity

The ratio of stress to strain in the elastic range of a stress-strain curve for concrete defines the modulus of elasticity (E) of that concrete (Figure 10-24). Normal-density concrete has a modulus of elasticity of 14,000 MPa to 41,000 MPa (2,000,000 psi to 6,000,000 psi), depending on factors such as compressive strength and aggregate type. For normal-density concrete with compressive strengths (f_c) between 20 MPa and 35 MPa (3000 psi and 5000 psi), the modulus of elasticity can be estimated as 5000 times the square root of f_c (57,000 times the square root of f_c in psi). The modulus of elasticity for structural lightweight concrete is between 7000 MPa and 17,000 MPa (1,000,000 psi and 2,500,000 psi). E for any particular concrete can be determined in accordance with ASTM C469, *Standard Test Method for Static Modulus of Elasticity and Poisson's Ratio of Concrete in Compression*. See Purl and Weiss (2006) and Weiss (2006) for stress-strain relationships.

Deflection

Deflection of concrete beams and slabs is one of the more common and obvious building movements. The deflections are the result of flexural strains that develop under dead and live loads. They may result in cracking in the tensile zone of concrete members. Reinforced concrete structural design anticipates these tension cracks. Concrete members are often cambered, that is, built with an upward bow, to compensate for the expected later deflection.

Poisson's Ratio

When a block of concrete is loaded in uniaxial compression, as in Figure 10-25, it will compress and simultaneously develop a lateral strain or bulging. The ratio of lateral to axial strain is called Poisson's ratio, μ. A common value used is 0.20 to 0.21, but the value may vary from 0.15 to 0.25 depending upon the aggregate, moisture content, concrete age, and compressive strength. Poisson's ratio (ASTM C469) is generally of no concern to the structural designer. It is used in advanced structural analysis of flat-plate floors, shell roofs, arch dams, and mat foundations.

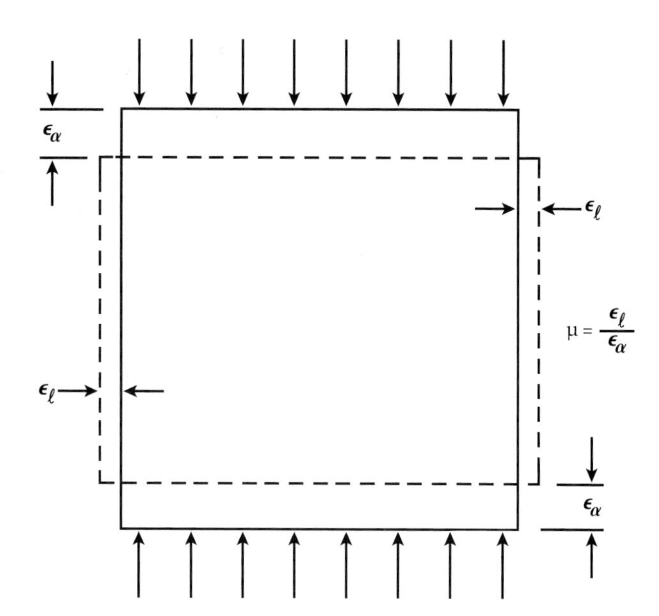

Figure 10-25. Ratio of lateral to axial strain is Poisson's ratio, μ.

Shear Strain

Concrete, like other materials, deforms under shear forces. The shear strain produced is important in determining the load paths or distribution of forces in indeterminate structures—for example where shear-walls and columns both participate in resisting horizontal forces in a concrete building frame. The amount of movement, while not large, is significant in short, stubby members. In larger members, flexural strains are much greater and more significant to designers. Calculation of the shear modulus (modulus of rigidity), G, is shown in Figure 10-26; G varies with the strength and temperature of the concrete.

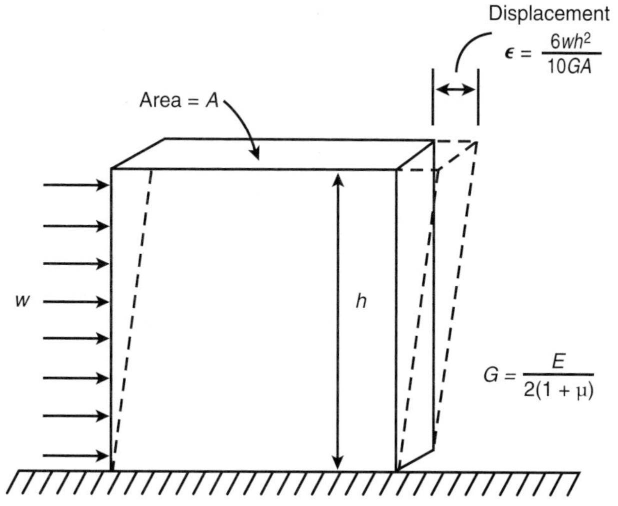

Figure 10-26. Strain that results from shear forces on a body. G = shear modulus. μ = Poisson's ratio. Strain resulting from flexure is not shown.

Torsional Strain

Plain rectangular concrete members can also fail in torsion, that is, a twisting action caused by bending about an axis parallel to the wider face and inclined at an angle of about 45 degrees to the longitudinal axis of a member. Microcracks develop at low torque. However, concrete behaves reasonably elastic up to the maximum limit of the elastic torque (Hsu 1968).

Creep

When concrete is loaded, the deformation caused by the load can be divided into two parts: a deformation that occurs immediately (elastic strain) and a time-dependent deformation that begins immediately but continues at a decreasing rate as long as the concrete is loaded. This latter deformation is called creep.

The amount of creep is dependent upon (1) the magnitude of the applied stress, (2) the age and strength of the concrete when stress is applied, and (3) the length of time the concrete is stressed. It is also affected by other factors related to the quality of the concrete and conditions of exposure, such as: (1) type, amount, and maximum size of aggregate; (2) type of cementing materials; (3) amount of cement paste; (4) size and shape of the concrete element; (5) volume to surface ratio of the concrete element; (6) amount of steel reinforcement; (7) prior curing conditions; and (8) the ambient temperature and humidity. See ACI Committee 209 (2005) for more on the factors that affect creep.

Within normal stress ranges, creep is proportional to stress. In relatively young concrete, the change in volume or length due to creep is largely unrecoverable; in older or drier concrete it is largely recoverable.

The creep curves shown in Figure 10-27 are based on tests conducted under laboratory conditions in accordance with ASTM C512, *Standard Test Method for Creep of Concrete in Compression* (Figure 10-28). Cylinders were loaded to almost 40% of their compressive strength. Companion cylinders not subject to load were used to measure drying shrinkage; this was then deducted from the total deformation of the loaded specimens to determine creep. Cylinders were allowed to dry while under load except for those marked "sealed." The two 28-day curves for each concrete strength in Figure 10-27 show that creep of concrete loaded under drying conditions is greater than creep of concrete sealed against drying. Concrete specimens loaded at a late age will creep less than those loaded at an early age. It can be seen that as concrete strength decreases, creep increases. Figure 10-29 illustrates recovery from the elastic and creep strains after load removal.

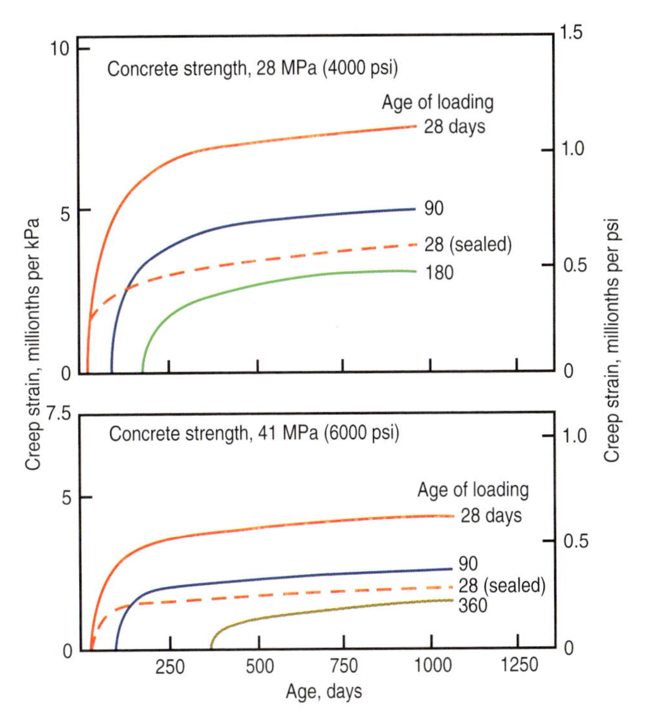

Figure 10-27. Relationship of time and age of loading to creep of two different strength concretes. Specimens were allowed to dry during loading, except for those labeled as sealed (Russell and Corley 1977).

Figure 10-28. ASTM C512 creep test frames loaded with both sealed and unsealed specimens.

A combination of strains occurring in a reinforced column is illustrated in Figure 10-30. The curves represent deformations and volume changes in a 14th-story column of a 76-story reinforced concrete building while under construction. The 400-mm x 1200-mm (16-in. x 48-in.) column contained 2.08% vertical reinforcement and was designed for 60-MPa (9000-psi) concrete.

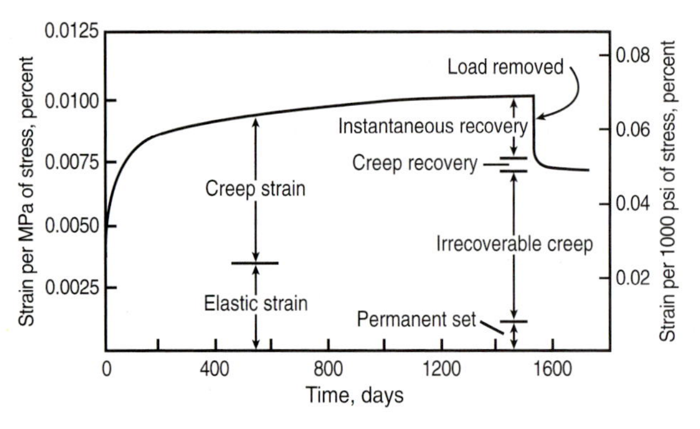

Figure 10-29. Combined curve of elastic and creep strains showing amount of recovery. Specimens (cylinders) were loaded at 8 days immediately after removal from fog curing room and then stored at 21°C (70°F) and 50% RH. The applied stress was 25% of the compressive strength at 8 days (Hansen and Mattock 1966).

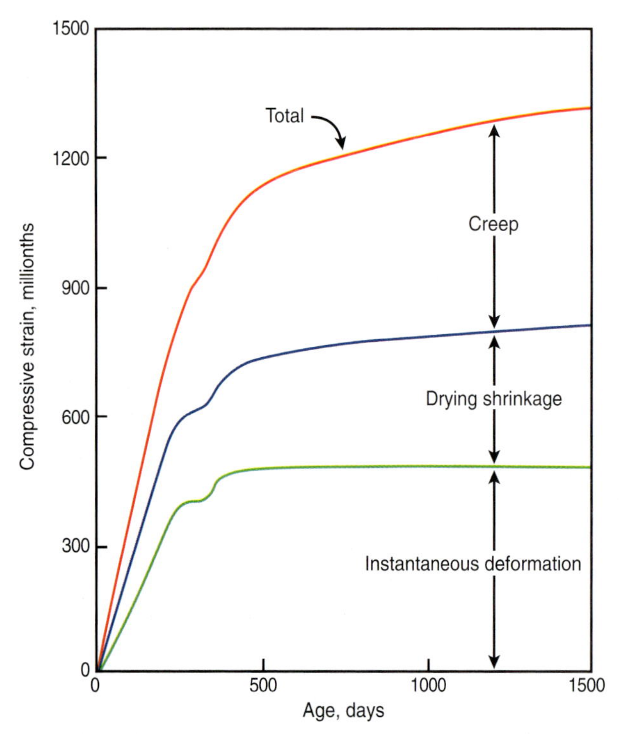

Figure 10-30. Summation of strains in a reinforced concrete column during construction of a tall building (Russell and Corley 1977).

The method of curing prior to loading has a marked effect on the amount of creep in concrete. The effects on creep of three different methods of curing are shown in Figure 10-31. Note that very little creep occurs in concrete that is cured by high-pressure steam (autoclaving). Note also that atmospheric steam-cured concrete has considerably less creep than 7-day moist-cured concrete. The two methods of steam curing shown in Figure 10-31 reduce drying shrinkage of concrete about half as much as they reduce creep.

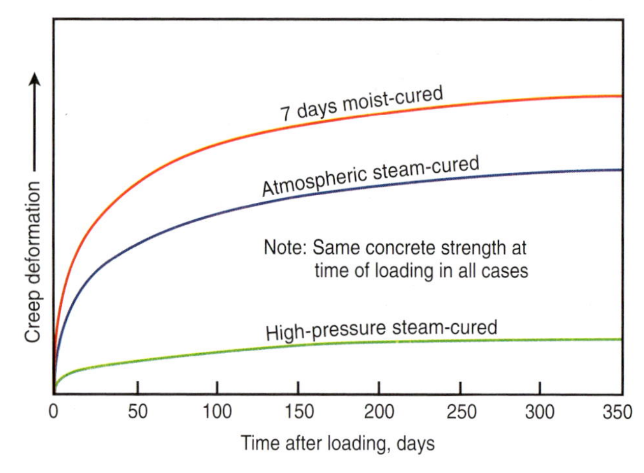

Figure 10-31. Effect of curing method on magnitude of creep for typical normal-density concrete (Hanson 1964).

Chemical Changes and Effects

Some volume changes of concrete result from chemical reactions; these may take place shortly after placing and finishing or later due to reactions within the hardened concrete in the presence of water or moisture.

Carbonation

Hardened concrete containing some moisture reacts with carbon dioxide present in air, a reaction that results in a slight shrinkage of the surface paste of the concrete. The effect, known as carbonation, is not destructive but actually increases the chemical stability and strength of the concrete. However, carbonation also reduces the pH of concrete. If steel is present in the carbonated area, steel corrosion can occur due to the absence of the protective oxide film provided by concrete's high pH. Rust is an expansive material and results in cracking and spalling of the concrete. The depth of carbonation is very shallow in dense, high-quality concrete, but can penetrate deeply in porous, poor-quality concrete. Because so little of a concrete element carbonates, carbonation shrinkage of cast-in-place concrete is insignificant and is usually not considered in engineering practice.

Carbonation of paste proceeds slowly and produces little direct shrinkage at relative humidities of 100% and 25%. Maximum carbonation and carbonation shrinkage occurs at about 50% relative humidity. Irreversible shrinkage and weight gain occurs during carbonation. And the carbonated product may show improved volume stability to subsequent moisture change and reduced permeability (Verbeck 1958).

During manufacture some concrete masonry units are deliberately exposed to carbon dioxide after reaching 80% of their rated strength. This introduction to carbonation shrinkage makes the units more dimensionally stable.

Future drying shrinkage is reduced 30% or more (Toennies and Shideler 1963).

One of the causes of surface crazing of concrete is the shrinkage that accompanies natural air carbonation of young concrete. More research is needed on the effect of early carbonation on deicer scaling resistance.

Carbonation of another kind also can occur in freshly placed, unhardened concrete. This carbonation causes a soft, chalky surface called dusting. It usually takes place during cold-weather concreting when there is an unusual amount of carbon dioxide in the air due to unvented heaters or gasoline-powered equipment operating in an enclosure.

Sulfate Attack

Sulfate attack of concrete can occur where soil and groundwater have a high sulfate content and measures to reduce sulfate attack, such as use of a low water to cementing materials ratio, have not been taken. The attack is greater in concrete that is exposed to wetting and drying, such as foundation walls and posts. Sulfate attack usually results in an expansion of the concrete because of the formation of solids from the chemical action or salt crystallization. The amount of expansion in severe circumstances has been significantly higher than 0.1%, and the disruptive effect within the concrete can result in extensive cracking and disintegration. The amount of expansion cannot be accurately predicted.

Alkali-Aggregate Reactions

Certain aggregates can react with alkali hydroxides in concrete, causing expansion and cracking over a period of years. The reaction is greater in those parts of a structure exposed to moisture. A knowledge of the characteristics of local aggregates is essential. There are two types of alkali-reactive aggregates, siliceous and carbonate. Alkali-aggregate reaction expansion may exceed 0.5% in concrete and can cause the concrete to fracture and break apart.

Structural design techniques cannot counter the effects of alkali-aggregate expansion, nor can the expansion be controlled by jointing. In areas where deleteriously reactive aggregates are known to exist, special measures must be taken to prevent the occurrence of alkali-aggregate reaction (see Chapter 11).

References

Abrams, M.S., *Performance of Concrete Structures Exposed to Fire*, Research and Development Bulletin RD060, Portland Cement Association, http://www.cement.org/pdf_files/RD060.pdf, 1977.

ACI Committee 209, *Prediction of Creep, Shrinkage, and Temperature Effects in Concrete Structures*, ACI 209R-92, reapproved 2008, American Concrete Institute, Farmington Hills, Michigan, 1992, 47 pages.

ACI Committee 209, *Report on Factors Affecting Shrinkage and Creep of Hardened Concrete*, ACI 209.1R-05, American Concrete Institute, Farmington Hills, Michigan, 2005, 12 pages.

ACI Committee 209, *Guide for Modeling and Calculating Shrinkage and Creep in Hardened Concrete*, ACI 209.2R-08, American Concrete Institute, Farmington Hills, Michigan, 2008, 45 pages.

ACI Committee 224, *Control of Cracking in Concrete Structures*, ACI 224R-01, reapproved 2008, American Concrete Institute, Farmington Hills, Michigan, 2001, 46 pages.

ACI Committee 231, *Report on Early-Age Cracking: Causes, Measurement, and Mitigation*, ACI 231R-10, American Concrete Institute, Farmington Hills, Michigan, 2010, 46 pages.

Aïtcin, Pierre-Claude, "Does Concrete Shrink or Does it Swell?," *Concrete International*, American Concrete Institute, Farmington Hills, Michigan, December 1999, pages 77 to 80.

Baker, W.F.; Korista, D.S.; Novak, L.C.; Pawlikowski, J.; and Young, B., "Creep and Shrinkage and the Design of Supertall Buildings – A Case Study: The Burj Dubai Tower," *Structural Implications of Shrinkage and Creep of Concrete*, SP-246CD, American Concrete Institute, Farmington Hills, Michigan, 2007, pages 133 to 148.

Bentur, A., ed., *Early Age Cracking in Cementitious Systems*, RILEM Report 25, RILEM Publications s.a.r.l., Bagneux, France, 2003, 351 pages.

Bentz, Dale P.; Irassar, Edgardo F.; Bucher, Brooks E.; and Weiss, W. Jason, "Limestone Fillers Conserve Cement – Part 2: Durability Issues and the Effects of Limestone Fineness on Mixtures," *Concrete International*, Vol. 31, No. 12, December 2009, pages 35 to 39.

Bucher, Brooks; Radlinska, Aleksandra; and Weiss, Jason, *Preliminary Comments on Shrinkage and Shrinkage Cracking Behavior of Cement Systems that Contain Limestone*, Concrete Technology Forum 2008, National Ready Mixed Concrete Association, Silver Spring, Maryland, 2008, 8 pages.

Bureau of Reclamation, "Long-Time Study of Cement Performance in Concrete—Tests of 28 Cements Used in the Parapet Wall of Green Mountain Dam," *Materials Laboratories Report No. C-345,* U.S. Department of the Interior, Bureau of Reclamation, Denver, 1947.

Carette, G.G.; Painter, K.E.; and Malhotra, V.M., "Sustained High Temperature Effect on Concretes Made with Normal Portland Cement, Normal Portland Cement and Slag, or Normal Portland Cement and Fly Ash," *Concrete International*, American Concrete Institute, Farmington Hills, Michigan, July 1982.

Copeland, L.E., and Bragg, R.H., *Self Desiccation in Portland Cement Pastes*, Research Department Bulletin RX052, Portland Cement Association, http://www.cement.org/pdf_files/RX052.pdf, 1955.

Davis, R.E., "A Summary of the Results of Investigations Having to Do with Volumetric Changes in Cements, Mortars, and Concretes Due to Causes Other Than Stress," *Proceedings of the American Concrete Institute*, American Concrete Institute, Farmington Hills, Michigan, Vol. 26, 1930, pages 407 to 443.

FHWA and Transtec, HIPERPAV, http://www.hiperpav.com, 2001.

Gebler, Steven H., and Klieger, Paul, *Effect of Fly Ash on Some of the Physical Properties of Concrete*, Research and Development Bulletin RD089, Portland Cement Association, http://www.cement.org/pdf_files/RD089.pdf, 1986.

Grasley, Zachary, *Internal Relative Humidity, Drying Stress Gradients, and Hygrothermal Dilation of Concrete*, Master's Thesis, University of Illinois at Urbana-Champaign, 2003, 81 pages. [PCA SN2625, www.cement.org/pdf_files/sn2625.pdf]

Hammer, T.A., "Test Methods for Linear Measurement of Autogenous Shrinkage Before Setting," *Autogenous Shrinkage of Concrete*, edited by E. Tazawa, E&FN Spon and Routledge, New York, 1999, pages 143 to 154. Also available through PCA as LT245.

Hansen, Torben C., and Mattock, Alan H., *Influence of Size and Shape of Member on the Shrinkage and Creep of Concrete*, Development Department Bulletin DX103, Portland Cement Association, http://www.cement.org/pdf_files/DX103.pdf, 1966.

Hanson, J.A., *Prestress Loss As Affected by Type of Curing*, Development Department Bulletin DX075, Portland Cement Association, http://www.cement.org/pdf_files/DX075.pdf, 1964.

Hanson, J.A., *Effects of Curing and Drying Environments on Splitting Tensile Strength of Concrete*, Development Department Bulletin DX141, Portland Cement Association, http://www.cement.org/pdf_files/DX141.pdf, 1968.

Hedenblad, Göran, *Drying of Construction Water in Concrete,* Byggforskningsradet, The Swedish Council for Building Research, Stockholm, 1997.

Hognestad, E.; Hanson, N.W.; and McHenry, D., *Concrete Stress Distribution in Ultimate Strength Design*, Development Department Bulletin DX006, Portland Cement Association, http://www.cement.org/pdf_files/DX006.pdf, 1955.

Holt, Erika E., and Janssen, Donald J., "Influence of Early Age Volume Changes on Long-Term Concrete Shrinkage," Transportation Research Record 1610, *Transportation Research Board*, National Research Council, Washington, D.C., 1998, pages 28 to 32.

Hsu, Thomas T.C., *Torsion of Structural Concrete—Plain Concrete Rectangular Sections*, Development Department Bulletin DX134, Portland Cement Association, http://www.cement.org/pdf_files/DX134.pdf, 1968.

Jackson, F.H., *Long-Time Study of Cement Performance in Concrete—Chapter 9. Correlation of the Results of Laboratory Tests with Field Performance Under Natural Freezing and Thawing Conditions*, Research Department Bulletin RX060, Portland Cement Association, http://www.cement.org/pdf_files/RX060.pdf, 1955.

Knaack, Adam; Kurama, Yahya; and Kirkner, David, *Stress-Strain Properties of Concrete at Elevated Temperatures*, Structural Engineering Research Report #NDSE-09-01, University of Notre Dame, April 2009, 153 pages. [PCA SN3111 www.cement.org/pdf_files/SN3111.pdf]

Le Chatelier, H., "Sur les Changements de Volume qui Accompagent le durcissement des Ciments," *Bulletin Societe de l'Encouragement pour l'Industrie Nationale*, 5eme serie, tome 5, Paris, 1900.

Lentz, A.E., and Monfore, G.E., *Thermal Conductivities of Portland Cement Paste, Aggregate, and Concrete Down to Very Low Temperatures*, Research Department Bulletin RX207, Portland Cement Association, http://www.cement.org/pdf_files/RX207.pdf, 1966.

Meininger, R.C., "Drying Shrinkage of Concrete," *Engineering Report*, No. RD3, National Ready Mixed Concrete Association, Silver Spring, Maryland, June 1966.

Monfore, G.E., and Lentz, A.E., *Physical Properties of Concrete at Very Low Temperatures*, Research Department Bulletin RX145, Portland Cement Association, http://www.cement.org/pdf_files/RX145.pdf, 1962.

PCA, *Building Movements and Joints*, EB086, Portland Cement Association, http://www.cement.org/pdf_files/EB086.pdf, 1982.

Philleo, Robert, *Some Physical Properties of Concrete at High Temperatures*, Research Department Bulletin RX097, Portland Cement Association, http://www.cement.org/pdf_files/RX097.pdf, 1958.

Pickett, Gerald, *The Effect of Change in Moisture Content on the Creep of Concrete Under a Sustained Load*, Research Department Bulletin RX020, Portland Cement Association, http://www.cement.org/pdf_files/RX020.pdf, 1947.

Powers, T.C., "Absorption of Water by Portland Cement Paste during the Hardening Process," *Industrial and Engineering Chemistry*, Vol. 27, No. 7, July 1935, pages 790 to 794.

Powers, Treval C., "Causes and Control of Volume Change," Volume 1, Number 1, *Journal of the PCA Research and Development Laboratories*, Portland Cement Association, January 1959.

Purl, S., and Weiss, W.J., "Assessment of Localized Damage in Concrete Under Compression Using Acoustic Emission," *ASCE Journal of Civil Engineering Materials*, Vol. 18, No. 3, 2006, pages 325 to 333.

Radlinska, A.; Rajabipour, F.; Bucher, B.; Henkensiefken, R.; Sant, G.; and Weiss, J., "Shrinkage Mitigation Strategies in Cementitious Systems: a Closer Look at Differences in Sealed and Unsealed Behavior," *Transportation Research Record*, Vol. 2070, 2008, pages 59 to 67.

Roper, Harold, *Volume Changes of Concrete Affected by Aggregate Type*, Research Department Bulletin RX123, Portland Cement Association, http://www.cement.org/pdf_files/RX123.pdf, 1960.

Russell, H.G., and Corley, W.G., *Time-Dependent Behavior of Columns in Water Tower Place*, Research and Development Bulletin RD052, Portland Cement Association, http://www.cement.org/pdf_files/RD052.pdf, 1977.

Suprenant, Bruce A., and Malisch, Ward R., "The Fiber Factor," *Concrete Construction*, Addison, Illinois, October 1999, pages 43 to 46.

Suprenant, Bruce A., and Malisch, Ward R., "A New Look at Water, Slump, and Shrinkage," *Concrete Construction*, Addison, Illinois, April 2000, pages 48 to 53.

Tarr, S.M., "Interior Cement Floors Don't Curl: But Concrete Floors Do Warp and Joints Suffer!" *L&M Concrete News*, L&M Construction Chemicals, Omaha, Nebraska, Vol. 5, No. 2, Fall 2004, page 5.

Tarr, Scott M., and Farny, James A., *Concrete Floors on Ground*, EB075, Fourth Edition, Portland Cement Association, Skokie, Illinois, 2008, 252 pages.

Tazawa, Ei-ichi, *Autogenous Shrinkage of Concrete*, E&FN Spon and Routledge, New York, 1999, 428 pages.

Toennies, H.T., and Shideler, J.J., *Plant Drying and Carbonation of Concrete Block—NCMA-PCA Cooperative Program*, Development Department Bulletin DX064, Portland Cement Association, http://www.cement.org/pdf_files/DX064.pdf, 1963.

Tremper, Bailey, and Spellman, D.L., "Shrinkage of Concrete—Comparison of Laboratory and Field Performance," *Highway Research Record Number 3, Properties of Concrete*, Transportation Research Board, National Research Council, Washington, D.C., 1963.

Verbeck, G.J., *Carbonation of Hydrated Portland Cement*, Research Department Bulletin RX087, Portland Cement Association, http://www.cement.org/pdf_files/RX087.pdf, 1958.

Weiss, W.J., "Chapter 19 – Elastic Properties, Creep, and Relaxation," ASTM 169D, Significance of Tests and Properties of Concrete and Concrete Making Materials, *ASTM International*, West Conshohocken, Pennsylvania, 2006, pages 194 to 206.

Whiting, D., and Dziedzic, W., *Effects of Conventional and High-Range Water Reducers on Concrete Properties*, Research and Development Bulletin RD107, Portland Cement Association, 1992.

Ytterberg, Robert F., "Shrinkage and Curling of Slabs on Grade, Part I—Drying Shrinkage, Part II—Warping and Curling, and Part III— Additional Suggestions," *Concrete International*, American Concrete Institute, Farmington Hills, Michigan, April 1987, pages 22 to 31; May 1987, pages 54 to 61; and June 1987, pages 72 to 81.

CHAPTER 11

Durability

Concrete is an inherently durable material, providing decades of service life. Concrete's durability improves the sustainability of our infrastructure, pavements, buildings, and other structures by conserving resources and reducing environmental impacts related to repair and replacement.

The durability of concrete may be defined as the ability of concrete to resist weathering action, chemical attack, and abrasion while maintaining its desired engineering properties for the expected service life of the structure. To ensure durability of concrete, care in materials selection, design, and construction is most important, with increased care necessary in more severe environments. When properly designed and produced, concrete structures are extremely durable making them sustainable, even under very severe conditions of use (Figure 11-1).

Figure 11-1. The Confederation Bridge, (shown here under construction), is a concrete bridge in a severe environment with a 100 year design life.

Concrete is resistant to most natural environments; however, concrete is sometimes exposed to various substances that can cause deterioration. Most concrete durability failure mechanisms are controlled by the availability of moisture. Deterioration processes typically involve two stages. Initially, aggressive fluids (water or solutions with dissolved solids or gases) penetrate or are transported through the capillary pore structure of the concrete to reaction sites (such as chlorides penetrating to steel reinforcement, or sulfates penetrating to aluminates) where they trigger chemical or physical deterioration mechanisms. Therefore, the resistance of concrete to the ingress of fluids (that is, low permeability) is fundamental to durability.

The design of concretes with low water content and a low w/cm is essential to durability. In addition, the use of supplementary cementitious materials (SCMs) and adequate curing are critical elements in the production of durable concrete.

Although a durable structure is expected to serve without deterioration or major repair before expiration of its design life, it must not be presumed that provision of durability is a substitution for maintenance. Even structures designed and constructed to a high durability standard require regular inspection and routine maintenance (Figure 11-2).

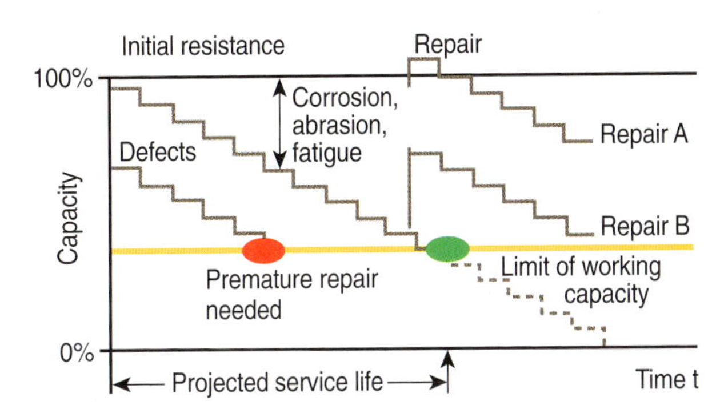

Figure 11-2. Service life of a durable structure.

Factors Affecting Durability

Permeability and Diffusion

Concrete subjected to severe exposure conditions should be highly impermeable. Permeability refers to the ease of fluid migration through concrete when the fluid is under pressure or to the ability of concrete to resist penetration by water or other substances (liquid, gas, or ions) as discussed in Chapter 9. Diffusivity refers to the ease with which dissolved ions move through concrete. Decreased permeability and diffusivity improves concrete's resistance to freezing-thawing cycles, resaturation, sulfate, and chloride-ion penetration, and other forms of chemical attack.

The size of the molecules or ions that are transported through the concrete, the viscosity of the fluid, and the valence of the ions can all affect the transport properties. Thus, permeability and diffusivity must be expressed in terms of the substance that is migrating through the concrete.

Permeability and diffusivity are influenced by porosity, but are distinct from porosity. Porosity is the volume of voids as a percent (or fraction) of the total volume. Permeability and diffusivity are affected by the connectivity of the voids. Figure 11-3 shows two hypothetical porous materials with approximately the same porosity. However, in one material the pores are discontinuous (as would be the case with entrained air bubbles), while in the other the pores are continuous. The latter material would be much more permeable than the former.

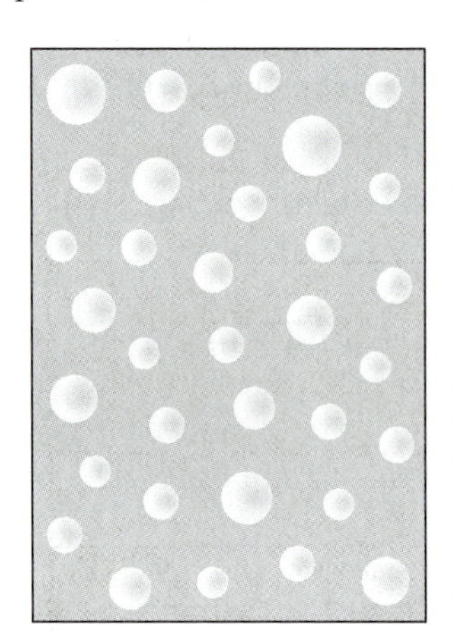

Figure 11-3. Porosity and permeability are related but distinct. The two hypothetical materials above have about the same porosity (total volume of pores), but very different permeabilities. Discrete pores have almost no effect on permeability, while interconnected pores increase permeability.

Figure 11-4 shows the relative sizes of the various pores and solids found in concrete. The capillary pores are primarily responsible for the transport properties. As a rough guide, Powers (1958) plotted the permeability versus capillary porosity for cement paste, as shown in Figure 11-5. It can be seen that as the porosity increases above about 30% (for example, due to a higher water-cement ratio), the permeability increases dramatically.

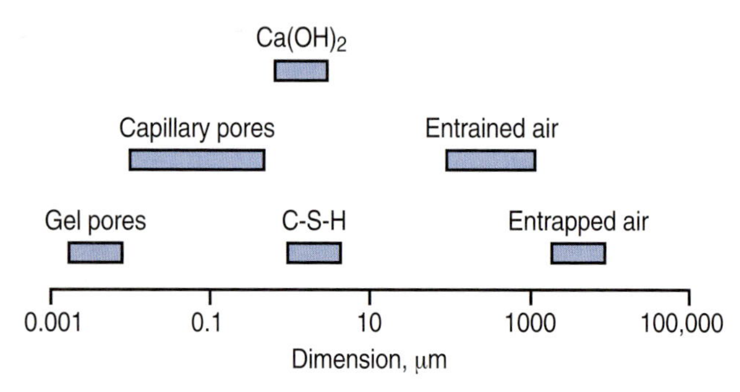

Figure 11-4. Relative sizes of different types of pores and other microstructural features.

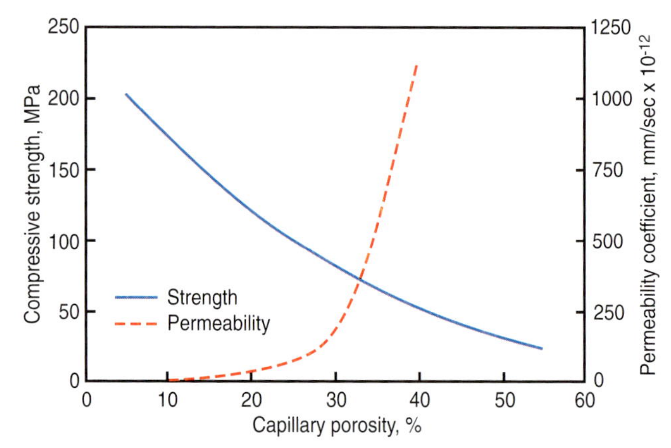

Figure 11-5. Both compressive strength and permeability are related to the capillary porosity of the cement paste, but in different ways. Above about 30% porosity, the permeability increases dramatically (adapted from Powers 1958).

Decreasing the porosity below 30% reduces the permeability, but any additional benefits obtained are relatively minor. The pore system of cement paste becomes discontinuous at about 30% porosity.

Powers and others (1959) calculated the time required for capillary pores to become discontinuous with increasing hydration of the cement, as shown in Table 11-1. It is notable that mixtures with a water-cement ratio greater than 0.7 will always have continuous pores. Figure 11-6 and Table 11-1 illustrate the relationship between water-cement ratio, permeability and curing. Observe the

Table 11-1. Approximate Age Required to Produce Maturity at Which Time Capillaries Become Discontinuous for Concrete Continuously Moist- Cured (Powers and others 1959)

Water-cement ratio by mass	Time required
0.40	3 days
0.45	7 days
0.50	14 days
0.60	6 months
0.70	1 year
Over 0.70	Impossible

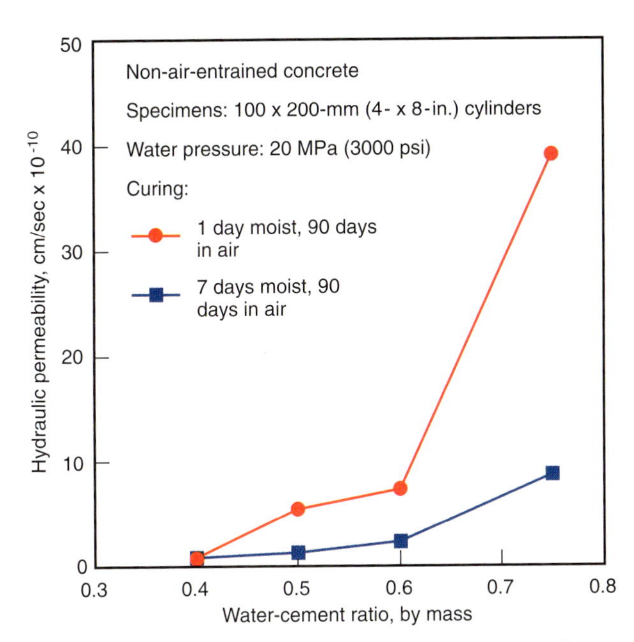

Figure 11-6. Relationship between hydraulic (water) permeability, water-cement ratio, and initial curing on concrete (Whiting 1989).

importance of using a low water-cement ratio. Typically, the water-to-cementitious materials ratio is limited to a maximum of 0.40 to 0.50 when concrete is being designed for durability, depending on the specific conditions of exposure (ACI 318-08). Favorable curing conditions are also necessary to ensure durability.

The use of SCMs can reduce the permeability and diffusivity of concrete. These materials may or may not reduce the total porosity to any great extent, but instead, act to refine and subdivide the capillary pores so that they become less continuous. Some other benefits are obtained by the improvements in fresh concrete properties afforded by many of these materials. For instance, the use of fly ash can reduce the water demand of the concrete, allowing the water-to-cementitious materials ratio to be reduced with equivalent workability.

Though fly ash benefits concrete from the standpoint of durability from the pozzolanic reaction – calcium hydroxide reacts with silica and water to form calcium silicate hydrate. Because calcium silicate hydrate has a greater volume than the calcium hydroxide and pozzolan from which it originates, the pozzolanic reaction results in a finer and less interconnected system of capillary pores. Slag cement is defined as a hydraulic material but may also exhibit minimal pozzolanic reaction; however, it contributes its own calcium silicate hydrate during hydration.

SCMs are highly effective in binding chloride ions so that they do not migrate further into the concrete. This binding effect is particularly important in applications, such as parking garages, bridge decks, and marine construction, where the reinforcing steel is vulnerable to chloride-induced corrosion. It should be noted that the indiscriminate use of SCMs is not necessarily beneficial. Because the pozzolanic reaction takes time, the engineer and contractor must take care to ensure that the properties of the concrete at early ages are satisfactory. Extended moist curing is necessary to achieve the best results.

Test Methods to Determine Permeability. Various methods (both direct and indirect) exist to determine permeability of concrete. Several methods are outlined in Table 11-2. Refer to Chapter 18 for more information on these test methods.

Table 11-2. Test Methods Used to Determine Various Permeability-Related Properties

Test method	Title	Comments
CRD 48	Standard Test Method for Water Permeability of Concrete	Only sensitive to concretes with low-cement contents, such as mass concrete in hydraulic structures
CRD 163	Test Method for Water Permeability of Concrete Using Triaxial Cell	Best for evaluating concretes with a w/c between 0.4 and 0.7
ASTM C1543 (AASHTO T 259)	Standard Test Method for Determining the Penetration of Chloride Ion into Concrete by Ponding	Used for years by highways agencies. The test requires between 6 and 12 months to complete. No clear way provided for interpretation of results.
ASTM C1556	Standard Test Method for Determining the Apparent Chloride Diffusion Coefficient of Cementitious Mixtures by Bulk Diffusion	Considered a useful method for prequalification of concrete mixtures, but takes about three months to complete
AASHTO TP 64	Predicting Chloride Penetration of Hydraulic Cement Concrete by the Rapid Migration Procedure	Less variable than ASTM C1202. Results are not influenced by corrosion inhibiting admixtures.
ASTM C1202 (AASHTO T 277)	Standard Test Method for Electrical Indication of Concrete's Ability to Resist Chloride Ion Penetration	Widely used in specifications for reinforced concrete exposed to chlorides
ASTM C1585	Standard Test Method for Measurement of Rate of Absorption of Water by Hydraulic-Cement Concretes	New test method. Initial ingress of aggressive ions by absorption into unsaturated concrete is much faster than by diffusion or permeability.
ASTM C642	Standard Test Method for Density, Absorption, and Voids in Hardened Concrete	Not a permeability test, but used as indicator of concrete quality.

Table 11-3. Exposure Categories for Durable Concrete (Adapted from ACI 318)

		Exposure Categories	Max. w/cm	Min. Design Compressive Strength, MPa (psi)
F	Freezing and thawing	**F0** (Not applicable) – for concrete not exposed to cycles of freezing and thawing	—	17 (2500)
		F1 (Moderate) – Concrete exposed to freezing and thawing cycles and occasional exposure to moisture (and no deicing chemicals)	0.45	31 (4500)
		F2 (Severe) – Concrete exposed to freezing and thawing cycles and in continuous contact with moisture	0.45	31 (4500)
		F3 (Very Severe) – Concrete exposed to freezing and thawing cycles that will be in continuous contact with moisture and exposure to deicing chemicals or seawater†	0.45	31 (4500)
S	Sulfates*** SO_4^{2-}	**S0** (Not applicable) – Soil: SO_4 <0.10% – Water: SO_4 <150 ppm	—	—
		S1 (Moderate) – Soil: 0.10% ≤ SO_4 < 0.20% – Water: 150 ppm ≤ SO_4 <1500 ppm – Seawater	0.50	28 (4000)
		S2 (Severe)‡ – Soil: 0.20% ≤ SO_4 < 2.0% – Water: 1500 ppm ≤ SO_4 <10,000 ppm	0.45	31 (4500)
		S3 (Very severe)‡ – Soil: SO_4 > 2.0% – Water: SO_4 >10,000 ppm	0.40*	35 (5000)**
C	Corrosion Cl^-	**C0** (Not applicable) – Concrete that will be dry or protected from moisture in service	—	17 (2500)
		C1 (Moderate) – Concrete exposed to moisture but not to an external source of chlorides in service	0.45	31 (4500)
		C2 (Severe) – Concrete exposed to moisture and an external source of chlorides in service	0.40	35 (5000)
P	Permeability	**P0** (Not applicable) – Concrete where low permeability to water is not required	—	17 (2500)
		P1 Concrete required to have low permeability to water	0.50	28 (4000)

*ACI 318 uses a w/cm of 0.45
**ACI 318 uses a min. strength of 31 (4500 psi)
*** ASTM C1580 method is recommended to determine sulfate content of soil and ASTM D516 for testing water
† See ACI 318 for the amount of SCMs allowed in concrete exposed to deicers
‡ Chloride admixtures are not allowed

Exposure Categories

Assessing the environment to which the concrete will be exposed is a fundamental part of designing durable concrete. This will influence both the design of concrete mixtures as well as troubleshooting and repair of distressed concrete. Table 11-3 shows the four main durability-related exposure categories covered in ACI 318. The specifier selects the relevant exposures for each component of the concrete structure, and determines the one that requires the greatest resistance in terms of:

- the lowest water-to-cementing materials ratio (w/cm)
- the highest minimum concrete strength

ACI 318 only addresses exposures to freezing and thawing, soluble sulfates in soil or water, conditions that need precautions to minimize corrosion of reinforcing steel, and conditions that will need low permeability for concrete members in contact with water. Additional durability concerns specific to a project (for example abrasion or alkali-aggregate reactivity) need to be separately addressed by the specifier.

Cracking and Durability

No matter how good the quality of concrete, if the concrete cracks extensively, moisture can enter the concrete and adversely affect its durability. Two basic causes of cracks in concrete are: (1) stress due to applied loads, and (2) stress due to volume changes when concrete is restrained. Random cracking can be avoided by the use of proper joint spacing to predetermine the location of the cracks, or by the use of properly sized and positioned reinforcing steel to reduce crack widths or increase joint spacing (see Chapter 10).

Protective Treatments

The first line of defense against chemical and mechanical attack is the use of good-quality concrete. This is enhanced through the application of protective treatments in especially severe environments to prevent aggressive substances from contacting the concrete. In addition to using concrete with a low permeability, surface treatments can be used to keep aggressive substances from coming into direct contact with concrete. *Effects of Substances on Concrete and Guide to Protective Treatments* (Kerkhoff 2007) discusses the effects of hundreds of chemicals on concrete and provides a list of treatments to help control chemical attack.

Deterioration Mechanisms and Mitigation

Abrasion and Erosion

Concrete surfaces that are exposed to strong mechanical stress require high abrasion and erosion resistance. Abrasion damage in concrete is defined by ACI as *"...wearing away of a surface by rubbing and friction."* Erosion is defined as *"...progressive disintegration of a solid by the abrasive or cavitation action of gases, fluids, or solids in motion"* (ACI 2010).

Wear on concrete surfaces can occur in the following situations:

- floors and slabs due to pedestrian and small-wheeled traffic
- pavements and slabs subject to vehicular traffic (in particular studded tire use and/or use of chains during winter weather events)
- sliding bulk material (for example in silos)
- impact stress of heavy objects (for example factory floors or loading ramps)
- erosion of hydraulic structures from the impact of objects transported by the fluid
- cavitation in hydraulic structures

Mechanism of Abrasion and Erosion. When concrete is exposed to sliding stress, fine material can be dislodged, depending on friction and roughness of the contact surfaces. This leads to abrasion of the surface. Revolving stress involves wear caused by rubber tires (soft) or plastic tires (hard). The hard tires in particular create an abrasive attack on the concrete surface. This leads to loss of paste around individual aggregate pieces and ultimate breaking away. Figure 11-7 shows aggregate exposed after decades of use on a concrete pavement

An impact stress is caused by an object hitting the concrete surface. The softer cement paste is attacked, gradually uncovering the coarser aggregates, and eventually removing them from their embedment.

Cavitation is the result of bubbles collapsing in a fast moving stream of water. Vapor bubbles are formed as the water moves over surface irregularities and later collapse explosively, causing damage to the concrete surface.

Figure 11-7. Picture of aggregate exposed after decades of use on a concrete pavement (Courtesy of R.D. Hooton).

In each of these cases, the form of loading is somewhat different, making it impossible to employ a single standardized test that covers all possible damage mechanisms for abrasion and erosion. Therefore, it is not surprising that there are a number of test methods for assessing the ability of a concrete sample to resist abrasion.

Abrasion and Erosion Testing. ASTM C779, *Standard Test Method for Abrasion Resistance of Horizontal Concrete Surfaces*, the most commonly referenced concrete abrasion test method, offers three loading regimes: revolving disks, dressing wheels, and ball bearings. There is little correlation between the different loading regimes in this test method, making it difficult to predict wear from one mechanism based on data from another test. The tests should be used for comparison purposes (within the same test method) to select the best concrete mixture for abrasion resistance. The heavier the abrasion, the more helpful the tests.

Other available tests for abrasion and erosion include:

- ASTM C418, *Standard Test Method for Abrasion Resistance of Concrete by Sandblasting*, which uses the depth of wear under sandblasting
- ASTM C944, *Standard Test Method for Abrasion Resistance of Concrete or Mortar Surfaces by the Rotating-Cutter Method*, which uses rotating cutters (more useful for smaller samples)
- ASTM C1138, *Standard Test Method for Abrasion Resistance of Concrete (Underwater Method)*, which simulates the effects of swirling water or cavitation
- clamping concrete slabs inside a rotating drum filled with steel shot or aggregate
- rotating wire brushes

Abrasion resistance of aggregates is commonly tested using the Los Angeles abrasion test (rattler method) performed in accordance with ASTM C131 (AASHTO T 96) or ASTM C535 (see Chapter 6).

Materials and Methods for Abrasion and Erosion Resistance. Test results indicate that abrasion resistance is closely related to the compressive strength of concrete. Higher compressive strength concrete has increased resistance to abrasion. Therefore, a low water-cement ratio and adequate curing are essential for abrasion resistance. The type of aggregate and the surface finish also have a strong influence on abrasion resistance. Hard dense aggregate is more wear resistant than soft porous aggregate. Figure 11-8 shows results of abrasion tests on concretes of different compressive strengths and different aggregate types. The total aggregate content in the concrete mixture should be reasonably high so that the thickness of the paste layer at the surface is kept to a minimum without compromising the surface finish.

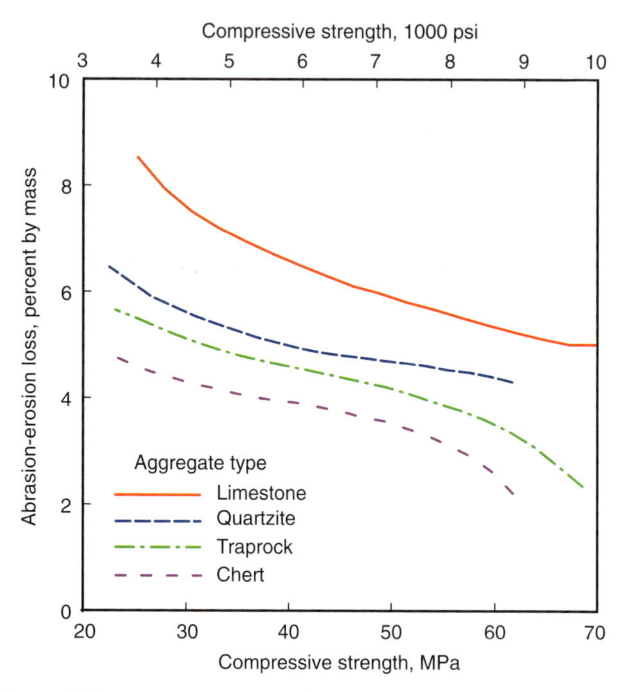

Figure 11-8. Effect of compressive strength and aggregate type on the abrasion resistance of concrete (ASTM C1138). High strength concrete made with a hard aggregate is highly resistant to abrasion (Liu 1981).

It is critical that the surface of the concrete be as durable as possible, which requires careful selection of finishing techniques for interior and exterior applications. A steel-troweled surface resists abrasion better than a surface that had not been troweled. Figure 11-9 illustrates the effect surface treatments, such as metallic or mineral aggregate surface hardeners, have on abrasion resistance of hard steel troweled surfaces. For the same reason, the mixture should be designed to minimize bleeding. ACI Committee 201 (2008) recommends that surfaces that are likely to be heavily abraded should be treated with a dry-shake surface hardener to provide additional protection.

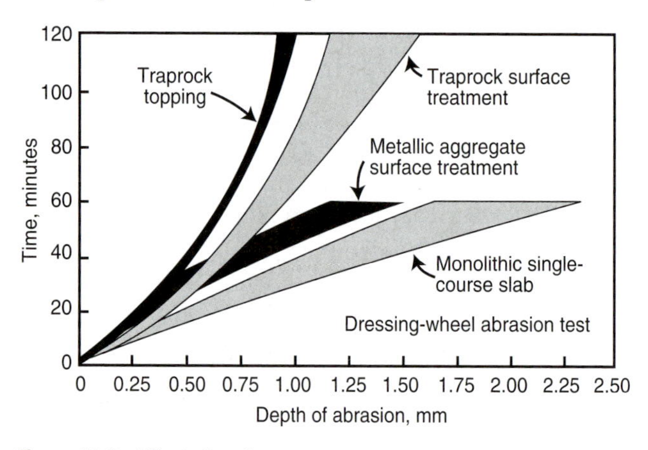

Figure 11-9. Effect of surface treatments on the abrasion resistance of concrete (ASTM C779). Base slab compressive strength was 40 MPa (6000 psi) at 28 days. All slabs were steel troweled (Brinkerhoff 1970).

Freezing and Thawing

Concrete elements exposed to weathering in colder climates need sufficient resistance to freezing and thawing cycles and deicer chemicals. Freeze-thaw exposures range from moderate (category F1), such as a facade element exposed to occasional moisture; to very severe (category F3), for example a pavement or bridge decks in continual contact with moisture and exposed to deicer chemicals or freezing seawater. Freeze-thaw damage in concrete can also be a consequence of the use of aggregates that are not resistant to freezing and thawing cycles (refer to Chapter 6), the use of non-frost resistant paste, or both. Concrete generally exhibits lower resistance to the combined effects of freezing and thawing cycles in conjunction with deicing chemicals as compared to freezing and thawing cycles alone.

Mechanism of Freeze-Thaw Damage. As the water in moist concrete freezes, it produces osmotic and hydraulic pressures in the capillaries and pores of the cement paste and aggregate. If the pressure exceeds the tensile strength of the surrounding paste or aggregate, the cavity will dilate and rupture. The accumulative effect of successive freeze-thaw cycles is the disruption of paste and aggregate eventually causing significant expansion and deterioration of the concrete. Deterioration is visible in the form of cracking, scaling, and disintegration (Figures 11-10 and 11-11).

Figure 11-10. Severe scaling of a concrete pavement.

The mechanisms of freeze-thaw damage and salt scaling in concrete are quite complex. Powers (1962) and Pigeon and Pleau (1995) conducted extensive reviews of the mechanisms of freeze-thaw action. Concrete damage due to freezing and thawing cycles is the result of complex microscopic and macroscopic interactions closely related to the freezing behavior of the pore solution.

Ice in capillary pores (or any ice in large voids or cracks) draws water from surrounding pores to advance its growth. Also, since most pores in cement paste and some aggregates are too small for ice crystals to form, water attempts to migrate to locations where it can freeze.

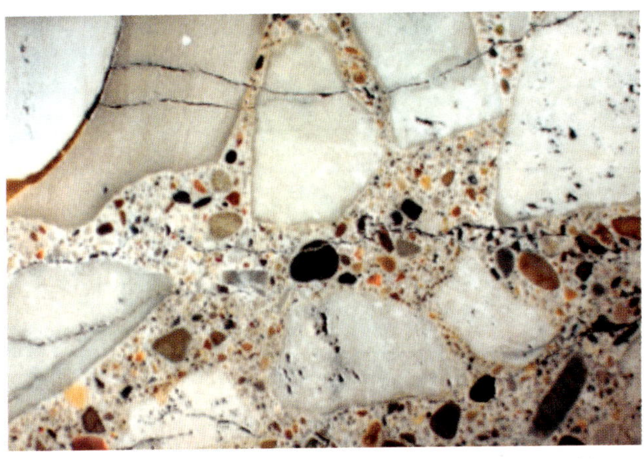

Figure 11-11. Sample of sawn and polished concrete damaged by freeze-thaw cycles.

The pore solution of concrete contains a high quantity of dissolved ions which lower the freezing point. Use of deicers will increase the ion concentration in the pore solution and therefore impact the freezing behavior. In addition, the freezing point of the pore solution is dependent on pore size; the smaller the pore, the lower the freezing point (Table 11-4). Pore volume and pore size distribution are therefore important variables for freeze-thaw resistance. As concrete ages, the pore size distribution changes due to continuing hydration and carbonation at the concrete surface.

Hydraulic pressures are caused by the 9% expansion of water upon freezing; in this process growing ice crystals displace unfrozen water. If a capillary is above critical saturation (91.7% filled with water), hydraulic pressures result as freezing progresses. At lower water contents, no hydraulic pressure should exist.

The moisture content of the pore system is a crucial detriment of the extent of freeze-thaw damage. Fagerlund (1975) introduced the term critical saturation. Critical saturation is influenced by pore volume and pore size distribution. It is reached when the moisture content is high enough to damage concrete during one or a few freeze-thaw cycle(s). Setzer (1999) found that pastes subjected to freeze-thaw cycles have the potential to acquire more water with each cycle due to a micropump effect. It is therefore likely to find concrete to be durable during initial freeze-thaw cycles, but then observe increasingly higher vulnerability with increased number of cycles. As well, additional water from the surrounding environment can enter the pores of concrete during thaw periods and raise the level of saturation as winter progresses (Fagerlund 1975)

Osmotic pressures develop from differential concentrations of alkali solutions in the paste (Powers 1965a). As pure water freezes, the alkali concentration increases in the adjacent unfrozen water. A high-alkali solution, through the mechanism of osmosis, draws water from

Table 11-4. Pore Size Distribution (Adapted from Setzer 1997)

Type	Size	Filling of pores	Temperature at freezing
Capillary pores	≥ 1 mm	empty	
Macro pores	< 1 mm	suction, immediately filling	
Meso capillaries	< 30 μm	suction, filling in minutes to weeks	Water freezes between 0°C (32°F) and -20°C (4°F)
Micro capillaries	< 1 μm	Complete filling through capillary suction not possible	
Meso gel pores	< 30 nm	Filling through condensation at relative humidity of 50% to 98%	Water freezes between -20°C (-4°F) and -40°C (-40°F)
Micro gel pores	< 1 nm	Filling through sorption at relative humidity of < 50%	Water freezes at about -90°C (-130°F)

lower-alkali solutions in the pores. This osmotic transfer of water continues until equilibrium in the fluids' alkali concentration is achieved. Osmotic pressure is considered a minor factor, if present at all, in aggregate frost action, whereas it may be dominant in the cement paste fraction of concrete. Osmotic pressures, as described above, are considered a major factor in "salt scaling."

Monteiro and others (2006) showed the presence of a transition layer surrounding the air void. This layer can have a significant effect on the ice formation inside the air voids. Stresses arise from differences in thermal expansion of ice and concrete and place the ice in tension as the temperature drops. Valenza and Scherer (2006) found cracking of the brine ice layer on the concrete surface to be the origin of salt scaling. Considering the mechanical and viscoelastic properties of ice, the researchers found that the differences will cause moderately concentrated brine solutions to crack. The stresses induced from crack formations in the brine ice will then result in cracks penetrating into the concrete surface, resulting in surface damage.

Exposure to Deicers and Anti-icers

Deicers are solid or liquid chemicals that are applied on concrete to melt ice or snow. Anti-icers are liquids applied before a precipitation event and work to keep water from freezing or refreezing. Deicing and anti-icing chemicals used for snow and ice removal can cause and aggravate surface scaling and joint deterioration (Figure 11-12). The damage is primarily a physical action. Deicer scaling and cracking of inadequately air-entrained or non-air-entrained concrete during freezing is believed to be caused by a buildup of osmotic and hydraulic pressures in excess of the normal hydraulic pressures produced when water in concrete freezes. These pressures become critical and result in scaling unless entrained air voids are both present at the surface and throughout the paste fraction of the concrete to relieve the pressure. The hygroscopic (moisture absorbing) properties of deicing salts also attract water and keep the concrete more saturated, increasing the potential for freeze-thaw deterioration. However, properly designed and placed air-entrained concrete will withstand deicers for many years.

Figure 11-12. V-shaped joints are a common sign of the effects of freeze-thaw damage in concrete pavements. Some joints exhibit an inverted V-shaped deterioration (Courtesy of D. Harrington).

Studies have also shown that, in the absence of freezing, the formation of salt crystals in concrete (from external sources of chloride, sulfate, and other salts) may contribute to concrete scaling and deterioration similar to the disintegration of rocks by salt weathering. The entrained air voids in concrete allow space for salt crystals to grow. This mechanism relieves internal stress similar to the way the voids relieve stress from freezing water in concrete (ASCE 1982 and Sayward 1984).

The extent of scaling depends upon the amount of deicer used and the frequency of application. Relatively low concentrations of deicer (on the order of 2% to 4% by mass) produce more surface scaling than higher concentrations or the absence of deicer (Verbeck and Klieger 1956). However, some highly concentrated anti-icer solutions can potentially include other surface damage due to chemical attack.

Deicers can reach concrete surfaces in ways other than direct application, such as splashing by vehicles and dripping from the undersides of vehicles. Scaling is more severe in poorly drained areas because more of the deicer solution remains on the concrete surface during freezing and thawing. Air entrainment is effective in preventing surface scaling and is recommended for all concretes that may come in contact with deicing or anti-icing chemicals (Figure 11-13).

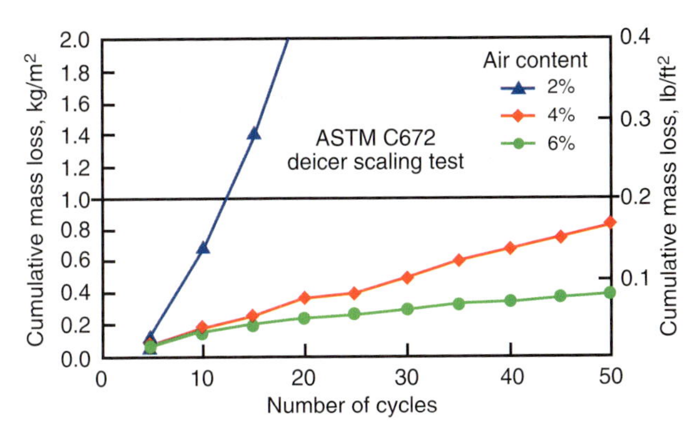

Figure 11-13. Cumulative mass loss for mixtures with a water to cement ratio of 0.45 and on-time finishing (Pinto and Hover 2001).

Table 11-5 summarizes the most commonly used snow and ice control materials (NCHRP 2007). Chloride-based salts containing sodium, calcium, magnesium, and potassium (NaCl, $CaCl_2$, $MgCl_2$, and KCl) comprise the majority of deicers used to melt snow and ice. These chemicals work well because they lower the freezing point of the precipitation that falls on concrete pavements.

In recent years, there has been considerable interest in enhancing the deicing effectiveness at low temperatures by substituting calcium chloride or magnesium chloride for a portion of the sodium chloride brine. Blending salt (sodium chloride) brine with these and other liquid ice control chemicals is considered an economical method to maintain low temperature performance properties. The most common chemical blended with salt brine is calcium chloride, but magnesium chloride is also used along with a myriad of other products that are mixtures of mostly calcium chloride and a number of carbohydrate chemicals. Deicers that contain ammonium nitrate or ammonium sulfate are not recommended for use on concrete because they severely attack concrete (Stark and others 2002).

Deicers can have many effects on concrete and the immediate environment. All deicers can aggravate scaling of concrete that is not properly air entrained. Sodium chloride (rock salt) (ASTM D632 or AASHTO M 143), and calcium chloride (ASTM D98 or AASHTO M 144) are the most frequently used deicers. In the absence of freezing, sodium chloride has little to no chemical effect on concrete but can damage plants and corrode metal. Calcium chloride in weak solutions generally has little chemical effect on concrete and vegetation but does corrode metal.

Studies have shown that concentrated calcium chloride solutions can chemically attack concrete (Brown and Cady 1975 and Sutter and others 2008). While there is disagreement in the laboratory studies about the effect of magnesium chloride on concrete deterioration (Leiser and Dombrowski 1967; Cody and others 1996; Lee and others 2000; Kozikowski and others 2007; and Sutter and others 2008), there has not been any observed damage of field concrete that can be attributed to magnesium chloride use (NCHRP 2007). Urea does not chemically damage concrete, vegetation, or metal. Nonchloride deicers are used to minimize corrosion of reinforcing steel and minimize groundwater chloride contamination. The use of deicers containing ammonium nitrate and ammonium sulfate should be strictly prohibited as they rapidly attack and disintegrate concrete.

The chemicals used for aircraft and airfield deicing are distinctly different from those commonly used for pavement deicing. For pavements at airports, only non-chloride deicing agents are used. These include: urea, potassium acetate, sodium acetate, sodium formate, calcium magnesium acetate, propylene glycol, and ethylene glycols (Mericas and Wagoner 2003). The latter two glycol deicers although not used for deicing pavement are commonly used for aircraft deicing. Propylene glycol and ethylene glycols make up 30% to 70% of the as-applied solution, with an increased use of propylene glycol because of the toxicity concerns related to ethylene glycol (Ritter 2001).

Freeze-Thaw and Deicer Scaling Testing. Nmai (2006) presents a historical evolution of accelerated tests to assess frost resistance of concrete. The most commonly used tests for freezing and thawing and deicer scaling are ASTM C666, *Standard Test Method for Resistance of Concrete to Rapid Freezing and Thawing*, and ASTM C672, *Standard Test Method for Scaling Resistance of Concrete Surfaces Exposed to Deicing Chemicals*. These test methods normally subject samples of concrete to a number of freezing and thawing cycles in order to obtain an accelerated indication of the degree of deterioration associated with long-term exposure. When these test methods are used to evaluate mass loss or surface degradation due to freezing and thawing, the results can be related to the frost resistance of the concrete surface.

Since the development of internal microcracking is anticipated, freeze-thaw resistance in accordance with ASTM C666 is indirectly evaluated by changes in the relative dynamic modulus of elasticity (as measured by resonant frequency), indicating the degree of internal microcrack formation.

A durability factor is then calculated according to the expression below:

$$DF = \frac{P_n N}{M} \qquad \text{(Eq. 1)}$$

where:

DF = durability factor of the specimen tested,

P_n = relative dynamic modulus at N cycles (%),

N = number of cycles at which the test specimen achieves the minimum specified value of P_c for discontinuing the test or the specified number of cycles of the test, whichever is less, and

M = specified number of cycles of the test (typically M = 300).

Table 11-5. Snow and Ice Control Materials (adapted from NCHRP 2007)*

Material type*		Forms sed	Practical temperature limit	Comment
Chlorides	Sodium NaCl	Solid and liquid brine	-10°C (15°F)	The most common ice melting salt. Has little chemical effect on concrete, but can damage lawns and shrubs and contribute to disintergration of low-quality, non-air-entrained concrete. Promotes corrosion of metal.
	Magnesium, $MgCl_2$	Liquid brine, some solid flake	-15°C (5°F)	Releases about 40 percent less chlorides into the environment than either rock salt or calcium chloride. Promotes corrosion of metal. Studies evaluating the effect of magnesium chloride deicing salts on concrete show conflicting results. (Leiser and Dombrowski 1967;Cody and others 1996; Lee and others 2000; Kozikowski and others 2007; Sutter and others 2008; and NCHRP 2007).
	Calcium $CaCl_2$	Liquid brine, some solid flake	-32°C (-25°F)	At low concentrations has little chemical effect on concrete, lawns, and shrubs, but does promote corrosion of metals and can contribute to damage of low-quality concrete. Can absorb moisture from the air, causing it to clump, harden, or even liquefy during storage. Calcium chloride can be hazardous to human health, can leave a slippery residue that is difficult to clean, and tends to refreeze quickly, which may require frequent reapplication (Peeples 1998). At high concentrations, it has been reported to damage concrete through the formation of calcium oxychlorides (Sutter and others 2008).
	Potassium, KCl	Solid	-4°C (25° F)	Common plant nutrient. It is not a skin irritant and is perceived to be less damaging to vegetation. Impractical as a deicer unless used in conjunction with other ingredients. For example, a natural product containing a blend of sodium, magnesium, and potassium chloride is used in some western states (NCHRP 2007).
Organics	Potassium acetate	Liquid	-32°C (-25°F)	Biodegradable deicer primarily used for airports. Environmentally friendly, Slightly corrosive, often mixed with a corrosion inhibitor (Peeples 1998).
	Potassium formate		-32°C (-25°F)	
	Sodium acetate	Solid	-18°C (0°F)	Questions as to possibly damaging interactions with concrete are currently under investigation, especially in relation to aggravating alkali-silica reactivity.
	Sodium formate		-18°C (0°F)	
	Calcium magnesium acetate, CMA	Mostly liquid, some solid	-4°C (20°F)	Made from limestone and acetic acid. Biodegradable, no toxic effects on terrestrial or aquatic animals or soils or vegetation (Wyatt and Fritzsche 1989). Can effectively prevent the formation of ice-surface bonds when applied prior to precipitation. The high cost has limited its practical use. CMA is not effective at very low temperatures.
	Agricultural by-products	Liquid	varies	For example alpha methyl glucoside (MG-104) a corn by-product that is most effective when combined with other ingredients.
	Manufactured organic materials (glycols, methanol)	Liquid	varies	
Ammonium	Ammonium nitrate			Should be strictly prohibited as they rapidly attack and disintegrate concrete.
	Ammonium sulfate			
Nitrogen products	Urea	Solid	-10°C (15°F)	A common fertilizer nutrient. In its pure form not corrosive. However, most of the commercially available products are not suitable for use in corrosion sensitive environments (Peeples 1998). Might damage low-quality concrete. Does not damage lawns, shrubs, or properly made air-entrained concrete.

*Sand and other abrasives are also used for snow and ice control.

As an indication of the degree of freeze-thaw resistance, Cordon (1966) suggested that a concrete of poor frost resistance would have a durability factor below 20%, while concrete with good frost resistance would have a durability factor greater than 80% (Figure 11-14). Natural freezing of concrete in service normally occurs at lower cooling rates, at later ages, and after some period of drying (Lin and Walker 1975, Pigeon and others 1985, and Vanderhost and Jansen 1990). It is generally concluded that the ASTM C666 test is more severe than most natural exposures.

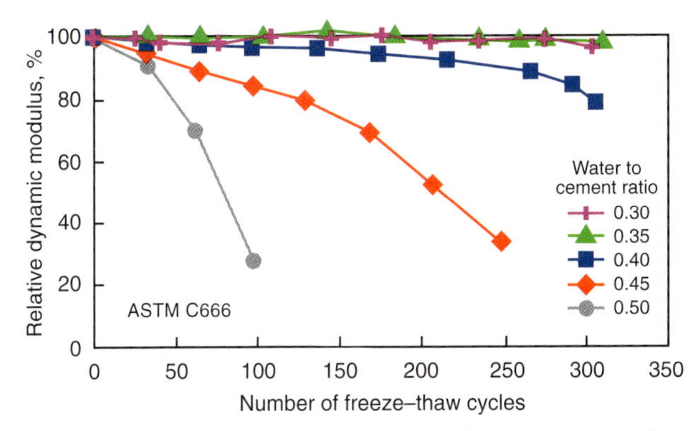

Figure 11-14. Durability factors vs. number of freeze-thaw cycles for selected non-air-entrained concretes (Pinto and Hover 2001).

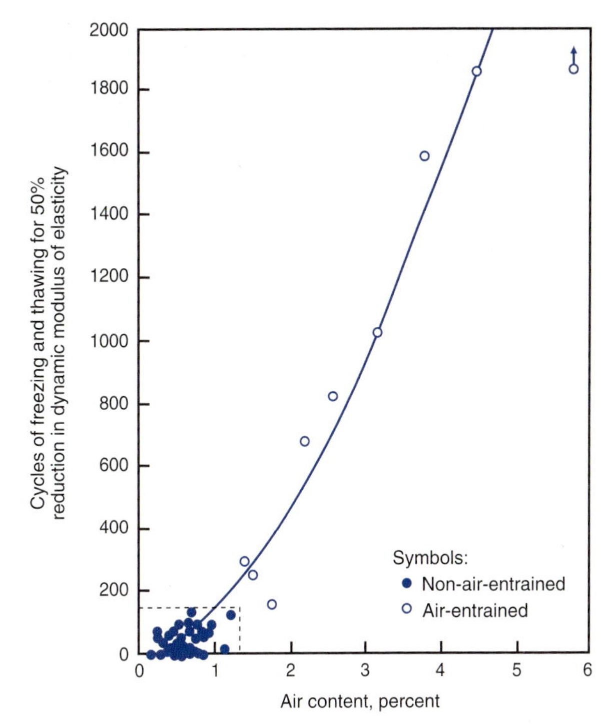

Figure 11-15. Effect of entrained air on the resistance of concrete to freezing and thawing in laboratory tests. Concretes were made with cements of different fineness and composition and with various cement contents and water-cement ratios (Bates and others 1952, and Lerch 1960).

In ASTM C672, small slabs of concrete are used to evaluate scaling resistance. Although ASTM C672 requires that only surface scaling be monitored using a visual damage rating, many practitioners also measure mass loss. Concrete mixtures that perform well in ASTM C666 (AASHTO T 161) do not always perform well in ASTM C672. ASTM C666 (AASHTO T 161) and ASTM C672 are often used to evaluate materials and mixture proportions to determine their effect on freeze-thaw and deicer resistance. The Swedish Standard SS 13 72 44, (from Marchand and others 1994, and Jacobsen and others 1996) correlates the scaling potential with the collected mass loss per unit area. It is suggested that scaling of less than 1.0 kg/m^2 (0.2 lbs/ft^2) after 50 freeze-thaw cycles in the presence of deicer salts indicates an acceptable scaling resistance (Gagné and Marchand 1993).

Materials and Methods to Control Freeze-Thaw and Deicer Damage. The resistance of hardened concrete to freezing and thawing and deicers in a moist condition is significantly improved by the use of intentionally entrained air. Convincing proof of the improvement in durability effected by air entrainment is shown in Figures 11-15 and 11-16.

Entrained air voids act as empty chambers in the paste where freezing (expanding) and migrating water can enter, thus relieving the pressures described above and preventing damage to the concrete. Upon thawing, most of the water returns to the capillaries as a result of capillary action and pressure from air compressed within the bubbles. Thus the bubbles protect the concrete from the next cycle of freezing and thawing (Powers 1955, Lerch 1960, and Powers 1965).

The pressure developed by water as it expands during freezing depends largely upon the distance the water must travel to the nearest air void for relief. Therefore, the voids must be spaced closely enough to reduce the pressure below that which would exceed the tensile strength of the concrete. The amount of hydraulic pressure is also related to the rate of freezing and the permeability of the paste.

The spacing and size of air voids are important factors contributing to the effectiveness of air entrainment in concrete. ASTM C457 describes a method of evaluating the air-void system in hardened concrete. Most authorities consider the following air-void characteristics as representative of a system with adequate freeze-thaw resistance (Powers 1949, Klieger 1952, Klieger 1956, Mielenz and others 1958, Powers 1965, Klieger 1966, Whiting and Nagi 1998, and Pinto and Hover 2001):

Figure 11-16. Effect of weathering on boxes and slabs on ground at the Long-Time Study outdoor test plot, PCA, Skokie, Illinois (Stark and others 2002). Specimens at top are air-entrained, specimens at bottom exhibiting severe crumbling and scaling are non air-entrained. All concretes were made with 335 kg (564 lb) of Type I portland cement per cubic meter (cubic yard). Periodically, calcium chloride deicer was applied to the slabs. Specimens were 40 years old when photographed (see Klieger 1963 for concrete mixture information).

1. Calculated spacing factor, $\bar{L}$, (an index related to the distance between bubbles but not the actual average spacing in the system)—less than 0.200 mm (0.008 in.) (Powers 1954 and 1965 and Mielenz and others 1958)
2. Specific surface, α, (surface area of the air voids) – 24 square millimeters per cubic millimeter (600 sq in. per cubic inch) of air-void volume, or greater.

Current U.S. field quality control practice usually involves only the measurement of total air volume in freshly mixed concrete. This practice does not distinguish air-void size in any way. In addition to total air volume, Canadian practice also requires attainment of spacing factors. CSA A23.1-09 requires the average spacing factor to be less than 0.230 mm with no single test value to exceed 0.260 mm. Figure 11-17 illustrates the relationship between spacing factor and total air content. Measurement of air volume alone does not permit full evaluation of the important characteristics of the air-void system; however, air-entrainment is generally considered effective for freeze-thaw resistance when the total volume of air in the mortar fraction of the concrete – material passing the 4.75-mm (No. 4) sieve – is about 9 ± 1% (Klieger 1952) or about 18% by paste volume. For equal admixture dosage rates per unit of cement, the air content of ASTM C185 (AASHTO T 137) mortar would be about 19% due to the standard aggregate's properties. The air content of concrete with 19-mm (¾-in.) maximum-size aggregate would be about 6% for effective freeze-thaw resistance.

The relationship between air content of standard mortar and concrete is illustrated by Taylor (1948). Pinto and Hover (2001) address paste air content versus frost resistance. The total required concrete air content for durability increases as the maximum-size aggregate is reduced (due to greater paste volume) and as the exposure conditions become more severe (see Chapter 12).

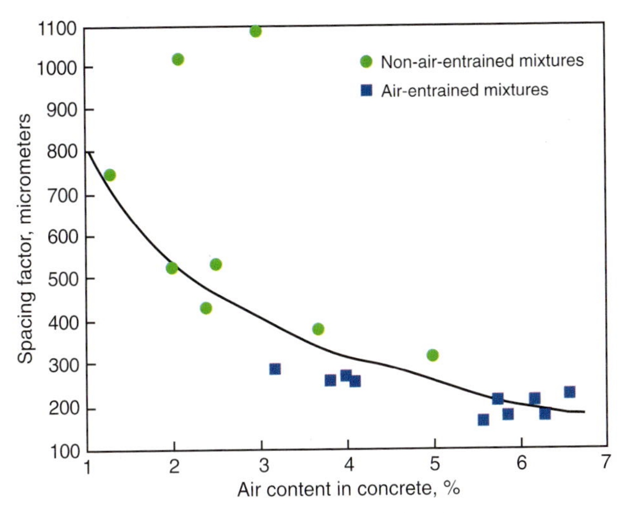

Figure 11-17. Spacing factor as a function of total air content in concrete (Pinto and Hover 2001).

A good air-void system with a low spacing factor (maximum of 200 micrometers) is perhaps more important to deicer environments than saturated frost environments without deicers. The relationship between spacing factor and deicer scaling is illustrated in Figure 11-18. A low water to portland cement ratio helps minimize scaling, but is not sufficient to control scaling at normal water-cement ratios. Figure 11-19 illustrates the overriding impact of air content over water-cement ratio in controlling scaling.

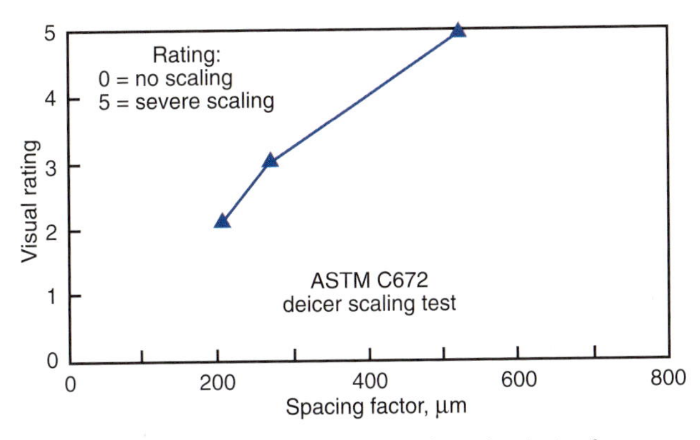

Figure 11-18. Visual rating as a function of spacing factor, for a concrete mixture with a water to cement ratio of 0.45 (Pinto and Hover 2001).

Field service and extensive laboratory testing have shown that when the proper spacing and volume of air voids are present, air-entrained concrete will have excellent resistance to freeze-thaw cycles and surface scaling due to freezing and thawing and the application of deicer chemicals, provided the concrete is properly proportioned, placed, finished, and cured.

Scaling resistance may decrease as the amount of certain SCMs increase. The ACI 318 building code limits the maximum dosage of fly ash, slag, and silica fume to 25%,

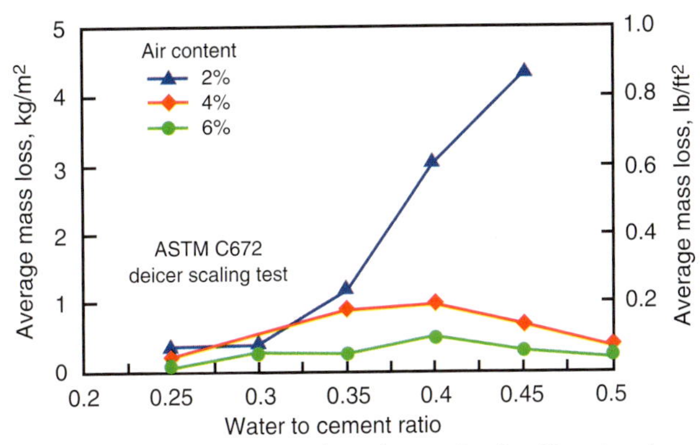

Figure 11-19. Measured mass loss of concrete after 40 cycles of deicer and frost exposure at various water to cement ratios (Pinto and Hover 2001).

50%, and 10%, respectively, by mass of cementing materials, for deicer exposures. Total SCM content should not exceed 50% by mass of the cementitious materials (ACI 318-08). Certain transportation departments in northern states allow the use of up to 30% fly ash in pavements. However, concretes that are properly designed, placed, and cured, have demonstrated good scaling resistance even when made with higher dosages of some of these materials. The selection of materials and dosages should be based on local experience and the durability should be demonstrated by field or laboratory performance.

When concrete in service will be exposed to cycles of freezing and thawing or deicing chemicals, consult local guidelines on allowable practices and use the following guidelines to ensure adequate concrete performance:

1. An adequate air content (a minimum of 5% to 8% for 19 mm [3/4 in.] nominal size aggregate) with a satisfactory air void system (having a spacing factor ≤ 0.200 mm (0.008 in.) and a specific surface area of 24 mm^2/mm^3 (600 $in.^2/in.^3$) or greater
2. A low water-to-cementitious materials ratio (≤ 0.45)
3. A minimum compressive strength of 31 MPa (4500 psi) for concrete exposed to freezing and thawing cycles that will be in continuous contact with moisture and exposure to deicing chemicals
4. Fly ash, slag cement, and silica fume dosages not exceeding 25%, 50%, and 10%, respectively with combinations not exceeding 50%, by mass of cementing materials, for deicer exposures, unless otherwise demonstrated by local practice or testing
5. Proper finishing after bleed water has evaporated from the surface
6. A minimum of 7 days moist curing at or above 10°C (50°F)
7. A minimum 30-day drying period after moist curing prior to exposure to freeze-thaw cycles and deicers when saturated

8. Adequate drainage (1% minimum slope, 2% preferred)
9. For additional protection, consider applying a breathable sealer after the initial drying period (see Chapter 15).

Air Drying. The resistance of air-entrained concrete to freeze-thaw cycles and deicers is greatly increased by an air drying period after initial moist curing. Air drying removes excess moisture from the concrete which in turn reduces the internal stress caused by freeze-thaw conditions and deicers. Water-saturated concrete will deteriorate faster than an air-dried concrete when exposed to moist freeze-thaw cycling and deicers. Concrete placed in the spring or summer usually has an adequate drying period. Concrete placed in the fall season, however, often does not dry out sufficiently enough before deicers are used. This is especially true of fall paving cured using membrane-forming compounds. These membranes remain intact until abraded by traffic. Thus, adequate drying may not occur before the onset of winter. Curing methods, such as use of plastic sheets that allow drying at the completion of the curing period are preferable for fall paving on all projects where deicers will be used. Concrete placed in the fall should be allowed at least 30 days for air drying after the moist-curing period. The exact length of time for sufficient drying to take place may vary with climate and weather conditions.

Treatment of Scaled Surfaces. If surface scaling (an indication of an inadequate air-void system or poor finishing practices) develops during the first frost season, or if the concrete is of poor quality, a breathable surface treatment can be applied to the dry concrete to help protect it against further damage. Treatment often consists of a penetrating sealer made with boiled linseed oil (ACPA 1996), breathable methacrylate, or other materials. Nonbreathable formulations should be avoided as they can trap water below the surface and cause delamination.

The effect of mix design, surface treatment, curing, or other variables on resistance to surface scaling can be evaluated by ASTM C672. In this test, concrete slabs are cured and then exposed to a salt solution at 28 days of age and subjected to 50 cycles of freezing and thawing. A visual rating of surface scaling is used to rank performance.

Alkali-Aggregate Reactivity

Aggregates containing certain constituents can react with alkali hydroxides in concrete (Farny and Kerkhoff 2007 and PCA 2007).

The reactivity is potentially harmful only when it produces significant expansion (Mather 1975). Alkali-aggregate reactivity (AAR) has two forms – alkali-silica reaction (ASR) and alkali-carbonate reaction (ACR). ASR is of greater concern than ACR because the occurrence of aggregates containing reactive silica minerals is more widespread. Alkali-reactive carbonate aggregates have a specific composition that is not common.

Alkali-Silica Reaction

Alkali-silica reactivity has been recognized as a potential source of distress in concrete since the late 1930s (Stanton 1940 and PCA 1940). Even though potentially reactive aggregates exist throughout North America, ASR distress in structural concrete is not common. There are a number of reasons for this condition:

- most aggregates are chemically stable in hydraulic cement concrete.
- aggregates with good service records are abundant in many areas.
- most concrete in service is dry enough to inhibit ASR.
- use of certain pozzolans or slags can control ASR.
- in many concrete mixtures, the alkali content of the concrete is low enough to control harmful ASR.
- some forms of ASR do not produce significant deleterious expansion.

The reduction of ASR potential requires understanding the ASR mechanism; properly using tests to identify potentially reactive aggregates; and, if needed, taking steps to minimize the potential for expansion and related cracking.

Typical indicators of ASR might be any of the following: a network of cracks (Figures 11-20 and 11-21); cracks with straining or exuding gel; closed or spalled joints; relative displacements of different parts of a structure; or fragments breaking out of the surface of the concrete (popouts) (Figure 11-22). Because ASR deterioration is slow, the risk of catastrophic failure is low. However, ASR can cause serviceability problems and can exacerbate other deterioration mechanisms such as those that occur in frost, deicer, or sulfate exposures.

Figure 11-20. Cracking of concrete from alkali-silica reactivity.

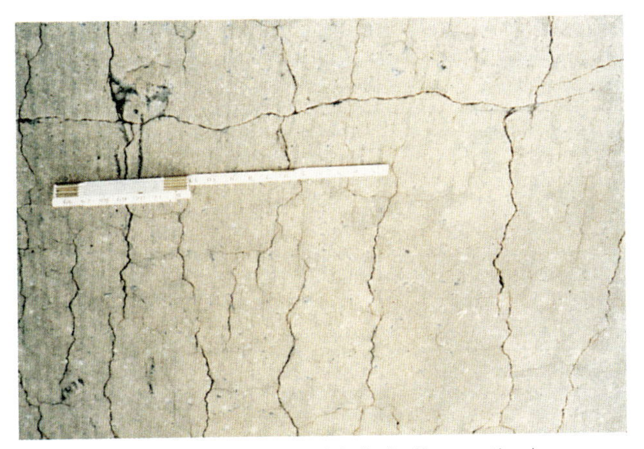

Figure 11-21. Indication of potential alkali-silica reaction is provided by the classic map cracks on the surface of the concrete. ASR should be verified by petrographic examination.

Figure 11-22. Popouts caused by ASR of sand-sized particles.

Mechanism of ASR. The alkali-silica reaction forms a gel that swells as it draws water from the surrounding cement paste. Reaction products from ASR have a great affinity for moisture. In absorbing water, these gels can induce pressure, expansion, and cracking of the aggregate and surrounding paste. The reaction can be visualized as a two-step process:

1. Alkali hydroxide + reactive silica gel → reaction product (alkali-silica gel)
2. Gel reaction product + moisture → expansion

The amount of gel formed in the concrete depends on the amount and type of silica, the alkali hydroxide concentration, and the availability of moisture. The presence of gel does not always coincide with distress, and thus, gel presence does not necessarily indicate destructive ASR. Conversely, certain aggregates produce relatively little gel, yet can lead to significant and deleterious expansion.

Factors Affecting ASR. For alkali-silica reaction to occur, the following three conditions must be present:

1. reactive forms of silica in the aggregate,
2. high-alkali (pH) pore solution, and
3. sufficient moisture.

If one of these conditions is absent, ASR cannot occur.

ASR Testing. Field performance history is the best method of evaluating the susceptibility of an aggregate to ASR. For the most definitive evaluation, the existing concrete should have been in service for at least 15 years. Comparisons should be made between the existing and proposed concrete's mix proportions, ingredients, and service environments. This process should indicate whether special requirements are needed, are not needed, or whether testing of the aggregate or job concrete is required.

Table 11-6 describes different test methods used to evaluate the potential for alkali-silica reactivity. The use of newer, faster test methods can be utilized for initial screening. Where uncertainties arise, lengthier tests can be used to confirm results. When SCMs or blended cements are used to control ASR, their effectiveness must be determined by tests such as ASTM C1567 or C1293. See AASHTO (2001), Farny and Kerkhoff (2007), and PCA Durability Committee (2007) for more information on tests to demonstrate the effectiveness of control measures below. These tests should not be used to disqualify use of potentially reactive aggregates. Reactive aggregates can often be safely used with the careful selection of cementitious materials as discussed below. Using the aggregate-reactivity class, the size of the concrete structure, and the exposure conditions, a level of ASR risk can be assigned. The level of prevention required is then determined considering the risk of ASR together with the class of structure (Safety, economic, or environmental consequences) (Thomas, Fournier, and Folliard 2008).

Materials and Methods to Control ASR. The best way to avoid ASR is to take appropriate precautions before concrete is placed. Standard concrete specifications may require modifications to address ASR. These modifications should be carefully tailored to avoid limiting the concrete producer's options. This permits careful analysis of cementitious materials and aggregates and choosing a control strategy that optimizes effectiveness and the economic selection of materials. If the aggregate is not reactive by historical identification or testing, no special requirements are needed.

The most effective way of controlling expansion due to ASR is to design mixtures specifically to control ASR, preferably using locally available materials. Current practices include the use of supplementary cementing materials or blended cement proven by testing to control ASR or by limiting the alkali content of the concrete (determined by summing up the alkali content (primarily from the cement). Limits of 3 kg/m^3 (5 lb/yd^3) or 1.8 kg/m^3 (3 lb/yd^3) have been used, depending on the reactivity of the aggregate (CSA A23.1 2009). Landgren and Hadley (2002) provide techniques to control popouts caused by ASR (Figure 11-22).

Table 11-6. Test Methods for Alkali-Silica Reactivity (Adapted from Farny and Kerkhoff 2007)

Test name	Purpose	Type of test	Type of sample
ASTM C227, Potential Alkali-Reactivity of Cement-Aggregate Combinations (Mortar-Bar Method)	To test the susceptibility of cement-aggregate combinations to expansive reactions involving alkalies.	Mortar bars stored over water at 37.8°C (100°F) and high relative humidity	At least 4 mortar bars; standard dimensions: 25 x 25 x 285 mm (1 x 1 x 11¼ in.)
ASTM C289, Potential Alkali-Silica Reactivity of Aggregates (Chemical Method)	To determine potential reactivity of siliceous aggregates	Aggregate sample reacted with alkaline solution at 80°C (176°F)	Four 25 gram samples of crushed and sieved aggregate
ASTM C294, Constituents of Natural Mineral Aggregates	To give descriptive nomenclature for the more common or important natural minerals – a good starting point to predict behavior	Visual identification	Varies, but should be representative of entire source
ASTM C295, Petrographic Examination of Aggregates for Concrete	To evaluate possible aggregate reactivity through petrographic examination	Visual and microscopic examination of prepared samples – sieve analysis, microscopy, scratch or acid tests	Varies with knowledge of quarry: cores 53 to 100 mm in diameter (2 1/8 to 4 in.), 45 kg (100 lb) or 300 pieces, or 2 kg (4 lb)
ASTM C441, Effectiveness of Pozzolans or GBFS in Preventing Excessive Expansion of Concrete Due to Alkali-Silica Reaction	To determine the effectiveness of pozzolans or slag in controlling expansion from ASR	Mortar bars – using Pyrex® glass as aggregate – stored over water at 37.8°C (100°F)	At least 3 mortar bars and also 3 mortar bars of control mixture
ASTM C856, Petrographic Examination of Hardened Concrete	To outline petrographic examination procedures of hardened concrete – useful in determining condition or performance	Visual (unmagnified) and microscopic examination of prepared samples	At least one core 150 mm diameter by 300 mm long (6 in. diameter by 12 in. long)
ASTM C856 Annex (AASHTO T 299), Uranyl-Acetate Treatment Procedure	To identify products of ASR in hardened concrete	Staining of a freshly-exposed concrete surface and viewing under UV light	Varies: core with lapped surface, core with broken surface
Los Alamos staining method (Powers 1999)	To identify products of ASR in hardened concrete	Staining of a freshly-exposed concrete surface with two different reagents	Varies: core with lapped surface, core with broken surface
ASTM C1260 (AASHTO T 303), Potential Alkali-Reactivity of Aggregates (Mortar-Bar Method)	To test the potential for deleterious alkali-silica reaction of aggregate in mortar bars	Immersion of mortar bars in alkaline solution at 80°C (176°F)	At least 3 mortar bars
ASTM C1293, Determination of Length Change of Concrete Due to Alkali-Silica Reaction (Concrete Prism Test)	To determine the potential ASR expansion of cement-aggregate combinations	Concrete prisms stored over water at 38°C (100.4°F)	3 prisms per cement-aggregate combination, standard dimensions: 75 x 75 x 285 mm (3 x 3 x 11¼ in.)
ASTM C1567, Potential Alkali-Silica Reactivity of Combinations of Cementitious Materials and Aggregates (Accelerated Mortar-Bar Method)	To test the potential for deleterious alkali-silica reaction of cementitious materials and aggregate combinations in mortar bars	Immersion of mortar bars in alkaline solution at 80°C (176°F)	At least 3 mortar bars for each cementitious materials and aggregate combination
CRD-C 662-10, Determining the Potential Alkali-Silica Reactivity of Combinations of Cementitious Materials, Lithium Nitrate Admixture and Aggregate (Accelerated Mortar-Bar Method)	To evaluate the use of pozzolans, slag cement, lithium admixtures, and combinations thereof to control deleterious expansion due to ASR	Immersion of mortar bars in alkaline solution at 80°C (176°F)	At least 3 mortar bars for each cementitious materials and aggregate combination

Duration of test	Measurement	Criteria	Comments
Varies: first measurement at 14 days, then 1, 2, 3, 4, 5, 6, 9, and 12 months and every 6 months after that as necessary	Length change	Per ASTM C33, maximum 0.10% expansion at 6 months, or if not available for a 6-month period maximum of 0.05% at 3 months	Test may not produce significant expansion, especially for carbonate aggregate Long test duration Expansions may not be from AAR
24 hours	Drop in alkalinity and amount of silica solubilized	Point plotted on graph falls in deleterious or potentially deleterious area	Quick results Not a reliable test method
Short duration – as long as it takes to visually examine the sample	Description of type and proportion of minerals in aggregate	Not applicable	These descriptions are used to characterize naturally-ocurring minerals that make up common aggregate sources.
Short duration – visual examination does not involve long test periods	Particle characteristics, like shape, size, texture, color, mineral composition, physical condition	Not applicable	Usually includes optical microscopy; may include XRD analysis, differential thermal analysis, or infrared spectroscopy – see C294 for descriptive nomenclature
Varies: first measurement at 14 days, then 1, 2, 3, 4, 5, 6, 9, and 12 months and every 6 months after that as necessary	Length change	Per ASTM C989, minimum 75% reduction in expansion or 0.020% maximum expansion or per C618, comparison against low-alkali control	Highly reactive artificial aggregate may not represent real aggregate conditions Pyrex® contains alkalies
Short duration – includes preparation of samples and visual and microscopic examination	Is the aggregate known to be reactive? Orientation and geometry of cracks Is there any gel present?	See measurement – this examination determines if ASR reactions have taken place and their effects upon the concrete. Used in conjunction with other tests	Specimens can be examined with stereo microscopes, polarizing microscopes, metallographic microscopes, and SEM
Immediate results	Intensity of fluorescence	Lack of fluorescence	Identifies small amounts of ASR gel whether they cause expansion or not. Opal, a natural aggregate, and carbonated paste can indicate – interpret results accordingly. Test must be supplemented by petrographic examination and physical tests for determining concrete expansion
Immediate results	Color of stain	Dark pink stain corresponds to ASR gel and indicates an advanced state of degredation	
16 days	Length change at 14 days after 2 day cure. Initial measurement after 2 days.	0.10% maximum expansion at 14 days	Accelerated test often used to assess aggregate reactivity. Test is quite aggressive and many aggregates will fail this test but pass the concrete prism test or perform well in the field. (Thomas and others 2006) There are various reported cases of coarse aggregates that pass this test but are actually reactive in concrete prism test or field. (Folliard and others 2006)
Varies: first measurement at 7 days, then 28 and 56 days, then 3, 6, 9, and 12 months and every 6 months after that as necessary	Length	Per Appendix X1, potentially deleteriously reactive if expansion equals or exceeds 0.04% at 1 year (or 0.04% at 2 years if SCMs are to be evaluated)	Preferred method of assessment Best represents the field Long test duration for meaningful results. Use as a supplement to C227, C295, C289, C1260, and C1567 Similar to CSA A23.2-14A
16 days	Length change	If greater than 0.10%, indicative of potential deleterious expansion; use C1293 (2-year test) to confirm	Very fast alternative to C1293 Allows for evaluation of SCMs and determination of optimum dosage
28 days	Length change		

Low-alkali portland cement (ASTM C150) with an alkali content of not more than 0.60% (equivalent sodium oxide) has been successful for ASR resistance with slightly reactive to moderately reactive aggregates. However, it is the total concrete alkalies that are of greatest importance.

Lithium-based admixtures are available to control ASR in fresh concrete (Stark 1992, Thomas and others 2007).

When pozzolans, slag cements, or blended cements are used to control ASR, their effectiveness must be determined by tests such as ASTM C1567 or C1293. ASTM C1567 with a 14 day expansion limit is the most common method used to evaluate effectiveness of control measures. Where applicable, different amounts of pozzolan or slag cement should be tested to determine the optimum dosage. Expansion usually decreases as the dosage of the pozzolan or slag cement increases (Figure 11-23).

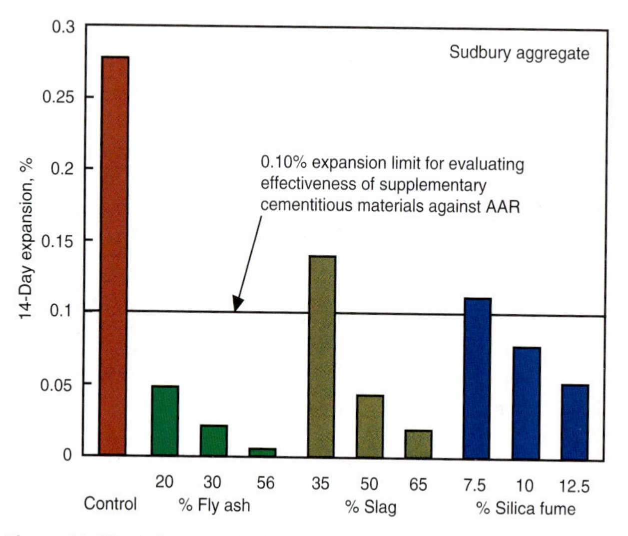

Figure 11-23. Influence of different amounts of fly ash, slag, and silica fume by mass of cementing material on mortar bar expansion (Modified ASTM C1260/C1567) after 14 days when using reactive aggregate (Fournier 1997).

Recommended methods to control ASR are given in Farny and Kerkhoff (2007) and PCA (2007), Thomas, Fournier, and Folliard (2008), and AASHTO PP65, *Standard Practice for Determining the Reactivity of Concrete Aggregates and Selecting Appropriate Measures for Preventing Deleterious Expansion in New Concrete Construction*. These methods include both prescriptive and performance approaches to controlling ASR. The PCA approach is illustrated in Figure 11-24.

Mitigation of ASR. It is important to distinguish between the reaction and the resulting damage from the reaction. In the diagnosis of concrete deterioration, it is most likely that a gel product will be identified. But, in some cases significant amounts of gel are formed without causing damage to concrete. To pinpoint ASR as the cause of damage, the presence of deleterious ASR gel must be verified. A site of expansive reaction can be defined as an aggregate particle that is recognizably reactive or potentially reactive and is at least partially replaced by gel. Gel can be present in cracks and voids and may also be present in a ring surrounding an aggregate particle at its edges. A network of internal cracks connecting reacted aggregate particles is an almost certain indication that ASR is responsible for cracking. A petrographic examination (ASTM C856) is the most conclusive method available for identifying ASR gel in concrete (Powers 1999). Petrography, when used to study a known reacted concrete, can confirm the presence of reaction products and verify ASR as an underlying cause of deterioration (Figure 11-25).

Structures that carry the potential for ASR damage should be monitored on a regular basis for signs of ASR. Typically, this process will begin with regular visual condition surveys and progress forward if any symptoms are found. Once ASR has been identified in a structure, the options for remediation are limited to either treating the cause or treating the symptoms. With the cause being throughout the structure, it is much more difficult to adequately treat the cause than it is to mitigate further damage. See Fournier and others (2010) for a detailed explanation of diagnosing and mitigating ASR.

Alkali-Carbonate Reaction

Reactions observed with certain dolomitic rocks are associated with alkali-carbonate reaction (ACR). Reactive rocks usually contain large crystals of dolomite scattered in, and surrounded by, a fine-grained matrix of calcite and clay. Calcite is one of the mineral forms of calcium carbonate; dolomite is the common name for calcium-magnesium carbonate. ACR is relatively rare because aggregates susceptible to this reaction are usually unsuitable for use in concrete for other reasons, such as strength potential. Argillaceous dolomitic limestone contains calcite and dolomite with appreciable amounts of clay and can also contain small amounts of reactive silica.

Mechanism of ACR. There is still some debate about the mechanisms of ACR, but some attribute the expansion to dedolomitization, or the breaking down of dolomite (Hadley 1961). Concrete that contains dolomite and has expanded also contains brucite (magnesium hydroxide, $Mg(OH)_2$), which is formed by dedolomitization. Dedolomitization proceeds according to the following equation (Ozol 2006):

$$CaMgCO_3 \text{ (dolomite)} + \text{alkali hydroxide solution} \rightarrow MgOH_2 \text{ (brucite)} + CaCO_3 \text{ (calcium carbonate)} + K_2CO_3 \text{ (potassium carbonate)} + \text{alkali hydroxide}$$

The dedolomitization reaction and subsequent crystallization of brucite may cause considerable expansion. Whether dedolomitization causes expansion directly or indirectly, it is usually a precursor to other expansive processes (Tang and others 1994).

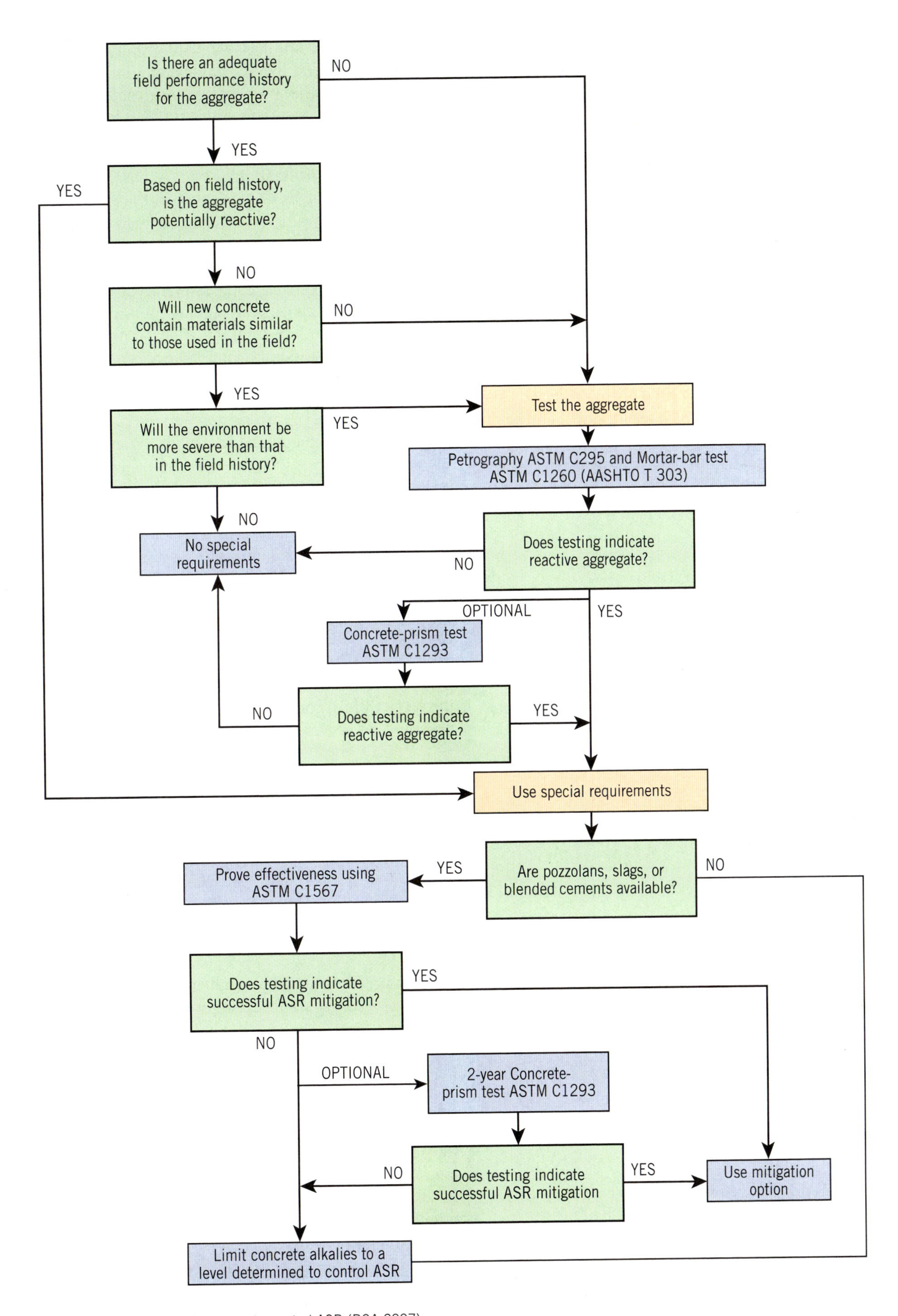

Figure 11-24. Flow chart showing steps to control ASR (PCA 2007).

Figure 11-25. Polished section view of an alkali reactive aggregate in concrete. Observe the alkali-silica reaction rim around the reactive aggregate and the crack formation.

Test Methods for Identifying the Potential for ACR Distress. The three test methods commonly used to identify potentially alkalicarbonate reactive aggregate are:

- petrographic examination (ASTM C295);
- rock cylinder method (ASTM C586); and
- concrete prism test (ASTM C1105).
- chemical composition (CSA A23.2-26A)

See Farny and Kerkhoff (2007) for detailed information.

Materials and Methods to Control ACR. The best and most practical preventative measure has been to avoid the use of these aggregates. Alkali reactivity of carbonate rocks is not usually dependent upon its clay mineral composition (Hadley 1961). ACR-susceptible aggregate has a specific composition that is readily identified by petrographic testing. If a rock indicates ACR susceptibility, one of the following measures should be taken to reduce the likelihood of damage:

- selective quarrying to completely avoid reactive aggregate;
- blend aggregate according to Appendix in ASTM C1105; or
- limit aggregate size to the smallest practical.

Low-alkali cement, SCMs, and lithium compounds are generally not found to be effective in controlling expansive ACR (Swenson and Gillot 1964, Rogus and Hooker 1992, Thomas and Innis 1998, Shehata and others 2009, and Wang and others 1994).

Carbonation

Carbonation of concrete is a process by which carbon dioxide typically in from the atmosphere penetrates the concrete and reacts with the various hydration products, such as calcium hydroxide, to form carbonates (Verbeck 1958). From the standpoint of durability, the most important effect of carbonation is a reduction in the pH of concrete. A high pH protects the reinforcement from corrosion by maintaining a passive layer on the surface of the steel. In addition, the ability of hydrated cement paste to bind chloride ions increases with pH (Page and Vennesland 1983).

Mechanism of Carbonation. Concrete may undergo carbonation caused by CO_2 from either the atmosphere or from carbonated water. The CO_2 reacts with calcium hydroxide in the concrete to form calcium carbonate, reducing the pH. The progress of carbonation depends on the following: 1) concrete composition (type and amount of cement and SCMs), 2) permeability (water-cement ratio, curing), 3) exposure conditions during carbonation (relative humidity, moisture content of the concrete element, CO_2 content of the air) and 4) duration of exposure.

The rate of ingress of carbonation into the concrete is most rapid at intermediate relative humidity, about 50%. Since the reaction involves the dissolution of CO_2 in water, some moisture must be present. However, in very wet concrete the transport of CO_2 is actually slower because its solubility in water is limited (Herholdt and others 1979); its rate of diffusion in water is four orders of magnitude slower than in air (Neville 1996). Therefore, the carbonation rate of water-saturated concrete is negligible.

Designers, specifiers, and users are particularly interested in estimating the time before the carbonation front reaches the reinforcement for a given concrete cover. This impacts the service life of concrete, if the moisture in the carbonated area is sufficient to trigger reinforcement corrosion.

The depth of carbonation, d_c, is approximately linear to the square root of the time of carbonation t_c:

$$d_c = d_0 + a\sqrt{t_c} \qquad \text{Eq. 2}$$

with:

d_0 = a parameter that depends on curing and early exposure. It becomes smaller with later start of carbonation t_0.

a = a factor that contains parameters resulting from concrete composition, curing, and exposure conditions.

The equation is conservative. Carbonation is slower than predicted by the $a\sqrt{t_c}$ -relationship, if the concrete element is at least occasionally exposed to moisture.

Carbonation Testing. Depth of carbonation is typically measured by applying phenolphthalein solution to a freshly fractured surface of concrete. The characteristic bright pink color appears when the pH is above 9.5, as shown in Figure 11-26. Where the pH has been reduced to below 9.5 (whether by carbonation or other causes), there is no color change of the indicator.

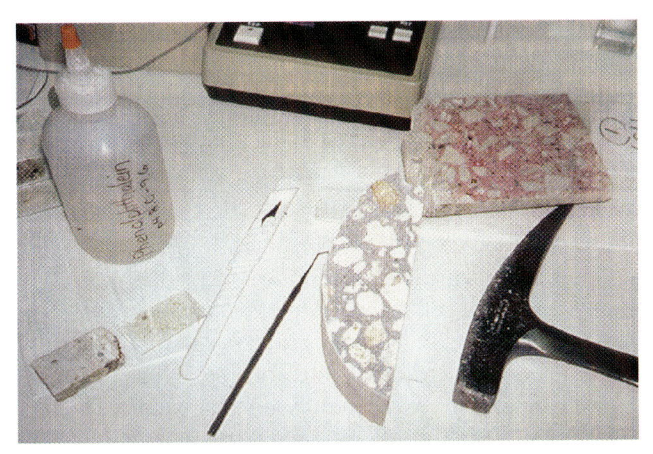

Figure 11-26. The depth of carbonation is determined by wetting a freshly fractured surface of concrete with phenolphthalein solution.

Materials and Methods to Resist Carbonation. The amount of carbonation is significantly increased in concrete with a high water to cement ratio, low cement content, short curing period, low strength, and highly permeable (porous) paste. Ensuring that the concrete exhibits sufficiently low permeability best reduces the rate of carbonation. To reduce early carbonation, concrete needs to be protected from drying for as long as possible. The concentration of CO_2 in the atmosphere varies from 0.03% by volume in rural environments to about 0.3% in urban environments; local concentrations are much higher in enclosed parking garages and tunnels (Neville 1996).

Concrete with portland cement resists rapid ingress of carbonation due to its high calcium hydroxide content. Carbonation of portland cement concrete converts the calcium hydroxide into calcium carbonate, which tends to reduce permeability and slightly increase strength (Taylor 1997). For concrete containing SCMs, this effect is reversed and the permeability of the surface layer increases somewhat after carbonation. Concrete containing moderate to high levels of fly ash or slag may carbonate more rapidly than portland cement concrete of the same w/cm, especially if the concrete is not properly cured (Osborne 1989 and Thomas and others 2000). Because concretes made with high volumes of SCMs typically gain strength more slowly, longer curing may be needed for sufficient resistance to carbonation.

The depth of carbonation in good-quality, well-cured concrete is generally of little practical significance provided the embedded steel has adequate concrete cover. Finished surfaces tend to experience less carbonation than formed surfaces. Carbonation of finished surfaces is often observed to a depth of 1 to 10 mm (0.04 to 0.4 in.) and for formed surfaces, between 2 and 20 mm (0.1 and 0.9 in.) after several years of exposure, depending on the concrete properties, ingredients, age, duration of curing, and environmental exposure (Campbell, Sturm, and Kosmatka 1991). ACI Committee 201 (ACI 201.2R-08) has more information on atmospheric and water carbonation and the ACI 318 building code (ACI 318-08), provides reinforcing steel cover requirements for different exposures.

Corrosion

Concrete protects embedded steel from corrosion through its highly alkaline nature. The high pH environment in concrete (usually greater than 13.0) causes a passive and non-corroding protective oxide film to form around the steel. However, the presence of chloride ions from deicers or seawater can destroy or penetrate this film leading to corrosion. Corrosion of steel, is an expansive process – the byproduct of corrosion, rust, induces significant internal stresses and eventual spalling of the concrete over reinforcing steel (Figure 11-27).

Figure 11-27. Corrosion of reinforcement results in cracking, spalling, and discoloration of the concrete (Courtesy of R.D. Hooton).

Mechanism of Corrosion. The corrosion of steel reinforcement is an electrochemical process. For corrosion to take place, all elements of a corrosion cell must be present: an anode, a cathode, an electrolyte, and an electrical connection. Once the chloride corrosion threshold of concrete (about 0.15% water-soluble chloride by mass of cement) is reached, an electric cell is formed along the steel or between steel bars and the electrochemical process of corrosion begins. Some steel areas along the bar act as the anode, discharging current in the electric cell; from there the iron goes into solution. Steel areas that receive current are the cathodes where hydroxide ions are formed (Figure 11-28). The iron and hydroxide ions form iron hydroxide. The iron hydroxide further oxidizes to form rust or other iron oxides as illustrated in Figure 11-29. The volume of the final product may be more than six times the volume of the original iron resulting in cracking and spalling of the concrete. The cross-sectional area of the steel can also be significantly reduced.

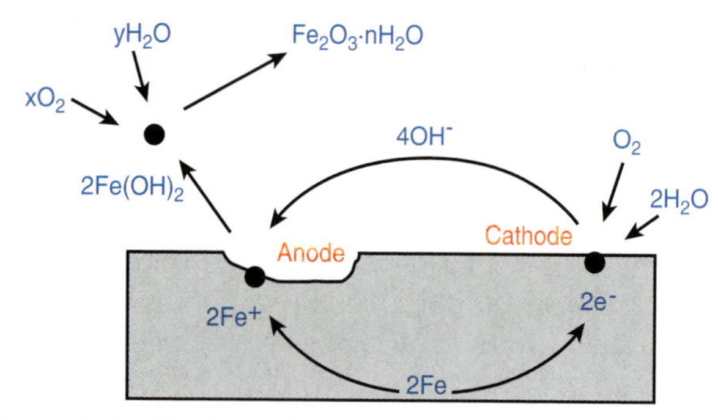

Figure 11-28. Dissolution of the iron takes place at the anode. The ferrous ions combine with hydroxyl ions, oxygen, and water to form various corrosion products (Source: Detwiler and Taylor 2005).

Once corrosion starts, the rate of steel corrosion is influenced by the concrete's electrical resistivity, moisture content, and the rate at which oxygen migrates through the concrete to the steel. Chloride ions alone can also penetrate the passive film on the reinforcement. They combine with iron ions to form a soluble iron chloride complex that carries the iron into the concrete for later oxidation (rust) (Whiting 1997, Taylor, Whiting, and Nagi 2000, and Whiting, Taylor and Nagi 2002).

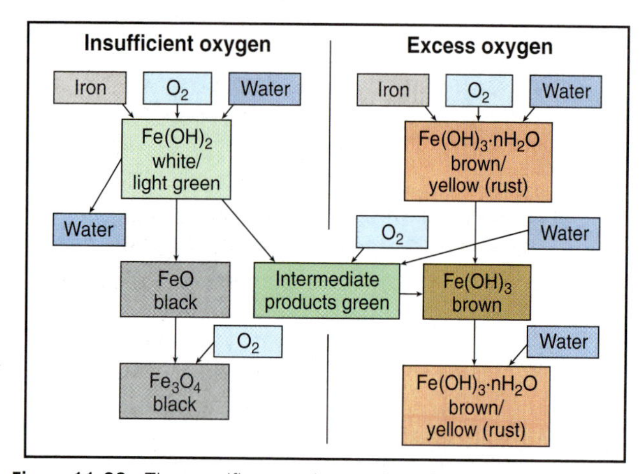

Figure 11-29. The specific corrosion products formed depend on the availability of oxygen (adapted from Herholdt and others 1979).

Various factors affect the rate of corrosion of steel. These include:

1. Water – Water both participates in the cathodic reaction and transports the ions.
2. Oxygen – Corrosion is two orders of magnitude slower when oxygen is not present.
3. The pH of the concrete – Illustrated in Figure 11-30, the pH affects the rate of corrosion. Below a pH of about 11, the passive layer of concrete breaks down allowing chloride ions to more rapidly penetrate the concrete. Below a pH of about 4, the protective film on the steel dissolves.
4. Chlorides – Chloride ions act as catalysts to the corrosion reaction. They break down the passive layer by a process illustrated in Figure 11-31. The localized microcell corrosion that takes place under these circumstances is called pitting.
5. Temperature – The rate of corrosion for a given concentration of oxygen approximately doubles for every 30°C (54°F) increase in temperature (Uhlig and Revie 1985).
6. Electrical resistivity of the concrete – In some installations, stray currents may induce corrosion in the steel, thereby artificially driving the reactions. The presence of dissimilar metals embedded in concrete can also form a corrosion cell. Concrete with a high electrical resistivity will reduce the rate of corrosion by reducing the rate of ion transport.
7. Permeability/Diffusivity of the concrete – As shown in Figure 11-31, oxygen and chloride ions must reach the anode through the concrete cover. Thus, the diffusivity of the concrete to these ions affects the rate of corrosion. In addition, the availability of oxygen at the cathode, and the availability of oxygen and water at the anode (which will affect the degree of expansion of the corrosion product) are a function of the permeability of the concrete.
8. Cathode-to-anode area – When the cathode is much larger than the anode, localized corrosion proceeds rapidly, resulting in loss of area of the steel cross section. When the areas of the cathode and anode are approximately equal, corrosion proceeds much more slowly and is distributed more evenly.

Nonferrous Metals. Nonferrous metals are also frequently used for construction in contact with portland cement concrete. Metals such as zinc, aluminum, and lead – and alloys containing these metals – may be subject to corrosion when embedded in, or in surface contact with, concrete. Galvanic corrosion will occur if aluminum and steel or other metals of dissimilar composition are both

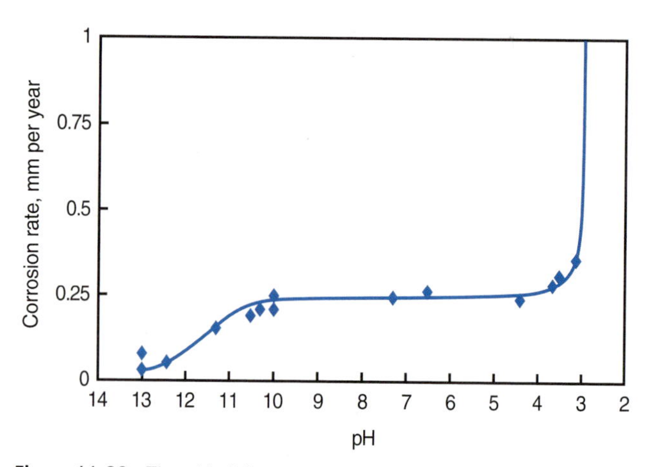

Figure 11-30. The pH of the surrounding medium affects the rate of corrosion. Below a pH of about 4, the passive oxide layer dissolves. The pH of concrete pore solution is normally greater than 13.0 (adapted from Uhlig and Revie 1985).

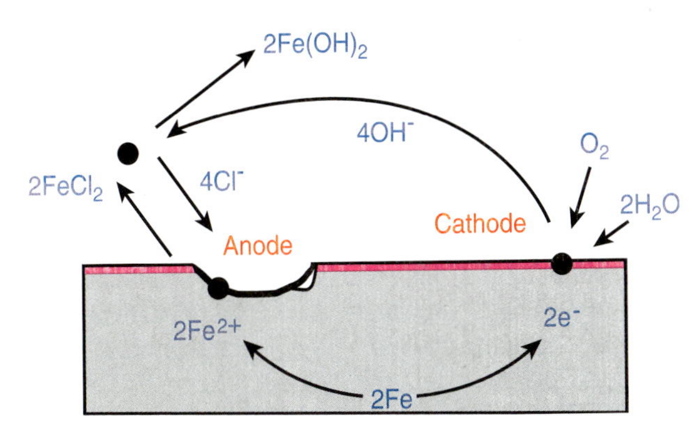

Figure 11-31. Chloride ions act as catalysts to the corrosion of iron. The localized corrosion that takes place is known as pitting corrosion (Source: Detwiler and Taylor 2005).

embedded in concrete and in contact with each other. See PCA (2002) and Kerkhoff (2007), and ACI Committee 201 (201.2R-08) for more information on dissimilar metal corrosion and nonferrous metals in concrete.

Corrosion Testing. Deductions concerning corrosion activity of embedded steel can be made using the information obtained from ASTM C876, *Test Method for Half-Cell Potentials of Uncoated Reinforcing Steel in Concrete*.

Acid-soluble chloride content of concrete is measured in accordance with ASTM C1152, *Test Method for Acid-Soluble Chloride in Mortar and Concrete*. ACI 318 bases the chloride limit on water-soluble chlorides, with maximum limits of 0.06% for prestressed concrete and 0.15% for reinforced concrete.

Testing to determine water-soluble chloride ion content should be performed in accordance with ASTM C1218, *Test Method for Water-Soluble Chloride in Mortar and Concrete*. ASTM C1524, *Test Method for Water-Extractable Chloride in Aggregate (Soxhlet Method)*, can be used to evaluate aggregates that contain a high amount of naturally occurring chloride. ASTM G109, *Standard Test Method for Determining Effects of Chemical Admixtures on Corrosion of Embedded Steel Reinforcement in Concrete Exposed to Chloride Environments*, can be used to determine the effects of admixtures on the corrosion of embedded steel reinforcement in concrete exposed to chloride environments.

Where stray currents are expected, the electrical conductivity of the concrete should be explicitly addressed by limiting the allowable charge passed as measured by ASTM C1202.

Materials and Methods to Control Corrosion. The concrete mixture proportions play an important role in preventing corrosion. In addition to good quality concrete, specific concrete materials and mixture proportions should be considered to lower corrosion activity and to optimize protection of embedded steel. To maximize chloride (corrosion) resistance, reduce permeability by specifying a maximum water-cement ratio of 0.40 or less (Stark 1989) and at least seven days of moist curing. Judicious use of one or more SCM, combined with extended moist curing, can effectively reduce the permeability, diffusivity, and electrical conductivity of concrete. Cements with high C_3A contents and/or slag cement or alumina-bearing pozzolans are also frequently used because of their effectiveness in binding chlorides.

The threshold level at which corrosion starts in normal concrete with no inhibiting admixture is about 0.15% water-soluble chloride ion (0.20% acid-soluble) by weight of cement. Admixtures, aggregate, and mixing water containing chlorides should be avoided. But in any case, the total acid-soluble chloride content of the concrete should be limited to a maximum of 0.08% and 0.20% (preferably less) by mass of cement for prestressed and reinforced concrete, according to ACI Committee 201 (ACI 201.2R-08) and ACI Committee 222 (ACI 222R-01).

Increasing the concrete cover over the steel also helps slow down the migration of chlorides. Sufficient concrete cover must be provided for reinforcement where the surface is to be exposed to corrosive substances. It is good practice to increase the concrete cover over the reinforcing steel above the amount specified in ACI 318 (2008). Additional cover extends the time corrosive chemicals, such as chlorides, need to reach the reinforcing steel. ACI committee 201 (ACI 201.2R-08) recommends a minimum cover of 40 mm (1½ in.) and preferably at least 50 mm (2 in.) for concrete in moderate-to-severe corrosion environments. Oesterle (1997) and Stark (1989) confirm the need for 65 mm to 75 mm (2½ in. to 3 in.) of cover over reinforcement to provide corrosion protection. Some engineers specify 90 mm (3½ in.) or more of concrete cover over steel in concrete exposed to chlorides or other corrosive solutions. However, large depths of cover on the tension side of concrete members can lead to excessive crack widths because they lose the benefit of the top mat of reinforcement in terms of controlling thermal and drying shrinkage.

Other methods of reducing steel corrosion include the use of epoxy-coated reinforcing steel (ASTM D3963 or AASHTO M 284), corrosion resistant steel, corrosion inhibiting admixtures (ASTM C1582), surface treatments, concrete overlays, and cathodic protection as discussed below.

Epoxy coated reinforcing steel. Fusion-bonded epoxy-coated reinforcing steel is very popular for the construction of marine structures and pavements, bridge decks, and parking garages exposed to deicer chemicals (Figure 11-32). The epoxy coating prevents chloride ions and other corrosive chemicals, moisture, and oxygen from reaching the steel. If the epoxy coating contains pinholes or is damaged during construction, its protection ability is effectively lost.

Figure 11-32. Epoxy-coated reinforcing steel used in bridge deck.

Epoxy repair kits are available and should be used to recoat the damaged portion of the bar prior to concrete placement. Epoxy-coated bars should conform to ASTM A775/A775M, *Specification for Epoxy- Coated Reinforcing Steel Bars*, and to ASTM D3963/D3963M, *Specification for Fabrication and Jobsite Handling of Epoxy-Coated Steel Reinforcing Bars*.

Stainless steel. Occasionally, selective use of stainless steel reinforcement in zones exposed to high chloride concentrations can ensure a long service life in that part of the structure, provided the concrete itself is made sufficiently resistant to avoid other types of deterioration. Stainless steel can be coupled with carbon steel reinforcement without causing galvanic corrosion (Gjørv 2009). Some stainless steels may be used in concretes containing chloride if test data support their performance.

Nickel-plated steel. Nickel-plated steel will not corrode when embedded in chloride-free concrete. The nickel plate will provide protection to steel as long as no discontinuities or pinholes are present in the coating. The coating should be about 0.1 mm thick to resist rough handling. Minor breaks in the coating may not be very detrimental in the case of embedment in chloride-free concrete. However, corrosion of the underlying steel could be strongly accelerated in the presence of chloride ions.

Galvanized steel should conform to ASTM A767/A767M, *Specification for Zinc-Coated (Galvanized) Steel Bars for Concrete Reinforcement*. Chloride-ions will cause corrosion of galvanized steel in concrete and may lead to severe cracking and spalling of the surrounding concrete. The use of chloride admixtures should be avoided in concrete containing galvanized steel exposed to corrosive or wet environments. Stark (1989) illustrates the effect of humidity and chloride content on corrosion of black (untreated) and galvanized steel bars.

Fiber-reinforced plastic (FRP) reinforcement. FRP can be occasionally be used to replace part or all of the steel reinforcement in portland cement concrete exposed to extremely corrosive chemicals. Plastic reinforcing bars are available in most conventional bar sizes. The lightweight, nonmagnetic, nonconductive, high strength (tensile strength greater than 690 MPa [100,000 psi]) bars are chemically resistant to many acids, salts, and gases and are unaffected by electrochemical attack. Commercially available FRP reinforcement is made of continuous aramid, carbon, or glass fibers embedded in a resin matrix. The resin allows the composite action of the fibers to work as a single element. Resins used in FRP reinforcement include polyester, vinyl ester, nylon, or polyethylene. Consult ACI 440.1R (ACI Committee 440 2006) for special design considerations.

Corrosion Inhibitors. Corrosion inhibitors such as calcium nitrite are used as an admixture to reduce corrosion. Organic-based corrosion inhibitors, based on amine and amine and fatty ester derivatives, are also available (Nmai and others 1992 and Berke and others 2003). See Chapter 7 for more on corrosion inhibitors.

Cathodic protection reverses the natural electron flow through concrete and reinforcing steel by inserting a nonstructural anode in the concrete. This forces the steel to act as the cathode by electrically charging the system. Since corrosion occurs where electrons leave the steel, the reinforcement cannot corrode, as it is receiving the electrons.

The methods noted above can be combined with other corrosion protection methods. Some additional protective strategies stop or reduce chloride-ion or chemical penetration at the concrete surface and include concrete surface sealers, water repellents, surfacings, and overlays. Materials commonly used include silanes, siloxanes, methyl methacrylates, epoxies, and other compounds. Latex-modified and polymer concrete are often used in overlays to reduce chloride-ion or chemical penetration. Impermeable interlayer membranes (primarily used on bridge decks), prestressing for crack control, or polymer impregnation are also available to help protect reinforcement.

Using more than one protection method simultaneously can result in significant savings in maintenance costs. For example, the advantages of using epoxy-coated reinforcement are obvious. Epoxy coating stops chloride at the reinforcing steel. However, damaged areas in the coating caused by handling during transportation and construction or coating imperfections can be a source of corrosion. An additional protection system, such as a corrosion-inhibitor or supplementary cementitious materials in the concrete mixture, can be used to further protect the steel at coating-damaged areas.

Sulfate Attack

High amounts of sulfates in soil or water can attack and destroy a concrete that is not properly designed. Sulfates (for example calcium sulfate, sodium sulfate, and magnesium sulfate) can attack concrete by reacting with hydrated compounds in the hardened cement paste. These reactions can induce sufficient pressure to disrupt the cement paste, resulting in disintegration of the concrete (loss of paste cohesion and strength) (Figure 11-33). Also see **Salt Crystallization or Physical Salt Attack** below for the related role of sulfates on physical salt attack.

Sulfate ions are sometimes found in soil and groundwater in North America, most notably in the Western United States and the Canadian Prairie provinces. They are also found in other locations throughout the world. Sulfates are also present in seawater, in some industrial environments, and in sewers.

Sulfate attack can lead to loss of strength, expansion, spalling of surface layers, and ultimately the disintegration of concrete (Taylor 1997). Flowing water is more aggressive than stagnant water, since new sulfate ions are constantly being transported to the concrete for chemical reaction.

Mechanisms of Sulfate Attack. Sulfate ions attack calcium hydroxide and the hydration products of C_3A, forming gypsum and ettringite in expansive reactions. Magnesium sulfate attacks in a manner similar to sodium sulfate but also forms brucite (magnesium hydroxide). Brucite forms primarily on the concrete surface; it consumes calcium hydroxide, lowers the pH of the pore solution, and then decomposes the calcium silicate hydrates (Santhanam and others 2001). Depending on the cation(s) involved, sulfate attack can be more or less aggressive; with magnesium sulfate attack being the most aggressive and calcium sulfate the least aggressive.

Although extremely rare, thaumasite may form during sulfate attack in moist conditions at temperatures usually between 0°C and 10°C (32°F to 50°F) occuring as a result of a reaction between calcium silicate hydrate, sulfate, carbonate, and water (Thaumasite Expert Group 1999). A small amount of thaumasite can be accommodated in concrete without causing distress; however deleterious thaumasite sulfate attack results from expansive stresses created by significant quantities of thaumasite forming. In concretes where deterioration is associated with excess thaumasite formation, cracks can be filled with thaumasite and haloes of white thaumasite are present around aggregate particles. At the concrete/soil interface the surface concrete layer can be soft with complete replacement of the cement paste by thaumasite (Figure 11-34) (Hobbs 2001).

Sulfate Testing. There is no standardized test for concrete sulfate resistance. ASTM C1012 is used in C595 (M240) and C1157 to define cements as moderately- or highly-sulfate resistant in specification, based on their performance in mortar specimens containing graded Ottawa sand and exposed to 5 percent sodium sulfate solution. (ASTM C150 uses prescriptive limits on C_3A content to define sulfate resistant cement types or an optional performance requirement, ASTM C452.) In ASTM C1012 standard mortar bars are exposed to a 5% sulfate solution and expansion is measured over time. In this aggressive environment, expansions of less than 0.10% at 6 months indicate moderate sulfate resistance and expansions less than 0.05% at 6 months (or less than 0.10% at 12 months), indicate high sulfate resistance. ACI 318 requires ASTM C1012 for testing for cements and use of SCMs for some sulfate exposure classes.

Figure 11-33. Concrete beams after seven years of exposure to sulfate-rich wet soil in a Sacramento, California, test plot. The beams in better condition have low water-cementitious materials ratios, and most have sulfate resistant cement. The inset shows two of the beams tipped on their side to expose decreasing levels of deterioration with depth and moisture level. Deterioration of concrete exposed to sulfates is often a combination of chemical sulfate attack and physical salt crystallization.

Figure 11-34. Thaumasite deterioration of concrete cubes (Courtesy of R.D. Hooton).

To assess the sulfate exposure condition for a specific location, soil testing is conducted using ASTM C1580, *Standard Test Method for Water-Soluble Sulfate in Soil.*

Materials and Methods to Control Sulfate Attack. The first step to designing a sulfate-resistant concrete is to determine the exposure class for sulfates and use the lowest water-cementing materials ratio required. ACI 318 defines 4 exposure classes, S0 to S3, for concrete, depending on the sulfate content in the soil or groundwater (S3 is the most severe).

Cements specially formulated to improve sulfate resistance, such as ASTM C150 (AASHTO M 85) Types II or V cements, ASTM C595 (AASHTO M 240) sulfate resistant cements, or ASTM C1157 Types MS or HS should be used. (See Chapter 3 for more information on cements for sulfate resistance).

Sulfate-resistant cements alone are not adequate to resist sulfate attack. It is essential to limit the ability of the sulfates to enter the concrete in the first place. This is accomplished by reducing the permeability of the concrete (minimizing the water-to-cementitious materials ratio and providing good curing) (Stark 2002).

In very severe sulfate exposures, a waterproof barrier protecting the concrete can be beneficial (Stark 2002 and Kanare 2008).

Sulfate attack can also occur when the sulfate is internally supplied to the concrete. ACI Committee 221 (1996) reported the presence of sulfates in a variety of aggregate types, either as an original component of the aggregate, or due to the oxidation of sulfides originally present. The most common form of sulfate in aggregates is gypsum ($CaSO_4 \cdot 2H_2O$). Gypsum occurs as a coating on sand and gravel, as a component of some sedimentary rock, or in weathered slags. Aggregates made from recycled building materials may contain sulfates in the form of contamination from plaster or gypsum wallboard. It is difficult to eliminate the gypsum present. On sieving, these particles break apart, becoming part of the sand fraction. In some quarries, it is necessary to reject the entire fraction of aggregate smaller than 2 mm (No. 8) in size because it contains 1% to 2% SO_3 by total mass of aggregate.

Salt Crystallization or Physical Salt Attack

Similar to natural rock formations like limestone, porous concretes are susceptible to weathering caused by salt crystallization. These salts may or may not contain sulfates and they may or may not react with the hydrated compounds in concrete. Examples of salts known to cause weathering of exposed concrete include sodium carbonate and sodium sulfate. (Laboratory studies have also related saturated solutions of calcium chloride and other salts to concrete deterioration.) The greatest damage occurs with drying of saturated solutions of these salts, often in an environment with specific cyclical changes in relative humidity and temperature that alter mineralogical phases. Both aggregate particles and cement paste can be attacked by salts.

Mechanism of Salt Attack. Haynes and others (1996) and Folliard and Sandberge (1994), described the mechanism of physical attack on concrete by salts present in groundwater. Groundwater enters the concrete by capillary action and diffusion. When pore water evaporates from above-ground concrete surfaces, the salt concentrates until it crystallizes, sometimes generating pressures large enough to cause cracking. Changes in ambient temperature and relative humidity cause some salts to undergo cycles of dissolution and crystallization, or hydration-dehydration. When crystallization or hydration is accompanied by volumetric expansion, repeated cycles can cause deterioration of concrete similar to that caused by cycles of freezing and thawing.

Physical attack by sulfate salts can be distinguished from conventional, chemical sulfate attack, for example, by evaluating the sulfate content of the concrete. Chemical sulfate attack can be evidenced by significant amounts of ettringite or gypsum, as well as the characteristic decalcification of the paste and cracking due to expansion. In physical sulfate attack, damage in the form of scaling is usually limited to the exterior surface of the concrete; the concrete is not affected below the surface. Damage due to salt crystallization can occur with a variety of salts; they need not contain sulfate ions. Concrete structures exposed to salt solutions should have a low water-cementing materials ratio (0.40 to 0.45 maximum depending on severity) to reduce permeability.

Materials and Methods to Control Salt Attack. The ideal way to prevent damage by salt crystallization is to prevent the salt-laden water from entering the concrete. Where conditions conducive to salt crystallization exist—that is, where the climate is arid but the local groundwater table is near the surface – a barrier on the upstream face is the surest way to prevent damage. Coating of the concrete surface, or placement of plastic sheeting beneath slabs (with compactable granular subbase under the plastic sheeting), helps keep the salt solution out (Kanare 2008). It should be noted that over-watering of lawns and plants located near the concrete can locally raise the elevation of the water table. Appropriate landscaping (use of native plants, locating lawns and plants away from the concrete) mitigates this problem. Any measure that reduces the permeability of the concrete – low water-cement ratio, use of SCMs, and good curing – also reduces the vulnerability of the concrete to salt crystallization damage.

Like sulfate attack, salt crystallization is more severe at locations where the concrete is exposed to wetting and drying cycles, than it is with continuously wet exposures. This is often seen in concrete posts where the concrete has deteriorated only a few centimeters above and below the soil line. The portion of concrete deep in the soil (where it is continuously wet) is in good condition (Figure 11-35).

Figure 11-35. Salt crystallization attack is often observed near the soil line. Here concrete posts have been attacked by sulfate salts near the soil line.

Delayed Ettringite Formation

Primary and Secondary Ettringite. Ettringite, a form of calcium sulfoaluminate, is found in all portland cement paste. Calcium sulfate sources, such as gypsum, are added to portland cement during final grinding at the cement mill to prevent rapid setting and improve strength development. Sulfate can be also be present in SCMs and some admixtures. Gypsum and other sulfate compounds react with tricalcium aluminate in cement to form ettringite within the first few hours after mixing with water. Most of the sulfate in cement is normally consumed to form ettringite or calcium monosulfate within 24 hours (Klemm and Miller 1997). At this stage ettringite is uniformly and discretely dispersed throughout the cement paste at a submicroscopic level (less than a micrometer in cross-section). This ettringite is often called primary ettringite.

If concrete is exposed to moisture for long periods (typically many years) ettringite can slowly dissolve and reform in less confined locations. Upon microscopic examination, harmless white needle-like crystals of ettringite can be observed lining air voids or cracks. This reformed ettringite is usually called secondary ettringite (Figure 11-36).

Figure 11-36. White secondary ettringite deposits in void. Field width 64 µm.

Concrete deterioration accelerates the rate at which ettringite leaves its original location in the paste to go into solution and then recrystallize in larger spaces such as air voids or cracks. Both water and sufficient space must be present for the crystals to form. Cracks can form due to damage caused by frost action, alkali-aggregate reactivity, drying shrinkage, thermal effects, strain due to excessive stress, or other mechanisms.

Ettringite crystals in air voids and cracks are typically two to four micrometers in cross section and 20 to 30 micrometers in length. Under conditions of extreme deterioration or long exposure to a moist environment, the white ettringite crystals can appear to completely fill voids or cracks. However, secondary ettringite, as large needle-like crystals, should not be interpreted as being harmful to the concrete (Detwiler and Powers-Couche 1997).

Mechanism of DEF. Delayed ettringite formation (DEF) refers to the delayed formation of ettringite, in which the normal early formation of ettringite that occurs in concrete cured at ambient temperature is interrupted as a result of exposure to high-temperatures (between 70°C and 100°C [158°F to 212°F] during placement or curing. If the ettringite forms at later ages and is exposed to moisture in service, this can lead to internal expansion and damage to the hardened concrete.

Since ettringite is not stable at elevated temperatures, monosulfoaluminate forms instead, even when sufficient sulfate is present. Later, after the concrete cools, ettringite forms from monosulfoaluminate, alumina, and sulfate initially trapped in hydration products (C-S-H). When later exposed to moisture these conditions create an expansive reaction in the paste (Famy and Taylor 2001; Scrivener, Damidot and Famy 1999; Taylor, Famy, and Scrivener 2001; and Ramlochan and others 2004). This process may take many years for the monosulfoaluminate to convert to ettringite. As a result of an increase in paste volume, separation of the paste from the aggregates is usually observed with DEF. It is characterized by the development of rims around the aggregates (Figure 11-37), sometimes filled with ettringite. An abundance of water (saturation or near saturation of the concrete) is necessary for the formation of ettringite as each mole of ettringite contains 32 moles of water.

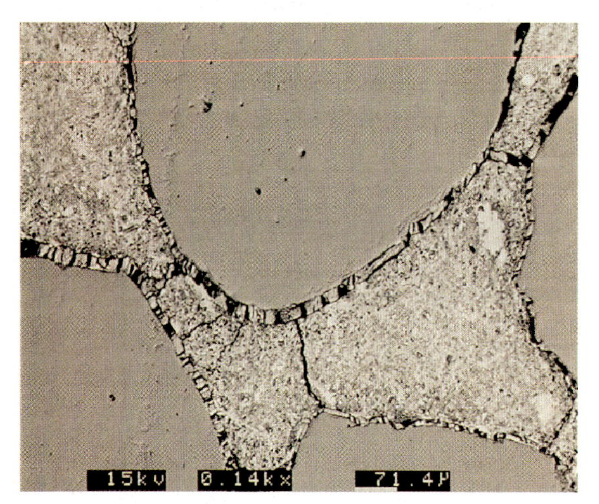

Figure 11-37. Heat induced delayed expansion (DEF) is characterized by expanding paste that becomes detached from nonexpansive components, such as aggregates, creating gaps at the paste-aggregate interface. The gap can subsequently be filled with larger opportunistic ettringite crystals as shown here (Courtesy of Z. Zhang and J. Olek).

The mechanism causing expansion in the paste is not fully understood at this time; the true influence of ettringite formation on expansion is still being investigated. Work by Shimada (2005) has proposed a possible explanation. DEF is a complex phenomenon that depends on the concrete temperature during curing as well as on both physical and chemical properties of the system. Several competing factors are at work, including: the amount of ettringite that decomposes, the rate at which ettringite reforms, the rate at which stress can be relieved through reprecipitation in larger spaces (Ostwald ripening), and the strength of the system to resist expansive stresses. Graf (2007) found that DEF-related expansion is influenced by the level of moisture exposure. A threshold level of relative humidity for expansion to occur seems to be slightly above 90%. The DEF-related expansion of a system will therefore be controlled by the curing temperature, the environment in which it exists, and the chemistry of the system.

DEF Testing. There are no commonly accepted test methods for DEF. The phenomenon is rarely observed in the field. The primary means of controlling DEF is to prevent the concrete from reaching excessive temperatures from heat of hydration or thermal treatments (as used in some precast operations).

Materials and Methods to Control DEF. Because of the risk of delayed ettringite formation, as well as the deleterious effects of elevated temperature on durability, concrete temperatures above 70°C (158°F) should be avoided (Shimada 2005 and Detwiler and Taylor 2005). The use of SCMs will help reduce the risk of DEF (Ramlochan and others 2005). The heat of hydration itself might raise the internal temperature of concrete above 70°C (158°F), even if no additional energy is applied to the concrete. Precautions must be taken to help minimize unacceptably high temperatures occurring during placement and curing. This is especially the case with mass concrete (typically concrete having elements with a minimum dimension of more than about 1 m [39 in.]), particularly if a high cement content is used. Refer to PCI for more information on DEF control measures.

Acid Attack

Most acidic solutions will disintegrate portland cement concrete. The rate of disintegration will be dependent on the type and concentration of acid. Certain acids, such as oxalic acid, are harmless. Weak solutions of some acids have insignificant effects.

Mechanism of Acid Attack. Acids attack concrete by dissolving both hydrated and unhydrated cement compounds as well as calcareous aggregate. Siliceous aggregates are resistant to most acids and other chemicals and are sometimes specified to improve the chemical resistance of concrete, especially with the use of chemically-resistant cement. Siliceous aggregate should be avoided when a strongly basic solution, like sodium hydroxide, is present as it attacks siliceous aggregate.

Materials and Methods to Control Acid Attack. In certain acidic solutions, it may be impossible to apply an adequate protective treatment to the concrete. The use of a sacrificial calcareous aggregate should be considered, particularly in those locations where the acidic solution may pond. Replacement of siliceous aggregate by limestone or dolomite having a minimum calcium oxide concentration of 50% will aid in neutralizing the acid. The acid will attack the entire exposed surface more uniformly, reducing the rate of attack on the paste and preventing loss of aggregate particles at the surface. Langelier Saturation Index values for a water solution and calcium absorption test data on a soil sample can be used to test for this condition (Hime and others 1986 and Steinour 1975). Negative Langelier Index values indicate a lime deficiency.

The use of calcareous aggregate may also retard expansion resulting from sulfate attack caused by some acid solutions. As practical, the paste content of the concrete should be minimized – primarily by reducing water content and using a well-graded aggregate – to reduce the area of paste exposed to attack and the concrete permeability. High cement contents are not necessary for acid resistance. Concrete deterioration increases as the pH of the acid decreases below about 6.5 (Kong and Orbison 1987 and Fattuhi and Hughes 1988).

Properly cured concrete with reduced calcium hydroxide contents, as occur with concretes using SCMs, may experience a slightly slower rate of attack from acids. This is because acid resistance is linked to the total quantity of calcium-containing phases, not just the calcium hydroxide content (Matthews 1992). Resistance to acid attack is primarily dependent on the concrete's permeability and cementitious materials ratio. Acid rain (often with a pH of 4 to 4.5) can slightly etch concrete surfaces, usually without affecting the performance of exposed concrete structures. Extreme acid rain or strong acids may warrant special concrete designs or precautions, especially in submerged areas.

The American Concrete Pressure Pipe Association (ACPPA) provides guidelines for granular soils with a pH below 5 and the total acidity of the soil exceeding 25 meq/100 gm and requires one of the following precautions to be used (ACPPA 2000):

- backfill in the pipe zone with consolidated clay material or calcareous material;
- acid resistant membrane on or around the pipe; or
- 8 to 10% silica fume in the mortar coating.

Where soil pH is below 4, the pipe should be installed in an acid resistant membrane or in an envelope of non-aggressive consolidated clay (ACPPA 2000). Natural waters usually have a pH of more than 7 and seldom less than 6. Waters with a pH greater than 6.5 may be aggressive if they contain bicarbonates. Water that contains bicarbonates also contains dissolved free carbon dioxide (CO_2) and carbonic acid (H_2CO_3) which can dissolve calcium carbonate unless it is saturated. This aggressive carbon dioxide acts by acid reaction and can attack concrete products regardless of whether they are carbonated. Methods are presented in Steinour (1975) for estimating the amount of aggressive carbon dioxide from an ordinary water analysis when the pH is between 4.5 and 8.6, and the temperature is between 0 °C (32 °F) and 65 °C (145 °F).

The oxidation of hydrogen sulfide (H_2S), present in sanitary sewers, forms sulfuric acid (H_2SO_4), a strong acid that dissolves the calcium, magnesium, and aluminum compounds in concrete. Certain species of bacteria feed on any form of sulfide, converting it to sulfuric acid. Thus odor complaints, which are often the first indication of the presence of H_2S, must be investigated so that the concrete does not undergo prolonged acid attack. Other anions in aggressive water may react with the components of cement paste to form insoluble calcium salts. Such reactions are not harmful provided the products are neither expansive nor removed by erosion.

The German Institute of Standardization Specification DIN 4030-2 (1991) includes criteria and a test method for assessing the potential of damage from carbonic acid-bearing water (German Institute of Standardization 1991).

Seawater Exposure

Concrete has been used in seawater exposures for decades with excellent performance. However, special care in mix design and material selection is necessary for these severe environments. A structure exposed to seawater or seawater spray is most vulnerable in the tidal or splash zone where there are repeated cycles of wetting and drying, freezing and thawing.

It is clear that all of the necessary ingredients are present for at least three different deterioration mechanisms:

1. Corrosion (chlorides, oxygen, water)
2. Sulfate attack (sulfates, water)
3. Magnesium ion substitution (magnesium, water)

Several additional deterioration mechanisms may also be presented:

1. Alkali-aggregate reaction (if reactive aggregates are used)
2. Abrasion/erosion (due to wave action, sand, gravel, icebergs)
3. Freeze-thaw damage
4. Ship and debris impact
5. Carbonation
6. Salt crystallization

Mechanisms of Seawater Attack. When several deterioration mechanisms take place in the same concrete element, they generally interact in a way such that the net effect is greater than the sum of individual effects. For example, cracking due to corrosion will allow carbon dioxide, sulfate, and magnesium to penetrate more readily into the concrete to attack the interior as well as the surface.

One notable exception is the effect of chloride ions on sulfate attack. Despite the high sulfate content of seawater, sulfate attack on marine concrete does not result in expansive formation of ettringite. Instead, deterioration takes the form of erosion or loss of the solid constituents. Ettringite expansion is suppressed in environments where OH^- ions are essentially replaced by Cl^- ions. Mindess and Young (1981) reported that gypsum and ettringite are both more soluble in chloride ion solutions, so that deleterious expansions are mitigated. Thus, seawater is considered a moderate sulfate exposure.

In concrete submerged in seawater, the rate of corrosion is limited by the supply of oxygen. The deeper the concrete, the slower the deterioration. The splash zone is the most severe exposure for several reasons: (1) both oxygen and seawater are in abundant supply; (2) cycles of wetting and drying concentrate the salts present; (3) cycles of freezing and thawing (if any) take place when the concrete is fully saturated; and (4) the action of waves, floating objects, and sand create the most severe conditions of abrasion and erosion. Floating ice may adhere to the concrete surface and later be pulled away by currents, taking some of the concrete with it.

Materials and Methods for a Marine Environment. Cement used in concrete for a marine environment (whether in or near the ocean) must balance the benefit of higher C_3A content to bind chlorides with the need for sulfate resistance (that is, low C_3A content). Thus, an intermediate C_3A content cement is desirable. A cement resistant to moderate sulfate exposure is helpful. Portland cements with tricalcium aluminate (C_3A) contents that range from 4% to 10% have been found to provide satisfactory protection against seawater sulfate attack, as well as protection against reinforcement corrosion by chlorides. Slag cement and alumina-bearing pozzolans are very effective in binding chlorides as well as providing sulfate resistance. For this reason, specially formulated marine cements generally contain high volumes (65% or more by mass) of slag cement. Other supplementary cementing materials can also be used to advantage in this environment. And, as always, for the best durability the concrete should be of high quality, that is, well compacted, well cured, and with a low water-to-cementitious materials ratio.

Sulfates and chlorides in seawater require the use of low permeability concrete to minimize steel corrosion and sulfate attack (Figure 11-38). Water-cementitious material ratios should not exceed 0.40. In northern climates, the concrete must be properly air entrained with at least 6% air. High-strength concrete should be considered where large ice formations abrade the structure. Proper concrete cover over reinforcing steel must be provided (see ACI 318). See Stark (1995 and 2001), Farny (1996), and Kerkhoff (2007).

Figure 11-38. Concrete exposed to a marine environment.

References

AASHTO, *Guide Specification For Highway Construction SECTION 56X Portland Cement Concrete Resistant to Excessive Expansion Caused by Alkali-Silica Reaction* (Appendix F to ASR Transition Plan), http://leadstates.transportation.org/ASR/library/gspec.stm, 2001.

ACI, *ACI Concrete Terminology*, American Concrete Institute, Farmington Hills, Michigan, http://www.concrete.org/Technical/CCT/FlashHelp/ACI_Concrete_Terminology.pdf, August 2010, 81 pages.

ACI Committee 201, *Guide to Durable Concrete*, ACI 201.2R-08, American Concrete Institute, Farmington Hills, Michigan, 2008, 49 pages.

ACI Committee 221, *Guide for Use of Normal Weight Aggregates in Concrete*, ACI 221R-96, American Concrete Institute, Farmington Hills, Michigan, 1996.

ACI Committee 222, *Protection of Metals in Concrete Against Corrosion*, ACI 222R-01, American Concrete Institute, Farmington Hills, Michigan, 2001, 41 pages.

ACI Committee 318, *Building Code Requirements for Structural Concrete (ACI 318-08) and Commentary*, ACI 318-08, American Concrete Institute, Farmington Hills, Michigan, 2008, 471 pages.

ACI Committee 440, *Guide for the Design and Construction of Structural Concrete Reinforced with FRP Bars*, ACI 440.1R-06, American Concrete Institute, Farmington Hills, Michigan, 2006, 44 pages.

ACPA, *Scale-Resistant Concrete Pavements*, IS117, American Concrete Pavement Association, Skokie, Illinois 1996.

ACPPA, *External Protection of Concrete Cylinder Pipe*, American Concrete Pressure Pipe Association, Reston, Virginia, http://www.acppa.org/documents/external.pdf, 2000, 12 pages.

ASCE, "Entrained Air Voids in Concrete Help Prevent Salt Damage," *Civil Engineering*, American Society of Civil Engineers, New York, May 1982.

Berke, N.S.; Aldykiewicz, A.J. Jr.; and Li, L., "What's New in Corrosion Inhibitors," *Structure*, July/August 2003, pages 10 to 12.

Brinkerhoff, C.H., "Report to ASTM C-9 Subcommittee III-M (Testing Concrete for Abrasion) Cooperative Abrasion Test Program," University of California and Portland Cement Association, 1970.

Bates, A.A.; Woods, H.; Tyler, I.L.; Verbeck, G.; and Powers, T.C., "Rigid-Type Pavement," Association of Highway Officials of the North Atlantic States, *28th Annual Convention Proceedings*, March 1952, pages 164 to 200.

Bouzoubaâ, Nabil; Bilodeau, Alain; Fournier, Benoit; Hooton, R. Doug; Gagné, Richard; and Jolin, Marc, "Deicing Salt Scaling Resistance of Concrete Incorporating Supplementary Cementing Materials: A Promising Laboratory Test", *Canadian Journal of Civil Engineering*, Vol. 35, November 2008, pages 1261 to 1275.

Boyd, A.J., and Hooton, R.D., "Long-Term Scaling Performance of Concretes Containing Supplementary Cementing Materials", *Journal of Materials in Civil Engineering*, ASCE, Vol. 16, No. 10, October 2007, pages 820 to 825.

Brown, F.P., and Cady, P.D., "Deicer Scaling Mechanisms in Concrete," *Durability of Concrete,* ACI SP-47, American Concrete Institute, Farmington Hills, Michigan, 1975, pages 101 to 119.

Campbell, D.H.; Sturm, R.D.; and Kosmatka, S.H., "Detecting Carbonation," *Concrete Technology Today*, PL911, http://www.cement.org/pdf_files/PL911.pdf, Portland Cement Association, March 1991, pages 1 to 5.

Cody, Rober D.; Cody, Anita M.; Spry, Paul G.; and Gan, Guo-Liang, "Concrete Deterioration by Deicing Salts: An Experimental Study," http://www.ctre.iastate.edu/pubs/semisesq/index.htm, *Semisequicentennial Transportation Conference Proceedings*, Center for Transportation Research and Education, Ames, Iowa, 1996.

Cordon, W.A., "Freezing and Thawing of Concrete – Mechanisms and Control," *ACI Monograph No. 3*, American Concrete Institute, Farmington Hills, Michigan, 1966.

CSA Standard A23.1-09/A23.2-09, *Concrete Materials and Methods of Concrete Construction/ Test Methods and Standard Practices for Concrete,* Canadian Standards Association, Toronto, Canada, 2009.

Detwiler, Rachel J., and Powers-Couche, Laura, "Effect of Ettringite on Frost Resistance," *Concrete Technology Today*, PL973, Portland Cement Association, http://www.cement.org/pdf_files/PL973.pdf, 1997, pages 1 to 4.

Detwiler, R.J., and Taylor, P.C., *Specifier's Guide to Durable Concrete,* EB221, Portland Cement Association, Skokie, Illinois, 2005, 68 pages.

DIN, *Assessment of Water, Soil and Gases for Their Aggressiveness to Concrete; collection and examination of water and soil samples*, DIN 4030-2, Deutsches Institut Für Normung E.V. (German National Standard), June 1991, 13 pages.

Fagerlund, G., "The Significance of Critical Degrees of Saturation at Freezing of Porous and Brittle Materials," *ACI Special Publication*, SP-47, American Concrete Institute, Farmington Hills, Michigan, 1975, pages 13 to 65.

Famy, Charlotte, and Taylor, Hal F.W., "Ettringite in Hydration of Portland Cement Concrete and its Occurrence in Mature Concretes," *ACI Materials Journal*, Vol. 98, No. 4, American Concrete Institute, Farmington Hills, Michigan, July-August 2001, pages 350 to 356.

Farny, Jamie, "Treat Island, Maine—The Army Corps' Outdoor Durability Test Facility," *Concrete Technology Today*, PL963, http://www.cement.org/pdf_files/PL963.pdf, December 1996, pages 1 to 3.

Farny, James A., and Kerkhoff, B., *Diagnosis and Control of Alkali-Aggregate Reactions*, IS413, Portland Cement Association, Skokie, Illinois, 2007, 26 pages.

Fattuhi, N.I., and Hughes, B.P., "Ordinary Portland Cement Mixes with Selected Admixtures Subjected to Sulfuric Acid Attack," *ACI Materials Journal*, American Concrete Institute, Farmington Hills, Michigan, November-December 1988, pages 512 to 518.

Folliard, J.J., and Sandberg, P., "Mechanisms of Concrete Deterioration by Sodium Sulfate Crystallization," ACI SP-145, *Third International ACI/CANMET Conference on Concrete Durability,* Nice, France, 1994, pages 993-946.

Folliard, K.J.; Barborak, R.; Drimalas, T.; Du, L.; Garber, S.; Ideker, J.; Ley, T.; Williams, S.; Juenger, M.; Thomas, M.D.A.; and Fournier, B., "Preventing ASR/DEF in New Concrete: Final Report," *The University of Texas at Austin, Center for Transportation Research (CTR),* CTR 4085-5, 2006.

Fournier, B., *CANMET/Industry Joint Research Program on Alkali-Aggregate Reaction—Fourth Progress Report,* Canada Centre for Mineral and Energy Technology, Ottawa, 1997.

Fournier, Benoit; Berube, Marc-Andre; Folliard, Kevin J.; and Thomas, Michael, *Report on the Diagnosis, Prognosis, and Mitigation of Alkali-Silica Reaction (ASR) in Transportation Structures*, FHWA-HIF-09-004, Federal Highway Administration, Washington, D.C., January 2010, 154 pages.

Gagné, R., and Marchand, J., "La résistance à l'écaillage des bétons à haute performance : état de la question," *Atelier international sur la résistance des bétons aux cycles de gel-dégel en présence de sels fondants, CRIB/RILEM*, 30-31 août, Sainte-Foy, 1993, pages 21 to 48.

Gjørv, Odd E., *Durability Design of Concrete Structures in Severe Environments*, Taylor and Francis, 2009, 232 pages.

Graf, Luis A., *Effect of Relative Humidity on Expansion and Microstructure of Heat-Cured Mortars*, RD139, Portland Cement Association, Skokie, Illinois, 2007, 50 pages.

Hadley, D.W., "Alkali Reactivity of Carbonate Rocks—Expansion and Dedolomitization," Research Department Bulletin RX139, Portland Cement Association, http://www.cement.org/pdf_files/RX139.pdf, 1961.

Haynes, Harvey; O'Neill, Robert; and Mehta, P. Kumar, "Concrete Deterioration from Physical Attack by Salts," *Concrete International*, Vol. 18, No. 1, American Concrete Institute, Farmington Hills, Michigan, January 1996, pages 63 to 68.

Herholdt, A.G.; Justesen, C.F.P.; Nepper-Christensen, P.; and Nielsen, A., editors, *Beton-Bogen*, Aalborg Portland, Aalborg, Denmark, 1979, 719 pages.

Hime, W.G.; Erlin, B.; and McOrmond, R.R., "Concrete Deterioration Through Leaching with Soil-Purified Water," *Cement, Concrete, and Aggregates*, American Society for Testing and Materials, Philadelphia, Summer 1986, pages 50 to 51.

Hobbs, D.W., "Concrete Deterioration: Causes, Diagnosis and Minimizing Risk," *International Materials Review*, February 2001, pages 117 to 144.

Jacobsen, S.; Marchand, J.; and Boisvert, L., "Effect of Cracking and Healing on Chloride Transport in OPC Concrete," *Cement and Concrete Research*, Vol. 26, No. 6, June 1996, pages 869 to 881.

Kanare, Howard M., *Concrete Floors and Moisture*, EB119, Portland Cement Association, Skokie, Illinois, and National Ready Mixed Concrete Association, Silver Spring, Maryland, 2008, 176 pages.

Kerkhoff, Beatrix, *Effects of Substances on Concrete and Guide to Protective Treatments*, IS001, Portland Cement Association, 2007, 36 pages.

Klemm, Waldemar A., and Miller, F. MacGregor, "Plausibility of Delayed Ettringite Formation as a Distress Mechanism—Considerations at Ambient and Elevated Temperatures," Paper 4iv059, *Proceedings of the 10th International Congress on the Chemistry of Cement*, Gothenburg, Sweden, June 1997, 10 pages.

Klieger, Paul, *Extensions to the Long-Time Study of Cement Performance in Concrete*, Research Department Bulletin RX199, Portland Cement Association, 1963.

Klieger, Paul, *Further Studies on the Effect of Entrained Air on Strength and Durability of Concrete with Various Sizes of Aggregate*, Research Department Bulletin RX077, Portland Cement Association, http://www.cement.org/pdf_files/RX077.pdf, 1956.

Klieger, Paul, *Studies of the Effect of Entrained Air on the Strength and Durability of Concretes Made with Various Maximum Sizes of Aggregates*, Research Department Bulletin RX040, Portland Cement Association, http://www.cement.org/pdf_files/RX040.pdf, 1952.

Klieger, Paul, *Air-Entraining Admixtures*, Research Department Bulletin RX199, Portland Cement Association, http://www.cement.org/pdf_files/RX199.pdf, 1966.

Kong, Hendrik, and Orbison, James G., "Concrete Deterioration Due to Acid Precipitation," *ACI Materials Journal*, American Concrete Institute, Farmington Hills, Michigan, March-April 1987.

Kozikowski, Ronald; Taylor, Peter; and Pyc, W. Agata, *Evaluation of Potential Concrete Deterioration Related to Magnesium Chloride ($MgCl_2$) Deicing Salts*, SN2770, Portland Cement Association, Skokie, Illinois, 2007, 30 pages.

Landgren, R., and Hadley, D.W., *Surface Popouts Caused by Alkali Aggregate Reaction*, RD121, Portland Cement Association, 2002, 20 pages.

Lee, H.; Cody, R.D.; Cody, A.M.; and Spry, P.G., "Effects of Various Deicing Chemicals on Pavement Concrete Deterioration," *Proceedings of the Mid-Continent Transportation Symposium*, Center for Transportation Research and Education, Iowa State University, 2000.

Leiser, K., and Dombrowski, G., "Research Work on Magnesium Chloride Solution Used in Winter Service on Roads," *Strasse*, Vol. 7, No. 5, Berlin, Germany, May 1967.

Lerch, William, *Basic Principles of Air-Entrained Concrete*, T-101, Portland Cement Association, 1960.

Lin, C., and Walker, R.D., "Effects of Cooling Rates on the Durability of Concrete," *Transportation Research Record 539*, Transportation Research Board, Washington, D.C., 1975, pages 8 to 19.

Liu, Tony C., "Abrasion Resistance of Concrete," Journal of the American Concrete Institute, Farmington Hills, Michigan, September-October 1981, pages 341 to 350.

Marchand, J.; Sellevold, E.J.; and Pigeon M., "The Deicer Salt Scaling Deterioration of Concrete—An Overview," *Proceedings of the Third CANMET/ACI International Conference on Durability of Concrete*, Nice, France, 1994, pages 1 to 46.

Mather, Bryant, *New Concern over Alkali-Aggregate Reaction*, Joint Technical Paper by National Aggregates Association and National Ready Mixed Concrete Association, NAA Circular No. 122 and NRMCA Publication No. 149, Silver Spring, Maryland, 1975.

Matthews, J.D., "The Resistance of PFA Concrete to Acid Groundwaters," 9th International Congress/India, Vol. V, 1992, Building Research Establishment, Watford, England, page 355.

Mercias, Dean, and Wagoner, Bryan, "Runway Deicers: A Varied Menu," *Airport Magazine*, Vol. 15, 2003.

Mielenz, R.C.; Wokodoff, V.E.; Blackstrom, J.E.; and Flack, H.L., "Origin, Evolutin and Effects of the Air-Void System in Concrete. Part 1–Entrained Air in Unhardened Concrete," July 1958, "Part 2–Influence of Type and Amount of Air-Entraining Agent," August 1958, "Part 3–Influence of Water-Cement Ratio and Compaction," September 1958, and "Part 4–The Air-Void System in Job Concrete," October 1958, *Journal of the American Concrete Institute,* Farmington Hills, Michigan, 1958.

Mindess, Sidney, and Young, J. Francis, *Concrete*, Prentice-Hall, Englewood Cliffs, New Jersey, 1981, 671 pages.

Monteiro, P.J.M.; Coussy, Oliver; and Silva, Denise A., "Effect of Cryo-Suction and Air Void Transition Layer on the Hydraulic Pressure Developed During Freezing of Concrete," *ACI Materials Journal*, Vol. 103, No. 2, 2006, pages 136 to 140.

NCHRP, *Guidelines for the Selection of Snow and Ice Control Materials to Mitigate Environmental Impacts*, NCHRP Report 577, National Cooperative Highway Research Program, Transportation Research Board, Washington D.C., 2007.

Neville, A.M., *Properties of Concrete*, 4th edition, John Wiley & Sons, New York, New York, USA, 1996, 844 pages.

Nmai, Charles K, "Freezing and Thawing," *Significance of Tests and Properties of Concrete and Concrete-Making Materials*, ASTM STP 169D, edited by Lamond, Joseph F. and Pielert, James H., American Society for Testing and Materials, Philadelphia, 2006, pages 154 to 163.

Nmai, C.K.; Farrington, S.; and Bobrowski, G.S., "Organic-Based Corrosion-Inhibiting Admixture for Reinforced Concrete," *Concrete International*, V. 14, No. 4, April 1992, pages 45 to 51.

Oesterle, R.G., *The Role of Concrete Cover in Crack Control Criteria and Corrosion Protection,* Serial No. 2054, Portland Cement Association, Skokie, Illinois, 1997, 88 pages.

Osborne, G.J., "Carbonation and Permeability of Blast Furnace Slag Cement Concretes from Field Structures," *Proceedings of the 3rd International Congress on the Use of Fly Ash, Silica Fume, Slag and Natural Pozzolans in Concrete,* ACI SP-114, Vol. 2, American Concrete Institute, Detroit, Michigan, 1989, pages 1209 to 1237.

Osborne, G.J., "Carbonation of Blast Furnace Slag Cement Concrete." *Durability of Building Materials*, Vol. 4, 1986, pages 81 to 96.

Ozol, Michael A., "Alkali-Carbonate Rock Reaction," *Significance of Tests and Properties of Concrete and Concrete-Making Materials*, ASTM STP 169D, edited by Lamond, Joseph, F. and Pielert, James H., American Society for Testing and Materials, Philadelphia, 2006, pages 410 to 424.

Page, C.L., and Vennesland, Ø., "Pore Solution Composition and Chloride Binding Capacity of Silica-Fume Cement Pastes," *Materials and Structures*, No. 91, January-February 1983, pages 19 to 25.

PCA, *Tests of Concrete Road Materials from California*, Major Series 285, Research Reports, Portland Cement Association, April 1940.

PCA, *Types and Causes of Concrete Deterioration*, IS536, Portland Cement Association, Skokie, Illinois, 2002, 16 pages.

PCA Durability Subcommittee, *Guide Specification for Concrete Subject to Alkali-Silica Reactions*, IS415, Portland Cement Association, 2007.

Peeples, Bob, "Re: USING SALT TO MELT ICE," MadSci Network, http://www.madsci.org/posts/archives/1998-11/910675052.Ch.r.html, 1998.

Pigeon, M., and Pleau, R., *Durability of Concrete In Cold Climates*, E&FN Spon Publishing House, London, UK, 1995, 244 pages.

Pigeon, M.; Prevost, J.; and Simard, J.-M., "Freeze-Thaw Durability versus Freezing Rate," *ACI Journal*, American Concrete Institute, Farmington Hills, Michigan, September-October 1985, pages 684 to 692.

Pinto, Roberto C.A., and Hover, Kenneth C., *Frost and Scaling Resistance of High-Strength Concrete,* Research and Development Bulletin RD122, Portland Cement Association, 2001, 70 pages.

Powers, Laura J., "Developments in Alkali-Silica Gel Detection," *Concrete Technology Today*, PL991, Portland Cement Association, http://www.cement.org/pdf_files/PL991.pdf, April 1999.

Powers, T.C., "Structure and Physical Properties of Hardened Portland Cement Paste," *Journal of the American Ceramic Society*, Vol. 41, No. 1, The American Ceramic Society, Westerville, Ohio, January 1, 1958, pages 1 to 5 (also available as PCA RX094).

Powers, T.C.; Copeland, H.E.; and Mann, H.M., *Capillary Continuity or Discontinuity in Cement Pastes*, Research Department Bulletin RX110, Reprinted from the Journal of the PCA Research and Development Laboratories, Vol. 1, No. 2, Portland Cement Association, Skokie, Illinois, May 1959, pages 38 to 48.

Powers, T.C., *Prevention of Frost Damage to Green Concrete*, Research Department Bulletin RX148, Portland Cement Association, Skokie, Illinois, 1962, 18 pages.

Powers, T.C., "The Mechanism of Frost Action in Concrete," *Stanton Walker Lecture Series on the Materials Sciences*, Lecture No. 3, National Sand and Gravel Association and National Ready Mixed Concrete Association, Silver Spring, Maryland, 1965a.

Powers, T.C., *Topics in Concrete Technology:...(3) Mixtures Containing Intentionally Entrained Air (4) Characteristics of Air-Void Systems*, Research Department Bulletin RX174, Portland Cement Association, http://www.cement.org/pdf_files/RX174.pdf, 1965.

Powers, T.C., *Basic Considerations Pertaining to Freezing and Thawing Tests*, Research Department Bulletin RX058, Portland Cement Association, http://www.cement.org/pdf_files/RX058.pdf, 1955.

Powers, T.C., *The Air Requirements of Frost-Resistant Concrete*, Research Department Bulletin RX033, Portland Cement Association, http://www.cement.org/pdf_files/RX033.pdf, 1949.

Powers, T.C., *Void Spacing as a Basis for Producing Air-Entrained Concrete*, Research Department Bulletin RX049, Portland Cement Association, http://www.cement.org/pdf_files/RX049.pdf, 1954.

Ramlochan, T.; Hooton, R.D.; and Thomas, M.D.A., "The Effect of Pozzolans and Slag on the Expansion of Mortars Cured at Elevated Temperature, Part II: Microstructural and Microchemical Investigations", *Cement and Concrete Research*, Vol. 34, No. 8, 2004, pages 1341 to 1356.

Ritter, Steve, "Aircraft Deicers," *Chemical and Engineering News*, Volume 79, No. 1, January 2001, page 30.

Rogers, Chris A.; Grattan-Bellew, Paddy E.; Hooton, R. Doug; Ryell, John; and Thomas, Michael D.A., "Alkali-Aggregate Reactions in Ontario", *Canadian Journal of Civil Engineering*, Vol. 27, No. 2, April 2000, pages 246 to 260.

Rogers, C.A., and Hooton, R.D., *Comparison between Laboratory and Field Expansion of Alkali-Carbonate Reactive Concrete*, Proceedings of the 9th International Conference on Alkali-Aggregate Reaction in Concrete, 1992, pages 877 to 884.

Santhanam, Manu; Cohen, Menahi D.; and Olek, Jan, "Sulfate Attack Research—Whither Now?," *Cement and Concrete Research*, 2001, pages 845 to 851.

Sayward, John M., *Salt Action on Concrete*, Special Report 84-25, U.S. Army Cold Regions Research and Engineering Laboratory, Hanover, New Hampshire, August 1984.

Scrivener, K.L.; Damidot, D.; and Famy, C., "Possible Mechanisms of Expansion of Concrete Exposed to Elevated Temperatures During Curing (Also Known as DEF) and Implications for Avoidance of Field Problems," *Cement, Concrete, and Aggregates*, CCAGDP, Vol. 21, No. 1, American Society for Testing and Materials, West Conshohocken, Pennsylvania, June 1999, pages 93 to 101.

Setzer, M.J., "Basis of Testing the Freeze-Thaw Resistance: Surface and Internal Deterioration," *Frost Resistance of Concrete: Proceedings of the International RILEM Workshop on Resistance of Concrete to Freezing and Thawing With or Without De-icing Chemicals*/ Setzer, M.J.; Auberg, R. (Editors), E & F.N. Spon, 1997. (RILEM Proceedings 4). ISBN 0-419-22900-0, pages 157 to 173.

Setzer, M.J., "Micro Ice Lens Formation and Frost Damage," *Frost Damage in Concrete*, Proceedings of the International RILEM Workshop (eds. D.J. Janssen, M.J. Setzer and M.B. Snyder), RILEM Publication PRO 25, Minneapolis, Minnesota, June 28 to 30, 1999, pages 1 to 15.

Shehata, M.H.; Jagdat, S.; Lachemi, M.; and Rogers, C., *Do Supplementary Cementing Materials Control Alkali-Carbonate Reaction?* Proceedings, 17th Annual Symposium, Ed. Fowler, D., and Allen, J., International Centre for Aggregate Research, University of Texas, Austin, 2009.

Shimada, Yukie E., *Chemical Path of Ettringite Formation in Heat Cured Mortar and Its Relationship to Expansion*, Ph.D. Thesis, Northwestern University, Evanston, Illinois, USA, 2005 [PCA SN2526].

Shimada, Y.; Johansen, V.C.; Miller, F.M.; and Mason, T.O., *Chemical Path of Ettringite Formation in Heat Cured Mortar and Its Relationship to Expansion: A Literature Review*, RD136, Portland Cement Association, 2005.

Stanton, Thomas E., "Expansion of Concrete through Reaction between Cement and Aggregate," *Proceedings*, American Society of Civil Engineers, Vol. 66, New York, 1940, pages 1781 to 1811.

Stark, D., *Durability of Concrete in Sulfate-Rich Soils*, RD097, Portland Cement Association, Skokie, Illinois, 1989.

Stark, D., "Lithium Salt Admixture—An Alternative Method to Prevent Expansive Alkali-Silica Reactivity," *Proceedings of the 9th International Conference on Alkali-Aggregate Reaction in Concrete*, The Concrete Society, London, July 1992. Also PCA Publication RP307.

Stark, David, *Long-Time Performance of Concrete in a Seawater Exposure*, RP337, Portland Cement Association, 1995, 58 pages.

Stark, David, *Long-Term Performance of Plain and Reinforced Concrete in Seawater Environments*, Research and Development Bulletin RD119, Portland Cement Association, 2001, 14 pages.

Stark, David, *Performance of Concrete in Sulfate Environments*, PCA Serial No. 2248, Portland Cement Association, Skokie, Illinois, 2002.

Stark, David C.; Kosmatka, Steven H.; Farny, James A.; and Tennis, Paul D., *Performance of Concrete Specimens in the PCA Outdoor Test Facility*, RD124, Portland Cement Association, Skokie, Illinois, 2002, 36 pages.

Steinour, H.H., *Estimation of Aggressive CO_2 and Comparison with Langelier Saturation Index*, Serial No. 1905, Portland Cement Association, Skokie, Illinois, 1975.

Sutter, L.; Petersen, K.; Julio-Betancourt, G.; Hooton, R.D.; VanDam, T.; and Smith, K., "The Deleterious Chemical Effects of Concentrated Deicing Solutions on Portland Cement Concrete", *Final Report, South Dakota Department of Transportation*, Research Report 2002-01-F, April 30, 2008, 198 pages.

Swenson, E.G., and Gillott, J.E., 1964. *Alkali-Carbonate Rock Reaction*, Highway Research Board Record 45, pages 21 to 40.

Tang, Mingshu; Deng, Min; Lon, Xianghui; and Han, Sufeng, "Studies on Alkali-Carbonate Reaction," *ACI Materials Journal*, American Concrete Institute, Farmington Hills, Michigan, January-February 1994, pages 26 to 29.

Taylor, H.F.W., *Cement Chemistry*, 2nd edition, Thomas Telford, London, UK, 1997.

Taylor, H.F.W.; Famy, C.; and Scrivener, K.L., "Delayed ettringite formation," *Cement and Concrete Research*, Vol. 31, Pergamon-Elsevier Science, Oxford, UK, 2001, pages 683 to 693.

Taylor, Peter C.; Whiting, David A.; and Nagi, Mohamad A., *Threshold Chloride Content for Corrosion of Steel in Concrete: A Literature Review*, SN2169, Portland Cement Association, http://www.cement.org/pdf_files/SN2169.pdf, 2000.

Taylor, Thomas G., *Effect of Carbon Black and Black Iron Oxide on Air Content and Durability of Concrete,* Research Department Bulletin RX023, Portland Cement Association, http://www.cement.org/pdf_files/RX923.pdf, 1948.

Thaumasite Expert Group, *The Thaumasite Form of Sulfate Attack: Risks, Diagnosis, Remedial Works and Guidance on New Construction*, Report of the Thaumasite Expert Working Group, Department of Environment, Transport, and the Regions, London, January 1999, 180 pages.

Thomas, M.D.A., and Innis, F.A., "Effect of Slag on Expansion Due to Alkali-Aggregate Reaction in Concrete." *ACI Materials Journal*, Vol. 95, No. 6, 1998.

Thomas, M.D.A.; Matthews, J.D.; and Haynes, C.A., "Carbonation of Fly Ash Concrete." *Proceedings of the 4th ACI/CANMET International Conference on the Durability of Concrete*, (Ed. V.M. Malhotra), ACI SP-192, Vol. 1, 2000, pages 539 to 556.

Thomas, M.D.A.; Matthews, J.D.; and Haynes, C.A., 2000. "Carbonation of fly ash concrete." *Proceedings of the 4th ACI/CANMET International Conference on the Durability of Concrete*, (Ed. V.M. Malhotra), ACI SP-192, Vol. 1, pages 539 to 556.

Thomas, Michael D.A.; Fournier, Benoit; and Folliard, Kevin J., *Report on Determining the Reactivity of Concrete Aggregates and Selecting Appropriate Measures for Preventing Deleterious Expansion in New Concrete Construction*, FHWA-HIF-09-001, Federal Highway Administration, Washington, D.C., April 2008, 28 pages.

Thomas, Michael D.A.; Fournier, Benoit; Folliard, Kevin J.; Ideker, Jason H.; and Resendez, Yadhira, *The Use of Lithium to Prevent or Mitigate Alkali-Silica Reaction in Concrete Pavements and Structures,* FHWA-HRT-06-133, Federal Highway Administration, U.S. Department of Transportation, Washington, D.C., March 2007, 62 pages.

Thomas, Michael D.A.; Fournier, Benoit; Folliard, Kevin J.; Ideker, Jason H.; and Shehata, Medhat, "Test Methods for Evaluating Preventive Measures for Controlling Expansion Due to Alkali-Silica Reaction in Concrete," *Cement and Concrete Research,* Vol. 36, No. 10, October 2006, pages 1842 to 1856.

Uhlig, Herbert H., and Revie, R. Winston, *Corrosion and Corrosion Control*, 3rd edition, John Wiley & Sons, New York, New York, USA, 1985.

Valenza II, J.J., and Scherer, G.W., "Mechanism for Salt Scaling," *Journal of the American Ceramic Society*, Vol. 89, No. 4, 2006, pages 1161 to 1179.

Vanderhost, N.M., and Jansen, D.J., "The Freezing and Thawing Environment: What is Severe?," *Paul Klieger Symposium on Performance of Concrete*, ACI SP – 122, David Whiting, editor, American Concrete Institute, Farmington Hills, Michigan, 1990, pages 181 to 200.

Verbeck, George, and Klieger, Paul, *Studies of "Salt" Scaling of Concrete,* Research Department Bulletin RX083, Portland Cement Association, http://www.cement.org/pdf_files/RX083.pdf, 1956.

Verbeck, G.J., *Carbonation of Hydrated Portland Cement*, Research Department Bulletin RX087, Portland Cement Association, http://www.cement.org/pdf_files/RX087.pdf, 1958.

Whiting, David A., *Origins of Chloride Limits for Reinforced Concrete*, R&D Serial No. 2153, Portland Cement Association, http://www.cement.org/pdf_files/SN2153.pdf, 1997, 18 pages.

Whiting, David A., and Nagi, Mohamad A., *Manual on Control of Air Content in Concrete,* EB116, National Ready Mixed Concrete Association and Portland Cement Association, 1998, 42 pages.

Whiting, David A.; Taylor, Peter C.; and Nagi, Mohamad, A., *Chloride Limits in Reinforced Concrete,* PCA Serial No. 2438, Portland Cement Association, http://www.cement.org/pdf_files/SN2438.pdf, 2002.

Wyatt, J., and Fritzsche, C., "The snow battle: salt vs. chemicals," *American City and County,* April 1989, pages 30 to 36.

CHAPTER 12

Designing and Proportioning Concrete Mixtures

There are three phases in the development of a concrete mixture for a particular project; specifying, designing, and proportioning.

Contract documents contain performance requirements for fresh and hardened concrete, typically along with criteria for acceptable concrete ingredients including the aggregates, admixtures, water, and cementitious materials. Such requirements differ in the level of detail concerning proportions or ratios of ingredients, varying from a fully prescriptive specification that defines all ingredients and batch weights to a more performance-oriented specification that lists end-result concrete properties with no limitations on ingredients or proportions. Contract documents may either fully state material and mixture requirements or they may incorporate such requirements by making reference to ASTM, AASHTO, ACI, or a variety of public-agency standards and specifications.

The terms "mixture design" (mix design) and "mixture proportioning" are often incorrectly used interchangeably (USDOT 2006). As far back as 1918, this distinction was recognized:

> *The term "design" is used in the title of this article as distinguished from "proportioning" since it is the intention to imply that each element of the problem is approached with a deliberate purpose in view which is guided by a rational method of accomplishment.* (Abrams 1918).

Mixture design is the process by which the concrete mixture performance characteristics are defined. It is in the mixture design process that parameters such as air content, workability, and required strength are established. These parameters are driven by factors such as the concrete's service environment, construction method, and structural requirements as discussed in Chapter 9. Once the mixture design parameters are established, the materials characteristics and production technology are identified and determined. Then, the concrete mixture proportions can be developed using relationships established either through research or from past experience.

The output from the mixture design process becomes part of the input for mixture proportioning. In turn, the output from the mixture proportioning process becomes the actual batch proportions of the various ingredients in the concrete mixture as shown in Figure 12-1.

Figure 12-2 outlines the process of mixture proportioning. Knowledge of the local materials and the relationships among ingredients combine to produce a set of mixture proportions. These proportions are designed to meet the requirements for both fresh and hardened concrete provided that the concrete is properly batched, mixed, transported, placed, consolidated, finished, textured, and cured. For a given concrete there may be many sets of mixture proportions that satisfy the mixture design requirements. Performance records from the lab or field are required to determine whether any given set of materials and proportions can reliably meet the design requirements. Concrete materials are themselves variable, and the relationships among proportions are mixture-and-materials-specific. Therefore, any set of proposed mixture proportions is referred to as a "trial mixture," until satisfactory performance has been demonstrated through laboratory or field testing.

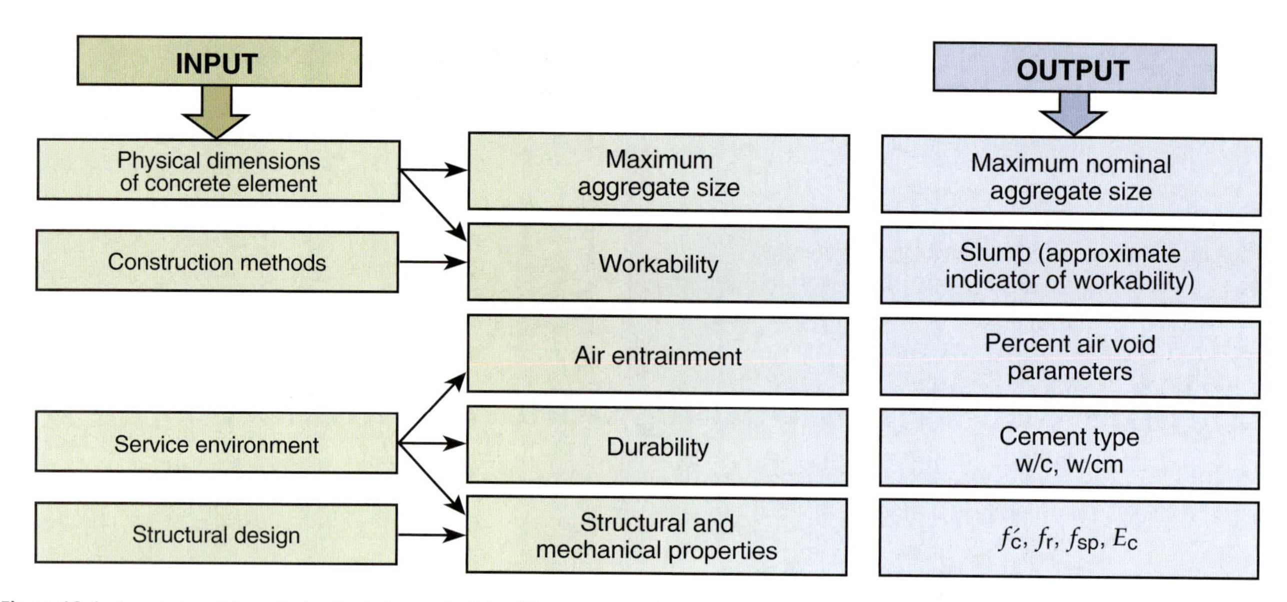

Figure 12-1. Inputs to mixture design include constraints of the concrete element and clear spacing between reinforcing, how the concrete will be batched and placed, the service environment with its associated durability requirements, and the required mechanical properties for the concrete. Output is the set of required characteristics for the mixture and each ingredient (Courtesy of K. Hover).

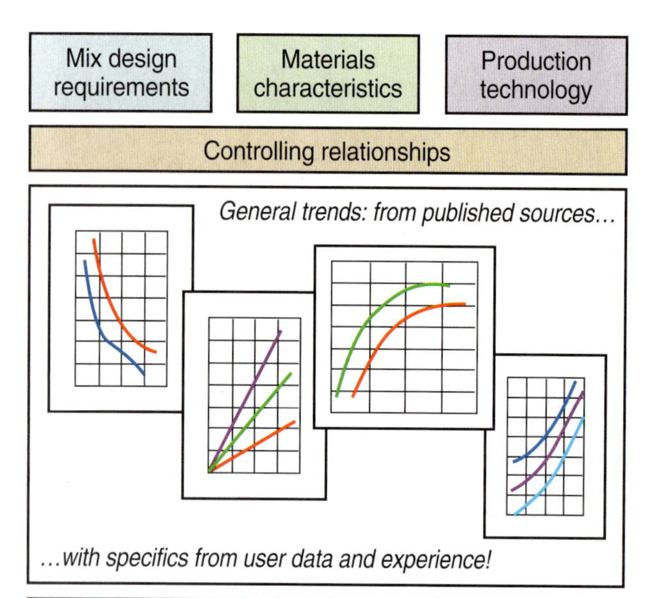

Mixture proportions	
Water	kg/m3 (lb/yd3)
Cement	kg/m3 (lb/yd3)
Fly ash	kg/m3 (lb/yd3)
Coarse aggregate	kg/m3 (lb/yd3)
Intermediate aggregate	kg/m3 (lb/yd3)
Fine aggregate	kg/m3 (lb/yd3)
Air content	%
Air-entraining admixture	ml (fl oz)
Water-reducing admixture	ml (fl oz)

Figure 12-2. Inputs to mixture proportioning include mixture design requirements, characteristics of the specific materials to be used, and requirements based on production technology. Output is a set of mixture proportions expressed as batch weights (Courtesy of K. Hover).

A trial batch should be produced with the mixer that will be used for the project. All relevant concrete properties (such as workability, density, air content, and strength) should be measured to ensure that the mixture performs as desired. Appropriate adjustments should be made, when necessary, and subsequent trial batches produced. This iterative process continues until a satisfactory combination of the constituent materials is identified. The entire process should establish the most economical and practical combination of readily available constituent materials that will meet the requirements of the mixture design.

Understanding the basic principles of mixture design is as important as the actual calculations used to establish mix proportions. Only with proper selection of materials and mixture characteristics can the above qualities be obtained in concrete construction (Abrams 1918, Hover 1998, and Shilstone 1990).

Mix Design (Selecting Characteristics)

Before a concrete mixture can be proportioned, mixture characteristics are selected based on the intended use of the concrete, the exposure conditions, the size and shape of building elements, and the physical properties of the concrete required for the structure. This is the mix design process. Although strength is an important characteristic, other properties such as durability, permeability, and wear resistance are just as important, especially when considering life-cycle design of structures. The mixture characteristics should reflect the needs of the structure; for example, resistance to chloride ions should be verifiable and the appropriate test methods specified.

Once the mix design characteristics are selected, the mixture can be proportioned from field or laboratory data. Since most of the desirable properties of hardened concrete depend primarily upon the quality of the cementitious paste, the first step in proportioning a concrete mixture is the selection of the appropriate water-cementing materials ratio for the durability and strength needed. Concrete mixtures should be kept as simple as possible. An excessive number of ingredients often makes a concrete mixture difficult to reliably produce. However, opportunities provided by modern concrete technology should not be overlooked.

Strength

Strength (compressive or flexural) is the most universally used performance measure for concrete quality. Within the normal range of strengths used in concrete construction, the compressive strength is inversely related to the water-cementitious materials ratio. For fully compacted concrete made with clean, sound aggregates, the strength and other desirable properties of concrete under given job conditions are governed by the quantity of mixing water used per unit of cement or cementing materials (Abrams 1918).

The strength of the cementitious paste binder in concrete depends on the quality and quantity of the reacting paste components and on the degree to which the hydration reaction has progressed. Concrete becomes stronger with time as long as there is moisture and a favorable temperature available. Therefore, the strength at any particular age is both a function of the original water cementitious material ratio and the degree to which the cementitious materials have hydrated. The importance of prompt and thorough curing is easily recognized.

Differences in concrete strength for a given water cementing materials ratio may result from: (1) changes in the aggregate size, grading, surface texture, shape, strength, and stiffness; (2) differences in types and sources of cementing materials; (3) entrained-air content; (4) the presence of admixtures; and (5) the length of curing time and temperature.

The specified compressive strength, f'_c, at 28 days is the strength that is expected to be equal to or exceeded by the average of any set of three consecutive strength tests. ACI 318 requires for f'_c to be at least 17.5 MPa (2500 psi). No individual test (average of two cylinders) can be more than 3.5 MPa (500 psi) below the specified strength. Specimens must be cured under laboratory conditions for an individual class of concrete (ACI 318). Some specifications allow alternative requirements.

The average strength should equal the specified strength plus an allowance to account for variations in materials; variations in methods of mixing, transporting, and placing the concrete; and variations in making, curing, and testing concrete cylinder specimens. The average strength, which is greater than f'_c, is called f'_{cr}; it is the strength required in the mix design. Requirements for f'_{cr} are discussed in detail under Proportioning later in this chapter. Tables 12-1 and 12-2 show strength requirements for various exposure conditions.

Flexural strength is commonly used for design of pavements. However, flexural strength testing is generally avoided due to greater testing variability. Strength tests of standard cured compressive strength cylinders can be correlated to the flexural strength of concrete (Kosmatka 1985). Because of the robust empirical relationship between these two strengths and the economics of testing cylinders instead of beams, most state departments of transportation are now using compression tests of cylinders to monitor concrete quality for their pavement and bridge projects. For more information on flexural strength, see Strength in Chapter 9 and Strength Specimens in Chapter 18.

Table 12-1. Maximum Water-Cementitious Material Ratios and Minimum Design Strengths for Various Exposure Conditions

Exposure category	Exposure condition	Maximum water-cementitious material ratio by mass for concrete	Minimum design compressive strength, f'_c, MPa (psi)
F0, S0, P0, C0	Concrete protected from exposure to freezing and thawing, application of deicing chemicals, or aggressive substances	Select water-cementitious ratio on basis of strength, workability, and finishing needs	Select strength based on structural requirements
P1	Concrete intended to have low permeability when exposed to water	0.50	28 (4000)
F1, F2, F3	Concrete exposed to freezing and thawing in a moist condition or deicers	0.45	31 (4500)
C2	For corrosion protection for reinforced concrete exposed to chlorides from deicing salts, salt water, brackish water, seawater, or spray from these sources	0.40	35 (5000)

Adapted from ACI 318-08. The following four exposure categories determine durability requirements for concrete: (1) F – Freezing and Thawing; (2) S – Sulfates; (3) P – Permeability; and (4) C – Corrosion. Increasing numerical values represent increasingly severe exposure conditions.

Table 12-2. Types of Cement Required for Concrete Exposed to Sulfates in Soil or Water*

Exposure class		Sulfate exposure**		Cementitious materials requirements†			Maximum w/cm
		Water-soluble sulfate in soil, % by mass	Dissolved sulfate in water, ppm	C150	C595	C1157	
S0	Negligible	$SO_4 < 0.10$	$SO_4 < 150$	NSR	NSR	NSR	None
S1	Moderate‡	$0.10 < SO_4 < 0.20$	$150 < SO_4 < 1500$	II or II(MH)	IP(MS) IS(< 70)(MS) IT(P ≥ S)(MS) IT(P < S < 70)(MS)	MS	0.50
S2	Severe	$0.20 < SO_4 < 2.00$	$1500 < SO_4 < 10,000$	V	IP(HS) IS(< 70)(HS) IT(P ≥ S)(HS) IT(P < S < 70)(HS)	HS	0.45
S3§	Very severe	$SO_4 > 2.00$	$SO_4 > 10,000$	V	IP(HS) IS(< 70)(HS) IT(P ≥ S)(HS) IT(P < S < 70)(HS)	HS	0.40

* Adapted from Bureau of Reclamation Concrete Manual, ACI 201 and ACI 318. "NSR" indicates no special requirements for sulfate resistance.

** Soil is tested per ASTM C1580 and water per ASTM D516.

† Pozzolans and slag that have been determined by testing according to ASTM C1012 or by service record to improve sulfate resistance may also be used in concrete. Maximum expansions when using ASTM C1012: Moderate exposure – 0.10% at 6 months; Severe exposure – 0.05% at 6 months or 0.10% at 12 months; Very Severe exposure – 0.10% at 18 months. Refer to ACI 201.2R for more guidance. ASTM C595 Type IT was adopted in 2009 and has not been reviewed by ACI at the time of printing.

‡ Includes seawater.

§ ACI 318 requires SCMs (tested to verify improved sulfate resistance) with Types V, IP(HS), IS(< 70)(HS) and HS cements for exposure class S3. ACI 318 requires a maximum water:cement ratio of 0.45 for exposure class S3.

Water-Cementitious Material Ratio

The water-cementitious material ratio is simply the mass of water divided by the mass of cementitious material (portland cement, blended cement, fly ash, slag cement, silica fume, and natural pozzolans). The water-cementitious material ratio selected for mix design must be the lowest value required to meet strength and anticipated exposure conditions. Tables 12-1 and 12-2 show w/cm requirements for various exposure conditions.

When durability is not critical, the water-cementitious materials ratio should be selected on the basis of concrete compressive strength. In such cases the water-cementitious materials ratio and mixture proportions for the required strength should be based on adequate field data or trial mixtures made with actual job materials to determine the relationship between the ratio and strength. Figure 12-3 or Table 12-3 can be used to select a water-cementitious materials ratio with respect to the required average strength, f'_{cr}, as a starting point for trial mixtures when no other data are available.

In mix design, the water to cementitious materials ratio, w/cm, is often used synonymously with water to cement ratio (w/c); however, some specifications differentiate between the two ratios.

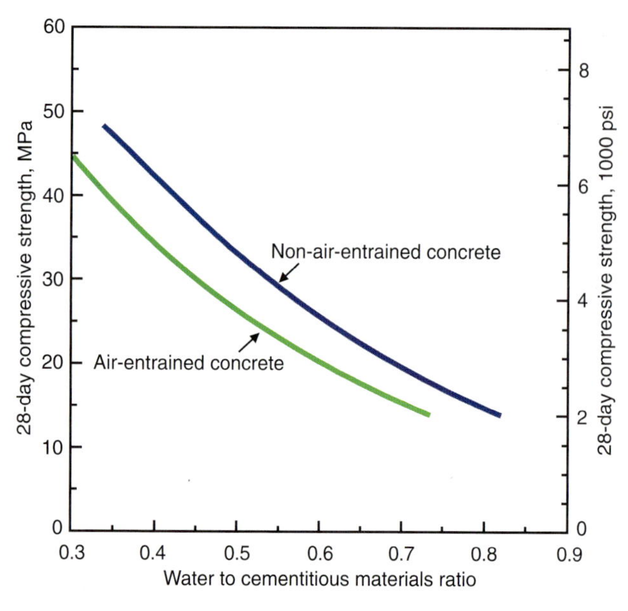

Figure 12-3. Approximate relationship between compressive strength and water to cementing materials ratio for concrete using 19-mm to 25-mm (3/4-in. to 1-in.) nominal maximum size coarse aggregate. Strength is based on cylinders moist cured 28 days per ASTM C31 (AASHTO T 23). Adapted from Table 9-3, ACI 211.1, ACI 211.3, and Hover 1995.

Table 12-3 (Metric). Relationship Between Water to Cementitious Material Ratio and Compressive Strength of Concrete

Compressive strength at 28 days, MPa	Water-cementitious materials ratio by mass	
	Non-air-entrained concrete	Air-entrained concrete
45	0.38	0.30
40	0.42	0.34
35	0.47	0.39
30	0.54	0.45
25	0.61	0.52
20	0.69	0.60
15	0.79	0.70

Strength is based on cylinders moist-cured 28 days in accordance with ASTM C31 (AASHTO T 23). Relationship assumes nominal maximum size aggregate of about 19 to 25 mm.
Adapted from ACI 211.1 and ACI 211.3.

Table 12-3 (Inch-Pound Units). Relationship Between Water to Cementitious Material Ratio and Compressive Strength of Concrete

Compressive strength at 28 days, psi	Water-cementitious materials ratio by mass	
	Non-air-entrained concrete	Air-entrained concrete
7000	0.33	—
6000	0.41	0.32
5000	0.48	0.40
4000	0.57	0.48
3000	0.68	0.59
2000	0.82	0.74

Strength is based on cylinders moist-cured 28 days in accordance with ASTM C31 (AASHTO T 23). Relationship assumes nominal maximum size aggregate of about ¾ in. to 1 in.
Adapted from ACI 211.1 and ACI 211.3.

Aggregates

The grading characteristics and nature of aggregate particles have an important influence on proportioning concrete mixtures because they affect the workability of the fresh concrete, as well as other concrete properties such as shrinkage, flexural strength, and to a certain extent, compressive strength.

Grading is important for attaining an economical mixture. It affects the amount of concrete that can be made with a given amount of cementitious materials and water. Coarse aggregates should be graded up to the largest size practical under job conditions. The maximum size that can be used depends on factors such as the size and shape of the concrete member to be cast, the amount and distribution of reinforcing steel in the member, and the thickness of slabs. Grading also influences the workability and placeability of the concrete. Sometimes mid-sized aggregate, around the 9.5 mm (⅜ in.) size, is lacking in an aggregate supply. This can result in a concrete with high shrinkage properties, high water demand, poor workability, and poor placeability. Durability may also be affected. Various options are available for obtaining optimal grading of aggregate (Shilstone 1990).

Requirements for limits on nominal maximum size of aggregate particles are covered by ACI 318 (ACI 318-08). The nominal maximum size of aggregate should not exceed:

1. One-fifth the narrowest dimension of a vertical concrete member;
2. Three-quarters the clear spacing between reinforcing bars and between the reinforcing bars and forms; and
3. One-third the depth of slabs.

These requirements may be waived if, in the judgment of the designer, the mixture possesses sufficient workability that the concrete can be properly placed without honeycomb or voids. Smaller sizes can be used when availability or economic consideration require them.

The amount of mixing water required to produce a unit volume of concrete of a given slump is dependent on the shape and the maximum size and amount of coarse aggregate. Larger sizes minimize the water requirement and thus potentially allow the cement content to be reduced. Also, rounded aggregate requires less mixing water than crushed aggregate in concretes of equal slump (see Water Content).

The maximum size of coarse aggregate that will produce concrete of maximum strength for a given cement content depends upon the aggregate source as well as its shape and grading. For high compressive-strength concrete (greater than 70 MPa or 10,000 psi), the maximum size is about 19 mm (¾ in.). Higher strengths can also sometimes be achieved through the use of crushed stone aggregate rather than rounded-gravel aggregate.

The most desirable fine-aggregate grading will depend upon the type of work, the paste content of the mixture, and the size of the coarse aggregate. For leaner mixtures, a fine grading (lower fineness modulus) is desirable for workability. For richer mixtures, a coarse grading (higher fineness modulus) is used for greater economy.

In some areas, the chemically bound chloride in aggregate may make it difficult for concrete to pass chloride limits set by ACI 318 or other specifications. However, some or all of the chloride in the aggregate may not be available for participation in corrosion of reinforcing steel, thus that chloride may be ignored. ASTM C1524, *Standard Test Method for Water-Extractable Chloride in Aggregate (Soxhlet Method)*, can be used to evaluate the amount of chloride available from aggregate. ACI 222.1 also provides guidance.

The bulk volume of coarse aggregate can be determined from Figure 12-4 or Table 12-4. These bulk volumes are based on aggregates in a dry-rodded condition as described in ASTM C29, *Standard Test Method for Bulk Density ("Unit Weight") and Voids in Aggregate* (AASHTO T 19). They are selected from empirical relationships to produce concrete with a degree of workability suitable for general reinforced concrete construction. For less workable concrete, such as required for concrete pavement construction, they may be increased about 10%. For more workable concrete, such as may be required when placement is by pump, they may be reduced up to 10%.

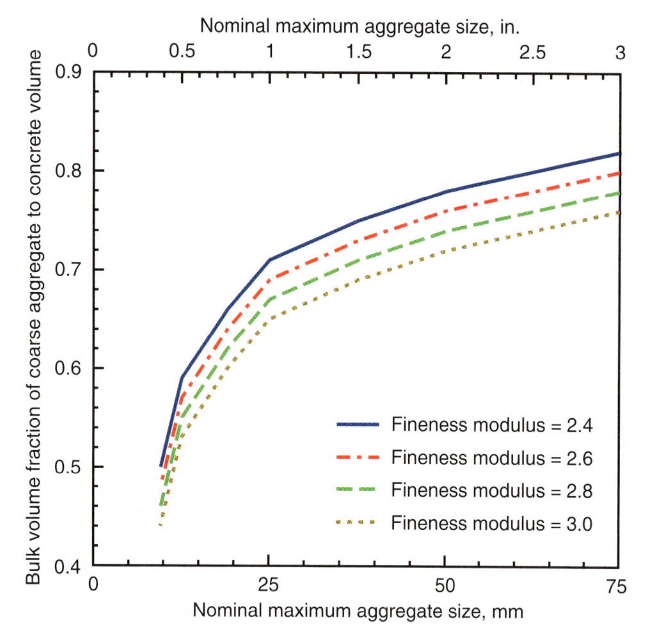

Figure 12-4. Bulk volume of coarse aggregate per unit volume of concrete. Bulk volumes are based on aggregates in a dry-rodded condition as described in ASTM C29 (AASHTO T 19). For more workable concrete, such as may be required when placement is by pump, they may be reduced up to 10%. Adapted from Table 12-4, ACI 211.1 and Hover (1995 and 1998).

Table 12-4. Bulk Volume of Coarse Aggregate Per Unit Volume of Concrete

Nominal maximum size of aggregate, mm (in.)	Bulk volume of dry-rodded coarse aggregate per unit volume of concrete for different fineness moduli of fine aggregate*			
	2.40	2.60	2.80	3.00
9.5 (⅜)	0.50	0.48	0.46	0.44
12.5 (½)	0.59	0.57	0.55	0.53
19 (¾)	0.66	0.64	0.62	0.60
25 (1)	0.71	0.69	0.67	0.65
37.5 (1½)	0.75	0.73	0.71	0.69
50 (2)	0.78	0.76	0.74	0.72
75 (3)	0.82	0.80	0.78	0.76
150 (6)	0.87	0.85	0.83	0.81

*Bulk volumes are based on aggregates in a dry-rodded condition as described in ASTM C29 (AASHTO T 19). Adapted from ACI 211.1.

Air Content

Entrained air should be used in all concrete that will be exposed to freezing and thawing and deicing chemicals. It can also be used to improve workability even where not required. Air entrainment is accomplished using an air-entraining portland cement or by adding an air-entraining admixture at the mixer. The amount of admixture should be adjusted to meet variations in concrete ingredients and job conditions. The amount recommended by the admixture manufacturer will, in most cases, produce the desired air content.

Recommended target air contents for air-entrained concrete are shown in Figure 12-5 and Table 12-5. Note that the amount of air required to provide adequate freeze-thaw resistance is dependent upon the nominal maximum size of aggregate and the level of exposure. In properly proportioned mixtures, the mortar content decreases as maximum aggregate size increases; decreasing the required concrete air content. This is evident in Figure 12-5. The levels of exposure are defined by ACI 318-08 as follows:

Mild Exposure, F0. Exposure Class F0 is assigned to concrete that will not be exposed to cycles of freezing and thawing or deicing agents. When air entrainment is desired for a beneficial effect other than durability, such as to improve workability or cohesion or in low cement content concrete to improve strength, air contents lower than those needed for durability can be used.

Moderate Exposure, F1. Exposure Class F1 is assigned to concrete exposed to cycles of freezing and thawing and that will be occasionally exposed to moisture before freezing and will not be exposed to deicing or other aggressive chemicals. Examples include exterior beams, columns, walls, girders, or slabs that are not in contact with wet soil and are located where they will not receive direct applications of deicing chemicals.

Severe Exposure, F2. Exposure Class F2 is assigned to concrete exposed to cycles of freezing and thawing that is in continuous contact with moisture before freezing where exposure to deicing salt is not anticipated. An example is an exterior water tank or vertical members in contact with soil.

Very Severe Exposure, F3. Exposure Class F3 is assigned to concrete exposed to cycles of freezing and thawing, in continuous contact with moisture, and where exposure to deicing chemicals is anticipated. Examples include pavements, bridge decks, curbs, gutters, sidewalks, canal linings, or exterior water tanks or sumps.

When mixing water is held constant, the entrainment of air will increase slump. When cement content and slump are held constant, the entrainment of air results in the need for less mixing water, particularly in leaner concrete mixtures. In batch adjustments, to maintain a constant slump while changing the air content, the water should be decreased by about 3 kg/m^3 (5 lb/yd^3) for each percentage point increase in air content or increased 3 kg/m^3 (5 lb/yd^3) for each percentage point decrease.

A specific air content may not be readily or repeatedly achieved because of the many variables affecting air content; therefore, a permissible range of air contents around a target value must be provided. Although a range of ±1% of the Figure 12-5 or Table 12-5 values is often used in project specifications, it is sometimes an impractical limit. The appropriate solution is to use a wider range, such as –1 to +2 percentage points of the target values. For example, for a target value of 6% air, the specified range for the concrete delivered to the jobsite could be 5% to 8%.

Slump

Concrete must always be made with a workability, consistency, and plasticity suitable for job conditions. Workability is a measure of how easy or difficult it is to place, consolidate, and finish concrete. Consistency is the ability of freshly mixed concrete to flow. Plasticity determines concrete's ease of molding. If more aggregate is used in a concrete mixture, or if less water is added, the mixture becomes stiff (less plastic and less workable) and difficult to mold. Neither very dry, crumbly mixtures nor very watery, fluid mixtures have sufficient plasticity.

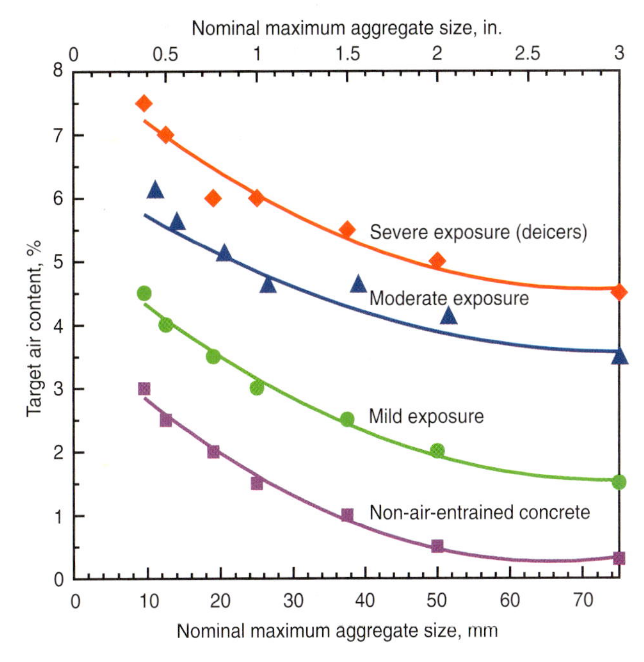

Figure 12-5. Target total air content requirements for concretes using different sizes of aggregate. The air content in job specifications should be specified to be delivered within –1 to +2 percentage points of the target value for moderate and severe exposures. Adapted from Table 12-5, ACI 211.1, and Hover (1995 and 1998).

Table 12-5 (Metric). Approximate Mixing Water and Target Air Content Requirements for Different Slumps and Nominal Maximum Sizes of Aggregate

Slump, mm	Water, kilograms per cubic meter of concrete, for indicated sizes of aggregate*							
	9.5 mm	12.5 mm	19 mm	25 mm	37.5 mm	50 mm**	75 mm**	150 mm**
	Non-air-entrained concrete							
25 to 50	207	199	190	179	166	154	130	113
75 to 100	228	216	205	193	181	169	145	124
150 to 175	243	228	216	202	190	178	160	—
Approximate amount of entrapped air in non-air-entrained concrete, percent	3	2.5	2	1.5	1	0.5	0.3	0.2
	Air-entrained concrete							
25 to 50	181	175	168	160	150	142	122	107
75 to 100	202	193	184	175	165	157	133	119
150 to 175	216	205	197	184	174	166	154	—
Recommended average total air content, percent, for level of exposure:†								
Mild exposure	4.5	4.0	3.5	3.0	2.5	2.0	1.5	1.0
Moderate exposure (Class F1)	6.0	5.5	5.0	4.5	4.5	4.0	3.5	3.0
Severe exposure (Class F2 and F3)	7.5	7.0	6.0	6.0	5.5	5.0	4.5	4.0

* These quantities of mixing water are for use in computing cementitious material contents for trial batches. They are maximums for reasonably well-shaped angular coarse aggregates graded within limits of accepted specifications.

** The slump values for concrete containing aggregates larger than 37.5 mm are based on slump tests made after removal of particles larger than 37.5 mm by wet screening.

† The air content in job specifications should be specified to be delivered within –1 to +2 percentage points of the table target value for moderate and severe exposures.

Adapted from ACI 211.1 and ACI 318. Hover (1995) presents this information in graphical form.

Table 12-5 (Inch-Pound Units). Approximate Mixing Water and Target Air Content Requirements for Different Slumps and Nominal Maximum Sizes of Aggregate

Slump, in.	Water, pounds per cubic yard of concrete, for indicated sizes of aggregate*							
	⅜ in.	½ in.	¾ in.	1 in.	1½ in.	2 in.**	3 in.**	6 in.**
	Non-air-entrained concrete							
1 to 2	350	335	315	300	275	260	220	190
3 to 4	385	365	340	325	300	285	245	210
6 to 7	410	385	360	340	315	300	270	—
Approximate amount of entrapped air in non-air-entrained concrete, percent	3	2.5	2	1.5	1	0.5	0.3	0.2
	Air-entrained concrete							
1 to 2	305	295	280	270	250	240	205	180
3 to 4	340	325	305	295	275	265	225	200
6 to 7	365	345	325	310	290	280	260	—
Recommended average total air content, percent, for level of exposure:†								
Mild exposure	4.5	4.0	3.5	3.0	2.5	2.0	1.5	1.0
Moderate exposure (Class F1)	6.0	5.5	5.0	4.5	4.5	3.5	3.5	3.0
Severe exposure (Class F2 and F3)	7.5	7.0	6.0	6.0	5.5	5.0	4.5	4.0

* These quantities of mixing water are for use in computing cement factors for trial batches. They are maximums for reasonably well-shaped angular coarse aggregates graded within limits of accepted specifications.

** The slump values for concrete containing aggregates larger than 1½ in. are based on slump tests made after removal of particles larger than 1½ in. by wet screening.

† The air content in job specifications should be specified to be delivered within −1 to +2 percentage points of the table target value for moderate and severe exposures.

Adapted from ACI 211.1. Hover (1995) presents this information in graphical form.

The slump test is used to measure concrete consistency. For a given proportion of cement and aggregate without admixtures, the higher the slump, the wetter the mixture. Slump is indicative of workability when assessing similar mixtures. However, slump should not be used to compare mixtures of different proportions, or make assumptions of hardened concrete properties. When used with different batches of the same mix design, a change in slump indicates a change in consistency and in the characteristics of materials, mixture proportions, water content, mixing, time of test, or the testing itself.

Different slumps are needed for various types of concrete construction. Slump is usually indicated in the job specifications as a range, such as 50 mm to 100 mm (2 in. to 4 in.), or as a maximum value not to be exceeded. Slump should be specified based on the method of placement. Slumps vary by application. For example: for pavements placed by mechanical paver, slump is typically 25 mm to 75 mm (1 in. to 3 in.); for floors placed by chute or pump, slump is typically 75 mm to 125 mm (3 in. to 5 in.); for walls and foundations placed by chute or pump, slumps typically range from 100 mm to 200 mm (4 in. to 8 in.). ASTM C94, *Standard Specification for Ready-Mixed Concrete*, addresses slump tolerances in detail. For minor batch adjustments, the slump can be increased by about 10 mm by adding 2 kilograms of water per cubic meter of concrete (1 in. by adding 10 lb of water per cubic yard of concrete). Slumps may also be modified with the use of a water-reducing admixture, plasticizer, or viscosity modifier.

Water Content

The water content of concrete is influenced by: aggregate size, aggregate shape, aggregate texture, slump, water to cementing materials ratio, air content, cementing materials type and content, admixtures, and environmental conditions. An increase in air content and aggregate size, a reduction in water-cementing materials ratio and slump, and the use of rounded aggregates, water-reducing admixtures, or fly ash will reduce water demand. On the other hand, increased temperatures, cement contents, slump, water-cement ratio, aggregate angularity, and a decrease in the proportion of coarse aggregate to fine aggregate will increase water demand.

The approximate water contents in Table 12-5 and Figure 12-6, used in proportioning, are for angular coarse aggregates (crushed stone). For some concretes and aggregates, the water estimates in Table 12-5 and Figure 12-6 can be reduced by approximately 10 kg (20 lb) for subangular aggregate, 20 kg (35 lb) for gravel with some crushed particles, and 25 kg (45 lb) for a rounded gravel. This illustrates the need for trial batch testing of local materials, as each aggregate source is different and can influence concrete properties differently.

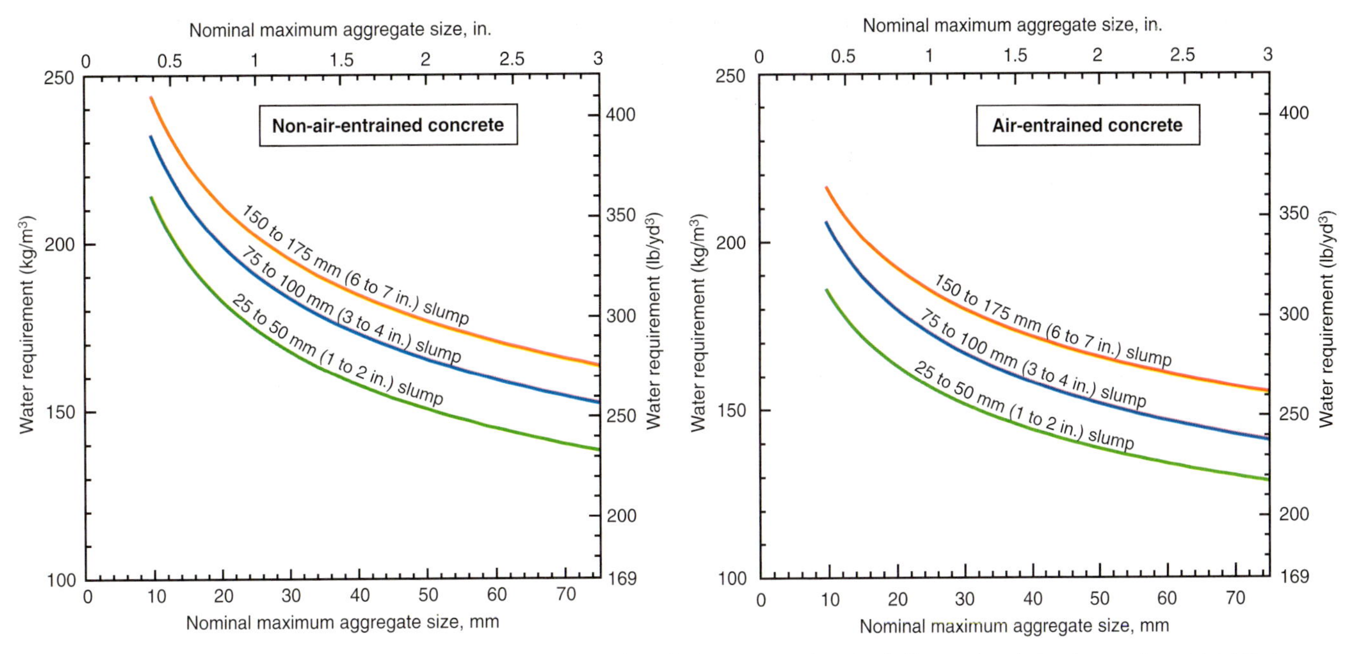

Figure 12-6. Approximate water requirement for various slumps and crushed aggregate sizes for (left) non-air-entrained concrete and (right) air-entrained concrete. Adapted from Table 12-5, ACI 211.1, and Hover (1995 and 1998).

Changing the amount of any single ingredient in a concrete mixture normally effects the proportions of other ingredients. The properties of the mixture are also altered. For example, the addition of 2 kg of water per cubic meter of concrete will increase the slump by approximately 10 mm (10 lb of water per cubic yard of concrete will increase the slump by approximately 1 in.); it will also increase the air content and paste volume, decrease the aggregate volume, and lower the density of the concrete. For the same slump, a decrease in air content by 1 percentage point will increase the water demand by about 3 kg/m^3 of concrete (5 lb/yd^3 of concrete).

Cementing Materials Content and Type

The cementing materials content is usually determined from the selected water-cementing materials ratio and water content. However, minimum cement content frequently is included in specifications in addition to a maximum water-cementing materials ratio. Minimum cement content requirements serve to ensure satisfactory durability and finishability, to improve wear resistance of slabs, and to guarantee a suitable appearance of vertical surfaces. This is important even though strength requirements may be met at lower cementing materials contents. However, excessively large amounts of cementing materials should be avoided to maintain economy in the mixture and to avoid adversely affecting workability and other properties.

For severe freeze-thaw, deicer, and sulfate exposures, it is desirable to specify only enough mixing water to achieve the desired consistency without exceeding the maximum water-cementitious materials ratios shown in Tables 12-1 and 12-2. For concrete placed underwater, usually not less than 390 kg of cementing materials per cubic meter of concrete (650 lb/yd^3) should be used with a maximum w/cm of 0.45. For workability, finishability, and abrasion resistance in flatwork, the quantity of cementing materials to be used should be not less than shown in Table 12-6. To obtain economy, proportioning should minimize the amount of cement required without sacrificing concrete quality.

Table 12-6. Minimum Requirements of Cementing Materials for Concrete Used in Flatwork

Nominal maximum size of aggregate, mm (in.)	Cementing materials, kg/m^3 (lb/yd^3)*
37.5 (1½)	280 (470)
25 (1)	310 (520)
19 (¾)	320 (540)
12.5 (½)	350 (590)
9.5 (⅜)	360 (610)

* Cementing materials quantities may need to be greater for severe exposure.

Adapted from ACI 302.

Since quality depends primarily on water-cementing materials ratio, the water content should be held to a minimum to reduce the cement requirement. Steps to minimize water and cement requirements include: (1) the stiffest practical mixture, (2) the largest practical maximum size of aggregate, and (3) the optimum ratio of fine-to-coarse aggregate. Concrete that will be exposed to sulfate conditions should be made with the cement type shown in Table 12-2.

Seawater contains significant amounts of sulfates and chlorides. Although sulfates in seawater are capable of attacking concrete, the presence of chlorides in seawater inhibits the expansive reaction that is characteristic of sulfate attack. This explains observations from a number of sources that the performance of concretes in seawater have shown satisfactory durability. This is despite the fact these concretes were made with portland cements having tricalcium aluminate (C_3A) contents as high as 10%, and sometimes greater. However, the permeability of these concretes was low, and the reinforcing steel had adequate cover. Portland cements meeting a C_3A requirement of not more than 10% or less than 4% (to ensure durability of reinforcement) are acceptable (ACI 357R).

Supplementary cementitious materials have varied effects on water demand and air contents. The addition of fly ash will generally reduce water demand and decrease the air content if no further adjustment in the amount of air-entraining admixture is made. Silica fume increases water demand and decreases air content. Slag and metakaolin have a minimal effect at normal dosages.

Table 12-7 shows limits on the amount of supplementary cementing materials in concrete exposed to deicers. Local practices should be consulted as dosages different than those shown in Table 12-7 may be used without jeopardizing scale-resistance, depending on the exposure severity.

Table 12-7. Cementitious Materials Requirements for Concrete Exposed to Deicing Chemicals (Exposure Class F3)

Cementitious materials*	Maximum percent of total cementitious materials by mass**
Fly ash and natural pozzolans	25
Slag	50
Silica fume	10
Total of fly ash, slag, silica fume and natural pozzolans	50†
Total of natural pozzolans and silica fume	35†

* Includes portion of supplementary cementing materials in blended cements.

** Total cementitious materials include the summation of portland cements, blended cements, fly ash, slag, silica fume and other pozzolans.

† Silica fume should not constitute more than 10% of total cementitious materials and fly ash or other pozzolans shall not constitute more than 25% of cementitious materials.

Adapted from ACI 318.

Chemical Admixtures

Water-reducing admixtures are added to concrete to reduce the water-cementing materials ratio, reduce cementing materials content, reduce water content, reduce paste content, or to improve the workability of a concrete without changing the water-cementing materials ratio. Water reducers will usually decrease water contents by 5% to 10% and some will also increase air contents by 0.5 to 1 percentage points. Retarders may also increase the air content.

High-range water reducers (plasticizers) reduce water contents between 12% and 30% and some can simultaneously increase the air content up to 1 percentage point; others can reduce or have no affect the air content.

Calcium chloride-based admixtures reduce water contents by about 3% and increase the air content by about 0.5 percentage point.

When using a chloride-based admixture, the risks of reinforcing steel corrosion should be considered. Table 12-8 provides recommended limits on the water-soluble chloride-ion content in reinforced and prestressed concrete for various conditions.

When using more than one admixture in concrete, the compatibility of intermixing admixtures should be assured by the admixture manufacturer or the combination of admixtures should be tested in trial batches. The water contained in admixtures should be considered part of the mixing water if the admixture's water content is sufficient to affect the water-cementing materials ratio by 0.01 or more.

Excessive use of multiple admixtures should be minimized to allow better control of the concrete mixture in production and to reduce the risk of admixture incompatibility.

Table 12-8. Maximum Chloride-Ion Content for Corrosion Protection

Type of member	Maximum water-soluble chloride ion (Cl^-) in concrete, percent by mass of cement*
Prestressed concrete (Classes C0, C1, C2)	0.06
Reinforced concrete exposed to chloride in service (Class C2)	0.15
Reinforced concrete that will be dry or protected from moisture in service (Class C0)	1.00
Other reinforced concrete construction (Class C1)	0.30

*ASTM C1218.

Adapted from ACI 318.

Proportioning

The design and proportioning of concrete mixtures involves: (1) the establishment of specific concrete characteristics and (2) the selection of proportions of available materials to produce concrete of required properties, with the greatest economy. Proportioning methods have evolved from the arbitrary volumetric method (1:2:3 – cement:sand: coarse aggregate) of the early 1900s (Abrams 1918) to the present-day weight and absolute-volume methods described in ACI's Committee 211, *Standard Practice for Selecting Proportions for Normal, Heavyweight and Mass Concrete* (ACI 211.1).

Weight proportioning methods are fairly simple and quick for estimating mixture proportions using an assumed or known weight of concrete per unit volume. A more accurate method, absolute volume, involves use of relative density (specific gravity) values for all the ingredients to calculate the absolute volume occupied by each other in a specific unit volume of concrete. The absolute volume method will be illustrated. A concrete mixture also can be proportioned from field experience (statistical data) or from trial mixtures.

Other valuable documents to help proportion concrete mixtures include the *Standard Practice for Selecting Proportions for Structural Lightweight Concrete* (ACI 211.2); *Guide for Selecting Proportions for No-Slump Concrete* (ACI 211.3); *Guide for Selecting Proportions for High-Strength Concrete with Portland Cement and Fly Ash* (ACI 211.4R); and *Guide for Submittal of Concrete Proportions* (ACI 211.5). Hover (1995 and 1998) provides a graphical process for designing concrete mixtures in accordance with ACI 211.1.

Proportioning from Field Data

Existing or previously used concrete mixture proportions can be used for a new project provided strength-test data and standard deviations show that the mixture is acceptable. Durability aspects previously presented must also be met. Standard deviation computations are outlined in ACI 318. The statistical data should essentially represent the same materials, proportions, and concreting conditions to be used in the new project. The data used for proportioning should also be from concrete with an f'_c that is within 7 MPa (1000 psi) of the strength required for the proposed work. Also, the data should represent at least 30 consecutive tests or two groups of consecutive tests totaling at least 30 tests (one test is the average strength of two cylinders from the same sample). If only 15 to 29 consecutive tests are available, an adjusted standard deviation can be obtained by multiplying the standard deviation (S) for the 15 to 29 tests and a modification factor from Table 12-9. The data must represent 45 or more days of tests.

Table 12-9. Modification Factor for Standard Deviation When Less Than 30 Tests Are Available

Number of tests*	Modification factor for standard deviation**
Less than 15	Use Table 9-11
15	1.16
20	1.08
25	1.03
30 or more	1.00

* Interpolate for intermediate numbers of tests.

** Modified standard deviation to be used to determine required average strength, f'_{cr}.

Adapted from ACI 318.

The standard or modified deviation is then used in Equations 12-1 to 12-3. The average compressive strength from the test record must equal or exceed the ACI 318 required average compressive strength, f'_{cr}, in order for the concrete proportions to be acceptable. The f'_{cr} for the selected mixture proportions is equal to the larger of Equations 12-1 and 12-2 (for $f'_c \leq 35$ MPa[5000 psi]) or Equations 12-1 and 12-3 (for $f'_c > 35$ MPa [5000 psi]).

$$f'_{cr} = f'_c + 1.34S \qquad \text{Eq. 12-1}$$

$$f'_{cr} = f'_c + 2.33S - 3.45 \text{ (MPa)} \qquad \text{Eq. 12-2}$$

$$f'_{cr} = f'_c + 2.33S - 500 \text{ (psi)} \qquad \text{Eq. 12-2}$$

$$f'_{cr} = 0.90\, f'_c + 2.33S \qquad \text{Eq. 12-3}$$

where

f'_{cr} = required average compressive strength of concrete used as the basis for selection of concrete proportions, MPa (psi)

f'_c = specified compressive strength of concrete, MPa (psi)

S = standard deviation, MPa (psi)

When field strength test records do not meet the previously discussed requirements, f'_{cr} can be obtained from Table 12-10. A field strength record, several strength test records, or tests from trial mixtures must be used for documentation showing that the average strength of the mixture is equal to or greater than f'_{cr}.

If less than 30, but not less than 10 tests are available, the tests may be used for average strength documentation if the time period is not less than 45 days. Mixture proportions may also be established by interpolating between two or more test records if each meets the above requirements and the project requirements. If a significant difference exists between the mixtures that are used in the interpolation, a trial mixture should be considered to check strength gain. If the test records meet the above requirements and limitations of ACI 318, the proportions for the mixture may then be considered acceptable for the proposed work.

If the average strength of the mixtures with the statistical data is less than f'_{cr}, or statistical data or test records are insufficient or not available, the mixture should be proportioned by the trial-mixture method. The approved mixture must have a compressive strength that meets or exceeds f'_{cr}. Three trial mixtures using three different water – cementing materials ratios or three different cementing materials contents should be tested. A water to cementing materials ratio to strength curve (similar to Figure 12-3) can then be plotted and the proportions interpolated from the data. It is also good practice to test the properties of the newly proportioned mixture in a trial batch.

ACI 214 provides statistical analysis methods for monitoring the strength of the concrete in the field to ensure that the mix properly meets or exceeds the design strength, f'_c.

Table 12-10 (Metric). Required Average Compressive Strength When Data Are Not Available to Establish a Standard Deviation

Specified compressive strength, f'_c, MPa	Required average compressive strength, f'_{cr}, MPa
Less than 21	$f'_c + 7.0$
21 to 35	$f'_c + 8.5$
Over 35	$1.10f'_c + 5.0$

Adapted from ACI 318.

Table 12-10 (Inch-Pound Units). Required Average Compressive Strength When Data Are Not Available to Establish a Standard Deviation

Specified compressive strength, f'_c, psi	Required average compressive strength, f'_{cr}, psi
Less than 3000	$f'_c + 1000$
3000 to 5000	$f'_c + 1200$
Over 5000	$1.10f'_c + 700$

Adapted from ACI 318.

Proportioning by Trial Mixtures

When field test records are not available or are insufficient for proportioning by field experience methods, the concrete proportions selected should be based on trial mixtures. The trial mixtures should use the same materials proposed for the work. Three mixtures with three different water-cementing materials ratios or cementing materials contents should be made to produce a range of strengths that encompass f'_{cr}. The trial mixtures should have a slump and air content within ±20 mm (±0.75 in.) and ±0.5 percentage points, respectively, of the maximum permitted. Three cylinders for each water-cementing materials ratio should be made and cured according to ASTM C192, *Standard Practice for Making and Curing Concrete Test Specimens in the Laboratory* (AASHTO R 39). At 28 days, or the designated test age, the compressive strength of the concrete should be determined by testing the cylinders in compression. The test results should be plotted to produce a strength versus water-cementing materials ratio curve (similar to Figure 12-3) that is used to proportion a mixture.

A number of different methods of proportioning concrete ingredients have been used including:

- Arbitrary assignment (1:2:3), volumetric
- Void ratio
- Fineness modulus
- Surface area of aggregates
- Cement content

Any one of these methods can produce approximately the same final mixture after adjustments are made in the field. The best approach, however, is to select proportions based on past experience and reliable test data with an established relationship between strength and water to cementing materials ratio for the materials to be used in the concrete. The trial mixtures can be relatively small batches made with laboratory precision or batches made to the quantity that will be yielded during the course of normal concrete production. A combination of these two approaches is often necessary to reach a satisfactory mixture for a given project.

The following parameters must first be selected: (1) required strength, (2) minimum cementing materials content or maximum water-cementing materials ratio, (3) nominal maximum size of aggregate, (4) air content, and (5) desired slump. Trial batches are then made varying the relative amounts of fine and coarse aggregates as well as other ingredients. Based on considerations of workability and economy, the proper mixture proportions are selected.

When the quality of the concrete mixture is specified by water-cementitious material ratio, the trial batch procedure consists essentially of combining a paste (water, cementing materials, and, generally, a chemical admixture) of the correct proportions with the necessary amounts of fine and coarse aggregates to produce the required slump and workability. Representative samples of the cementing materials, water, aggregates, and admixtures must be used.

Quantities per cubic meter (cubic yard) are then calculated. To simplify calculations and eliminate error caused by variations in aggregate moisture content, the aggregates should be prewetted then dried to a saturated surface dry (SSD) condition. Properly store the aggregates in this SSD condition until they are used. The moisture content of the aggregates should be determined and the batch weights corrected accordingly.

The size of the trial batch is dependent on the equipment available and on the number and size of test specimens to be made. Larger batches will produce more accurate data. Machine mixing is recommended since it more nearly represents job conditions. This is mandatory if the concrete is to contain entrained air. The mixing procedures outlined in ASTM C192 (AASHTO R 39) should be used.

Measurements and Calculations

Tests for slump, air content, and temperature should be made on the trial mixture, and the following measurements and calculations should also be performed:

Density (Unit Weight) and Yield. The density (unit weight) of freshly mixed concrete is expressed in kilograms per cubic meter (pounds per cubic foot). The yield is the volume of fresh concrete produced in a batch, usually expressed in cubic meters (cubic feet). The yield is calculated by dividing the total mass of the materials batched by the density of the freshly mixed concrete. Density and yield are determined in accordance with ASTM C138, *Standard Test Method for Density (Unit Weight), Yield, and Air Content (Gravimetric) of Concrete* (AASHTO T 12).

Absolute Volume. The absolute volume of a granular material (such as cement and aggregates) is the volume of the solid matter in the particles; it does not include the volume of air spaces between particles. The volume (yield) of freshly mixed concrete is equal to the sum of the absolute volumes of the concrete ingredients including: cementing materials, water (exclusive of that absorbed in the aggregate), aggregates, admixtures when applicable, and air. The absolute volume is computed from a material's mass and relative density (specific gravity) as follows:

$$\text{Absolute volume} = \frac{\text{mass of loose material}}{\text{relative density of a material x density of water}}$$

A value of 3.15 can be used for the relative density (specific gravity) of portland cement. Blended cements have relative densities ranging from 2.90 to 3.15. The relative density of fly ash varies from 1.9 to 2.8, slag from 2.85 to 2.95, and silica fume from 2.20 to 2.25. The relative density of water is 1.0 and the density of water is 1000 kg/m^3 (62.4 lb/ft^3) at 4°C (39°F) (accurate enough for mix calculations at room temperature). More accurate water density values are given in Table 12-11. Relative density of normal aggregate usually ranges between 2.4 and 2.9.

The relative density of aggregate as used in mix-design calculations is the relative density of either SSD material or ovendry material. Relative densities of admixtures, such as water reducers, can also be considered if needed. Absolute volume is usually expressed in cubic meters (cubic feet).

The absolute volume of air in concrete, expressed as cubic meters per cubic meter (cubic feet per cubic yard), is equal to the total air content in percent divided by 100 (for example, 7% ÷ 100) and then multiplied by the volume of the concrete batch.

The volume of concrete in a batch can be determined by either of two methods: (1) if the relative densities of the aggregates and cementing materials are known, these can be used to calculate concrete volume; or (2) if relative densities are unknown, or they vary, the volume can be computed by dividing the total mass of materials in the mixer by the density of concrete. In some cases, both determinations are made, one serving as a check on the other.

Examples of Mixture Proportioning

Example 1. Absolute Volume Method (Metric)

Conditions and Specifications. Concrete is required for a pavement that will be exposed to moisture in a severe freeze-thaw environment. A specified compressive strength, f'_c, of 35 MPa is required at 28 days. Air entrainment is required. Slump should be between 25 mm and 75 mm. A nominal maximum size aggregate of 25 mm is required. No statistical data on previous mixtures are available. The materials available are as follows:

Cement:
Type GU (ASTM C1157, *Standard Performance Specification for Hydraulic Cement*) with a relative density of 3.0.

Coarse aggregate:
Well-graded, 25-mm nominal maximum-size rounded gravel (ASTM C33, *Standard Specification for Concrete*

Table 12-11. Density of Water Versus Temperature

Temperature, °C	Density, kg/m³	Temperature, °F	Density, lb/ft³
16	998.93	60	62.368
18	998.58	65	62.337
20	998.19	70	62.302
22	997.75	75	62.261
24	997.27	80	62.216
26	996.75	85	62.166
28	996.20		
30	995.61		

Aggregates [AASHTO M 80]) with an ovendry relative density of 2.68, absorption of 0.5% (SSD), and ovendry rodded bulk density (unit weight) of 1600 kg/m³. The laboratory sample for trial batching has a moisture content of 2%.

Fine aggregate:
Natural sand (ASTM C33 [AASHTO M 6]) with an oven-dry relative density of 2.64 and absorption of 0.7%. The laboratory sample moisture content is 6%. The fineness modulus is 2.80.

Air-entraining admixture:
Wood-resin type (ASTM C260, *Standard Specification for Air-Entraining Admixtures for Concrete* [AASHTO M 154]).

Water reducer:
ASTM C494, *Standard Specification for Chemical Admixtures for Concrete* (AASHTO M 194). This particular admixture is known to reduce water demand by 10% when used at a dosage rate of 3 g/kg (or 3 mL/kg) of cement. Assume that the chemical admixtures have a density close to that of water, meaning that 1 mL of admixture has a mass of 1 g.

From this information, the task is to proportion a trial mixture that will meet the above conditions and specifications.

Strength. The design strength of 35 MPa is greater than the 31 MPa required in Table 12-1 for the exposure condition. Since no statistical data is available, f'_{cr} (required compressive strength for proportioning) from Table 12-11 is equal to f'_c + 8.5 MPa. Therefore, $f'_{cr} = 35 + 8.5 = 43.5$ MPa.

Water-Cementitious Materials Ratio. For an environment with moist freezing and thawing, the maximum w/cm should be 0.45. The recommended water-cementitious material ratio for an f'_{cr} of 43.5 MPa is 0.31 from Figure 12-3 or interpolated from Table 12-3 [{(45 – 43.5)(0.34 – 0.30)/(45 – 40)} + 0.30 = 0.31]. Since the lower w/cm governs, the mixture must be designed for 0.31. If a plot from trial batches or field tests had been available, the w/cm could have been extrapolated from that data.

Air Content. For a severe freeze-thaw exposure, Table 12-5 recommends a target air content of 6.0% for a 25-mm aggregate. Therefore, design the mixture for 5% to 8% air and use 8% (or the maximum allowable) for batch proportions. The trial-batch air content must be within ±0.5 percentage points of the maximum allowable air content.

Slump. The slump is specified at 25 mm to 75 mm. Use 75 mm ±20 mm for proportioning purposes.

Water Content. Table 12-5 and Figure 12-6 recommend that a 75-mm slump, air-entrained concrete made with 25-mm nominal maximum-size aggregate should have a water content of about 175 kg/m³. However, rounded gravel should reduce the water content of the table value by about 25 kg/m³. Therefore, the water content can be estimated to be about 150 kg/m³ (175 kg/m³ minus 25 kg/m³). In addition, the water reducer will reduce water demand by 10% resulting in an estimated water demand of 135 kg/m³.

Cement Content. The cement content is based on the maximum water-cement ratio and the water content. Therefore, 135 kg/m³ of water divided by a water-cement ratio of 0.31 requires a cement content of 435 kg/m³.

Coarse-Aggregate Content. The quantity of 25-mm nominal maximum-size coarse aggregate can be estimated from Figure 12-4 or Table 12-4. The bulk volume of coarse aggregate recommended when using sand with a fineness modulus of 2.80 is 0.67. Since it has a bulk density of 1600 kg/m³, the ovendry mass of coarse aggregate for a cubic meter of concrete is

$$1600 \times 0.67 = 1072 \text{ kg}$$

Admixture Content. For an 8% air content, the air-entraining admixture manufacturer recommends a dosage rate of 0.5 g per kg of cement. From this information, the amount of air-entraining admixture per cubic meter of concrete is

$$0.5 \times 435 = 218 \text{ g or } 0.218 \text{ kg}$$

The water reducer dosage rate of 3 g per kg of cement results in

$$3 \times 435 = 1305 \text{ g or } 1.305 \text{ kg of water reducer per cubic meter of concrete}$$

Fine-Aggregate Content. At this point, the amounts of all ingredients except the fine aggregate are known. In the absolute volume method, the volume of fine aggregate is determined by subtracting the absolute volumes of the known ingredients from 1 cubic meter. The absolute volume of the water, cement, admixtures and coarse aggregate is calculated by dividing the known mass of each by the product of their relative density and the density of water. Volume computations are as follows:

$$\text{Water} = \frac{135}{1 \times 1000} = 0.135 \text{ m}^3$$

$$\text{Cement} = \frac{435}{3.0 \times 1000} = 0.145 \text{ m}^3$$

$$\text{Air} = \frac{8.0}{100} = 0.080 \text{ m}^3$$

$$\text{Coarse aggregate} = \frac{1072}{2.68 \times 1000} = 0.400 \text{ m}^3$$

Total volume of known ingredients 0.760 m³

The calculated absolute volume of fine aggregate is then

$$1 - 0.76 = 0.24 \text{ m}^3$$

The mass of dry fine aggregate is

$$0.24 \times 2.64 \times 1000 = 634 \text{ kg}$$

The mixture then has the following proportions before trial mixing for one cubic meter of concrete:

Water	135 kg
Cement	435 kg
Coarse aggregate (dry)	1072 kg
Fine aggregate (dry)	634 kg
Total mass	2276 kg
Air-entraining admixture	0.218 kg
Water reducer	1.305 kg

Slump 75 mm (±20 mm for trial batch)

Air content 8% (±0.5% for trial batch)

Estimated concrete density (using SSD aggregate) = 135 + 435 + (1072 x 1.005*) + (634 x 1.007*) = 2286 kg/m³

The liquid admixture volume is generally too insignificant to include in the water calculations. However, certain admixtures, such as shrinkage reducers, plasticizers, and corrosion inhibitors are exceptions due to their relatively large dosage rates; their volumes should be included.

Moisture. Corrections are needed to compensate for moisture in and on the aggregates. In practice, aggregates will contain some measurable amount of moisture. The dry-batch weights of aggregates, therefore, have to be increased to compensate for the moisture that is absorbed in and contained on the surface of each particle and between particles. The mixing water added to the batch must be reduced by the amount of free moisture contributed by the aggregates. Tests indicate that for this example, coarse-aggregate moisture content is 2% and fine-aggregate moisture content is 6%.

With the aggregate moisture contents (MC) indicated, the trial batch aggregate proportions become

Coarse aggregate (2% MC) = 1072 x 1.02 = 1093 kg

Fine aggregate (6% MC) = 634 x 1.06 = 672 kg

Water absorbed by the aggregates does not become part of the mixing water and must be excluded from the water adjustment. Surface moisture contributed by the coarse aggregate amounts to 2% – 0.5% = 1.5%; that contributed by the fine aggregate is, 6% – 0.7% = 5.3%. The estimated requirement for added water becomes

135 – (1072 x 0.015) – (634 x 0.053) = 85 kg

The estimated batch weights for one cubic meter of concrete are revised to include aggregate moisture as follows:

Water (to be added)	85 kg
Cement	435 kg
Coarse aggregate (2% MC, wet)	1093 kg
Fine aggregate (6% MC, wet)	672 kg
Total	2285 kg
Air-entraining admixture	0.218 kg
Water reducer	1.305 kg

Trial Batch. At this stage, the estimated batch weights should be checked by means of trial batches or by full-size field batches. Enough concrete must be mixed for appropriate air and slump tests and for casting the three cylinders required for 28-day compressive-strength tests, plus beams for flexural tests if necessary. For a laboratory trial batch it is convenient, in this case, to scale down the weights to produce 0.1 m³ of concrete as follows:

Water	85 x 0.1 =	8.5 kg
Cement	435 x 0.1 =	43.5 kg
Coarse aggregate (wet)	1093 x 0.1 =	109.3 kg
Fine aggregate (wet)	672 x 0.1 =	67.2 kg
Total		228.5 kg
Air-entraining admixture	218 g x 0.1 = 21.8 g or 21.8 mL	
Water reducer	1305 g x 0.1 = 130 g or 130 mL	

The above concrete, when mixed, had a measured slump of 100 mm, an air content of 9%, and a density of 2274 kg/m³. During mixing, some of the premeasured water may remain unused or additional water may be added to approach the required slump. In this example, although 8.5 kg of water was calculated to be added, the trial batch actually used only 8.0 kg. The mixture excluding admixtures therefore becomes

Water	8.0 kg
Cement	43.5 kg
Coarse aggregate (2% MC)	109.3 kg
Fine aggregate (6% MC)	67.2 kg
Total	228.0 kg

The yield of the trial batch is

$$\frac{228.0 \text{ kg}}{2274 \text{ kg/m}^3} = 0.10026 \text{ m}^3$$

The mixing water content is determined from the added water plus the free water on the aggregates and is calculated as follows:

Water added 8.0 kg

Free water on coarse aggregate

$$= \frac{109.3}{1.02} \times 0.015^{**} = 1.61 \text{ kg}$$

Free water on fine aggregate

$$= \frac{67.2}{1.06} \times 0.053^{**} = 3.36 \text{ kg}$$

Total water 12.97 kg

* (0.5% absorption ÷ 100) + 1 = 1.005
(0.7% absorption ÷ 100) + 1 = 1.007

**(2% MC – 0.5% absorption) ÷ 100 = 0.015
(6% MC – 0.7% absorption) ÷ 100 = 0.053

The mixing water required for a cubic meter of the same slump concrete as the trial batch is

$$\frac{12.97}{0.10026} = 129 \text{ kg}$$

Batch Adjustments. The measured 100-mm slump of the trial batch is unacceptable (above 75 mm ±20 mm), the yield was slightly high, and the 9.0% air content as measured in this example is also too high (more than 0.5% above 8.5%). Adjust the yield, reestimate the amount of air-entraining admixture required for 8% air content, and adjust the water to obtain a 75-mm slump. Increase the mixing water content by 3 kg/m³ for each 1% by which the air content is decreased from that of the trial batch and reduce the water content by 2 kg/m³ for each 10 mm reduction in slump. The adjusted mixture water for the reduced slump and air content is

(3 kg water x 1 percentage point difference for air) – (2 kg water x 25/10 for slump change) + 129 = 127 kg of water

With less mixing water needed in the trial batch, less cement also is needed to maintain the desired water-cement ratio of 0.31. The new cement content is

$$\frac{127}{0.31} = 410 \text{ kg}$$

The amount of coarse aggregate remains unchanged because workability is satisfactory. The new adjusted batch weights based on the new cement and water contents are calculated after the following volume computations:

Water	=	$\frac{127}{1 \times 1000}$	= 0.127 m³
Cement	=	$\frac{410}{3.0 \times 1000}$	= 0.137 m³
Coarse aggregate (dry)	=	$\frac{1072}{2.68 \times 1000}$	= 0.400 m³
Air	=	$\frac{8}{100}$	= 0.080 m³
Total			0.744 m³

Fine aggregate volume = 1 – 0.744 = 0.256 m³

The weight of dry fine aggregate required is
0.256 x 2.64 x 1000 = 676 kg

Air-entraining admixture (the manufacturer suggests reducing the dosage by 0.1 g to reduce air 1 percentage point) = 0.4 x 410 = 164 g or mL

Water reducer = 3.0 x 410 = 1230 g or mL

Adjusted batch weights per cubic meter of concrete are

Water	127 kg
Cement	410 kg
Coarse aggregate (dry)	1072 kg
Fine aggregate (dry)	676 kg
Total	2285 kg
Air-entraining admixture	164 g or mL
Water reducer	1230 g or mL
Estimated concrete density (aggregates at SSD)	= 127 + 410 + (1072 x 1.005) + (676 x 1.007) = 2295 kg/m³

After checking these adjusted proportions in a trial batch, it was found that the concrete had the desired slump, air content, and yield. The 28-day test cylinders had an average compressive strength of 48 MPa, which exceeds the f'_{cr} of 43.5 MPa. Due to fluctuations in moisture content, absorption rates, and relative density (specific gravity) of the aggregate, the density determined by volume calculations may not always equal the density determined by ASTM C138 (AASHTO T 121). Occasionally, the proportion of fine-to-coarse aggregate is kept constant in adjusting the batch weights to maintain workability or other properties obtained in the first trial batch. After adjustments to the cementitious materials, water, and air content have been made, the volume remaining for aggregate is appropriately proportioned between the fine and coarse aggregates.

Additional trial concrete mixtures with water-cement ratios above and below 0.31 should also be tested to develop a strength to water-cement ratio relationship. From that data, a new more economical mixture with a compressive strength closer to f'_{cr} and a lower cement content can be proportioned and tested. The final mixture would probably look similar to the above mixture with a slump range of 25 mm to 75 mm and an air content of 5% to 8%. The amount of air-entraining admixture must be adjusted to field conditions to maintain the specified air content.

Example 2. Absolute Volume Method (Inch-Pound Units)

Conditions and Specifications. Concrete is required for a building foundation. A specified compressive strength, f'_c, of 3500 psi is required at 28 days using a Type I cement. The design calls for a minimum of 3 in. of concrete cover over the reinforcing steel. The minimum distance between reinforcing bars is 4 in. The only admixture allowed is for air entrainment. No statistical data on previous mixes are available. The materials available are as follows:

Cement:
Type I, ASTM C150, with a relative density of 3.15.

Coarse aggregate:
Well-graded ¾-in. nominal maximum-size gravel containing some crushed particles (ASTM C33) with an ovendry relative density (specific gravity) of 2.68, absorption of 0.5% (SSD), and ovendry rodded bulk density (unit weight) of 100 lb/ft³. The laboratory sample for trial batching has a moisture content of 2%.

Fine aggregate:
Natural sand (ASTM C33) with an ovendry relative density (specific gravity) of 2.64 and absorption of 0.7%. The laboratory sample moisture content is 6%. The fineness modulus is 2.80.

Air-entraining admixture:
Wood-resin type, ASTM C260.

From this information, the task is to proportion a trial mixture that will meet the above conditions and specifications.

Strength. Since no statistical data is available, f'_{cr} from Table 12-11 is equal to $f'_c + 1200$. Therefore, $f'_{cr} = 3500 + 1200 = 4700$ psi.

Water-Cementitious Materials Ratio. Table 12-1 requires no maximum w/cm. The recommended water-cementing materials ratio for an f'_{cr} of 4700 psi is 0.42 interpolated from Figure 12-3 or Table 12-3 [w/cm = {(5000 – 4700) (0.48 – 0.40)/(5000 – 4000)} + 0.40 = 0.42].

Coarse-Aggregate Size. From the specified information, a ¾-in. nominal maximum-size aggregate is adequate as it is less than ¾ of the distance between reinforcing bars and between the reinforcement and forms (cover).

Air Content. A target air content of 6.0% is specified in this instance not for exposure conditions but to improve workability and reduce bleeding. Therefore, design the mix for 6% ±1.0% air and use 7% (or the maximum allowable) for batch proportions. The trial batch air content must be within ±0.5 percentage points of the maximum allowable air content.

Slump. As no slump was specified, a slump of 3 in. is selected for proportioning purposes.

Water Content. Figure 12-6 and Table 12-5 recommend that a 3-in. slump, air-entrained concrete made with ¾-in. nominal maximum-size aggregate should have a water content of about 305 lb/yd³. However, gravel with some crushed particles should reduce the water content of the table value by about 35 lb/yd³. Therefore, the water content can be estimated to be about 305 lb minus 35 lb, which is 270 lb.

Cement Content. The cement content is based on the maximum water-cement ratio and the water content. Therefore, 270 lb of water divided by a water-cement ratio of 0.42 requires a cement content of 643 lb.

Coarse-Aggregate Content. The quantity of ¾-in. nominal maximum-size coarse aggregate can be estimated from Figure 12-4 or Table 12-4. The bulk volume of coarse aggregate recommended when using sand with a fineness modulus of 2.80 is 0.62. Since it weighs 100 lb/ft³, the ovendry weight of coarse aggregate for a yd³ of concrete (27 ft³) is

$$100 \times 27 \times 0.62 = 1674 \text{ lb/yd}^3 \text{ of concrete}$$

Admixture Content. For a 7% air content, the air-entraining admixture manufacturer recommends a dosage rate of 0.9 fl oz per 100 lb of cement. From this information, the amount of air-entraining admixture is

$$0.9 \times \frac{643}{100} = 5.8 \text{ fl oz/yd}^3$$

Fine-Aggregate Content. At this point, the amounts of all ingredients except the fine aggregate are known. In the absolute volume method, the volume of fine aggregate is determined by subtracting the absolute volumes of the known ingredients from 27 ft³ (1 yd³). The absolute volume of the water, cement, and coarse aggregate is calculated by dividing the known weight of each by the product of their relative density (specific gravity) and the density of water. Volume computations are as follows:

$$\text{Water} = \frac{270}{1 \times 62.4} = 4.33 \text{ ft}^3$$

$$\text{Cement} = \frac{643}{3.15 \times 62.4} = 3.27 \text{ ft}^3$$

$$\text{Air} = \frac{7.0}{100} \times 27 = 1.89 \text{ ft}^3$$

$$\text{Coarse aggregate} = \frac{1674}{2.68 \times 62.4} = 10.01 \text{ ft}^3$$

Total volume of known ingredients 19.50 ft³

The liquid admixture volume is generally too insignificant to include in these calculations. However, certain admixtures such as shrinkage reducers, plasticizers, and corrosion inhibitors are exceptions due to their relatively large dosage rates; their volumes should be included.

The calculated absolute volume of fine aggregate is then

$$27 - 19.50 = 7.50 \text{ ft}^3$$

The weight of dry fine aggregate is

$$7.50 \times 2.64 \times 62.4 = 1236 \text{ lb}$$

The mixture then has the following proportions before trial mixing for one cubic yard of concrete:

Water	270 lb
Cement	643 lb
Coarse aggregate (dry)	1674 lb
Fine aggregate (dry)	1236 lb
Total weight	3823 lb

Air-entraining admixture 5.8 fl oz

Slump	3 in. (±¾ in. for trial batch)
Air content	7% (±0.5% for trial batch)
Estimated density (using SSD aggregate)	= [270 + 643 + (1674 x 1.005*) + (1236 x 1.007*)] ÷ 27 = 142.22 lb/ft³

Moisture. Corrections are needed to compensate for moisture in the aggregates. In practice, aggregates will contain some measurable amount of moisture. The dry-batch weights of aggregates, therefore, have to be increased to compensate for the moisture that is absorbed in and contained on the surface of each particle and between particles. The mixing water added to the batch must be reduced by the amount of free moisture contributed by the aggregates. Tests indicate that for this example, coarse-aggregate moisture content is 2% and fine-aggregate moisture content is 6%.

With the aggregate moisture contents (MC) indicated, the trial batch aggregate proportions become

Coarse aggregate (2% MC) = 1674 x 1.02 = 1707 lb

Fine aggregate (6% MC) = 1236 x 1.06 = 1310 lb

Water absorbed by the aggregates does not become part of the mixing water and must be excluded from the water adjustment. Surface moisture contributed by the coarse aggregate amounts to 2% – 0.5% = 1.5%; that contributed by the fine aggregate is 6% – 0.7% = 5.3%. The estimated requirement for added water becomes

$$270 - (1674 \times 0.015) - (1236 \times 0.053) = 179 \text{ lb}$$

The estimated batch weights for one cubic yard of concrete are revised to include aggregate moisture as follows:

Water (to be added)	179 lb
Cement	643 lb
Coarse aggregate (2% MC, wet)	1707 lb
Fine aggregate (6% MC, wet)	1310 lb
Total	3839 lb
Air-entraining admixture	5.8 fl oz

Trial Batch. At this stage, the estimated batch weights should be checked by means of trial batches or by full-size field batches. Enough concrete must be mixed for appropriate air and slump tests and for casting the three cylinders required for compressive strength tests at 28 days. For a laboratory trial batch it is convenient, in this case, to scale down the weights to produce 2.0 ft^3 of concrete or 2/27 yd^3.

$$\text{Water} \quad 179 \times \frac{2}{27} = 13.26 \text{ lb}$$

$$\text{Cement} \quad 643 \times \frac{2}{27} = 47.63 \text{ lb}$$

$$\text{Coarse aggregate (wet)} \quad 1707 \times \frac{2}{27} = 126.44 \text{ lb}$$

$$\text{Fine aggregate (wet)} \quad 1310 \times \frac{2}{27} = 97.04 \text{ lb}$$

$$\text{Total} \quad 284.37 \text{ lb}$$

$$\text{Air-entraining admixture} \quad 5.8 \times \frac{2}{27} = 0.43 \text{ fl oz}$$

(Laboratories often convert fluid ounces to milliliters by multiplying fluid ounces by 29.57353 to improve measurement accuracy. Also, most laboratory pipettes used for measuring fluids are graduated in milliliter units.)

The above concrete, when mixed, had a measured slump of 4 in., an air content of 8%, and a density (unit weight) of 141.49 lb/ ft^3. During mixing, some of the premeasured water may remain unused or additional water may be added to approach the required slump. In this example, although 13.26 lb of water was calculated to be added, the trial batch actually used only 13.12 lb. The mixture excluding admixture therefore becomes:

Water	13.12 lb
Cement	47.63 lb
Coarse aggregate (2% MC)	126.44 lb
Fine aggregate (6% MC)	97.04 lb
Total	284.23 lb

The yield of the trial batch is

$$\frac{284.23}{141.49} = 2.009 \text{ ft}^3$$

The mixing water content is determined from the added water plus the free water on the aggregates and is calculated as follows:

$$\text{Water added} = 13.12 \text{ lb}$$

$$\text{Free water on coarse aggregate} = \frac{126.44}{1.02^{**}} \times 0.015^{\dagger} = 1.86 \text{ lb}$$

$$\text{Free water on fine aggregate} = \frac{97.04}{1.06^{**}} \times 0.053^{\dagger} = 4.85 \text{ lb}$$

$$\text{Total} = 19.83 \text{ lb}$$

The mixing water required for a cubic yard of the same slump concrete as the trial batch is

$$\frac{19.83 \times 27}{2.009} = 267 \text{ lb}$$

Batch Adjustments. The measured 4-in. slump of the trial batch is unacceptable (more than 0.75 in. above 3 in.), the yield was slightly high, and the 8.0% air content as measured in this example is also too high (more than 0.5% above 7%). Adjust the yield, reestimate the amount of air-entraining admixture required for 7% air content, and adjust the water to obtain a 3-in. slump. Increase the mixing water content by 5 lb for each 1% by which the air content is decreased from that of the trial batch and reduce the water content by 10 lb for each 1-in. reduction in

* (0.5% absorption ÷ 100) + 1 = 1.005;
(0.7% absorption ÷ 100) + 1 = 1.007

** 1 + (2% MC/100) = 1.02; 1 + (6% MC/100) = 1.06;

† (2% MC – 0.5% absorption)/100 = 0.015; (6% MC – 0.7% absorption)/100 = 0.053

slump. The adjusted mixture water for the reduced slump and air content is

$$(5 \times 1) - (10 \times 1) + 267 = 262 \text{ lb/yd}^3$$

With less mixing water needed in the trial batch, less cement also is needed to maintain the desired water-cement ratio of 0.42. The new cement content is

$$\frac{262}{0.42} = 624 \text{ lb/yd}^3$$

The amount of coarse aggregate remains unchanged because workability is satisfactory. The new adjusted batch weights based on the new cement and water contents are calculated after the following volume computations:

$$\text{Water} = \frac{262}{1 \times 62.4} = 4.20 \text{ ft}^3$$

$$\text{Cement} = \frac{624}{3.15 \times 62.4} = 3.17 \text{ ft}^3$$

$$\text{Coarse aggregate} = \frac{1674}{2.68 \times 62.4} = 10.01 \text{ ft}^3$$

$$\text{Air} = \frac{7.0}{100} \times 27 = 1.89 \text{ ft}^3$$

$$\text{Total} \quad 19.27 \text{ ft}^3$$

$$\text{Fine aggregate volume} = 27 - 19.27 = 7.73 \text{ ft}^3$$

The weight of dry fine aggregate required is

$$7.73 \times 2.64 \times 62.4 = 1273 \text{ lb}$$

An air-entraining admixture dosage of 0.8 fluid ounces per 100 pounds of cement is expected to achieve the 7% air content in this example. Therefore, the amount of air-entraining admixture required is:

$$= \frac{0.8 \times 624}{100} = 5.0 \text{ fl oz}$$

Adjusted batch weights per cubic yard of concrete are

Water	262 lb	262 lb
Cement	624 lb	624 lb
Coarse aggregate	1674 lb (dry)	1682 lb (SSD)
Fine aggregate	1273 lb (dry)	1282 lb (SSD)
Total	3833 lb	3850 lb
Air-entraining admixture	5.0 fl oz	

Estimated concrete density (unit weight) with the aggregates at SSD:

$$= \frac{[262 + 624 + (1674 \times 1.005) + (1273 \times 1.007)]}{27}$$

$$= 142.60 \text{ lb/ft}^3$$

Upon completion of checking these adjusted proportions in a trial batch, it was found that the proportions were adequate for the desired slump, air content, and yield. The 28-day test cylinders had an average compressive strength of 4900 psi, which exceeds the f'_{cr} of 4700 psi. Due to fluctuations in moisture content, absorption rates, and specific gravity of the aggregate, the density determined by volume calculations may not always equal the unit weight determined by ASTM C138 (AASHTO T 121). Occasionally, the proportion of fine to coarse aggregate is kept constant in adjusting the batch weights to maintain workability or other properties obtained in the first trial batch. After adjustments to the cement, water, and air content have been made, the volume remaining for aggregate is appropriately proportioned between the fine and coarse aggregates.

Additional trial concrete mixtures with water-cement ratios above and below 0.42 should also be tested to develop a strength curve. From the curve, a new more economical mixture with a compressive strength closer to f'_{cr}, can be proportioned and tested. The final mixture would probably look similar to the above mixture with a slump range of 1 in. to 3 in. and an air content of 5% to 7%. The amount of air-entraining admixture must be adjusted to field conditions to maintain the specified air content.

Water Reducers. Water reducers are used to increase workability without the addition of water or to reduce the w/cm of a concrete mixture to improve permeability or other properties.

Using the final mixture developed in the last example, assume that the project engineer approves the use of a water reducer to increase the slump to 5 in. to improve workability for a difficult placement area. Assuming that the water reducer has a manufacturer's recommended dosage rate of 4 oz per 100 lb of cement to increase slump 2 in., the admixture amount becomes

$$\frac{624}{100} \times 4 = 25.0 \text{ oz/yd}^3$$

The amount of air-entraining agent may also need to be reduced (up to 50%), as many water reducers also entrain air. If a water reducer was used to reduce the water-cement ratio, the water and sand content would also need adjustment.

Pozzolans and Slag. Pozzolans and slag are sometimes added in addition to or as a partial replacement of cement to aid in workability and resistance to sulfate attack and alkali reactivity. If a pozzolan or slag were required for the above example mixture, it would have been entered in the first volume calculation used in determining fine aggregate content. For example:

Assume that 75 lb of fly ash with a relative density (specific gravity) of 2.5 were to be used in addition to the originally derived cement content. The ash volume would be

$$\frac{75}{2.5 \times 62.4} = 0.48 \text{ ft}^3$$

The water to cementing materials ratio would be

$$\frac{W}{C + P} = \frac{270}{643 + 75} = 0.38 \text{ by weight}$$

The water to portland cement only ratio would still be

$$\frac{W}{C} = \frac{270}{643} = 0.42 \text{ by weight}$$

The fine aggregate volume would have to be reduced by 0.48 ft^3 to allow for the volume of ash.

The pozzolan amount and volume computation could also have been derived in conjunction with the first cement content calculation using a water to cementing materials ratio of 0.42 (or equivalent). For example, assume 15% of the cementitious material is specified to be a pozzolan and

$$W/CM \text{ or } W/(C + P) = 0.42.$$

Then with $W = 270 \text{ lb and } C + P = 643 \text{ lb},$

$$P = 643 \times \frac{15}{100} = 96 \text{ lb}$$

and $C = 643 - 96 = 547 \text{ lb}$

Appropriate proportioning computations for these and other mix ingredients would follow.

Example 3. Laboratory Trial Mixture Using the PCA Water-Cement Ratio Method (Metric)

With the following method, the mix designer develops the concrete proportions directly from the laboratory trial batch rather than the absolute volume of the constituent ingredients.

Conditions and Specifications. Concrete is required for a plain concrete pavement to be constructed in North Dakota. The pavement specified compressive strength is 35 MPa at 28 days. The standard deviation of the concrete producer is 2.0 MPa. Type IP cement and 19-mm nominal maximum-size coarse aggregate is locally available. Proportion a concrete mixture for these conditions and check it by trial batch. Enter all data in the blank spaces on a trial mixture data sheet (Figure 12-7).

Durability Requirements. The pavement will be exposed to freezing, thawing, and deicers and therefore should have a maximum water to cementitious material ratio of 0.45 (Table 12-1) and at least 335 kg of cement per cubic meter of concrete as required by the owner.

Strength Requirements. For a standard deviation of 2.0 MPa, the f'_{cr} (required compressive strength for proportioning) must be the larger of

$$f'_{cr} = f'_c + 1.34S = 35 + 1.34(2.0) = 37.7 \text{ MPa}$$

or

$$f'_{cr} = f'_c + 2.33S - 3.45 = 35 + 2.33(2.0) - 3.45 = 36.2 \text{ MPa}$$

Therefore the required average compressive strength = 37.7 MPa.

Aggregate Size. The 19-mm maximum-size coarse aggregate and the fine aggregate are in saturated-surface dry condition for the trial mixtures.

Air Content. The target air content should be 6% (Table 12-5) and the range is set at 5% to 8%.

Slump. The specified target slump for this project is 40 (±20) mm.

Batch Quantities. For convenience, a batch containing 10 kg of cement is to be made. The quantity of mixing water required is 10 x 0.45 = 4.5 kg. Representative samples of fine and coarse aggregates are measured in suitable containers. The values are entered as initial mass in Column 2 of the trial-batch data sheet (Figure 12-7).

All of the measured quantities of cement, water, and air-entraining admixture are used and added to the mixer. Fine and coarse aggregates, previously brought to a saturated, surface-dry condition, are added until a workable concrete mixture with a slump deemed adequate for placement is produced. The relative proportions of fine and coarse aggregate for workability can readily be judged by an experienced concrete technician or engineer.

Workability. Results of tests for slump, air content, density, and a description of the appearance and workability are noted in the data sheet and Table 12-13.

The amounts of fine and coarse aggregates not used are recorded on the data sheet in Column 3, and mass of aggregates used (Column 2 minus Column 3) are noted in Column 4. If the slump when tested had been greater than that required, additional fine or coarse aggregates (or both) would have been added to reduce slump. Had the slump been less than required, water and cement in the appropriate ratio (0.45) would have been added to increase slump. It is important that any additional quantities be measured accurately and recorded on the data sheet.

Mixture Proportions. Mixture proportions for a cubic meter of concrete are calculated in Column 5 of Figure 12-7 by using the batch yield (volume) and density (unit weight). For example, the number of kilograms of cement per cubic meter is determined by dividing one cubic meter by the volume of concrete in the batch and multiplying the result by the number of kilograms of cement in the batch. The percentage of fine aggregate by mass of total aggregate is also calculated. In this trial batch, the cement content was 341 kg/m³ and the fine aggregate made up 38% of the total aggregate by mass. The air content and slump were acceptable. The 28-day strength was 39.1 MPa, greater than f'_{cr}. The mixture in Column 5, along with slump and air content limits of 40 (±20) mm and 5% to 8%, respectively, is now ready for submission to the project engineer.

Data and Calculations for Trial Batch (saturated surface-dry aggregates)

Batch size: 10 kg ✓ 20 kg ______ 40 kg ______ of cement

Note: Complete Columns 1 through 4, fill in items below, the complete 5 and 6.

1 Material	2 Initial mass, kg	3 Final mass, kg	4 Mass. used, (Col. 2 minus Col. 3)	5 Mass per m³ No. of batches (C) x Col. 4	6 Remarks
Cement	10.0	0	10.0	341	
Water	4.5	0	4.5	153	
Fine aggregate	37.6	17.3	20.3	691 (a)	% F.A.* = $\frac{a}{a+b}$ x 100 = 38%
Coarse aggregate	44.1	11.0	33.1	1128 (b)	
Air-entraining admixture	10 ml	Total (T) =	67.9	2313	
		T x C = 67.9 x 34.0648 = 2313			Math check

Measured slump: 45 mm Measured air content 7.5 %

Appearance: Sandy ______ Good ✓ Rocky ______

Workability: Good ✓ Fair ______ Poor ______

Mass of container + concrete = 42.7 kg

Mass of container = 8.0 kg

Mass of concrete (A) = 34.7 kg

Volume of container (B) = 0.015 m³

Density of concrete (D) = $\frac{A}{B}$ = 34.7/0.015 = 2313 kg/m³

Volume of concrete produced = $\frac{\text{Total mass of material per batch}}{\text{Density}} = \frac{T}{D}$

= 67.9/2313 = 0.0293558 m³

Number of 67.9 kg batches per m³ (C) = $\frac{1.0\text{ m}^3}{\text{Volume}} = \frac{1.0}{0.0293558}$ = 34.0648 batches

*Percentage fine aggregate of total aggregates = $\frac{\text{Mass of fine aggregate}}{\text{Total mass of aggregates}}$ x 100

Figure 12-7. Trial mixture data sheet (metric).

Table 12-13. Example of Results of Laboratory Trial Mixtures (Metric)*

Batch no.	Slump, mm	Air content, percent	Density, kg/m³	Cement content, kg/m³	Fine aggregate, percent of total aggregate	Workability
1	50	5.7	2341	346	28.6	Harsh
2	40	6.2	2332	337	33.3	Fair
3	45	7.5	2313	341	38.0	Good
4	36	6.8	2324	348	40.2	Good

*Water-cement ratio was 0.45.

Example 4. Laboratory Trial Mixture Using the PCA Water-Cement Ratio Method (Inch-Pound Units)

With the following method, the mix designer develops the concrete proportions directly from a laboratory trial batch, rather than the absolute volume of the constituent ingredients as in Example 2.

Conditions and Specifications. Air-entrained concrete is required for a foundation wall that will be exposed to moderate sulfate soils. A compressive strength, f'_c, of 4000 psi at 28 days using Type II cement is specified. Minimum thickness of the wall is 10 in. and concrete cover over ½-in.-diameter reinforcing bars is 3 in. The clear distance between reinforcing bars is 3 in. The w/cm versus compressive strength relationship based on field and previous laboratory data for the example ingredients is illustrated by Figure 12-8. Based on the test records of the materials to be used, the standard deviation is 300 psi. Proportion and evaluate by trial batch a mixture meeting the above conditions and specifications. Enter all data in the appropriate blanks on a trial-mixture data sheet (Figure 12-9).

Water-Cementitious Materials Ratio. For these exposure conditions, Table 12-2 indicates that concrete with a maximum w/cm of 0.50 should be used and the minimum design strength should be 4000 psi.

The w/cm for strength is selected from a graph plotted to show the relationship between the w/cm and compressive strength for these specific concrete materials (Figure 12-8).

For a standard deviation of 300 psi, f'_{cr} must be the larger of

$$f'_{cr} = f'_c + 1.34S = 4000 + 1.34(300) = 4402 \text{ psi}$$

or

$$f'_{cr} = f'_c + 2.33S - 500 = 4000 + 2.33(300) - 500 = 4199 \text{ psi}$$

Therefore, $f'_{cr} = 4400$ psi

From Figure 12-8, the w/cm for air-entrained concrete is 0.55 for an f'_{cr} of 4400 psi. This is greater than the 0.50 permitted for the exposure conditions; therefore, the exposure requirements govern. A w/cm of 0.50 must be used, even though this may produce strengths higher than needed to satisfy structural requirements.

Aggregate Size. Assuming it is economically available, 1½-in. maximum-size aggregate is satisfactory; it is less than ⅕ the wall thickness and less than ¾ the clear distance between reinforcing bars and between reinforcing bars and the form. If this size were not available, the next smaller available size would be used. Aggregates are to be in a saturated surface-dry condition for these trial mixtures.

Air Content. Because of the exposure conditions and to improve workability, a moderate level of entrained air is needed. From Table 12-5, the target air content for concrete with 1½-in. aggregate in a moderate exposure is 4.5%. Therefore, proportion the mixture with an air content range of 4.5% ±1% and aim for 5.5% ±0.5% in the trial batch.

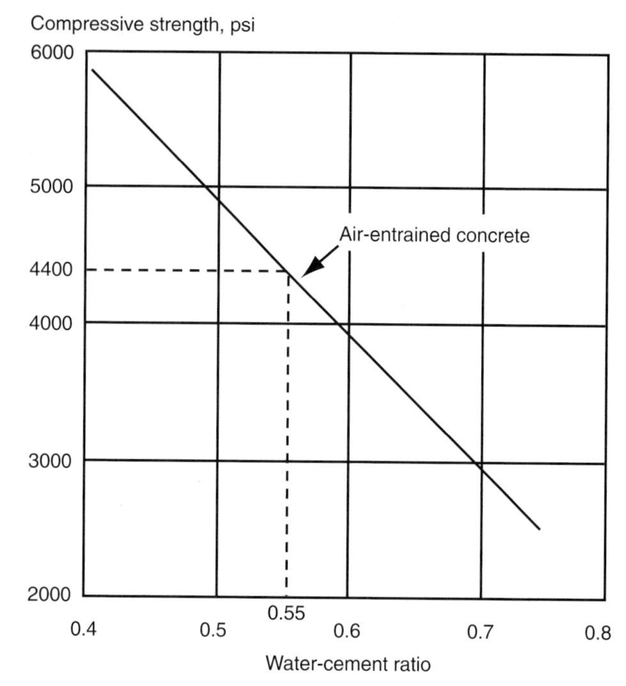

Figure 12-8. Relationship between strength and water to cement ratio based on field and laboratory data for specific concrete ingredients.

Slump. The recommended slump range for placing a reinforced concrete foundation wall is 1 in. to 3 in., assuming that the concrete will be consolidated by vibration (Table 12-6). Batch for 3 in. ±0.75 in.

Batch Quantities. For convenience, a batch containing 20 lb of cement is to be made. The quantity of mixing water required is 20 x 0.50 = 10 lb. Representative samples of fine and coarse aggregates are weighed into suitable containers. The values are entered as initial weights in Column 2 of the trial-batch data sheet (Figure 12-9).

All of the measured quantities of cement, water, and air-entraining admixture are used and added to the mixer. Fine and coarse aggregates, previously brought to a saturated surface-dry condition, are added in proportions similar to those used in mixtures from which Figure 12-8 was developed. Mixing continues until a workable concrete with a 3-in. slump deemed adequate for placement is produced. The relative proportions of fine and coarse aggregate for workability can readily be judged by an experienced concrete technician or engineer.

Workability. Results of tests for slump, air content, unit weight, and a description of the appearance and workability ("Good" for this example) are noted on the data sheet.

The amounts of fine and coarse aggregates not used are recorded on the data sheet in Column 3, and masses of aggregates used (Column 2 minus Column 3) are noted in Column 4. If the slump when tested had been greater than that required, additional fine or coarse aggregates (or both) would have been added to reduce slump. Had the slump been less than required, water and cement in the appropriate ratio (0.50) would have been added to increase slump. It is important that any additional quantities be measured accurately and recorded on the data sheet.

Mixture Proportions. Mixture proportions for a cubic yard of concrete are calculated in Column 5 of Figure 12-9 by using the batch yield (volume) and density (unit weight). For example, the number of pounds of cement per cubic yard is determined by dividing 27 ft^3 (1 yd^3) by the volume of concrete in the batch and multiplying the result by the number of pounds of cement in the batch. The percentage of fine aggregate by weight of total aggregate is also calculated. In this trial batch, the cement content was 539 lb/yd^3 and the fine aggregate made up 33.5% of the total aggregate by weight. The air content and slump were acceptable. The 28-day strength was 4950 psi (greater than f'_{cr}). The mixture in Column 5, along with slump and air content limits of 1 in. to 3 in. and 3.5% to 5.5%, is now ready for submission to the project engineer.

Data and Calculations for Trial Batch
(saturated surface-dry aggregates)

Batch size: 10 lb ______ 20 lb ✓ 40 lb ______ of cement

Note: Complete Columns 1 through 4, fill in items below, the complete 5 and 6.

1 Material	2 Initial mass, lb	3 Final mass, lb	4 Mass. used, (Col. 2 minus Col. 3)	5 Mass per cubic yard (C) x Col. 4	6 Remarks
Cement	20.0	0	20.0	539	
Water	10.0	0	10.0	269	
Fine aggregate	66.2	27.9	38.3	1032 (a)	% fine aggregate = $\frac{a}{a+b}$ x 100
Coarse aggregate	89.8	13.8	76.0	2048 (b)	= 33.5 %
Air-entraining admixture	0.3 oz	Total (T) =	144.3	3888	
		T x C =	144.3 x 26.943	= 3888	Math check

Measured slump: 3 in. Measured air content 5.4 %

Appearance: Sandy ______ Good ✓ Rocky ______

Workability: Good ✓ Fair ______ Poor ______

Weight of container + concrete = 93.4 lb

Weight of container = 21.4 lb

Weight of concrete (A) = 72.0 lb

Volume of container (B) = 0.50 cu ft

Density of concrete (D) = $\frac{A}{B}$ = $\frac{72.0}{0.50}$ = 144.0 lb/cu ft

Yield (volume of concrete produced) = $\frac{\text{Total weight of material per batch}}{\text{Density of concrete}}$ = $\frac{144.3}{144.0}$ = 1.0021 cu ft

Number of 144.3 lb batches per cu yd (C) = $\frac{\text{27 cu ft*}}{\text{Yield}}$ = $\frac{27}{1.0021}$ = 26.943 batches

*One cubic yard has 27 cu ft.

Figure 12-9. Trial mixture data sheet (inch-pound units).

Table 12-14. Example of Results of Laboratory Trial Mixtures (Inch-Pound Units)*

Batch no.	Slump, in.	Air content, percent	Density, lb/ft³	Cement content, lb/yd³	Fine aggregate, percent of total aggregate	Workability
1	3	5.4	144	539	33.5	Good
2	2¾	4.9	144	555	27.4	Harsh
3	2½	5.1	144	549	35.5	Excellent
4	3	4.7	145	540	30.5	Excellent

*Water-cement ratio was 0.50.

Table 12-15 (Metric). Example: 1 m³ Trial Mixtures for Air-Entrained Concrete of Medium Consistency, 75-mm to 100-mm slump

w/cm, kg per kg	Nominal maximum size of aggregate, mm	Air content, percent	Water, kg	Cement, kg	With fine sand, fineness modulus = 2.50			With coarse sand, fineness modulus = 2.90		
					Fine aggregate, percent of total aggregate	Fine aggregate, kg	Coarse aggregate, kg	Fine aggregate, percent of total aggregate	Fine aggregate, kg	Coarse aggregate, kg
0.40	9.5	7.5	202	505	50	744	750	54	809	684
	12.5	7.5	194	485	41	630	904	46	702	833
	19.0	6	178	446	35	577	1071	39	648	1000
	25.0	6	169	424	32	534	1151	36	599	1086
	37.5	5	158	395	29	518	1255	33	589	1184
0.45	9.5	7.5	202	450	51	791	750	56	858	684
	12.5	7.5	194	387	43	678	904	47	750	833
	19.0	6	178	395	37	619	1071	41	690	1000
	25.0	6	169	377	33	576	1151	37	641	1086
	37.5	5	158	351	31	553	1225	35	625	1184
0.50	9.5	7.5	202	406	53	833	750	57	898	684
	12.5	7.5	194	387	44	714	904	49	785	833
	19.0	6	178	357	38	654	1071	42	726	1000
	25.0	6	169	338	34	605	1151	38	670	1086
	37.5	5	158	315	32	583	1225	36	654	1184
0.55	9.5	7.5	202	369	54	862	750	58	928	684
	12.5	7.5	194	351	45	744	904	49	815	833
	19.0	6	178	324	39	678	1071	43	750	1000
	25.0	6	169	309	35	629	1151	39	694	1086
	37.5	5	158	286	33	613	1225	37	684	1184
0.60	9.5	7.5	202	336	54	886	750	58	952	684
	12.5	7.5	194	321	46	768	904	50	839	833
	19.0	6	178	298	40	702	1071	44	773	1000
	25.0	6	169	282	36	653	1151	40	718	1086
	37.5	5	158	262	33	631	1225	37	702	1184
0.65	9.5	7.5	202	312	55	910	750	59	976	684
	12.5	7.5	194	298	47	791	904	51	863	833
	19.0	6	178	274	40	720	1071	44	791	1000
	25.0	6	169	261	37	670	1151	40	736	1086
	37.5	5	158	244	34	649	1225	38	720	1184
0.70	9.5	7.5	202	288	55	928	750	59	994	684
	12.5	7.5	194	277	47	809	904	51	880	833
	19.0	6	178	256	41	738	1071	45	809	1000
	25.0	6	169	240	37	688	1151	41	753	1086
	37.5	5	158	226	34	660	1225	38	732	1184

Mixture Adjustments. To determine the most workable and economical proportions, additional trial batches could be made varying the percentage of fine aggregate. In each batch the water-cement ratio, aggregate gradation, air content, and slump should remain about the same. Results of four such trial batches are summarized in Table 12-14.

Figure 12-10 and Table 12-15 illustrate the change in mix proportions relative to a w/cm for various types of concrete mixtures using a particular aggregate source. Information for concrete mixtures using particular ingredients can be plotted in several ways to illustrate the relationship between ingredients and properties. This is especially useful when optimizing concrete mixtures for best economy or to adjust to specification or material changes.

Table 12-15 (Inch-Pound Units). Example: 1 yd^3 Trial Mixtures for Air-Entrained Concrete of Medium Consistency, 3-in. to 4-in. slump

					With fine sand, fineness modulus = 2.50			With coarse sand, fineness modulus = 2.90		
w/cm lb per lb	Nominal maximum size of aggregate, in.	Air content, percent	Water, lb	Cement, lb	Fine aggregate, percent of total aggregate	Fine aggregate, lb	Coarse aggregate, lb	Fine aggregate, percent of total aggregate	Fine aggregate, lb	Coarse aggregate, lb
0.40	⅜	7.5	340	850	50	1250	1260	54	1360	1150
	½	7.5	325	815	41	1060	1520	46	1180	1400
	¾	6	300	750	35	970	1800	39	1090	1680
	1	6	285	715	32	900	1940	36	1010	1830
	1½	5	265	665	29	870	2110	33	990	1990
0.45	⅜	7.5	340	755	51	1330	1260	56	1440	1150
	½	7.5	325	720	43	1140	1520	47	1260	1400
	¾	6	300	665	37	1040	1800	41	1160	1680
	1	6	285	635	33	970	1940	37	1080	1830
	1½	5	265	590	31	930	2110	35	1050	1990
0.50	⅜	7.5	340	680	53	1400	1260	57	1510	1150
	½	7.5	325	650	44	1200	1520	49	1320	1400
	¾	6	300	600	38	1100	1800	42	1220	1680
	1	6	285	570	34	1020	1940	38	1130	1830
	1½	5	265	530	32	980	2110	36	1100	1990
0.55	⅜	7.5	340	620	54	1450	1260	58	1560	1150
	½	7.5	325	590	45	1250	1520	49	1370	1400
	¾	6	300	545	39	1140	1800	43	1260	1680
	1	6	285	520	35	1060	1940	39	1170	1830
	1½	5	265	480	33	1030	2110	37	1150	1990
0.60	⅜	7.5	340	565	54	1490	1260	58	1600	1150
	½	7.5	325	540	46	1290	1520	50	1410	1400
	¾	6	300	500	40	1180	1800	44	1300	1680
	1	6	285	475	36	1100	1940	40	1210	1830
	1½	5	265	440	33	1060	2110	37	1180	1990
0.65	⅜	7.5	340	525	55	1530	1260	59	1640	1150
	½	7.5	325	500	47	1330	1520	51	1450	1400
	¾	6	300	460	40	1210	1800	44	1330	1680
	1	6	285	440	37	1130	1940	40	1240	1830
	1½	5	265	410	34	1090	2110	38	1210	1990
0.70	⅜	7.5	340	485	55	1560	1260	59	1670	1150
	½	7.5	325	465	47	1360	1520	51	1480	1400
	¾	6	300	430	41	1240	1800	45	1360	1680
	1	6	285	405	37	1160	1940	41	1270	1830
	1½	5	265	380	34	1110	2110	38	1230	1990

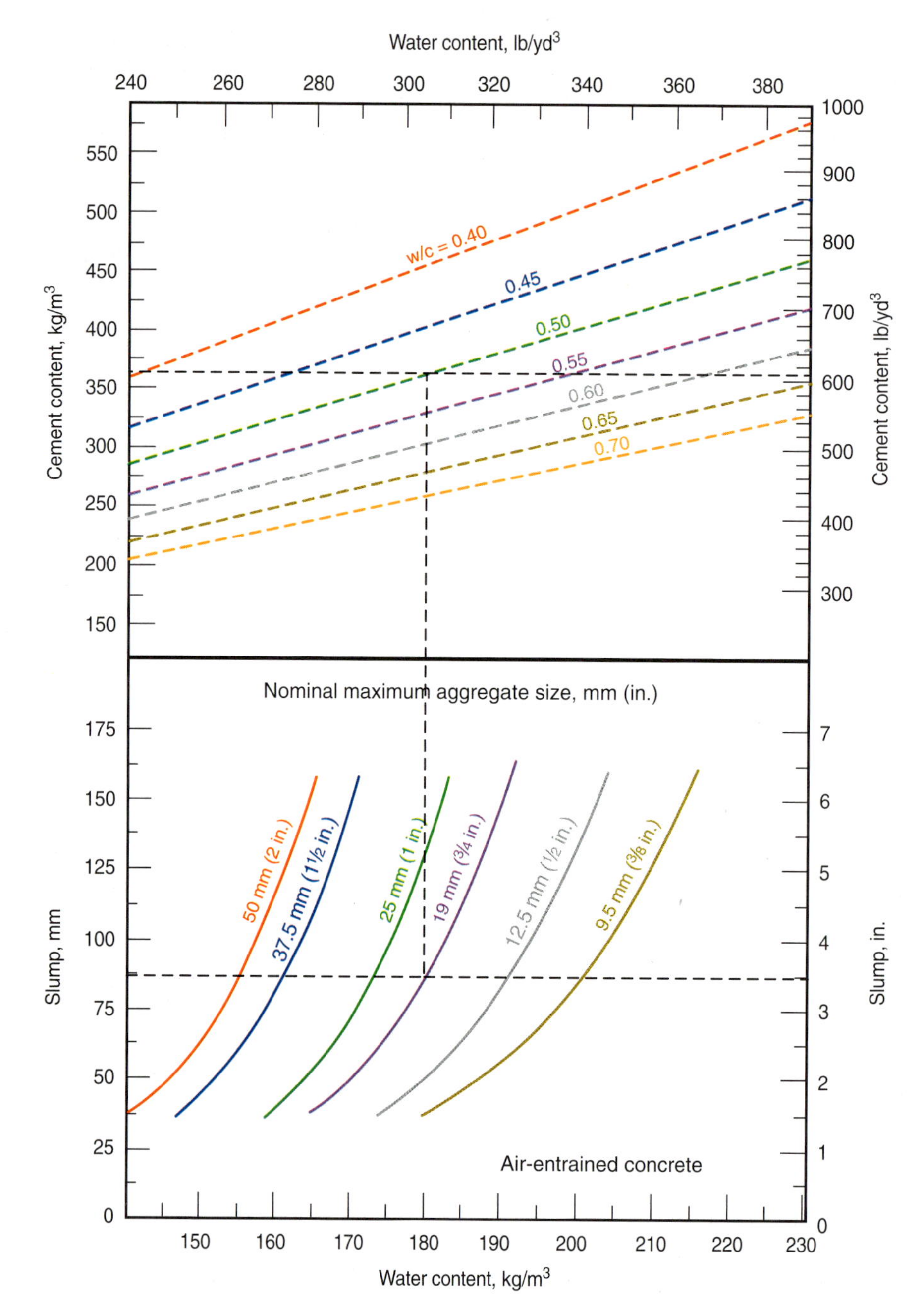

Figure 12-10. Example graphical relationship for a particular aggregate source demonstrating the relationship between slump, aggregate size, water to cement ratio, and cement content (Hover 1995).

Example 5. Absolute Volume Method Using Multiple Cementing Materials and Admixtures (Metric)

The following example illustrates how to develop a mixture using the absolute volume method when more than one cementing material and admixture are used.

Conditions and Specifications. Concrete with a structural design strength of 40 MPa is required for a bridge to be exposed to freezing and thawing, deicers, and very severe sulfate soils. A value not exceeding 1500 coulombs is required to minimize permeability to chlorides. Water reducers, air entrainers, and plasticizers are allowed. A shrinkage reducer is requested to keep shrinkage under 300 millionths. Some structural elements exceed a thickness of 1 meter, requiring control of heat development. The concrete producer has a standard deviation of 2 MPa for similar mixtures to that required here. For difficult placement areas, a slump of 200 mm to 250 mm is required. The following materials are available:

Cement:
Type HS, silica fume modified portland cement, ASTM C1157. Relative density of 3.14. Silica fume content of 5%.

Fly ash:
Class F, ASTM C618, *Standard Specification for Coal Fly Ash and Raw or Calcined Natural Pozzolan for Use in Concrete* (AASHTO M 295). Relative density of 2.60.

Slag:
Grade 120, ASTM C989, *Standard Specification for Slag Cement for Use in Concrete and Mortars* (AASHTO M 302). Relative density of 2.90.

Coarse aggregate:
Well-graded 19-mm nominal maximum-size crushed rock (ASTM C33 or AASHTO M 80) with an ovendry relative density of 2.68, absorption of 0.5%, and ovendry density of 1600 kg/m^3. The laboratory sample has a moisture content of 2.0%. This aggregate has a history of alkali-silica reactivity in the field.

Fine aggregate:
Natural sand with some crushed particles (ASTM C33 or AASHTO M 6) with an ovendry relative density of 2.64 and an absorption of 0.7%. The laboratory sample has a moisture content of 6%. The fineness modulus is 2.80.

Air entrainer:
Synthetic, ASTM C260 (AASHTO M 154).

Retarding water reducer:
Type D, ASTM C494 (AASHTO M 194). Dosage of 3 g per kg of cementing materials.

Plasticizer:
Type 1, ASTM C1017, *Standard Specification for Chemical Admixtures for use in Flowing Concrete*. Dosage of 30 g/kg of cementing materials.

Shrinkage reducer:
Dosage of 15 g/kg of cementing materials.

Strength. For a standard deviation of 2.0 MPa, the f'_{cr} must be the greater of

$$f'_{cr} = f'_c + 1.34S = 40 + 1.34(2) = 42.7$$

or

$$f'_{cr} = 0.9\ f'_c + 2.33S = 36 + 2.33(2) = 40.7$$

therefore $f'_{cr} = 42.7$ MPa

Water-Cementing Materials Ratio. Past field records using these materials indicate that a w/cm of 0.35 is required to provide a strength of 42.7 MPa.

For a deicer environment and to protect embedded steel from corrosion, Table 12-1 requires a maximum water-cementing materials ratio of 0.40 and a strength of at least 35 MPa. For a severe sulfate environment, Table 12-2 requires a maximum water to cementing materials ratio of 0.40 and a strength of at least 35 MPa. Both the water to cementing materials ratio requirements and strength requirements are met and exceeded using the above determined 0.35 water-cementing materials ratio and 40 MPa design strength.

Air Content. For a severe exposure, Figure 12-5 suggests a target air content of 6% for 19-mm aggregate. Therefore, design the mixture for 5% to 8% and use 8% for batch proportions. The trial batch air content must be within ±0.5 percentage points of the maximum allowable air content.

Slump. Assume a slump of 50 mm without the plasticizer and a maximum of 200 mm to 250 mm after the plasticizer is added. Use 250 ±20 mm for proportioning purposes.

Water Content. Figure 12-6 recommends that a 50-mm slump, air-entrained concrete with 19-mm aggregate should have a water content of about 168 kg/m^3. Assume the retarding water reducer and plasticizer will jointly reduce water demand by 15% in this case, resulting in an estimated water demand of 143 kg/m^3, while achieving the 250-mm slump.

Cementing Materials Content. The amount of cementing materials is based on the maximum water-cementing materials ratio and water content. Therefore, 143 kg of water divided by a water-cementing materials ratio of 0.35 requires a cement content of 409 kg. Fly ash and slag cement will be used to help control alkali-silica reactivity and control temperature rise. Local use has shown that a fly ash dosage of 15% and a slag dosage of 30% by mass of cementing materials are adequate. Therefore, the suggested cementing materials for one cubic meter of concrete are as follows:

Cement:	55% of 409	= 225 kg
Fly ash:	15% of 409	= 61 kg
Slag:	30% of 409	= 123 kg

These dosages meet the requirements of Table 12-7 (2.8% silica fume from the cement + 15% fly ash + 30% slag = 47.8% which is less than the 50% maximum allowed).

Coarse-Aggregate Content. The quantity of 19-mm nominal maximum-size coarse aggregate can be estimated from Figure 12-4. The bulk volume of coarse aggregate recommended when using sand with a fineness modulus of 2.80 is 0.62. Since the coarse aggregate has a bulk density of 1600 kg/m^3, the ovendry mass of coarse aggregate for a cubic meter of concrete is

$$1600 \times 0.62 = 992\ kg/m^3$$

Admixture Content. For an 8% air content, the air-entraining admixture manufacturer recommends a dosage of 0.5 g per kg of cementing materials. The amount of air entrainer is then

$$0.5 \times 409 = 205\ g = 0.205\ kg$$

The retarding water reducer dosage rate is 3 g per kg of cementing materials.

This results in 3 x 409 = 1227 g or 1.227 kg of water reducer per cubic meter of concrete.

The plasticizer dosage rate is 30 g per kg of cementing materials.

This results in 30 x 409 = 12,270 g or 12.270 kg of plasticizer per cubic meter of concrete.

The shrinkage reducer dosage rate is 15 g per kg of cementing materials.

This results in 15 x 409 = 6135 g or 6.135 kg of shrinkage reducer per cubic meter of concrete.

Fine-Aggregate Content. At this point, the amounts of all ingredients except the fine aggregate are known. The volume of fine aggregate is determined by subtracting the absolute volumes of all known ingredients from 1 cubic meter. The absolute volumes of the ingredients are calculated by dividing the known mass of each by the product of their relative density and the density of water. Assume a relative density of 1.0 for the chemical admixtures. Assume a density of water of 997.75 kg/m³ as all materials in the laboratory are maintained at a room temperature of 22°C (Table 12-12). Volumetric computations are as follows:

Water (including chemical admixtures) $= \dfrac{143}{1.0 \times 997.75} = 0.143\ m^3$

Cement $= \dfrac{225}{3.14 \times 997.75} = 0.072\ m^3$

Fly ash $= \dfrac{61}{2.60 \times 997.75} = 0.024\ m^3$

Slag $= \dfrac{123}{2.90 \times 997.75} = 0.043\ m^3$

Air $= \dfrac{8.0}{100} = 0.080\ m^3$

Coarse aggregate $= \dfrac{992}{2.68 \times 997.75} = 0.371\ m^3$

Total $= 0.733\ m^3$

The calculated absolute volume of fine aggregate is then
1 – 0.733 = 0.267 m³

The mass of dry fine aggregate is
0.267 x 2.64 x 997.75 = 703 kg

The admixture volumes are

Air entrainer $= \dfrac{0.205}{(1.0 \times 997.75)} = 0.0002\ m^3$

Water reducer $= \dfrac{1.227}{(1.0 \times 997.75)} = 0.0012\ m^3$

Plasticizer $= \dfrac{12.270}{(1.0 \times 997.75)} = 0.0123\ m^3$

Shrinkage reducer $= \dfrac{6.135}{(1.0 \times 997.75)} = 0.0062\ m^3$

Total = 19.84 kg of admixture with a volume of 0.0199 m³

Consider the admixtures part of the mixing water

Mixing water minus admixtures = 143 – 19.84 = 123 kg

The mixture then has the following proportions before trial mixing for 1 cubic meter of concrete:

Water	123 kg
Cement	225 kg
Fly ash	61 kg
Slag	123 kg
Coarse aggregate (dry)	992 kg
Fine aggregate (dry)	703 kg
Air entrainer	0.205 kg
Water reducer	1.227 kg
Plasticizer	12.27 kg
Shrinkage reducer	6.135 kg
Total	= 2247 kg

Slump = 250 mm (±20 mm for trial batch)
Air content = 8% (±0.5% for trial batch)

Estimated concrete density using SSD aggregate (adding absorbed water)

= 123 + 225 + 61 + 123 + (992 x 1.005) + (703 x 1.007) + 20 (admixtures) = 2257 kg/m³

Moisture. The dry batch weights of aggregates have to be increased to compensate for the moisture on and in the aggregates and the mixing water reduced accordingly. The coarse aggregate and fine aggregate have moisture contents of 2% and 6%, respectively. With the moisture contents indicated, the trial batch aggregate proportions become

Coarse aggregate (2% MC) = 992 x 1.02 = 1012 kg

Fine aggregate (6% MC) = 703 x 1.06 = 745 kg

Absorbed water does not become part of the mixing water and must be excluded from the water adjustment. Surface moisture contributed by the coarse aggregate amounts to 2% – 0.5% = 1.5% and that contributed by the fine aggregate, 6% – 0.7% = 5.3%. The estimated added water becomes

123 – (992 x 0.015) – (703 x 0.053) = 71 kg

The batch quantities for one cubic meter of concrete are revised to include aggregate moisture as follows:

Water (to be added)	71 kg
Cement	225 kg
Fly ash	61 kg
Slag	123 kg

Coarse aggregate (2% MC)	1012 kg
Fine aggregate (6% MC)	745 kg
Air entrainer	0.205 kg
Water reducer	1.227 kg
Plasticizer	12.27 kg
Shrinkage reducer	6.14 kg

Trial Batch. The above mixture is tested in a 0.1 m^3 batch in the laboratory (multiply above quantities by 0.1 to obtain batch quantities). The mixture had an air content of 7.8%, a slump of 240 mm, a density of 2257 kg/m^3, a yield of 0.1 m^3, and a compressive strength of 44 MPa. Rapid chloride testing resulted in a value of 990 coulombs (ASTM C1202, *Standard Test Method for Electrical Indication of Concrete's Ability to Resist Chloride Ion Penetration* [AASHTO T 277]). ASTM C1567, *Standard Test Method for Determining the Potential Alkali-Silica Reactivity of Combinations of Cementitious Materials and Aggregate (Accelerated Mortar-Bar Method)*, was used to evaluate the potential of the mix for alkali-silica reactivity, resulting in an acceptable expansion of 0.02%. Temperature rise was acceptable and shrinkage was within specifications. The water-soluble chloride content was 0.06%, meeting the requirements of Table 12-9. The following mix proportions meet all applicable requirements and are ready for submission to the project engineer for approval:

Water	123 kg (143 kg total including admixtures)
Cement, Type HS	225 kg
Fly ash, Class F	61 kg
Slag, Grade 120	123 kg
Coarse aggregate	992 kg (ovendry) or 997 kg (SSD)
Fine aggregate	703 kg (ovendry) or 708 kg (SSD)
Air entrainer*	0.205 kg
Water reducer*	1.227 kg
Plasticizer*	12.27 kg
Shrinkage reducer*	6.14 kg
Slump	200 mm to 250 mm
Air content	5% to 8%
Density (SSD agg.)	2257 kg/m^3
Yield	1 m^3
w/cm	0.35

*Liquid admixture dosages are often provided in liters or milliliters in mix proportion documents.

Concrete for Small Jobs

Although well-established ready mixed concrete mixtures are used for most construction, ready mix is not always practical for small jobs, especially those requiring one cubic meter (yard) or less. Small batches of concrete mixed at the site are required for such jobs.

If mixture proportions or mixture specifications are not available, Tables 12-16 and 12-17 can be used to select proportions for concrete for small jobs. Recommendations with respect to exposure conditions discussed earlier should be followed.

The proportions in Tables 12-16 and 12-17 are only a guide and may need adjustments to obtain a workable mixture with locally available aggregates (PCA 1988). Packaged, combined, dry concrete ingredients (ASTM C387, *Standard Specification for Packaged, Dry, Combined Materials for Mortar and Concrete*) are also available.

Mixture Review

In practice, concrete mixture proportions will be governed by the limits of data available on the properties of materials, the degree of control exercised over the production of concrete at the plant, and the amount of supervision at the jobsite. It should not be expected that field results will be an exact duplicate of laboratory trial batches. An adjustment of the selected trial mixture is usually necessary on the job.

The mixture design and proportioning procedures presented here and summarized in Figure 12-11 are applicable to normal-weight concrete. For concrete requiring some special property, using special admixtures or materials—lightweight aggregates, for example—different proportioning principles may be involved.

Internet web sites and computer models also provide assistance with designing and proportioning concrete mixtures (Bentz 2001 and COMPASS 2004). Use caution when using computer models to design concrete mixtures, as the locally available materials may not perform exactly as modeled by the program. Therefore, trial batching and field verification are still the best method to determine the performance of a concrete mixture.

Table 12-16 (Metric). Proportions by Mass to Make One Tenth Cubic Meter of Concrete for Small Jobs

Nominal maximum size coarse aggregate, mm	Air-entrained concrete				Non-air-entrained concrete			
	Cement, kg	Wet fine aggregate, kg	Wet coarse aggregate, kg*	Water, kg	Cement, kg	Wet fine aggregate, kg	Wet coarse aggregate, kg	Water, kg
9.5	46	85	74	16	46	94	74	18
12.5	43	74	88	16	43	85	88	18
19.0	40	67	104	16	40	75	104	16
25.0	38	62	112	15	38	72	112	15
37.5	37	61	120	14	37	69	120	14

*If crushed stone is used, decrease coarse aggregate by 5 kg and increase fine aggregate by 5 kg.

Table 12-16 (Inch-Pound). Proportions by Mass to Make One Cubic Foot of Concrete for Small Jobs

Nominal maximum size coarse aggregate, in.	Air-entrained concrete				Non-air-entrained concrete			
	Cement, lb	Wet fine aggregate, lb	Wet coarse aggregate, lb*	Water, lb	Cement, lb	Wet fine aggregate, lb	Wet coarse aggregate, lb	Water, lb
3/8	29	53	46	10	29	59	46	11
1/2	27	46	55	10	27	53	55	11
3/4	25	42	65	10	25	47	65	10
1	24	39	70	9	24	45	70	10
1 1/2	23	38	75	9	23	43	75	9

*If crushed stone is used, decrease coarse aggregate by 3 lb and increase fine aggregate by 3 lb.

Table 12-17. Proportions by Bulk Volume* of Concrete for Small Jobs

Nominal maximum size coarse aggregate, mm (in.)	Air-entrained concrete				Non-air-entrained concrete			
	Cement	Wet fine aggregate	Wet coarse aggregate	Water	Cement	Wet fine aggregate	Wet coarse aggregate	Water
9.5 (3/8)	1	2 1/4	1 1/2	1/2	1	2 1/2	1 1/2	1/2
12.5 (1/2)	1	2 1/4	2	1/2	1	2 1/2	2	1/2
19.0 (3/4)	1	2 1/4	2 1/2	1/2	1	2 1/2	2 1/2	1/2
25.0 (1)	1	2 1/4	2 3/4	1/2	1	2 1/2	2 3/4	1/2
37.5 (1 1/2)	1	2 1/4	3	1/2	1	2 1/2	3	1/2

*The combined volume is approximately 2/3 of the sum of the original bulk volumes.

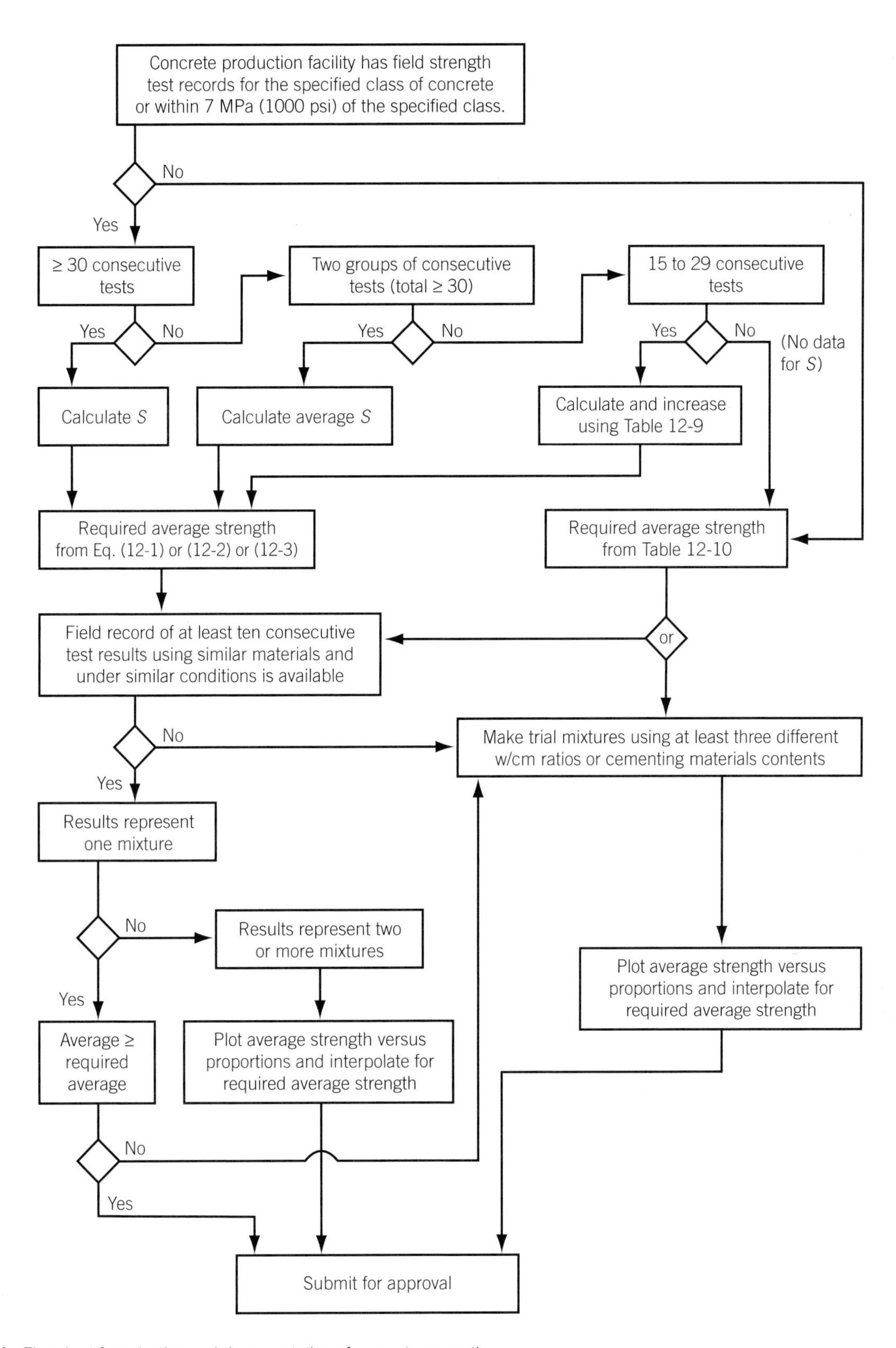

Figure 12-11. Flowchart for selection and documentation of concrete proportions.

References

Abrams, D.A., *Design of Concrete Mixtures, Lewis Institute, Structural Materials Research Laboratory*, Bulletin No. 1, PCA LS001, Chicago, http://www.cement.org/pdf_files/LS001.pdf, 1918, 20 pages.

ACI Committee 201, *Guide to Durable Concrete*, ACI 201.2R-08, American Concrete Institute, Farmington Hills, Michigan, 2008, 49 pages.

ACI Committee 211, *Standard Practice for Selecting Proportions for Normal, Heavyweight and Mass Concrete*, ACI 211.1-91, reapproved 2009, American Concrete Institute, Farmington Hills, Michigan, 1991, 38 pages.

ACI Committee 211, *Guide for Selecting Proportions for High-Strength Concrete Using Portland Cement and Other Cementitious Material*, ACI 211.4R-08, American Concrete Institute, Farmington Hills, Michigan, 2008, 25 pages.

ACI Committee 211, *Guide for Submittal of Concrete Proportions*, ACI 211.5R-01, reapproved 2009, American Concrete Institute, Farmington Hills, Michigan, 2001, 12 pages.

ACI Committee 211, *Guide for Selecting Proportions for No-Slump Concrete*, ACI 211.3R-02, reapproved 2009, American Concrete Institute, Farmington Hills, Michigan, 2002, 26 pages.

ACI Committee 211, *Standard Practice for Selecting Proportions for Structural Lightweight Concrete*, ACI 211.2- 98, reapproved 2004, American Concrete Institute, Farmington Hills, Michigan, 1998, 20 pages.

ACI Committee 214, *Evaluation of Strength Test Results of Concrete*, ACI 214R-02, American Concrete Institute, Farmington Hills, Michigan, 2002, 20 pages.

ACI Committee 222, *Provisional Standard Test Method for Water Soluble Chloride Available for Corrosion of Embedded Steel in Mortar and Concrete Using the Soxhlet Extractor*, ACI 222.1-96, withdrawn, American Concrete Institute, Farmington Hills, Michigan, 1996, 3 pages.

ACI Committee 302, *Guide for Concrete Floor and Slab Construction*, ACI 302.1R-04, American Concrete Institute, Farmington Hills, Michigan, 2004, 77 pages.

ACI Committee 318, *Building Code Requirements for Structural Concrete and Commentary*, ACI 318-08/318R-08, American Concrete Institute, Farmington Hills, Michigan, 2008, 467 pages.

ACI Committee 357, *Guide for the Design and Construction of Fixed Offshore Concrete Structures*, ACI 357R-84, reapproved 1997, American Concrete Institute, Farmington Hills, Michigan, 1984, 23 pages.

Bentz, Dale, *Concrete Optimization Software Tool*, http://ciks.cbt.nist.gov/bentz/fhwa, National Institute of Standards and Technology, 2001.

Bureau of Reclamation, *Concrete Manual*, 8th ed., Denver, revised 1981.

COMPASS, The Transtec Group, http://www.pccmix.com/, 2004.

Hover, Ken, "Graphical Approach to Mixture Proportioning byACI 211.1-91," *Concrete International*, American Concrete Institute, Farmington Hills, Michigan, September 1995, pages 49 to 53.

Hover, Kenneth C., "Concrete Design: Part 1, Finding Your Perfect Mix,", *CE News*, September 1998, and "Concrete Design: Part 2, Proportioning Water, Cement, and Air," *CE News*, October 1998, and "Concrete Design: Part 3, Proportioning Aggregate to Finish the Process," *CE News*, November 1998.

Kosmatka, Steven H., "Compressive versus Flexural Strength for Quality Control of Pavements," *Concrete Technology Today*, PL854, Portland Cement Association, http://www.cement.org/pdf_files/PL854.pdf, 1985, pages 4 and 5.

PCA, *Concrete for Small Jobs*, IS174, Portland Cement Association, http://www.cement.org/pdf_files/IS174.pdf, 1988.

Shilstone, James M., Sr., "Concrete Mixture Optimization," *Concrete International*, American Concrete Institute, Farmington Hills, Michigan, June 1990, pages 33 to 39.

U.S. Department of Transportation (USDOT), Federal Highway Administration, *Integrated Materials and Construction Practices for Concrete Pavement: A State-of-the-Practice Manual*, HIF-07-004, Washington DC, 2006.

CHAPTER 13

Batching, Mixing, Transporting, and Handling Concrete

The production and delivery of concrete are achieved in different ways. The basic processes and common techniques are explained in this chapter. ASTM C94/C94M, *Standard Specification for Ready-Mixed Concrete*, provides standard specifications for the manufacture and delivery of freshly mixed concrete. Guidelines to improve concrete specifications are in NRMCA (2009), Daniel and Lobo (2005) and Lamond and Pielert (2006). Standards of the *Concrete Plant Manufacturers Bureau, Truck Mixer Manufacturers Bureau*, and *Volumetric Mixer Manufacturers Bureau* can be found on the National Ready Mixed Concrete Association's (NRMCA) website at http://www.nrmca.org.

Ordering Concrete

ASTM C94 provides three options for ordering concrete:

1. Option A is a performance based order. The purchaser designates the compressive strength of the concrete, while the concrete producer selects the mixture proportions needed to obtain the required compressive strength.
2. Option B is a prescription based order. The purchaser selects mixture proportions, including cement, water, and admixture contents.
3. Option C is a combined option. The purchaser designates the compressive strength and minimum cement content. The concrete producer selects the mixture proportions to comply with the requirements of the purchaser.

For all options, the purchaser should include requirements for slump and nominal maximum size of coarse aggregate.

Prescription Versus Performance Based Specifications

Traditionally, the owner/agency develops the design requirements and establishes prescriptive provisions for the proportions of the concrete. These may include minimum cement content, and performance requirements for the concrete, such as the air content and strength. Contractors are then directed by the specifications to order the concrete mixture from the producer in accordance with the prescriptive provisions provided by the specifications. This procedure allows the owner to control the concrete mixture design and proportions, while any risk associated with the performance of the mixture resides with the owner as well. This method may lead to confusion over responsibility and authority if a problem with the concrete performance arises.

With the emergence of value engineering, design-build, performance specifications, and warranties, more owners are requiring the producer to develop the concrete mixture proportions based on performance criteria. This affords more flexibility and encourages innovation, but it also transfers more responsibility onto the concrete producer.

Prescriptive specifications are not a guarantee of performance. Significant reductions in project cost may be realized by allowing concrete suppliers to optimize mixtures for performance properties (Obla and Lobo 2006). See Detwiler and Taylor (2005), Calderone and others (2005), Hover, Bickley, and Hooton (2008), and ACI (2010) for guidelines on specifying for concrete performance.

As trends continue to move away from prescriptive specifications toward performance specifications and warranties, concrete producers will become increasingly responsible for concrete mixture design and proportioning within the scope of their projects (see Chapter 12).

Batching

Batching is the process of measuring quantities of concrete mixture ingredients by either mass or volume and introducing them into the mixer. To produce concrete of uniform quality, the ingredients must be measured accurately for each batch. Most specifications require that batching be done by mass rather than by volume (ASTM C94 and AASHTO M 157). Water and liquid admixtures can be measured either by volume or mass. ASTM C685, *Standard Specification for Concrete Made by Volumetric Batching and Continuous Mixing* and AASHTO M 241 cover volumetric batching in equipment designed for continuous mixing.

Specifications generally require that materials be measured for individual batches within the following percentages of accuracy: cementitious material ±1% of each intermediate weighing in cumulative weigh batchers, aggregates ±2% of each intermediate weighing in cumulative weigh batchers (note that individual scales for aggregates are rare but they require 1%), batched water to ±1% of the total mixing water, and admixtures ±3% the desired quantity.

Scales and volumetric devices for measuring quantities of concrete ingredient materials should be accurate. The accuracy of scales and batching equipment should be checked periodically and adjusted when necessary (Figure 13-1). Scales are checked for accuracy by using a combination of certified test weights and product substitute loading. ASTM C94 requires that scales should be accurate to the larger of ±0.15% of the scale capacity of ±0.4% of the applied load in all quarters of the scale capacity.

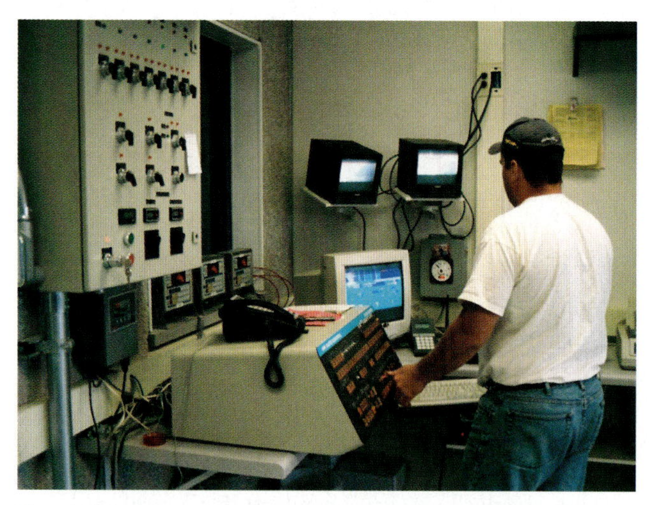

Figure 13-1. Batch control room in a typical ready mixed concrete plant.

Water is typically measured through water meters, in volumetric tanks, or in scales that measure the mass.

Chemical admixtures are typically charged into the mixture as aqueous solutions. Admixtures that cannot be added in solution can be either batched by mass or volume (generally in bag quantities) as directed by the manufacturer. Admixture dispensers should be checked frequently since errors in dispensing admixtures, particularly overdoses, can lead to problems in both fresh and hardened concrete.

Mixing Concrete

All concrete should be mixed thoroughly until its ingredients are uniformly distributed. Mixers should not be loaded above their rated mixing capacities and should be operated at the mixing speed and for the period, either based on revolutions or time, recommended by the manufacturer. The rated mixing capacity of revolving drum truck mixers is limited to 63% of the gross volume of the mixer. For stationary plant mixers, the mixing capacity varies depending on the design. Increased output should be obtained by using a larger mixer or additional mixers, rather than by speeding up or overloading the equipment on hand. If the blades of a mixer become worn or coated with hardened concrete, mixing action will be less efficient. These conditions should be corrected.

If concrete has been adequately mixed, samples taken from different portions of a batch should have essentially the same strength, density, air content, slump, and coarse-aggregate content, with some allowance for testing variability. Maximum allowable differences to evaluate mixing uniformity within a batch of freshly mixed concrete are given in ASTM C94 (AASHTO M 157).

Structural lightweight concrete can be mixed the same way as normal-density concrete when the aggregates have less than 10% total absorption by mass or when the absorption is less than 2% by mass during the first hour after immersion in water. For aggregates not meeting these limits, mixing procedures are described by Bohan and Ries (2008).

Stationary Mixing

Concrete may be mixed at the jobsite in a stationary mixer or a paving mixer (Figure 13-2). Stationary mixers include both onsite mixers and central mixers in ready mix plants. They are available in sizes up to 9.0 m^3 (12 yd^3) and can be of the revolving drum tilting or nontilting type, reversing drum, or the horizontal shaft revolving blade or paddle type. All types may be equipped with loading skips and some are equipped with a swinging discharge chute. Stationary mixers have timing devices, some of which can be set for a given mixing time and locked so that the batch cannot be discharged until the designated mixing time has elapsed.

Careful attention should be paid to the required mixing time. Many specifications require a minimum mixing time of one minute plus 15 seconds for every cubic meter (yard), unless mixer performance tests demonstrate that shorter periods are acceptable and will provide a uniform concrete mixture. The standards of the Concrete Plant Mixers

Figure 13-2. Concrete can be mixed at the jobsite in a stationary mixer (Courtesy of Baker Concrete).

Manufacturer's Bureau, CPMB 100, qualify mixers for a shorter mixing time between 45 seconds and 90 seconds. Short mixing times, less than what the mixer is designated for, can result in non-homogenous mixtures, poor distribution of air voids (resulting in poor freeze-thaw resistance), poor strength gain, and early stiffening problems. The mixing period should be measured from the time all cement and aggregates are in the mixer drum, provided all the water is added before one-fourth of the mixing time has elapsed (ACI 304R).

Under usual conditions, most of the mixing water should be charged in the drum before the solid materials are added. Coarse aggregate should be charged initially to avoid head packs, or cement balls. Water then should be added uniformly with the solid materials, leaving about 10% to be added after all other materials are in the drum. When heated water is used in cold weather, this order of charging may require some modification to prevent possible rapid stiffening when hot water contacts the cement. In this case, addition of the cementitious materials should be delayed until most of the aggregate and water have intermingled in the drum. If supplementary cementing materials are used, they should be added with the cement.

If retarding or water-reducing admixtures are used, they should be added in the same sequence in the charging cycle each time. If not, significant variations in the time of initial setting and percentage of entrained air may result. Addition of the admixture should be completed not later than one minute after addition of water to the cement has been completed or prior to the start of the last three-fourths of the mixing cycle, whichever occurs first. If two or more admixtures are used in the same batch of concrete, they should be added separately to avoid any interaction that might interfere with the efficiency of the admixtures and adversely affect the concrete properties. The sequence in which they are added to the mixture is also important.

Ready Mixed Concrete

Ready mixed concrete is batched and mixed at a concrete plant and delivered to the project in a freshly mixed and unhardened state. Figure 13-3 illustrates a central mix ready mix plant.Ready mixed concrete can be manufactured by any of the following methods:

1. Central-mixed concrete is mixed completely in a stationary mixer and is delivered either in a truck mixer operating at agitating speed (Figure 13-4), a truck agitator (Figure 13-5 top), or a nonagitating truck (Figure 13-5 bottom).
2. Shrink-mixed concrete is mixed partially in a stationary mixer and completed in a truck mixer.
3. Truck-mixed concrete is mixed completely in a truck mixer (Figure 13-6).

ASTM C94 (AASHTO M 157) notes that when a truck mixer is used for complete mixing, 70 to 100 revolutions of the drum or blades at mixing speed are usually required to produce the uniformly mixed concrete. The homogeneity of concrete is maintained after mixing and during delivery by turning the drum at agitating speed. Agitating speed is usually about 2 rpm to 6 rpm, and mixing speed is generally about 12 rpm to 18 rpm. Mixing at high speeds for long periods (one or more hours), along with the addition of water to maintain slump, can result in concrete strength loss, temperature rise, excessive loss of entrained air, and accelerated slump loss.

When truck mixers are used, ASTM C94 (AASHTO M 157) also limits the time between batching and complete discharge of the concrete at the job site to 1½ hours. ASTM C94 (AASHTO M 157) also limits the number of drum revolutions to 300 times after introduction of water to the cement and aggregates, or the cement to the aggregates. With the use of specialized concrete mixtures, the limit on time and number of revolutions may be exceeded. Mixers and agitators should always be operated within the limits for volume and speed of rotation designated by the equipment manufacturer.

Mobile Batcher Mixed Concrete (Continuous Mixer)

Mobile volumetric mixers (Figure 13-7) batch concrete by volume and continuously mix concrete as the dry ingredients, water, and admixtures are fed into a mixing trough that is typically an auger system. The concrete must

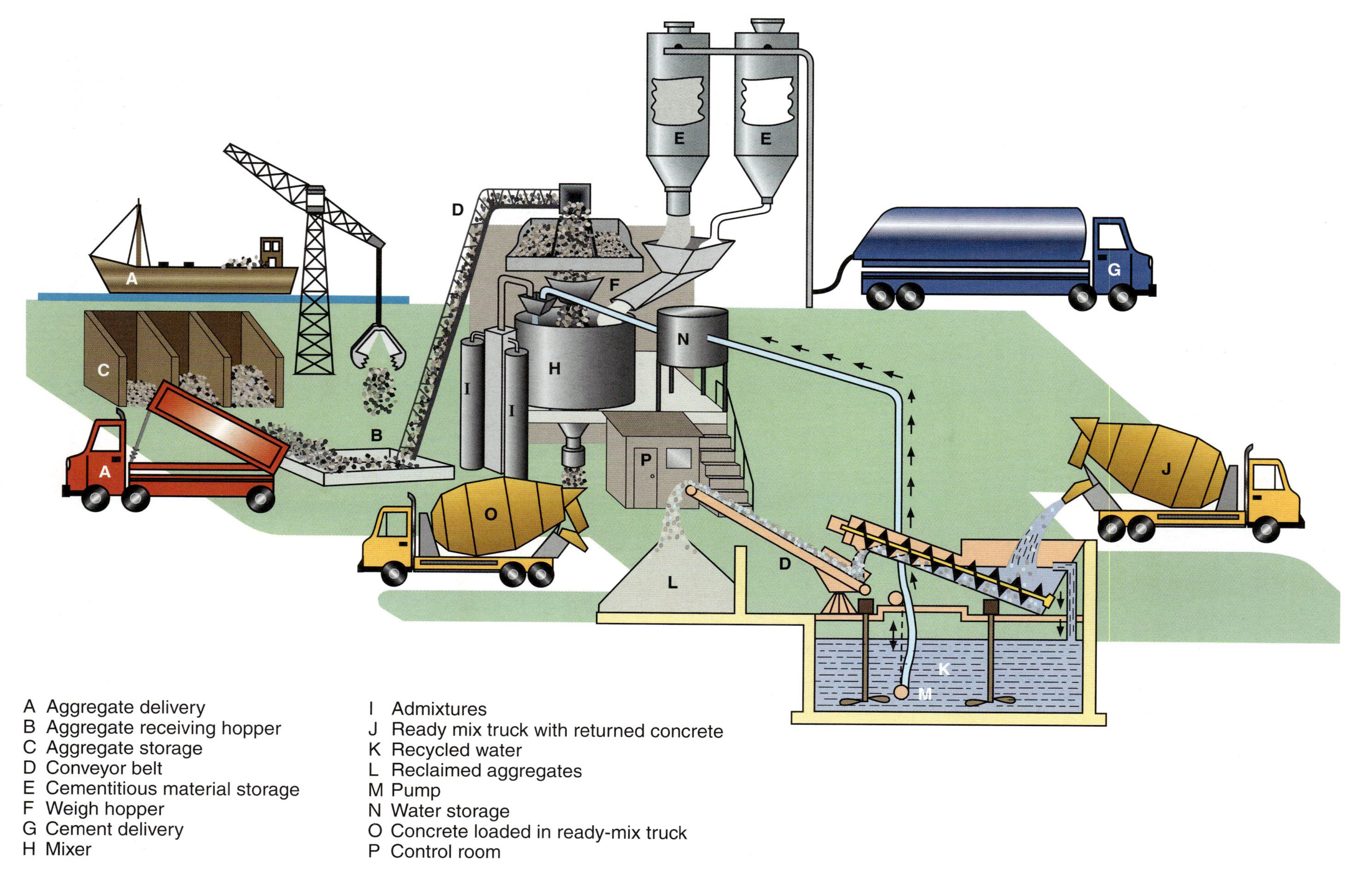

Figure 13-3. Schematic of a ready mix plant.

Figure 13-4. Central mixing in a stationary mixer of the tilting drum type with delivery by a truck mixer operating at agitating speed.

Figure 13-5. (top) Truck agitators are also used with central-mix batch plants. Agitation mixing capabilities allow truck agitators to supply concrete to projects with slow rates of concrete placement and at distances greater than nonagitating trucks. (bottom) Nonagitating trucks are used with central-mix batch plants where short hauls and quick concrete discharge (by conveyor, in this example) allows the rapid placement of large volumes of concrete (Courtesy of Gomaco).

Figure 13-6. Truck-mixed concrete is mixed completely in a truck mixer.

Figure 13-7. Mobile batcher measures materials by volume and continuously mixes concrete as the dry ingredients, water, and admixtures are fed into a mixing trough at the rear of the vehicle.

conform to ASTM C685 (AASHTO M 241). The concrete mixture is easily adjusted for quantities needed during project placement and varying weather conditions. These mixers are typically used to produce smaller quantities of concrete or to produce concrete that are avoided in truck mixers – such as rapid setting or latex modified concrete.

Retempering (Remixing) Concrete

Fresh concrete that is left to agitate in the mixer drum tends to stiffen before initial set develops. Such concrete may be used if upon remixing it becomes sufficiently plastic to be placed and compacted in the forms. ASTM C94 (AASHTO M 157) allows water to be added to the concrete when the truck arrives on the jobsite and the slump is less than specified providing the following conditions are met: (1) maximum allowable water-cement ratio is not exceeded; (2) maximum allowable slump is not exceeded; (3) maximum allowable mixing and agitating time (or drum revolutions) are not exceeded; and (4) concrete is remixed for a minimum of 30 revolutions at mixing speed or until the uniformity of the concrete is within the limits described in ASTM C94 (AASHTO M 157). Water should not be added to a partial load. Indiscriminate addition of water to make concrete more fluid should not be allowed because this increases the w/cm and lowers the quality of concrete. The later addition of water and remixing to retemper the mixture can result in marked strength reduction.

If early setting becomes a persistent problem, a retarder may be used to control early hydration. Mixture adjustments are permitted at the jobsite. These adjustments are for air entrainment and the addition of other admixtures. For example, using a water reducing admixture to increase slump, followed by sufficient mixing.

Transporting and Handling Concrete

Good advanced planning can help select the appropriate transporting method for concrete. Delays, early stiffening and drying, and segregation can all seriously affect the quality of the finished work and must be taken into consideration during transporting and handling of concrete.

Delays. The objective in planning any work schedule is to produce the fastest work using the best labor force and the proper equipment for the work at hand. Machines for transporting and handling concrete are being improved all the time. The greatest productivity will be achieved if the work is planned to optimize the productivity of personnel and equipment. Additionally, the equipment should be selected to reduce the delay time during concrete placement.

Early Stiffening and Drying Out. Concrete begins to stiffen as soon as the cementitious materials and water are mixed. The degree of stiffening that occurs in the first 30 minutes is not usually a problem. Concrete that is kept agitated generally can be placed and compacted within 90 minutes after mixing unless hot concrete temperatures or high cement contents speed up hydration excessively. Admixture technology can considerably extend the delivery time to discharge if the project requires it. Proper planning should eliminate or minimize any variables that would allow the concrete to stiffen to the extent that full consolidation is not achieved and finishing becomes difficult. Less time is available during conditions that hasten the stiffening process, such as hot and dry weather, use of accelerators, and use of heated concrete. In turn, low concrete temperatures and low ambient temperatures may extend setting time well beyond the 90 minute time limit.

Segregation. Segregation is the tendency for coarse aggregate to separate from the sand-cement mortar. This results in part of the concrete having too little coarse aggregate and the remainder having too much. The former is likely to shrink more, crack, and have poor resistance to abrasion. The latter may be too harsh for full consolidation and finishing and is a frequent cause of honeycombing. The method and equipment used to transport and handle the concrete must not result in segregation of the concrete materials.

Methods and Equipment for Transporting and Handling Concrete

Table 13-1 summarizes the most common methods and equipment for moving concrete to the point where it is needed.

Table 13-1. Methods and Equipment for Transporting and Handling Concrete

Equipment	Type and range of work for which equipment is best suited	Advantages	Points to watch for
Belt conveyors	For conveying concrete horizontally or to a higher or lower level. Usually positioned between main discharge point and secondary discharge point.	Belt conveyors have adjustable reach, traveling diverter, and variable speed both forward and reverse. Can place large volumes of concrete quickly when access is limited.	End-discharge arrangements needed to prevent segregation and leave no mortar on return belt. In adverse weather (hot, windy) long reaches of belt need cover.
Belt conveyors mounted on truck mixers	For conveying concrete to a lower, horizontal, or higher level.	Conveying equipment arrives with the concrete. Adjustable reach and variable speed.	End-discharge arrangements needed to prevent segregation and leave no mortar on return belt.
Buckets	Used with cranes, cableways, and helicopters for construction of buildings and dams. Convey concrete directly from central discharge point to formwork or to secondary discharge point.	Enables full versatility of cranes, cableways, and helicopters to be exploited. Clean discharge. Wide range of capacities.	Select bucket capacity to conform to size of the concrete batch and capacity of placing equipment. Discharge should be controllable.
Chutes on truck mixers	For conveying concrete to a lower level, usually below ground level, on all types of construction.	Low cost and easy to maneuver. No power required; gravity does most of the work.	Slopes should range between 1 to 2 and 1 to 3 and chutes must be adequately supported in all positions. End-discharge arrangements (downpipe) needed to prevent segregation.

Table 13-1. Methods and Equipment for Transporting and Handling Concrete (Continue)

Equipment	Type and range of work for which equipment is best suited	Advantages	Points to watch for
Cranes and buckets	Useful for work above ground level.	Can handle concrete, reinforcing steel, formwork, and sundry items in bridges and concrete framed buildings.	Limited to one left at a time. Careful scheduling between trades and operations is needed to keep crane busy.
Dropchutes	Used for placing concrete in vertical forms of all kinds. Some chutes are one piece tubes made of flexible rubberized canvas or plastic, others are assembled from articulated metal cylinders (elephant trunks).	Dropchutes direct concrete into formwork and carry it to bottom of forms without segregation. Their use avoids spillage of grout and concrete on reinforcing steel and form sides, which is harmful when off-the-form surfaces are specified. They also will prevent segregation of coarse particles.	Dropchutes should have sufficiently large, splayed-top openings into which concrete can be discharged without spillage. The cross section of a dropchute should be chosen to permit inserting into the formwork without interfering with reinforcing steel.
Mobile batcher mixers	Used for intermittent production of concrete at jobsite, or where only small quantities are required.	A combined materials transporter and mobile batching and mixing system for quick, precise proportioning of specified concrete. One-person operation.	Trouble-free operation requires good preventive maintenance program on equipment. Materials must be identical to those in original mix design.
Nonagitating trucks	Used to transport concrete on short hauls over smooth roadways.	Capital cost of nonagitating equipment is lower than that of truck agitators or mixers.	Concrete slump should be limited. Possibility of segregation. Clearance is needed for high lift of truck body upon discharge.
Pneumatic guns (shotcrete)	Used where concrete is to be placed in difficult locations and where thin sections and large areas are needed.	Ideal for placing concrete in freeform shapes, for repairing structures, for protective coatings, thin linings, and building walls with one-sided forms.	Quality of work depends on skill of those using equipment. Only experienced nozzlemen should be employed.
Pumps	Used to convey concrete directly from central discharge point at jobsite to formwork or to secondary discharge point.	Pipelines take up little space and can be readily extended. Delivers concrete in continuous stream. Pump can move concrete both vertically and horizontally. Truck-mounted pumps can be delivered when necessary to small or large projects. Tower-crane mounted pump booms provide continuous concrete for tall building construction.	Constant supply of freshly-mixed concrete is needed with average consistency and without any tendency to segregate. Care must be taken in operating pipeline to ensure an even flow and to clean out at conclusion of each operation. Pumping vertically, around bends, and through flexible hose will considerably reduce the maximum pumping distance.
Screw spreaders	Used for spreading concrete over large flat areas, such as in pavements and bridge decks.	With a screw spreader a batch of concrete discharged from a bucket or truck can be quickly spread over a wide area to a uniform depth. The spread concrete has good uniformity of compaction before vibration is used for final compaction.	Screw spreaders are normally used as part of paving train. They should be used for spreading before vibration is applied.
Tremies	For placing concrete through congested rebar, in slurry applications, and under water.	Can be used to funnel concrete through slurry or water into a foundation.	Precautions are needed to ensure that the tremie discharge end is always buried in fresh concrete, so that the seal is preserved between water and concrete mass. Diameter should be 250 to 300 mm (10 to 12 in.) unless pressure is available. Concrete mixture needs more cement, 390 kg/m^3 (658 lb/yd^3), and greater slump, 150 to 230 mm (6 to 9 in.), because concrete must flow and consolidate without any vibration.
Truck agitators	Used to transport concrete for all uses in pavements, structures, and buildings. Haul distances must allow discharge of concrete within 1½ hours, but limit may be waived under certain circumstances.	Truck agitators usually operate from central mixing plants where quality concrete is produced under controlled conditions. Discharge from agitators is well controlled. There is uniformity and homogeneity of concrete on discharge.	Timing of deliveries should suit job organization. Concrete crew and equipment must be ready onsite to handle concrete.
Truck mixers	Used to transport concrete for uses in pavements, structures, and buildings. Haul distances must allow discharge of concrete within 1½ hours, but limit may be waived under certain circumstances.	No central mixing plant needed, only a batching plant, since concrete is completely mixed in truck mixer. Discharge is same as for truck agitator.	Timing of deliveries should suit job organization. Concrete crew and equipment must be ready onsite to handle concrete. Control of concrete quality is not as good as with central mixing.
Wheelbarrows and buggies	For short flat hauls on all types of onsite concrete construction, especially where accessibility to work area is restricted.	Very versatile and therefore ideal inside and on jobsites where placing conditions are constantly changing.	Slow and labor intensive.

There have been few major changes in the principles of conveying concrete over the last 75 years. What has changed is the technology that led to development of more efficient machinery. The wheelbarrow and buggy, although still used, have evolved to become the power buggy (Figure 13-8); the bucket hauled over a pulley wheel evolved into the bucket and crane (Figure 13-9); and the horse-drawn wagon is now the ready mixed concrete truck (Figures 13-10 and 13-11).

Figure 13-8. Versatile power buggy can move all types of concrete over short distances.

Figure 13-9. Concrete is easily lifted to its final location by bucket and crane.

Figure 13-10. Ready mixed concrete is often placed in its final location by direct chute discharge from a truck mixer.

Figure 13-11. In comparison to conventional rear-discharge trucks, front-discharge truck mixers provide the driver with more mobility and control for direct discharge into place.

Figure 13-12. The tower crane and bucket can easily handle concrete for tall-building construction (Courtesy of Baker Concrete).

As concrete-framed buildings became taller, the need to hoist reinforcement and formwork as well as concrete to higher levels led to the development of the tower crane – a familiar sight on the building skyline today (Figure 13-12). It is fast and versatile, but its capacity is limited to one lifting point.

The first mechanical concrete pump was developed and used in the 1930s and the hydraulic pump was developed in the 1950s. The advanced mobile pump with hydraulic placing boom (Figure 13-13) is probably the single most important innovation in concrete handling equipment. It is economical to use in placing both large and small quantities of concrete, depending on jobsite conditions. For small to medium size projects, a combination of truck mixer and boom pump can be used to transport and place concrete.

Figure 13-13. (top) A truck-mounted pump and boom can conveniently move concrete vertically or horizontally to the desired location. (bottom) View of concrete discharging from flexible hose connected to rigid pipeline leading from the pump. Rigid pipe is used in pump booms and in pipelines to move concrete over relatively long distances. Up to 8 m (25 ft) of flexible hose may be attached to the end of a rigid line to increase placement mobility (Bottom photo Courtesy of Baker Concrete).

The conveyor belt is an efficient, portable method of handling and transporting concrete. A dropchute prevents concrete from segregating as it leaves the belt; a scraper prevents loss of mortar. Conveyor belts can be operated in series and on extendable booms of hydraulic cranes (Figure 13-14). Truck-mixer-mounted conveyor belts are also available (Figure 13-15).

Figure 13-14. The conveyor belt is an efficient, portable method of transporting concrete.

Figure 13-15. A conveyor belt mounted on a truck mixer places concrete up to about 12 meters (40 feet) without the need for additional transporting equipment.

The screw spreader (Figure 13-16) has been very effective in placing and distributing concrete for pavements. Screw spreaders can place a uniform depth of concrete quickly and efficiently.

Shotcrete is concrete that is pneumatically projected onto a surface at high velocity (Figure 13-17). It may also be known as "gunite" and "sprayed concrete," Shotcrete is used for both new construction and repair work. It is especially suited for curved or thin concrete structures and shallow repairs (see Chapter 20).

See ACI Committee 304 and Panarese (1987) for extensive information on methods to transport concrete.

Figure 13-16. The screw spreader quickly spreads concrete over a wide area to a uniform depth. Screw spreaders are used primarily in pavement construction.

Figure 13-17. Shotcrete is pneumatically applied concrete.

Choosing the Best Method of Concrete Placement

When choosing the best method for concrete placement, the initial consideration is the type of job, its physical size, the total amount of concrete to be placed, and the placement schedule. Further consideration will identify the amount of work that is below, at, or above ground level. This aids in selecting the concrete transporting equipment necessary for placing concrete at the required levels.

Concrete must be moved from the mixer to the point of placement as rapidly as possible without segregation or loss of ingredients. The transporting and handling equipment must have the capacity to move sufficient concrete so that cold joints are eliminated.

Work At and Below Ground Level

The largest concrete volume placements on a typical job are usually either below or at ground level and therefore can be placed by methods different from those employed on a superstructure. Concrete work below ground can vary enormously – from filling large-diameter bored piles or massive mat foundations to the intricate work involved in basement and subbasement walls. A crane can be used to handle formwork, reinforcing steel, and concrete. The concrete may be chuted directly from the truck mixer to the point needed. They must not slope greater than 1 vertical to 2 horizontal or less than 1 vertical to 3 horizontal. Long chutes, over 6 meters (20 ft), or those not meeting slope standards must discharge into a hopper before distribution to point of need.

Belt conveyors are very useful for work near ground level. Since placing concrete below ground is frequently a matter of horizontal movement assisted by gravity, lightweight portable conveyors can be used for high output at relatively low cost.

Alternatively, a concrete pump can move the concrete to its final position. Pumps must be of adequate capacity and must be capable of moving concrete without segregation. The loss of slump caused by pressure forcing mix water into the aggregates as the mix travels from pump hopper to discharge at the end of the pipeline must be minimal – not greater than 50 mm (2 in). The air content generally should not be reduced by more than 2 percentage points during pumping. Air loss greater than this may be caused by a boom configuration that allows the concrete to freefall. In view of this, specifications for both slump and air content may be met at the discharge end of the pump (ACI 301). Pipelines must not be made of aluminum or aluminum alloys. These cause excessive entrainment of air; aluminum reacts with cement alkali hydroxides to form hydrogen gas. This can result in high air content and serious reduction in concrete strength. For more information on pumping concrete see the American Concrete Pumping Association (www.concretepumpers.com).

Work Above Ground Level

Conveyor belt, crane and bucket, hoist, pump, or the ultimate sky-hook, the helicopter, can be used for lifting concrete to locations above ground level (Figure 13-18). The tower crane and pumping boom (Figure 13-19) are the right tools for tall buildings. The volume of concrete needed per floor as well as boom placement and length affect the use of a pump; large volumes minimize pipeline movement in relation to output.

The specifications and performance of transporting and handling equipment are being continuously improved. The best results and lowest costs will be realized if the work is planned to get the most out of the equipment. Panarese (1987) is very helpful in deciding which method to use based on capacity and range information for various methods and equipment.

Figure 13-18. For work aboveground or at inaccessible sites, a concrete bucket can be lifted by helicopter (Courtesy of Paschal).

Figure 13-19. A pump boom mounted on a mast and located near the center of a structure can frequently reach all points of placement. It is especially applicable to tall buildings where tower cranes cannot be tied up with placing concrete. Concrete is supplied to the boom through a pipeline from a ground-level pump. Concrete can be pumped hundreds of meters (feet) vertically with these pumping methods.

References

ACI Committee 301, *Specifications for Structural Concrete*, ACI 301-05, ACI Committee 301 Report, American Concrete Institute, Farmington Hills, Michigan, 2005, 49 pages.

ACI Committee 304, *Guide for Measuring, Mixing, Transporting, and Placing Concrete*, ACI 304R-00, ACI Committee 304 Report, American Concrete Institute, Farmington Hills, Michigan, 2009.

ACI Committee 304, *Placing Concrete by Pumping Methods*, ACI 304.2R-96, ACI Committee 304 Report, American Concrete Institute, Farmington Hills, Michigan, 1996.

ACI Committee 304, *Placing Concrete with Belt Conveyors*, ACI 304.4R-95, ACI Committee 304 Report, American Concrete Institute, Farmington Hills, Michigan, 1995.

ACI, *Guide to Performance-Based Requirements for Concrete*, ITG-8R-10, American Concrete Institute, Farmington Hills, Michigan, 2010.

Bohan, Richard P., and Ries, John, *Structural Lightweight Aggregate Concrete*, IS032, Portland Cement Association, Skokie, Illinois, USA, 2008, 8 pages.

Caldarone, Michael A.; Taylor, Peter C.; Detwiler, Rachel J.; and Bhidé, Shrinivas B., *Guide Specification for High-Performance Concrete for Bridges*, EB233, Portland Cement Association, Skokie, Illinois, 2005, 64 pages.

Daniel, D. Gene, and Lobo, Colin L., *User's Guide to ASTM Specification C94 for Ready-Mixed Concrete*, MNL 49, ASTM International, West Conshohocken, Pennsylvania, 2005, 130 pages.

Detwiler, R.J., and Taylor, P.C., *Specifier's Guide to Durable Concrete*, EB221, 2nd edition, Portland Cement Association, Skokie, Illinois, 2005, 72 pages.

Hover, Kenneth C.; Bickley, John; and Hooton, R. Doug, *Guide to Specifying Concrete Performance: Phase II Report of Preparation of a Performance-Based Specification For Cast-in-Place Concrete*, RMC Research and Education Foundation, Silver Spring, Maryland, USA, March 2008, 53 pages.

Lamond, J., and Pielert, J., *Significance of Tests and Properties of Concrete and Concrete-Making Materials*, STP169D, ASTM, West Conshohocken, Pennsylvania, 2006, 645 pages.

NRMCA, *Guide to Improving Specifications for Ready Mixed Concrete*, 2PE003, National Ready Mixed Concrete Association, Silver Spring, Maryland, 2009, 27 pages.

Obla, Karthik, and Lobo, Colin, *Experimental Case Study Demonstrating Advantages of Performance Specifications*, Report to the RMC Research Foundation, Project 04-02, RMC Research Foundation, Silver Spring, Maryland, USA, January 2006, 39 pages.

Panarese, William C., *Transporting and Handling Concrete*, IS178, Portland Cement Association, 1987.

CHAPTER 14

Placing and Finishing Concrete

Preparation Before Placing

Preparation prior to placing concrete for pavements or slabs on ground includes compacting, trimming, and moistening the subgrade (Figures 14-1, 14-2, and 14-3); erecting the forms; and setting reinforcing steel and other embedded items securely in place. To effectively produce the desired structure, every construction project requires good planning,well-executed project documents, construction practices that follow accepted standards, and a high level of communication among team members.

Pre-construction meetings are of prime importance in planning concrete construction work because many potential problems can be avoided well before the start of the project. The National Ready Mixed Concrete Association (NRMCA) and the American Society of Concrete Contractors (ASCC) developed a guide *Checklist for the Concrete Pre-Construction Conference* (1999) to assist decision makers and participants with planning quality concrete construction work. The checklist allocates responsibilities and establishes procedures related to concrete construction including: subgrade preparation, forming, concrete mixture proportioning, necessary equipment, ordering and scheduling materials and operations, placing, consolidating, finishing, jointing, curing and protection, testing and acceptance as well safety and environmental issues (NRMCA and ASCC 1999).

Figure 14-1. A base course foundation for concrete pavement is shaped by an auto-trimmer to design grades, cross section and alignment using automatic sensors that follow string lines.

Figure 14-2. Water trucks with spray-bars are used to moisten subgrades and base course layers to achieve adequate compaction and to reduce the amount of water drawn out of concrete as it's placed.

Figure 14-3. (top) Adequate compaction of a base course foundation for concrete pavement can be achieved by using a vibratory roller. (bottom) Vibratory plate compactors are also used to prepare subgrades under slabs.

Subgrade Preparation

Cracks, slab settlement, and structural failure can often be traced to an inadequately prepared and poorly compacted subgrade. The subgrade on which concrete is to be placed should be well drained, of uniform bearing capacity, proper moisture content (clays should not be too wet or too dry), level or properly sloped, and free of sod, organic matter, and frost. The three major causes of nonuniform support are: (1) the presence of soft unstable saturated soils or hard rocky soils, (2) backfilling without adequate compaction, and (3) expansive soils. Uniform support cannot be achieved by merely dumping granular material on a soft area. To prevent bridging and settlement cracking, soft or saturated soil areas and hard spots (rocks) should be dug out and replaced with soils similar to the surrounding subgrade or, if a similar soil is not available, with granular material such as crushed stone. All fill materials must be compacted to provide the same uniform support as the rest of the subgrade. Proof rolling the subgrade using a fully-loaded dump truck or similar heavy equipment is commonly used to identify areas of unstable soils requiring additional attention.

The strength, or bearing capacity, of the subgrade should be adequate to support anticipated structural loads, and the material should be prepared in such a manner as to assure the absence of hard or soft spots leading to inconsistent support conditions (see Tarr and Farny 2008). In general, undisturbed soil is superior to compacted material for supporting concrete slabs. Expansive, compressible, and potentially troublesome soils should be evaluated by a geotechnical engineer; a special slab design may be required.

The subgrade should be moistened with water in advance of placing concrete, but should not contain puddles or wet, soft, muddy spots when the concrete is placed. Moistening the subgrade is important, especially in hot, dry weather. This keeps the dry subgrade from drawing too much water from the concrete; it also increases the immediate air-moisture level thereby decreasing the amount of evaporation from the concrete surface.

In cold weather, concrete must not be placed on frozen subgrade. Snow, ice, and other debris must be removed from within forms before concrete is placed.

Where concrete is to be deposited on and bonded to rock or hardened concrete, all loose material must be removed. Where concrete is intended to act independently, granular fill, sand, sheet goods, or coating materials may be used as bond breakers. Cut faces defining the boundaries of concrete placements should be nearly vertical or horizontal rather than sloping.

Subbase

Concrete can be built without a subbase. However, a subbase is frequently placed on the subgrade to improve stability during construction, serve as a leveling course to correct minor surface irregularities, enhance uniformity of support, bring the site to the desired grade, and serve as a capillary break between a concrete slab on ground and the subgrade.

Where a subbase is used, the contractor should place and compact to near maximum density a 100-mm (4-in.) thick layer of granular material such as limestone trimmings, gravel, or crushed stone. If a thicker subbase is needed for achieving the desired grade, the material should be compacted in thin layers about 100 mm (4 in.) deep unless tests determine compaction of a thicker lift is possible (Figure 14-4). Subgrades and subbases can be compacted with small plate vibrators, vibratory rollers, or hand tampers. If a subbase is used, it must be well compacted.

Figure 14-4. Nuclear gauges containing radioactive sources used to measure soil density and moisture can determine if a subbase has been adequately compacted.

Moisture Control and Vapor Retarders

Many of the moisture problems associated with enclosed slabs on ground (floors) can be minimized or eliminated by (1) sloping the landscape away from buildings, (2) using a 100-mm (4-in.) thick granular subbase to form a capillary break between the soil and the slab, (3) providing drainage for the granular subbase to prevent water from collecting under the slab, (4) installing foundation drain tile, and (5) installing a vapor retarder directly beneath a concrete slab.

A vapor retarder is not a vapor barrier. A vapor retarder slows the movement of water vapor by use of a 0.25 mm to 0.40 mm (10 mil to 15 mil) polyethylene film or polyolefin sheet that is overlapped approximately 150 mm (6 in.) at the edges and taped (Figure 14-5). A vapor retarder does not stop 100% of vapor migration; a vapor barrier stops nearly 100%. Vapor barriers are thick, rugged multiple-ply-reinforced membranes that are sealed at the edges. Vapor retarders are more commonly used. However, many of the following principles apply to vapor barriers as well. Newer polyolefin products are very low permeance and highly tear and puncture resistant (Figure 14-6).

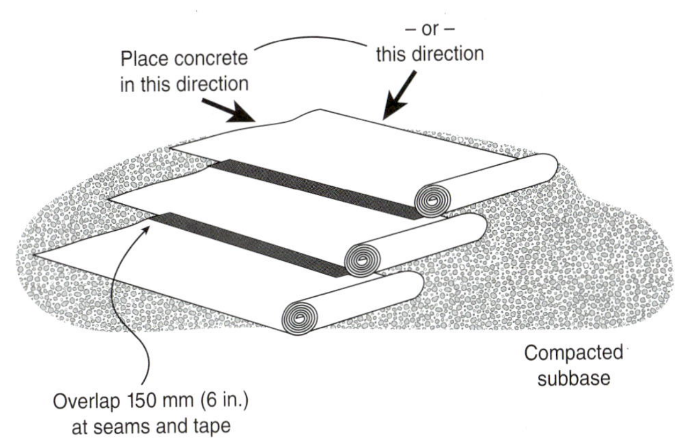

Figure 14-5. Install vapor retarder sheets so that concrete cannot work its way between adjacent sheets during placement (Kanare 2008).

Figure 14-6. Vapor retarders and vapor barriers can be made from many types of materials, but all serve the same purpose: to reduce or virtually eliminate the passage of water vapor through concrete.

A vapor retarder should be placed under all concrete floors on ground that are likely to receive an impermeable floor covering such as sheet vinyl tile. They should also be used for any purpose where the passage of water vapor through the floor might damage moisture-sensitive equipment or materials in contact with the floor (Kanare 2008).

Insulation is sometimes installed over the vapor retarder to assist in keeping the temperature of a concrete floor above the dew point; this helps prevent moisture in the air from condensing on the slab surface. This practice also creates a warm floor for thermal comfort. Codes and specifications often require insulation at the perimeter of a floor slab. Placing insulation under the entire slab on ground for energy conservation alone is usually not economical.

Vapor retarders placed directly under concrete slabs may increase the time delay before final finishing due to longer bleeding times, particularly in cold weather. To minimize this effect, a minimum 75-mm (3-in.) thick layer of approved granular, self-draining compactible subbase material can be placed over the vapor barrier (or insulation if present) (ACI Committee 302). However, slabs that will receive moisture-sensitive flooring should be placed directly on the vapor retarder (Kanare 2008). Sand over polyethylene sheeting is slippery, somewhat dangerous, and difficult to keep in place while concreting. It often results in sand lenses protruding into the concrete. A 50 mm to 100 mm (2 in. to 4 in.)-thick subbase will alleviate this problem in areas where moisture vapor will not damage flooring or products stored on the slab. The subbase over a vapor retarder must be protected from rain, construction activities, or other external sources of moisture to prevent excessive vapor migration after the concrete slab is placed.

If concrete is placed directly on a vapor retarder, the water-cementitious materials ratio should be kept low (0.45 or less) because excess mix water can only escape to the surface as bleed water. Because of a longer bleeding period, settlement cracking over reinforcement and shrinkage cracking is more likely. For more information see ACI (2001) and ACI Committee 302.

Good quality, well-consolidated concrete at least 100-mm (4-in.) thick is practically impermeable to the passage of liquid water unless the water is under considerable pressure. However, such concrete – even concrete several times as thick – is not impermeable to the passage of water vapor.

Water vapor that passes through a concrete slab transports soluble salts which can increase the surface pH and form a deposit (efflorescence) as the moisture evaporates. Floor coverings such as linoleum, vinyl tile, carpeting, wood, and synthetic surfacing effectively seal the moisture within the slab; eventually sufficiently high pH moisture condensate may deteriorate the adhesives causing the floor covering to loosen, buckle, or blister.

To prevent problems with floor covering materials caused by moisture within the concrete, the following steps should be taken: (1) use a low water-cement ratio concrete, (2) moist-cure the slab for 7 days, (3) allow the slab a drying period of at least 2 months (Hedenblad 1997 and 1998), and (4) test the slab moisture condition before installing the floor covering. Two commonly used tests are ASTM F1869, *Standard Test Method for Measuring Moisture Vapor Emission Rate of Concrete Subfloor Using Anhydrous Calcium Chloride*, and ASTM F2170, *Standard Test Method for Determining Relative Humidity in Concrete Floor Slabs Using in situ Probes*. These tests measure the moisture vapor emission rate (MVER) from a concrete slab and the internal relative humidity of a concrete slab, respectively.

ASTM F1869 involves the use of a desiccant beneath a sealed plastic dome (Figure 14-7) while ASTM F2170 incorporates a relative humidity sensor sealed in a hole extending into the slab interior. Both of these tests require the building to be enclosed and maintained at operating ambient temperature and relative humidity. Flooring-material manufacturers publish the recommended test and specified moisture limits for installing their product. For more information and additional tests for water vapor transmission, see Chapter 18, Kosmatka (1985), PCA (2000), Tarr and Farny (2008), and Kanare (2008).

Figure 14-7. Moisture vapor emission rate is determined according to ASTM F1869 using commercially available calcium chloride kits that absorb moisture from a specific test area over a known length of time.

Formwork

Forms should be properly aligned, clean, tight, adequately braced, and constructed of materials that will impart the desired off-the-form finish to the hardened concrete. Edge forms and intermediate screeds should be set accurately and firmly to the specified elevation and contour for the finished surface. Slab edge forms are usually metal or wood braced firmly with wood or steel stakes to keep them in horizontal and vertical alignment. The forms should be straight and free from warping and have sufficient strength to resist concrete pressure without deforming. They should also be strong enough to support any mechanical placing and finishing equipment used. Wood forms, unless oiled or otherwise treated with a form-release agent, should be moistened before placing concrete, otherwise they will absorb water from the concrete and may swell. Forms should be made for easy removal so as to minimize damage to the concrete. Avoid the use of large nails with wood forms, and minimize the quantity to facilitate removal and reduce damage to concrete. For architectural concrete, the form-release agent should be a nonstaining material. See Hurd (2005) and ACI committee 347 and 303 reports for more information on formwork.

Reinforcement

Reinforcing steel should be clean of debris and free of excessive rust or mill scale when concrete is placed. Thin films of surface discoloration are common and not deemed significant. Unlike subgrades, reinforcing steel ≥ 650 mm^3 (1 in.^{2}) can be colder than 0°C (32°F) when placing concrete (ACI 306). Mortar splattered on reinforcing bars from previous placements need not be removed from steel and other embedded items if the next lift is to be completed within a few hours. However, loose, dried mortar must be removed from items that will be encased by later lifts of concrete. See Chapter 8 for more details on fabrication of reinforcement for concrete.

Depositing the Concrete

Concrete should be deposited continuously as near as possible to its final position without objectionable segregation (Figures 14-8 and 14-9). In slab construction, placing should start along the perimeter at one end of the slab with each batch discharged against previously placed concrete (Figure 14-10). The concrete should not be dumped in separate piles and then leveled and worked together; nor should the concrete be deposited in large piles and moved horizontally into final position. These practices result in segregation because mortar tends to flow ahead of the coarser material.

Figure 14-8. Concrete should be placed as near as possible to its final position.

Figure 14-9. The swing arm on a conveyor belt allows fresh concrete to be placed evenly across a deck.

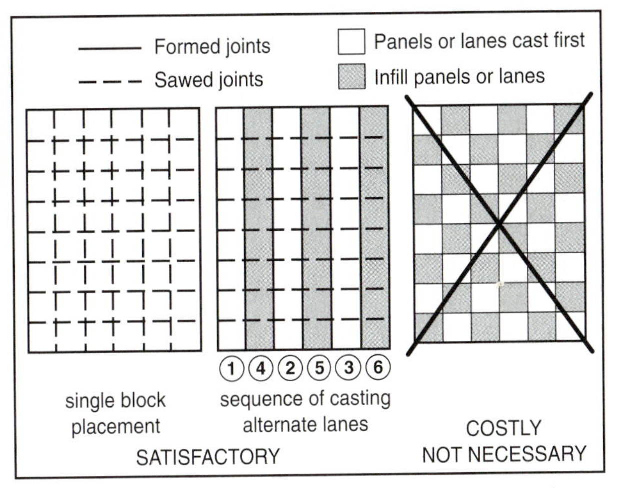

Figure 14-10. Sequence for casting concrete floors on ground.

In general, concrete should be placed in walls, thick slabs, or foundations in layers of uniform thickness and thoroughly consolidated before the next layer is placed. Lift heights are governed by formwork pressures and the economics of the form. (Refer to ACI 347 for formwork formulas for calculating formwork pressures). The depth will depend on the width between forms, size of coarse aggregate used, the volume of steel reinforcement and consistency of the concrete mixture. In general, layers should be about 150 mm to 500 mm (6 in. to 20 in.) deep for reinforced members and 375 mm to 500 mm (15 in. to 20 in.) thick for mass work using large aggregates (> 25 mm [1 in.]) or stiff consistency concrete mixtures [slump ≤ 75 mm (3 in.)].

Walls (< 300 mm [12 in.]) using smaller aggregates and flowing concrete have been placed successfully with lifts of 900 mm to 1200 mm (36 in. to 48 in.); with 900 mm (36 in.) typically being the maximum lift height for cast-in-place architectural concrete. However, with the use of chemical admixtures, vertical placements for walls and columns in lifts can clearly exceed 500 mm (20 in.). The rate of placement should be rapid enough that previously placed concrete has not yet set when the next layer of concrete is placed upon it. Timely placement and adequate consolidation will prevent flow lines, seams, and planes of weakness (cold joints) that result from placing freshly mixed concrete on top of concrete that is beyond its initial set. Typically, the formwork supplier will provide the allowable formwork pressure, and the contractor will then determine the rate of placement.

To avoid segregation, concrete should not be moved horizontally over long distances as it is being placed in forms or slabs. In some work, such as placing concrete in sloping wingwalls or beneath window openings in walls, it is necessary to move the concrete horizontally within the forms, but this should be kept to a minimum.

Where standing water is present, concrete should be placed in a manner that displaces the water ahead of the concrete and does not allow the water to be mixed into the concrete. In all cases, water should be prevented from collecting at the ends, in corners, and along faces of forms. Care should be taken to avoid disturbing saturated subgrade soils so that they maintain sufficient bearing capacity to support structural loads.

Chutes on the ready mix truck provide an efficient means to deposit concrete near grade or where truck access is available. Dropchutes are used to move concrete to lower elevations (usually in wall forms) without segregation and spattering of mortar on reinforcement and forms (Figure 14-11). Field studies indicate that free fall of concrete from heights of up to 46 m (150 ft) directly over reinforcing steel or at a high slump, so long as the material is confined, does not result in segregation of the concrete ingredients nor reduce compressive strength (Suprenant 2001). However, if a baffle is not used to control the flow of concrete onto sloped surfaces at the end of an inclined chute, segregation can occur.

Figure 14-11. Dropchutes minimize the risk of segregation.

Concrete is sometimes placed through openings (called windows) in the sides of tall, narrow forms. There is danger of segregation when a chute discharges directly through the opening without controlling concrete flow at the end of the chute. To decrease the tendency of segregation, a collecting hopper should be used outside the opening. This permits the concrete to flow more smoothly through the opening.

When concrete is placed in tall forms at a fairly rapid rate, some bleed water may collect on the top surface, especially with non-air-entrained concrete. Bleeding can be reduced by placing concrete more slowly and at a stiffer consistency.

In monolithic placement of deep beams, walls, or columns, to avoid cracks between structural elements, concrete placement should pause (about 1 hr) to allow settlement of the deep element before concreting is continued in any slabs, beams, or girders framing into them. The delay should be short enough to allow the next layer of concrete to knit with the previous layer by vibration, thus preventing cold joints and honeycombing (ACI Committee 304). Haunches and column capitals are considered part of the floor or roof slab and should be placed integrally with them.

The use of self-consolidating concrete (SCC) may allow for continuous placements if precautions are taken to account for form pressure increases due to the fluid behavior of SCC and for the heat of hydration in mass placements due to the high cementitious content of SCC (see Chapter 19).

Placing Concrete Underwater

When concrete must be placed underwater, the work should be done under experienced supervision. The basic principles for normal concrete work apply to underwater concreting.

It is important that the concrete flow without segregation; therefore, the aim in proportioning should be to obtain a cohesive mixture with high workability. The slump of the concrete should be specified at 150 mm to 230 mm (6 to 9 in.) and the mixture should have a maximum w/cm of 0.45. Generally, the concrete mixture will have a minimum cementitious materials content of 390 kg/m^3 (600 lb/yd^3). Antiwashout admixtures can be used to make concrete cohesive enough to be placed in limited depths of water, even without tremies. Using rounded aggregates, a higher percentage of fines, and entrained air may help to obtain the desired consistency.

Methods for placing concrete underwater include the following: tremie, concrete pump, bottom-dump buckets, grouting preplaced aggregate, toggle bags, bagwork, and the diving bell. A tremie is a smooth, straight pipe long enough to reach the lowest point to be concreted from a working platform above the water (Figure 14-12). The diameter of the tremie pipe should be at least 8 times the diameter of the maximum size of aggregate. A hopper to receive the concrete is attached to the top of the pipe. The lower end of the tremie should be kept buried in the fresh concrete to maintain a seal below the rising top surface and to force the concrete to flow in beneath it by pressure. Placing should be continuous with as little disturbance to the previously placed concrete as possible. The top surface should be kept as level as possible. See ACI Committee 304 for additional information.

Mobile concrete pumps with a variable radius boom make easy work of placing concrete underwater. Because the flexible hose on a concrete pump is similar to a tremie, the same placement techniques apply.

With the grouting preplaced aggregate method, the forms are first filled with clean coarse aggregate, then grouted. Grouting preplaced aggregate is advantageous when placing concrete in flowing water. Concrete can be placed more quickly and economically than by conventional placement methods. However, the method is very specialized and must be performed by qualified experienced personnel.

The current in the water through which the concrete is deposited should not exceed 3 m (10 ft) per minute.

Figure 14-12. A tremie pipe allows for placement of concrete under water.

Placing on Hardened Concrete

Concrete requiring an overlay is usually roughened to produce a better mechanical bond with the next placement. As long as no laitance (a weak layer of concrete), dirt, or loose particles are present, newly placed concrete requires little preparation prior to placing freshly mixed concrete on it. When in service, hardened concrete usually requires mechanical cleaning and roughening prior to placement of new concrete.

Bonded Construction Joints in Structural Concrete

A bonded construction joint is required between two structural concrete placements. When freshly mixed concrete is placed in contact with existing hardened concrete, a high-quality, watertight bond to minimize water infiltration at the joint is required. Poorly bonded construction joints are usually the result of (1) lack of bond between old and new concrete or (2) a weak porous surface layer in the hardened concrete at the joint. The quality of a bonded joint therefore depends on the quality of the hardened concrete and on the preparation of its surface.

In columns and walls, the concrete near the top surface of a lift is often of inferior quality to the concrete beneath it. This may be due to poor consolidation or excessive laitance, bleeding, and segregation. Even in well-proportioned and carefully consolidated mixtures, some aggregate particle settlement and water gain (bleeding) at the top surface is unavoidable; this is particularly true with rapid placement rates and with tall vertical structures. Also, the encasing formwork prevents the escape of moisture from the fresh concrete. While formwork provides adequate curing as long as it remains in place, where there is no encasing formwork, the top surface may dry out too rapidly; this may result in a weak porous layer unless protection and curing are provided.

Preparing Hardened Concrete

When freshly mixed concrete is placed on recently hardened concrete, certain precautions must be taken to secure a well-bonded, watertight joint. The hardened concrete must be clean, sound, and reasonably rough with some coarse aggregate particles exposed. Any laitance, soft mortar, dirt, wood chips, form oil, or other foreign materials must be removed since they could interfere with proper bonding of the subsequent placement.

The existing concrete surface upon which fresh concrete is to be placed must be thoroughly roughened and cleaned of all dust, surface films, deposits, loose particles, grease, oil, and other foreign material. In most cases it will be necessary to remove the entire surface down to sound concrete. Roughening and cleaning with lightweight chipping hammers, waterblasting, scarifiers, sandblasting (Figure 14-13), shotblasting, and hydrojetting are some of the satisfactory methods for exposing sound concrete. Care must be taken to avoid contamination of the clean surface before a bonding grout and overlay concrete are placed.

Figure 14-13. Sandblasting can clean any size or shape surface – horizontal, vertical, or overhead. Consult local environmental regulations regarding sandblasting.

Partially set or recently hardened concrete may only require stiff-wire brushing. In some types of construction such as dams, the surface of each concrete lift is cut with a high-velocity air-water jet to expose clean, sound concrete just before final set. This is usually done 4 to 12 hours after placing. The surface must then be protected and continuously cured until concreting is resumed for the next lift.

For two-course floors, the top surface of the base slab can be roughened with a steel or stiff fiber broom just before it sets. The surface should be level, heavily scored, and free of laitance. It should be protected until it is thoroughly cleaned just before the grout coat and topping mix are placed. When placing a bonded topping on a floor slab, the base slab should be cleaned of all curing compounds, laitance, dust, debris, grease, or other foreign substances using one of the following methods:

a. Wet- or dry-grit sandblasting
b. High-pressure water blasting
c. Mechanical removal by scabblers or grinding wheels
d. Power brooming and vacuuming

Hardened concrete may be left dry or moistened before new concrete is placed on it; however, the surface should not be wet or have any free-standing water. Laboratory studies indicate a slightly better bond is obtained on a dry surface than on a damp surface; however, the increased moisture level in the hardened concrete and in the environment around the concrete reduces water loss from the concrete mixture. This can be very beneficial on hot, dry days.

Bonding New to Previously Hardened Concrete

Care must be used when making horizontal construction joints in wall sections where freshly-mixed concrete is to be placed on hardened concrete. A good bond can be obtained by placing a rich concrete (higher cement and sand content than normal) in the bottom 150 mm (6 in.) of the new lift and thoroughly vibrating the joint interface. Alternatively, a cement-sand grout can be scrubbed into a clean surface immediately ahead of concreting.

A topping concrete mixture for slabs can be bonded to the previously prepared base slab by one of the following procedures:

1. Portland cement-sand grouting: A 1:1 cement-sand grout mixture having a water-cement ratio of not greater than 0.45, mixed to a creamlike consistency, is scrubbed into the prepared dry or damp (no free water) base slab surface. Grout is placed just a short distance ahead of the overlay or top-course concrete (Figure 14-14). This method may also be applicable to horizontal joints in walls. The grout should not be allowed to dry out prior to the overlay placement; otherwise, the dry grout may act as a poor surface for bonding.

2. Latex. A latex-bonding agent is added to the cement-sand grout and is spread in accordance with the latex manufacturer's direction.
3. Epoxy. An approved epoxy-bonding agent placed on the base concrete, prepared in accordance with the epoxy manufacturer's direction.

The surface of the base slab should have been prepared by one of the methods discussed previously. The bonding procedure should produce tensile bond strength with the base concrete in excess of 1.0 MPa (150 psi). Guidance on placing concrete overlays on pavements is provided by Harrington (2008).

Figure 14-14. Application of a bonding grout just ahead of the overlay concrete. The grout must not dry out before the concrete is placed.

Special Placing Techniques

Concrete may be placed by methods other than the usual cast-in-place method. Special concrete methods, such as extrusion (Figure 14-15) and shotcreting, are described further in Chapter 20. No matter what method is used, the basics of mixing, placing, consolidating, and curing apply to most concrete applications.

Figure 14-15. Curb machines continuously extrude low-slump concrete into a shape that immediately stands without support of formwork.

Consolidation

Consolidation is the process of compacting fresh concrete; to mold it within the forms and around embedded items and reinforcement; and to eliminate stone pockets, honeycombs, and entrapped air (Figure 14-16). Consolidation should not remove significant amounts of intentionally entrained air.

Consolidation is accomplished by hand or by mechanical methods. The method chosen depends on the consistency of the mixture and the placing conditions, such as complexity of the formwork and amount and spacing of reinforcement. Generally, mechanical methods using either internal or external vibration are preferred.

Workable, flowing mixtures can be consolidated by hand rodding; inserting a tamping rod or other suitable tool repeatedly into the concrete. The tamping rod should be long enough to reach the bottom of the form or lift and thin enough to easily pass between the reinforcing steel and the forms. For easier consolidation, low-slump concrete can be transformed into flowing concrete through the use of superplasticizers and without the addition of water to the concrete mixture (ASTM C1017, *Standard Specification for Chemical Admixtures for Use in Producing Flowing Concrete*).

Spading can be used to improve the appearance of formed surfaces. A flat, spadelike tool should be repeatedly inserted adjacent to the form. This forces the larger coarse aggregates away from form faces and assists entrapped air voids in their upward movement toward the top surface where they can escape. A mixture designed to be readily consolidated by hand methods should not be mechanically consolidated; otherwise, the concrete is likely to segregate under intense mechanical action.

Figure 14-16. Honeycombs and rock pockets are the results of inadequate consolidation.

Even in highly reinforced elements, proper mechanical consolidation makes possible the placement of stiff mixtures with low water-cementitious materials ratios and high coarse-aggregate contents. Mechanical methods of consolidation include:

1. Centrifugation – used to consolidate moderate-to-high-slump concrete in making pipes, poles, and piles;
2. Shock or Drop Tables – used to compact very stiff low-slump concrete in the manufacture of architectural precast units; and
3. Vibration – internal and external.

Vibration

Vibration, either internal or external, is the most widely used method for consolidating concrete. When concrete is vibrated, the internal friction between the aggregate particles is temporarily disrupted and the concrete behaves like a liquid; it settles in the forms under the action of gravity and the large entrapped air voids rise more easily to the surface. Internal friction is reestablished as soon as vibration stops.

Vibrators, whether internal or external, are usually characterized by their frequency of vibration, expressed as the number of vibrations per second (Hertz), or vibrations per minute (vpm); they are also designated by the amplitude of vibration, which is the deviation in millimeters (inches) from the point of rest. The frequency of vibration can be measured using a vibrating reed tachometer.

When vibration is used to consolidate concrete, a standby vibrator should be on hand at all times in the event of a mechanical breakdown.

Internal Vibration. Internal or immersion-type vibrators, often called spud or poker vibrators (Figures 14-17 and 14-18), are commonly used to consolidate concrete in walls, columns, beams, and slabs. Flexible-shaft vibrators consist of a vibrating head connected to a driving motor by a flexible shaft. Inside the head, an eccentric weight connected to the shaft rotates at high speed, causing the head to revolve in a circular orbit. The motor can be powered by electricity, gasoline, or air. The vibrating head is usually cylindrical with a diameter ranging from 19 to 175 mm (¾ to 7 in.). Some vibrators have an electric motor built directly into the head, which is generally at least 50 mm (2 in.) in diameter. The dimensions of the vibrator head as well as its frequency and amplitude in conjunction with the workability of the mixture affect the performance of a vibrator.

Small-diameter vibrators have high frequencies ranging from 160 Hz to 250 Hz (10,000 vpm to 15,000 vpm) and low amplitudes ranging between 0.4 mm and 0.8 mm (0.016 in. and 0.03 in.). As the diameter of the head increases, the frequency decreases and the amplitude increases. The effective radius of action of a vibrator increases with increasing diameter. Vibrators with a diameter of 19 mm to 38 mm (¾ in. to 1½ in.) have a radius of action in freshly mixed concrete ranging between 75 mm and 150 mm (3 in. and 6 in.), whereas the radius of action for vibrators of 50-mm to 75-mm (2-in. to 3-in.) diameter ranges between 175 mm and 350 mm (7 in. and 14 in.).

Figure 14-17. Proper vibration makes possible the placement of stiff concrete mixtures, even in heavily-reinforced concrete members.

Figure 14-18. Internal vibrators are commonly used to consolidate concrete in walls, columns, beams, and slabs (Courtesy of Baker Concrete).

Table 14-1 shows the range of characteristics and applications for internal vibrators for various applications.

Proper use of internal vibrators is critical for best results. Vibrators should not be used to move concrete horizontally since this causes segregation. Whenever possible, the vibrator should be lowered vertically into the concrete at regularly spaced intervals and allowed to descend by gravity. It should penetrate to the bottom of the layer being placed and at least 150 mm (6 in.) into any previously placed layer. The vibrator is then withdrawn at roughly the same pace as the gravity insertion. The height of each layer will depend on the type of application and concrete properties (see **Depositing Concrete**).

In thin slabs, the vibrator should be inserted at an angle or horizontally in order to keep the vibrator head completely

Table 14-1. Range of Characteristics, Performance, and Applications of Internal* Vibrators

Group	Diameter of head, mm (in.)	Recommended frequency, vibrations per minute**	Suggested values of: Eccentric moment, mm-kg in.-lb (10^{-3})	Average amplitude, mm (in.)	Centrifugal force, kg (lb)	Approximate values of: Radius of action,† mm (in.)	Rate of concrete placement, m^3/h (yd^3/h)‡	Application
1	20-40 (¾-1½)	9000-15,000	3.5-12 (0.03-0.10)	0.4-0.8 (0.015-0.03)	45-180 (100-400)	80-150 (3-6)	0.8-4 (1-5)	Plastic and flowing concrete in very thin members and confined places. May be used to supplement larger vibrators, especially in prestressed work where cables and ducts cause congestion in forms. Also used for fabricating laboratory test specimens.
2	30-60 (1¼-2½)	8500-12,500	9-29 (0.08-0.25)	0.5-1.0 (0.02-0.04)	140-400 (300-900)	130-250 (5-10)	2.3-8 (3-10)	Plastic concrete in thin walls, columns, beams, precast piles, thin slabs, and along construction joints. May be used to supplement larger vibrators in confined areas.
3	50-90 (2-3½)	8000-12,000	23-81 (0.20-0.70)	0.6-1.3 (0.025-0.05)	320-900 (700-2000)	180-360 (7-14)	4.6-15 (6-20)	Stiff plastic concrete (less than 80-mm [3-in.] slump) in general construction such as walls, columns, beams, prestressed piles, and heavy slabs. Auxiliary vibration adjacent to forms of mass concrete and pavements. May be gang mounted to provide full-width internal vibration of pavement slabs.
4	80-150 (3-6)	7000-10,500	8-290 (0.70-2.5)	0.8-1.5 (0.03-0.06)	680-1800 (1500-4000)	300-510 (12-20)	11-31 (15-40)	Mass and structural concrete up to 50-mm (2-in.) slump deposited in quantities up to 3 m^3 (4 yd^3) in relatively open forms of heavy construction (powerhouses, heavy bridge piers, and foundations). Also used for auxiliary vibration in dam construction near forms and around embedded items and reinforcing steel.
5	130-150 (5-6)	5500-8500	260-400 (2.25-3.50)	1.0-2.0 (0.04-0.08)	1100-2700 (2500-6000)	400-610 (16-24)	19-38 (25-50)	Mass concrete in gravity dams, large piers, massive walls, etc. Two or more vibrators will be required to operate simultaneously to mix and consolidate quantities of concrete of 3 m^3 (4 yd^3) or more deposited at one time into the form.

* Generally, extremely dry or very stiff concrete does not respond well to internal vibrators.
** While vibrator is operating in concrete.
† Distance over which concrete is fully consolidated.
‡ Assumes the insertion spacing is 1½ times the radius of action, and that vibrator operates two-thirds of time concrete is being placed. These ranges reflect not only the capability of the vibrator but also differences in workability of the mixture, degree of deaeration desired, and other conditions experienced in construction.
Adapted from ACI 309.

immersed. However, the vibrator should never be dragged around randomly in the slab. For slabs on ground, the vibrator should not make contact with the subgrade. The distance between insertions should be about 1½ times the radius of action so that the area visibly affected by the vibrator overlaps the adjacent previously vibrated area by a few centimeters (inches).

The vibrator should be held stationary until adequate consolidation is attained and then slowly withdrawn. An insertion time of 5 to 15 seconds will usually provide adequate consolidation. The concrete should move to fill the hole left by the vibrator on withdrawal. If the hole does not refill, reinsertion of the vibrator at a nearby point should solve the problem.

Adequacy of internal vibration is judged by changes in the surface appearance of the concrete. Changes to watch for are the embedment of large aggregate particles, general batch leveling, the appearance of a thin film of mortar on the top surface, and the cessation of large bubbles of entrapped air escaping at the surface. Internal vibration may significantly affect the entrained-air-void system in concrete (Stark 1986 and Hover 2001). Detailed guidance for proper vibration should be followed (ACI Committee 309).

Allowing a vibrator to remain immersed in concrete after paste accumulates over the head can result in nonuniformity. The length of time that a vibrator should remain in the concrete will depend on the workability of the concrete, the power of the vibrator, and the nature of the section being consolidated.

In heavily-reinforced sections where an internal vibrator cannot be inserted, it is sometimes helpful to vibrate the reinforcing bars by attaching a form vibrator to their exposed portions. This practice eliminates air and water trapped under the reinforcing bars and increases bond between the bars and surrounding concrete. This method should only be used if the concrete is still workable under the action of vibration. Internal vibrators should not be attached to reinforcing bars for this purpose because the vibrators may be damaged.

External Vibration. External vibrators can be form vibrators, vibrating tables, or surface vibrators such as vibratory screeds, plate vibrators, vibratory roller screeds, or vibratory hand floats or trowels. Form vibrators, designed to be securely attached to the outside of the forms, are especially useful for the following: consolidating concrete in members that are very thin or congested with reinforcement, stiff mixtures where internal vibrators cannot be used, and to supplement internal vibration.

Attaching a form vibrator directly to the concrete form generally is unsatisfactory. Rather, the vibrator should be attached to a steel plate that in turn is attached to steel I-beams or channels passing through the form stiffeners. Loose attachments can result in significant vibration energy losses and inadequate consolidation.

Form vibrators can be either electrically or pneumatically operated. They should be spaced to distribute the intensity of vibration uniformly over the form; optimum spacing is typically determined by experimentation. Sometimes it may be necessary to operate some of the form vibrators at a different frequency for better results; therefore, it is recommended that form vibrators be equipped to regulate their frequency and amplitude. The necessary duration of external vibration of the concrete is considerably longer than for internal vibration – generally between 1 and 2 minutes. A reed tachometer can not only determine frequency of vibration, but also give a rough estimate of amplitude of vibration by noting the oscillation of the reed at various points along the forms. This will assist in identifying dead spots or weak areas of vibration. A vibrograph could be used if more reliable measurements of frequency and amplitude are needed.

Form vibrators should not be applied within the top meter (yard) of vertical forms. Vibration of the top of the form, particularly if the form is thin or inadequately stiffened, causes an in-and-out movement that can create a gap between the concrete and the form.

Vibrating tables are used in precasting plants. They should be equipped with controls so that the frequency and amplitude can be varied according to the size of the element to be cast and the consistency of the concrete. Stiffer mixtures generally require lower frequencies (below 6000 vpm) and higher amplitudes (over 0.13 mm [0.005 in.]) than more workable mixtures. Increasing the frequency and decreasing the amplitude as vibration progresses will improve consolidation.

Surface vibrators, such as vibratory screeds (Figures 14-19, 14-20, and 14-21), are used to consolidate concrete in floors and other flatwork. Vibratory screeds give positive control of the strikeoff operation and save a great deal of labor. When using this equipment, concrete should not have slumps in excess of 75 mm (3 in.). For slumps greater than 75 mm (3 in.), care should be taken because surface vibration of such concrete will result in an excessive accumulation of mortar and fine material on the surface; this may reduce wear resistance. For the same reason, surface vibrators should not be operated after the concrete has been adequately consolidated.

Figure 14-19. Vibratory screeds such as this truss-type unit reduce the work of strikeoff while consolidating the concrete.

Figure 14-20. Where floor tolerances are not critical, an experienced operator using this vibratory screed does not need screed poles supported by chairs to guide the screed. Instead, the operator visually matches elevations to forms or previous passes. The process is called wet screeding.

Because surface vibration of concrete slabs is least effective along the edges, a spud or poker-type vibrator should be used along the edge forms immediately before the vibratory screed is applied.

Vibratory screeds are used for consolidating slabs up to 250 mm (10 in.) thick, provided such slabs are unreinforced or only lightly reinforced (welded-wire fabric). Internal vibration or a combination of internal and surface vibration is recommended for reinforced slabs. More detailed information regarding internal and external vibration of concrete can be obtained from ACI Committee 309.

Figure 14-21. A laser level striking the sensors on this screed guides the operator as he strikes off the concrete. Screed poles and chairs are not needed and fewer workers are required to place concrete. Laser screeds interfaced with total station surveying equipment can also strike off sloped concrete surfaces.

Consequences of Improper Vibration. Undervibration may cause honeycombing; excessive amount of entrapped air voids, often called bugholes; sand streaks; cold joints; placement lines; and subsidence cracking.

Honeycombing results when the spaces between coarse aggregate particles are not filled with mortar. Faulty equipment, improper placement procedures, too much coarse aggregate, or congested reinforcement can cause honeycombing.

Entrapped air voids at the surface of concrete are often called bugholes. Vibratory equipment and operating procedures are the primary causes of excessive entrapped air voids, but the other causes of honeycombing can also apply.

Sand streaks result when heavy bleeding washes mortar out from along the form. A wet, harsh mixture that lacks workability because of an insufficient amount of mortar or fine aggregate may cause sand streaking. Segregation caused by aggregate striking reinforcing steel may also contribute to streaking.

Cold joints are a discontinuity resulting from a delay in placement that allowed one layer to harden before the adjacent concrete was placed. The discontinuity can reduce the structural integrity of a concrete member if the successive lifts did not properly bond together.

Placement lines or "pour" lines are dark lines between adjacent placements of concrete batches. They may occur when the vibrator did not penetrate the underlying layer enough to knit the two layers together.

Subsidence cracking may occur at or near the initial setting time as concrete settles over reinforcing steel in relatively deep elements that have not been adequately vibrated. Revibration while vibrator is still able to sink into the concrete under its own weight may eliminate these cracks.

Defects from overvibration include: (1) segregation as vibration and gravity causes heavier aggregates to settle while lighter aggregates rise; (2) sand streaks; (3) loss of entrained air in air-entrained concrete; (4) excessive form deflections or form damage; and (5) form failure caused by excessive pressure from vibrating the same location too long or placing concrete more quickly than the designed rate of placement. Undervibration is much more prevalent than overvibration.

Finishing

Concrete can be finished in many ways, depending on the intended service use. Various colors and textures, such as exposed-aggregate or a pattern-stamped surface, may be specified. Some surfaces may require only strikeoff and screeding to proper contour and elevation, while other surfaces may require a broomed, floated, or troweled finish may be specified. Details are given in ACI Committee 302, Kosmatka and Collins (2004), Collins; Panarese; and Bradley (2006), PCA (1980a), and Tarr and Farny (2008).

The mixing, transporting, and handling of concrete for slabs should be carefully coordinated with the finishing operations. Concrete should not be placed on the subgrade or into forms more rapidly than it can be spread, struck off, consolidated, and bullfloated or darbied. Concrete should not be spread over too large an area before strikeoff, nor should a large area be struck off and bleed water allowed to accumulate before bullfloating or darbying.

Finishing crews should be large enough to correctly place, finish, and cure concrete slabs with due regard for the effects of concrete temperature and atmospheric conditions on the setting time of the concrete and the size of the placement to be completed.

Screeding (Strikeoff)

Screeding or strikeoff is the process of cutting off excess concrete to bring the top surface of a slab to proper grade. The manually used template is called a straightedge, although the lower edge may be straight or slightly curved, depending on the surface specified. It should be moved across the concrete with a sawing motion while advancing forward a short distance with each movement. There should be a surplus (surcharge) of concrete against the front face of the straightedge to fill in low areas as the straightedge passes over the slab. A 150-mm (6-in.) slab needs a surcharge of about 25 mm (1 in.). Straightedges are sometimes equipped with vibrators that consolidate the concrete and assist in reducing the strikeoff work. This combination straightedge and vibrator is called a vibratory screed. Vibratory screeds are discussed earlier in this chapter under **Consolidation**. Screeding, consolidation, and bullfloating must be completed before bleed water collects on the surface.

Bullfloating or Darbying

To eliminate high and low spots and to embed large aggregate particles, a bullfloat or darby (Figure 14-22 top) should be used immediately after strikeoff. The long-handle bullfloat (Figure 14-22 bottom) is used on areas too large to reach with a short-handle darby. Highway straightedges are often used to obtain very flat surfaces (Figure 14-23). For non-air-entrained concrete, these tools can be made of wood, aluminum, or magnesium alloy; for air-entrained concrete they should be of aluminum or magnesium alloy. It is important that the surface not be closed during this finishing phase as bleeding is typically still in progress and trapped bleedwater or bleed air can lead to surface delaminations.

Bullfloating or darbying must be completed before bleed water accumulates on the surface. Care must be taken not to overwork the concrete as this could result in a less durable surface.

Figure 14-22. (top) Darbying brings the surface to the specified level and is done in tight places where a bullfloat cannot reach. (bottom) Bullfloating must be completed before any bleed water accumulates on the surface.

The preceeding operations should level, shape, and smooth the surface and work up a slight amount of cement paste. Although further finishing may not be required in all situations, on most slabs bullfloating or darbying is followed by finishing operations. These include: edging, jointing, floating, troweling, and brooming. A slight hardening of the concrete is necessary before the start of any of these finishing operations. When the bleed water sheen has evaporated the concrete may then receive additional finishing passes. These can be made

Figure 14-23. Highway straightedges are used on highway pavement and floor construction where very flat surfaces are desired (Courtesy of Baker Concrete).

using hand tool methods or tools with handles applied to the slab surface from outside of the slab area, bullfloats, fresnos, walking edgers, and walking jointers. When the concrete will sustain foot pressure with only about 6-mm (¼-in.) indentation, the surface is ready for continued finishing operations using mechanical means such as power floats and power trowels.

Warning: *One of the principal causes of surface defects in concrete slabs is finishing while bleed water is present on the surface. If bleed water is worked into the surface, the water-cement ratio is significantly increased. This reduces strength, entrained-air content, and watertightness of the surface. Any finishing operation performed on the surface of a concrete slab while bleed water is present can cause crazing, dusting, or scaling (PCA 2001).* Floating and troweling the concrete (discussed later) before the bleeding process is completed may also trap bleed water or air under the finished surface producing a weakened zone or void under the finished surface; this occasionally results in delaminations. The use of low-slump concrete with an adequate cement content and properly graded fine aggregate will minimize bleeding and help ensure maintenance-free slabs. Air-entrained slabs should never be overworked or hard-troweled. However, for exterior slabs to be bullfloated and broomed, air entrainment also reduces bleeding. ACI Committee 302 and Tarr and Farny (2008) present placing and finishing techniques in more detail and PCA (2001) discusses slab surface defects.

Edging and Jointing

Edging is required along all edge forms and isolation and construction joints in floors and outdoor slabs such as walks, drives, and patios. Edging densifies and compacts concrete next to the form where floating and troweling are less effective, making it more durable and less vulnerable to scaling, chipping, and popouts.

Edging tools should be held flat across their width and the leading face of the edger slightly inclined to provide a flat edge and prevent the edger from catching on the concrete surface which can causes tearing and defects. First pass edging operations should be completed before the onset of bleeding; otherwise, the concrete should be cut away from the forms to a depth of 25 mm (1 in.) using a pointed mason trowel or a margin trowel to push coarse aggregate particles away from the form edge. Edging may be required after each subsequent finishing operation for exterior and interior slabs.

Proper jointing practices can eliminate unsightly random cracks. Contraction joints, sometimes called control joints, can be formed with a hand groover or by inserting strips of plastic, wood, metal, or preformed joint material into the fresh concrete. When hand methods are used to form control joints in exterior concrete slabs, mark the forms to accurately locate the joints. Prior to bullfloating, the edge of a thin strip of wood or metal may be used to knock down the coarse aggregate where the joint will be hand tooled. The slab should then be jointed immediately after bullfloating or in conjunction with the edging operation. If insert strips are used, they should be kept vertical. Hand tooling and insert strips are not recommended for jointing industrial floors. For these slabs, contraction joints should be sawn into hardened concrete. Jointing is discussed further in this chapter under **Jointing Concrete.**

Floating

After the concrete has been hand-edged and hand-jointed, it should be floated with a hand float or with a finishing machine using float blades (Figure 14-24).The purpose of floating is to embed aggregate particles just beneath the surface; remove slight imperfections, humps, and voids; compact the mortar at the surface in preparation for additional finishing operations; and reestablish the moisture content of the paste at the near surface where evaporation has its greatest impact. The concrete should not be overworked as this may bring an excess of water and fine material to the surface and result in subsequent surface defects.

Hand floats usually are made of fiberglass, magnesium, or wood. The metal float reduces the amount of work required because drag is reduced as the float slides more readily over the concrete surface. A magnesium float is essential for hand-floating air-entrained concrete because a wood float tends to stick to and tear the concrete surface. The light metal float also produces a smoother surface than the wood float.

The hand float should be held flat on the concrete surface and moved with a slight sawing motion in a sweeping arc

to fill in holes, cut off lumps, and smooth ridges. When finishing large slabs, power floats can be used to reduce finishing time.

Floating produces a relatively even (but not smooth) texture that has good slip resistance and is often used as a final finish, especially for exterior slabs. Where a float finish is the desired final finish, it may be necessary to float the surface a second time after additional stiffening of the concrete surface.

Marks left by hand edgers and groovers are normally removed during floating unless the marks are desired for decorative purposes; in such cases the edger and groover should be used again after final floating.

Figure 14-24. Power floating using walk-behind (top) and ride-on equipment (bottom). Footprints indicate proper timing. When the bleedwater sheen has evaporated and the concrete will sustain foot pressure with only slight indentation, the surface is ready for floating and final finishing operations (Courtesy of Baker Concrete).

Troweling

Where a smooth, hard, dense surface is desired, floating should be followed by steel troweling (Figure 14-25). Troweling should not be done on a surface that has not been floated. Troweling after only bullfloating or darbying is not considered adequate finishing.

It is customary when hand-finishing large slabs to float and immediately trowel an area before moving the kneeboards. However power equipment enables the proper sequence for finishing operations. That is, to make an initial pass with floats only; a second pass after the concrete has been allowed to stiffen slightly using a float and steel trowel: and a third pass using only steel trowels (again after a delay to allow for additional stiffening of the slab surface). These operations should be delayed until the concrete has hardened sufficiently so that water and fine material are not brought to the surface. Too long a delay, of course, will result in a surface that is too hard to float and trowel. The tendency, however, is to float and trowel the surface too soon. Premature floating and troweling can cause scaling, crazing, or dusting and produce a surface with reduced wear resistance.

Figure 14-25. Hand floating (left hand) the surface with a hand float held flat on the concrete surface and moved in a sweeping arc with a slight sawing motion. Troweling (right hand) with blade tilted is performed before moving the kneeboards (Courtesy of Baker Concrete).

Spreading dry cement on a wet surface to absorb excess water causes surface defects such as dusting, crazing, and mortar flaking. These wet spots should be avoided, if possible, by adjustments in aggregate gradation, mix proportions, and consistency. When wet spots do occur, finishing operations should be delayed until the water either evaporates or is removed with a rubber floor squeegee or by dragging a soft rubber garden hose to push the standing water off of the slab surface (ACI 302). If a squeegee or hose is used, extreme care must be taken so that excess cement paste is not removed with the water, and so that the surface is not damaged.

Initial troweling may produce the desired surface free of defects. However, surface smoothness, density, and wear resistance can all be improved by additional trowelings. There should be a lapse of time between successive trowelings to permit the concrete to become harder. As the surface stiffens, each successive troweling should be made with smaller trowels, using progressively more tilt and pressure on the trowel blade. The final pass should make a ringing sound as the trowel moves over the hardening surface.

A power trowel is similar to a power float, except that the machine is fitted with smaller, individual steel trowel blades that are adjustable for tilt and pressure on the concrete surface. If necessary, tooled edges and joints should be rerun after troweling to maintain uniformity and true lines.

Exterior concrete should not be troweled for several reasons: (1) because it can lead to a loss of entrained air caused by overworking the surface, and (2) troweled surfaces are slippery when wet leading to slip and fall accidents. For safety reasons, floating and brooming are the preferred textures for safety for outdoor concrete. Traditionally, bullfloats have been used successfully without sealing the surface or removing substantial air from the surface. The risk of these occurrences increases with the use of power floating equipment.

Brooming

Brooming or tining should be performed before the concrete has thoroughly hardened, but it should be sufficiently hard to retain the scoring impression to produce a slip-resistant surface (Figure 14-26 and 14-27). Rough scoring or tining, can be achieved with a rake, a steel-wire broom, or a stiff, coarse, fiber broom; such coarse-textured brooming usually follows floating. If a finer texture is desired, the concrete should be floated to a smooth surface and then brushed with a soft-bristled broom.

Figure 14-26. Brooming provides a slip-resistant surface mainly used on exterior concrete.

Figure 14-27. (top) This machine is tining the surface of fresh concrete. (bottom) Tining of pavements improves tire traction and reduces hydroplaning.

Interior concrete could also be troweled before brooming. Slabs are usually broomed transversely to the main direction of traffic.

The magnitude of the texture is dependent on the degree of hard troweling – a light trowel will texture more easily than a burnished finish. Best results are obtained with equipment that is specially made for texturing concrete.

Finishing Formed Surfaces

Many off-the-form concrete surfaces require little or no additional treatment when they are carefully constructed with the proper forming materials. In accordance with ACI 301 (2010), these surfaces are divided into several classes; as-cast finishes, surface finish 1.0, surface finish 2.0, and surface finish 3.0. Table 14-2 provides more detail on each finish type.

A smooth, rubbed finish is produced on a newly hardened concrete surface no later than the day following form removal. The forms are removed and necessary patching completed as soon as possible. The surface is then wetted and rubbed with a carborundum brick or other abrasive until a satisfactory uniform color and texture are produced.

Table 14-2. Specified Surface Finishes (ACI 301-10)

As-cast finishes— Produce as-cast formed finishes in accordance with Contract Documents.
Surface finish-1.0 (SF-1.0): • No formwork facing material is specified; • Patch voids larger than 1½ in. wide or ½ in. deep; • Remove projections larger than 1 in.; • Tie holes need not be patched; • Surface tolerance Class D as specified in ACI 117; and • Mockup not required.
Surface finish-2.0 (SF-2.0): • Patch voids larger than ¾ in. wide or ½ in. deep; • Remove projections larger than ¼ in.; • Patch tie holes; • Surface tolerance Class B as specified in ACI 117; and • Unless otherwise specified, provide mockup of concrete surface appearance and texture.
Surface finish-3.0 (SF-3.0): • Patch voids larger than ¾ in. wide or ½ in. deep; • Remove projections larger than ⅛ in.; • Patch tie holes; • Surface tolerance Class A as specified in ACI 117; and • Provide mockup of concrete surface appearance and texture.

A sand-floated finish can also be produced on newly hardened concrete surfaces. No later than 5 to 6 hours following form removal, the surface should be thoroughly wetted and rubbed with a wood or rubber float in a circular motion, working fine sand into the surface until the resulting finish is even and uniform in texture and color.

A grout cleandown (sack-rubbed finish) can be used to impart a uniform color and appearance to a smooth surface. After defects have been repaired, the surface should be saturated thoroughly with water and kept wet at least one hour before finishing operations begin. Next a grout of 1 part cement, 1½ to 2 parts of fine sand passing a 600 µm (No. 30) sieve, and sufficient water for a thick, creamy consistency should be prepared. It should be preshrunk by mixing at least 15 minutes before use and then remixed without the addition of water and applied uniformly by brush, plasterer's trowel, or rubber float to completely fill all air bubbles and holes. The surface should be vigorously floated with a wood, rubber, or cork float immediately after applying the grout to fill any small air holes (bugholes) that are left; any remaining excess grout should be scraped off with a rubber float. If the float pulls grout from the holes, a sawing motion of the tool should correct the difficulty. Any grout remaining on the surface should be allowed to stand undisturbed until it loses some of its plasticity but not its damp appearance. Finally, the surface should be rubbed with clean, dry burlap to remove all excess grout. All air holes should remain filled, but no visible film of grout should remain after the rubbing. Any section cleaned with grout must be completed in one day as grout remaining on the surface overnight is difficult to remove.

If possible, work should be done in the shade and preferably during cool, damp weather. During hot or dry weather, the concrete can be kept moist with a fine atomized fog spray.

The completed rubbed surface should be moist-cured by keeping the area wet for 36 hours following the clean down. When completely dry, the surface should have a uniform color and texture.

Special Surface Finishes

Patterns and Textures

A variety of patterns and textures can be used to produce decorative finishes. Patterns can be formed with divider strips or by scoring or stamping the surface just before the concrete hardens. Textures can be produced with little effort and expense with floats, trowels, and brooms; more elaborate textures can be achieved with special techniques (Figure 14-28). See Kosmatka and Collins (2004).

Exposed-Aggregate Concrete

An exposed-aggregate finish provides a rugged, attractive surface in a wide range of textures and colors. Select aggregates are carefully chosen to avoid deleterious substances; they are usually of uniform size such as 9.5 mm to 12.5 mm (3⁄8 in. to 1⁄2 in.) or larger. They should be washed thoroughly before use to assure satisfactory bond. Flat or elongated aggregate particles should not be used since they are easily dislodged when the aggregate is exposed. Caution should be exercised when using crushed stone; it not only has a greater tendency to stack during the seeding operation (requiring more labor), but the sharp angular edges may be undesirable in some applications (pool decks, for example).

The aggregate should be evenly distributed or seeded in one layer onto the concrete surface immediately after the slab has been bullfloated or darbied. The particles must be completely embedded in the concrete. This can be done by lightly tapping with a hand float, a darby, or the broad side of a piece of lumber. Then, when the concrete can support a finisher on kneeboards, the surface should be hand-floated with a magnesium float or darby until the mortar completely surrounds and slightly covers all the aggregate particles.

Methods of exposing the aggregate usually include washing and brushing, using retarders, and scrubbing. When the concrete has hardened sufficiently, simultaneously brushing and flushing with water should expose the aggregate. In washing and brushing, the surface layer of mortar should be carefully washed away with a light spray of water and brushed until the desired exposure is achieved.

Mock-up. Test panels should be made to determine the correct time for exposing the aggregate without dislodging the particles. On large surface areas, a water-insoluble retarder can be sprayed or brushed on the surface immediately after floating, but on small jobs this may not be necessary. When the concrete becomes too hard to produce the required finish with normal washing and brushing, a dilute hydrochloric acid could be used. Surface preparation should be minimized and applicable local environmental laws should be followed.

Other methods for obtaining an exposed aggregate surface include: 1) the monolithic technique where a select aggregate, usually gap-graded, is mixed throughout the batch of concrete, and 2) the topping technique in which the select exposed-aggregate is mixed into a topping that is placed over a base slab of conventional concrete.

Figure 14-28. Patterned, textured, and colored concretes are very attractive.

The following techniques expose the aggregate after the concrete has hardened to a compressive strength of around 28 MPa (4000 psi):

Abrasive blasting is best applied to a gap-graded aggregate concrete. The nozzle should be held perpendicular to the surface and the concrete removed to a maximum depth of about one-third the diameter of the coarse aggregate.

Waterblasting can also be used to texture the surface of hardened concrete, especially where local ordinances prohibit the use of sandblasting for environmental reasons. High-pressure water jets are used on surfaces with or without retarders.

In tooling or bushhammering, a layer of hardened concrete is removed and the aggregate is fractured at the surface. Resulting finishes can vary from a light scaling to a deep, bold texture obtained by jackhammering with a single-pointed chisel. Combs and multiple points can be used to produce finishes similar to those sometimes used on cut stone.

Grinding and polishing will produce an exposed-aggregate concrete such as terrazzo, which is primarily used indoors. This technique is accomplished in several successive steps using either a stone grinder or diamond-disk grinder. Each successive step is done with progressively finer grits. A polishing compound and buffer can then be used for a honed finish.

Regardless of the method employed, it is wise for the contractor to make a preconstruction mock-up (field sample) for each finish to determine the timing and steps involved; in addition, the mock-up is used to obtain aesthetic approval from the architect and owner. For more information see Kosmatka and Collins (2004), PCA (1972), and PCA (1995).

Colored Finishes

Colored concrete finishes for decorative effects in both interior and exterior applications can be achieved by four different methods: (1) the one-course or integral method, (2) the two-course method, (3) the dry-shake method, and (4) stains and paints (Kosmatka and Collins 2004).

Color pigments added to the concrete in the mixer to produce a uniform color is the basis for the one-course method. Both natural and synthetic pigments are satisfactory if they are: (1) insoluble in water, (2) free from soluble salts and acids, (3) resistant to sunlight, (4) resistant to alkali and weak acids, (5) limited to small amounts of calcium sulfate, and (6) ground fine enough so that 90% passes a 45 micron screen. Use only the minimum amount necessary to produce the desired color and not more than 10% by weight of the cement. White cements improve the efficiency of pigment dosage and allow for the widest range of color choices, especially the lighter and pastel colors.

In the two-course method, a base slab is placed and left with a rough texture to bond better to a colored topping layer. As soon as the base slab can support a cement mason's weight, the topping course can be placed. If the base slab has hardened, prepare a bonding grout for the base slab prior to placing the topping mix. The topping mix is normally 13 mm (½ in.) to 25 mm (1 in.) thick, with a ratio of cement to sand of 1:3 or 1:4. The mix is floated and troweled in the prescribed manner. The two-course method is commonly used because it is more economical than the one-course method.

In the dry-shake method, a prepackaged dry-color material is cast onto the surface of a concrete slab. The dry-shake material is applied after the concrete has been screeded and darbied or bullfloated, excess moisture has evaporated from the surface, and preliminary floating is complete. Two-thirds of the dry material is distributed evenly by hand over the surface and thoroughly floated into the surface. Immediately, the remaining material is cast onto the surface at right angles to the initial application and then floated to assure uniform color distribution. The surface can then be troweled at the same time as a typical slab. For exterior surfaces that will be exposed to freezing and thawing, little or no troweling followed by brooming with a soft bristle concrete broom is recommended.

Stains, Paints, and Clear Coatings

Many types of stains, paints, and clear coatings can be applied to concrete surfaces. Among the principal paints used are portland cement base, latex-modified portland cement, and latex (acrylic and polyvinyl acetate) paints (PCA 1992). However, stains and paints are used only when it is necessary to color existing concrete. It is difficult to obtain a uniform color with dyes or stains; therefore, the manufacturer's directions should be closely followed. Stains and dyes are often used to even out colors on existing slabs or to create mottled or variegated effects.

Portland cement based paints can be used on either interior or exterior exposures. The surface of the concrete should be damp at the time of application and each coat should be dampened as soon as possible without disturbing the paint. Damp curing of conventional portland cement paint is essential. On open-textured surfaces, such as concrete masonry, the paint should be applied with stiff-bristle brushes (scrub brushes). Paint should be worked well into the surface. For concrete with a smooth or sandy finish, whitewash or Dutch-type calcimine brushes are best.

The latex materials used in latex-modified portland cement paints retard evaporation, thereby retaining the necessary water for hydration of the portland cement. When using latex-modified paints, moist curing is typically not required. Most latex paints are resistant to alkali and can be applied to new concrete after 10 days of good drying weather. The preferred method of application is by long-fiber, tapered nylon brushes 100 to 150 mm (4 to 6 in.) wide; however, roller or spray methods can also be used. The paints may be applied to damp, but not wet surfaces. If the surface is moderately porous, or if extremely dry conditions prevail, prewetting the surface is advisable.

Clear coatings are frequently used on concrete surfaces to (1) prevent soiling or discoloration of the concrete by air pollution, (2) to facilitate later cleaning the surface, (3) to brighten the color of the aggregates, and (4) to render the surface water-repellent and thus prevent color change due to rain and water absorption. The better coatings often consist of methyl methacrylate forms of acrylic resin, as indicated by a laboratory evaluation of commercial clear coatings (Litvin 1968). The methyl methacrylate coatings should have a higher viscosity and solids content when used on smooth concrete, since the original appearance of smooth concrete is more difficult to maintain than the original appearance of exposed-aggregate concrete.

When a change of color would be objectionable, other materials, such as silane and siloxane penetrating sealers, are commonly used as water repellents for many exterior concrete applications.

Curing and Protection

All newly placed and finished concrete slabs should be cured and protected from drying, from extreme changes in temperature, and from damage by subsequent construction activities and traffic (see Chapter 15).

Initial curing should begin immediately after strike-off operations and continue until finishing operations are complete (Figure 14-29). Final curing is needed to ensure continued hydration of the cement, assure proper strength gain and durability of the concrete, and to minimize early drying shrinkage.

Special precautions are necessary when concrete work continues during periods of adverse weather. In cold weather, arrangements should be made in advance for heating, covering, insulating, or enclosing the concrete (see Chapter 16). Hot-weather work may require special precautions against rapid evaporation and drying and high temperatures (see Chapter 17).

Rain Protection

Prior to commencing placement of concrete, the owner and contractor should be aware of procedures to be followed in the event of rain during the placing operation. Protective coverings such as polyethylene sheets or tarpaulins should be available and onsite at all times.

Figure 14-29. An excellent method of wet curing is to completely cover the surface with wet burlap and keep it continuously wet during the curing period.

When rain occurs, if practical, all batching and placing operations should stop and the fresh concrete should be covered to the extent that the rain does not indent the surface of the concrete or wash away the cement paste. When rain ceases, the covering should be removed and remedial measures taken such as surface retexturing or reworking in-place plastic concrete, before concrete placing resumes.

Jointing Concrete

The following three types of joints are common in concrete construction: isolation joints, contraction joints, and construction joints.

Isolation Joints

Isolation joints permit both horizontal and vertical differential movements at adjoining parts of a structure (Figure 14-30). They are used, for example, around the perimeter of a floor on ground, around columns; around machine foundations to separate the slab from the more rigid parts of the structure; and to isolate slabs-on ground from fixed foundation members.

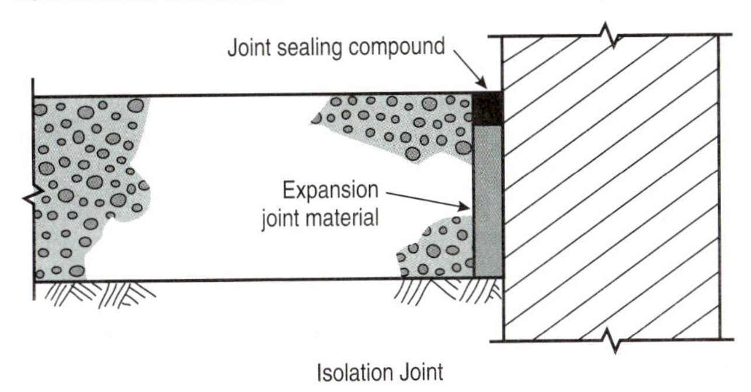

Figure 14-30. Isolation joints permit horizontal and vertical movements between abutting faces of a slab and fixed parts of a structure.

Isolation-joint material (often called expansion-joint material) can be as thin as 6 mm (¼ in.) or less, but 13-mm (½-in.) material is commonly used. Care should be taken to ensure that all the edges for the full depth of the slab are isolated from adjoining construction to minimize cracking stress potential.

Columns on separate footings are isolated from the floor slab either with a circular or square-shaped isolation joint. The square shape should be rotated to align its corners with control and construction joints.

Contraction Joints

Contraction joints provide for movement in the plane of a slab or wall. Joints induce controlled cracking caused by drying and thermal shrinkage at preselected locations (Figure 14-31). Contraction joints (also sometimes called control joints) should be constructed to permit transfer of loads perpendicular to the plane of a slab or wall. If no contraction joints are used, or if they are too widely spaced in slabs on ground or in lightly reinforced walls, random cracks may occur; cracks are most likely when drying and thermal shrinkage produce tensile stresses in excess of the concrete's tensile strength.

Contraction joints in slabs on ground can be created using several techniques. One of the most common methods is to saw a continuous straight slot in the top of the slab (Figure 14-32). When installed to a depth of ¼ the slab thickness, these joints create planes of weakness at which cracks are expected to form. Vertical loads are transmitted across a contraction joint by aggregate interlock between the opposite faces of the crack providing the crack is not too wide and the spacing between joints is not too great.

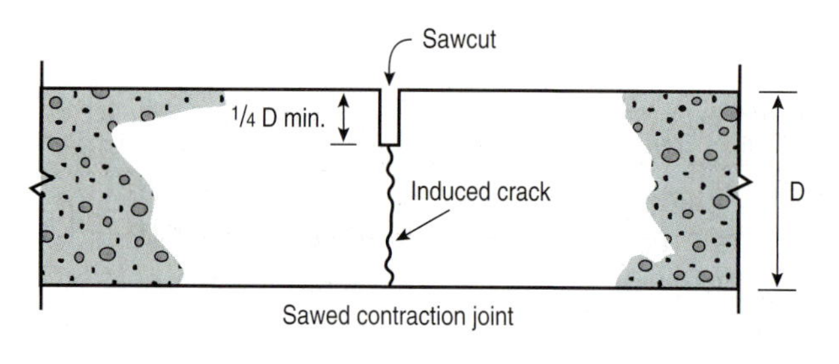

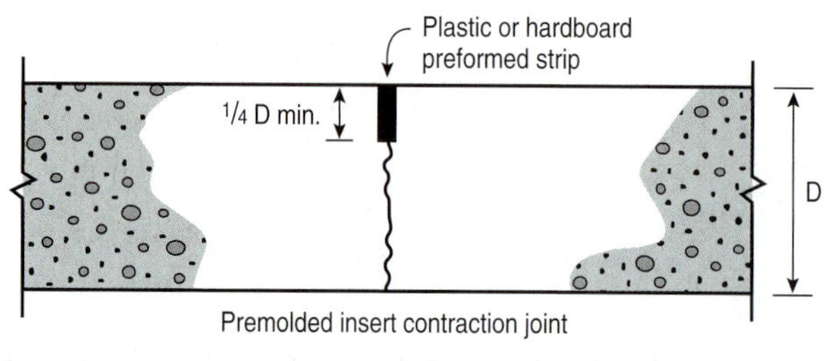

Figure 14-31. Contraction joints provide for horizontal movement in the plane of a slab or wall and induce controlled cracking caused by drying and thermal shrinkage.

Crack widths at saw-cut contraction joints that exceed 0.9 mm (0.035 in.) through the depth of the slab do not reliably transfer loads. However, the width of the crack at the suface cannot be used to evaluate potential load transfer capability as cracks are typically much wider at the drying surface than at the bottom. In addition, the effectiveness of load transfer by aggregate interlock depends on other variables than crack width. Other factors include: slab thickness, subgrade support, load magnitude, repetitions of load, and aggregate angularity. Smooth steel dowels may be used to increase load transfer at contraction joints when heavy wheel loads are anticipated. Dowels may be round, square, or plate shapes. Sizes and spacing of dowels, which are placed at the center of the slab depth, are shown in Tarr and Farny (2008). See ACI Committee 302 and PCA (1982) for further discussions on doweled joints.

Figure 14-32. Sawing a continuous cut in the top of a slab is one of the most economical methods for making a contraction joint.

Sawing must be coordinated with the setting time of the concrete. It should be started as soon as the concrete has hardened sufficiently to prevent aggregates from being dislodged by the saw (usually within 4 to 12 hours after the concrete hardens). Sawing should be completed before drying shrinkage stresses become large enough to produce cracking. The timing depends on factors such as mix proportions, ambient conditions, and type and hardness of aggregates. New dry-cut sawing techniques allow saw cutting to take place shortly after final finishing is completed. Generally, the slab should be cut before the concrete cools, after the concrete sets enough to prevent raveling or tearing while saw cutting, and before drying-shrinkage cracks start to develop.

Contraction joints can also be formed in the fresh concrete with hand groovers or by placing strips of wood, metal, plastic, or preformed joint material at the joint locations. The top of the strips should be flush with the concrete surface. Contraction joints, whether sawed, grooved, or preformed, should extend vertically into the slab to a depth of at least one-fourth the slab thickness or a minimum of 25 mm (1 in.) deep. It is recommended that the joint depth not exceed one-third the slab thickness if load transfer from aggregate interlock is important. While industrial slabs subjected to hard-wheeled traffic should be sawcut, contraction joints may be formed using a combination of either grooving or inserts and sawing to achieve the final joint depth.

Contraction joints in walls are also planes of weakness that permit differential movements in the plane of the wall. The thickness of the wall at a contraction joint should be reduced by a minimum of 25%, preferably 30%. Under the guidance of the design engineer, in lightly reinforced walls, half of the horizontal steel reinforcement should be cut at the joint. Care must be taken to cut alternate reinforcement bars precisely at the joint. A gap created by two cuts in the bar can lower the risk that the cut is not precisely located at the joint. At the corners of openings in walls where contraction joints are located, extra diagonal or vertical and horizontal reinforcement should be provided to control crack width. Contraction joints in walls should be spaced not more than about 6 meters (20 ft) apart. In addition, contraction joints should be placed where abrupt changes in wall thickness or height occur, and near corners – if possible, within 3 meters to 4 meters (10 ft to 15 ft). Depending on the structure, these joints may need to be caulked to prevent the passage of water through the wall. Instead of caulking, a waterstop can be used to prevent water from leaking through the crack that occurs in the joint.

The spacing of contraction joints in floors on ground depends on (1) slab thickness, (2) shrinkage potential of the concrete, (3) subgrade friction, (4) service environment, and (5) the absence or presence of steel reinforcement. Unless reliable data indicate that more widely spaced joints are feasible, the suggested intervals given in Table 14-3 should be used for well-proportioned concrete with aggregates having normal shrinkage characteristics. However, the risk of random cracking is reduced if the joint spacing is held to a maximum of 4.5 m (15 ft) regardless of the slab thickness. Joint spacing should certainly be decreased for concrete that may have high shrinkage characteristics. The panels created by contraction joints should be approximately square. Panels with excessive length-to-width ratio (more than 1½ to 1) are likely to crack near the center of the long dimension of the panel. In joint layout design it is also important to remember that contraction (control) joints should only terminate at a free edge or at an isolation joint. Contraction joints should never terminate at another contraction joint (T-intersection) as cracking will be induced from the end of the terminated joint into the adjacent panel. This is sometimes referred to as sympathetic cracking. Refer to Figure 14-36, which illustrates one possible joint layout solution to eliminate the potential for induced sympathetic cracking.

Table 14-3. Spacing of Contraction Joints in Meters (Feet)*

Slab thickness mm (in.)	Maximum-size aggregate less than 19 mm (¾ in.)	Maximum-size aggregate 19 mm (¾ in.) and larger
125 (5)	3.0 (10)	3.75 (13)
150 (6)	3.75 (12)	4.5 (15)
175 (7)	4.25 (14)	5.25 (18)**
200 (8)	5.0 (16)**	6.0 (20)**
225 (9)	5.5 (18)**	6.75 (23)**
250 (10)	6.0 (20)**	7.5 (25)**

* If concrete cools at an early age, shorter spacings may be needed to control random cracking. A temperature difference of only 6°C (10°F) may be critical. For slump less than 100 mm (4 in.), joint spacing can be increased by 20%.

**When spacings exceed 4.5 m (15 ft), load transfer by aggregate interlock decreases markedly. If shrinkage is high or unknown, joints should not exceed 4.5 m (15 ft).

Construction Joints

Construction joints (Figure 14-33) are stopping places in the process of construction. In structural building systems, a true construction joint should bond new concrete to existing concrete and permit no movement. Deformed

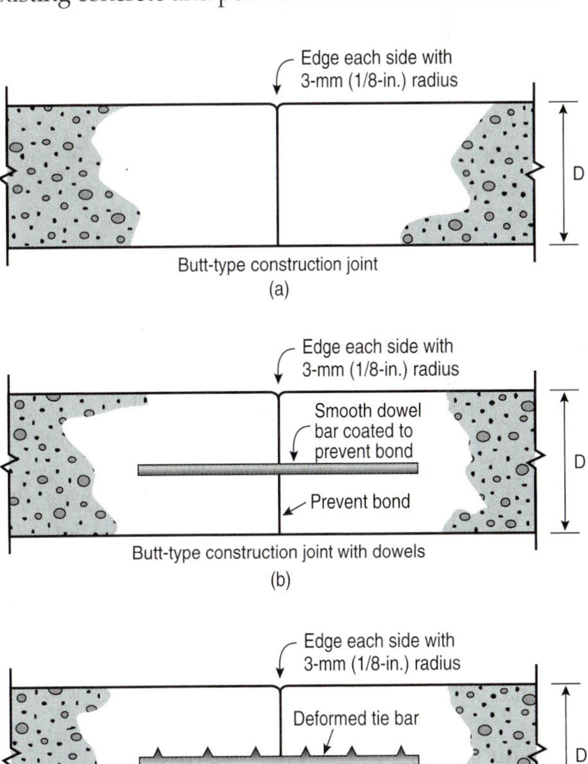

Figure 14-33. Construction joints are stopping points in the construction process. Construction-joint types (a) and (b) are also used as contraction joints.

tiebars are often used in construction joints to restrict movement. For slabs on ground, construction joints are designed and built to function as contraction or isolation joints. For example, in a floor on ground the construction joints align with columns and function as contraction joints and therefore are purposely made unbonded. The structural designer of suspended slabs should decide on the location of construction joints. Oils, form-release agents, and paints are used as debonding materials. In thick, heavily-loaded floors, unbonded doweled construction joints are commonly used. For thin slabs that are not loaded by heavy moving loads or subject to differential soil movement, the flat-faced butt-type joint will suffice.

On most structures it is desirable to have wall joints that will not detract from appearance. When properly made, wall joints can become inconspicuous or hidden by rustication strips. They can become an architectural as well as a functional feature of the structure. However, if rustication strips are used in structures exposed to deicing salts, such as bridge columns and abutments, care should be taken to ensure that the reinforcing steel has the required depth of concrete cover to prevent corrosion.

Horizontal Construction Joints

For making a horizontal construction joint in reinforced concrete wall construction, good results have been obtained by constructing the forms to the level of the joint, overfilling the forms a few centimeters (inches), and then removing the excess concrete just before hardening occurs; the top surface then can be manually roughened with stiff brushes. The procedure is illustrated in Figure 14-34.

In the case of vertical construction joints cast against a bulkhead, the concrete surface generally is too smooth to permit proper bonding. So, particular care should be given to removal of the smooth surface finish before re-erecting the forms for newer placements against the joint. Stiff-wire brushing may be sufficient if the concrete is less than three days old; otherwise, bushhammering or sandblasting may be needed. This should be followed by washing with clean water to remove all dust and loose particles.

Horizontal joints in walls should be made straight, horizontal, and should be placed at suitable locations. A straight horizontal construction joint can be made by nailing a 25 mm (1 in.) wood strip to the inside face of the form near the top (Figure 14-34). Concrete should then be placed to a level slightly above the bottom of the strip. After the concrete has settled and before it becomes too hard, any laitance that has formed on the top surface should be removed. The strip can then be removed and any irregularities in the joint leveled off. The forms are removed and then reerected above the construction joint for the next lift of concrete. To prevent concrete leakage from staining the wall below, gaskets should be used where forms contact previously placed hardened concrete.

A variation of this procedure makes use of a rustication strip instead of the 25 mm (1 in.) wood strip to form a groove in the concrete for architectural effect (Figure 14-35). Rustication strips can be V-shaped, rectangular, or slightly beveled. If a V-shaped strip is used, the joint should be located at the point of the V. If rectangular or beveled, the joint should be made at the top edge of the inner face of the strip.

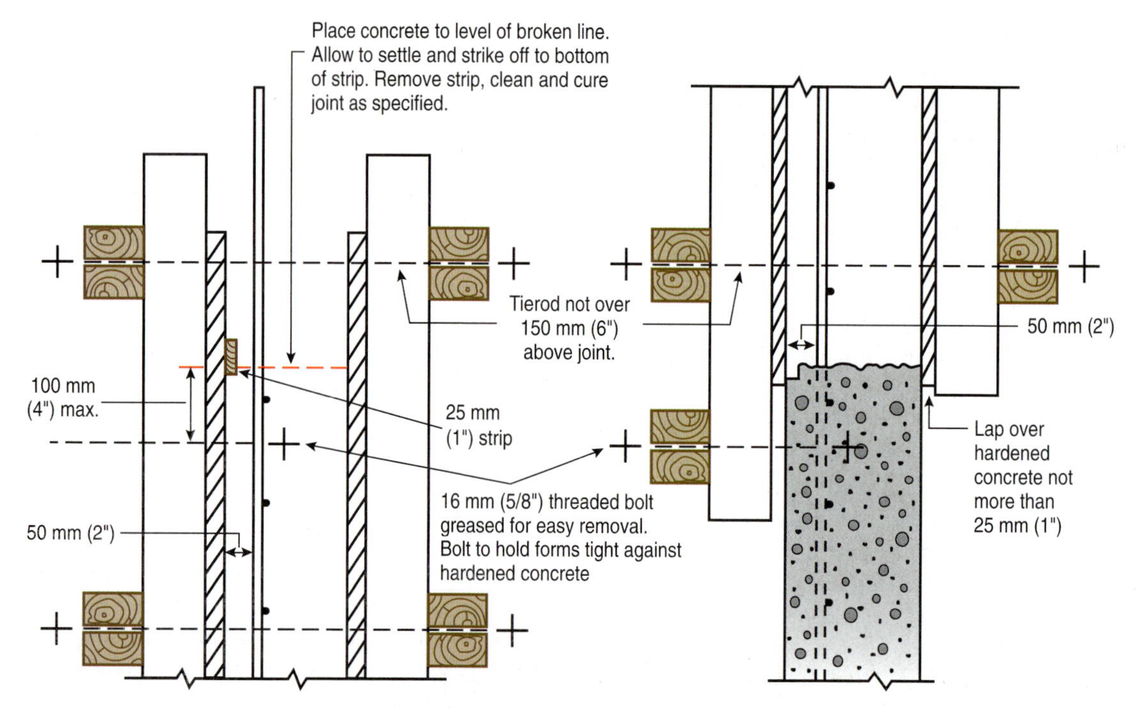

Figure 14-34. A straight, horizontal construction joint can be built using this detail.

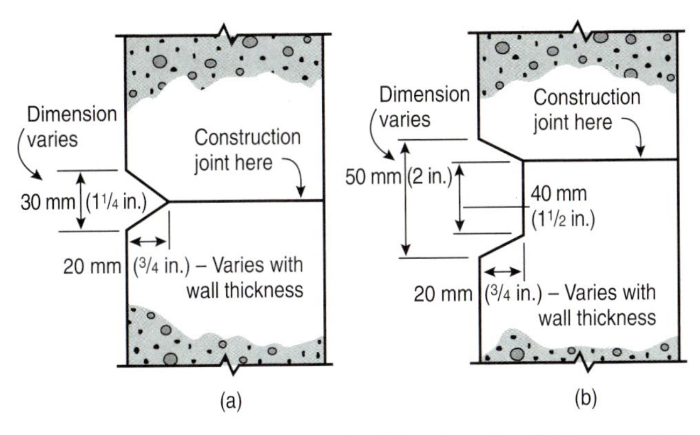

Figure 14-35. Horizontal construction joints in walls with V-shaped (a) and beveled (b) rustication strips.

Joint Layout For Floors

A typical joint layout for all three joint types – isolation, contraction, and construction – is illustrated in Figure 14-36. Isolation joints are provided around the perimeter of the floor where it abuts the walls and around all fixed elements that may restrain movement of the slab. This includes columns and machinery bases that penetrate the floor slab. These isolation joints will not be loaded by moving wheel loads. With the slab isolated from other building elements, the remaining task is to locate and correctly space contraction joints to eliminate random cracking. Construction joint locations are coordinated with the floor contractor to accommodate work schedules and crew size. Unbonded construction joints should coincide with the contraction joint pattern and act as contraction joints. Construction joints should be planned to provide long-strips for each placement rather than a checker-board pattern. Contraction joints are then placed to divide the long-strips into relatively square panels, with panel length not exceeding 1.5 times the width. Contraction joints should stop at free edges or isolation joints. Avoid reentrant corners in joint layout whenever possible. For more information on joints, see ACI Committee 302 (2004), PCA (1982), and Tarr and Farny (2008). For joints in walls, see PCA (1982), PCA (1982a), PCA (1984), PCA (1984a), and PCA (1984b).

Filling Floor Joints

There are three options for treating joints: they can be filled, sealed, or left open. The movement at contraction joints in a floor is generally very small. For some industrial and commercial uses, these joints can be left unfilled or unsealed. Where there are wet conditions or hygienic and dust-control requirements, joints should be sealed to prevent moisture penetration through joints or infiltration of dirt and debris. Joint sealers are flexible and can accommodate substantial joint movement. Exterior pavement

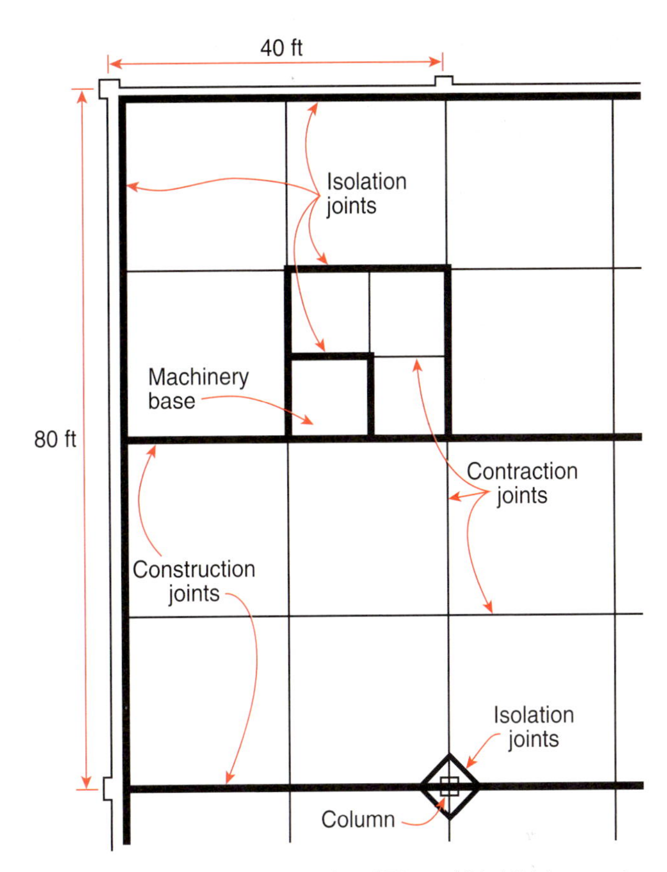

Figure 14-36. Typical joint layout for a 150-mm (6-in.) thick concrete floor on ground. Note, due to the isolated machinery base the full adjoining slab panel should be bordered with an isolation joint and load transfer devices (plates or dowels) should be provided, or the adjoining panel should be bordered with contraction joints, the interior contraction joints should be eliminated and steel reinforcement provided at the re-entrant corner to hold random cracks tightly closed.

joints are often sealed because they are only loaded by pneumatic tires. This type of loading imparts relatively low stresses that typically do not cause spalling. However, when traffic by small, hard-wheeled vehicles will cross joints, such as for interior industrial slabs, filling is necessary. Joint fillers are semi-rigid so they can provide some compressive lateral support to the joint walls which protects the joints from spalling.

The difference between a filler and a sealer is the hardness of the material. Fillers are more rigid than sealers and provide support to joint edges. In many places where traffic loading is light, a resilient material such as a polyurethane elastomeric sealant is satisfactory. However, heavy-traffic areas require support for joint edges to prevent spalling at saw-cuts; in such cases a good quality, semi-rigid epoxy or polyurea filler with a Shore Hardness of A-80 or D-50 (ASTM D2240, *Standard Test Method for Rubber Property – Durometer Hardness*) should be used. The material should be installed full depth in the saw cut, without a backer rod, and trimmed flush with the floor surface.

Isolation joints are intended to accommodate movement; thus a flexible, elastomeric sealant should be used to keep foreign materials out of the joint.

Unjointed Floors

An unjointed floor, or one with a limited number of joints, can be constructed when joints are unacceptable. Three unjointed floor methods are suggested:

1. A prestressed floor can be built through the use of post-tensioning. With this method, steel strands in ducts are tensioned after the concrete hardens to produce compressive stress in the concrete. This compressive stress will counteract the development of tensile stresses in the concrete and provide a crack-free floor. Large areas, 1000 m^2 (10,000 ft^2) and greater, can be constructed in this manner without intermediate joints.
2. Large areas – a single day of slab placement, usually 800 m^2 to 1000 m^2 (8000 ft^2 to 10,000 ft^2) – can be cast without contraction joints when the amount of distributed steel in the floor is about one-half of one percent of the cross-sectional area of the slab. Special effort should be made to reduce subgrade friction in floors without contraction joints. Tarr and Farny (2008) discuss use of distributed steel in floors.
3. Concrete made with expansive cement can be used to offset the amount of drying shrinkage to be anticipated after curing. Contraction joints are not needed when construction joints are used at intervals of 10 meters to 35 meters (40 ft to 120 ft). Large areas without joints, up to 2000 m^2 (20,000 ft^2), have been cast in this manner. Steel reinforcement is needed in order to produce compressive stresses during andafter the expansion period – this is a form of prestressing.

Removing Forms

It is advantageous to leave forms in place as long as possible to continue the curing period. However, there are times when it is necessary to remove forms as soon as possible. Vertical forms are typically removed at the start of the next day provided that concrete has a minimum strength of 0.4 f'_c, or at least 11 MPa (1600 psi). Where a rubbed finish is specified, forms must be removed early to permit the first rubbing before the concrete becomes too hard. Furthermore, it is often necessary to remove forms quickly to permit their immediate reuse.

In any case, shoring should never be removed until the concrete is strong enough to satisfactorily carry the stresses from both the dead load of the structure and any imposed construction loads. The concrete should be strong enough so that the surfaces will not be damaged in any way when reasonable care is used in removing forms. In general, for concrete temperatures above 10°C (50°F), the side forms of reasonably thick, supported sections can usually be removed 24 hours after concreting. Beam and floor slab forms and supports (shoring) may be removed between 3 and 21 days, depending on the size of the member and the strength gain of the concrete. For most conditions, it is better to rely on the strength of the concrete as determined by in situ or field-cured specimen testing rather than arbitrarily selecting an age at which forms may be removed. Advice on reshoring is provided by ACI Committee 347.

For form removal, the designer should specify the minimum strength requirements for various members. The age-strength relationship should be determined from representative samples of concrete used in the structure and field-cured under job conditions. It should be remembered, however, that strengths are affected by the materials used, concrete temperature, curing temperature and other conditions. The time required for form removal, therefore, will vary from job to job.

Stripping should be started some distance away from and then move toward a projection. This relieves pressure against projecting corners and reduces the chance of edges breaking off. If it is necessary to wedge between the concrete and the form, only wooden wedges should be used. A pinch bar or other metal tool should not be placed against the concrete to pry away formwork.

Recessed forms require special attention. Wooden wedges should be gradually driven behind the form and the form should be tapped lightly to break it away from the concrete. Forms should not be pulled off rapidly after wedging has been started at one end; this is almost certain to break the edges of the concrete.

Patching and Cleaning Concrete

After forms are removed, all undesired bulges, fins, and small projections can be removed by chipping or tooling. Undesired bolts, nails, ties, or other embedded metal can be removed or cut back to a depth of 13 mm (½ in.) from the concrete surface. When required, the surface can be rubbed or ground to provide a uniform appearance (see **Finishing Formed Surfaces**). Any cavities such as tierod holes should be filled unless they are intended for decorative purposes. Honeycombed areas must be repaired and stains removed to present a concrete surface that is uniform in color. All of these operations can be minimized by exercising care in constructing the formwork and placing the concrete. In general, repairs are easier and more successful if they are made as soon as practical, preferably as soon as the forms are removed. However, the procedures discussed below apply to both new and old hardened concrete.

Holes, Defects, and Overlays

Patches usually appear darker than the surrounding concrete. White cement can be used in mortar or concrete for patching where appearance is important. Samples should be applied and cured in an inconspicuous location to determine the most suitable proportions of white and gray cements. Steel troweling should be avoided since this may darken the patch.

Bolt holes, tierod holes, and other cavities that are small in area but relatively deep should be filled with a dry-pack mortar. The mortar should be mixed as stiff as is practical: use 1 part cement, 2½ parts sand passing a 1.25 mm (No. 16) sieve, and just enough water to form a ball when the mortar is squeezed gently in the hand. The cavity should be clean with no oil or loose material and kept damp with water for several hours. A neat-cement paste should be scrubbed onto the void surfaces, but not allowed to dry before the mortar is placed. The mortar should be tamped into place in layers about 13 mm (½ in.) thick. Vigorous tamping and adequate curing will ensure good bond and minimum shrinkage of the patch.

Concrete used to fill large patches and thin-bonded overlays should have a low water-cement ratio, often with a cement content equal to or greater than the concrete to be repaired. Cement contents range from 360 kg/m^3 to 500 kg/m^3 (600 lb/ft^3 to 850 lb/ft^3) and the water-cement ratio is usually 0.45 or less. The aggregate size should be no more than ⅓ the patch or overlay thickness. A 9.5-mm (⅜-in.) nominal maximum size coarse aggregate is commonly used. The sand proportion can be higher than usual, often equal to the amount of coarse aggregate, depending on the desired properties and application.

Before the patching concrete is applied, the surrounding concrete should be clean and sound (Figure 14-37). Abrasive methods of cleaning (sandblasting, hydrojetting, waterblasting, scarification, or shotblasting) are usually required. For overlays, a cement-sand grout, a cement-sand-latex grout, or an epoxy bonding agent should be applied to the prepared surface with a brush or broom (see **Bonding New to Previously Hardened Concrete**). Typical grout mix proportions are 1 part cement and 1 part fine sand and latex or epoxy admixtures. The grout should be applied immediately before the new concrete is placed. The grout should not be allowed to dry before the freshly mixed concrete is placed; otherwise bond may be impaired. The existing concrete may be dry or damp when the grout is applied but not wet with free-standing water. The minimum thickness for most patches and overlays is 20 mm (¾ in.). Some structures, like bridge decks, should have a minimum repair thickness of 40 mm (1½ in.). A superplasticizer is one of many admixtures often added to overlay or repair concrete to reduce the water-cement ratio, to improve workability and ease consolidation (Kosmatka 1985a).

Figure 14-37. Concrete prepared for patch installation.

Honeycombed and other defective concrete should be cut out to expose sound material. If defective concrete is left adjacent to a patch, moisture may get into the voids; in time, weathering action will cause the patch to spall. The edges of the defective area should be cut or chipped straight and at right angles to the surface, or slightly undercut to provide a key at the edge of the patch. No featheredges should be permitted (Figure 14-38). Based on the size of the patch, either a mortar or a concrete patching mixture should be used.

(a) Incorrectly installed patch. The feathered edges will break down under traffic or weather away.

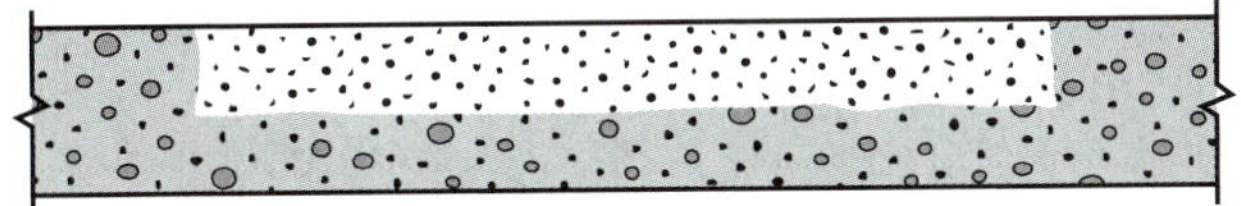

(b) Correctly installed patch. The chipped area should be at least 20 mm (¾ in.) deep with the edges at right angles or undercut to the surface.

Figure 14-38. Concrete patch installation.

Shallow patches can be filled with a dry-pack mortar as described earlier. This should be placed in layers not more than 13 mm (½ in.) thick, with each layer given a scratch finish to improve bond with the subsequent layer. The final layer can be finished to match the surrounding concrete by floating, rubbing, or tooling. Formed surfaces are matched by pressing a section of form material against the patch while still plastic.

Deep patches can be filled with concrete held in place by forms. Such patches should be reinforced and doweled to the hardened concrete (Bureau of Reclamation 1981). Large, shallow vertical or overhead repairs may best be accomplished by shotcreting. Several proprietary low-shrinkage cementitious repair products are also available.

Curing Patches

Following patching, good curing is essential (Figure 14-39). Curing should be started immediately to avoid early drying. Wet burlap, wet sand, plastic sheets, curing paper, tarpaulins, or a combination of these can be used. In locations where it is difficult to hold these materials in place, an application of two coats of membrane-curing compound is often the preferred method.

Figure 14-39. Good curing is essential to successful patching. This patch is covered with polyethylene sheeting plus rigid insulation to retain moisture and heat for rapid hydration and strength gain.

Cleaning Concrete Surfaces

Concrete surfaces are not always uniform in color when forms are removed; they may have a somewhat blotchy appearance and there may be a slight film of form-release agent in certain areas. There may be mortar stains from leaky forms or there may be rust stains. Flatwork can also become discolored during construction. Where appearance is important, all surfaces should be cleaned after construction has progressed to the stage where there will be no discoloration from subsequent construction activities.

There are three techniques for cleaning concrete surfaces: water, chemical, and mechanical (abrasion). Water dissolves dirt and rinses it from the surface. Chemical cleaners, usually mixed with water, react with dirt to separate it from the surface, and then the dirt and chemicals are rinsed off with clean water. Mechanical methods – sandblasting is most common – remove dirt by abrasion.

Before selecting a cleaning method, it should be tested on an inconspicuous area to be certain it will be helpful and not harmful. If possible, identify the characteristics of the discoloration because some treatments are more effective than others in removing certain materials.

Water cleaning methods include low-pressure washes, moderate-to-high-pressure waterblasting, and steam. Low-pressure washing is the simplest method, requiring only that water run gently down the concrete surface for a day or two. The softened dirt is then flushed using a slightly higher pressure rinse. Stubborn areas can be scrubbed with a nonmetallic-bristle brush and rinsed again. High-pressure waterblasting is used effectively by experienced operators. Steam cleaning must be performed by skilled operators using special equipment. Water methods are the least harmful to concrete, but they are not without potential problems. Serious damage may occur if the concrete surface is subjected to freezing temperatures while it is still wet; and water can bring soluble salts to the surface, forming a chalky, white deposit called efflorescence.

Chemical cleaning is usually done with water-based mixtures formulated for specific materials such as brick, stone, and concrete. An organic compound called a surfactant (surface-active agent), which acts as a detergent to wet the surface more readily, is included in most chemical cleaners. A small amount of acid or alkali is included to separate the dirt from the surface. For example, hydrochloric (muriatic) acid is commonly used to clean masonry walls and remove efflorescence. There can be problems related to the use of chemical cleaners. Their acid or alkaline properties can lead to reaction between cleaner and concrete as well as mortar, painted surfaces, glass, metals, and other building materials. Since chemical cleaners are used in the form of water-diluted solutions, they too can liberate soluble salts from within the concrete to form efflorescence. Some chemicals can also expose the aggregate in concrete. Chemicals commonly used to clean concrete surfaces and remove discoloration include weak solutions (1% to 10% concentration) of hydrochloric, acetic, or phosphoric acid. Diammonium citrate (20% to 30% water solution) is especially useful in removing discoloration stains and efflorescence on formed and flatwork surfaces. Chemical cleaners should be used by skilled operators taking suitable safety precautions. See Greening and Landgren (1966) and PCA (1988) for more information.

Mechanical cleaning includes sandblasting, shotblasting, scarification, power chipping, and grinding. These methods wear the dirt off the surface rather than separate it from the surface. They, in fact, wear away both the dirt and some of the concrete surface; it is inevitable that there will be some loss of decorative detail, increased surface roughness, and rounding of sharp corners. Abrasive methods may also reveal defects (voids) hidden just beneath the formed surface.

Chemical and mechanical cleaning can each have an abrading effect on the concrete surface that may change the appearance of a surface compared to that of an adjacent uncleaned surface.

Precautions

Protect Your Head and Eyes. Construction equipment and tools present constant potential hazards to busy construction personnel. That's why hard hats are required on construction projects.

Proper eye protection is essential when working with cement or concrete. Eyes are particularly vulnerable to blowing dust, splattering concrete, and other foreign objects. On some jobs it may be advisable to wear full-cover goggles or safety glasses with side shields. Actions that cause dust to become airborne should be avoided. Local or general ventilation can control exposures below applicable exposure limits. Respirators may be used in poorly ventilated areas, where exposure limits are exceeded, or when dust causes discomfort or irritation.

Protect Your Back. All materials used to make concrete – portland cement, sand, coarse aggregate, and water – can be quite heavy, even in small quantities. When lifting heavy materials, the back should be straight, legs bent, and the weight between the legs as close to the body as possible. Mechanical equipment should be used to place concrete as close as possible to its final position. After the concrete is deposited in the desired area by chute, pump, or wheelbarrow, it should be pushed – not lifted – into final position with a shovel. A short-handled, square-end shovel is an effective tool for spreading concrete, but special concrete rakes or come-alongs can also be used. Excessive horizontal movement of the concrete should be avoided. It not only requires extra effort, but may also lead to segregation of the concrete ingredients.

Working Safely with Concrete. Fresh, moist concrete is alkaline in nature and can be extremely caustic. It is, therefore, important to avoid prolonged, direct contact between the skin and wet concrete or clothing saturated with wet concrete. Such contact can cause skin irritation and severe chemical burns (PCA 1998). Following contact, use fresh water to thoroughly wash the skin areas exposed to the fresh concrete.

A chemical burn from concrete occurs with very little warning because little or no heat is sensed by the skin. Seek **immediate** medical attention if irritation or inflammation begins on exposed skin areas or if you have persistent or severe discomfort.

If clothing areas become saturated from contact with fresh concrete, mortar, or grout, promptly rinse them with clean water to prevent continued contact with skin surfaces. Indirect contact through clothing can be as serious as direct contact. When working with fresh concrete, begin each workday by wearing clean clothing. To insure all concrete has been removed, it is prudent to shower or bathe after the workday is complete.

Wear waterproof gloves, long-sleeve shirts, full length trousers, and waterproof leather or rubber boots when working with fresh concrete. Boots must be high enough to prevent fresh concrete from flowing into them. When finishing concrete, use waterproof pads between fresh concrete surfaces and knees. Wear gloves to protect hands. Also, wear proper eye protection when working with fresh concrete, mortar, or grout. If contact with the eyes is made, flush them immediately with fresh water and seek **prompt** medical attention.

When handling dry cement, or when working in an environment where cement dust is present, wear proper eye protection and a NIOSH-approved dust respirator. As with fresh concrete, if dry cement gets in your eyes, flush the eyes immediately and repeatedly with water and seek **prompt** medical attention.

References

ACI Committee 117, *Specifications for Tolerances for Concrete Construction and Materials (ACI 117-10) and Commentary*, ACI 117-10, ACI Committee 117 Report, American Concrete Institute, Farmington Hills, Michigan, 2010, 76 pages.

ACI Committee 301, *Specifications for Structural Concrete*, ACI 301-10, ACI Committee 301 Report, American Concrete Institute, Farmington Hills, Michigan, 2010.

ACI Committee 302, *Guide for Concrete Floor and Slab Construction*, ACI 302.1R-04, American Concrete Institute, Farmington Hills, Michigan, 2004, 77 pages.

ACI Committee 303, *Guide to Cast-in-Place Architectural Concrete Practice*, ACI 303R-04, American Concrete Institute, Farmington Hills, Michigan, 2004, 32 pages.

ACI Committee 304, *Guide for Measuring, Mixing, Transporting and Placing Concrete*, ACI 304R-00, American Concrete Institute, Farmington Hills, Michigan, 2000, 41 pages.

ACI Committee 306, *Cold-Weather Concreting*, ACI 306R-10, American Concrete Institute, Farmington Hills, Michigan, 2010, 30 pages.

ACI Committee 309, *Guide for Consolidation of Concrete*, ACI 309R-96, American Concrete Institute, Farmington Hills, Michigan, 1996, 39 pages.

ACI Committee 347, *Guide to Formwork for Concrete*, ACI 347-04, American Concrete Institute, Farmington Hills, Michigan, 2004, 32 pages.

ACI, "Vapor Retarder Location," *Concrete International*, American Concrete Institute, Farmington Hills, Michigan, April 2001, pages 72 and 73.

Bureau of Reclamation, *Concrete Manual*, 8th ed., U.S. Bureau of Reclamation, Denver, revised 1981.

Caldarone, Michael A.; Taylor, Peter C.; Detwiler, Rachel J.; and Bhidé, Shrinivas B., *Guide Specification for High-Performance Concrete for Bridges*, EB233, Portland Cement Association, Skokie, Illinois, 2005, 64 pages.

Collins, Terry C.; Panarese, William C.; and Bradley, Bentley J., *Concrete Finisher's Guide*, EB122, Portland Cement Association, 2006, 88 pages.

Daniel, D. Gene, and Lobo, Colin L., *User's Guide to ASTM Specification C94 for Ready-Mixed Concrete,* MNL 49, ASTM International, West Conshohocken, Pennsylvania, 2005, 130 pages.

Detwiler, R.J., and Taylor, P.C., *Specifier's Guide to Durable Concrete,* EB221, 2nd edition, Portland Cement Association, Skokie, Ilinois, 2005, 72 pages.

Greening, N.R., and Landgren, R., *Surface Discoloration of Concrete Flatwork,* Research Department Bulletin RX203, Portland Cement Association, http://www.cement.org/pdf_files/RX203.pdf, 1966.

Harrington, Dale, *Guide to Concrete Overlays,* ACPA TB021, National Concrete Pavement Technology Center, Ames, Iowa, www.cptechcenter.org/publications/overlays/guide_concrete_overlays_2nd_ed.pdf, 2008, 86 pages.

Hedenblad, Göran, *Drying of Construction Water in Concrete-Drying Times and Moisture Measurement*, LT229, Swedish Council for Building Research, Stockholm, 1997, 54 pages.

Hedenblad, Göran, "Concrete Drying Time," *Concrete Technology Today*, PL982, Portland Cement Association, http://www.cement.org/pdf_files/PL982.pdf, 1998, pages 4 and 5.

Hover, K.C., "Vibration Tune-up," *Concrete International*, American Concrete Institute, Farmington Hills, Michigan, September 2001, pages 31 to 35.

Hurd, M.K., *Formwork for Concrete*, SP-4, 7th edition, American Concrete Institute, Farmington Hills, Michigan, 2005, 500 pages.

Kanare, Howard M., *Concrete Floors and Moisture*, EB119, Portland Cement Association, Skokie, Illinois, and National Ready Mixed Concrete Association, Silver Spring, Maryland, USA, 2008, 176 pages.

Kosmatka, Steven H., and Collins, Terry C., *Finishing Concrete with Color and Texture*, PA124, Portland Cement Association, 2004, 72 pages.

Kosmatka, Steven H., "Floor-Covering Materials and Moisture in Concrete," *Concrete Technology Today*, PL853, Portland Cement Association, http://www.cement.org/pdf_files/PL853.pdf, September 1985.

Kosmatka, Steven H., "Repair with Thin-Bonded Overlay," *Concrete Technology Today*, PL851, Portland Cement Association, http://www.cement.org/pdf_files/PL851.pdf, March 1985a.

Litvin, Albert, *Clear Coatings for Exposed Architectural Concrete,* Development Department Bulletin DX137, Portland Cement Association, http://www.cement.org/pdf_files/DX137.pdf, 1968.

NRMCA/ASCC, *Checklist for the Concrete Pre-Construction Conference*, 1999.

PCA, *Building Movements and Joints*, EB086, Portland Cement Association, http://www.cement.org/pdf_files/EB086.pdf, 1982, 68 pages.

PCA, *Bushhammering of Concrete Surfaces*, IS051, Portland Cement Association, 1972.

PCA, *Color and Texture in Architectural Concrete*, SP021, Portland Cement Association, 1995, 36 pages.

PCA, *Concrete Basements for Residential and Light Building Construction*, IS208, Portland Cement Association, 1980a.

PCA, *Concrete Slab Surface Defects: Causes, Prevention, Repair*, IS177, Portland Cement Association, 2001, 12 pages.

PCA, *Joints in Walls Below Ground*, CR059, Portland Cement Association, 1982a.

PCA, "Joints to Control Cracking in Walls," *Concrete Technology Today*, PL843, Portland Cement Association, http://www.cement.org/pdf_files/PL843.pdf, September 1984.

PCA, *Painting Concrete*, IS134, Portland Cement Association, 1992, 8 pages.

PCA, *Removing Stains and Cleaning Concrete Surfaces*, IS214, Portland Cement Association, http://www.cement.org/pdf_files/IS214.pdf, 1988, 16 pages.

PCA, "Sealants for Joints in Walls," *Concrete Technology Today*, PL844, Portland Cement Association, http://www.cement.org/pdf_files/PL844.pdf, December 1984a.

PCA, "Why Concrete Walls Crack," *Concrete Technology Today*, PL842, Portland Cement Association, http://www.cement.org/pdf_files/PL842.pdf, June 1984b.

PCA, *Working Safely with Concrete*, MS271, Portland Cement Association, 1998, 6 pages.

Stark, David C., *Effect of Vibration on the Air-Void System and Freeze-Thaw Durability of Concrete*, Research and Development Bulletin RD092, Portland Cement Association, http://www.cement.org/pdf_files/RD092.pdf, 1986.

Suprenant, Bruce A., "Free Fall of Concrete," *Concrete International*, American Concrete Institute, Farmington Hills, Michigan, June 2001, pages 44 and 45.

Tarr, Scott M., and Farny, James A., *Concrete Floors on Ground*, EB075, Portland Cement Association, 2008, 256 pages.

CHAPTER 15

Curing Concrete

Curing is the maintenance of a satisfactory moisture content and temperature in concrete for a sufficient period of time during and immediately following placing so that the desired properties may develop (Figure 15-1). The need for adequate curing of concrete cannot be overemphasized. Curing has a strong influence on the properties of hardened concrete. Curing improves strength, volume stability, permeability resistance, and durability (including abrasion resistance and resistance to freezing and thawing and deicer scaling). Exposed slab surfaces are especially sensitive to curing as strength development and durability of the top surface of a slab can be reduced significantly when curing is neglected.

Figure 15-1. Curing should begin as soon as the concrete stiffens enough to prevent marring or erosion of the surface. Burlap sprayed with water is an effective method for moist curing.

Proper curing promotes continued hydration of cementitious materials. The extent to which hydration is completed influences the strength and durability of the concrete. Freshly mixed concrete normally contains more water than is required for hydration of the cement; however, excessive loss of water by evaporation can delay or prevent adequate hydration. The surface is particularly susceptible to insufficient hydration because it dries first. If temperatures are favorable, hydration is relatively rapid the first few days after concrete is placed. However, it is important for water to be retained in the concrete during this period, that is, for evaporation to be prevented or substantially reduced.

With proper curing, concrete becomes stronger. The strength improvement is rapid at early ages but continues more slowly thereafter for an indefinite period. Figure 15-2 shows the strength gain of concrete with age for different moist curing periods while Figure 15-3 shows the relative strength gain of concrete cured at different temperatures.

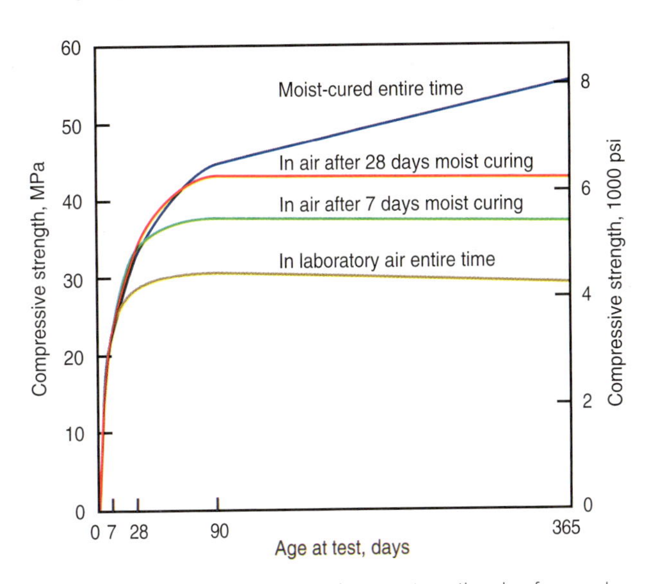

Figure 15-2. Effect of moist curing time on strength gain of concrete (Gonnerman and Shuman 1928).

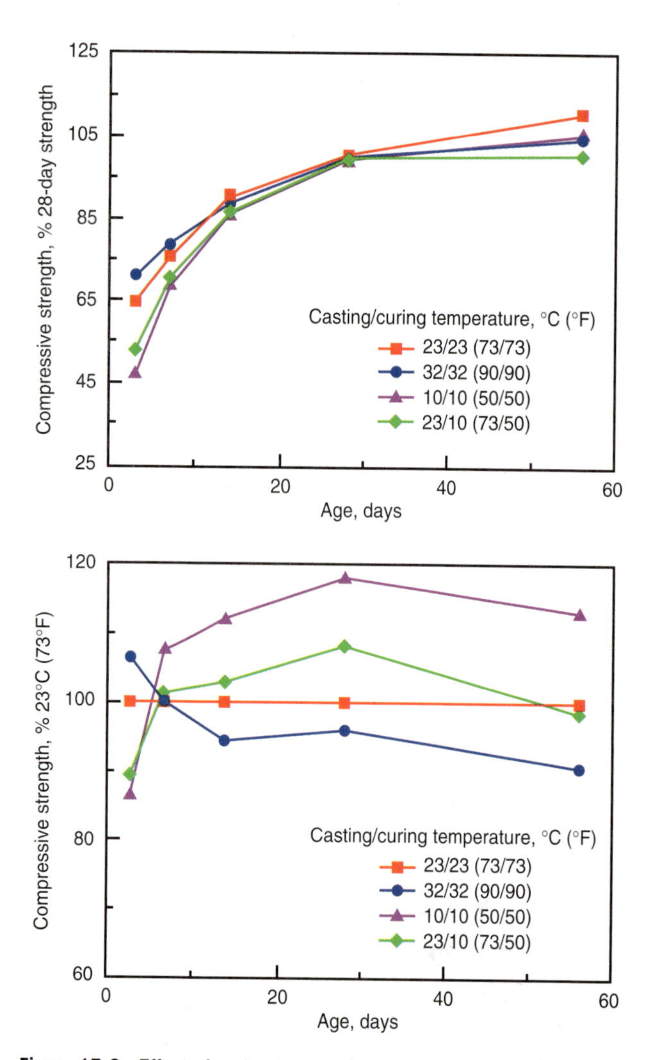

Figure 15-3. Effect of curing temperature on strength gain (top) relative to 28-day strength and (bottom) relative to the strength of concrete at 23°C (73°F) (Burg 1996).

The most effective method for curing concrete depends on the materials used, method of construction, temperature during placement, and the intended use of the hardened concrete. For most jobs, curing generally involves applying curing compounds, or covering the freshly placed and finished concrete with impermeable sheets or wet burlap. In some cases, such as in hot and cold weather, special care using other precautions is needed.

Concrete mixtures with high cement contents and low water-cementing materials ratios (less than 0.40) may require special curing needs. As cement hydrates the internal relative humidity decreases causing the paste to self-desiccate if no external water is provided. The paste can self-desiccate to a level where hydration stops. This may influence desired concrete properties, especially if the internal relative humidity drops below 80% within the first seven days. In view of this, membrane-forming curing compounds may not retain enough water in the concrete. Therefore, fogging and wet curing may become necessary to maximize hydration (Copeland and Bragg 1955). Fogging during and after placing and finishing also helps minimize plastic cracking in concretes with very low water-cement ratios.

When moist curing is interrupted, the development of strength continues for a short period and then stops after the concrete's internal relative humidity drops to about 80%. However, if moist curing is resumed, strength development will be reactivated, but the original potential strength may not be achieved. Thus, it is best to moist-cure the concrete continuously from the time it is placed and finished until it has gained sufficient strength, impermeability, and durability.

Loss of water will also cause the concrete to shrink, thus creating tensile stresses within the concrete. If these stresses develop before the concrete has attained adequate tensile strength, surface cracking can result. All exposed surfaces, especially exposed edges and joints, must be protected against moisture evaporation.

Hydration proceeds at a much slower rate when the concrete temperature is low. Temperatures below 10°C (50°F) are unfavorable for the development of early strength; below 4°C (40°F) the development of early strength is greatly retarded; and at or below freezing temperatures, down to -10°C (14°F), little or no strength develops.

A maturity concept is available to estimate the development of strength when there is variation in the curing temperature of the concrete. Maturity is the cumulative product of the age of the concrete and its average curing temperature above a certain base temperature. Refer to Chapter 17 for more information on the maturity concept. It follows that concrete should be protected so that its temperature remains favorable for hydration and moisture is not lost during the early hardening period.

Curing Methods and Materials

Concrete can be kept moist (and in some cases at a favorable temperature) by three curing methods:

Supplying additional moisture. Methods that provide additional sources of moisture replace moisture lost through evaporation or hydration and maintain the mixing water in the concrete during the early hardening period. These include ponding or immersion, spraying or fogging, and saturated wet coverings. These methods afford some cooling through evaporation, which is beneficial in hot weather.

Sealing in the mix water. Covering the concrete with impervious paper or plastic sheets, or applying membrane-forming curing compounds reduces the loss of mixing water from the surface of the concrete.

Accelerated curing. Supplying heat and additional moisture to the concrete accelerates strength gain. This is usually accomplished with live steam, heating coils, or electrically heated forms or pads.

The method or combination of methods chosen depends on factors such as availability of curing materials, size, shape, and age of concrete, production facilities (in place or in a plant), aesthetic appearance, and economics. As a result, curing often involves a series of procedures used at a particular time as the concrete ages. For example, fog spraying or plastic covered wet burlap can precede application of a curing compound. The timing of each procedure depends on the ambient evaporative conditions and the degree of hardening of the concrete needed to prevent the particular procedure from damaging the concrete surface (ACI 308 2009).

Ponding and Immersion

On horizontal surfaces, such as pavements and floors, concrete can be cured by ponding. Earth or sand dikes around the perimeter of the concrete surface retain a pond of water. Ponding is an ideal method for preventing loss of moisture from the concrete; it is also effective for maintaining a uniform temperature in the concrete. The curing water should not be more than about 11°C (20°F) cooler than the concrete to prevent thermal stresses that result in cracking. Since ponding requires considerable labor and supervision, the method is typically used on small jobs. Occasional checks to maintain continuous ponding are suggested especially in arid climates.

The most thorough method of curing with water consists of total immersion of the finished concrete element. This method is commonly used in the laboratory for curing concrete test specimens. Where appearance of the concrete is important, the water used for curing by ponding or immersion must be free of substances that will stain or discolor the concrete. The material used for dikes may also discolor the concrete.

Fogging and Sprinkling

Fogging (Figure 15-4) and sprinkling with water are excellent methods of curing when the ambient temperature is well above freezing and the humidity is low. A fine fog mist is frequently applied through a system of nozzles or sprayers to raise the relative humidity of the air over flatwork, thus slowing evaporation from the surface. Fogging is applied to minimize plastic shrinkage cracking until finishing operations are complete. Once the concrete has set sufficiently to prevent water erosion, ordinary lawn sprinklers are effective if good coverage is provided and water runoff is not an issue. Soaker hoses are useful on surfaces that are vertical or nearly so.

Figure 15-4. Fogging minimizes moisture loss during and after placing and finishing of concrete.

The cost of sprinkling may be a disadvantage. The method requires an ample water supply and careful supervision. If sprinkling is done at intervals, the concrete must be prevented from drying between applications of water by using burlap or similar materials; otherwise alternate cycles of wetting and drying can cause surface crazing or cracking.

Wet Coverings

Fabric coverings saturated with water, such as burlap, cotton mats, rugs, or other moisture-retaining fabrics, are commonly used for curing (Figure 15-5). Treated burlaps that reflect light and are resistant to rot and fire are also available. The requirements for burlap are described in AASHTO M 182, *Specification for Burlap Cloths Made from Jute or Kenaf* and those for white burlap-polyethylene sheeting are described in ASTM C171, *Standard Specification for Sheet Materials for Curing Concrete* (AASHTO M 171).

Figure 15-5. Lawn sprinklers saturating burlap with water keep the concrete continuously moist. Intermittent sprinkling is acceptable if no drying of the concrete surface occurs.

Burlap must be free of any substance that is harmful to concrete or causes discoloration. New burlap should be thoroughly rinsed in water to remove soluble substances and to make the burlap more absorbent.

Wet, moisture-retaining fabric coverings should be placed as soon as the concrete has hardened sufficiently to prevent surface damage. During the waiting period other curing methods are used, such as fogging or the use of membrane forming finishing aids. Care should be taken to cover the entire surface with wet fabric, especially at the edges of slabs where drying occurs on two or more adjacent surfaces. The coverings should be kept continuously moist so that a film of water remains on the concrete surface throughout the curing period. Use of polyethylene film over wet burlap is a good practice; it will eliminate the need for continuous watering of the covering. Periodically rewetting the fabric under the plastic before it dries out should be sufficient. Alternate cycles of wetting and drying during the early curing period may cause crazing of the surface.

Wet coverings of earth, sand, or sawdust are effective for curing and are often useful on small jobs. Sawdust from most woods is acceptable, but oak and other woods that contain tannic acid should not be used since deterioration of the concrete may occur. A layer about 50 mm (2 in.) thick should be evenly distributed over the previously moistened surface of the concrete and kept continuously wet.

Wet hay or straw can be used to cure flat surfaces. If used, it should be placed in a layer at least 150 mm (6 in.) thick and held down with wire screen, burlap, or tarpaulins to prevent its being blown off by wind.

A major disadvantage of moist earth, sand, sawdust, hay, or straw coverings is the possibility of discoloring the concrete and the difficulty in removal.

Impervious Paper

Impervious paper for curing concrete consists of two sheets of kraft paper cemented together by a bituminous adhesive with fiber reinforcement. Such paper, conforming to ASTM C171 (AASHTO M 171), is an efficient means of curing horizontal surfaces and structural concrete of relatively simple shapes. An important advantage of this method is that periodic additions of water are not required. Curing with impervious paper enhances the hydration of cement by preventing loss of moisture from the concrete (Figure 15-6).

As soon as the concrete has hardened sufficiently to prevent surface damage, it should be thoroughly wetted and the widest paper available applied. Edges of adjacent sheets should be overlapped about 150 mm (6 in.) and tightly sealed with sand, wood planks, pressure-sensitive tape, mastic, or glue. The sheets must be weighted to maintain close contact with the concrete surface during the entire curing period.

Figure 15-6. Impervious curing paper is an efficient means of curing horizontal surfaces.

Impervious paper can be reused if it effectively retains moisture. Tears and holes can easily be repaired with curing-paper patches. When the condition of the paper is questionable, additional use can be obtained by using it in double thickness.

In addition to curing, impervious paper provides some protection to the concrete against damage from subsequent construction activity as well as protection from the direct sun. It should be light in color and nonstaining to the concrete. Paper with a white upper surface is preferable for curing exterior concrete during hot weather.

Plastic Sheets

Plastic sheet materials, such as polyethylene film, can be used to cure concrete (Figure 15-7). Polyethylene film is a lightweight, effective moisture retarder and is easily applied to complex as well as simple shapes. Its application is the same as described for impervious paper.

Figure 15-7. Polyethylene film is an effective moisture barrier for curing concrete and easily applied to complex as well as simple shapes. To minimize discoloration, the film should be kept as flat as possible on the concrete surface.

Curing with polyethylene film (or impervious paper) can cause patchy discoloration, especially if the concrete contains calcium chloride and has been finished by hard-steel troweling. This discoloration is more pronounced when the film becomes wrinkled, but it is difficult and time consuming on a large project to place sheet materials without wrinkles. Flooding the surface under the covering may prevent discoloration, but other means of curing should be used when uniform color is important.

Polyethylene film should conform to ASTM C171 (AASHTO M 171), which specifies a 0.10-mm (4-mil) thickness for curing concrete, but lists only clear and white opaque film. However, black film is available and is satisfactory under some conditions. White film should be used for curing exterior concrete during hot weather to reflect the sun's rays. Black film can be used during cool weather or for interior locations. Clear film has little effect on heat absorption.

ASTM C171 (AASHTO M 171) also includes a sheet material consisting of burlap impregnated on one side with white opaque polyethylene film. Combinations of polyethylene film bonded to an absorbent fabric such as burlap help retain moisture on the concrete surface.

Polyethylene film may also be placed over wet burlap or other wet covering materials to retain the water in the wet covering material. This procedure reduces the labor-intensive need to re-wet covering materials. There are single-use plastic coverings available for use that help eliminate potential staining from re-use of coverings.

Membrane-Forming Curing Compounds

Liquid membrane-forming compounds consisting of waxes, resins, chlorinated rubber, and other materials can be used to retard or reduce evaporation of moisture from concrete. They are the most practical and most widely used method for curing not only for freshly placed concrete but also for extending curing of concrete after removal of forms or after initial moist curing. However, the most effective methods of curing concrete are wet coverings or water spraying that keeps the concrete continually damp. Curing compounds should be able to maintain the relative humidity of the concrete surface above 80% for seven days to sustain cement hydration.

Membrane-forming curing compounds are of two general types: clear, or translucent; and white pigmented. Clear or translucent compounds may contain a fugitive dye that makes it easier to check visually for complete coverage of the concrete surface when the compound is applied. The dye fades away soon after application. White-pigmented compounds are recommended on hot, sunny days as they reduce solar-heat gain, thus reducing the concrete temperature. Pigmented compounds should be agitated in the container prior to application to prevent pigment from settling out resulting in non-uniform coverage and ineffective curing.

Curing compounds should be applied immediately after final finishing of the concrete (Figure 15-8). The concrete surface should be damp when the coating is applied. On dry, windy days, or during periods when adverse weather conditions could result in plastic shrinkage cracking, protection may be required until curing operations can be initiated without damaging surfaces. Application of a curing compound immediately after final finishing and before all free water on the surface has evaporated will help prevent the formation of cracks. Power-driven spray equipment is recommended for uniform application of curing compounds on large paving projects. Spray nozzles recommended by the product manufacturer or use of windshields should be arranged to prevent wind-blown loss of curing compound. Otherwise proper coverage application rates will not be achieved.

Figure 15-8. Liquid membrane-forming curing compounds should be applied with uniform and adequate coverage over the entire surface and edges for effective, extended curing of concrete.

Normally only one smooth, even coat is applied at a typical rate of 3 m^2/liter to 5 m^2 /liter (150 ft^2/gallon to 200 ft^2/gallon); but products may vary, so manufacturer's recommended application rates should be followed. If two coats are necessary to ensure complete coverage, for effective protection the second coat should be applied at right angles to the first. Complete coverage of the surface must be attained because even small pinholes in the membrane will result in loss of moisture from the concrete.

Note that bonding of subsequent materials might be inhibited by the presence of a curing compound even after the moisture retention characteristics of the compound have diminished. Most curing compounds are not compatible with adhesives used with floor covering materials. Consequently, they should either be tested for compatibility, or not used when bonding of overlying materials is necessary. For example, a curing compound should generally not be applied to the base slab of a two-course floor. Similarly, some curing compounds may affect the adhesion of paint to concrete floors. Curing compound and floor covering manufacturers should be consulted to determine if their products are suitable for the intended application (Kanare 2008).

Curing compounds should be thoroughly mixed and uniformly applied. They should not sag, run off peaks, or collect in grooves. They should form a tough film to withstand early construction traffic without damage, be non-yellowing, and have good moisture-retention properties.

Caution is necessary when using curing compounds containing solvents of high volatility in confined spaces or near sensitive occupied spaces such as hospitals because evaporating volatiles may cause respiratory problems. Applicable local environmental laws concerning volatile organic compound (VOC) emissions should be followed.

Curing compounds should conform to ASTM C309 (AASHTO M 148), *Standard Specification for Liquid Membrane-Forming Compounds for Curing Concrete*. A method for determining the efficiency of curing compounds, waterproof paper, and plastic sheets is described in ASTM C156 (AASHTO T 155), *Standard Test Method for Water Loss [from a Mortar Specimen] Through Liquid Membrane-Forming Curing Compounds for Concrete*. Curing compounds with sealing properties are specified under ASTM C1315, *Standard Specification for Liquid Membrane-Forming Compounds Having Special Properties for Curing and Sealing Concrete*.

Internal Curing

Fully prewetted lightweight concrete aggregates and superabsorbent polymers provide a source of moisture for internal curing. Internal curing refers to the process by which hydration of cement and pozzolanic reactions can continue because of an internal water supply that is available in addition to the mixing water (ACI 213 2003 and Lam 2005). Internal moist curing must be accompanied by external curing methods to assure that surface or exposed concrete surfaces are properly cured.

This internal curing process allows the concrete to gain additional strength and also results in a reduction of permeability due to a significant extension in the time of curing (Holm and Ries 2006). Internal curing can also help to avoid early age cracking of concrete with high cementitious materials content that is typical of many high strength concretes. Likewise, early shrinkage of concrete caused by rapid drying can also be avoided through internal curing.

For concretes with low water to cement ratios, 60 kg/m^3 to 180 kg/m^3 (100 lb/yd^3 to 300 lb/yd^3) of saturated lightweight fine aggregate can provide the additional moisture for internal curing. All of the fine aggregate in a mixture can be replaced with saturated lightweight fine aggregate to maximize internal curing. Superabsorbent polymer does not replace any aggregate, but instead is used an admixture dosed by weight of cementitious materials. The superabsorbent polymers should be batched dry, due to their tendency to clump when wet (Lam 2005).

In most cases, any excess moisture will eventually diffuse out of the concrete. The time it takes for the concrete to dry must be taken into consideration when it is to be covered with a moisture-sensitive flooring system.

Forms Left in Place

Forms provide satisfactory protection against loss of moisture if the top exposed concrete surfaces are kept wet. A soaker hose is excellent for this. The forms should be left on the concrete as long as practical.

Wood forms left in place should be kept moist by sprinkling, especially during hot, dry weather. If this cannot be done, they should be removed as soon as practical and another curing method started without delay. Color variations may occur from formwork and uneven water curing of walls.

Steam Curing

Steam curing is advantageous where early strength gain in concrete is important or where additional heat is required to accomplish hydration, as in cold weather.

Two methods of steam curing are used: live steam at atmospheric pressure (for enclosed cast-in-place structures and large precast concrete units) and high-pressure steam in autoclaves (for small manufactured units). Only live steam at atmospheric pressure will be discussed here.

A typical steam-curing cycle consists of (1) an initial delay prior to steaming, (2) a period for increasing the temperature, (3) a period for holding the maximum temperature constant, and (4) a period for decreasing the temperature. A typical atmospheric steam-curing cycle is shown in Figure 15-9.

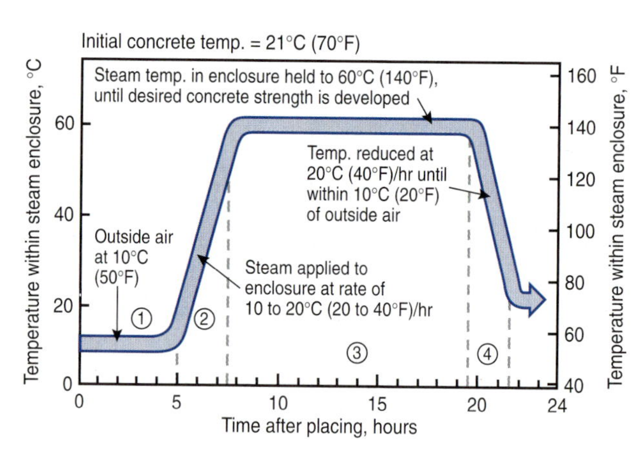

Figure 15-9. A typical atmospheric steam-curing cycle.

Steam curing at atmospheric pressure is generally done in an enclosure to minimize moisture and heat losses. Tarpaulins are frequently used to form the enclosure. Application of steam to the enclosure should be delayed until initial set occurs or delayed at least 3 hours after final placement of concrete to allow for some hardening of the concrete. However, a 3- to 5-hour delay period prior to steaming will achieve maximum early strength, as shown in Figure 15-10.

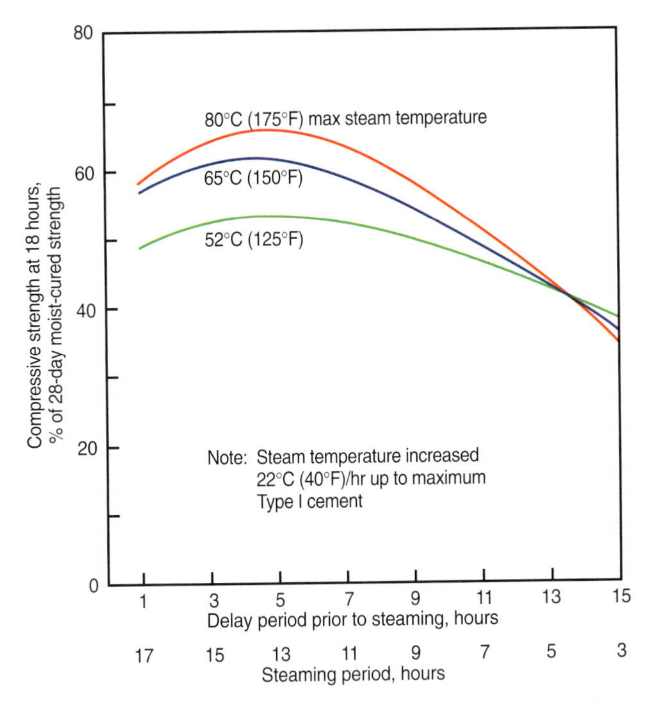

Figure 15-10. Relationship between strength at 18 hours and delay period prior to steaming. In each case, the delay period plus the steaming period totaled 18 hours (Hanson 1963).

Steam temperature in the enclosure should be kept at about 60°C (140°F) until the desired concrete strength has developed. Strength will not increase significantly if the maximum steam temperature is raised from 60°C to 70°C (140°F to 158°F). Steam-curing temperatures above 70°C (158°F) should be avoided whenever possible as it may result in damage (see Chapter 11 on delayed ettrignite formation). It is recommended that the internal temperature of concrete not exceed 70°C (158°F) to avoid heat induced delayed expansion and undue reduction in ultimate strength. Use of concrete temperatures above 70°C (158°F) should be demonstrated to be safe by test or historic field data.

Concrete temperatures are commonly monitored at the exposed ends of the concrete element. Monitoring air temperatures alone is not sufficient because the heat of hydration may cause the internal temperature of the concrete to exceed 70°C (158°F). Besides early strength gain, there are other advantages of curing concrete at temperatures of around 60°C (140°F); for example, there is reduced drying shrinkage and creep as compared to concrete cured at 23°C (73°F) for 28 days (Klieger 1960 and Tepponen and Eriksson 1987).

To prevent damaging volume changes excessive rates of heating and cooling should be avoided. Temperatures in the enclosure surrounding the concrete should not be increased or decreased more than 22°C to 33°C (40°F to 60°F) per hour depending on the size and shape of the concrete element.

The curing temperature in the enclosure should be held until the concrete has reached the desired strength. The time required will depend on the concrete mixture and steam temperature in the enclosure.

Insulating Blankets or Covers

Layers of dry, porous material such as straw or hay can be used to provide insulation against freezing of concrete when temperatures fall below 0°C (32°F).

Formwork can be economically insulated with commercial blanket or batt insulation that has a tough moisture-proof covering. Suitable insulating blankets are manufactured of fiberglass, sponge rubber, cellulose fibers, mineral wool, vinyl foam, and open-cell polyurethane foam. When insulated formwork is used, care should be taken to ensure that concrete temperatures do not become excessive.

Framed enclosures of canvas tarpaulins, reinforced polyethylene film, or other materials can be placed around the structure and heated by indirect fired or properly vented space heaters or steam. Portable hydronic heaters are used to thaw subgrades as well as heat concrete without the use of an enclosure

Curing concrete in cold weather should follow the recommendations in Chapter 17 and ACI Committee 306, *Cold-Weather Concreting*. Recommendations for curing concrete in hot weather can be found in Chapter 16 and ACI Committee 305, *Hot-Weather Concreting*.

Electrical, Oil, Microwave, and Infrared Curing

Electrical, hot oil, microwave, and infrared curing methods have been available for accelerated and normal curing of concrete for many years. Electrical curing methods include a variety of techniques: (1) use of the concrete itself as the electrical conductor, (2) use of reinforcing steel as the heating element, (3) use of a special wire as the heating element, (4) electric blankets, and (5) the use of electrically heated steel forms (presently the most popular method). Electrical heating is especially useful in cold-weather concreting. Hot oil, hot water, or closed loop steam may be circulated through pipes surrounding the steel forms to heat the concrete. Infrared rays and microwave have had limited use in accelerated curing of concrete. Concrete that is cured by infrared methods is usually under a covering or enclosed in steel forms. Electrical, oil, and infrared curing methods are used primarily in the precast concrete industry.

Curing Period and Temperature

The period of time that concrete should be protected from freezing, abnormally high temperatures, and against loss of moisture depends upon a number of factors: the type and quality of cementing materials used; mixture proportions; required strength, size and shape of the concrete member; ambient conditions; and future exposure conditions. The curing period may be 3 weeks or longer for lean concrete mixtures used in massive structures such as dams; conversely, it may be only a few days for rich mixtures, especially if Type III or HE cement is used. Steam-curing periods are normally much shorter, ranging from a few hours to 3 days; but generally 24-hour cycles are used. Since all the desirable properties of concrete are improved by curing, the curing period should be as long as necessary and reasonable.

For concrete slabs on ground (floors, pavements, canal linings, parking lots, driveways, sidewalks) and for structural concrete (cast-in-place walls, columns, slabs, beams, small footings, piers, retaining walls, bridge decks), the length of the curing period for ambient temperatures above 5°C (40°F) should be a minimum of 7 days (ACI 301). For high-early-strength concretes, the curing period may be shortened to 3 days (ACI 301). In cold weather concreting, additional time may be needed to attain 70% of the specified compressive or flexural strength. When the daily mean ambient temperature is 5°C (40°F) or lower, ACI Committee 306 recommendations for curing and protection period should be followed to prevent damage by freezing.

A higher curing temperature provides earlier strength gain in concrete than a lower temperature but it may decrease 28-day strength as shown in Figure 15-11.

Field Cured Cylinders. If strength tests are made to establish the time when curing can cease or forms can be removed, representative concrete test cylinders or beams should be fabricated in the field, located adjacent to the structure or pavement they represent, and cured using the same methods.

Match Curing. Equipment is available that can monitor internal concrete temperatures and match that temperature in the concrete cylinder curing box; this is the most accurate means of representing in-place concrete strengths. Cores, cast-in-place removable cylinders, and nondestructive testing methods may also be used to determine the strength of a concrete member.

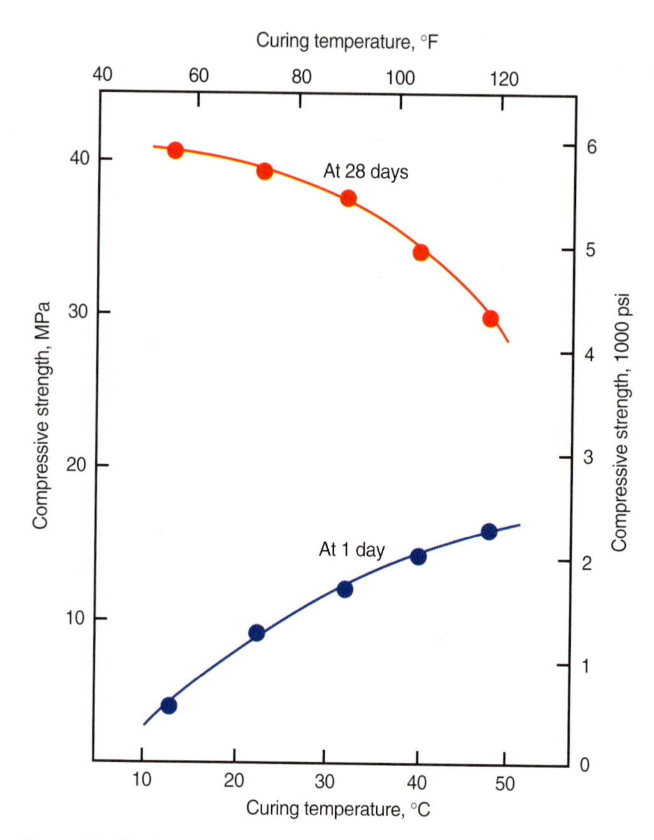

Figure 15-11. One-day strength increases with increasing curing temperature, but 28-day strength decreases with increasing curing temperature (Verbeck and Helmuth 1968).

Since the rate of hydration is influenced by cement type and the presence of supplementary cementing materials, the curing period should be prolonged for concretes made with cementing materials possessing slow-strength-gain characteristics, especially in cold weather. For mass concrete (large piers, locks, abutments, dams, heavy footings, and massive columns and transfer girders) in which no pozzolan is used as part of the cementitious material, curing of unreinforced sections should continue for at least 2 weeks. If the mass concrete contains a pozzolan, minimum curing time for unreinforced sections should be extended to 3 weeks. Heavily-reinforced mass concrete sections should be cured for a minimum of 7 days.

During cold weather, additional heat is often required to maintain favorable curing temperatures of 10°C to 20°C (50°F to 70°F). Vented gas or oil-fired heaters, heating coils, portable hydronic heaters, or live steam can be used to supply the required heat. In all cases, care must be taken to avoid loss of moisture from the concrete. Exposure of fresh concrete to heater or engine exhaust gases must be avoided as this can result in surface deterioration and dusting (rapid carbonation).

High-early-strength concrete can be used in cold weather to accelerate setting time and strength development. This can reduce the curing period, but a minimum temperature of 10°C (50°F) must be maintained.

For adequate deicer scale resistance of concrete, the minimum curing period generally corresponds to the time required to develop the design strength of the concrete at the surface. A period of air-drying after curing will enhance resistance to scaling. This drying period should be at least 1 month of relatively dry weather followed by the optional application of a sealing compound prior to exposure to freezing and thawing cycles and preferably three months before the application of deicing salts.

Sealing Compounds

Sealing compounds (sealers) are liquids applied to the surface of hardened concrete to reduce the penetration of liquids or gases such as water, deicing solutions, and carbon dioxide to protect concrete from freeze-thaw damage, corrosion of reinforcing steel, and acid attack. In addition, sealing compounds used on interior floor slabs reduce dusting and the absorption of spills while making the surface easier to clean.

Sealing compounds are fundamentally different from curing compounds. The primary purpose of a curing compound is to retard the loss of water from newly placed concrete and it is applied immediately after finishing. Surface sealing compounds on the other hand retard the penetration of harmful substances into hardened concrete and are typically not applied until the concrete is 28 days old and sufficiently surface dry to allow sealers to penetrate into surfaces of the concrete. Surface sealers are generally classified as either film-forming or penetrating.

Sealing exterior concrete is an optional procedure generally performed to aid in protection of concrete from freeze-thaw damage and chloride penetration from deicers. Curing is an absolute must when using a sealer; curing is necessary to produce properties needed for concrete to perform adequately for its intended purpose. Satisfactory freeze-thaw durability of exterior concrete still primarily depends on an adequate air-void system, sufficient strength, and the use of proper placing, finishing and curing techniques. However, not all concrete placed meets all of these recommended criteria; surface sealers can help improve the durability of these concretes (Figure 15-12).

Film-forming sealing compounds remain primarily on the surface. Only a slight amount of the material actually penetrates the concrete. The relatively large molecular structure of these compounds limits their ability to penetrate the surface. Thinning them with solvents will not improve their penetrating capability. These materials not only reduce the penetration of water, they also protect against mild chemicals; furthermore, they prevent the absorption of grease and oil as well as reduce dusting under pedestrian traffic.

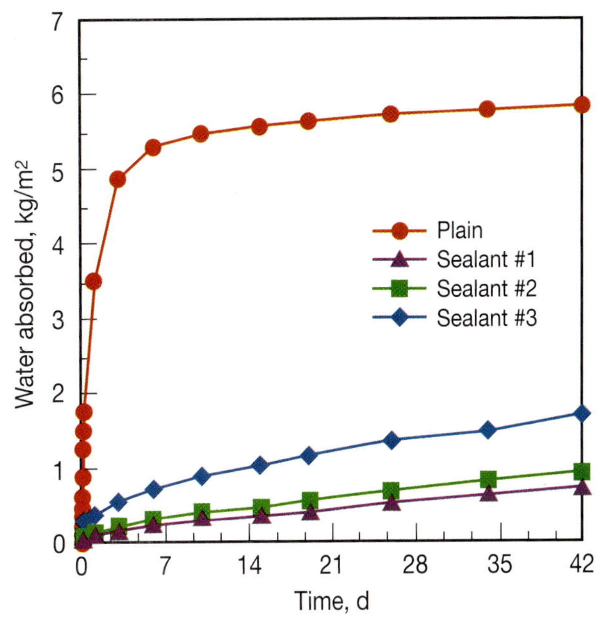

Figure 15-12. Sealants shown are effective at reducing the amount of water absorbed into concrete (Golias 2010).

Surface sealers consist of acrylic resins, chlorinated rubber, urethanes, epoxies, and alpha methyl styrene. The effectiveness of film-forming sealers depends on the continuity of the layer formed. Abrasive grit and heavy traffic can damage the layer requiring the reapplication of the material. Consult manufacturers' application recommendations because some of these materials are intended for interior use only and may yellow and deteriorate under exposure to ultraviolet light.

The penetrating sealer used most extensively historically is a mixture of 50 percent boiled linseed oil and 50 percent mineral spirits (AASHTO M 233, *Standard Specification for Boiled Linseed Oil Mixture for Treatment of Portland Cement Concrete*). Although this mixture is an effective sealer, it has two main disadvantages: it darkens the concrete, and periodic reapplication is necessary for long-term protection.

A new generation of water-repellent penetrating sealers have a very small molecular size that allows penetration and saturation of the concrete as deep as 3 mm (⅛ in.). The two most common are silane and siloxane, compounds which are derived from the silicone family. These sealers allow the concrete to breath, thus preventing a buildup of vapor pressure between the concrete and sealer that can occur with some film-forming materials. Because the sealer is embedded within the concrete, making it more durable to abrasive forces or ultraviolet deterioration, it can provide longer lasting protection than film-forming sealers. However, periodic retreatment is recommended. In northern states and coastal areas silanes and siloxanes are popular for protecting bridge decks and other exterior structures from corrosion of reinforcing steel caused by chloride infiltration from deicing chemicals or sea spray (Figure 15-13).

Figure 15-13. Penetrating sealers help protect reinforcing steel in bridge decks from corrosion due to chloride infiltration without reducing surface friction.

At least 28 days should be allowed to elapse before applying sealers to new concrete. Application of any sealer should only be done on concrete that is clean and allowed to dry for at least 24 hours at temperatures above 16°C (60°F). Penetrating sealers cannot fill surface voids if they are filled with water. Some surface preparation may be necessary if the concrete is old and dirty. Concrete placed in the late fall should not be sealed until spring because the sealer may cause the concrete to retain water that may exacerbate freeze-thaw damage.

The precautions outlined earlier regarding volatile solvents in curing compounds also apply to sealing compounds. The effectiveness of water-based surface sealers is still being determined. The scale resistance provided by concrete sealers should be evaluated based on criteria established in ASTM C672. For more information on surface sealing compounds, see AASHTO M 224, ACI Committee 330, and ACI Committee 362.

References

ACI Committee 213, *Guide for Structural Lightweight-Aggregate Concrete*, ACI 213R-03, American Concrete Institute, Farmington Hills, Michigan, 2003.

ACI Committee 301, *Specification for Structural Concrete*, ACI 301-10, American Concrete Institute, Farmington Hills, Michigan, 2010.

ACI Committee 305, *Hot-Weather Concreting*, ACI 305R-10, American Concrete Institute, Farmington Hills, Michigan, 2010, 17 pages.

ACI Committee 306, *Cold-Weather Concreting*, ACI 306R-10, American Concrete Institute, Farmington Hills, Michigan, 2010, 23 pages.

ACI Committee 308, *Guide to Curing Concrete*, ACI 308-09, American Concrete Institute, Farmington Hills, Michigan, 2009, 31 pages.

ACI Committee 330, *Guide for Design and Construction of Concrete Parking Lots*, ACI 330R-01, 1997, American Concrete Institute, Farmington Hills, Michigan, 2001, 32 pages.

ACI Committee 362, *Guide for the Design of Durable Parking Structures*, ACI 362.1R-97, reapproved 2002, American Concrete Institute, Farmington Hills, Michigan, 1997, 40 pages.

Burg, Ronald G., *The Influence of Casting and Curing Temperature on the Properties of Fresh and Hardened Concrete*, Research and Development Bulletin RD113, Portland Cement Association, 1996, 20 pages.

Copeland, L.E., and Bragg, R.H., *Self Desiccation in Portland Cement Pastes*, Research Department Bulletin RX052, Portland Cement Association, http://www.cement.org/pdf_files/RX052.pdf, 1955, 13 pages.

Gonnerman, H.F., and Shuman, E.C., "Flexure and Tension Tests of Plain Concrete," Major Series 171, 209, and 210, *Report of the Director of Research*, Portland Cement Association, November 1928, pages 149 and 163.

Hanson, J.A., Optimum Steam Curing Procedure in Precasting Plants, with discussion, Development Department Bulletins DX062 and DX062A, Portland Cement Association, http://www.cement.org/pdf_files/DX062.pdf, and http://www.cement.org/pdf_files/DX062A.pdf, 1963, 28 pages and 19 pages, respectively.

Holm, T., and Ries, J., "Lightweight Concrete and Aggregates," *Significance of Tests and Properties of Concrete and Concrete-Making Materials*, STP169D, ASTM International, West Conshohocken, Pennsylvania, 2006.

Kanare, Howard M., *Concrete Floors and Moisture*, EB119, Portland Cement Association, Skokie, Illinois, and National Ready Mixed Concrete Association, Silver Spring, Maryland, USA, 2008, 176 pages.

Klieger, Paul, *Some Aspects of Durability and Volume Change of Concrete for Prestressing*, Research Department Bulletin RX118, Portland Cement Association, http://www.cement.org/pdf_files/RX118.pdf, 1960, 15 pages.

Lam, H., *Effects of Internal Curing Methods on Restrained Shrinkage and Permeability*, SN2620, Portland Cement Association, Skokie, Illinois, 2005, 134 pages.

Tepponen, Pirjo, and Eriksson, Bo-Erik, "Damages in Concrete Railway Sleepers in Finland," *Nordic Concrete Research*, Publication No. 6, The Nordic Concrete Federation, Oslo, 1987.

Verbeck, George J., and Helmuth, R.A., "Structures and Physical Properties of Cement Pastes," *Proceedings, Fifth International Symposium on the Chemistry of Cement*, Vol. III, The Cement Association of Japan, Tokyo, 1968, page 9.

CHAPTER 16

Hot Weather Concreting

Weather conditions at a jobsite may be vastly different from optimum concrete placement conditions assumed at the time a concrete mixture is specified, designed, or selected. They rarely coincide with laboratory conditions in which concrete specimens are stored and tested. Longer duration projects will require changes to the concrete mixture as the seasonal weather changes. Hot weather conditions can adversely influence concrete quality primarily by accelerating the rate of evaporation/moisture loss and rate of cement hydration. Detrimental hot weather conditions include:

- high ambient air temperature
- high concrete temperature
- low relative humidity
- wind
- solar radiation

Hot weather conditions can create difficulties in fresh concrete, such as:

- increased water demand
- accelerated slump loss
- increased rate of setting
- increased tendency for plastic shrinkage cracking
- increased potential for cold joints
- difficulties in controlling entrained air content
- increased concrete temperature
- increased potential for thermal cracking

Many difficulties with hot weather concreting including increased water demand and accelerated slump loss lead to the addition of water on site. Adding water to the concrete at the jobsite can adversely affect properties and serviceability of the hardened concrete, resulting in:

- decreased strength
- decreased durability
- increased permeability
- nonuniform surface appearance
- increased tendency for drying shrinkage
- reduced abrasion resistance

For more information on the above topics, see ACI Committee 305 (2010).

When to Take Precautions

During hot weather the most favorable temperature for achieving high quality freshly mixed concrete is usually lower than can be obtained without artificial cooling. A concrete temperature of 10°C to 15°C (50°F to 60°F) is most desirable to maximize beneficial concrete properties, but such temperatures are not always practical. Many specifications require only that concrete when placed should have a temperature of less than 29°C to 35°C (85°F to 95°F). ASTM C94 (AASHTO M 157), *Standard Specification for Ready Mixed Concrete*, notes in some situations difficulty may be encountered when concrete temperatures approach 32°C (90°F). However, this specification does not mandate a maximum concrete temperature unless heated aggregates or heated water are used. ACI 301 (2010) requires that the temperature of concrete as delivered shall not exceed 35°C (95°F), unless otherwise specified or permitted.

Advanced planning is required for concrete placed in ambient conditions that are somewhere between 24°C and 38°C (75°F and 100°F). Last-minute attempts to mitigate hot-weather concreting are rarely performed soon enough to prevent damaging effects. If acceptable field data is not available, the maximum temperature limit should be established for conditions at the jobsite; this should be based on trial-batch tests at the temperature and for the typical concrete section thickness anticipated, rather than on ideal temperatures of 20°C to 30°C (68°F to 86°F) cited in ASTM C192 (AASHTO T 126), *Standard Practice for Making and Curing Concrete Test Specimens in the Laboratory*. If possible, large trial batches should be made to establish the relationship for the property of interest as a function of time at various concrete temperatures. This process will establish the maximum allowable time to discharge concrete after batching for various concrete temperatures.

Setting a maximum concrete temperature is not a guarantee of strength or durability. For most work it is too complex to simply limit only the maximum temperature of concrete as placed; circumstances and concrete requirements vary too widely. For example, a temperature limit that would serve successfully at one jobsite (such as in a cooler climate) could be highly restrictive at another (such as in a warmer climate). The inverse could occur if concrete mixtures that were designed for hot-weather conditions were placed in cooler ambient conditions.

Atmospheric conditions, including air temperature, relative humidity and wind speed, in conjunction with site conditions influence the precautions needed. For example, flatwork done under a roof that blocks solar radiation with exterior walls in place that screen the wind could be successfully completed using a concrete with a high temperature. However, this concrete would cause difficulty if placed outdoors on the same day where it would be exposed to direct sun and wind. Additionally, placement of concretes with a vulnerable surface to volume ratio such as thin overlays with relatively high cementitious contents require special care.

Figure 16-1. Liquid nitrogen added directly into a truck mixer at the ready mix plant is an effective method of reducing concrete temperature for mass concrete placements or during hot-weather concreting.

Which precautions to use and when to use them will depend on: the type of member or construction; characteristics of the materials being used; and the experience of the placing and finishing crew in dealing with the atmospheric conditions on the site. The following list of precautions will reduce or avoid the potential problems of hot-weather concreting:

- organize a preconstruction conference to discuss the precautions required for the project
- use materials and mix proportions that have proven performance in hot-weather conditions (lower cement contents, and set retarding admixtures)
- cool the concrete or one or more of its ingredients (Figure 16-1)
- use a concrete consistency (slump) that allows rapid placement and consolidation
- reduce the time of transport, placing, and finishing as much as possible so as to reduce and minimize waiting time
- schedule concrete placements to limit exposure to harsh atmospheric conditions. Consider night or more favorable weather conditions
- consider methods to limit moisture loss during placing and finishing, such as sunshades, windscreens, fogging, or spraying
- apply temporary moisture-retaining films to control evaporation after strike-off and prior to finishing concrete

The above precautions are discussed in further detail throughout this chapter.

Effects of High Concrete Temperatures

As concrete temperature increases there is a loss in slump that is often compensated for by adding water to the concrete at the jobsite. At higher temperatures a greater amount of water is required to hold slump constant than is needed at lower temperatures. The addition of water results in a higher water-cement ratio, thereby lowering the strength at all ages and adversely affecting other desirable properties of the hardened concrete. This is in addition to the adverse effect on strength at later ages due to the higher temperature, even without the addition of water.

As shown in Figure 16-2, if the temperature of freshly mixed concrete is increased from 10°C to 38°C (50°F to 100°F), about 20 kg/m^3 (33 lb/yd^3) of additional water is needed to maintain the same 75-mm (3-in.) slump. This additional water could reduce strength by 12% to 15% and produce compressive strength cylinder test results that may not comply with specifications. Adjusting mixture proportions for the higher water demand while maintaining w/cm will improve the concrete strength, however the durability and resistance to cracking (due to volume change) will still be impacted by the higher water content. Also, the higher cement content necessary to maintain the w/cm with additional mix water will further increase the concrete temperature and water demand.

High temperatures of freshly mixed concrete increase the rate of setting and shorten the length of time within which the concrete can be transported, placed, and finished. As a general rule of thumb, the setting time changes by about

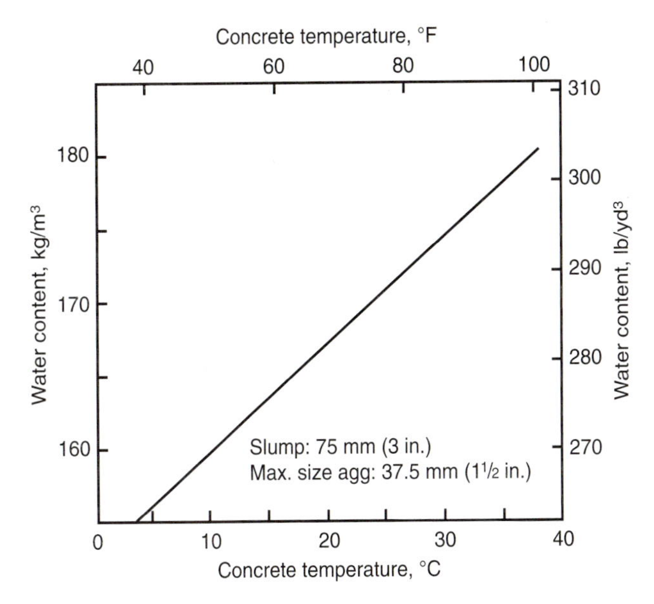

Figure 16-2. The water requirement of a concrete mixture increases with an increase in concrete temperature Bureau of Reclamation (1981).

33% for every 5°C (10°F) change in the initial concrete temperature. Different cements behave differently and don't always follow this generalization, as can be seen in Figure 16-3. This shows that the setting time can be reduced by 2 or more hours with a 10°C (18°F) increase in concrete temperature. Concrete should remain plastic long enough so that each layer can be placed without development of cold joints in the concrete. Set retarding admixtures, ASTM C494, *Standard Specification for Chemical Admixtures for Concrete* (AASHTO M 194) Type B, and hydration control admixtures can be beneficial in offsetting the accelerating effects of high temperature.

In hot weather, there is an increased tendency for cracks to form in both the fresh and hardened concrete. Rapid evaporation of water from freshly placed concrete can cause plastic-shrinkage cracks before the surface has hardened (see **Plastic Shrinkage Cracking**). Cracks may also develop in the hardened concrete because of increased drying shrinkage due to higher water contents or thermal volume changes as the concrete cools (see Chapter 10).

Air entrainment is also affected in hot weather. At elevated temperatures, an increase in the amount of air-entraining admixture dosage is generally required to produce a given entrained air content.

Figure 16-4 shows the effect of high initial concrete temperatures on compressive strength. The concrete temperatures at the time of mixing, casting, and curing were 23°C (73°F), 32°C (90°F), 41°C (105°F), and 49°C (120°F). After 28 days, the specimens were all moist-cured at 23°C (73°F) until the 90-day and one-year test ages. The tests, using identical concretes of the same water-cement ratio, show that while higher concrete temperatures give higher early strength than concrete at 23°C (73°F), at later ages concrete

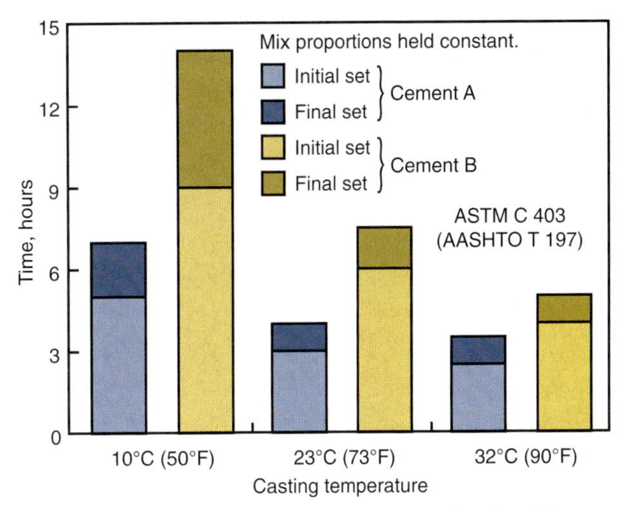

Figure 16-3. Effect of concrete temperature on setting time (Burg 1996).

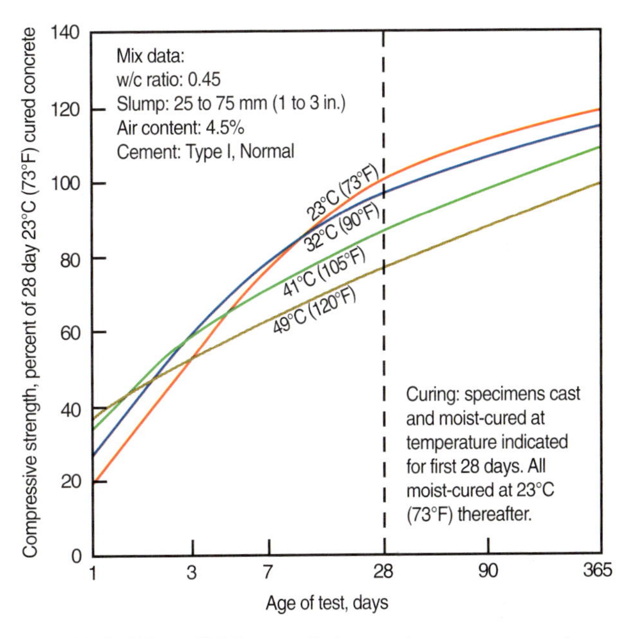

Figure 16-4. Effect of high concrete temperatures on compressive strength at various ages (Klieger 1958).

strengths are lower. If the water content had been increased to maintain the same slump (without increasing cement content), the reduction in strength may have been even greater than shown.

The proper fabrication, curing, and testing of compression test specimens during hot weather is critical. Steps should be taken to assure ASTM C31, *Standard Practice for Making and Curing Concrete Test Specimens in the Field* (AASHTO T 23) procedures are followed regarding initial curing of strength specimens for acceptance or quality control testing at 16°C to 27°C (60°F to 80°F). If the initial 24 hour curing is at 38°C (100°F), the 28-day compressive strength of the test specimens may be 10% to 15% lower than if cured at the required ASTM C31 (AASHTO T 23) curing temperatures (Gaynor, Meininger, and Khan 1985).

Specimens cured in the field in the same manner as the structure more closely represent the actual strength of concrete in the structure at the time of testing. However, test specimens give little indication of whether a deficiency is due to the quality of the concrete as delivered or to improper handling and curing. On some projects, field-cured specimens are made in addition to those destined for controlled laboratory curing. These are especially useful when the weather is unfavorable, to determine when forms can be removed, or when the structure can be put into use. For more information see Chapter 18.

Cooling Concrete Materials

Because of the detrimental effects of high concrete temperatures, all operations in hot weather should be directed toward keeping the concrete below specified temperature limits. This is accomplished by cooling the concrete. The usual method of cooling concrete is to lower the temperature of the concrete materials before mixing. One or more of the ingredients can be cooled. The contribution of each ingredient in a concrete mixture to the temperature of the freshly mixed concrete is related to the temperature, specific heat, and quantity of each material.

The aggregates and mixing water have a greater influence on concrete temperature after mixing than other ingredients. Therefore these materials should be kept as cool as practical in hot weather conditions. Figure 16-5 graphically shows the effect of material temperature on the temperature of fresh concrete. It is evident that although concrete temperature is primarily dependent upon the aggregate temperature (due to quantity of material in the mixture), cooling the mixing water can also be effective.

The approximate temperature of concrete can be calculated from the temperatures of its ingredients using the following equation (NRMCA 1962):

$$T = \frac{0.22(T_aM_a + T_cM_c) + T_wM_w + T_{wa}M_{wa}}{0.22(M_a + M_c) + M_w + M_{wa}}$$

Where:

T = temperature of the freshly mixed concrete, °C (°F)

T_a, T_c, T_w, and T_{wa} = temperature in °C (°F) of aggregates, cement, added mixing water, and free water on aggregates

M_a, M_c, M_w, and M_{wa} = mass, kg (lb), of aggregates, cementing materials, added mixing water, and free water on aggregates

Example calculations for initial concrete temperature are shown in Table 16-1.

Aggregates

Aggregates have a pronounced effect on the fresh concrete temperature because they represent 70% to 85% of the total mass of concrete. To lower the temperature of concrete 0.5°C (1°F) requires only a 0.8°C to 1.1°C (1.5°F to 2°F) reduction in the temperature of the coarse aggregate.

There are several simple methods of keeping aggregates cool. Cooling effects are realized when stockpiles are shaded from the sun and kept moist by sprinkling. Since evaporation is a cooling process, sprinkling provides effective cooling, especially when the relative humidity is low. Do not cool aggregate stockpiles with seawater. Using seawater can contribute to acceleration of concrete setting time and corrosion of steel reinforcement.

Sprinkling of coarse aggregates should be controlled to avoid excessive variations in the surface moisture content which may impact consistency.

Refrigeration is another method of cooling materials. Aggregates can be immersed in cold-water tanks, or cooled air can be circulated through storage bins. Vacuum cooling can reduce aggregate temperatures to as low as 7°C (45°F).

Table 16-1A. (Metric). Effect of Temperature of Materials on Initial Concrete Temperatures

Material	Mass, *M*, kg	Specific heat kJ/kg • K	Kilojoules to vary temperature, 1°C	Initial temperature of material, *T*, °C	Total kilojoules in material*
	(1)	(2)	(3) Col. 1 x Col. 2	(4)	(5) Col. 3 x Col. 4
Cement	335 (M_c)	0.92	308	66 (T_c)	20,328
Water	123 (M_w)	4.184	515	27 (T_w)	13,905
Total aggregate	1839 (M_a)	0.92	1692	27 (T_a)	45,684
			2515		79,917

Initial concrete temperature $= \frac{79{,}917}{2515} = 31.8°C$

To achieve 1°C reduction in initial concrete temperature:

Cement temperature must be lowered $= \frac{2515}{308} = 8.2°C$

Or water temperature dropped $= \frac{2515}{515} = 4.9°C$

Or aggregate temperature cooled $= \frac{2515}{1692} = 1.5°C$

* Total kilojoules are relative to a baseline of 0°C.

Table 16-1B. (Inch-Pound). Effect of Temperature of Materials on Initial Concrete Temperatures

Material	Mass, *M*, lb	Specific heat	Btu to vary temperature, 1°F	Initial temperature of material, *T*, °F	Total Btu in material*
	(1)	(2)	(3) Col. 1 x Col. 2	(4)	(5) Col. 3 x Col. 4
Cement	564 (M_c)	0.22	124	150 (T_c)	18,600
Water	282 (M_w)	1.00	282	80 (T_w)	22,560
Total aggregate	3100 (M_a)	0.22	682	80 (T_a)	54,560
			1088		95,720

Initial concrete temperature = $\frac{95,720}{1088}$ = 88.0°F

To achieve 1°F reduction in initial concrete temperature:

Cement temperature must be lowered = $\frac{1088}{124}$ = 8.8°F

Or water temperature dropped = $\frac{1088}{282}$ = 3.9°F

Or aggregate temperature cooled = $\frac{1088}{682}$ = 1.6°F

* Total Btu are relative to a baseline of 0°F.

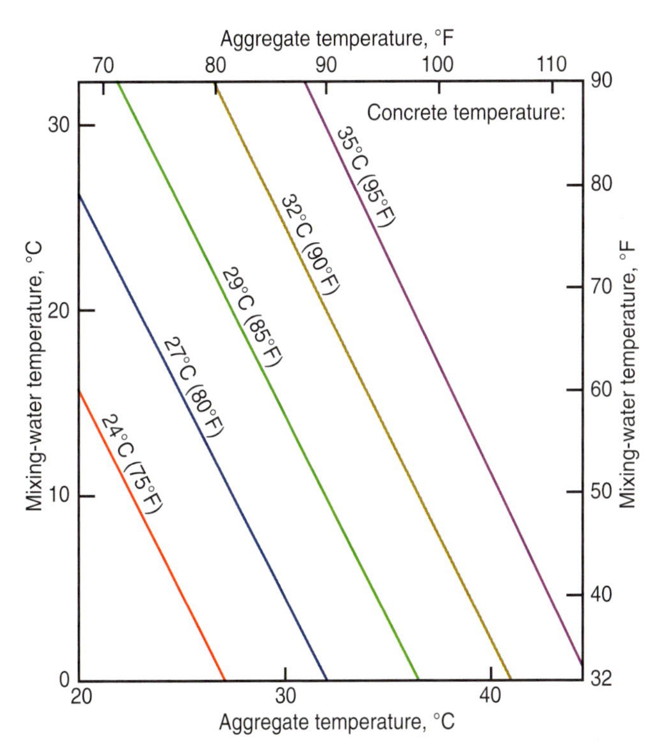

Figure 16-5. Temperature of freshly mixed concrete as affected by temperature of its ingredients. Although the chart is based on the following mixture, it is reasonably accurate for other typical mixtures: Aggregate 1360 kg (3000 lb). Moisture in aggregate 27 kg (60 lb), Added mixing water 109 kg (240 lb), Cement at 66°C (150°F) 256 kg (564 lb).

Water

Although aggregates provide the most influence on concrete temperature, water is the easiest ingredient to cool. Even though it is used in smaller quantities than the other ingredients, cold water produces a moderate reduction in the concrete temperature. Mixing water should be stored in tanks that are not exposed to the direct rays of the sun. Tanks and pipelines carrying mixing water should be buried, insulated, shaded, or painted white to keep water as cool as practical. Water can be cooled by refrigeration, liquid nitrogen, or ice. Cooling the mix water temperature 2.0°C to 2.2°C (3.5°F to 4°F) will usually lower the concrete temperature about 0.5°C (1°F). However, because mix water is such a small percentage of the total mixture, it is difficult to lower concrete temperatures more than about 4°C (8°F) by cooling the batch water alone (Gajda, Kaufman, and Sumodjo 2005).

Ice

Substituting ice for batch water is more effective than using chilled batch water for cooling concrete. Ice both lowers the batch water temperature and lowers the mix temperature by extracting heat during the phase change that occurs as ice melts into liquid water (Gajda, Kaufman, and Sumodjo 2005).

When ice is used, the ice must be completely melted by the time mixing is completed and the concrete is discharged. When using crushed ice, care must be taken to store it at a temperature that will prevent the formation of lumps.

The approximate temperature of concrete can be calculated from the temperatures of its individual ingredients by using the following equation. This equation accounts for the heat of fusion of ice:

$$T\,(°C) = \frac{0.22\,(T_a M_a + T_c M_c) + T_w M_w + T_{wa} M_{wa} - 80 M_i}{0.22(M_a + M_c) + M_w + M_{wa} + M_i}$$

$$T\,(°F) = \frac{0.22\,(T_a M_a + T_c M_c) + T_w M_w + T_{wa} M_{wa} - 112 M_i}{0.22(M_a + M_c) + M_w + M_{wa} + M_i}$$

Where M_i is the mass in kg (lb) of ice (NRMCA 1962).

The heat of fusion of ice in metric units is 335 kJ per kg (in British thermal units, 144 Btu per pound). Calculations in Table 16-2 show the effect of 44 kg (75 lb) of ice in reducing the temperature of concrete. Crushed or flaked ice is more effective than chilled water in reducing concrete temperature. The amount of water and ice must not exceed the total mixing-water requirements.

Table 16-2A. (Metric). Effect of Ice (44 kg) on Temperature of Concrete

Material	Mass, M, kg	Specific heat kJ/kg • K	Kilojoules to vary temperature, 1°C	Initial temperature of material, T, °C	Total kilojoules in material*
	(1)	(2)	(3) Col. 1 x Col. 2	(4)	(5) Col. 3 x Col. 4
Cement	335 (M_c)	0.92	308	66 (T_c)	20,328
Water	123 (M_w)	4.184	515	27 (T_w)	13,905
Total aggregate	1839 (M_a)	0.92	1692	27 (T_a)	45,684
Ice	44 (M_i)	4.184	184	0 (T_i)	79,917
			2699		
minus	44 (M_i) x heat of fusion, (335 kJ/kg) =				−14,740
					65,177

Concrete temperature = $\frac{65,177}{2699}$ = 24.1°C

* Total kilojoules are relative to a baseline of 0°C.

Table 16-2B. (Inch-Pound Units). Effect of Ice (75 lb) on Temperature of Concrete

Material	Mass, M, kg	Specific heat Btu/lb • R	Btu to vary temperature, 1°F	Initial temperature of material, T, °F	Total Btu in material**
	(1)	(2)	(3) Col. 1 x Col. 2	(4)	(5) Col. 3 x Col. 4
Cement	564 (M_c)	0.22	124	150 (T_c)	18,600
Water	207 (M_w)	1.00	207	80 (T_w)	16,560
Total aggregate	3100 (M_a)	0.22	682	80 (T_a)	54,560
Ice*	75 (M_i)	1.00	75	32 (T_i)	2,400
			1088		
minus	75 (M_i) x heat of fusion, (144 Btu/lb) =				−10,800
					81,320

Concrete temperature = $\frac{81,320}{1088}$ = 74.7°F

*32 M_i – 144 M_i = – 112 M_i

**Total Btu are relative to a baseline of 0°F.

Figure 16-6 shows crushed ice being charged into a truck mixer prior to the addition of other materials. Mixing time should be long enough to completely melt the ice. The volume of ice generally should not replace more than approximately 75% of the total batch water. The maximum temperature reduction from the use of ice is limited to about 11°C (20°F).

Figure 16-6. Substituting ice for part of the mixing water will substantially lower concrete temperature. A crusher delivers finely crushed ice to a truck mixer reliably and quickly.

Liquid Nitrogen

If a greater temperature reduction is required, the injection of liquid nitrogen into the mixer may be the best alternative method. Liquid nitrogen can be added directly into a central mixer drum or the drum of a truck mixer to lower concrete temperature. Figure 16-1 shows liquid nitrogen added directly into a truck mixer near a ready mix plant. This may also be added at the jobsite.

Liquid nitrogen precooling should be performed by trained professionals since liquid nitrogen vapors can displace oxygen from air, cause localized fog, and liquid nitrogen is a super-cold liquid. Care should be taken to prevent the liquid nitrogen from contacting the metal drum; the super-cold temperature of the liquid nitrogen can crack the drum. The addition of liquid nitrogen does not in itself influence the amount of mix water required except that lowering the concrete temperature can reduce water demand.

In general, approximately 6 liters (1 ½ gallons) of liquid nitrogen are needed to cool 1 cubic meter (yard) of concrete by 0.5°C (1°F). Liquid nitrogen has been successfully used to precool concrete to temperatures as low as 2°C (35°F) for specialized applications. The temperature of the concrete during injection can be determined through the use of an infrared thermometer aimed at the bottom of the drum.

Cement

Cement temperature has only a minor effect on the temperature of the freshly mixed concrete because of cement's low specific heat and the relatively small amount of cement in a concrete mixture (ACI Committee 305). A cement temperature change of 4°C (8°F) generally will change the concrete temperature by only 0.5°C (1°F). Because cement loses heat slowly during storage, it may still be warm when delivered (this heat is produced in grinding the cement clinker during manufacture). Cement temperatures in storage silos at ready-mix plants are often greater than 50°C (120°F) even in cold weather conditions.

Since the temperature of cement affects the temperature of the fresh concrete to some extent, some specifications place a limit on its temperature at the time of use. This limit varies from 66°C to 82°C (150°F to 180°F). However, it is more practical to specify a maximum temperature for freshly mixed concrete rather than place a temperature limit on individual ingredients (Lerch 1955).

Supplementary Cementitious Materials

Many concrete producers consider the use of supplementary cementitious materials (SCMs) to be essential in hot weather conditions. The materials of choice are fly ash and other pozzolans (ASTM C618, *Standard Specification for Coal Fly Ash and Raw or Calcined Natural Pozzolan for Use in Concrete* or AASHTO M 295) and slag cement (ASTM C989, *Standard Specification for Slag Cement for Use in Concrete and Mortars* or AASHTO M 302). Chapter 4 discusses the impact of each SCM on the water demand and heat of hydration. These materials generally slow both the rate of setting and the rate of slump loss. However, some caution regarding finishing is needed. The rate of bleeding can be slower than the rate of evaporation and plastic shrinkage cracking or crazing may result. This is discussed in greater detail under **Plastic Shrinkage Cracking.**

Chemical Admixtures

A set retarding admixture may be beneficial in delaying the setting time in hot weather concreting, despite the potential for increased rate of slump loss resulting from their use. A hydration control admixture or set stabilizer can be used to stop cement hydration and setting. Hydration is resumed, when desired, with the addition of a special accelerator (reactivator).

Set retarding admixtures should conform to the requirements of ASTM C494 (AASHTO M 194) Type B. Admixtures should be tested under job conditions before construction begins; this will determine their compatibility with the other concrete ingredients. Trial mixtures that simulate the anticipated temperatures as well as haul and placement times while measuring slump loss over time have been helpful in verifying that certain dosages of admixtures will perform properly in the field.

Preparation Before Concreting

Before concrete is placed, certain precautions should be taken during hot weather to maintain or reduce concrete temperature. Mixers, chutes, conveyor belts, hoppers, pump lines, and other equipment for handling concrete should be shaded, painted white, or covered with wet burlap to reduce the effect of solar heating.

Forms, reinforcing steel, and subgrade should be wetted with cool water just before the concrete is placed. During placing and finishing operations, fogging can be directed over the concrete surface. This not only cools the contact surfaces and surrounding air but also increases its relative humidity. The increase in relative humidity minimizes the rate of evaporation of water from the concrete after placement. For slabs on ground, it is a good practice to moisten the subgrade the evening before concreting. There should be no standing water or puddles in the forms or on the subgrade when concrete is placed.

During extremely hot periods, improvements may be obtained by restricting concrete placements to early morning, late evening, or nighttime hours, especially in arid climates. This practice has resulted in substantially less thermal shrinkage and cracking of thick slabs and pavements.

Transporting, Placing, and Finishing

Transporting and placing concrete should be completed as quickly as practical during hot weather. Delays contribute to slump loss and an increase in concrete temperatures. Sufficient labor and equipment must be available at the jobsite to handle and place concrete immediately upon delivery.

Prolonged mixing, even at agitating speed, should be avoided. If delays occur, stopping the mixer and then agitating intermittently can minimize the heat generated by mixing. ASTM C94 (AASHTO M 157) requires that discharge of concrete be completed within 90 minutes or before the drum has completed 300 revolutions, whichever occurs first. However, these restrictions may be extended under certain conditions (ACI 301-10). During hot weather the time limit may be reduced to 60 minutes or even 45 minutes. If specific time limitations on the completion of discharge of the concrete are desired, they should be included in the project specifications. It is also reasonable to obtain test data from a trial batch simulating the batch to placement time, mixing conditions, and anticipated concrete temperatures to document, if necessary, a reduction in the time limit.

Since the setting of concrete is more rapid in hot weather, extra care must be taken with placement techniques to avoid cold joints. For placement of walls, shallower layers can be specified to assure enough time for consolidation with the previous lift. Temporary sunshades and windbreaks help to minimize cold joints.

While proper consolidation is critical to concrete placement, overvibration with internal vibrators should be avoided. Internal vibrators produce significant amounts of heat and can raise the temperature of the concrete locally by 10°C (18°F) or more during a 10 second insertion (Figure 16-7) (Burlingame 2004).

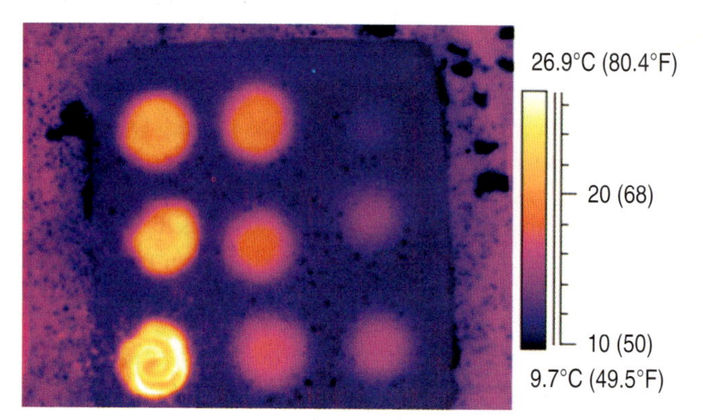

Figure 16-7. Infrared image of concrete surface after completed internal vibration (Burlingame 2004).

Floating of slabs should be done promptly after the water sheen disappears from the surface or when the concrete can support the weight of a finisher with no more than a 5-mm (¼-in.) indentation. Finishing on dry and windy days requires additional precautions to minimize the potential for plastic shrinkage cracking.

Plastic Shrinkage Cracking

Plastic shrinkage cracks sometimes occur in the surface of freshly mixed concrete soon after placement, during finishing or shortly thereafter (Figure 16-8). These cracks which appear mostly on horizontal surfaces can be substantially eliminated using preventive measures.

Figure 16-8. Typical plastic shrinkage cracks.

Plastic shrinkage cracking is usually associated with hot-weather concreting; however, it can occur any time ambient conditions produce rapid evaporation of moisture from the concrete surface. These cracks occur when water evaporates from the surface faster than it can travel to the surface during the bleeding process. This condition creates rapid drying shrinkage and tensile stresses in the surface that often result in short, irregular cracks. The following conditions, individually or collectively, increase evaporation of surface moisture and also increase the possibility of plastic shrinkage cracking:

1. High cementitious materials content
2. Low w/cm
3. High concrete temperature
4. High air temperature
5. Low humidity
6. Wind

The crack length is generally 50 mm to 1000 mm (a few inches to 3 ft) in length and they are usually spaced in a somewhat regular pattern (perpendicular to the wind, or a random pattern on where winds are swirling) with an irregular spacing from 50 mm to 700 mm (a few inches to 2 ft) apart.

The nomograph in Figure 16-9 is a graphical solution borrowed directly from hydrologic studies sponsored by the U.S. Navy and performed by Kohler, Nordenson, and Fox (Kohler 1952, Kohler 1954, and Kohler, Nordenson, and Fox 1955) on the shores of Lake Hefner in Oklahoma. Menzel (1954) sought to estimate the rate of evaporation of bleed water from a water-covered concrete surface. The most recent and most comprehensive entry in the hydrological literature available at that time was Kohler's Lake Hefner work (Kohler 1952, Kohler 1954).

Figure 16-9 is useful for determining when precautionary measures should be taken. However, there is no sure or absolute predictor for plastic shrinkage cracking. Al-Fadhala and Hover (2001) reinforced the principle that the nomograph evaluates the evaporative potential of the environment, not the rate of water loss from the concrete. However, the difference between the two may be small when the concrete surface is covered with bleed water.

Menzel (1954) adopted the Kohler (1952) equations, simply converting the units used to express vapor pressure and wind speed. The values for the saturation vapor pressure of water are themselves temperature dependent:

$W = 0.315\,(e_o - e_a)(0.253 + 0.060V)$ (for pressure in SI unit of kPa)

$W = 0.44\,(e_o - e_a)(0.253 + 0.096V)$ (in.-lb units)

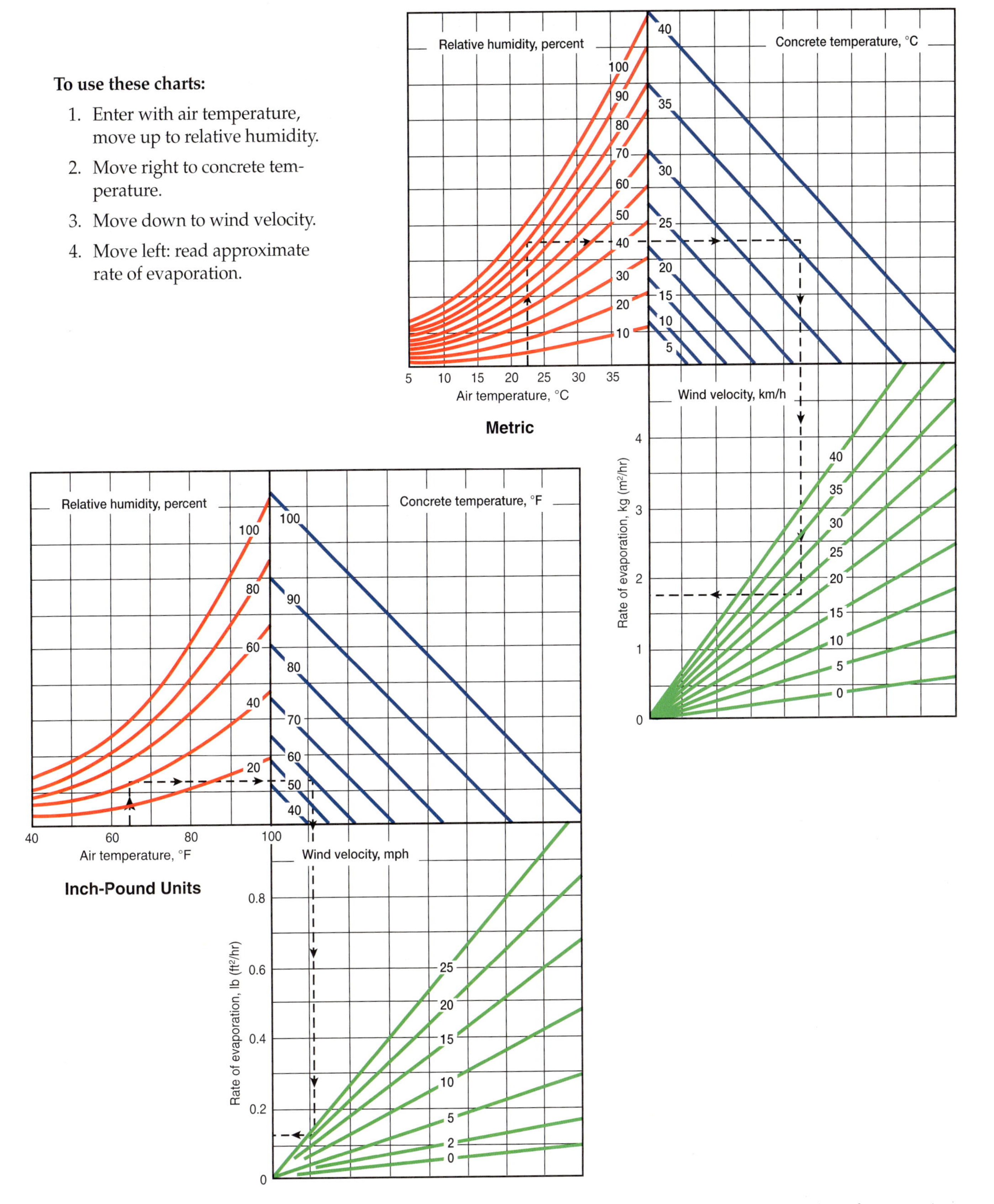

Figure 16-9. Effect of concrete and air temperatures, relative humidity, and wind velocity on rate of evaporation of surface moisture from concrete. Wind speed is the average horizontal air or wind speed in km/h (mph) measured at 500 mm (20 in.) above the evaporating surface. Air temperature and relative humidity should be measured at a level approximately 1.2 to 1.8 m (4 to 6 ft) above the evaporating surface and on the windward side shielded from the sun's rays (Kohler 1952, Menzel 1954, and NRMCA 1960).

Uno (1998) built-in regression equations for saturation vapor pressure and combined them with the Kohler/Menzel equation to produce a unified equation that takes vapor pressure into account:

$$E = 5([T_c + 18]^{2.5} - r \cdot [T_a + 18]^{2.5})(V + 4) \times 10^{-6} \text{ (SI units)}$$

$$E = (T_c^{2.5} - r \cdot T_a^{2.5})(1 + 0.4V) \times 10^{-6} \text{ (in.-lb units)}$$

Where:

W = mass of water evaporated in kg (lb) per m^2 (ft^2) of water-covered surface per hour

e_0 = saturation water vapor pressure in mm Hg (psi) in the air immediately over the concrete surface, at the concrete temperature. Obtain e_0 from Weast (1986) or ACI 305 (2010).

e_a = water vapor pressure in mmhg (psi) in the air surrounding the concrete obtained by multiplying the saturation vapor pressure at the temperature of the air surrounding the concrete by the relative humidity of the air. Air temperature and relative humidity are measured approximately 1.2 m to 1.8 m (4 ft to 6 ft) above the concrete surface on the windward side and shielded from the sun's rays.

V = average wind speed in km/h (mph), measured at 0.5 m (20 in.) above the concrete surface.

E = evaporation rate, lb/ft²/h (kg/m²/h)

T_c = concrete (water surface) temperature, °F (°C)

r = (relative humidity percent)/100

T_a = air temperature, °F (°C)

When the rate of evaporation of bleed water exceeds 1 kg/m² (0.2 lb/ft²) per hour, precautionary measures such as windscreens or fogging are desired. With some concrete mixtures, such as those containing certain pozzolans, cracking is possible if the rate of evaporation exceeds 0.5 kg/m² (0.1 lb/ft²) per hour. Concrete containing silica fume is particularly prone to plastic shrinkage because bleeding rates are commonly as low as 0.25 kg/m² (0.05 lb/ft²) per hour. Therefore, protection from premature drying is essential even at lower evaporation rates.

One or more of the precautions listed below can minimize the occurrence of plastic shrinkage cracking. They should be considered while planning for hot-weather concrete construction or while dealing with the problem after construction has started.

1. Keep the concrete temperature low by cooling aggregates and mixing water.
2. Add fibers to the concrete mixture
3. Moisten concrete aggregates that are dry and absorptive.
4. Dampen the subgrade (Figure 16-10) and fog forms prior to placing concrete.
5. Erect temporary windbreaks to reduce wind velocity over the concrete surface.
6. Erect temporary sunshades to reduce concrete surface temperatures.
7. Fog the slab immediately after placing and before finishing, taking care to prevent the accumulation of water that may increase the w/cm at the surface and reduce the quality of the cement paste in the slab surface. Tooling a wet surface will supply the mixing energy to increase the water-cement ratio.
8. Protect the concrete with temporary coverings, such as reflective (white) polyethylene sheeting, during any appreciable delay between placing and finishing.

Figure 16-10. Dampening the subgrade, yet keeping it free of standing water will lessen drying of the concrete and reduce problems from hot weather conditions.

Fogging the concrete surface before and after final finishing is the most effective way to minimize evaporation and reduce plastic shrinkage cracking (Figure 16-11). Use of a fog spray directed at a level approximately 1.5 m (5 ft) above the concrete surface will raise the relative humidity of the ambient air over the slab, thus reducing evaporation from the concrete. Fog nozzles atomize water using air pressure (Figure 16-12) to create a fog blanket. They should not be confused with garden-hose nozzles, which leave an excess amount of water on the slab. Fogging should be continued until a suitable curing material such as a curing compound, wet burlap, or curing paper can be applied.

Other methods to prevent the rapid loss of moisture from the concrete surface include a temporary application of plastic sheeting to the surface, spray applications, and reducing placement time. The plastic sheeting, when used, is typically installed after strike-off and removed immediately before finishing operations commence. Temporary moisture-retaining films (usually polymers) can be applied immediately after screeding to reduce water evaporation before final finishing operations and curing commence. These materials are floated and troweled into the surface during finishing and should have no adverse effect on the concrete or inhibit the adhesion of membrane-curing compounds. Repeated applications of moisture-retaining films followed by finishing are not recommended since these materials typically contain water contents of 90%. Reduce time between placing and the start of curing by eliminating delays during construction.

Figure 16-11. Fogging cools the air and raises the relative humidity above flatwork to lessen rapid evaporation from the concrete surface, thus reducing cracking and improving surface durability.

Figure 16-12. Fog nozzle.

Curing and Protection

The need for moist curing is greatest during placement and the first few hours after finishing. Initial curing includes procedures implemented anytime between placement and final finishing to reduce moisture loss from the concrete surface. Examples of initial curing include fogging and the use of evaporation reducers (see Chapter 15 and ACI 308).

Curing and protection are more critical in hot weather than in temperate periods. On hardened concrete and on flat concrete surfaces in particular, curing water should not be more than about 11°C (20°F) cooler than the concrete. This will minimize cracking caused by thermal stresses due to temperature differentials between the concrete and curing water.

To prevent the drying of exposed concrete surfaces, moist curing should commence as soon as the surfaces are finished and continue for at least 24 hours. In hot weather, continuous moist curing for the entire curing period is preferred. Avoid wetting and drying of a surface.

At lower w/cm concrete mixtures, the permeability of the paste is normally so low that externally applied curing water will not penetrate far beyond the surface layer (ACI 308 and Meeks and Carino 1999). Therefore, bulk properties such as compressive strength can be considerably less sensitive to surface moisture conditions at lower w/cm; however other surface properties such as abrasion and scaling resistance can be markedly improved by wet-curing low w/cm concrete.

However, if moist curing cannot be continued beyond 24 hours, while the surfaces are still damp, the concrete should be protected from drying with curing paper, heat-reflecting plastic sheets, or membrane-forming curing compounds. White-pigmented curing compounds can be used on horizontal surfaces. Application of a curing compound during hot weather should be preceded by 24 hours of moist curing. If this is not practical, the compound should be applied immediately after final finishing to keep the concrete surface moist.

Concrete surfaces should dry out slowly after the curing period to reduce the possibility of surface crazing and cracking. Crazing, a network pattern of fine cracks that do not penetrate much below the surface, is caused by minor surface shrinkage. Crazing cracks are very fine and barely visible except when the concrete is drying after the surface has been wet. The cracks form a chicken-wire like pattern (Figure 16-13).

Figure 16-13. Crazing cracks are a network of fine cracks, compared to a flexural or tensile stress crack.

Heat of Hydration

Heat generated during cement hydration raises the temperature of concrete to a greater or lesser extent depending on the size of the concrete placement, its surrounding environment, and the amount of cement in the concrete. As a general rule a total temperature rise of 2°C to 9°C (5°F to 15°F) per 45 kg (100 lb) of portland cement can be expected for thinner slab-type placements from the heat of hydration. There may be instances in hot-weather concrete work and massive concrete placements when measures must be taken to cope with the generation of heat from cement hydration and attendant thermal volume changes to control cracking (see Chapters 10 and 20).

References

ACI Committee 301, *Specifications for Structural Concrete*, ACI 301-10, American Concrete Institute, Farmington Hills, Michigan, 2010, 74 pages.

ACI Committee 305, *Hot-Weather Concreting*, ACI 305R-10, American Concrete Institute, Farmington Hills, Michigan, 2010, 23 pages.

ACI Committee 308, *Guide to Curing Concrete*, ACI 308-09, American Concrete Institute, Farmington Hills, Michigan, 2009, 31 pages.

Al-Fadhala, M., and Hover, K.C., "Rapid Evaporation from Freshly Cast Concrete and the Gulf Environment," *Construction and Building Materials*, Vol. 15, No. 1, January 2001, pages 1 to 7.

Burg, Ronald G., *The Influence of Casting and Curing Temperature on the Properties of Fresh and Hardened Concrete*, Research and Development Bulletin RD113, Portland Cement Association, 1996, 13 pages.

Burlingame, Scott, *Application of Infrared Imaging to Fresh Concrete: Monitoring Internal Vibration*, Master's Thesis, Cornell University, 2004, 465 pages. [Also PCA SN2806]

Bureau of Reclamation, *Concrete Manual*, 8th ed., Denver, revised 1981.

Gajda, J.; Kaufman, A.; and Sumodjo, F., "Precooling Mass Concrete", *Concrete Construction*, August 2005, pages 36 to 38.

Gaynor, Richard D.; Meininger, Richard C.; and Khan, Tarek S., *Effect of Temperature and Delivery Time on Concrete Proportions*, NRMCA Publication No. 171, National Ready Mixed Concrete Association, Silver Spring, Maryland, June 1985.

Haynes, W.M., ed., *CRC Handbook of Chemistry and Physics*, 91st edition, CRC Press, Boca Raton, Florida, 2010.

Klieger, Paul, *Effect of Mixing and Curing Temperature on Concrete Strength*, Research Department Bulletin RX103, Portland Cement Association, http://www.cement.org/pdf_files/RX103.pdf, 1958.

Kohler, M.A., "Lake and Pan Evaporation," Water Loss Investigations: Lake Hefner Studies, Geological Survey Circular 229, U.S. Government Printing Office, Washington, D.C., 1952, pages 127 to 158.

Kohler, M.A., "Lake and Pan Evaporation," Water Loss Investigations: Lake Hefner Studies, Technical Report, Geological Survey Professional Paper 269, U.S. Government Printing Office, Washington, D.C., 1954, pages 127 to 148.

Kohler, M.A.; Nordenson, T.J.; and Fox, W.E., "Evaporation from Pans and Lakes," Research Paper No. 38, U.S. Department of Commerce, Washington, D.C., May 1955, 21 pages.

Lerch, William, *Hot Cement and Hot Weather Concrete Tests*, IS015, Portland Cement Association, http://www.cement.org/pdf_files/IS015.pdf, 1955.

Meeks, K.W., and Carino, N.J., *Curing of High Performance Concrete: Report of the State-of-the-Art*, NISTR 6295, National Institute of Standards and Technology, Building and Fire Research Laboratory, Gaithersburg, Maryland, March 1999, 191 pages.

Menzel, Carl A., "Causes and Prevention of Crack Development in Plastic Concrete," *Proceedings of the Portland Cement Association*, 1954, pages 130 to 136.

NRMCA, *Cooling Ready Mixed Concrete*, NRMCA Publication No. 106, National Ready Mixed Concrete Association, Silver Spring, Maryland, 1962.

National Ready Mixed Concrete Association, "Plastic Cracking of Concrete," *Engineering Information*, NRMCA, Silver Spring, Maryland, July 1960, 2 pages.

Uno, P.J., "Plastic Shrinkage Cracking and Evaporation Formulas," *ACI Materials Journal*, Vol. 95, No. 4, July-August 1998, pages 365 to 375.

CHAPTER 17

Cold Weather Concreting

Concrete can be placed safely without damage from freezing in cold climates if certain precautions are taken. Cold weather is defined by ACI Committee 306 as existing when the air temperature has fallen to, or is expected to fall below 4°C (40°F) during the protection period. Under these circumstances, all materials and equipment needed for adequate protection and curing must be on hand and ready for use before concrete placement is started.

During cold weather, the concrete mixture and its temperature should be adapted to the construction procedure and ambient weather conditions. Preparations should be made to protect the concrete from excessively low temperatures using: enclosures, windbreaks, portable heaters, insulated forms, and blankets to maintain a suitable concrete temperature (Figure 17-1). Concrete must be delivered at the proper temperature and the temperature of forms, reinforcing steel, the ground, or other concrete on which the fresh concrete is cast must also be considered. Concrete should not be cast on frozen concrete or on frozen ground. Forms, reinforcing steel, and embedded fixtures must be free of snow and ice at the time concrete is placed. Thermometers and proper storage facilities for test cylinders should be available to verify that precautions are adequate.

Effect of Freezing on Fresh Concrete

Concrete gains very little strength at low temperatures. Freshly mixed concrete must be protected against the disruptive effects of freezing (Figure 17-2) until the degree of saturation of the concrete has been sufficiently reduced by the process of hydration. The time at which this reduction is accomplished corresponds roughly to the time required for the concrete to attain a compressive strength of 3.5 MPa (500 psi) (Powers 1962). At normal temperatures and water-cement ratios less than 0.60, this occurs within the first 24 hours after placement. Significant ultimate strength reductions, up to about 50%, can occur if concrete is frozen within a few hours after placement or before it attains a compressive strength of 3.5 MPa

Figure 17-1. When suitable preparations to build enclosures and insulate equipment have been made, cold weather is no obstacle to concrete construction.

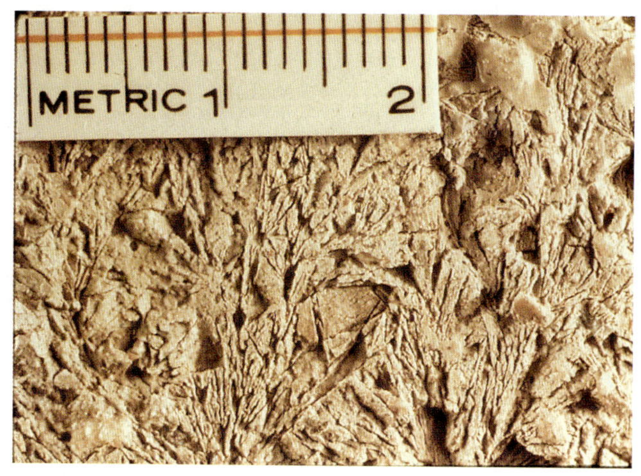

Figure 17-2. Closeup view of ice impressions in paste of frozen fresh concrete. The ice crystal formations occur as unhardened concrete freezes. They do not occur in adequately hardened concrete. The disruption of the paste matrix by freezing can cause reduced strength gain and increased porosity.

(500 psi) (McNeese 1952). Concrete exposed to deicers should attain a compressive strength of 35 MPa (4,500 psi) prior to repeated cycles of freezing and thawing (Klieger 1957 and Gebler and Klieger 1986).

Concrete that has been frozen just once at an early age can be restored to nearly normal strength by providing favorable subsequent curing conditions. Such concrete, however, will not be as resistant to weathering nor as impermeable. The critical period after which concrete is not seriously damaged by one or two freezing cycles is dependent upon the concrete ingredients and conditions of mixing, placing, curing, and subsequent drying. For example, air-entrained concrete is less susceptible to damage by early freezing than non-air-entrained concrete. See Chapter 11 for more information on freeze-thaw resistance.

Strength Gain of Concrete at Low Temperatures

Temperature affects the rate at which hydration of cement occurs – low temperatures retard hydration and consequently retard the hardening and strength gain of concrete.

If hardened concrete is frozen and kept frozen above about minus 10°C (14°F), it will still be able to gain strength slowly. However, below that temperature, cement hydration and concrete strength gain cease. Figure 17-3 illustrates the effect of cool temperatures on setting time of concrete. Figure 17-4 illustrates the effects of casting temperature on slump. Figures 17-5 and 17-6 show the age-compressive strength relationship for concrete that has been cast and cured at various temperatures. Note in Figure 17-6 that concrete cast and cured at 4°C (40°F) and 13°C (55°F) had relatively low strengths for the first week; but after 28 days—when all specimens were moist-cured at 23°C (73°F)—strengths for the 4°C (40°F) and 13°C (55°F) concretes grew faster than the 23°C (73°F) concrete and at one year they were slightly higher.

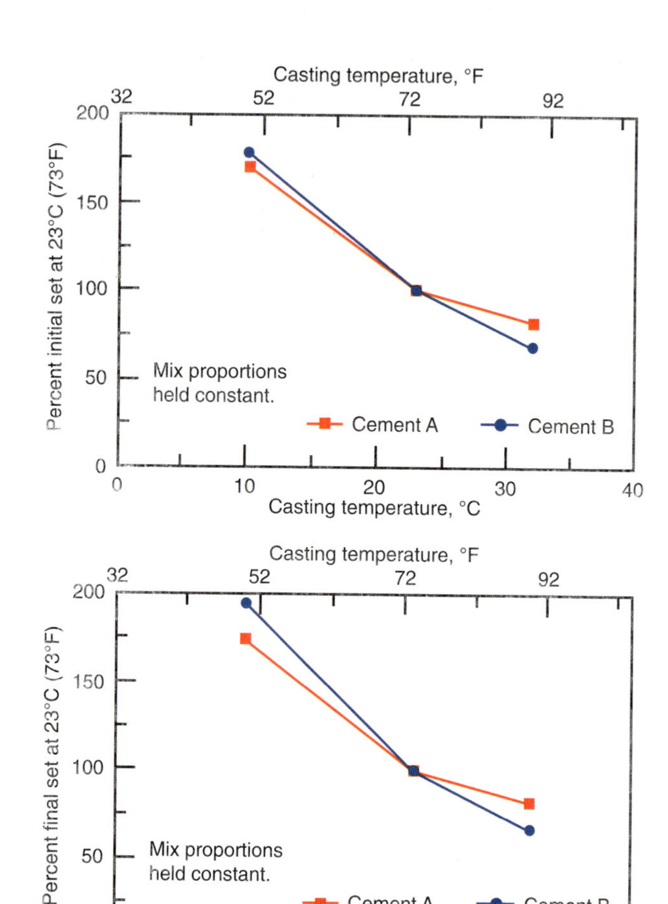

Figure 17-3. Initial set characteristics as a function of casting temperature (top), and final set characteristics as a function of casting temperature (bottom) (Burg 1996).

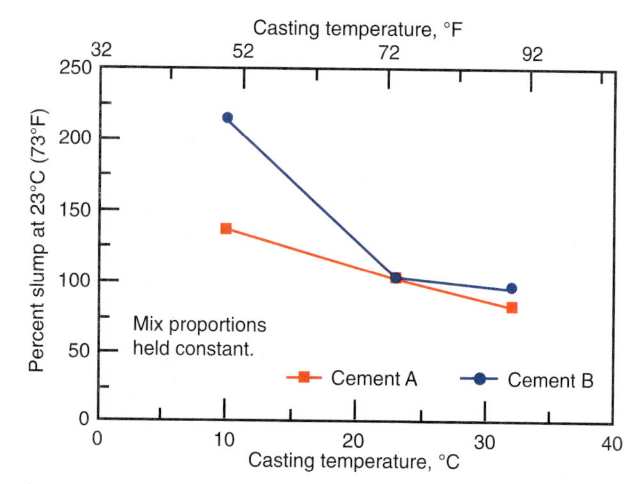

Figure 17-4. Slump characteristics as a function of casting temperature (Burg 1996).

Heat of Hydration

Concrete generates heat during hardening as a result of the chemical reaction by which cement reacts with water to form a hard, stable paste. The heat generated is called heat of hydration; it varies in both amount and rate for

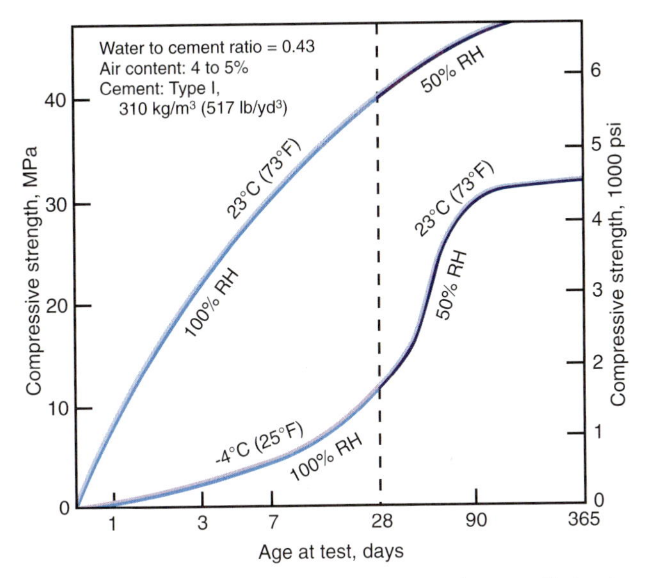

Figure 17-5. Effect of temperature conditions on the strength development of concrete. Concrete for the lower curve was cast at 4°C (40°F) and placed immediately in a curing room at -4°C (25°F). Both concretes received 100% relative humidity curing for first 28 days followed by 50% relative humidity curing (Klieger 1958).

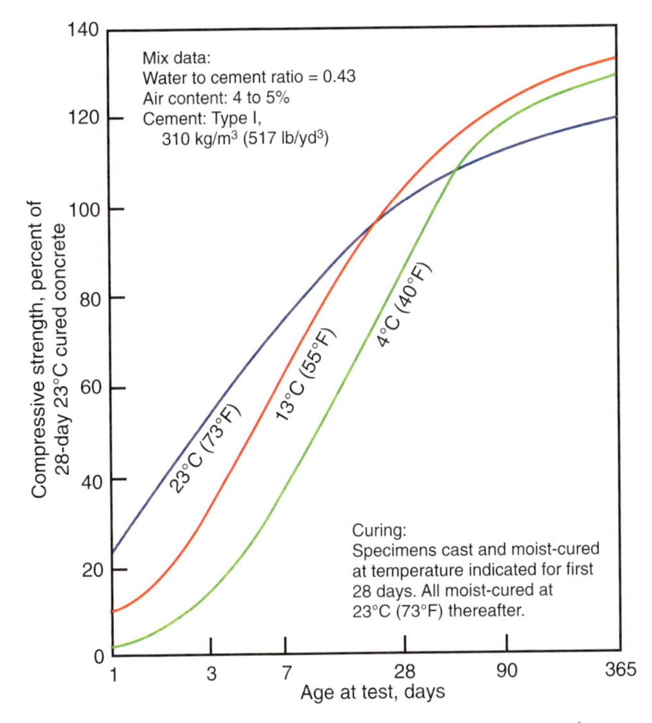

Figure 17-6. Effect of low temperatures on concrete compressive strength at various ages. Note that for this particular mixture made with Type I cement, the best temperature for long-term strength (1 year) was 13°C (55°F) (Klieger 1958).

different cements. Dimensions of the concrete placement, ambient air temperature, initial concrete temperature, water-cement ratio, admixtures, and the composition, fineness, and amount of cementitious material all affect heat generation and buildup.

Heat of hydration is useful in winter concreting as it contributes to the heat needed to provide a satisfactory curing temperature. This is especially true even without other temporary heat sources, particularly in more massive elements.

Figure 17-7 shows a concrete pedestal being covered with a tarpaulin just after the concrete was placed. Tarpaulins and insulated blankets are often necessary to retain the heat of hydration more efficiently and keep the concrete as warm as possible. Thermometer readings of the concrete's temperature will indicate whether the covering is adequate. The heat liberated during hydration will offset to a considerable degree the loss of heat during placing, finishing, and early curing operations. As the heat of hydration subsides, the need to cover the concrete becomes even more important.

Figure 17-7. Concrete footing pedestal being covered with a tarpaulin to retain the heat of hydration.

Special Concrete Mixtures

High strength at an early age is desirable in cold weather construction to reduce the length of time temporary protection is required. The additional cost of high-early-strength concrete is often offset by earlier reuse of forms and shores, savings in the shorter duration of temporary heating, earlier setting times that allows the finishing of flatwork to begin sooner, and earlier use of the structure. High-early-strength concrete can be obtained by using oneor a combination of the following:

1. Type III or HE high-early-strength cement
2. Additional portland cement [60 kg/m³ to 120 kg/m³ (100 lb/yd³ to 200 lb/yd³)]
3. Set accelerating admixtures

Principal advantages occur during the first 7 days. At a 4°C (40°F) curing temperature, the advantages of Type III cement are more pronounced and persist longer than at the higher temperature (Figure 17-8).

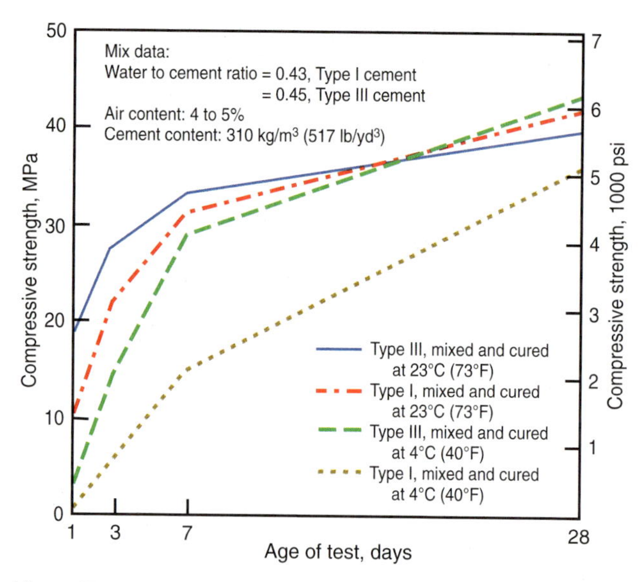

Figure 17-8. Early-age compressive-strength relationships for Type I and Type III portland cement concretes mixed and cured at 4°C (40°F) compared to 23°C (73°F) (Klieger 1958).

Figure 17-9. Finishing this concrete flatwork can proceed because a windbreak has been provided, there is adequate heat under the slab, and the concrete has low slump.

Small amounts of a set accelerating admixture can be used to accelerate the setting and early-age strength development of concrete in cold weather. Set accelerators containing chlorides should not be used where there is an in-service potential for corrosion, such as in concrete members containing steel reinforcement or where aluminum or galvanized inserts will be used. Chlorides are not recommended for concretes exposed to soil or water containing sulfates or for concretes susceptible to alkali-aggregate reaction.

Set accelerators must not be used as a substitute for proper curing and frost protection. Specially designed accelerating admixtures allow concrete to be placed at temperatures down to -7°C (20°F). The purpose of these admixtures is to reduce the time of initial setting, but not necessarily to speed up strength gain. Covering concrete to keep out moisture and to retain heat of hydration is still necessary. Furthermore, traditional antifreeze solutions, as used in automobiles, should never be used. The quantity of these materials needed to appreciably lower the freezing point of concrete is so great that strength and other properties can be seriously affected.

Since the goal of using special concrete mixtures during cold weather concreting is to reduce the time of setting, a low water-cement ratio, low-slump concrete is particularly desirable, especially for cold-weather flatwork. In addition, bleed water is minimized so that finishing can be accomplished more quickly (Figure 17-9). Concrete mixtures with higher slumps provided by water or retarding water-reducing admixtures usually take longer to set.

Air-Entrained Concrete

Entrained air is particularly desirable in any concrete that will be exposed to freezing weather while in service. Concrete that is not air entrained can suffer strength loss and internal as well as surface damage as a result of freezing and thawing (Figure 17-10). Air entrainment provides the capacity to absorb stresses due to ice formation within the concrete.

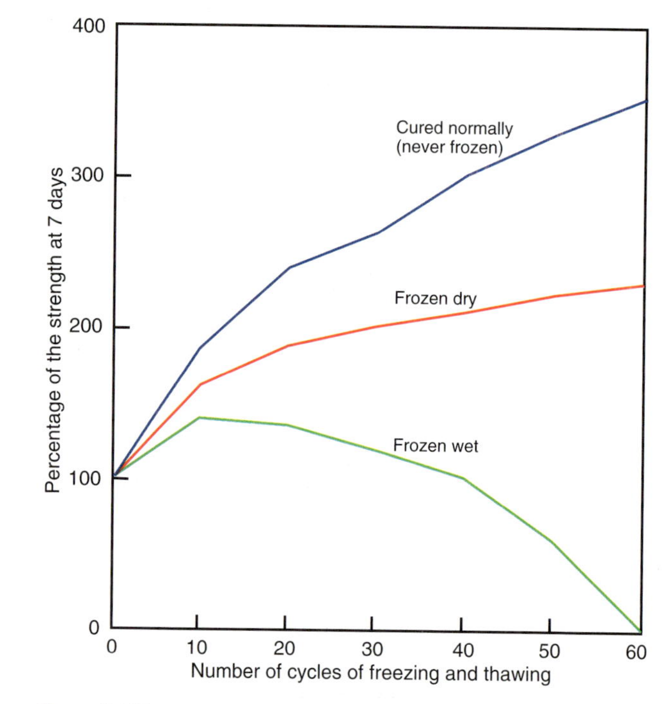

Figure 17-10. Effect of freezing and thawing on strength of concrete that does not contain entrained air (cured 7 days before first freeze) (Powers 1956).

Air entrainment should always be used for construction during the freezing months. The exception is concrete work done under roof where there is no chance that rain, snow, or water from other sources can saturate the concrete and where there is no chance of freezing. See Chapters 7 and 11 for more information on air-entraining admixtures and freeze-thaw resistance of concrete due to air-entrainment.

There is a likelihood of water saturating a concrete floor during construction. Figure 17-11 shows conditions in the upper story of an apartment building during winter construction. Snow accumulated on the top deck and when heaters were used below to warm the deck below, then the snow melted. Water ran through floor openings down to a level that was not being heated. The water-saturated concrete froze, which caused a strength loss, particularly at the floor surface. This could also result in greater deflection of the floor and a surface that is less wear-resistant than originally designed. Steel trowelled (burnished) floor finishes should not be used when entrained air content is specified for slabs. Entrained air may promote blistering and delamination of the slab surface. In addition, the intense energy applied to the slab surface densifies that surface, removing the entrained air content leaving the surface vulnerable to surface scaling deterioration in cold weather conditions.

Figure 17-11. Example of a concrete floor that was saturated with rain, snow, or water and then frozen, showing the need for air entrainment.

Temperature of Concrete

Temperature of Concrete as Mixed

The temperature of fresh concrete as mixed should not be less than shown in Lines 1, 2, or 3 of Table 17-1 for the respective thickness of section. Note that lower concrete temperatures are recommended for more massive concrete sections because heat generated during hydration is dissipated less rapidly in heavier sections. Also note that at lower ambient air temperatures more heat is lost from concrete during transporting and placing. Therefore, the recommended concrete temperatures as mixed are higher for colder weather.

There is little advantage in using fresh concrete at a temperature much above 21°C (70°F). Higher concrete temperatures do not afford proportionately longer protection from freezing because the rate of heat loss is greater. Also, high concrete temperatures are undesirable since they increase thermal shrinkage after hardening, require more mixing water for the same slump, and contribute to the possibility of plastic shrinkage cracking (caused by rapid moisture loss through evaporation). Therefore, the temperature of the concrete as mixed should not be more than 8°C (15°F) above the minimum recommended in Table 17-1.

Aggregate Temperature. The temperature of aggregates varies with weather and type of storage. Aggregates usually contain frozen lumps and ice when the temperature is below freezing. Frozen aggregates must be thawed to avoid aggregate pockets in the concrete after batching, mixing, and placing. If thawing takes place in the mixer, excessively high water contents in conjunction with the cooling effect caused by melting ice must be considered.

At air temperatures consistently above -4°C (25°F) it is seldom necessary to heat aggregates, the desired concrete temperature can usually be obtained by heating only the mixing water. At temperatures below freezing, in addition to heating the mixing water, often only the fine aggregate needs to be heated to produce concrete of the required temperature, provided the coarse aggregate is free of frozen lumps.

Three of the most common methods for heating aggregates are: (1) storing in bins or weigh hoppers heated by steam coils or live steam; (2) storing in silos heated by hot air or steam coils; and (3) stockpiling over heated slabs, steam vents, or pipes. Although heating aggregates stored in bins or weigh hoppers is most commonly used, the volume of aggregate that can be heated is often limited and quickly consumed during production. Circulating steam through pipes over which aggregates are stockpiled is a recommended method for heating aggregates. Stockpiles can be covered with tarpaulins to retain and distribute heat and to prevent formation of ice. Live steam, preferably at pressures of 500 kPa to 900 kPa (75 psi to 125 psi), can be injected directly into the aggregate pile to heat it, but the resultant variable moisture content in aggregates might result in erratic mixing-water control and must be accurately measured and accounted for in batching. Avoid heating methods that may promote hot spots in the aggregates of greater than 100°C (212°F) or average temperatures greater than 65°C (150°F).

Table 17-1. Recommended Concrete Temperature for Cold-Weather Construction – Air-Entrained Concrete*

Line	Condition		Thickness of sections, mm (in.)			
			Less than 300 (12)	300 to 900 (12 to 36)	900 to 1800 (36 TO 72)	Over 1800 (72)
1	Minimum temperature of fresh concrete as *mixed* for weather indicated.	Above -1°C (30°F)	16°C (60°F)	13°C (55°F)	10°C (50°F)	7°C (45°F)
2		-18°C to -1°C (0°F to 30°F)	18°C (65°F)	16°C (60°F)	13°C (55°F)	10°C (50°F)
3		Below -18°C (0°F)	21°C (70°F)	18°C (65°F)	16°C (60°F)	13°C (55°F)
4	Minimum temperature of fresh concrete *as placed and maintained.***		13°C (55°F)	10°C (50°F)	7°C (45°F)	5°C (40°F)

* Adapter from Table 5.1 of ACI 306R-10.
** Placement temperatures listed are for normal-weight concrete. Lower temperatures cn be used for lightweight concrete if justified by tests. For recommended duration of temperatures in Line 4, see Table 17-3.

On small jobs aggregates can be heated by stockpiling over metal culvert pipes in which fires are maintained. Care should be taken to prevent scorching the aggregates.

Mixing-Water Temperature. Of the ingredients used to make concrete, mixing water is the easiest and most practical to heat. The mass of aggregates and cement in concrete is much greater than the mass of water. However, water can store about five times as much heat per unit weight as can cement and aggregate. For cement and aggregates, the average specific heat (that is, heat units required to raise the temperature 1°C (1°F) per kg (lb) of material) can be assumed as 0.925 kJ (0.22 Btu) compared to 4.187 kJ (1.0 Btu) for water.

Figure 17-12 shows the effect of temperature of materials on temperature of fresh concrete. The chart is based on equation 1.

$$T = \frac{[0.22(T_aM_a+T_SM_S+T_CM_C)+T_wM_w+T_{ws}M_{ws}+T_{wa}M_{wa}}{[0.22(M_a+M_S+M_C)+M_w+M_{ws}+M_{wa}]} \quad \text{(Eq. 1)}$$

Where:

T = temperature in degrees Celsius (Fahrenheit) of the fresh concrete

T_a, T_s, T_c, T_w, and T_{ws} and T_{wa} = temperature in degrees Celsius (Fahrenheit) of the coarse aggregate, sand, cement, added mixing water, and free moisture on sand and free moisture on coarse aggregate;
generally $T_s = T_{ws}$ and $T_a = T_{wa}$

M_a, M_s, M_c, M_w, M_{ws} and M_{wa} = mass in kilograms (pounds) of the coarse aggregate, sand, cement, mixing water, free moisture on sand, and free moisture on coarse aggregate

If the weighted average temperature of aggregates and cement is above 0°C (32°F), the proper mixing water temperature for the required concrete temperature can be selected from Figure 17-12. The range of concrete temperatures in the chart corresponds with the recommended values given in Lines 1, 2, and 3 of Table 17-1.

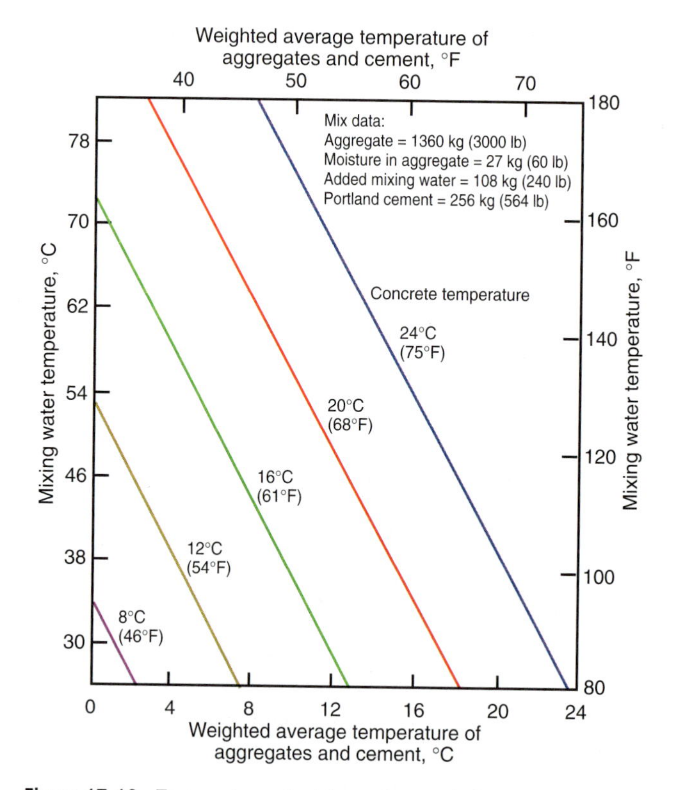

Figure 17-12. Temperature of mixing water needed to produce heated concrete of required temperature. Temperatures are based on the mixture shown but are reasonably accurate for other typical mixtures.

When the temperature of one or more of the aggregates is below 0°C (32°F) the free moisture on the aggregates will freeze, equation 1 can be modified to account for the additional heat required to return the frozen free moisture to a thawed moist state.

Si units – substitute $M_{ws}(0.5T_s - 80)$ for T_sM_{ws}

substitute $M_{wa}(0.5T_a - 80)$ for T_aM_{wa}

Inch pounds – substitute $M_{ws}(0.5T_s - 128)$ for T_sM_{ws}

substitute $M_{wa}(0.5T_a - 128)$ for T_aM_{wa}

To avoid the possibility of a quick or flash set of the concrete when either water or aggregates are heated to above 38°C (100°F), they should be combined in the mixer first, before the cement is added. If this mixer-loading sequence is followed, water temperatures up to the boiling point can be used, provided the aggregates are cold enough to reduce the final temperature of the aggregates and water mixture to appreciably less than 38°C (100°F).

Fluctuations in mixing-water temperature from batch to batch should be avoided. The temperature of the mixing water can be adjusted by blending hot and cold water.

Temperature Loss During Delivery

Temperature loss during haul time may be an issue when delivery times approaching or greater than 1 hour are anticipated. The following equations may be used to estimate temperature loss. These equations accommodate the adjustments to initial temperature to assure minimum temperature requirements at delivery. Adjust the values proportionally for times less or greater than 1 hour (ACI 306-10).

For revolving drum mixers $T = 0.25\,(t_r - t_a)$

For covered-dump body $T = 0.10\,(t_r - t_a)$

For open-dump $T = 0.15\,(t_r - t_a)$

Where:

t_r – Is required delivery temperature in degrees Celsius or Fahrenheit

t_a – Is required air temperature in degrees Celsius or Fahrenheit

Temperature of Concrete as Placed and Maintained

The concrete should be placed in the forms before its temperature drops below that given on Line 4 of Table 17-1. That temperature should be maintained for the duration of the protection period given in Chapter 15 under **Curing Period and Temperature.**

Cooling After Protection

To avoid cracking of the concrete due to sudden temperature change at the end of the curing period, ACI Committee 306 requires that the source of heat and cover protection be slowly removed. The maximum allowable temperature drop during the first 24 hours after the end of the protection is given in Table 17-2. The temperature drops apply to surface temperatures. Notice that the cooling rates for surfaces of mass concrete (thick sections) are lower than they are for thinner members.

Table 17-2. Maximum Allowable Temperature Drop During First 24 Hours After End of Protection Period*

Section size, minimum dimensions, mm (in.)			
Less than 300 (12)	300 to 900 (12 to 36)	900 to 1800 (36 to 72)	Over 1800 (72)
27°C (50°F)	22°C (40°F)	17°C (30°F)	11°C (20°F)

*Adapted from Table 5.1 of ACI 306R-10.

Monitoring Concrete Temperature

Calibrated thermometers are needed to check the concrete temperatures as delivered, as placed, and as maintained. An inexpensive pocket thermometer is shown in Figure 17-13.

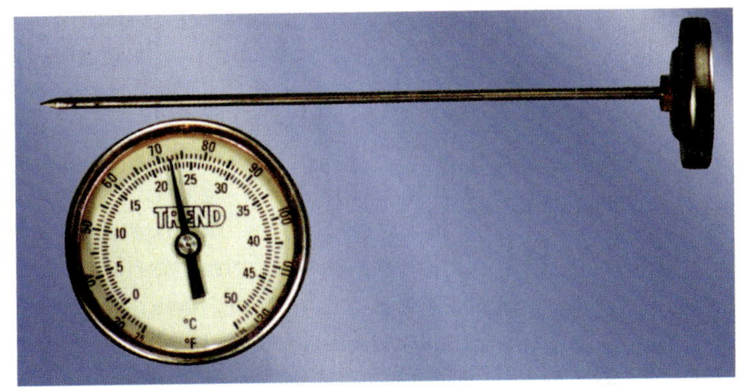

Figure 17-13. A bimetallic pocket thermometer with a metal sensor suitable for checking fresh concrete temperatures.

After the concrete has hardened, temperatures can be checked with special surface thermometers or with an ordinary thermometer that is kept covered with insulating blankets. A simple way to check temperature below the concrete surface is shown in Figure 17-14. Instead of filling the hole shown in Figure 17-14 with a fluid, it can be fitted with insulation. In that application the bulb would remain exposed.

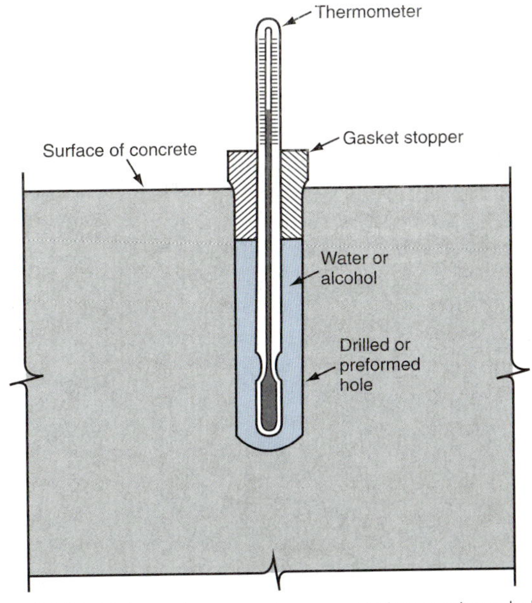

Figure 17-14. Scheme for measuring concrete temperatures below the surface with a glass thermometer.

Concrete test cylinders must be maintained at a temperature between 16°C (60°F) and 27°C (80°F) at the jobsite for up to 48 hours until they are taken to a laboratory for curing (ASTM C31, *Standard Practice for Making and Curing Concrete Test Specimens in the Field* or AASHTO T 23). For concrete mixtures with a specified strength of 40 MPa (6,000 psi) or greater, the initial curing temperature shall be between 20°C and 26°C (68°F and 78°F). During this period, cylinders should be kept in a curing box and covered with a nonabsorptive, nonreactive plate or impervious plastic bag. The temperature in the box should be accurately controlled by a thermostat (Figure 17-15). When stored in an insulated curing box outdoors, cylinders are less likely to be disturbed by vibrations than if left on the floor of a trailer. If kept in a trailer the heat may be inadvertently turned off at night or over a weekend or holiday in cold weather conditions. In that case, the cylinders would not be at the prescribed curing temperatures during this critical period.

Cylinders stripped of molds after the first 24±8 hours must be wrapped tightly in plastic bags or laboratory curing started immediately. When cylinders with strengths less than 40 MPa (6000 psi) are picked up for delivery to the laboratory, they must be maintained at a temperature of 16°C (60°F) to 27°C (80°F) until they are placed in the laboatory curing room. Cylinders with strengths of 40 MPa (6000 psi) or greater must be maintained at a temperature of 20°C (68°F) to 26°C (78°F) (ASTM C31).

Figure 17-15. Insulated curing box with thermostat for curing test cylinders. Heat is supplied by electric rubber heating mats on the bottom. A wide variety of designs are possible for curing boxes.

In addition to laboratory cured cylinders, it is useful to field cure some test cylinders in order to monitor actual jobsite curing conditions in cold weather. It is sometimes difficult to find the right locations for field curing. Differences in the surface to volume ratios between cylinders and the structure, in conjunction with differences in mass, make correlating field cured cylinder strengths to in-place strengths difficult. A preferred location is in a boxout in a floor slab or wall with thermal insulation for cover. When placed on a formwork ledge just below a heated, suspended floor, possible high temperatures there will not duplicate the average temperature in the slab, or the lowest temperature on top of the slab. Still, field cured cylinders are more indicative of actual concrete strength than laboratory-cured cylinders. Particular care should be taken to protect compressive strength test cylinders from freezing; their small mass may not generate enough heat of hydration to protect them.

Cast-in-place cylinders (ASTM C873, *Standard Test Method for Compressive Strength of Concrete Cylinders Cast in Place in Cylindrical Molds*) and nondestructive testing methods (see Chapter 18) as well as maturity techniques (discussed later in this chapter) are helpful in monitoring in place concrete strength.

Concreting on Ground During Cold Weather

Concreting on ground during cold weather involves some extra preparation. Placing concrete on the ground involves different procedures than those used at an upper level: (1) the ground must be thawed before placing concrete; (2) cement hydration will furnish some of the curing heat; (3) construction of enclosures is much simpler and use of insulating blankets may be sufficient; (4) in the case of a floor slab, a vented heater is required if the area is enclosed; and (5) hydronic heaters can be used to thaw subgrades using insulated blankets or to heat enclosures without concern for carbonation. For more on hydronic heaters, see **Heaters** later in this chapter.

Once cast, footings should be backfilled as soon as possible with unfrozen fill. Concrete should never be placed on a frozen subgrade or backfilled with frozen fill. Once these frozen materials thaw, uneven settlements may occur and cause cracking.

ACI Committee 306 requires that concrete not be placed on any surface that would lower the temperature of the concrete in place below the minimum values shown on Line 4 in Table 17-1. In addition, concrete placement temperatures should not be higher than these minimum values by more than 11°C (20°F) to reduce rapid moisture loss and the potential development of plastic shrinkage cracks.

When the subgrade is frozen to a depth of approximately 80 mm (3 inches), the surface region can be thawed by (1) steaming; (2) spreading a layer of hot sand, gravel, or other granular material where the grade elevations allow it; (3) removing and replacing with unfrozen fill; (4) covering the subgrade with insulation for a few days; or (5) using hydronic heaters under insulated blankets. Placing concrete for floor slabs and exposed footings should be delayed until the ground thaws and warms sufficiently to ensure that it will not freeze again during the protection and curing period.

Slabs can be cast on ground at ambient temperatures as low as 2°C (35°F) as long as the minimum concrete temperature as placed is not less than shown on Line 4 of Table 17-1. Although surface temperatures need not be higher than a few degrees above freezing, they also should preferably not be more than 5°C (10°F) higher than the minimum placement temperature. The duration of curing should not be less than that described in Chapter 15 for the appropriate exposure classification. Because of the risk of surface imperfections that might occur on exterior concrete placed in late fall and winter, many concrete contractors choose to delay concrete placement until spring. By waiting until spring, temperatures will be more favorable for cement hydration; this will help generate adequate strengths along with sufficient drying so the concrete can resist freeze-thaw damage.

Concreting Above Ground During Cold Weather

Working above ground in cold weather usually involves several different approaches in comparison to work at ground level:

1. The concrete mixture may not need to be changed to generate more heat because portable heaters can be used to heat the undersides of floor and roof slabs. However, there are advantages to having a mix that will produce a high strength at an early age; for example, artificial heat can be cut off sooner (see Table 17-3), and forms can be recycled faster.
2. Enclosures must be constructed to retain the heat under floor and roof slabs.
3. Portable heaters used to warm the underside of formed concrete can be direct fired heating units (without venting).

Before placing concrete, the heaters under a formed deck should be turned on to preheat the forms and melt any snow or ice remaining on top. Temperature requirements for surfaces in contact with fresh concrete are the same as those outlined in **Concreting on Ground During Cold Weather.** Metallic embedments at temperatures below the freezing point may result in local freezing that decreases the bond between concrete and steel reinforcement. ACI Committee 306 suggests that a reinforcing bar having a cross-sectional area of about 650 mm² (1 in.²) should have a temperature of at least -12°C (10°F) immediately before being surrounded by fresh concrete at a temperature of at least 13°C (55°F). Caution and additional study are required before definitive recommendations can be formulated. Good concrete placement practice suggests heating surfaces of formwork and large embedments to no more than 5°C (10°F) above and no less than 8°C (15°F) below the temperature of the concrete mixture. See ACI 306 for additional information.

Table 17-3A. Recommended Duration of Concrete Protection in Cold Weather – Air-Entrained Concrete*

Service category	Conventional concrete,** days	High-early-strength concrete,† days
No load, not exposed‡ favorable moist-curing	2	1
No load, exposed, but later has favorable moist-curing	3	2
Partial load, exposed	6	4
Fully stressed, exposed	See Table B below	

Table 17-3B. Recommended Duration of Concrete Protection for Fully Stressed, Exposed, Air-Entrained Concrete

Required percentage of standard-cured 28-day strength	Days at 10°C (50°F)			Days at 21°C (70°F)		
	Type of hydraulic cement			Type of hydraulic cement		
	I or GU	II or MH	III or HE	I or GU	II or MH	III or HE
50	6	9	3	4	6	3
65	11	14	5	8	10	4
85	21	28	16	16	18	12
95	29	35	26	23	24	20

* Adapted from Table 7.1 of ACI 306-10. Cold weather is defined as when the temperature has, or is expected to, fall below 4°C (40°F). For recommended concrete temperatures, see Table 17-1. For concrete that is not air entrained, ACI Committee 306 states that protection for durability should be at least twice the number of days listed in Table A.

Part B was adapted from Table 8.2 of ACI 306R-10. The values shown are approximations and will vary according to the thickness of concrete, mix proportions, and so on. They are intended to represent the ages at which supporting forms can be removed. For recommended concrete temperatures, see Table 17-1.

** Made with ASTM C150 Type I, II, or C1157 GU, or MH hydraulic cement.

† Made with ASTM C150Type III or C1157 HE hydraulic cement, an accelerator, or an extra 60 kg/m³ (100 lb/yd³) of cement.

‡ "Exposed" means subject to freezing and thawing.

When slab finishing is completed, insulating blankets or other insulation must be placed on top of the slab to ensure that proper curing temperatures are maintained. The insulation value (R) necessary to maintain the concrete surface temperature of walls and slabs above ground at 10°C (50°F) or above for 7 days may be estimated from Figure 17-16. To maintain a temperature for longer periods, more insulation is required. ACI 306 has additional graphs and tables for slabs placed on ground at a temperature of 2°C (35°F). Insulation can be selected based on R values provided by insulation manufacturers or by using the information in Table 17-4.

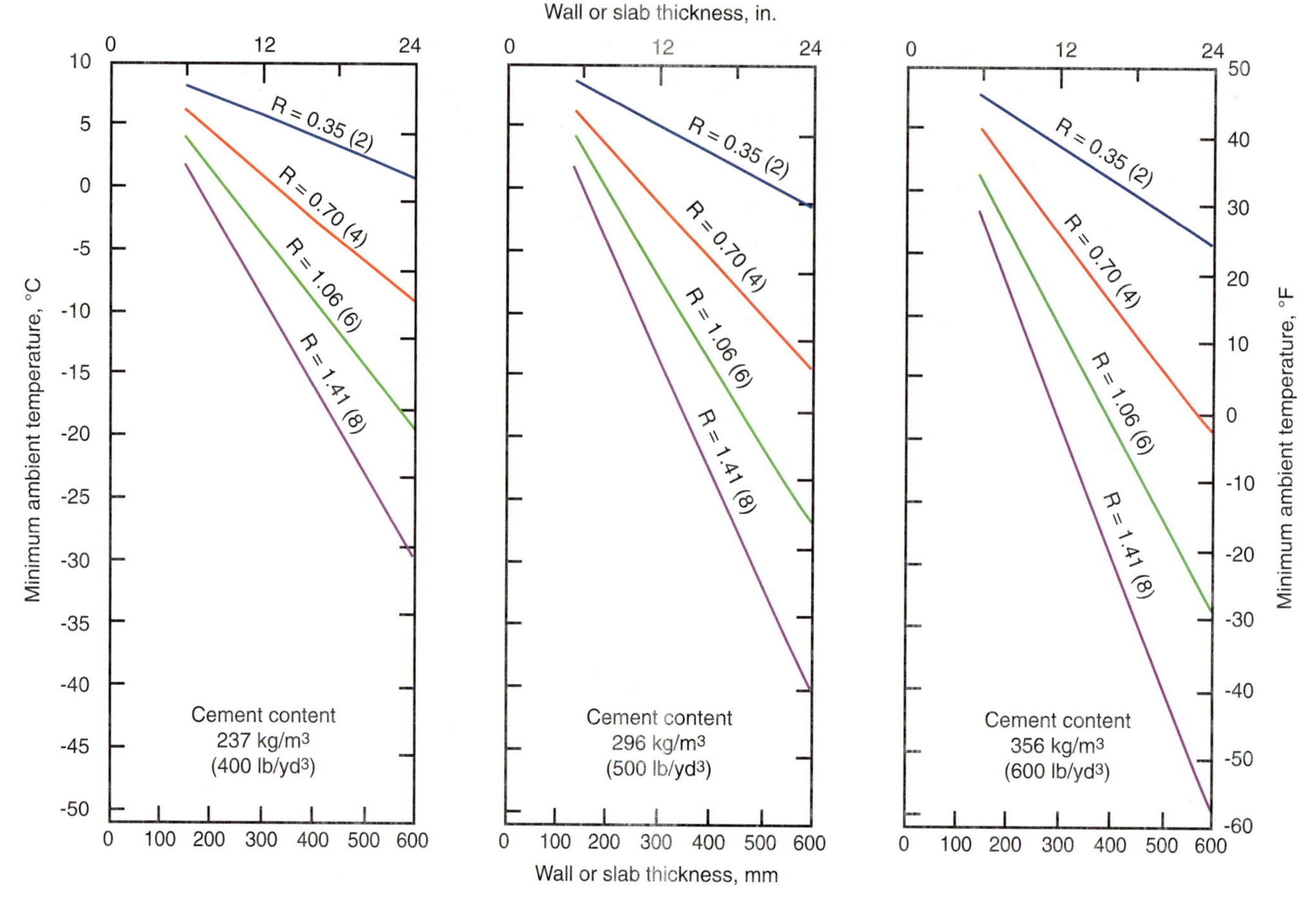

Figure 17-16. Thermal resistance (R) of insulation required to maintain the concrete surface temperature of walls and slabs aboveground at 10°C (50°F) or above for 7 days. Concrete temperature as placed: 10°C (50°F). Maximum wind velocity: 24 km/h (15 mph). Note that in order to maintain a certain minimum temperature for a longer period of time, more insulation or a higher R value is required (adapted from ACI 306).

When concrete strength development is not determined, a conservative estimate can be made if adequate protection at the recommended temperature is provided for the duration of time found in Table 17-3. However, the actual amount of insulation and length of the protection period should be determined from the monitored in place concrete temperature and the desired strength. A correlation between curing temperature, curing time, and compressive strength can be determined from laboratory testing of the particular concrete mixture used in the field (see **Maturity Concept**). Corners and edges are particularly vulnerable during cold weather. As a result, the thickness of insulation for these areas, especially on columns, should be about three times the thickness that is required for walls or slabs. On the other hand, if the ambient temperature rises much above the temperature assumed in selecting insulation values, the temperature of the concrete may become excessive. This increases the probability of thermal shock and cracking when forms are removed. Temperature readings of insulated concrete should therefore be taken at regular intervals and should not vary from ambient air temperatures by more than the values given in ACI 306. The maximum temperature differential between the concrete interior and the concrete surface should be about 20°C (35°F) to minimize cracking. The weather forecast should be checked and appropriate action taken for expected temperature changes.

Columns and walls should not be cast on foundations at temperatures below 0°C (32°F) because chilling of concrete in the bottom of the column or wall will retard strength development. Concrete should not be placed on any surface that would lower the temperature of the as-placed concrete below the minimum values shown on Line 4 in Table 17-1.

Enclosures

Heated enclosures are very effective for protecting concrete in cold weather, but are expensive (Figure 17-17). Enclosures can be of wood, canvas tarpaulins, or polyethylene film (Figure 17-18). Prefabricated, rigid plastic enclosures are also available. Plastic enclosures that admit daylight are the most popular but temporary heat in these enclosures can prove expensive.

Table 17-4. Insulation Values of Various Materials

Material	Density kg/m^3 (lb/ft^3)	Thermal resistance, *R*, for 10-mm (1-in.) thickness of material,* (m^2 • K)/W ([°F • hr • ft^2]/Btu)
Board and slabs		
Expanded polyurethane	24 (1.5)	0.437 (6.25)
Expanded polystyrene, extruded smooth-silk surface	29 to 56 (1.8 to 3.5)	0.347 (5.0)
Expasnded polystyrene, extruded cut-cell surface	29 (1.8)	0.277 (4.0)
Glass fiber, organic bonded	64 to 144 (4 to 9)	0.277 (4.0)
Expanded polystyrene, molded beads	16 (1)	0.247 (3.85)
Mineral fiber with resin binder	240 (15)	0.239 (3.45)
Mineral fiberboard, wet felted	256 to 272 (16 to 17)	0.204 (2.94)
Vegetable fiberboard sheathing	288 (18)	0.182 (2.64)
Cellular glass	136 (8.5)	0.201 (2.86)
Laminated paperboard	480 (30)	0.139 (2.00)
Particle board (low density)	590 (37)	0.128 (1.85)
Plywood	545 (34)	0.087 (1.24)
Loose fill		
Wood fiber, soft woods	32 to 56 (2.0 to 3.5)	0.231 (3.33)
Perlite (expanded)	80 to 128 (5.0 to 8.0)	0.187 (2.70)
Vermiculite (exfoliated)	64 to 96 (4.0 to 6.0)	0.157 (2.27)
Vermiculite (exfoliated)	112 to 131 (7.0 to 8.2)	0.148 (2.13)
Sawdust or shavings	128 to 240 (8.0 to 15.0)	0.154 (2.22)

Material	Density kg/m^3 (lb/ft^3)	Thermal resistance, *R*, for 10-mm (1-in.) thickness of material,* (m^2 • K)/W ([°F • hr • ft^2]/Btu)
Mineral fiber blanket, fibrous form (rock, slag, or glass) 5 to 32 kg/m^3 (0.3 to 2 lb/ft^3)	50 to 70 (2 to 2.75)	1.23 (7)
	75 to 85 (3 to 3.5)	1.90 (11)
	90 to 165 (5.5 to 6.5)	3.34 (19)
Mineral fiber loose fill (rock, slag, or glass) 10 to 32 kg/m^3 (0.6 to 2 lb/ft^3)	95 to 125 (3.75 to 5)	1.90 (11)
	165 to 220 (6.5 to 8.75)	3.34 (19)
	190 to 250 (7.5 to 10)	3.87 (22)
	260 to 350 (10.25 to 13.75)	5.28 (30)

* Values are from *ASHRAE Handbook of Fundamentals,* American Society of Heating, Refrigerating, and Air-conditioning Enginers, Inc., New York, 1977 and 1981.
R values are the reciprocal of U values (conductivity).

Figure 17-17. Even in the winter, an outdoor swimming pool can be constructed if a heated enclosure is used.

Figure 17-18. (top) Tarpaulin heated enclosure maintains an adequate temperature for proper curing and protection during severe and prolonged winter weather. (bottom) Polyethylene plastic sheets admitting daylight are used to fully enclose a building frame. The temperature inside is maintained at 10°C (50°F) with space heaters.

When enclosures are being constructed below a deck, the framework can be extended above the deck to serve as a windbreak. Typically, a height of 2 m (6 ft) will protect concrete and construction personnel against biting winds that cause temperature drops and excessive evaporation. Wind breaks may be taller or shorter depending on anticipated wind velocities, ambient temperatures, relative humidity, and concrete placement temperatures.

Enclosures can be quickly transported using flying forms; more often, though, they must be removed so that the wind will not interfere with maneuvering the forms into position. Similarly, enclosures can be built in large panels with the windbreak included; much like gang forms (Figure 17-1).

Insulating Materials

Heat and moisture can be retained in the concrete by covering it with commercial insulating blankets or batt insulation (Figure 17-19). The effectiveness of insulation can be determined by placing a thermometer under it and in contact with the concrete. If the temperature falls below the minimum required on Line 4 in Table 17-1, additional insulating material, or material with a higher R value, should be applied. Corners and edges of concrete are most vulnerable to freezing. In view of this, temperatures at these locations should be checked more often.

Figure 17-19. Stack of insulating blankets. These blankets trap heat and moisture in the concrete, providing beneficial curing.

The thermal resistance (R) values for common insulating materials are given in Table 17-4. For maximum efficiency, insulating materials should be kept dry and in close contact with concrete or formwork.

Concrete pavements can be protected from cold weather by spreading 300 mm (1 ft) or more of dry straw or hay on the surface for insulation. Tarpaulins, polyethylene film, or waterproof paper should be used as a protective cover over the straw or hay to make the insulation more effective and prevent it from blowing away. The straw or hay should be kept dry or its insulation value will drop considerably.

Insulating blankets for construction are made of fiberglass, sponge rubber, open cell polyurethane foam, vinyl foam, mineral wool, or cellulose fibers. The outer covers are made of canvas, woven polyethylene, or other tough fabrics that will withstand rough handling. The R value for a typical insulating blanket is about 1.2 $m^2 \cdot °C/W$ for 50 to 70 mm thickness, ($7°F \cdot hr \cdot ft^2$)/Btu, but since R values are not marked on the blankets, their effectiveness should be checked with a thermometer. If necessary, they can be used in multiple layers to attain the desired insulation.

Stay-in-place insulating concrete forms (ICF) became popular for cold-weather construction in the 1990s (Figure 17-20). Gajda (2002) showed that ICFs can be used to successfully place concrete in ambient temperatures as low as -29°C (-20°F). Forms built for repeated use often can be economically insulated with commercial blanket or batt insulation. The insulation should have a tough moisture proof covering to withstand handling abuse and exposure to the weather. Rigid insulation can also be used (Figure 17-21).

Figure 17-20. Insulating concrete forms (ICF) permit concreting in cold weather.

Heaters

Three types of heaters are used in cold weather concrete construction: direct fired, indirect fired, and hydronic systems (Figures 17-22 to 17-25). Indirect fired heaters are vented to remove the products of combustion. Where heat is supplied to the top surface of fresh concrete – for example, a floor slab – vented heaters are required. Carbon dioxide (CO_2) in the exhaust must be vented to the outside and prevented from reacting with the fresh concrete (Figure 17-23). Direct fired units can be used to heat the enclosed space beneath concrete placed for a floor or a roof deck (Figure 17-24).

Figure 17-21. With air temperatures down to -23°C (-10°F), concrete was cast in this insulated column form made of 19-mm (¾-in.) high-density plywood inside, 25-mm (1-in.) rigid polystyrene in the middle, and 13-mm (½-in.) rough plywood outside. R value: 1.0 $m^2 \cdot °C/W$ (5.6 [$°F \cdot hr \cdot ft^2$]/Btu).

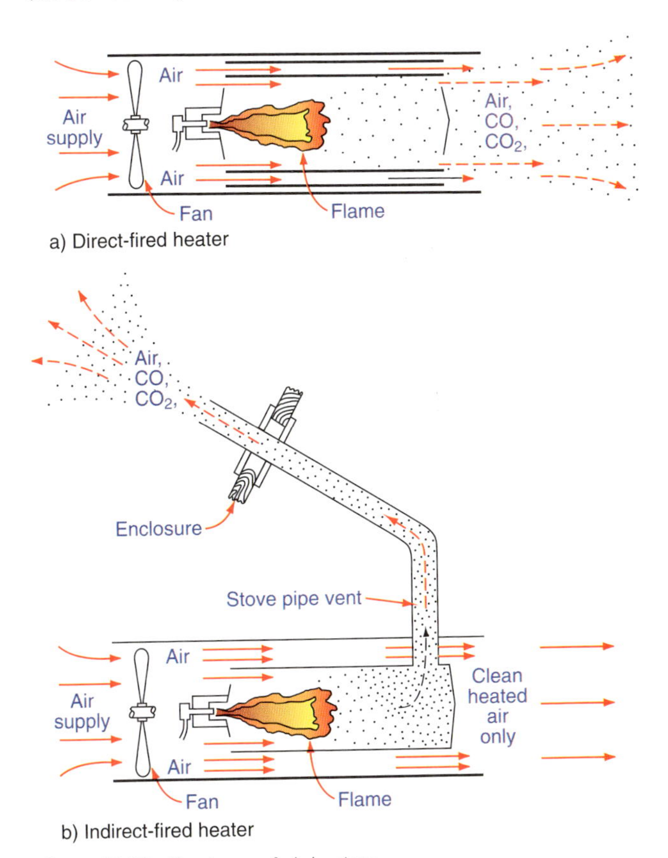

Figure 17-22. Two types of air heaters.

Figure 17-23. An indirect-fired heater. Notice vent pipe that carries combustion gases outside the enclosure.

Figure 17-24. A direct-fired heater installed through the enclosure, thus using a fresh air supply.

Any heater burning a fossil fuel produces carbon dioxide (CO_2); this gas will combine with calcium hydroxide on the surface of fresh concrete to form a weak layer of calcium carbonate that interferes with cement hydration (Kauer and Freeman 1955). The result is a soft, chalky surface that will dust under traffic. Depth and degree of carbonation depend on concentration of CO_2, curing temperature, humidity, porosity of the concrete, length of exposure, and method of curing. Direct-fired heaters, therefore, should not be permitted to heat the air over concreting operations – at least until 24 hours have elapsed. In addition, the use of gasoline-powered construction equipment should be restricted in enclosures during that time. If unvented heaters are used, immediate wet curing or the use of a curing compound will minimize carbonation.

Carbon monoxide (CO), another product of combustion, is not usually a problem unless the heater is using recirculated air. Four hours of exposure to 200 parts per million of CO will produce headaches and nausea. Three hours of exposure to 600 ppm can be fatal. The American National Standard Safety Requirements for Temporary and Portable Space Heating Devices and Equipment Used in the Construction Industry (ANSI A10.10) limits concentrations of CO to 50 ppm at worker breathing levels. The standard also establishes safety rules for ventilation and the stability, operation, fueling, and maintenance of heaters.

A salamander is an inexpensive combustion heater without a fan that discharges its combustion products directly into the surrounding air; heating is accomplished by radiation from its metal casing. Salamanders are fueled by coke, oil, wood, or liquid propane. They are one form of a direct-fired heater. A primary disadvantage of salamanders is the high temperature of their metal casing; a definite fire hazard. Salamanders should be placed so that they will not overheat formwork or enclosure materials. When placed on floor slabs, they should be elevated to avoid scorching the concrete.

Some heaters burn more than one type of fuel. The approximate heat values of fuels are as follows:

No. 1 fuel oil	37,700 kJ/L (135,000 Btu/gal)
Kerosene	37,400 kJ/L (134,000 Btu/gal)
Gasoline	35,725 kJ/L (128,000 Btu/gal)
Liquid-propane gas	25,500 kJ/L (91,500 Btu/gal)
Natural gas	37,200 kJ/m³ (1,000 Btu/ft³)

The output rating of a portable heater is usually the heat content of the fuel consumed per hour. A rule of thumb is that about 134,000 kJ are required for each 100 m³ (36,000 Btu for 10,000 ft³) of air to develop a 10°C (20°F) temperature rise.

Hydronic systems transfer heat by circulating a glycol/water solution in a closed system of pipes or hoses (see Figure 17-25). These systems transfer heat more efficiently than forced air systems without the negative effects of exhaust gases and drying of the concrete from air movement. The specific heat of water/glycol solutions is more than six times greater than air. As a result, hydronic heaters can deliver very large quantities of heat at low temperature differentials of 5°C (10°F) or less between the heat transfer hose and the concrete. Cracking and curling induced by temperature gradients within the concrete are nearly eliminated as is the danger of accidentally overheating the concrete and in potentially damaging long term strength gain.

Typical applications for hydronic systems include thawing and preheating subgrades. They are also used to cure elevated and on-grade slabs, walls, foundations, and columns. To heat a concrete element, hydronic heating hoses are usually laid on or hung adjacent to the structure and covered with insulated blankets and sometimes plastic sheets. Usually, construction of temporary enclosures is not necessary. Hydronic systems can be used over areas much larger than would be practical to enclose. If a heated enclosure is necessary for other work, hydronic hoses can be sacrificed (left under a slab on grade) to make the slab a radiant heater for the structure built above (Grochoski 2000).

Figure 17-25. Hydronic system showing hoses (top) laying on soil to defrost subgrade and (bottom) warming the forms while fresh concrete is pumped in.

Electricity can also be used to cure concrete in winter. The use of large electric blankets equipped with thermostats is one method. The blankets can also be used to thaw subgrades or concrete foundations.

Use of electrical resistance wires that are cast into the concrete is another method. The power supplied is under 50 volts, and from 7.0 to 23.5 MJ (1.5 to 5 kilowatt-hours) of electricity per cubic meter (cubic yard) of concrete is required, depending on the circumstances. Where electrical resistance wires are used, insulation should be included during the initial setting period. If insulation is removed before the recommended time, the concrete should be covered with an impervious sheet and the power continued for the required time.

Steam is another source of heat for winter concreting. Live steam can be piped into an enclosure or supplied through radiant heating units. In choosing a heat source, it must be remembered that the concrete itself supplies heat through hydration of cement; this is often enough for curing needs provided the heat can be retained within the concrete using insulation.

Duration of Heating

After concrete is in place, it should be protected and kept at the recommended temperatures listed on Line 4 of Table 17-1. These curing temperatures should be maintained until sufficient strength is gained to withstand exposure to low temperatures, anticipated environment, and construction and service loads. The length of protection required to accomplish this will depend on the cement type and amount, whether accelerating admixtures were used, and the loads that must be carried. Recommended minimum periods of protection are given in Table 17-3. The duration of heating structural concrete requiring full service loading before forms and shores are removed should be based on the adequacy of in-place compressive strengths rather than an arbitrary time period. If no data are available, a conservative estimate of the length of time for heating and protection can be made using Table 17-3.

Moist Curing

Strength gain stops when moisture required for hydration is no longer available. Concrete retained in forms or covered with insulation seldom loses enough moisture at 5°C to 13°C (40°F to 55°F) to impair curing. However, a positive means of providing moist curing is needed to offset drying from low wintertime humidity and from the dry air produced by heaters used in enclosures during cold weather.

Live steam exhausted into an enclosure around the concrete is an excellent method of curing because it provides both heat and moisture. Steam is especially practical in extremely cold weather because the moisture provided offsets the rapid drying that occurs when very cold air is heated.

Liquid membrane forming compounds can be used for early curing of concrete surfaces within heated enclosures.

Terminating the Heating Period

Rapid cooling of concrete at the end of the heating period should be avoided. Sudden cooling of the concrete surface while the interior is still warm may cause thermal cracking, especially in massive sections such as bridge piers, abutments, dams, and large structural members; thus cooling should be gradual. A safe temperature differential between a concrete wall and the ambient air temperature can be obtained from ACI 306R-10. The maximum uniform drop in temperature throughout the first 24 hours after the end of protection should not be greater than the amounts given in Table 17-2. Gradual cooling can be accomplished by lowering the heat or by simply shutting off the heat and allowing the heat to dissipate.

Form Removal and Reshoring

It is good practice in cold weather to leave forms in place as long as possible. Even within heated enclosures, forms serve to distribute heat more evenly and help prevent drying and local overheating.

If the curing temperatures listed on Line 4 of Table 17-1 are maintained, Table 17-3A can be used to determine the minimum time in days that vertical support for forms should remain in place. Before shores and forms are removed, fully stressed structural concrete should be tested to determine if in-place strengths are adequate. In-place strengths can be monitored using one of the following: (1) field-cured cylinders (ASTM C31 or AASHTO T 23); (2) probe penetration tests (ASTM C803, *Standard Test Method for Penetration Resistance of Hardened Concrete*); (3) cast-in-place cylinders (ASTM C873, *Standard Test Method for Compressive Strength of Concrete Cylinders Cast in Place in Cylindrical Molds*); (4) pullout testing (ASTM C900, *Standard Test Method for Pullout Strength of Hardened Concrete*); or (5) maturity testing (ASTM C1074, *Standard Practice for Estimating Concrete Strength by the Maturity Method*). Many of these tests are indirect methods of measuring compressive strength; they require correlation in advance with standard cylinders before estimates of in-place strengths can be made.

If in-place compressive strengths are not documented, Table 17-3B lists conservative time periods in days to achieve various percentages of the standard laboratory cured 28-day strength. The engineer issuing project drawings and specifications in cooperation with the formwork contractor must determine what percentage of the design strength is required (see ACI Committee 306R-10). Vertical forms can be removed sooner than shoring and temporary falsework (ACI Committee 347).

Maturity Concept

The maturity concept is based on the principle that strength gain in concrete is a function of curing time and temperature. The maturity concept, as described in ACI 306R-10 and ASTM C1074 can be used to evaluate strength development. Two maturity methods to estimate the in-place concrete strength are shown in Table 17-5. The first method is based on the Nurse-Saul function, also called Time-Temperature Factor method. This method is simple and very popular. However, it fails to recognize the fact that maturity increases disproportionately at elevated temperatures and that this increase depends on the type(s) of cementitious materials used and the water-to-cementitious materials ratio. Therefore, the Time-Temperature Factor method typically underestimates the strength development at elevated temperatures. The "Equivalent Age" maturity function is based on the Arrhenius equation; this function presents maturity in terms of equivalent age of curing at a specified temperature.

Table 17-5. Time-Temperature Factor and Equivalent Age Maturity Equations

Time-Temperature Factor	M = maturity index, °C-hours
$M = \sum_{0}^{t} (T - T_0)\Delta t$	T = average concrete temperature, °C, during the time interval Δt T_0 = datum temperature (usually taken to be 0 °C) t = elapsed time, hours Δt = time intervals, hours
Equivalent Age	t_e = equivalent age at the reference temperature
$t_e = \sum_{0}^{t} e^{\frac{-E}{R}\left(\frac{1}{T_r} - \frac{1}{T}\right)} \Delta t$	E = apparent activation energy, J/mol (see ASTM C1074 for typical values) R = universal gas constant, 8.314 J/mol-K T = average concrete temperature, Kelvin, during the time interval Δt T_r = absolute reference temperature, Kelvin Δt = time intervals, hours

The Time-Temperature Factor method presents maturity in terms of °C•hr. Most maturity equipment uses a datum temperature of 0°C, which further simplifies the calculation. Given this simplification, maturity is typically calculated using metric units.

To monitor the strength development of concrete in place using the maturity concept, the following information must be available:

1. The strength-maturity relationship of the concrete used in the structure. The results of compressive strength tests at various ages on a series of cylinders made of a concrete similar to that used in the structure; this must be done to develop a strength-maturity curve. These cylinders are cured in a laboratory at 23°C ± 2°C (73°F ± 3°F).
2. A time-temperature record of the concrete in place. Temperature readings are obtained by placing expendable thermistors or thermocouples at varying depths in the concrete. The location giving the lowest values provides the series of temperature readings to be used in the computation (Figure 17-26).

See Figure 17-27 for an example using the maturity concept. Before construction begins, a calibration curve is drawn plotting the relationship between compressive strength and the maturity factor for a series of test cylinders (of the particular concrete mixture proportions) cured in a laboratory and tested for strength at successive ages.

The maturity concept is imprecise and somewhat limited. However, the concept is useful in checking the curing of concrete and in estimating strength in relation to time and temperature. It presumes that all other factors affecting concrete strength have been properly controlled. With these limitations in mind, the maturity method has gained greater acceptance as a surrogate for measuring the compressive strength of the concrete for removal of shoring or opening a pavement to traffic. It is no substitute for quality control and proper concreting practices (Gajda 2007, Malhotra 1974, and ACI Committee 347).

Figure 17-26. Downloading data from a maturity sensor embedded in a pavement.

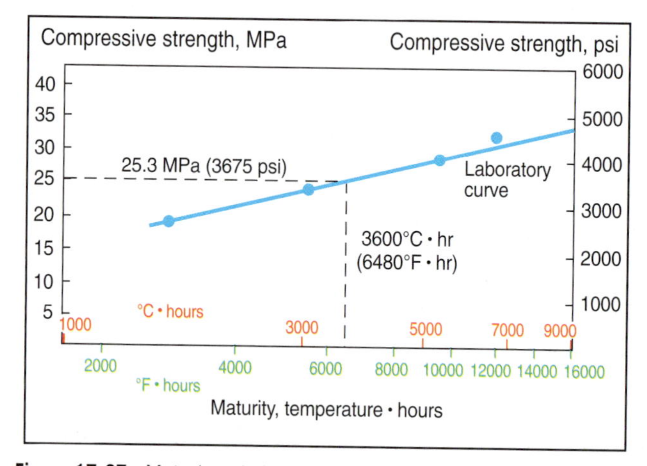

Figure 17-27. Maturity relation example.

References

ACI Committee 306, *Cold-Weather Concreting*, ACI 306R-10, American Concrete Institute, Farmington Hills, Michigan, 2010, 30 pages.

ACI Committee 347, *Guide to Formwork for Concrete*, ACI 347-04, American Concrete Institute, Farmington Hills, Michigan, 2004, 32 pages.

Burg, Ronald G., *The Influence of Casting and Curing Temperature on the Properties of Fresh and Hardened Concrete*, Research and Development Bulletin RD113, Portland Cement Association, 1996, 20 pages.

Gajda, John, *Cold-Weather Construction of ICF Walls*, R&D Serial No. 2615, Portland Cement Association, 2002, 36 pages.

Gajda, John, *Mass Concrete for Buildings and Bridges*, EB547, Portland Cement Association, Skokie, Illinois, USA, 2007, 44 pages.

Gebler, Steven H., and Klieger, Paul, *Effect of Fly Ash on Durability of Air-Entrained Concrete*, Research and Development Bulletin RD090, Portland Cement Association, http://www.cement.org/pdf_files/RD090.pdf, 1986.

Grochoski, Chet, "Cold-Weather Concreting with Hydronic Heaters," *Concrete International*, American Concrete Institute, Farmington Hills, Michigan, April 2000, pages 51 to 55.

Kauer, J.A., and Freeman, R.L., "Effect of Carbon Dioxide on Fresh Concrete," *Journal of the American Concrete Institute Proceedings*, Vol. 52, December 1955, pages 447 to 454. ,American Concrete Institute, Farmington Hills, Michigan. Discussion: December 1955, Part II, pages 1299 to 1304.

Klieger, Paul, *Curing Requirements for Scale Resistance of Concrete*, Research Department Bulletin RX082, Portland Cement Association, http://www.cement.org/pdf_files/RX082.pdf, 1957.

Klieger, Paul, *Effect of Mixing and Curing Temperature on Concrete Strength*, Research Department Bulletin RX103, Portland Cement Association, http://www.cement.org/pdf_files/RX103.pdf, 1958.

Malhotra, V.M., "Maturity Concept and the Estimation of Concrete Strength: A Review," Parts I and II, *Indian Concrete Journal*, Vol. 48, Associated Cement Companies, Ltd., Bombay, April and May 1974.

McNeese, D.C., "Early Freezing of Non-Air-Entrained Concrete," *Journal of the American Concrete Institute Proceedings*, Vol. 49, American Concrete Institute, Farmington Hills, Michigan, December 1952, pages 293 to 300.

Powers, T.C., *Resistance of Concrete to Frost at Early Ages*, Research Department Bulletin RX071, Portland Cement Association, http://www.cement.org/pdf_files/RX071.pdf, 1956.

Powers, T.C., *Prevention of Frost Damage to Green Concrete*, Research Department Bulletin RX148, Portland Cement Association, http://www.cement.org/pdf_files/RX148.pdf, 1962, 18 pages.

CHAPTER 18

Test Methods

Quality control and acceptance testing are indispensable parts of the construction process. Test results provide important feedback on compliance with project specifications and also may be used to base decisions regarding any necessary adjustments to the concrete mixture. Past experience and sound judgment must be relied on in evaluating test results.

Most specifications today are still a combination of prescriptive and performance requirements (Parry 2000). Specifiers are, however, moving toward performance-based specifications (also called end-result or end-property specifications) that are concerned with the final performance of concrete rather than the process used to achieve the performance (Hover, Bickley, and Hooton 2008). Such specifications may not have acceptance limits for process control tests (such as slump or limits on the quantities of concrete ingredients) as with prescriptive specifications. Instead, physical tests are used to measure in-place performance. These tests then become the basis for acceptance. Of course, even though process control tests may not be specified, a producer may use them to guide the product to a successful end result.

Classes of Tests

Project specifications may require specific characteristics of the concrete mixture (such as maximum size of aggregate, aggregate proportions, and minimum amount of cementitous materials) and certain properties of the freshly mixed and hardened concrete (such as temperature, slump, density, air content, and compressive or flexural strength).

Cementitious materials are tested for their compliance with ASTM or AASHTO standard specifications to avoid abnormal performance such as early stiffening, delayed setting, or low strength in concrete.

Aggregate are tested to determine their suitability for use in concrete (including tests for abrasion, resistance to cyclic freezing and thawing, harmful materials by petrographic examination, and potential alkali-aggregate reactivity), and to assure uniformity (such as tests for moisture control, (relative density or specific gravity), and gradation). Some tests are used for both purposes.

Fresh concrete is tested to evaluate the performance of available materials, establish mixture proportions, and control concrete quality during construction. ASTM C94, *Standard Specification for Ready-Mixed Concrete* (or AASHTO M 157), specifies that slump, air-content, density, and temperature tests be performed when strength test specimens are made. Following is a discussion of frequency of testing and descriptions of the major control tests to ensure uniformity of materials, desired properties of freshly mixed concrete, and required strength of hardened concrete. Special test methods are also described.

ASTM (2009) and Lamond and Pielert (2006) provide extensive discussions of test methods for concrete and concrete ingredients.

Computational Software: to provide an easier method for recording test data and calculating test results, NRMCA (2001) provides a CD with spread sheets for a variety of concrete and aggregate tests.

Frequency of Testing

The frequency of testing is a significant factor in the effectiveness of quality control of concrete. Specified testing frequencies are intended for acceptance of concrete or its components. Tests should be conducted at random locations within the quantity or time period represented by the test. Occasionally, specified testing frequencies may be insufficient to effectively control materials within specified limits during production. Therefore, process control tests (nonrandom tests) are often performed in addition to acceptance tests to document trends so that adjustments can be made to the concrete mixture before performing the required acceptance tests.

The frequency of testing aggregates and concrete for typical batch-plant procedures depends largely upon the uniformity of materials, including the moisture content of aggregates, and the production process. Initially, it is advisable to make process control tests several times a day, but as work progresses and materials become more predictable, the testing frequency often can be reduced. ASTM C1451, *Standard Practice for Determining Uniformity of Ingredients of Concrete From a Single Source*, provides a standard practice for determining the uniformity of cementitious materials, aggregates, and chemical admixtures used in concrete.

Usually, aggregate moisture tests are made once or twice a day. The first batch of fine aggregate in the morning is often overly wet because moisture will migrate overnight to the bottom of the storage bin. As fine aggregate is drawn from the bottom of the bin and additional aggregate is added, the moisture content should stabilize at a lower level and the first moisture test can be conducted. It is important to obtain moisture samples representative of the aggregates being batched; a 1% change in moisture content of fine aggregate corresponds to approximately 8 kg/m^3 (13 lb/yd^3) of mix water.

Slump, air content, density (unit weight), and temperature tests should be made for the first batch of concrete each day, whenever consistency of concrete appears to vary, and whenever strength-test specimens are made at the jobsite. Air-content tests should be made often enough at the point of delivery to ensure proper air content, particularly if temperature and aggregate grading change.

The number of strength tests will depend on the job specifications and the occurrence of variations in the concrete mixture. The ACI 318 building code and ASTM C94 require that strength tests for each class of concrete placed each day be made at least once a day, at least once for each 115 m^3 (150 yd^3) of concrete, and if applicable, at least once for each 500 m^2 (5000 ft^2) of surface area for slabs or walls. In ACI 318, a strength test is defined as the average strength of two 150-mm x 300-mm (6-in. x 12-in.) or three 100-mm x 200-mm (4-in. x 8-in.) cylinders tested at 28-days or other age designated for f'_c. A 7-day test cylinder, along with the two or three 28-day test cylinders, is often made and tested to provide an early indication of strength development. As a rule of thumb, the 7-day strength is about 60% to 75% of the 28-day compressive strength, depending upon the type and amount of cementitious materials, water-cement ratio, initial curing temperature, and other variables. Additional specimens may be required when high-strength concrete is involved or where structural requirements are critical. Specimens for strength tests are subjected to standard curing as defined in ASTM C31/C31M, *Standard Practice for Making and Curing Concrete Test Specimens in the Field*, (or AASHTO T23). Strength of standard-cured specimens should not be used as an indication of in-place concrete strengths (ACI 318).

In-place concrete strengths are typically estimated by testing specimens that have been field-cured in the same manner (as nearly as practical) as concrete in the structure. ASTM C31/C31M (or AASHTO T 23) provides requirements for the handling and testing of field-cured specimens. Tests of field-cured specimens are commonly used to decide when forms and shores under a structural slab might be removed or to determine when traffic will be allowed on new pavement. Although field-cured specimens may be tested at any age, 7-day tests are often made for comparison with tests of standard-cured specimens at the same age. These are useful to judge if curing and protection during cold weather concreting is adequate.

Testing Aggregates

Sampling Aggregates

Methods for obtaining representative samples of aggregates are given in ASTM D75, *Standard Practice for Sampling Aggregates* (or AASHTO T 2). Accurate sampling is important. The location in the production process where samples will be obtained must be carefully planned. Sampling from a conveyor belt, stockpile, or aggregate bin may require special sampling equipment; caution must be exercised to obtain a sample free from segregation of different particle sizes. The sample must be large enough to meet ASTM minimum sample size requirements. Samples obtained for moisture content testing should be placed in a sealed container or plastic bag as soon as possible to retain moisture until testing.

Reducing large field samples to small quantities for individual tests must be done in accordance with ASTM C702, *Standard Practice for Reducing Samples of Aggregate to Testing Size* (or AASHTO T 248) so that the final samples will be truly representative. For coarse aggregate, this is done by the quartering method. The sample is thoroughly mixed and formed into a conical pile. The pile is flattened into a layer of uniform thickness and diameter (four to eight times the thickness). The flattened mass is divided into four equal parts, and two opposite quarters are discarded. This process is repeated until the desired size of sample remains. A similar procedure is sometimes used for moist, fine aggregate. Sample splitters are desirable for dry aggregate (Figure 18-1) but should not be used for samples that are more moist than saturated surface dry. A sample splitter comprises of chutes that empty into alternating directions so that one-half of the sample introduced into a hopper is diverted into one receptacle and the other half into another receptacle. The sample from one receptacle is reintroduced into the splitter as many times as necessary to obtain the required sample size.

Figure 18-1. Sample splitter commonly used to reduce coarse aggregate samples.

Organic Impurities

Organic impurities in fine aggregate should be determined in accordance with ASTM C40, *Standard Test Method for Organic Impurities in Fine Aggregates for Concrete* (or AASHTO T 21). A sample of fine aggregate is placed in a sodium hydroxide solution in a colorless glass bottle and shaken. The next day the color of the sodium hydroxide solution is compared with a glass color standard or standard color solution. If the color of the solution containing the sample is darker than the standard color solution or Organic Glass Plate No. 3, the fine aggregate should not be used without further investigation.

Some fine aggregates contain small quantities of coal or lignite that give the solution a dark color. The quantity may be insufficient to reduce the strength of the concrete appreciably. If surface appearance of the concrete is not important, ASTM C33, *Standard Specification for Concrete Aggregates* (AASHTO M 6), states that fine aggregate is acceptable if the amount of coal and lignite does not exceed 1.0% of the total fine aggregate mass. A fine aggregate failing this ASTM C33 (AASHTO M 6) limit may be used if, when tested in accordance with ASTM C87, *Standard Test Method for Effect of Organic Impurities in Fine Aggregate on Strength of Mortar* (or AASHTO T 71), the 7-day strengths of mortar cubes made with the fine aggregate following ASTM C109, *Standard Test Method for Compressive Strength of Hydraulic Cement Mortars (Using 2-in. or [50-mm] Cube Specimens)* (or AASHTO T 106), are at least 95% of the 7-day strengths of mortar made with the same fine aggregate, but washed in a 3% solution of sodium hydroxide and then thoroughly rinsed in water. It should be realized that appreciable quantities of coal or lignite in aggregates can cause popouts and staining of the concrete and can reduce durability when concrete is exposed to weathering. Local experience is often the best indication of the potential durability of concrete made with such aggregates.

Objectionable Fine Material

Large amounts of clay and silt in aggregates can adversely affect durability, increase water requirements, and increase shrinkage. ASTM C33 (AASHTO M 6/M 80) limits the amount of material passing the 75 µm (No. 200) sieve to 3% (abrasion exposure) or 5% in fine aggregate and to 1% or less in coarse aggregate. Testing for material finer than the 75-µm (No.-200) sieve should be done in accordance with ASTM C117, *Standard Test Method for Materials Finer than 75-µm (No. 200) Sieve in Mineral Aggregates by Washing* (AASHTO T 11). Testing for clay lumps should be performed in accordance with ASTM C142, *Standard Test Method for Clay Lumps and Friable Particles in Aggregates* (AASHTO T 112).

Grading

The particle size distribution, or grading, of an aggregate significantly affects concrete mixture proportioning and workability and are an important element in the assurance of concrete quality. The grading of an aggregate is determined by a sieve analysis in which the particles are divided into their various sizes as the sample passes through a stack of standard sieves. The sieve analysis should be made in accordance with ASTM C136, *Standard Test Method for Sieve Analysis of Fine and Coarse Aggregates* (AASHTO T 27).

Results of sieve analyses are used: (1) to determine whether or not the materials meet specifications; (2) to select the most suitable material if several aggregates are available; and (3) to detect variations in grading that are sufficient to warrant blending selected sizes or an adjustment of concrete mixture proportions.

The grading requirements for concrete aggregate are shown in Chapter 6 and ASTM C33 (AASHTO M 6/M 80). Materials containing too much or too little of any one size should be avoided. Some specifications require that mixture proportions be adjusted if the average fineness modulus of fine aggregate changes by more than 0.20. Other specifications require an adjustment in mixture proportions if the amount retained on any two consecutive sieves changes by more than 10% by mass of the total fine-aggregate sample. A small quantity of clean particles that pass a 150-µm (No.-100) sieve but are retained on a 75-µm (No.-200) sieve is desirable for workability. Most specifications permit up to 10% of this finely divided material in fine aggregate.

Well-graded (uniform) aggregates contain particles on each sieve size. Well-graded aggregates enhance numerous characteristics and result in greater workability and durability. The more well-graded an aggregate is, the more it will pack together efficiently, thus reducing the volume between aggregate particles that must be filled by paste. On the other hand, gap-graded aggregates – those having either a large quantity or a deficiency of one or more sieve sizes – can result in reduced workability during mixing, pumping, placing, consolidation and finishing. Durability can suffer too as a result of using more fine aggregate and water to produce a workable mixture. See Chapter 6 and Graves (2006) for additional information on aggregate grading.

Moisture Content of Aggregates

Several methods are used for determining the amount of moisture in aggregate samples. The total moisture content for fine or coarse aggregate can be measured in accordance with ASTM C566, *Standard Test Method for Total Evaporable Moisture Content of Aggregate by Drying*, (AASHTO T 255). In this method a measured sample of damp aggregate is dried either in a ventilated conventional oven, microwave oven, or over an electric or gas hotplate. Using the mass measured before and after drying, the total moisture content can be calculated as follows:

$$P = 100(M - D)/D$$

Where:

P = moisture content of sample, percent

M = mass of original sample

D = mass of dried sample

The surface (free) moisture can be calculated if the aggregate absorption is known. Absorption refers to the increase in aggregate mass due to filling of permeable pores following a standard procedure. It is expressed as a percentage of the dry mass. The surface moisture content is equal to the total moisture content minus the absorbed moisture. Historic information for an aggregate source can be used to obtain absorption data if the mineral composition of the pit or quarry has not changed significantly. However, if recent data are not available, they can be determined using methods outlined in ASTM C127, *Standard Test Method for Density, Relative Density (Specific Gravity), and Absorption of Coarse Aggregate* (AASHTO T 85), for coarse aggregate and ASTM C128, *Standard Test Method for Density, Relative Density (Specific Gravity), and Absorption of Fine Aggregate* (AASHTO T 84) for fine aggregate.

Only the surface moisture, not the absorbed moisture, becomes part of the mixing water in concrete. Surface moisture percentages are used to calculate the amount of water in the aggregates to reduce the amount of mix water added to the batch as discussed in Chapter 12. In addition, the batch weight of aggregates should be increased by the percentage of surface moisture present in each type of aggregate. If adjustments are not made during batching, surface water will replace a portion of the aggregate mass and the batch will not have the correct yield. Table 18-1 illustrates a method of adjusting batch weights for moisture in aggregates.

When drying equipment is not available a field or plant determination of surface (free) moisture in fine aggregate can be made in accordance with ASTM C70, *Standard Test Method for Surface Moisture in Fine Aggregate*. The same procedure can be used for coarse aggregate with appropriate changes in the size of sample and dimensions of the container. This test depends on displacement of water by a known mass of moist aggregate. Therefore, the relative density (specific gravity) of the aggregate must be known accurately.

Electrical moisture meters are used in many concrete batching plants primarily to monitor the moisture content of fine aggregates, but some plants also use them to check coarse aggregates. They operate on the principle that the electrical resistance of damp aggregate decreases as

Table 18-1. Example of Adjustment in Batch Weights for Moisture in Aggregates

Aggregate data	Absorption, %	Moisture content, %
Fine aggregate	1.2	5.8
Coarse aggregate	0.4	0.8

Concrete ingredients	Mix design mass (aggregates in dry BOD condition),* kg/m³ (lb/yd³)	Aggregate mass (SSD condition),** kg/m³ (lb/yd³) $BOD \cdot \frac{(Absorbed\ \%)}{100}$	Aggregate mass (in moist condition), kg/m³ (lb/yd³) $BOD \cdot \frac{(Moisture\ \%)}{100}$	Mix water correction for surface moisture in aggregates, kg/m³ (lb/yd³) $BOD \cdot \frac{(Moist\%-Absorb\%)}{100}$	Adjusted batch weight, kg/m³ (lb/yd³)
Cement	355 (598)				355 (598)
Fine aggregate	695 (1171)	703 (1185)	735 (1239)	32 (54)	735 (1239)
Coarse aggregate	1060 (1787)	1064 (1793)	1068 (1800)	4 (7)	1068 (1800)
Water	200 (337)				164 (276)
Total	2310 (3893)			36 (61)	2322 (3913)†

* An aggregate in a bulk-oven dry (BOD) condition is one with its permeable voids completely dry so that it is fully absorbent.

** An aggregate in a saturated, surface-dry (SSD) condition is one with its permeable voids filled with water and with no surface moisture on it. Concrete suppliers often request mix design proportions on a SSD basis because of batching software requirements.

† Total adjusted batch weight is higher than total mix design weight by the amount of water absorbed in the aggregate.

moisture content increases, within the range of dampness normally encountered. The meters measure the electrical resistance of the aggregate between electrodes protruding into the batch hopper or bin. Moisture meters based on the microwave-absorption method are gaining popularity because they are more accurate than the electrical resistance meters. However, both methods measure moisture contents accurately and rapidly, but only at the level of the probes. These meters require frequent calibration and must be properly maintained. The variable nature of moisture contents in aggregates cause difficulty in obtaining representative samples for comparison with moisture meter readings. Several oven-dried moisture content tests should be performed to verify the calibration of these meters before trends in accuracy can be established.

Testing Freshly Mixed Concrete

Sampling Freshly Mixed Concrete

The importance of obtaining truly representative samples of freshly mixed concrete for control tests is critical. Unless the sample is representative, test results will be misleading. Samples should be obtained and handled in accordance with ASTM C172, *Standard Practice for Sampling Freshly Mixed Concrete* (AASHTO T 141). Except for routine slump and air-content tests performed for process control, ASTM C172 (AASHTO T 141) requires that sample size used for acceptance purposes be at least 28 L (1 ft^3) and be obtained within 15 minutes between the first and final portions of the sample. The composite sample, made of two or more portions, should not be taken from the very first or last portion of the batch discharge. The sample should be protected from sunlight, wind, contamination, and other sources of rapid evaporation during sampling and testing.

Consistency

The slump test described by ASTM C143, *Standard Test Method for Slump of Hydraulic-Cement Concrete* (AASHTO T 119), is the most generally accepted method used to measure the consistency of concrete (Figure 18-2). In this context, the term consistency refers to the relative fluidity of fresh concrete. The test equipment consists of a slump cone (a metal conical mold 300 mm [12 in.] high, with a 200-mm [8-in.] diameter base and 100-mm [4-in.] diameter top) and a steel rod 16 mm (5⁄8 in.) in diameter and 600 mm (24 in.) long with hemispherically shaped tips. The dampened slump cone, placed upright on a flat, nonabsorbent rigid surface, should be filled in three layers of approximately equal volume. Therefore, the cone should be filled to a depth of about 70 mm (2.5 in.) for the first layer, a depth of about 160 mm (6 in.) for the second layer, and overfilled for the third layer. Each layer is rodded 25 times. Following rodding, the last layer is struck off and the cone is slowly raised vertically 300 mm (12 in.) in 5 ± 2 seconds. As the concrete subsides or settles to a new height, the empty slump cone is then inverted and gently placed next to the settled concrete. The tamping rod is placed on the inverted cone to provide a reference for the original height. The slump is the vertical distance the concrete settles, measured to the nearest 5 mm (1⁄4 in.); a ruler is used to measure from the top of the slump cone (mold) to the displaced original center of the subsided concrete (see Figure 18-2).

A higher slump value is indicative of a more fluid concrete. The entire test through removal of the cone should be completed in 2½ minutes, as concrete will lose slump with time. If a portion of the concrete falls away or shears off while performing the slump test, another test should be run on a different portion of the sample. Shearing of the concrete mass may indicate that the mixture lacks cohesion.

Figure 18-2. Slump test for consistency of concrete. Figure A illustrates lower slump, Figure B a higher slump.

Another test method for flow of fresh concrete involves the use of the K-Slump Tester (ASTM C1362, *Standard Test Method for Flow of Freshly Mixed Hydraulic Cement Concrete*). This is a probe-type instrument that is inserted into the fresh concrete in any location where there is a minimum depth of 175 mm (7 in.) of concrete and a 75-mm (3-in.) radius of concrete around the tester. The height of the mortar that has flowed through the openings into the tester provides a measure of fluidity.

Additional consistency tests include: the FHWA vibrating slope apparatus (Wong and others 2001 and Saucier 1966); British compacting factor test (BS 1881); Powers remolding test (Powers 1932); German flow table test (Mor and Ravina 1986); ASTM C1170, *Standard Test Method for Determining Consistency and Density of Roller-Compacted Concrete Using a Vibrating Table*; Kelly ball penetration test (ASTM C360-92 withdrawn) (Daniel 2006); Thaulow tester; Powers and Wiler plastometer (Powers and Wiler 1941); Tattersall (1971) workability device; Colebrand test; BML viscometer (Wallevik 1996); BTRHEOM rheometer for fluid concrete (de Larrard and others 1993); ICAR rheometer (Koehler and Fowler 2004); free-orifice rheometer (Bartos 1978); delivery chute torque meter (US patent 4,332,158 [1982]); delivery-chute vane (US patent 4,578,989 [1986]); Angles flow box (Angles 1974); ring penetration test (Teranishs and others 1994); and the Wigmore consistometer(1948). The Vebe test and the Thaulow test are especially applicable to stiff and extremely dry mixes while the flow table is especially applicable to flowing concrete (Daniel 2006).

For self-consolidating concrete, ASTM C1611/C1611M, *Standard Test Method for Flow of Self-Consolidating Concrete*, can be used to evaluate consistency. The slump cone mold is filled with fresh concrete without rodding. The mold is raised and the concrete is allowed to spread. After spreading has ceased, the average diameter of the concrete mass is measured and reported as the slump flow.

Temperature Measurement

Concrete temperature is measured in accordance with ASTM C1064, *Standard Test Method for Temperature of Freshly Mixed Hydraulic-Cement Concrete* (AASHTO T 309). Because of the important influence concrete temperature has on the properties of freshly mixed and hardened concrete, many specifications place limits on the temperature of fresh concrete. Glass or armored thermometers are available (Figure 18-3). The thermometer should be accurate to plus or minus 0.5°C (±1°F) and should remain in a representative sample of concrete for a minimum of 2 minutes or until the reading stabilizes. At least 75 mm (3 in.) of concrete should surround the sensing portion of the thermometer. Electronic temperature meters with precise digital readouts are also available. The temperature test should be completed within 5 minutes after obtaining the sample.

Figure 18-3. A thermometer is used to take the temperature of fresh concrete.

Density and Yield

The density (unit weight) and yield of freshly mixed concrete (Figure 18-4) are determined in accordance with ASTM C138, *Standard Test Method for Density (Unit Weight), Yield, and Air Content (Gravimetric) of Concrete* (AASHTO T 121). The results may be used to determine the volumetric quantity (yield) of concrete produced per batch (see Chapter 12). The test also can give indications of air content provided the relative densities of the ingredients are known. A balance or scale sensitive to 0.3% of the anticipated mass of the sample and container is required. For example, a 7-L (0.25-ft^3) density container requires a scale sensitive to 50 g (0.1 lb). The size of the container used to determine density and yield varies with the size of aggregate; the 7-L (0.25-ft^3) air meter container is commonly used with aggregates up to 25 mm (1 in.); a14-L (0.5-ft^3)

Figure 18-4. Fresh concrete is measured in a container of known volume to determine density (unit weight).

container is used with aggregates up to 50 mm (2 in.). The volume of the container should be determined at least annually in accordance with ASTM C29/29M, *Test Method for Bulk Density ("Unit Weight") and Voids in Aggregate*. Care is needed to consolidate the concrete adequately by either rodding or internal vibration. Strike off the top surface using a flat plate so that the container is filled to a flat smooth finish. The density is expressed in kilograms per cubic meter (pounds per cubic foot) and the yield in cubic meters (cubic feet). Yield is determined by dividing the total batch weight by the density.

The density of unhardened as well as hardened concrete can also be determined by nuclear methods as described in ASTM C1040/C1040M, *Standard Test Methods for In-Place Density of Unhardened and Hardened Concrete, Including Roller Compacted Concrete, By Nuclear Methods* (AASHTO T 271).

Air Content

A number of methods for measuring air content of freshly mixed concrete can be used. ASTM test methods include: ASTM C231/C231M, *Standard Test Method for Air Content of Freshly Mixed Concrete by the Pressure Method* (AASHTO T 152); ASTM C173/173M, *Standard Test Method for Air Content of Freshly Mixed Concrete by the Volumetric Method* (AASHTO T 196); and ASTM C138/138M (AASHTO T 121). Although they measure only total air volume and not air-void characteristics, laboratory tests have demonstrated that total air content is indicative of the adequacy of the air-void system. With any of the above methods, air-content tests should be started within 5 minutes after the final portion of the composite sample has been obtained.

The pressure method, ACTM C231 (AASHTO T 152), is based on Boyle's law, which relates pressure to volume. Many commercial air meters of this type are calibrated to read air content directly when a predetermined pressure is applied (Figure 18-5). The applied pressure compresses the air within the concrete sample, including the air in the pores of aggregates. For this reason, the pressure method is not suitable for determining the air content of concretes made with some lightweight aggregates or other very porous materials. Aggregate correction factors that compensate for air trapped in normal-weight aggregates are relatively constant and, though small, should be subtracted from the pressure meter gauge reading to obtain the correct air content. The instrument should be calibrated for various elevations above sea level if it is to be used in localities having considerable differences in elevation. Some meters are based on the change in pressure of a known volume of air and are not affected by changes in elevation. Pressure meters are widely used because the mixture proportions and specific gravities of the concrete ingredients need not be known. Also, a test can be conducted in less time than is required for other methods.

Figure 18-5. Pressure-type meter for determining air content.

The volumetric method (Figure 18-6) described in ASTM C173/C173M (AASHTO T 196) is based on the removal of air from a known volume of concrete by agitating the concrete in a fixed volume of water-isopropyl alcohol mixture. This method can be used for concrete containing any type of aggregate, including lightweight or porous materials. An aggregate correction factor is not necessary with this test. The volumetric test is not affected by atmospheric pressure, and the specific gravity of the concrete ingredients need not be known. Care must be taken to agitate the sample sufficiently to remove all air. The addition of 500 mL (1 pt) or more of alcohol accelerates the removal of air, thus shortening test times; it also dispels most of the foam and increases the accuracy of the test, including tests performed on high-air-content or high-cement-content concretes.

Figure 18-6. Volumetric air meter.

The gravimetric method, ASTM C138/C138M or AASHTO T 121, uses the same test equipment used that is for determining the density (unit weight) of fresh concrete. The measured density of concrete is subtracted from the theoretical density as determined from the absolute volumes of the ingredients, assuming no air is present. This difference, expressed as a percentage of the theoretical density, is the air content. Mixture proportions and specific gravities of the ingredients must be accurately known; otherwise results may be in error. Consequently, this method is suitable only where laboratory-type control is exercised. Significant changes in density can be a convenient way to detect variability in air content.

AASHTO T 199, *Standard Method of Test for Air Content of Freshly Mixed Concrete by the Chace Indicator*, can be used as a quick check for the presence of low, medium, or high levels of air in concrete. It is not a substitute for the other more accurate methods. A representative sample of mortar from the concrete is placed in a cup and introduced into a graduated glass container (Figure 18-7). The container is then filled with alcohol to the zero mark on the stem. A thumb is placed over the stem opening and the container is rotated repeatedly from vertical to horizontal end to remove the air from the mortar. The drop in the alcohol level and the mortar content are used to estimate the air content of concrete.

Figure 18-7. Chace air indicator (AASHTO T 199).

Studies into the effect of fly ash on the air-void stability of concrete resulted in the development of the foam-index test. The test can be used to measure the relative air-entraining admixture requirements for concrete mixtures containing fly ash. The fly ash is placed in a wide mouth jar along with the air-entraining admixture and shaken vigorously. Following a waiting period of 45 seconds, a visual determination of the stability of the foam or bubbles is made (Gebler and Klieger 1983).

Air-Void Analysis of Fresh Concrete

The conventional methods for analyzing air in fresh concrete, such as the pressure method noted above, only measure the total air content; consequently, they provide no information about the parameters that determine the quality of the air-void system. These parameters – the size and number of voids and spacing between them – can be measured on polished specimens of hardened concrete (see **Testing Hardened Concrete, Air Content**); but the result of such analysis will only be available several days after the concrete has hardened. A test method has been developed to determine the key air-void parameters in samples of fresh air-entrained concrete. The method uses an apparatus known as an air-void analyzer (AVA) (Figure 18-8). The test apparatus determines the volume and size distributions of entrained air bubbles. The measured data are used to estimate the spacing factor, specific surface, and total volume of entrained air.

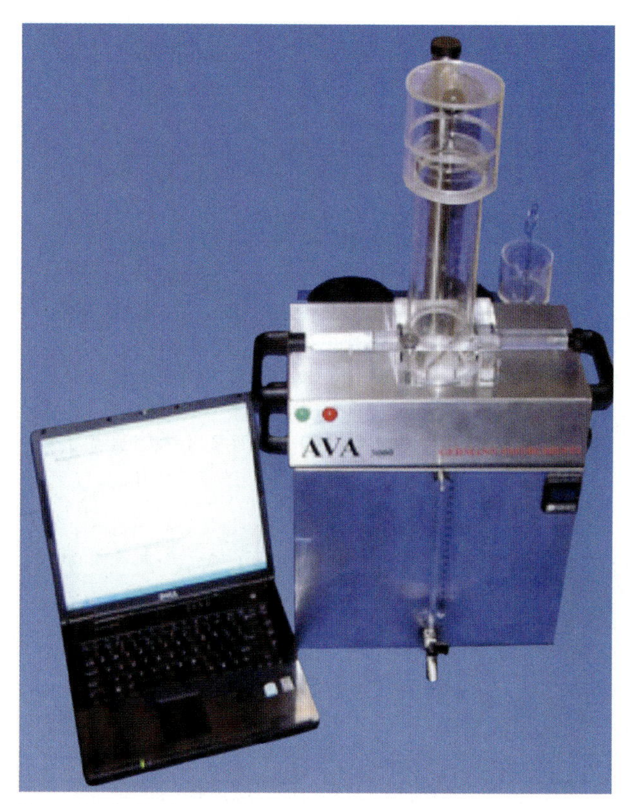

Figure 18-8. Equipment for the air-void analyzer.

In this test method, air bubbles from a sample of fresh concrete rise through a viscous liquid, enter a column of water above it, then rise through the water and collect under a submerged pan that is attached to a sensitive balance (Figure 18-9). The viscous liquid retains the original bubble sizes. Large bubbles rise faster than small ones through the liquids. As air bubbles accumulate under the pan, the buoyancy of the pan increases. The balance measures this change in buoyancy, which is recorded as a function of time and can be related to the number of bubbles of different size.

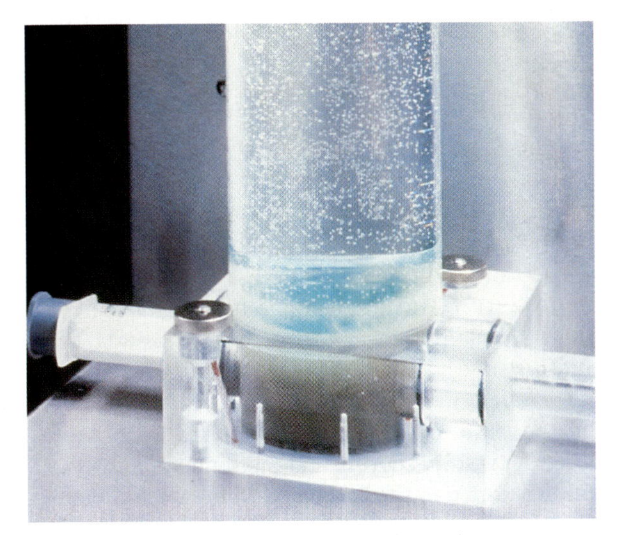

Figure 18-9. Air bubbles rising through liquids in column.

Fresh concrete samples can be taken at the ready mix plant and on the jobsite. Testing concrete before and after placement into forms can verify how the applied methods of transporting, placing, and consolidation affect the airvoid system. Because the samples are taken on fresh concrete, the air content and air-void system can be adjusted during production.

In 2008, AASHTO adopted a provisional test method for the AVA (AASHTO TP 75-08, *Method of Test for Air-Void Characteristics of Freshly Mixed Concrete by Buoyancy Change*). The AVA was not developed for measuring the total air-content of concrete, and because of the small sample size, may not give accurate results for this quantity. However, this method may be useful in assessing the quality of the air-void system; it gives good results in conjunction with traditional methods for measuring air content (Aarre 1998, Distlehorst and Kurgan 2007, and Peterson 2009).

Strength Specimens

Specimens molded for strength tests should be made and cured in accordance with ASTM C31/C31M or AASHTO T 23, and laboratory-molded specimens according to ASTM C192/C192M, *Standard Practice for Making and Curing Concrete Test Specimens in the Laboratory* (or AASHTO R39). Molding of strength specimens should be started within 15 minutes after the composite sample is obtained.

Traditionally, the standard test specimen for compressive strength of concrete with a nominal maximum aggregate size of 50 mm (2 in.) or smaller was a cylinder 150 mm (6 in.) in diameter by 300 mm (12 in.) high (Figure 18-10). In 2008, ACI 318 was revised to permit 100 mm (4 in.) in diameter by 200 mm (8 in.) high cylinders. The smaller cylinders can only be used for nominal maximum aggregate size of 25 mm (1 in.) or less. For larger aggregates, the diameter of the cylinder should be at least three times the nominal maximum size of aggregate and the height should be twice the diameter. Alternatively, it is permitted to wet-sieve fresh concrete with large aggregate using a 50 mm (2 in.) sieve in accordance with ASTM C172. While rigid metal molds are preferred, paraffin-coated cardboard, plastic, or other types of single-use molds conforming to ASTM C470, *Standard Specification for Molds for Forming Concrete Test Cylinders Vertically*, can be used. They should be placed on a smooth, level, rigid surface and filled carefully to avoid distortion of their shape.

Figure 18-10. Preparing standard test specimens for compressive strength of concrete.

The smaller 100-mm (4-in.) diameter by 200-mm (8-in.) high cylinders have been commonly used with high strength concrete containing up to 19 mm (¾ in.) maximum nominal size aggregate (Burg and Ost 1994, Forstie and Schnormeier 1981, and Date and Schnormeier 1984). The 100-mm x 200-mm (4-in. x 8-in.) cylinders are easier to cast, requires less material, weigh considerably less than 150-mm x 300-mm (6-in. x 12-in.) cylinders and requires less storage space for curing. In addition, the smaller cross-sectional area allows the use of smaller capacity testing machines to test high-strength concrete cylinders. The difference in indicated strength between the two cylinder sizes is insignificant as illustrated in Figure 18-11. The standard deviation and coefficient of variation of 100-mm (4-in.) cylinders is slightly higher or similar to that for 150-mm (6-in.) cylinders (Detwiler and others 2001, Burg and others 1999, Pistilli and Willems 1993, and Carino and others 2004). The predominant size used in Canada is the 100-mm (4-in.) diameter cylinder (CSA A23.1). Consult job specifications for allowable cylinder sizes.

Beams for the flexural strength test should be 150 mm x 150 mm (6 in. x 6 in.) in cross section for nominal maximum size of aggregates up to 50 mm (2 in.). For larger aggregates, the minimum cross-sectional dimension should be at least three times the nominal maximum size of aggregate. The length of beams should be at least three times the depth of the beam plus 50 mm (2 in.), or a total length of at least 500 mm (20 in.) for a 150-mm x 150-mm (6-in. x 6-in.) beam.

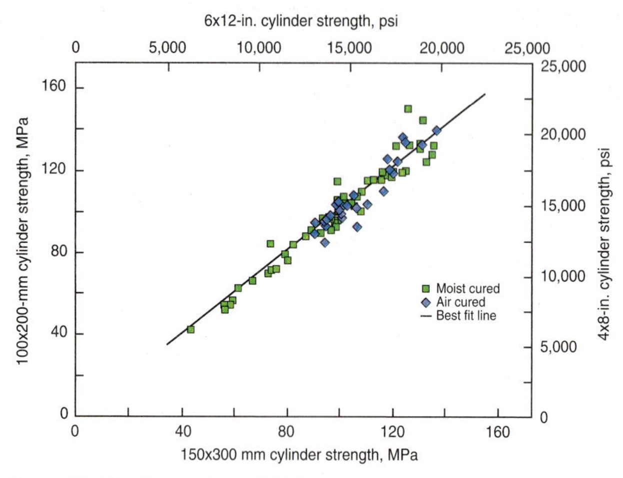

Figure 18-11. Comparison of 100 x 200-mm (4 x 8-in.) and 150 x 300-mm (6 x 12-in.) cylinder strengths (Burg and Ost 1994).

ASTM C31/C31M prescribes the method of consolidation and the number of layers to be used in making test specimens. For concrete with slump less than 25 mm (1 in.), consolidation is by vibration. For slump of at least 25 mm (1 in.), consolidation is by vibration or rodding. For cylinders to be consolidated by rodding, the molds are filled in two layers for 100-mm (4-in.) diameter cylinders and three layers for 150-mm (6-in.) diameter cylinders. Each layer is rodded 25 times and a smaller diameter rod is used for the smaller cylinders. If the rodding leaves holes, the sides of the mold should be lightly tapped with a mallet or open hand. Cylinders to be vibrated should be filled in two layers with one insertion per layer for 100-mm (4-in.) diameter cylinders and two insertions per layer for 150-mm (6-in.) cylinders. Care should be used to avoid over-vibration. Usually vibration is sufficient when the top surface becomes relatively smooth and large air bubbles no longer break through the surface.

Beams up to 200 mm (8 in.) deep are molded using two layers if consolidated by rodding and using one layer if consolidated by vibration. Each layer is rodded once for each 1400 mm² (2 in.²) of top surface area. If vibration is used, the vibrator should be inserted at intervals less than 150 mm (6 in.) along the center of the beam. For beams wider than 150 mm (6 in.), alternate insertions of the vibrator should be along two lines. Internal vibrators should have a maximum width not more than ⅓ the width of beams or ¼ the diameter of cylinders. Immediately after casting, the tops of the specimens should be sealed with plastic caps or plastic bags or covered with an oiled glass or steel plates.

The strength of a test specimen can be greatly affected by jostling, changes in temperature, and exposure to drying, particularly within the first 24 hours after casting. Thus, test specimens should be cast in locations where subsequent movement is unnecessary and where protection is available. Cylinders and test beams should be protected from rough handling at all ages. Identify specimens on the exterior of the mold to prevent confusion and errors in reporting.

Standard testing procedures require that specimens be cured under controlled conditions, either in the laboratory (Figure 18-12) or in the field. Standard curing gives an accurate indication of the quality of the concrete as delivered. After specimens are molded in the field, they are subjected to initial curing for up to 48 h in accordance with ASTM C31/C31M. The temperature surrounding the specimens should be between 16 and 27°C (60 and 80°F) and moisture loss should be prevented. For concrete with a specified strength greater than 40 MPa (6000 psi), the storage temperature should be between 20 and 26°C (68 and 78°F). After initial curing and mold removal, specimens are subjected to final curing with free water maintained on their surfaces and at a temperature of 23.0± 2.0°C (73.5 ± 3.5°F). Specimens can be submerged in limewater or stored in a moist room. To prevent leaching of calcium hydroxide from concrete specimens, limewater must be saturated with hydrated lime, not agricultural lime (limestone), in accordance with ASTM C511, *Standard Specification for Mixing Rooms, Moist Cabinets, Moist Rooms, and Water Storage Tanks Used in the Testing of Hydraulic Cements and Concretes* (AASHTO M 201).

Figure 18-12. Controlled moist curing in the laboratory for standard test specimens at a relative humidity of 95% to 100% and temperature of 23±2°C (73.5±3.5°F) (ASTM C511 or AASHTO M 201).

Specimens cured in the field in the same manner as the structure more closely represent the actual strength of concrete in the structure at the time of testing. However, they give little indication of whether a low strength test result is due to the quality of the concrete as delivered or to improper handling and curing. On some projects, field-cured specimens are made in addition to those destined for standard curing; these are especially useful during cold weather, to determine when forms can be removed, or to determine when the structure can be put into use. For more information see **Strength Tests of Hardened Concrete** and ASTM (2009).

In-place concrete strength development can also be evaluated by maturity testing (ACI Committee 228 [2003] and ASTM C1074, *Standard Practice for Estimating Concrete Strength by the Maturity Method*), which is discussed in Chapter 17.

Time of Setting

ASTM C403/C403M, *Standard Test Method for Time of Setting of Concrete Mixtures by Penetration Resistance* (AASHTO T 197), is used to determine the time of setting of concrete by means of penetration resistance measurements made at regular time intervals on mortar sieved from the concrete mixture (Figure 18-13). The initial and final time of setting are determined as the times when the penetration resistance equals 3.4 MPa (500 psi) and 27.6 MPa (4000 psi). Typically, initial setting occurs between 2 and 6 hours after batching and final setting occurs between 4 and 12 hours. Temperature, water cementitious materials ratio, and admixtures all affect setting time.

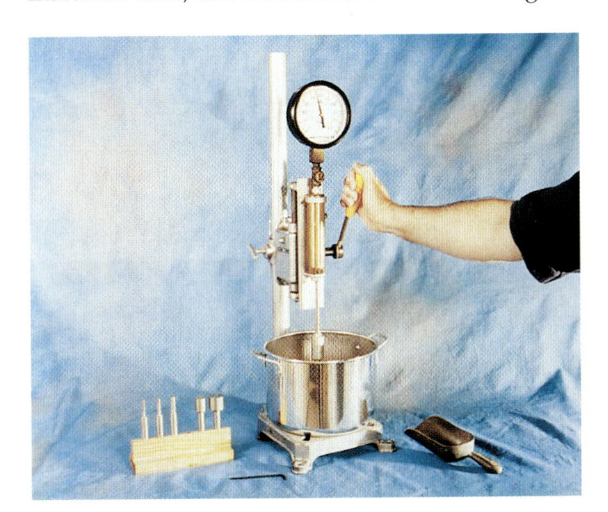

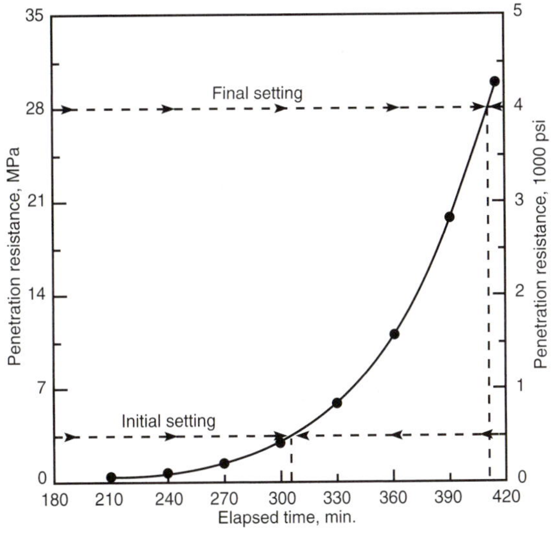

Figure 18-13. (top) Time of setting equipment. (bottom) Plot of test results.

Accelerated Compression Tests to Project Later-Age Strength

ASTM C684, *Standard Test Method for Making, Accelerated Curing, and Testing Concrete Compression Test Specimens*, uses accelerated strength tests to expedite quality control of concrete. Strength development of test specimens is accelerated using one of four curing procedures: warm water at 35°C ± 3°C (95°F ± 5°F), in boiling water for 3.5 h, autogenous curing in an insulated container, or 5 hours at a high temperature of 150°C ± 3°C (300°F ± 5°F) and a pressure of 10.3 ± 0.02 MPa (1500 ± 25 psi). Accelerated strength tests are performed at ages ranging between 5 and 49 hours, depending on the curing procedure used. Later-age strengths are estimated using previously established relationships between accelerated strength and standard 28-day compressive strength tests (Carino 2006).

ASTM C918/C918M, *Standard Test Method for Measuring Early-Age Compressive Strength and Projecting Later-Age Strength*, uses the maturity method of monitoring temperature of cylinders cured in accordance with ASTM C31/C31M (AASHTO T 23). Cylinders are tested at early ages beyond 24 hours, and the concrete temperature history is used to compute the maturity index at the time of test. To use this method, a prediction equation relating strength to maturity index, is developed from laboratory or field data in accordance with ASTM C918/C918M. The prediction equation is used to project the strength at later ages based on the maturity index and measured strength of the specimens tested at early-age. See Carino (2006).

Chloride Content

The chloride content of fresh concrete should be checked to make sure it is below the specified limits, such as those given in ACI 318, to avoid corrosion of reinforcing steel. An approximation of the water-soluble chloride content of aggregates, admixtures, and freshly mixed concrete can be made using a method developed by the National Ready Mixed Concrete Association (NRMCA 1986). The total chloride content of freshly mixed concrete may be estimated by summing up the chloride contents of all of the individual constituents of the mixture. The NRMCA method provides only a quick approximation and should not be used to determine compliance.

The water extractable chloride content of aggregate may be determined using ASTM C1524, *Standard Test Method for Water-Extractable Chloride in Aggregate (Soxhlet Method)*. In this method, the aggregate is not pulverized so that chloride ions within the aggregate particles are not extracted. These chlorides would not be available to initiate or contribute toward steel corrosion (ACI 222R). See also **Testing Hardened Concrete, Chloride Content**.

Portland Cement Content, Water Content, and Water-Cement Ratio

Test methods have been developed for estimating the portland cement and water content of freshly mixed concrete. Due to their complexity, however, they are used rarely for routine quality control. Nevertheless, results of these tests can assist in determining the strength and durability potential of concrete prior to setting and hardening and can indicate whether or not the desired cement and water contents were obtained. While not currently in issue, ASTM C1078, *Test Methods for Determining the Cement Content of Freshly Mixed Concrete* (Withdrawn 1998), and ASTM C1079, *Test Methods for Determining the Water Content of Freshly Mixed Concrete* (Withdrawn 1998), based on the Kelly-Vail method, can be used to determine cement content and water content. Experimental methods using microwave absorption have also been developed to estimate the water-cement ratio. The disadvantage of these test methods is that they require sophisticated equipment and special operator skills, which may not be readily available.

Other tests for determining cement or water contents can be classified into four categories: chemical determination, separation by settling and decanting, nuclear methods, and electrical methods. The Rapid Analysis Machine (RAM) and nuclear cement gage have been used to measure cement contents (Forester, Black, and Lees 1974 and PCA 1983). The microwave oven drying method (AASHTO T 318, *Standard Method of Test for Water Content of Freshly Mixed Concrete Using Microwave Oven*) and neutron-scattering methods have been used to measure water contents. For an overview of these and other tests from all four categories, see Hime (2006). A combination of these tests can be run independently to determine either cement content or water content to calculate the water-cement ratio. None of these test methods, however, have the level of reliability required for use as acceptance tests.

Supplementary Cementitious Materials Content

Standard test methods are not available for determining the supplementary cementitious materials content of freshly mixed concrete. However, the presence of certain supplementary cementitious materials, such as fly ash, can be determined by washing a sample of the concrete's mortar over a 45 µm (No. 325) sieve and using a stereo microscope (150 to 250 X) to view the residue retained (Figure 18-14). Fly ash particles appear as spheres of various colors. Sieving the mortar through a 150- or 75-µm (No. 100 or 200) sieve is helpful in removing sand grains.

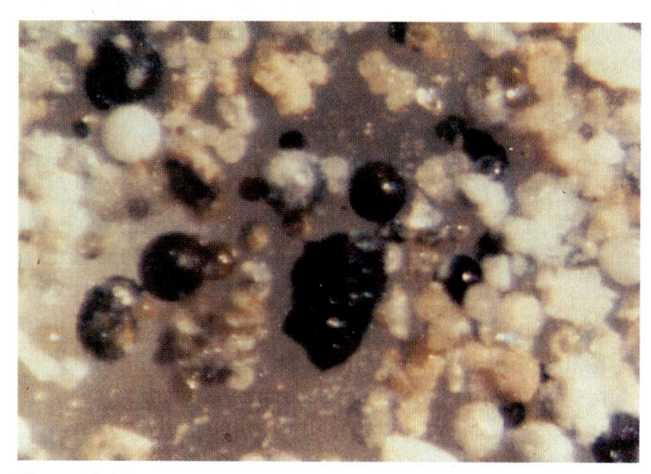

Figure 18-14. Fly ash particles retained on a 45µm sieve after washing, as viewed through a microscope at 200x.

Bleeding of Concrete

The bleeding tendency of fresh concrete can be determined by two methods described in ASTM C232, *Standard Test Methods for Bleeding of Concrete* (or AASHTO T 158). One method consolidates the specimen by tamping without further disturbance; the other method consolidates the specimen by vibration after which the specimen is vibrated intermittently throughout the test. The bleeding tendency is expressed as the volume of bleed water at the surface per unit area of exposed concrete, or as a percentage of the net mixing water in the test specimen. Typical values range from 0.01 to 0.08 mL/cm^2 or 0.1% to 2.5% of mixing water. The bleeding test is rarely used in the field, but it is useful for evaluating alternative mixtures in the laboratory (Figure 18-15).

Figure 18-15. ASTM C232 (AASHTO T 158) test for bleeding of concrete; Method A without vibration. The container has an inside diameter of about 255 mm (10 in.) and a height of about 280 mm (11 in.). The container is filled to a height of about 255 mm (10 in.). Bleed water is drawn off the concrete surface and recorded in regular intervals until cessation of bleeding. Inset: The container needs to be covered during the test to prevent evaporation.

Testing Hardened Concrete

Molded specimens prepared as described in the previous section **Strength Specimens** (ASTM C31/C31M [AASHTO T 23], ASTM C192/C192M [AASHTO R39], or field specimens obtained in accordance with ASTM C873, *Standard Test Method for Compressive Strength of Concrete Cylinders Cast in Place in Cylindrical Molds*), or samples of hardened concrete obtained from construction (ASTM C42/C42M, *Standard Test Method for Obtaining and Testing Drilled Cores and Sawed Beams of Concrete* [AASHTO T 24]), can be used for tests of hardened concrete. Separate specimens should be obtained for each test performed because specimen preconditioning for certain tests can make the specimen unusable for other tests.

Strength Tests of Hardened Concrete

Strength tests of hardened concrete can be performed on the following: (1) specimens molded from samples of freshly mixed concrete and cured in accordance with ASTM C31/C31M or ASTM C192/C192M (AASHTO T 23 and R 39) (Figure 18-16); (2) in-situ specimens cored or sawed from hardened concrete in accordance with ASTM C42/C42M (AASHTO T 24); or (3) cast-in-place specimens made using special cylinder molds and cured in the structure in accordance with ASTM C873 (Figure 18-17).

Figure 18-16. Compressive strength test cylinders and flexural strength specimens made in the field in accordance with ASTM C31/C31M.

Cast-in-place cylinders can be used in concrete that is 125 mm to 300 mm (5 in. to 12 in.) in depth. The mold is filled in the normal course of concrete placement. The specimen is then cured in place and in the same manner as the rest of the concrete section. The specimen is removed from the concrete and mold immediately prior to testing to determine the in-place concrete strength. If cast-in-place specimens have length-diameter ratios (L/D) less than 1.75, the strength correction factors given in ASTM C42/C42M are applied. This method is particularly applicable in cold-weather concreting, post-tensioning work, slabs, or any concrete work where a minimum in-place strength must be achieved before construction can continue.

Figure 18-17. Concrete cylinders cast in place in cylindrical molds provide a means for determining the in-place compressive strength of concrete (ASTM C873/C873M).

For all methods, cylindrical specimens should have a diameter at least three times the nominal maximum size of coarse aggregate in the concrete and a length as close to twice the diameter as possible. Correction factors are available in ASTM C42/C42M (AASHTO T 24) for specimens with length-diameter ratios between 1.0 and 1.75. Cores and cylinders with a height of less than 95% of the diameter before or after capping should not be tested. A minimum core diameter of at least 94 mm (3.70 in.) should be used if a length to diameter (L/D) ratio greater than one is possible.

Cores should not be taken until the concrete can be drilled without disturbing the bond between the mortar and the coarse aggregate. For horizontal surfaces, cores should be taken vertically away from formed joints or edges. For vertical or sloped faces, cores should be taken perpendicular to the central portion of the concrete placement. Diamond-studded coring bits can cut through reinforcing steel. This should be avoided if possible when obtaining compression test specimens. The presence of bar reinforcement perpendicular to the core axis may reduce the measured compressive strength. There are, however, insufficient research data to develop appropriate correction factors to account for the presence of steel. If the specifier permits testing of cores with embedded reinforcement, engineering judgment is required to evaluate test results. A covermeter (electromagnetic device) or a surveyor's magnetic locator can be used to locate reinforcing steel.

Length of a core drilled from a concrete structure should be determined in accordance with ASTM C1542/C1542M, *Standard Test Method for Measuring Length of Concrete Cores*. For measuring member thickness, the use of ASTM C174/C174 M, *Standard Test Method for Measuring Thickness of Concrete Elements Using Drilled Concrete Cores*, may be stipulated.

Cores taken from structures should be tested in a moisture condition as near as that of the in-place concrete as possible. Procedures for moisture conditioning of cores are described in ASTM C42/C42M (AASHTO T 24).

It has been observed that the presence of a moisture gradient has a detrimental effect on the measured core strength. Figure 18-18 shows the effects of core conditioning on the strength of drilled cores. Forty-eight hour water immersion of the specimens prior to testing yields significantly lower test results than air-drying specimens for seven days prior to testing. The work by Fiorato and others (2000) showed that measured strengths varied by up to 25%, depending upon the time and type of conditioning prior to testing. Before testing, ASTM C42/C42M requires that cores be kept in sealed plastic bags for at least 5 days after last being wetted. This is intended to provide reproducible moisture conditions and reduce the effects of moisture gradients introduced during specimen preparation.

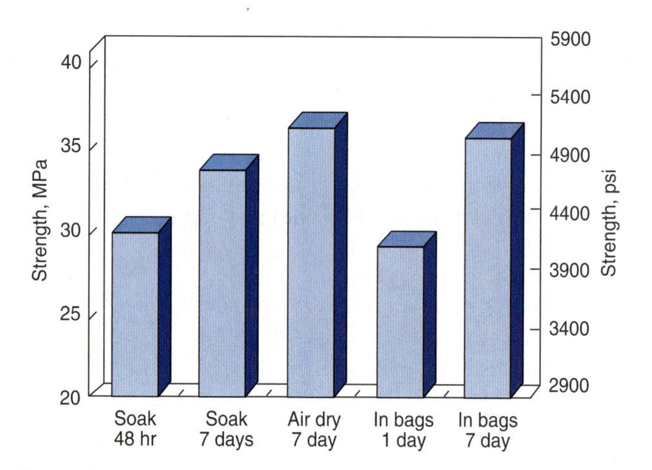

Figure 18-18. Effect of core conditioning on strength of drilled cores (Fiorato, Burg, and Gaynor 2000).

Flexure test specimens that are saw-cut from in-place concrete are always immersed in lime-saturated water at 23.0°C ± 2.0°C (73.5°F ± 3.5°F) for at least 40 hours immediately prior to testing.

Compressive strength test results are greatly influenced by the condition of the ends of cylinders and cores. For compression testing, specimens should be ground or capped in accordance with the requirements of ASTM C617, *Standard Practice for Capping Cylindrical Concrete Specimens* (AASHTO T 231), or ASTM C1231/C1231M, *Standard Practice for Use of Unbonded Caps in Determination of Compressive Strength of Hardened Concrete Cylinders* (AASHTO T22). Various commercially available materials can be used to cap compressive test specimens. ASTM C617 (AASHTO T 231) outlines methods for using bonded caps made of sulfur mortar, neat cement paste, and high-strength gypsum paste. Sulfur mortar caps must be allowed to harden at least two hours before the specimens are tested. For concrete strength of 35 MPa (5000 psi) or greater, hardening time should be at least 16 hr unless data shows that shorter times are suitable. Sulfur mortar caps should be made as thin as is practical to provide accurate test results (Lobo and others 1994). ASTM C617 requires that the average cap thickness not exceed 6 mm (¼ in.) for concrete compressive strengths less than 50 MPa (7000 psi) and not exceed 3 mm (⅛ in.) for higher concrete strength.

As an alternative to bonded caps, ASTM C1231/C1231M describes the use of neoprene caps that are not bonded to the ends of the specimen. This method of capping uses a disk-shaped 13 ± 2-mm (½ ± 1⁄16-in.) thick neoprene pad that is approximately the diameter of the specimen. The pad is placed in a cylindrical steel retainer with a cavity approximately 25 mm (1 in.) deep and slightly smaller than the diameter of the pad. A cap is placed on one or both ends of the cylinder; the specimen is then tested in accordance with ASTM C39/C39M, *Standard Test Method for Compressive Strength of Cylindrical Concrete Specimens* (AASHTO T 22), with the exception that the test is stopped at 10% of the anticipated ultimate load to check that the axis of the cylinder is vertical within a tolerance of 0.5 degrees. The end of the specimen to receive an unbounded cap should not depart by more than 0.5 degrees from perpendicularity with the cylinder axis. If either the perpendicularity of the cylinder end, or the vertical alignment during loading are not met, the load applied to the cylinder may be concentrated on one side of the specimen. This can cause a short shear fracture in which the failure plane intersects the end of the cylinder. This type of fracture usually indicates the cylinder failed prematurely, yielding results lower than the actual strength of the concrete. If end perpendicularity requirements are not met, the cylinder can be saw-cut, ground, or capped with a sulfur mortar compound in accordance with ASTM C617 (AASHTO T 231).

Short shear fractures can also be reduced by: dusting the pad and end of cylinder with corn starch or talcum powder, preventing excess water from cylinders or burlap from draining into the retainer and below the pad, and checking bearing surfaces of retainers for planeness and indentations. In addition, annually clean and lubricate the spherically seated block and adjacent socket on the compression machine.

Testing of specimens for strength (Figure 18-19) should be done in accordance with (1) ASTM C39/C39M (AASHTO T 22) for compressive strength, (2) ASTM C78, *Standard Test Method for Flexural Stenth of Cocnrete (Using Simple Beam with Third-Point Loading)* (AASHTO T 97), (3) ASTM C293, *Standard Test Method for Flexural Stength of Concrete (Using Simple Beam With Center-Point Loading)* (AASHTO T 177), and (4) ASTM C496/C496M, *Standard Test Method for Splitting Tensile Strength of Cylindrical Concrete Specimens* (AASHTO T 198).

Figure 18-19. Testing hardened concrete specimens: (left) cylinder, (right) beam.

For both pavement thickness design and pavement mixture proportioning, the modulus of rupture (flexural strength) should be determined by ASTM C78 (AASHTO T 97). However, ASTM C293 (AASHTO T 177) can be used for job control if empirical relationships to third-point loading test results are determined before construction starts.

The moisture content of the specimen has considerable effect on the resulting strength. Beams for flexural tests are especially vulnerable to moisture gradient effects. A saturated specimen will show lower compressive strength and higher flexural strength than those for companion specimens tested dry. This is important to consider when cores taken from hardened concrete in service are compared with molded specimens tested as taken from the moist-curing room or water storage tank. Cylinders used for acceptance testing for a specified strength must be cured in accordance with ASTM C31/C31M (AASHTO T 23), to accurately represent the quality of the concrete. However, cores are subject to workmanship, variable environmental site conditions, and variable conditioning after extraction. Cores are tested usually in the as-received condition, but rarely in a moist condition similar to standard-cured cylinders. Because cores and cylinders are handled in very different manners, they cannot be expected to yield the same results at the same test age.

The amount of variation in compressive-strength testing is far less than for flexural-strength testing. To avoid the extreme care needed in field flexural-strength testing, compressive-strength tests can be used to monitor concrete quality in projects designed on the basis of modulus of rupture. However, a laboratory-determined empirical relationship (Figure 18-20) must be developed between the compressive and flexural strength of the concrete used (Kosmatka 1985a). Because of the robust empirical relationship between these two strengths and the economics of testing cylinders instead of beams, most state departments of transportation are now using compression tests of cylinders to monitor concrete quality for their pavement and bridge projects.

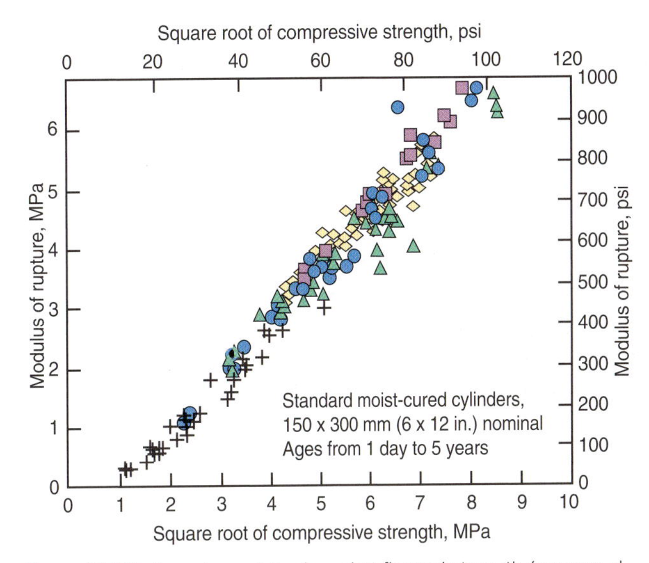

Figure 18-20. Long-term data show that flexural strength (measured by third-point loading) is proportional to the square root of compressive strength over a wide range of strength levels (Wood 1992).

Evaluation of Compression Test Results. The ACI 318 building code states that the compressive strength of concrete can be considered satisfactory if the following conditions are met: the averages of all sets of three consecutive strength tests equal or exceed the specified strength f'_c and no individual strength test (average of two or three cylinders) is more than 3.5 MPa (500 psi) below the specified strength. If the results of the cylinder tests do not meet these criteria, steps should be taken to increase the measured compressive strength. In addition, if the second condition is not met, steps have to be taken to ensure that the load capacity of the structure is not jeopardized. If the likelihood of low strength concrete is confirmed and the load carrying capacity is in question, the strength of the in-place concrete may be evaluated by drilled cores as discussed below.

In addition to the cylinders for acceptance testing, job specifications often require one or two 7-day cylinders and one or more "hold" cylinders. The 7-day cylinders monitor early strength gain. Hold cylinders are commonly used to provide additional information in case the cylinders tested for acceptance are damaged or do not meet the required compressive strength. For low 28-day test results, the hold cylinders are typically tested at 56 days.

Protection and curing procedures for the structure should also be evaluated to judge if they are adequate when field-cured cylinders tested at the age designated for f'_c have a strength of less than 85% of that of companion standard-cured cylinders. The 85% requirement may be waived if the field-cured strength exceeds f'_c by more than 3.5 MPa (500 psi).

If, as a result of low strength test results, it becomes necessary, the in-place concrete strength should be determined by testing three cores from each portion of the structure where the standard-cured cylinders did not meet acceptance criteria. Moisture conditioning of cores prior to compression testing should be in accordance with ASTM C42/C42M (AASHTO T 24).

If the average strength of three cores is at least 85% of f'_c, and if no single core is less than 75% of f'_c, the concrete in the area represented by the cores is considered structurally adequate. If the results of properly conducted core tests are so low as to leave structural integrity in doubt, load tests as outlined in ACI 318 may be performed. Nondestructive test methods are not a substitute for core tests, but they can be used to confirm the likelihood of low strength concrete prior to drilling cores. Refer to Chapter 12 and NRMCA (1979), ACI Committee 214 (2002), and ACI Committee 318 (2008).

Air Content

The air-content and air-void-system parameters of hardened concrete can be determined using ASTM C457 to assure that the air-void system is adequate to resist damage from a freezing and thawing environment. The test is also used to determine the effects of different admixtures and methods of placement and consolidation on the air-void system. The test can be performed on molded specimens or samples removed from the structure. Using a polished section of a concrete sample, the air-void system is characterized with measurements from a microscope. The information obtained from this test method includes the volume of entrained and entrapped air, its specific surface (surface area of the air voids per unit volume of air), the spacing factor, and the number of voids per linear distance (Figure 18-21). See Chapter 11 for more information.

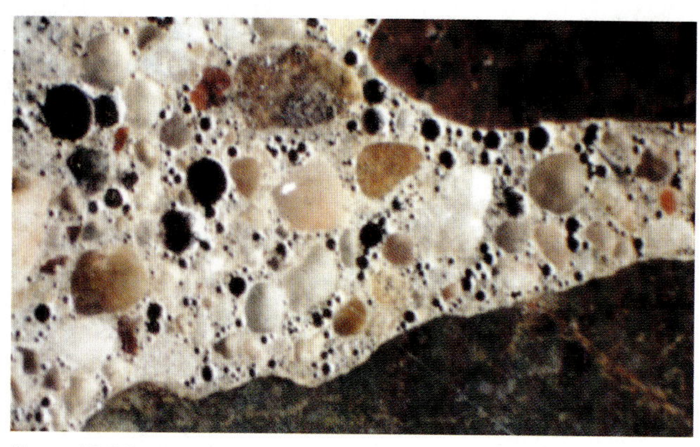

Figure 18-21. View of concrete air-void system under a microscope.

Density, Absorption, and Voids

The density, absorption, and voids content of hardened concrete can be determined in accordance with ASTM C642, *Standard Test Method for Density, Absorption, and Voids in Hardened Concrete* (Table 18-2). It should be noted that the boiling procedure in ASTM C642 can render the specimens useless for other tests, especially strength tests.

Saturated, surface-dry density (SSD) is often required for specimens prior to many other tests. In this case, the density can be determined by soaking the specimen in water for 48 hours and then determining its mass in air (in the SSD condition) and immersed in water. The SSD density is calculated as follows:

$$D_{SSD} = \frac{M_1 \rho}{M_1 - M_2}$$

Where:

D_{SSD} is density in the SSD condition

M_1 is the SSD mass in air, kg (lb)

M_2 is the apparent mass immersed in water, kg (lb)

ρ is the density of water, 1000 kg/m^3 (62.4 lb/ft^3)

The SSD density provides an approximation of the density of freshly mixed concrete. The density of hardened concrete can also be determined using ASTM C1040/C1040M (AASHTO T 271). For this test, a relationship is established between gravimetric density measurements of concrete mixtures and nuclear gauge measurements.

The rate of absorption (sorptivity) of water by hardened concrete can be determined using ASTM C1585, *Standard Test Method for Measurement of Rate of Absorption of Water by Hydraulic-Cement Concretes*. In this test, a concrete disk with a diameter of 100 mm (4 in.) and a height of 50 mm (2 in.) is preconditioned to control the internal relative humidity at the start of the test. The side of the disk and one end is sealed. The bare end is placed in water and the

Table 18-2. Permeability and Absorption of Concretes Moist Cured 7 Days and Tested After 90 Days

Mix No.	Cement, kg/m³ (lb/yd³)	w/cm	Compressive strength at 90 days, MPa (psi)	Permeability				Porosity, %†	Vol. of permeable voids, %	Absorption after immersion, %	Absorption after immersion and boiling, %
				RCPT, coulombs	90 days ponding, % Cl	Water, m/s**	Air, m/s**				
			ASTM C39 (AASHTO T 22)	ASTM C1202 (AASHTO T 277)	ASTM C1543 (AASHTO T 259)	API RP 27	API RP 27		ASTM C642	ASTM C642	ASTM C642
1	445 (750)	0.26*	104.1 (15100)	65	0.013	—	2.81×10^{-10}	7.5	6.2	2.43	2.56
2	445 (750)	0.29*	76.7 (11130)	852	0.022	—	3.19×10^{-10}	8.8	8.0	3.13	3.27
3	381 (642)	0.40*	46.1 (6690)	3242	0.058	2.61×10^{-13}	1.16×10^{-9}	11.3	12.2	4.96	5.19
4	327 (550)	0.50	38.2 (5540)	4315	0.076	1.94×10^{-12}	1.65×10^{-9}	12.5	12.7	5.45	5.56
5	297 (500)	0.60	39.0 (5660)	4526	0.077	2.23×10^{-12}	1.45×10^{-9}	12.7	12.5	5.37	5.49
6	245 (413)	0.75	28.4 (4120)	5915	0.085	8.32×10^{-12}	1.45×10^{-9}	13.0	13.3	5.81	5.90

* Admixtures: 59.4 kg/m³ (100 lb/yd³) silica fume and 25.4 ml/kg of cement (30 fl.oz/cwt) HRWR (Mix 1); 13.0 ml/kg (20 fl.oz/cwt) HRWR (Mix 2); 2.2 ml/kg (3.4 fl. oz/cwt) WR (Mix 3).

** To convert from m/s to Darcy, multiply by 1.03×10^5, from m/s to m², multiply by 1.02×10^{-7}.

† Measured with helium porosimetry.

Adapted from Whiting (1988).

uptake of water is measured as a function of time. The water gain is plotted as a function of the square root of time, and the initial rate of water absorption is calculated through regression analysis.

Portland Cement Content

The portland cement content of hardened concrete can be determined by ASTM C1084, *Standard Test Method for Portland-Cement Content of Hardened Hydraulic-Cement Concrete* (or AASHTO T 178). The test method includes two independent procedures: an oxide analysis procedure and a maleic acid extraction procedure. Each procedure requires substantial expertise in chemical testing and elaborate apparatus. Although not frequently performed, the cement content tests are valuable in determining the cause of lack of strength gain or poor durability of concrete. The user of these test methods should be aware of certain admixtures and aggregate types that can interfere with test results. The presence of supplementary cementitious materials would also be reflected in the test results.

Supplementary Cementitious Material and Organic Admixture Content

The presence and amount of certain supplementary cementitious materials, such as fly ash and slag cement, can be determined by petrographic techniques (ASTM C856) as discussed below. A sample of the supplementary cementitious material used in the concrete is usually necessary as a reference to determine the type and amount of the supplementary cementitious material present in the concrete. The presence and possibly the amount of an organic admixture (such as a water reducer) can be determined by infrared spectrophotometry (Hime, Mivelaz, and Connolly 1966).

Chloride Content

The water-soluble chloride-ion content of hardened concrete can be determined in accordance with ASTM C1218/C1218M, *Standard Test Method for Water-Soluble Chloride in Mortar and Concrete*. In addition, ASTM C1152/C1152M, *Standard Test Method for Acid-Soluble Chloride in Mortar and Concrete*, can be used to determine the acid-soluble chloride content of concrete which in most cases is equivalent to total chloride. ACI 318 places limits on the water-soluble chloride-ion content of concrete for different exposure categories and whether the concrete is prestressed. These limits are based on testing samples of harded concrete at ages between 28 and 42 days.

The above tests for chloride-ion content use pulverized samples of concrete. Thus, they also extract chloride ions from within fine and coarse aggregates particles that generally are not available to contribute to corrosion of reinforcing steel. ASTM C1524 can be used to investigate the amount of water-extractable chloride ions from aggregate particles. Because ASTM C1524 does not involve pulverizing the aggregate particles, it provides a more realistic measure of the chloride ions available for corrosion. ACI 222.1 is also a Soxhlet extraction procedure that tests chunks of concrete for water-extractable chloride. The accuracy and interpretation of information obtained from the Soxhlet procedure is still a matter of debate.

Petrographic Analysis

Petrographic analysis uses microscopical techniques described in ASTM C856, *Standard Practice for Petrographic Examination of Hardened Concrete* (or AASHTO T 299), to determine the constituents, characteristics, and distress mechanisms. The future performance of concrete elements can also be estimated. Some of the features that can be analyzed by a petrographic examination include paste, aggregate, fly ash, and air content; frost and sulfate attack; alkali-aggregate reactivity; degree of hydration and carbonation; water-cement ratio; bleeding characteristics; fire damage; scaling; popouts; effect of admixture; and other aspects. Almost any kind of concrete failure can be analyzed by petrography (St. John, Poole, and Sims 1998). However, a standard petrographic analysis is sometimes accompanied by wet chemical analyses, infrared spectroscopy, X-ray diffractometry, scanning electron microscopy with attendant elemental analysis, differential thermal analysis, and other analytical tools. Some results from a petrographic examination depend on the subjective judgment of the petrographer. ASTM C856 provides criteria for the qualifications of petrographers and technicians.

The Annex to ASTM C856 (AASHTO T 299) describes a technique for field and laboratory detection of gel resulting from alkali-silica reactivity (ASR) (See Chapter 6 and 11). In this method, a uranyl-acetate solution is applied to a broken or roughened concrete surface that has been dampened with distilled or deionized water. After one minute, the solution is rinsed off and the treated surface is viewed under ultraviolet light. Areas of gel fluoresce bright yellow-green. It must be recognized, however, that several materials not related to ASR in concrete can also fluoresce and interfere with an accurate indication of ASR gel. Materials that fluoresce like gel include: naturally fluorescent minerals, carbonated paste, opal, some other rock ingredients, and reaction products from fly ash, silica fume, and other pozzolans. ASTM C856 includes a prescreening procedure that gives a visual impression of the background fluorescence to compensate for the effects of these materials. However, this test is considered ancillary to more definitive petrographic examinations and other tests. In addition, the toxicity and radioactivity of uranyl acetate warrants special handling and disposal procedures regarding the solution and treated concrete. Potential eye damage from ultraviolet light also requires special safety considerations.

The Los Alamos method is a staining technique for detecting ASR gel that does not require ultraviolet light or uranyl-acetate solution. Instead, solutions of sodium cobaltinitrite and rhodamine B are used to condition the specimen and produce a dark pink stain that corresponds to calcium-rich ASR gel. These rapid visual methods can identify evidence of ASR gel that has not caused damage to concrete. That is, ASR gel can be present when other mechanisms such as freeze-thaw action, sulfate attack, and other deterioration mechanisms have caused the damage. These rapid methods for detecting the presence of ASR gel are useful but their limitations must be understood. Neither of the rapid procedures is a viable substitute for petrographic examination coupled with proper field inspection (Powers 1999).

Volume and Length Change

Volume or length change (shrinkage) limits are sometimes specified for certain concrete applications. ASTM C157/C157M, *Standard Test Method for Length Change of Hardened Hydraulic-Cement Mortar and Concrete* (AASHTO T 160), determines length change in concrete due to drying shrinkage, chemical shrinkage, and causes other than externally applied forces and temperature changes. Early volume change of concrete before final setting can be determined using ASTM C827, *Standard Test Method for Change in Height at Early Ages of Cylindrical Specimens of Cementitious Mixtures*. ASTM C1698, *Standard Test Method for Autogenous Strain of Cement Paste and Mortar*, can be used to measure the autogenous strain of paste or mortar specimens from the time of final setting up to a specified age. Creep can be determined in accordance with ASTM C512, *Standard Test Method for Creep of Concrete in Compression*. The static modulus of elasticity and Poisson's ratio of concrete can be determined using ASTM C469, *Standard Test Method for Static Modulus of Elasticity and Poisson's Ratio of Concrete in Compression*, and dynamic values of these parameters can be determined by using ASTM C215, *Standard Test Method for Fundamental Transverse, Longitudinal, and Torsional Frequencies of Concrete Specimens*. Age at cracking due to the combined effects of both drying shrinkage, autogenous shrinkage, heat of hydration, and stress relaxation can be determined in accordance with ASTM C1581/C1581M, *Standard Test Method for Determining Age at Cracking and Induced Tensile Stress Characteristics of Mortar and Concrete under Restrained Shrinkage*. ASTM E1155, *Standard Test Method for Determining F_F Floor Flatness and F_L Floor Levelness Numbers*, can be used to investigate the deflections of a concrete slab surface caused by curling and warping. Volume change is discussed further in Chapter 10.

Durability

In addition to tests for air content and chloride content previously described, the following tests are used to measure the durability performance of concrete (also see Chapter 11):

Resistance to Freezing-and-Thawing. The freeze-thaw resistance of concrete is usually determined in accordance with ASTM C666/C666M, *Standard Test Method for Resistance of Concrete to Rapid Freezing and Thawing* (AASHTO T 161). Prismatic specimens are monitored for changes in the dynamic modulus of elasticity, mass, and specimen length over a period of 300 or more cycles of freezing and thawing. ASTM C1646/1646M, *Standard Practice for Making and Curing Test Specimens for Evaluating the Resistance of Coarse Aggregate to Freezing and Thawing in Air-Entrained Concrete*, can be used to prepare test specimens to evaluate the performance of coarse aggregates in air-entrained concrete when tested using ASTM C666/C666M.

Concrete that will be exposed to deicers as well as freezing in a saturated condition should be tested for deicer-scaling resistance using ASTM C672/C672M, *Standard Test Method for Scaling Resistance of Concrete Surfaces Exposed to Deicing Chemicals*. Although ASTM C672/C672M requires that only surface scaling be monitored, many practitioners also measure mass loss, as is the current practice in Canada (Figure 18-22). Concrete mixtures that perform well in ASTM C666/C666M (AASHTO T 161) do not always perform well in ASTM C672/C672M. ASTM C666/C666M (AASHTO T 161) and ASTM C672/C672M are often used to evaluate innovative designs of concrete mixtures, or new materials such as chemical admixtures, supplementary cementitious materials, and aggregates to determine their effect on resistance to freezing and thawing and deicers.

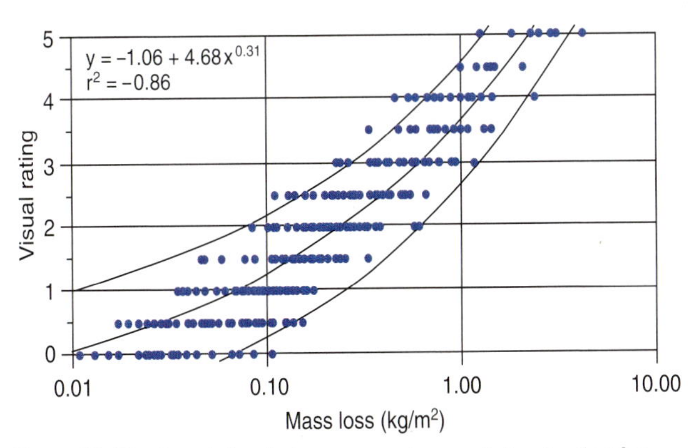

Figure 18-22. Correlation between mass loss and visual rating for each specimen tested according to ASTM C672/C672M (Pinto and Hover 2001).

Sulfate Resistance. The sulfate resistance of cementitious materials can be evaluated using a mortar bar test in accordance with ASTM C1012, *Standard Test Method for Length Change of Hydraulic-Cement Mortars Exposed to a Sulfate Solution*. Mortar bar specimens are immersed in a sodium sulfate solution and length change is measured as a function of time. This test is valuable in assessing the sulfate resistance of concrete that will be continuously wet. It does not evaluate the more aggressive wetting and drying environment. The test can be modified to include wet-dry cycling or the U.S. Bureau of Reclamation (USBR) wet-dry concrete prism test for sulfate attack can be used. ASTM C1580, *Standard Test Method for Water-Soluble Sulfate in Soil*, and ASTM D516, *Standard Test Method for Sulfate Ion in Water* (AASHTO T 290), or the USBR method (Bureau of Reclamation 1975) can be used to test soil and water for sulfate ion content to determine the severity of the sulfate exposure.

Alkali-Silica Reactivity. Aggregate can be evaluated for potential alkali-silica reactivity using the ASTM C227, *Standard Test Method for Potential Alkali Reactivity of Cement-Aggregate Combinations (Mortar-Bar Method)*, ASTM C289, *Standard Test Method for Potential Alkali-Silica Reactivity of Aggregates (Chemical Method)*, ASTM C295, *Standard Guide for Petrographic Examination of Aggregates for Concrete*, ASTM C1260, *Standard Test Method for Potential Alkali Reactivity of Aggregates (Mortar-Bar Method)* (AASHTO T 303), and ASTM C1293, *Standard Test Method for Determination of Length Change of Concrete Due to Alkali-Silica Reaction*.

SCMs, such as fly ash and slag should be evaluated by tests such as ASTM C227, ASTM C441, *Standard Test Method for Effectiveness of Pozzolans or Ground Blast-Furnace Slag in Preventing Excessive Expansion of Concrete Due to the Alkali-Silica Reaction*, ASTM C1567, *Standard Test Method for Determining the Potential Alkali-Silica Reactivity of Combinations of Cementitious Materials and Aggregate (Accelerated Mortar-Bar Method)* or ASTM C1293. Additionally, lithium admixtures to control ASR can be evaluated using a modified ASTM C1293 (Thomas, Fournier, and Folliard 2008), CSA A23.2-28A, or CRD-C 662, *Determining the Potential Alkali-Silica Reactivity of Combinations of Cementitious Materials, Lithium Nitrate Admixture and Aggregate (Accelerated Moratar-Bar Method)*. Discussion on the effectiveness of each of these ASR test methods can be found in Chapter 11.

A sequence of tests to evaluate aggregate reactivity, developed by Thomas, Fournier, and Folliard (2008), has been adopted by AASHTO (AASHTO PP65, *Standard Practice for Determining the Reactivity of Concrete Aggregates and Selecting Appropriate Measures for Preventing Deleterious Expansion in New Concrete Construction*). The aggregates are first evaluated based on field history, then through petrographic examination (ASTM C295). Following the petrographic examination, the aggregates are tested according to ASTM C1260. If the expansion exceeds 0.10%, the aggregates are then tested according to ASTM C1293. Once the aggregates have been evaluated for reactivity, appropriate preventive measures can be prescribed (see Chapter 11). A drawback to this approach is that ASTM 1293 requires one year to complete the testing.

An alternate to testing aggregate separately for potential reactivity is to test the concrete mixture using ASTM C1567 or ASTM C1293. A rapid 3-month version of ASTM C1293 is under development at the University of Texas at Austin through the International Center for Aggregate Research (Thomas and others 2006). Existing concrete structures can be evaluated for alkali-silica reaction using ASTM C856.

Alkali-Carbonate Reactivity. Alkali-carbonate reactivity is relatively rare. Potential reactivity of aggregates can be evaluated by using ASTM C295, ASTM C586, *Standard Test Method for Potential Alkali Reactivity of Carbonate Rocks as Concrete Aggregates (Rock-Cylinder Method)*, and ASTM C1105, *Standard Test Method for Length Change of Concrete Due to Alkali-Carbonate Rock Reaction*. Existing concrete structures can be evaluated for alkali-carbonate reaction using ASTM C856.

Corrosion Activity. The corrosion activity of reinforcing steel in concrete is rarely evaluated unless unusual materials are used, or concrete will be used in a very severe environment, or there is a need to evaluate the potential for in-place corrosion. Corrosion activity can be evaluated using ASTM C876, *Standard Test Method for Half-Cell Potentials of Uncoated Reinforcing Steel in Concrete*. The half-cell potential test provides an indication of the likelihood that steel corrosion is active, but it does not indicate the rate of corrosion. The linear polarization resistance (LPR) method described in ACI 228.2R can be used to estimate corrosion rate. LPR provides an estimate of the corrosion rate at the time of the test. This rate can change with changes in temperature, moisture content, and other factors. Thus LPR measurements need to be made several times a year to obtain a valid estimate of the mean corrosion rate.

Abrasion Resistance. Abrasion resistance can be determined using ASTM C418, *Standard Test Method for Abrasion Resistance of Concrete by Sandblasting*, ASTM C779/C779M, *Standard Test Method for Abrasion Resistance of Horizontal Concrete Surfaces*, ASTM C944/C944M, *Standard Test Method for Abrasion Resistance of Concrete or Mortar Surfaces by the Rotating-Cutter Method*, and ASTM C1138M, *Standard Test Method for Abrasion Resistance of Concrete by Sandblasting (Underwater Method)*.

Moisture Testing

The in-place moisture content, water vapor emission rate, and relative humidity of hardened concrete are useful indicators in determining if concrete is dry enough for the application of floor-covering materials and coatings. The moisture content of concrete should be low enough to avoid spalling when exposed to temperatures above the boiling point of water. Moisture related test methods fall into two general categories: qualitative or quantitative. Qualitative tests provide a gross indication of the presence or absence of moisture while quantitative tests measure the amount of moisture. Qualitative tests may give a strong indication that excessive moisture is present and the floor is not ready for floor-covering materials. Quantitative tests are performed to verify that the floor is dry enough for these materials.

Qualitative moisture tests include: plastic sheet, mat bond, electrical resistance, electrical impedance, and nuclear moisture gauge tests. ASTM D4263, *Standard Test Method for Indicating Moisture in Concrete by the Plastic Sheet Method*, uses a square sheet of clear plastic film that is taped to the slab surface and left for 24 hours to provide an indication of moisture. The plastic sheet test is unreliable. In the mat bond test, a 1-m^2 (9-ft^2) sheet of floor covering is glued to the floor with the edges taped to the concrete for 72 hours. The force needed to remove the flooring is an indication of the slab moisture condition. Electrical resistance is measured using a moisture meter through two probes placed in contact with the concrete. Electrical impedance uses an electronic signal that is influenced by the moisture in the concrete. Nuclear moisture gauges use a source of high-speed neutrons that are slowed by the hydrogen atoms in water. The effect of these collisions provides a measure of the moisture content of the concrete. Although the last three tests each yield a numeric test result, their value is quite limited. A common limitation is that the volume of concrete that contributes to the gauge reading is not known with certainty. Experience and skill are needed to judge the reliability of the devices and the test results produced by them.

Quantitative test methods include: gravimetric moisture content, moisture vapor emission rate, and relative humidity probe tests. The most direct method for determining moisture content is to dry cut a specimen from the concrete element in question, place it in a moisture proof container, and transport it to a laboratory for testing. After obtaining the specimen's initial mass, dry the specimen in an oven at about 105°C (220°F) for 24 hours or until constant mass is achieved. The difference between the two masses divided by the dry mass, multiplied by 100, provides the moisture content in percent. ASTM F1869, *Standard Test Method for Measuring Moisture Vapor Emission Rate of Concrete Subfloor Using Anhydrous Calcium Chloride*, is a commonly used test for measuring the readiness of concrete for application of floor coverings (Figure

Figure 18-23. The calcium chloride test kit (ASTM F1869).

18-23). The emission rate is expressed according to the standard as pounds of moisture emitted from 1000 ft^2 in 24 hours. Converted to SI units, the rate is expressed as micrograms per square meter per second (µg/m^2s). See Kosmatka (1985) and Kanare (2008) for more information.

ASTM F2420, *Standard Test Method for Determining Relative Humidity on the Surface of Concrete Floor Slabs Using Relative Humidity Probe Measurement and Insulated Hood*, uses a hygrometer or relative humidity probe sealed under an insulated, impermeable box to trap moisture in an air pocket above the floor (Figure 18-24). The probe is allowed to equilibrate for at least 72 hours or until two consecutive readings at 24-hour intervals are within the accuracy of the instrument (typically ± 3% RH). Acceptable relative humidity limits for the installation of floor coverings range from a maximum of 60% to 90%. It can require several months of air-drying to achieve the desired relative humidity. A method for estimating drying time to reach a specified relative humidity based on water-cement ratio, thickness of structure, number of exposed sides, relative humidity, temperature, and curing conditions can be found in Hedenblad (1997), Hedenblad (1998), Tarr and Farny (2008), and Kanare (2008).

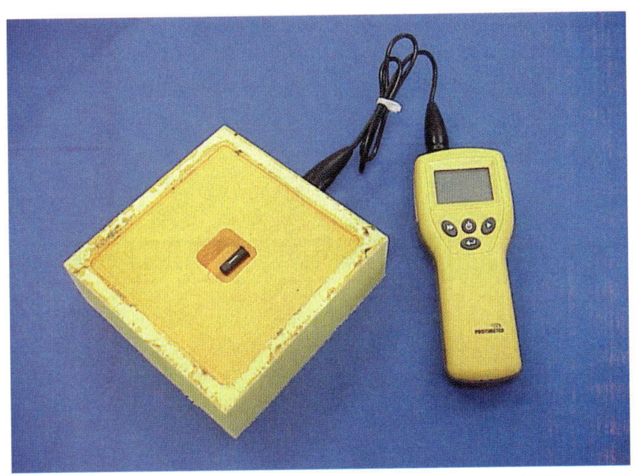

Figure 18-24. Relative humidity hood test (ASTM F2420).

Carbonation

The depth or degree of carbonation can be determined by petrographic techniques (ASTM C856) through the observation of calcium carbonate – the primary chemical product of carbonation. In addition, a pH stain (phenolphthalein) can be used to estimate the depth of carbonation by testing the pH of concrete (carbonation reduces pH). For example, upon application of a phenolphthalein solution to a freshly fractured or freshly cut surface of concrete, noncarbonated areas turn red or purple while carbonated areas remain colorless. (Figure 18-25). The phenolphthalein indicator when observed against hardened paste changes color at a pH of 9.0 to 9.5. The pH of good quality noncarbonated concrete without admixtures is usually greater than 12.5. For more information, see **pH Testing Methods**, and Verbeck (1958), Steinour (1964), and Campbell, Sturm, and Kosmatka (1991).

Figure 18-25. The depth of carbonation is determined by spraying phenolphthalein solution on a freshly broken concrete surface. Noncarbonated areas turn red or purple, carbonated areas stay colorless.

pH Testing Methods

There are three practical methods for measuring the surface pH of hardened concrete in the field. The first uses litmus paper designed for the alkaline range of pH readings. Place a few drops of distilled water on the concrete, wait 60 ± 5 seconds and immerse an indicator strip (litmus paper) in the water for 2 to 3 seconds. After removing the strip, compare it to the standard pH color scale supplied with the indicator strips. A second method uses a pH pencil. The pencil is used to make a 25 mm (1 in.) long mark after which 2 to 3 drops of distilled water are placed on the mark. After waiting 20 seconds, the color is compared to a standard color chart to judge the pH of the concrete. Finally, the third method uses a wide-range liquid pH indicator on a freshly fractured surface of the concrete or a core obtained from the concrete. After several minutes, the resulting color is compared to a color chart to determine the pH of the concrete. This method is also effective for measuring the depth of carbonation present on the concrete surface. See Kanare (2008) for more information.

Permeability and Diffusion

Both direct and indirect methods of measuring permeability are used. Table 18-2 shows typical values of indicators of permeability. Resistance to chloride-ion penetration, for example, can be determined by ponding chloride solution on a concrete surface and, at a later age, determining the chloride content of the concrete at particular depths using ASTM C1543, *Standard Test Method for Determining the Penetration of Chloride Ion into Concrete by Ponding* (or AASHTO T 259). ASTM C1202, *Standard Test Method for Electrical Indication of Concrete's Ability to Resist Chloride Ion Penetration* (or AASHTO T 277), also called the Coulomb or rapid chloride permeability test (RCPT), is often specified for concrete bridge decks. While ASTM C1202 is referred to as a "permeability" test, it is actually a test of electrical conductivity. Conductivity is a good indicator of permeability because the same factors that affect conductivity also affect permeability. A more rapid test for electrical conductivity (or its inverse resistivity) than the rapid chloride penetrability test was developed by the Florida Department of Transportation (FDOT 2004). This procedure uses the Wenner probe array method (Malhotra and Carino 2004) for measuring the resistivity of 100 mm x 200 mm (4 in. x 8 in.) cylinders. ACI Committee 222 recommends using this method for assessing the resistivity of in-place concrete. Because dry concrete has a high resistivity, the concrete needs to be in a saturated condition to obtain a reliable indicator of the permeability of concrete. The results of the electrical resistivity test have been correlated to the RCPT permeability rating system (Smith 2006).

The apparent chloride diffusion coefficient of hardened cementitious mixtures can be determined in accordance with ASTM C1556, *Standard Test Method for Determining the Apparent Chloride Diffusion Coefficient of Cementitious Mixtures by Bulk Diffusion*. Various absorption methods, including ASTM C642 and ASTM C1585, can be used as indicators of the resistance of concrete to the ingress of fluids. Direct water permeability data can be obtained using CRD-C 163, *Test Method for Water Permeability of Concrete Using a Triaxial Cell*. A test method recommended by the American Petroleum Institute for determining the permeability of rock is also available. All these methods have limitations. Direct permeability testing using applied pressure is impractical for testing high quality concrete because of the time required to force measurable quantities of water through a specimen. For more information, see American Petroleum Institute (1956), Tyler and Erlin (1961), Whiting (1981), Pfeifer and Scali (1981), and Whiting (1988).

Nondestructive Test Methods

Nondestructive tests (NDT) can be used to evaluate the relative strength and other properties of hardened concrete. The most widely used methods are the rebound hammer, probe penetration, pullout test, and a variety of tests based on stress-wave propagation. Other techniques include: X-rays, gamma radiography, neutron moisture gages, magnetic cover meters, eddy current, microwave absorption, and acoustic emissions. When using methods to estimate in-place strength, caution should be exercised against acceptance of nondestructive test results assuming a unique correlation to the compression strength. Empirical correlations must be developed for the specific instrument and concrete mixture prior to use (Malhotra 1976, NRMCA 1979, Malhotra 1984, Clifton 1985, Malhotra and Carino 2004). For more information on test methods used to estimate quality of concrete, see ACI Committee 228. Table 18-3 lists several nondestructive test methods along with main applications.

Table 18-3. Nondestructive Test Methods for Concrete

Property or condition	Primary NDT method	Secondary NDT method
In-place strength	Pullout test Pull-off test (for bond strength) Probe penetration Maturity method (new construction)	Rebound hammer
General Quality and Uniformity	Visual inspection Rebound hammer Pulse velocity	Probe penetration
Thickness	Impact-echo Ground penetrating radar Eddy-current thickness gage	Ultrasonic-echo
Dynamic modulus of elasticity	Resonant frequency (on small specimens)	Pulse velocity Impact-echo Spectral analysis of surface waves
Density	Gamma radiometry (nuclear gauge)	
Reinforcement location	Covermeter Ground penetrating radar	Radiography Ultrasonic echo
Reinforcement bar size	Covermeter	Radiography
Corrosion state of steel reinforcement	Half-cell potential Polarization resistance	Radiography
Presence of near-surface defects	Sounding Infrared thermography	Ground penetrating radar Radiography
Presence of internal defects	Impact-echo Ultrasonic echo Impulse response	Pulse velocity Spectral analysis of surface waves Radiography

An NDT program may be undertaken for a variety of purposes regarding the strength or condition of hardened concrete, including:

- determination of in-place concrete strength
- monitoring rate of concrete strength gain
- location of defects, such as voids or honeycombing in concrete
- determination of relative strength of comparable members
- evaluation of concrete cracking and delaminations
- evaluation of damage from mechanical or chemical actions
- steel reinforcement location, size, and corrosion activity
- member dimensions

Irrespective of the type of NDT test used to estimate in-place strength, adequate and reliable correlation data with compressive strength data is necessary for a reliable estimate of in-place strength. In addition, correlation to in-place compressive strengths using drilled cores from one or two locations can provide guidance in interpreting NDT test results. NDT methods can then be used to survey larger portions of the structure. Care should be taken to consider the influence that varying sizes and locations of structural elements can have on the NDT test performed.

Rebound Hammer Tests. ASTM C805/C805M, *Standard Test Method for Rebound Number of Hardened Concrete*, is essentially a surface-hardness tester that provides a quick, simple means of checking concrete uniformity (Figure 18-26). It measures the rebound of a spring-loaded mass after it has struck a steel rod in contact with a smooth concrete surface. The rebound number reading gives an indication of the relative compressive strength and elastic modulus of the concrete. Two different concrete mixtures

Figure 18-26. The rebound hammer gives an indication of the uniformity of concrete.

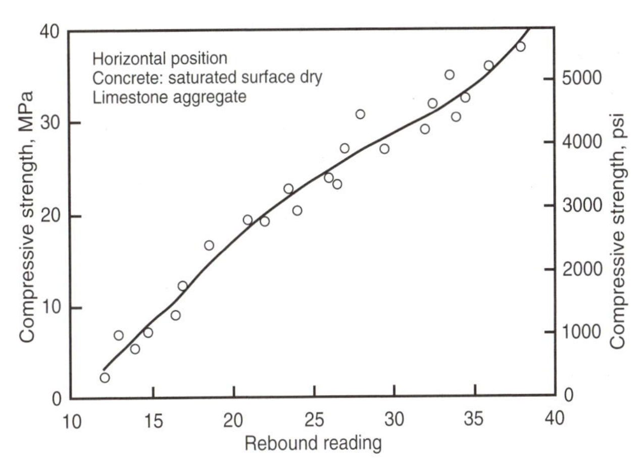

Figure 18-27. Example of a correlation curve for rebound hammer tests.

having the same strength but different elastic modulus will yield different readings. In view of this, an understanding of the factors influencing the rebound number is required.

Probe Penetration Tests. ASTM C803/C803M, *Standard Test Method for Penetration Resistance of Hardened Concrete*, is also called the Windsor Probe test. The Windsor probe, like the rebound hammer, is basically a hardness tester that provides a quick means of determining the relative strength of the concrete. The equipment consists of a powder-actuated gun that drives a hardened alloy probe into the concrete (Figure 18-28). The exposed length of the probe is measured and related to the compressive strength of the concrete by using a previously-established correlation curve.

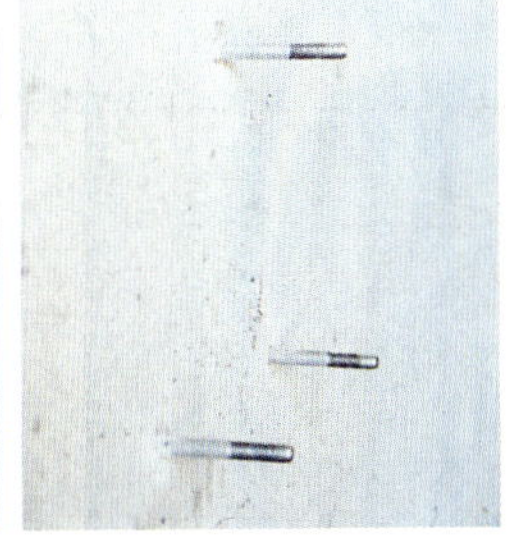

Figure 18-28. The Windsor-probe technique for determining the relative compressive strength of concrete. (left) Powder-actuated gun drives hardened alloy probe into concrete. (right) Exposed length of probe is measured and relative compressive strength of the concrete then determined from a calibration table.

The results of the Windsor-probe test will be influenced greatly by the type of coarse aggregate used in the concrete. Therefore, to improve accuracy of the estimated in-place strength, a correlation curve for the particular concrete to be tested should be developed. This can be done using a cast slab for probe penetration tests and companion cores or cast cylinders for compressive strength.

Both the rebound hammer and the probe damage the concrete surface to some extent. The rebound hammer leaves a small indentation on the surface; the probe leaves a small hole and may cause minor cracking and small craters similar to popouts.

Maturity Tests. The maturity principle is that strength gain of concrete is a function of time and temperature and samples of the same concrete will have equal strength if they have the same maturity index irrespective of their actual temperature histories. ASTM C1074, *Standard Practice for Estimating Concrete Strength by the Maturity Method*, provides procedures for measuring the maturity index of concrete based on one of two temperature-time (maturity) functions. ASTM C1074 also provides the procedure for developing the strength-maturity relationship for the specific concrete mixture to be used. Maturity meters include temperature probes embedded in the fresh concrete and control units that monitor and record the concrete temperature at regular intervals and calculate the maturity index. When a strength estimate is desired, the maturity index is read from the maturity meter and the strength is estimated form the strength-maturity relationship. ASTM C1074 also includes a procedure for establishing the most appropriate maturity function for the cementitious materials and admixtures in the concrete.

Pullout Tests. ASTM C900, *Standard Test Method for Pullout Strength of Hardened Concrete*, involves casting the enlarged end of a steel insert in the concrete to be tested and then measuring the force required to pull it out (Figure 18-29). The tensile load reacts against a ring bearing on the surface of the concrete, which constrains the failure along a well-defined surface. The pullout strength has a strong correlation to compressive strength, and the relationship is affected little by factors such w/cm and type of materials used (Peterson 1997) (Figure 18-30). ASTM C900 also describes a procedure for performing pullout tests in existing concrete. In this case, special hardware is used to insert a ring into an undercut created in the concrete by diamond studded router.

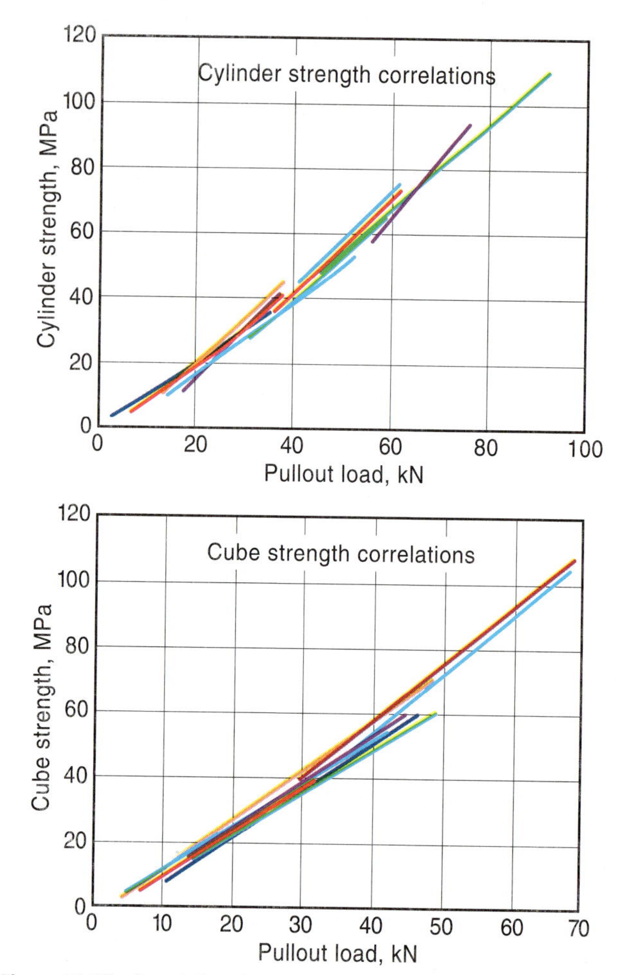

Figure 18-30. Correlations between pullout test (ASTM C900) and cylinder (ASTM C39/C39M) or cube strength (ASTM C109/109M) (Petersen 1997).

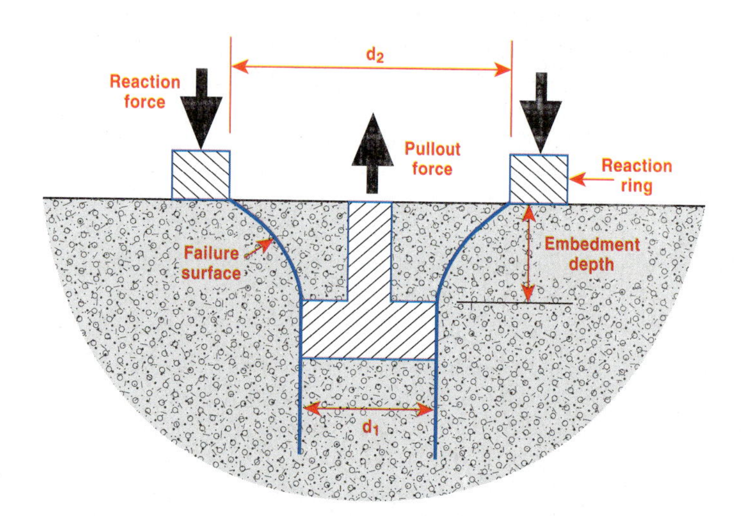

Figure 18-29. (right) Pullout test equipment being used to measure the in-place strength of concrete. (left) An illustration of the resulting failure surface (photo and illustration courtesy of N. Carino).

Pull-Off Tests. The pull-off test in accordance with ASTM C1583/1583M, *Standard Test Method for Tensile Strength of Concrete Surfaces and the Bond Strength or Tensile Strength of Concrete Repair and Overlay Materials by Direct Tension (Pull-off Method)*, can be used to evaluate the strength of the concrete surface before applying a repair material and for evaluating the bond strength of an overlay bonded to concrete. A metal disc is bonded to the surface using a fast-setting adhesive, a partial depth core is drilled to a specified depth, and the disc is pulled off by applying direct tensile force (Figure 18-31). Failure can occur in one of several potential failure planes: in the overlay, at the interface, or in the substrate. The failure load is divided by the cross sectional area of the core and reported as the pull-off strength in MPa (psi).

Figure 18-31. Demonstration of pull-off test to measure the strength of the concrete substrate before application of an overlay or repair material (photo courtesy of N. Carino).

Stress Wave and Vibration Tests. ASTM C597, *Standard Test Method for Pulse Velocity Through Concrete*, is based on the principle that velocity of sound (or stress waves) in a solid can be measured by measuring the travel time of short pulses of compressional waves through a test object. The pulse is introduced on one side of the test object by a transmitting transducer. The arrival of the pulse is measured by another transducer on the opposite side of the object. A control unit measures the time it takes for the pulse to travel between the two transducers. The distance between the transducers is measured and divided by the pulse travel time. The result is reported as the ultrasonic pulse velocity in m/s (ft/s), which in concrete typically ranges between 3500 m/s and 4500 m/s (11,500 ft/s to 15,000 ft/s). In general, high velocities are indicative of sound concrete and low velocities are indicative of poor concrete. The pulse velocity, however, is affected greatly by the amount and type of aggregate and to a minor degree by the w/cm. Therefore, care must be used when comparing the pulse velocities of different concrete. If internal defects are present in the test object, the actual pulse path length and pulse travel time will be increased. This results in a lower calculated pulse velocity. This test is very useful for assessing the uniformity of the concrete in a structure.

ASTM C215, *Standard Test Method for Fundamental Transverse, Longitudinal, and Torsional Frequencies of Concrete Specimens*, covers the procedures to determine the fundamental resonant frequencies of prismatic and cylindrical test specimens. Resonant frequency is a function of the dynamic modulus of elasticity, Poisson's ratio, density, and geometry of the test specimen. Thus if the geometry and mass of the specimen are known, the measured resonant frequencies can be used to calculate the dynamic modulus of elasticity and dynamic Poisson's ratio. In ASTM C215, two methods are presented to determine resonant frequency. In the *forced resonance method*, the specimen is forced to vibrate at continuously increasing frequencies until a maximum vibrational response is recorded. In the *impact resonance method*, the specimen is struck with a small impactor and the resulting vibrational response is subjected to a frequency analysis. The fundamental transverse, longitudinal, and torsional frequencies of the test specimens are determined by changing the locations of the excitation and where the response is measured. Because the test method is nondestructive, it can be used to monitor changes in vibrational response as a function of time or as a function of the number of cycles of a disruptive action. Hence, resonant frequency testing is frequently used in laboratory durability tests such as freezing and thawing (ASTM C666/C666M or AASHTO T 161).

A number of other test methods based on stress waves (or sound) have been developed. ASTM D4580, *Standard Practice for Measuring Delaminations in Concrete Bridge Decks by Sounding*, employs simple hammer and chain drag soundings that are low-cost accurate tests used to identify delaminated areas of concrete. Hammer soundings can be used on either vertical or horizontal surfaces, but are usually limited to small areas of delaminations. These areas are identified by striking the surface of the concrete with a hammer while listening for either a high pitched ringing or low pitched hollow sound. Dragging either a single chain, in small areas, or for larger areas, a T-bar with four or more chains attached can be used to identify delaminated concrete. The sound emitted as the chains are dragged over the surface indicates whether the concrete is delaminated. Chain drag soundings are usually limited to horizontal surfaces that have a relatively rough texture. Smooth concrete may not bounce the chain links enough to generate adequate sound to detect delaminated areas. Note that corrosion of reinforcing bars in the area of delaminated concrete will probably extend beyond the boundary identified as delaminated.

A versatile technique based on stress-wave propagation is the impact-echo method. In this method, the surface of the test object is struck with a small impactor and a transducer located near the impact point monitors the arrival of the stress pulse after undergoing reflections from within the test object. The method is capable of detecting voids, cracks, and other defects within structural elements (Sansalone and Street 1997). ASTM C1383, *Standard Test Method for Measuring the P-Wave Speed and the Thickness of Concrete Plates Using the Impact-Echo Method*, describes the use of the method to measure the thickness of plate-like concrete elements such as slabs, pavements, bridge decks and walls. ASTM C1383 includes two procedures: Procedure A is used to measure the P-wave, and Procedure B uses that wave speed to measure the plate thickness. The advantage of the impact-echo method is that it's not only nondestructive, but only requires access to one side of the structure.

Other stress-wave methods not covered in this review include: ultrasonic-echo, impulse response, and spectral analysis of surface waves (SASW). Information on these other methods may be found in Malhotra and Carino (2004) and ACI 228.2R.

Acoustic Emission Tests. Acoustic emission (AE) refers to the stress waves emitted from inside an object during a sudden release of stress. In concrete members, the most common source of acoustic emission is the formation of microcracks when the tensile capacity is reached. In acoustic emission monitoring, transducers are placed on the surface of the member, and these transducers record the arrival of stress waves caused by sudden release of stress. Compared with other stress-wave methods discussed previously, AE is a passive monitoring method as opposed to an active method for probing the interior conditions of a member. Common applications of AE include monitoring the response of a structure during load testing and detecting breaks in post-tensioning tendons. By using triangulation methods, it may be possible to estimate the location of the AE source within the member. Additional information on acoustic emission may be found in Malhotra and Carino (2004).

Nuclear (Radioactive) Tests. Nuclear (or radioactive) methods involve the use of high-energy electromagnetic radiation to gain information about the density or internal structure of the test object. These involve a source of penetrating electromagnetic radiation (X-rays or gamma rays) and a sensor to measure the intensity of the radiation after it has traveled through the object. If the sensor is in the form of photographic plate, the technique is called *radiography*. If the sensor is an electronic device that converts the incident radiation into electrical pulses, the technique is called *radiometry*. The use of X-rays for evaluating concrete members is limited due to the costly and potentially dangerous high-voltage equipment required as well as radiation hazards.

Gamma-radiography equipment can be used in the field to determine the location of reinforcement, honeycombing, and voids in structural concrete members (Mariscotti and others 2009). ASTM C1040, (AASHTO T 271) uses gamma radiometry to determine the density of fresh and hardened concrete in place. Users of nuclear-based methods require special training and certifications to ensure radiological safety.

Covermeters. Battery-operated devices, known as covermeters, are available to measure the depth of reinforcement in concrete and to estimate bar size. There are two types of covermeters: those based on the principle of magnetic reluctance and those based on the principles of eddy currents (ACI 228.2R). Magnetic covermeters will only detect embedded magnetic materials, such as reinforcing steel. Eddy-current meters will detect any type of electrically conductive metal. Sophisticated instruments and software have been developed that permit reconstructing two-dimensional or three-dimensional images of the layout of embedded reinforcement. An eddy-current device has been developed to measure the thickness of concrete pavement slabs. This is done by placing a metal plate on the subgrade before placing the concrete. After the concrete has gained sufficient strength, as sensing head is passed over the plate and the measured response is converted to a slab thickness.

Ground Penetrating Radar. Ground-penetrating (or short-pulse) radar (GPR) is analogous to sonar except that a pulse of electromagnetic energy is used instead of sound. An antenna placed on the surface sends out a short pulse of electromagnetic radiation in the microwave. The pulse travels through the test object and a portion of the incident radiation is reflected when the pulse encounters and interfaces between materials with different electrical properties. At a concrete-metal interface there is 100 % reflection of energy in the electromagnetic pulse. At a concrete-air interface only about 50 % of the incident energy is reflected. GPR is very effective in locating embedded metal and can probe deeper than electrical covermeters. It is also useful for evaluating grouting in reinforced concrete masonry construction. There are two test methods for evaluating bridge decks and pavements. ASTM D4748, *Standard Test Method for Determining the Thickness of Bound Pavement Layers Using Short-Pulse Radar*, can be used to measure the thickness of the upper layer of a multilayer pavement system. ASTM D6087, *Standard Test Method for Evaluating Asphalt-Covered Concrete Bridge Decks Using Ground Penetrating Radar*, can be used to evaluate the presence of delaminations due to corrosion of the top layer of reinforcement in concrete bridge decks.

Infrared Thermography. Infrared thermography (IRT) is a technique for locating near-surface defects by measuring surface temperature. It is based on two principles: 1) a surface emits infrared radiation with an intensity that depends on its temperature, and 2) the presence of a near-surface air gap will interfere with heat flow and alter the surface-temperature distribution. Thus, by measuring the surface temperature distribution using an infrared camera, the presence of a near surface defect can be inferred. This technique requires heat flow either into or out of the surface of the test object. This can be accomplished by artificial heating with lamps, by solar radiation, or night time cooling. ASTM D4788, *Standard Test Method for Detecting Delaminations in Bridge Decks Using Infrared Thermography*, describes the application of IRT for performing a delamination survey of bridge decks. Another potential application is in the evaluation of structural repairs made using bonded fiber-reinforced polymer (FRP) sheets. It has been shown that active infrared thermography, in which heating lamps are used to create a transient heat flow condition, is effective in locating voids that may exist between FRP plies or at the FRP-concrete interface (Starnes and others 2003 and Levar and Hamilton 2003).

References

Aarre, Tine, "Air-Void Analyzer," *Concrete Technology Today*, PL981, Portland Cement Association, http://www.cement.org/pdf_files/PL981.pdf, April 1998, page 4.

ACI Committee 214, *Evaluation of Strength Test Results of Concrete*, ACI 214-02, American Concrete Institute, Farmington Hills, Michigan, 2002, 20 pages.

ACI Committee 222, *Provisional Standard Test Method for Water Soluble Chloride Available for Corrosion of Embedded Steel in Mortar and Concrete Using the Soxhlet Extractor*, ACI 222.1-96, American Concrete Institute, Farmington Hills, Michigan, 1996, 3 pages.

ACI Committee 222, *Protection of Metals in Concrete*, ACI 222R-01, American Concrete Institute, Farmington Hills, Michigan, 2001, 41 pages.

ACI Committee 228, *In-Place Methods to Estimate Concrete Strength*, ACI 228.1R-03, American Concrete Institute, Farmington Hills, Michigan, 2003, 44 pages.

ACI Committee 228, *Nondestructive Test Methods for Evaluation of Concrete in Structures*, ACI 228.2R-98, reapproved 2004, American Concrete Institute, Farmington Hills, Michigan, 1998, 62 pages.

ACI Committee 318, *Building Code Requirements for Structural Concrete and Commentary*, ACI 318-08, American Concrete Institute, Farmington Hills, Michigan, 2008.

American Petroleum Institute, *Recommended Practice for Determining Permeability of Porous Media*, API RP 27, American Petroleum Institute, Washington, D.C., 1956.

Angles, J., "Measuring Workability," 8(12), *Concrete*, 1974, page 26.

ASTM, *Manual of Aggregate and Concrete Testing*, American Society for Testing and Materials, West Conshohocken, Pennsylvania, 2009.

Bartos, P., "Workability of Flowing Concrete—Assessment by a Free Orifice Rheometer," 12(10), *Concrete*, 1978, pages 28 to 30.

Bureau of Reclamation, *Concrete Manual*, 8th Edition, Denver, 1975, page 11.

Burg, R.G.; Caldarone, M.A.; Detwiler, G.; Jansen, D.C.; and Willems, T.J., "Compression Testing of HSC: Latest Technology." *Concrete International*, American Concrete Institute, Farmington Hills, Michigan, August 1999, pages 67 to 76.

Burg, Ron G., and Ost, Borje W., *Engineering Properties of Commercially Available High-Strength Concrete (Including Three-Year Data)*, Research and Development Bulletin RD104, Portland Cement Association, 1994, 58 pages.

Campbell, D.H.; Sturm, R.D.; and Kosmatka, S.H., "Detecting Carbonation," *Concrete Technology Today*, PL911, http://www.cement.org/pdf_files/PL911.pdf, Portland Cement Association, March 1991, pages 1 to 5.

Carino, Nicholas J., "Prediction of Potential Concrete Strength at Later Ages," *Significance of Tests and Properties of Concrete and Concrete-Making Materials*, STP 169D, American Society for Testing and Materials, West Conshohocken, Pennsylvania, 2006, pages 141 to 153.

Carino, N.J.; Guthrie, W.F.; and Lagergren, E.S., "Effects of Testing Variables on the Measured Compressive Strength of High-Strength (90 MPa) Concrete," ACI Special Publication SP-149, American Concrete Institute, Farmington Hills, Michigan, 2004, pages 589 to 632.

Clifton, James R., "Nondestructive Evaluation in Rehabilitation and Preservation of Concrete and Masonry Materials," SP-85-2, *Rehabilitation, Renovation, and Preservation of Concrete and Masonry Structures*, SP-85, American Concrete Institute, Farmington Hills, Michigan, 1985, pages 19 to 29.

CSA Standard A23.1-09/A23.2-09, *Concrete Materials and Methods of Concrete Construction/Test methods and Standard Practices for Concrete*, Canadian Standards Association, Toronto, Canada, 2009.

Daniel, D. Gene, "Factors Influencing Concrete Workability," *Significance of Tests and Properties of Concrete and Concrete-Making Materials*, STP 169D, American Society for Testing and Materials, West Conshohocken, Pennsylvania, 2006, pages 59 to 72.

Date, Chetan G., and Schnormeier, Russell H., "Day-to-Day Comparison of 4 and 6 Inch Diameter Concrete Cylinder Strengths," *Concrete International*, American Concrete Institute, Farmington Hills, Michigan, August 1984, pages 24 to 26.

de Larrard, F.; Szitkar, J.; Hu, C.; and Joly, M., "Design and Rheometer for Fluid Concretes," *Proceedings, International RILEM Workshop*, Paisley, Scotland, March 2-3, 1993, pages 201 to 208.

Detwiler, Rachel J.; Thomas, Wendy; Stangebye, Thorlief; and Urahn, Michael, "Variability of 4x8 in. Cylinder Tests," *Concrete International*, American Concrete Institute, May 2001.

Distlehorst, Jennifer A., and Kurgan, Geoffrey J., "Development of Precision Statement for Determining Air Void Characteristics of Fresh Concrete with Use of Air Void Analyzer" Transportation Research Record Number 2020, *Transportation Research Board*, 2007.

FDOT, *Florida Method of Test for Concrete Resistivity as an Electrical Indicator of its Permeability*, FM 5-578, Florida Department of Transportation, January 27, 2004.

Fiorato, A.E.; Burg, R.G.; and Gaynor, R.D., "Effects of Conditioning on Measured Compressive Strength of Concrete Cores" *Concrete Technology Today*, CT003, Portland Cement Association, http://www.cement.org/pdf_files/CT003.pdf, 2000, pages 1 to 3.

Forester, J.A.; Black, B.F.; and Lees, T.P., *An Apparatus for Rapid Analysis Machine, Technical Report, Cement and Concrete Association*, Wexham Springs, Slough, England, April 1974.

Forstie, Douglas A., and Schnormeier, Russell, "Development and Use of 4 by 8 Inch Concrete Cylinders in Arizona," *Concrete International*, American Concrete Institute, Farmington Hills, Michigan, July 1981, pages 42 to 45.

Gebler, S.H., and Klieger, P., *Effects of Fly Ash on the Air-Void Stability of Concrete*, Research and Development Bulletin RD085T, Portland Cement Association, http://www.cement.org/pdf_files/RD085.pdf, 1983.

Graves, Robin E., "Grading, Shape, and Surface Texture" *Significance of Tests and Properties of Concrete and Concrete-Making Materials*, STP 169D, American Society for Testing and Materials, West Conshohocken, Pennsylvania, 2006, pages 337 to 345.

Hedenblad, Göran, *Drying of Construction Water in Concrete—Drying Times and Moisture Measurement*, T9, Swedish Council for Building Research, Stockholm, 1997, 54 pages. [Available from Portland Cement Association as LT229.]

Hedenblad, Göran, "Concrete Drying Time," *Concrete Technology Today*, PL982, Portland Cement Association, http://www.cement.org/pdf_files/PL982.pdf, 1998, pages 4 and 5.

Hime, William G., "Analyses for Cement and Other Materials in Hardened Concrete," *Significance of Tests and Properties of Concrete and Concrete-Making Materials*, STP 169D, American Society for Testing and Materials, West Conshohocken, Pennsylvania, 2006, pages 309 to 313.

Hime, W.G.; Mivelaz, W.F.; and Connolly, J.D., *Use of Infrared Spectrophotometry for the Detection and Identification of Organic Additions in Cement and Admixtures in Hardened Concrete*, Research Department Bulletin RX194, Portland Cement Association, http://www.cement.org/pdf_files/RX194.pdf, 1966, 22 pages.

Hover, Kenneth C.; Bickley, John; and Hooton, R. Doug, *Guide to Specifying Concrete Performance: Phase II Report of Preparation of a Performance-Based Specification For Cast-in-Place Concrete*, RMC Research and Education Foundation, Silver Spring, Maryland, USA, March 2008, 53 pages.

Kanare, Howard M., *Concrete Floors and Moisture*, EB119, Portland Cement Association, Skokie, Illinois, and National Ready Mixed Concrete Association, Silver Spring, Maryland, 2008, 176 pages.

Koehler, E.P., and Fowler, D.W., "Development of a Portable Rheometer for Fresh Portland Cement concrete," *International Center for Aggregate Research*, Report, ICAR 105-3F, Austin, Texas, 321 pages http://www.icar.utexas.edu/publications/105/105-3F.pdf, 2004, 302 pages.

Kosmatka, Steven H., "Compressive versus Flexural Strength for Quality Control of Pavements," *Concrete Technology Today*, PL854, Portland Cement Association, http://www.cement.org/pdf_files/PL854.pdf, 1985a, pages 4 and 5.

Kosmatka, Steven H., "Floor-Covering Materials and Moisture in Concrete," *Concrete Technology Today*, PL853, Portland Cement Association, http://www.cement.org/pdf_files/PL853.pdf, September 1985, pages 4 and 5.

Lamond, Joseph F., and Pielert, James H., *Significance of Tests and Properties of Concrete and Concrete-Making Materials*, STP 169D, American Society for Testing and Materials, West Conshohocken, Pennsylvania, 2006, 664 pages.

Levar, J.M., and Hamilton, H.R., "Nondestructive Evaluation of Carbon Fiber-Reinforced Polymer-Concrete Bond Using Infrared Thermography," *ACI Materials Journal*, Vol. 100, No. 1, January-February 2003, pages 63 to 72.

Lobo, C.L.; Mullings, G.M.; and Gaynor, R.D., "Effects of Cappings Materials and Procedures on the Compressive Strength of 100 x 200 mm (4 x 8 in.) High Strength Concrete Cylinders," *Cement, Concrete, and Aggregates*, Vol. 16, No. 2, December 1994, pages 173 to 180.

Malhotra, V.M., *Testing Hardened Concrete*, Nondestructive Methods, ACI Monograph No. 9, American Concrete Institute-Iowa State University Press, Farmington Hills, Michigan, 1976.

Malhotra, V.M., *In Situ/Nondestructive Testing of Concrete*, SP-82, American Concrete Institute, Farmington Hills, Michigan, 1984.

Malhotra, V.M., and Carino, N.J., eds. *Handbook on Nondestructive Testing of Concrete*, 2nd Edition, ISBN 0-8031-2099-0, CRC Press, Boca Raton, Florida, and ASTM International, West Conshohocken, Pennsylvania, 2004.

Mariscotti, M.A.J.; Jalinoos, F.; Frigerio, T.; Ruffolo, M.; and Thieberger, P., "Gamma-Ray Imaging for Void and Corrosion Assessment," *Concrete International*, Vol. 31, No. 11, November 2009, pages 48 to 53.

Mor, Avi, and Ravina, Dan, "The DIN Flow Table," *Concrete International*, American Concrete Institute, Farmington Hills, Michigan, December 1986.

NRMCA, *Concrete Tool Box*, Version 4.0.2, National Ready Mixed Concrete Association, Silver Spring, Maryland, 2001.

NRMCA, *In-Place Concrete Strength Evaluation—A Recommended Practice*, NRMCA Publication 133, revised 1979, National Ready Mixed Concrete Association, Silver Spring, Maryland, 1979.

NRMCA, "Standard Practice for Rapid Determination of Water Soluble Chloride in Freshly Mixed Concrete, Aggregate and Liquid Admixtures," *NRMCA Technical Information Letter No. 437*, National Ready Mixed Concrete Association, March 1986.

Parry, James M., *Wisconsin Department of Transportation QC/QA Concept*, Portland Cement Concrete Technician I/IA Course Manual, Wisconsin Highway Technician Certification Program, Platteville, Wisconsin, 2000, pages B-2 to B-4.

PCA, "Rapid Analysis of Fresh Concrete," *Concrete Technology Today*, PL832, Portland Cement Association, http://www.cement.org/pdf_files/PL832.pdf, June 1983, pages 3 and 4.

Petersen, C.G., "Air Void Analyser (AVA) for fresh concrete, latest Advances", *Ninth ACI International Conference on Superplasticizers and Other Chemical Admixtures in Concrete*, Seville, Spain, October 2009.

Petersen, C.G., "LOK-Test and CAPO-Test Pullout Testing: Twenty Years Experience," *Conference on Non-Destructive Testing in Civil Engineering*, Liverpool, UK, April 1997, British Institute of Non-Destructive Testing.

Pfeifer, D.W., and Scali, M.J., *Concrete Sealers for Protection of Bridge Structures*, NCHRP Report 244, Transportation Research Board, National Research Council, 1981.

Pinto, Roberto C.A., and Hover, Kenneth C., *Frost and Scaling Resistance of High-Strength Concrete*, Research and Development Bulletin RD122, Portland Cement Association, 2001, 70 pages.

Pistilli, Michael F., and Willems, Terry, "Evaluation of Cylinder Size and Capping Method in Compression Testing of Concrete," *Cement, Concrete and Aggregates*, American Society for Testing and Materials, West Conshohocken, Pennsylvania, Summer 1993.

Powers, Laura J., "Developments in Alkali-Silica Gel Detection," *Concrete Technology Today*, PL991, Portland Cement Association, http://www.cement.org/pdf_files files/PL991.pdf, April 1999, pages 5 to 7.

Powers, T.C., "Studies of Workability of Concrete," *Journal of the American Concrete Institute*, Volume 28, American Concrete Institute, Farmington Hills, Michigan, 1932, page 419.

Powers, T.C., and Wiler, E.M., "A Device for Studying the Workability of Concrete," *Proceedings of ASTM*, Vol. 41, American Society for Testing and Materials, West Conshohocken, Pennsylvania, 1941.

Saucier, K.L., *Investigation of a Vibrating Slope Method for Measuring Concrete Workability*, Miscellaneous Paper 6-849, U.S. Army Engineer Waterways Experiment Station, Vicksburg, Mississippi, 1966.

Sansalone, M., and Streett, W.B., *Impact-Echo: Nondestructive Testing of Concrete and Masonry*, Bullbrier Press, Jersey Shore, Pennsylvania, 1997.

Smith, David, *The Development of a Rapid Test for Determining the Transport Properties of Concrete*, SN2821, Portland Cement Association, Skokie, Illinois, 2006, 125 pages.

St. John, Donald A.; Poole, Alan W.; and Sims, Ian, *Concrete Petrography—A handbook of investigative techniques*, Arnold, London, http://www.arnoldpublishers.com, PCA LT226, 485 pages.

Starnes, M.A.; Carino, N.J.; and Kausel, E.A., "Preliminary Thermography Studies for Quality Control of Concrete Structures Strengthened with FRP Composites," *ASCE Journal of Materials in Civil Engineering*, Vol. 15, No. 3, May/June 2003, pages 266 to 273.

Steinour, Harold H., *Influence of the Cement on Corrosion Behavior of Steel in Concrete*, Research Department Bulletin RX168, Portland Cement Association, http://www.cement.org/pdf_files/RX168.pdf, 1964, 22 pages.

Tarr, Scott M., and Farny, James A., *Concrete Floors on Ground*, EB075, Portland Cement Association, 2008, 256 pages.

Tattersall, G.H., *Measurement of Workability of Concrete*, East Midlands Region of the Concrete Society of Nottingham, England, 1971.

Teranishs, K.; Watanabe, K.; Kurodawa, Y.; Mori, H.; and Tanigawa, Y., "Evaluation of Possibility of High-Fluidity Concrete," *Transactions of the Japan Concrete Institute (16)*, Japan Concrete Institute, Tokyo, 1994, pages 17 to 24.

Thomas, Michael; Fournier, Benoit; Folliard, Kevin; Ideker, Jason; and Shehata, Medhat, *Test Methods for Evaluating Preventive Measures for Controlling Expansion Due to Alkali-Silica Reaction in Concrete*, ICAR 302-1, International Center for Aggregates Research, 2006, 62 pages.

Thomas, Michael D.A.; Fournier, Benoit; and Folliard, Kevin J., *Report on Determining the Reactivity of Concrete Aggregates and Selecting Appropriate Measures for Preventing Deleterious Expansion in New Concrete Construction*, FHWA-HIF-09-001, Federal Highway Administration, Washington, D.C., April 2008, 28 pages.

Tyler, I.L., and Erlin, Bernard, *A Proposed Simple Test Method for Determining the Permeability of Concrete*, Research Department Bulletin RX133, Portland Cement Association, http://www.cement.org/pdf_files/RX133.pdf, 1961.

Verbeck, G.J., *Carbonation of Hydrated Portland Cement*, Research Department Bulletin RX087, Portland Cement Association, http://www.cement.org/pdf_files/RX087.pdf, 1958.

Wallevik, O., "The Use of BML Viscometer for Quality Control of Concrete," *Concrete Research*, Nordic Concrete Research, Espoo, Finland, 1996, pages 235 to 236.

Whiting, David, *Rapid Determination of the Chloride Permeability of Concrete*, FHWA-RD-81-119, Federal Highway Administration, Washington, D.C., 1981.

Whiting, David, "Permeability of Selected Concretes," *Permeability of Concrete*, SP-108, American Concrete Institute, Farmington Hills, Michigan, 1988.

Wigmore, V.S., *Consistometer*, Civil Engineering (London), Vol. 43, No. 510, December 1948, pages 628 to 629.

Wong, G. Sam; Alexander, A. Michel; Haskins, Richard; Poole, Toy S.; Malone, Phillip G.; and Wakeley, Lillian, *Portland-Cement Concrete Rheology and Workability: Final Report*, FHWA—RD-00-025, Federal Highway Administration, Washington, D.C., 2001, 117 pages.

Wood, Sharon L., *Evaluation of the Long-Term Properties of Concrete*, Research and Development Bulletin RD102, Portland Cement Association, 1992, 99 pages.

CHAPTER 19

High-Performance Concrete

High-performance concrete (HPC) exceeds the properties and constructability of normal concrete. Normal and special materials are used to make these specially designed concretes that must meet a combination of performance requirements. Special mixing, placing, and curing practices may be needed to produce and handle high-performance concrete. Performance tests are usually required to demonstrate compliance with specific project needs (ASCE 1993, Russell 1999, and Bickley and Mitchell 2001). High-performance concrete has been primarily used in bridges, and tall buildings for its durability, strength, and high modulus of elasticity (Figure 19-1). It has also been used in shotcrete repair, poles, tunnels, parking garages, and agricultural applications.

High-performance concrete characteristics are defined, categorized, or developed for particular applications and environments (Goodspeed, Vanikar, and Cook 1996 and Russell and Ozyildirim 2006); some of the characteristics that may be required include:

- Enhanced Durability
  - High abrasion resistance
  - Low permeability and diffusion
  - Resistance to chemical attack
  - High resistance to freeze-thaw, and deicer scaling damage
  - Resistance to alkali silica reactivity
- Enhanced Engineering Properties
  - High strength
  - High early strength
  - High modulus of elasticity
  - Toughness and impact resistance
  - Volume stability

Figure 19-1. High-performance concrete is often used in bridges and tall buildings. (left) I-35W St. Anthony Falls bridge in Minneapolis, Minnesota. (right) 311 W. Wacker Drive, Chicago, Illinois.

- Other Enhanced Properties
  - Ease of placement
  - Temperature control
  - Compaction without segregation
  - Inhibition of bacterial and mold growth

High-performance concretes are made with carefully selected high-quality ingredients and optimized mixture designs; these are batched, mixed, placed, compacted and cured to the highest industry standards. Typically, they will have low water-cementing materials ratios of 0.20 to 0.45. High-range water reducers (or superplasticizers) are usually used to make these concretes fluid and workable at lower w/cm.

High-performance concrete almost always has greater durability than normal concrete. This greater durability may be accompanied by normal strength or it may be partnered with high strength. Note that strength is not always the primary required property. For bridge decks, a normal strength concrete with very high durability and very low permeability is considered high performance concrete (Lane 2010). Table 19-1 lists materials often used in high-performance concrete and their selection criteria. Table 19-2 lists properties that can be selected for high-performance concrete. Typical mix designs and member properties for many bridges can be found in the FHWA report compiling results for HPC bridges (Russell and others 2006). Not all properties can be achieved concurrently.

High-performance concrete specifications should be performance oriented. However, many specifications are a combination of performance requirements (such as permeability or strength limits) and prescriptive requirements (such as air content limits or dosage of SCMs) (Ferraris and Lobo 1998 and Caldarone and others 2005). Table 19-3 provides examples of high-performance concrete mixtures used in a variety of structures. Selected high-performance concretes are presented in this chapter.

Table 19-1. Materals Used in High-Performance Concrete

Material	Primary contribution/desired property
Portland cement	Cementing material/durability
Blended cement	Cementing material/durability/high strength
Fly ash	Cementing material/durability/high strength
Slag cement	Cementing material/durability/high strength
Silica fume	Cementing material/durability/high strength
Calcined clay	Cementing material/durability/high strength
Metakaolin	Cementing material/durability/high strength
Calcined shale	Cementing material/durability/high strength
Expanded shale, clay, and/or slate	Lightweight
Superplasticizers	Flowability
High-range water reducers	Reduce water to cement ratio
Hydration control admixtures	Control setting
Retarders	Control setting
Accelerators	Accelerate setting
Corrosion inhibitors	Control steel corrosion
Water reducers	Reduce cement and water content
Shrinkage reducers	Reduce shrinkage
ASR inhibitors	Control alkali-silica reactivity
Polymer/latex modifiers	Durability
Optimally graded aggregate	Improve workability and reduce paste demand

Table 19-2. Selected Properties of High-Performance Concrete

Property	Test method	Criteria that may be specified
High compressive strength	ASTM C39 (AASHTO T 22)	55 to 140 MPa (8000 to 20,000 psi) at 28 to 91 days
High-early compressive strength	ASTM C39 (AASHTO T 22)	20 to 41 MPa (3000 to 6000 psi) at 3 to 18 hours, or 1 to 3 days
High-early tensile strength	ASTM C78 (AASHTO T 97)	2 to 4 MPa (300 to 600 psi) at 3 to 12 hours, or 1 to 3 days
Abrasion resistance	ASTM C944	0 to less than 2 mm depth of wear
Low permeability	ASTM C1202 (AASHTO T 277)	500 to 2500 coulombs
Reduced chloride penetration	ASTM C1543 (AASHTO T 259 and AASHTO T 260)	Less than 0.07% Cl at 6 months
High resistivity	ASTM G59	
Low absorption	ASTM C642	2% to 5%
Low diffusion coefficient	ASTM C1556	1000 x 10^{-13} m/s
Resistance to chemical attack	Expose concrete to saturated solution in wet/dry environment	No deterioration after 1 year
Resistance to sulfate attack	ASTM C1012	Mild Exposure: 0.10% max expansion at 6 months; Moderate Exposure: 0.10% max. expansion at 12 months; Severe Exposure: 0.10% max expansion at 18 months;
High modulus of elasticity	ASTM C469	34 to more than 48 GPa (5 to more than 7 million psi)
High resistance to freezing and thawing damage	ASTM C666, Procedure A (AASHTO T 161)	Relative dynamic modulus of elasticity after 300 cycles of 70% to more than 90%
High resistance to deicer scaling	ASTM C672	Visual rating of the surface after 50 cycles of 0 to 3
Low shrinkage	ASTM C157	Less than 800 millionths (microstrain) to less than 400 millionths (microstrain)
Low creep	ASTM C512	70 microstrain/MPa to less than 30 microstrain/MPa (0.52 microstrain/psi to less than 0.21 microstrain/psi)
Increased workability	ASTM C143 (AASHTO T 119)	Slump more than 190 mm (7.5 in)
Increased workability for SCC	ASTM C1611	Slump flow ≤ 600 mm (24 in)
Resistance to alkali silica reactivity	ASTM C441	Expansion at 56 days of 0.20% to less than 0.10%
Resistance to delayed ettringite formation	Maximum internal curing temperature (within concrete)	Less than 70°C (158°F)

Table 19-3 (Metric). High-Performance Concrete Mixtures Used in Various Structures

Mixture Number/Mixture Ingredient	1	2	3	4	5	6	7	8	9
Water, kg/m3	151	145	135	145	130	130	119	157	151
Cement, kg/m3	311	398*	500	335*	513	315	530	387	371
Fly ash, kg/m3	31	45	—	—	—	40	—	68	59
Slag, kg/m3	47	—	—	125	—	—	—	—	—
Silica fume, kg/m3	16	32*	30	40*	43	23	—	—	30
Coarse aggregate, kg/m3	1068	1030	1100	1130	1080	1140	949	973	997
Fine aggregate, kg/m3	676	705	700	695	685	710	766	652	801
Water reducer, L/m3	1.6	1.7	—	1.0	—	1.5	—	—	—
Retarder, L/m3	—	—	1.8	—	—	—	—	—	—
Water reducing/Retarding Admixture, L/m3	—	—	—	—	—	—	—	0.6	—
Shrinkage-Reducing Admixture, L/m3	—	—	—	—	—	—	—	4.8	—
Hydration Stabilizer, L/m3	—	—	—	—	—	—	—	0.9	—
Air Entraining Admixture, L/m3	—	—	—	—	—	—	1.4	0.3	—
Air, %	7 ± 1.5	5 – 8	—	—	—	5.5	—	—	—
HRWR or plasticizer, L/m3	2.1	3	14	6.5	15.7	5.0	2.4	0.7	0 to 3090
Water to cementitious materials ratio	0.37	0.30	0.27	0.29	0.25	0.34	0.22	0.35	0.33
Comp. strength at 28 days, MPa	59	—	93	99	119	—	61	—	76
Comp. strength at 91 days MPa	—	60	107	104	145	—	—	36	—
Permeability at 56 days, coulombs	—	—	—	—	—	—	—	—	Less than 800

1. Wacker Drive bi-level roadway, Chicago, 2001.
2. Confederation Bridge, Northumberland Strait, Prince Edward Island/New Brunswick, 1997.
3. La Laurentienne Building, Montreal, 1984.
4. BCE Place Phase 2, Toronto, 1993.
5. Two Union Square, Seattle, 1988.
6. Great Belt Link, East Bridge, Denmark, 1996.
7. Girders, Angeles Crest Bridge (Higareda 2010).
8. Deck, Angeles Crest Bridge.
9. Pontoon, Hood Canal Floating Bridge (Gaines and Tragesser 2008).

* Originally used a blended cement containing silica fume. Portland cement and silica fume quantities have been separated for comparison purposes.

Table 19-3 (Inch-Pound Units). High-Performance Concrete Mixtures Used in Various Structures

Mixture Number/Mixture Ingredient	1	2	3	4	5	6	7	8	9
Water, lb/yd^3	254	244	227	244	219	219	200	265	255
Cement, lb/yd^3	525	671*	843	565*	865	531	893	652	625
Fly ash, lb/yd^3	53	76	—	—	—	67	—	115	100
Slag, lb/yd^3	79	—	—	211	—	—	—	—	—
Silica fume, lb/yd^3	27	54*	51	67	72	39	—	—	50
Coarse aggregate, lb/yd^3	1800	1736	1854	1905	1820	1921	1600	1640	1680
Fine aggregate, lb/yd^3	1140	1188	1180	1171	1155	1197	1292	1099	1350
Water reducer, oz/yd^3	41	47	—	27	—	38	—	—	—
Retarder, oz/yd^3	—	—	48	—	—	—	—	—	—
Water reducing/Retarding Admixture, oz/yd^3	—	—	—	—	—	—	—	15	—
Shrinkage-Reducing Admixture, oz/yd^3	—	—	—	—	—	—	—	123	—
Hydration Stabilizer, oz/yd^3	—	—	—	—	—	—	—	23	—
Air Entraining Admixture, oz/yd^3	—	—	—	—	—	—	37	8.5	
Air, %	7 ± 1.5	5 – 8	—	—	—	5.5			
HRWR or plasticizer, oz/yd^3	55	83	975	175	420	131	63	18	0 to 80
Water to cementitious materials ratio	0.37	0.30	0.27	0.29	0.25	0.34	0.22	0.35	0.33
Comp. strength at 28 days, psi	8590	—	13,500	14,360	17,250	—	8750		11,000
Comp. strength at 91 days psi	—	8700	15,300	15,080	21,000	—		5190	
Permeability at 56 days, coulombs	—	—	—	—	—	—			Less than 800

1. Wacker Drive bi-level roadway, Chicago, 2001.
2. Confederation Bridge, Northumberland Strait, Prince Edward Island/New Brunswick, 1997.
3. La Laurentienne Building, Montreal, 1984.
4. BCE Place Phase 2, Toronto, 1993.
5. Two Union Square, Seattle, 1988.
6. Great Belt Link, East Bridge, Denmark, 1996.
7. Girders, Angeles Crest Bridge (Higareda 2010).
8. Deck, Angeles Crest Bridge.
9. Pontoon, Hood Canal Floating Bridge (Gaines and Tragesser 2008).

* Originally used a blended cement containing silica fume. Portland cement and silica fume quantities have been separated for comparison purposes.

High-Durability Concrete

HPC concrete today is more on concretes with high durability in mild, moderate, or severe environments. For example, the Confederation Bridge across the Northumberland Strait between Prince Edward Island and New Brunswick, Canada has a 100-year design life (see Mix No. 2 in Table 19-3). This bridge contains HPC designed to efficiently protect the embedded reinforcement. The concrete had a diffusion coefficient of 4.8×10^{-13} at six months (a value 10 to 30 times lower than that of conventional concrete). The electrical resistivity was measured at 470 Ω-m to 530 Ω-m, compared to 50 Ω-m for conventional concrete. The design required that the concrete be rated at less than 1000 coulombs. The high concrete resistivity in itself will result in a rate of corrosion that is potentially less than 10% of the corrosion rate for conventional concrete (Dunaszegi 1999). The following sections review durability issues that high-performance concrete can address.

Abrasion Resistance

Abrasion resistance is related to the strength of concrete. This makes high strength HPC ideal for abrasive environments. The abrasion resistance of HPC incorporating silica fume is especially high. This makes silica fume concrete particularly useful for spillways and stilling basins, bridge decks subject to studded tires or tires with chains, and concrete pavements or concrete pavement overlays subjected to heavy or abrasive traffic.

Holland and others (1986) describe how severe abrasion-erosion had occurred in the stilling basin of a dam. Repairs using fiber-reinforced concrete were not durable. The new HPC mixtures used to repair the structure the second time contained 386 kg/m^3 (650 lb/yd^3) of cement, 70 kg/m^3 (118 lb/yd^3) of silica fume, and admixtures; had a water-to-cementitious materials ratio of 0.28; and had a 90-day compressive strength exceeding 103 MPa (15,000 psi).

Berra, Ferrara, and Tavano (1989) studied the addition of fibers to silica fume mortars to optimize abrasion resistance. The best results were obtained with a mixture using slag cement, steel fibers, and silica fume. Mortar strengths ranged from 75 MPa to 100 MPa (11,000 psi to 14,500 psi). In addition to better erosion resistance; less drying shrinkage, high freeze-thaw resistance, and good bond to the substrate were achieved.

In Norway steel studs are allowed in tires. This causes severe abrasion wear on pavement surfaces, with resurfacing required within one to two years. Tests using an accelerated road-wear simulator showed that in the range of 100 MPa to 120 MPa (14,500 psi to 17,000 psi), concrete had the same abrasion resistance as granite (Helland 1990). Abrasion-resistant highway mixtures usually contain between 320 kg/m^3 and 450 kg/m^3 (539 lb/yd^3 and 758 lb/yd^3) of cement, plus silica fume or fly ash. They have water to cementing materials ratios of 0.22 to 0.36 and compressive strengths in the range of 85 MPa to 130 MPa (12,000 psi to 19,000 psi). Applications have included new pavements and overlays to existing pavements.

Blast Resistance

High-performance concrete can be designed to have excellent blast resistance properties. These concretes often have a compressive strength exceeding 120 MPa (14,500 psi) and contain steel fibers. Blast-resistant concretes are often used in bank vaults, military applications, and some transportation structures.

Permeability

The durability and service life of steel reinforced concrete exposed to weather is related to the permeability of the concrete cover protecting the reinforcement. HPC typically has very low permeability to air, water, and chloride ions. Low permeability is often specified through the use of a coulomb value, such as a maximum of 1000 coulombs.

Test results obtained on specimens from a concrete column specified to be 70 MPa (10,000 psi) at 91 days and which had not been subjected to any wet curing were as follows (Bickley and others 1994):

Water permeability of vacuum-saturated specimens:

Age at test:	7 years
Applied water pressure:	0.69 MPa
Permeability:	7.6×10^{-13} cm/s

Rapid chloride permeability (ASTM C1202):

Age at test, years	Coulombs
1	303
2	258
7	417

The dense pore structure of high-performance concrete, gives it characteristics that make it eminently suitable for uses where a high quality concrete would not normally be considered. Ternary mixtures achieve lower permeabilities more easily for a given w/cm (Bouzoubaa and others 2004). Latex-modified HPC is able to achieve these same low levels of permeability at normal strength levels without the use of supplementary cementing materials.

Diffusion

Aggressive ions, such as chloride, in contact with the surface of concrete will diffuse through the concrete until a state of equilibrium is achieved. If the concentration of ions at the surface is high, diffusion may result in corrosion-inducing concentrations at the level of the reinforcement.

The lower the water-cementing materials ratio the lower the diffusion coefficient will be for any given set of materials. Supplementary cementing materials, particularly silica fume, further reduce the diffusion coefficient. Typical values for diffusion for HPC are as follows:

Type of Concrete	Diffusion Coefficient (ASTM C1556)
Portland cement-fly-ash silica fume mix:	1000×10^{-15} m²/s
Portland cement-fly ash mix:	1600×10^{-15} m²/s

Carbonation

HPC has a very good resistance to carbonation due to its low permeability. It was determined that after 17 years the concrete in the CN Tower in Toronto had carbonated to an average depth of 6 mm (0.24 in.) (Bickley, Sarkar, and Langlois 1992). The concrete mixture in the CN Tower had a water-cement ratio of 0.42. For a cover to the reinforcement of 35 mm (1.4 in.), this concrete would provide corrosion protection for 500 years. For the lower water-cementing materials ratios common to HPC, significantly longer times to corrosion would result, assuming a crack free structure. In practical terms, uncracked HPC cover concrete is immune to carbonation to a depth that would cause corrosion.

Freeze-Thaw Resistance

Because of its very low water-cementing materials ratio (0.20 to 0.45), it is widely believed that HPC should be highly resistant to both scaling and physical breakup due to freezing and thawing. There is ample evidence that properly air-entrained high performance concretes are highly resistant to freezing and thawing and to scaling.

Gagne, Pigeon, and Aïtcin (1990) tested 27 mixtures using cement and silica fume with water-cementing materials ratios of 0.30, 0.26, and 0.23 and a wide range of quality in air-voids systems. All specimens performed exceptionally well in salt-scaling tests, confirming the durability of high-performance concrete, and suggesting that air-entrainment is not needed. Tachitana and others (1990) conducted ASTM C666 (Procedure A) tests on non-air-entrained high performance concretes with water-cementing materials ratios between 0.22 and 0.31. All were found to be extremely resistant to freeze-thaw damage and again it was suggested that air-entrainment is not needed.

Pinto and Hover (2001) found that non-air-entrained concrete with a w/c of 0.25 was deicer-scaling resistant with no supplementary cementing materials present. They found that higher strength portland cement concretes needed less air than normal concrete to be frost and scale resistant.

Burg and Ost (1994) found that of the six mixtures tested in Table 19-5 using ASTM C666, only the silica fume concrete (Mix 4) with a water to cementing materials ratio of 0.22 was frost resistant.

Sidewalks constructed in Chicago in the 1920s used 25-mm (1-in.) thick toppings made of no-slump dry-pack mortar rammed into place. The concrete contained no air entrainment. Many of these sidewalks are still in use today; they are in good condition (minus some surface weathering exposing fine aggregate) after 90 plus years of exposure to frost and deicers. No documentation exists on the water to cement ratio; however, it can be assumed that the water-to-cement ratio was comparable to that of modern HPC.

While the above experiences prove the excellent durability of certain high-performance concretes to freeze-thaw damage and salt scaling, it is considered prudent to use air-entrainment. No well-documented field experiments have been made to prove that air-entrainment is not needed. Until such data are available, current practice for air-entrainment should be followed. It has been shown that the prime requirement of an air-void system for HPC is a preponderance of air bubbles of 200 μm size and smaller. If the correct air bubble size and spacing can be assured, then a moderate air content will ensure durability and minimize strength loss. The best measure of air-entrainment is the spacing factor.

Chemical Attack

For resistance to chemical attack on most structures, HPC offers a much improved performance. Resistance to various sulfates is achieved primarily by the use of a dense, strong concrete of very low permeability and low water-to-cementing materials ratio; these are all characteristics of HPC. Similarly, as discussed by Gagne, Chagnon, and Parizeau (1994), resistance to acid from wastes is also much improved.

Alkali-Silica Reactivity

Reactivity between certain siliceous aggregates and alkali hydroxides can affect the long-term performance of concrete. Two characteristics of HPC that help combat alkali-silica reactivity are:

1. HPC concretes at very low water to cement ratios can self desiccate (dry out) to a level that does not allow ASR to occur (relative humidity less than 80%). Burg and Ost (1994) observed relative humidity values ranging from 62% to 72% for their six mixtures in Table 19-5. The low permeability of HPC also minimizes external moisture from entering the concrete.
2. HPC concretes can use significant amounts of supplementary cementing materials that may have the ability to control alkali-silica reactivity. However, this must be demonstrated by test. HPC concretes can also use ASR inhibiting admixtures to control ASR.

HPC concretes are not immune to alkali-silica reactivity and appropriate precautions must be taken. Procedures are available for determining the potential reactivity of aggregates and limiting ASR (Thomas, Fournier, and Folliard, 2008).

Resistivity

HPC, particularly that formulated with silica fume, has very high resistivity, up to 20 to 25 times that of normal concrete. This increases resistance to the flow of electrical current and reduces corrosion rates. Particularly if dry, HPC acts as an effective dielectric. Where cracking occurs in HPC, the corrosion is localized and minor; this is due to the high resistivity of the concrete which suppresses the development of a macro corrosion cell.

High-Early-Strength Concrete

High-early-strength concrete, also called fast-track concrete, achieves its specified strength at an earlier age than normal concrete. The time period in which a specified strength should be achieved may range from a few hours (or even minutes) to several days. High-early-strength can be attained using traditional concrete ingredients and concreting practices, although sometimes special materials or techniques are needed.

High-early-strength can be obtained using one or a combination of the following, depending on the age at which the specified strength must be achieved and on job conditions:

1. Type III or HE high-early-strength cement
2. High cement content (400 kg/m^3 to 600 kg/m^3 [675 lb/yd^3 to 1000 lb/yd^3])
3. Low water-cementing materials ratio (0.20 to 0.45 by mass)
4. Higher freshly mixed concrete temperature
5. Higher curing temperature (Note: Keep internal member temperature under 70°C [158°F] to help prevent delayed ettringite formation)
6. Chemical admixtures
7. Silica fume (or other supplementary cementing materials)
8. Steam or autoclave curing (see note on 5)
9. Insulation to retain heat of hydration (see note on 5)
10. Special rapid hardening cements

High-early-strength concrete is used for prestressed concrete to allow for early stressing, precast concrete for rapid production of elements, high-speed cast-in-place construction, rapid form reuse, cold-weather construction, rapid repair of pavements (to reduce traffic downtime), fast-track paving, and several other uses.

In fast-track paving, use of high-early-strength mixtures allows traffic to open just hours after concrete is placed. An example of a fast-track concrete mixture used for a bonded concrete highway overlay consisted of 380 kg (640 lb) of Type III cement, 42 kg (70 lb) of Type C fly ash, 6½% air, a water reducer, and a water-to-cementing materials ratio of 0.4. Strength data for this 40-mm (1½ in.) slump concrete are given in Table 19-4. Figures 19-2 and 19-3 illustrate early strength development of concretes designed to open to traffic within 4 hours after placement. Figure 19-4 illustrates the benefits of blanket curing to develop early strength for patching or fast-track applications.

Table 19-4. Strength Data for Fast-Track Bonded Overlay

Age	Compressive strength, MPa (psi)	Flexural strength, MPa (psi)	Bond strength, MPa (psi)
4 hours	1.7 (252)	0.9 (126)	0.9 (120)
6 hours	7.0 (1020)	2.0 (287)	1.1 (160)
8 hours	13.0 (1883)	2.7 (393)	1.4 (200)
12 hours	17.6 (2546)	3.4 (494)	1.6 (225)
18 hours	20.1 (2920)	4.0 (574)	1.7 (250)
24 hours	23.9 (3467)	4.2 (604)	2.1 (302)
7 days	34.2 (4960)	5.0 (722)	2.1 (309)
14 days	36.5 (5295)	5.7 (825)	2.3 (328)
28 days	40.7 (5900)	5.7 (830)	2.5 (359)

Adapted from Knutson and Riley 1987.

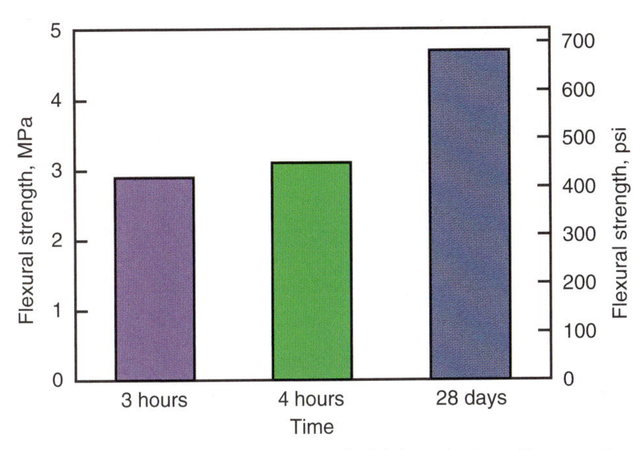

Figure 19-2. Strength development of a high-early strength concrete mixture using 390 kg/m³ (657 lb/yd³) of rapid hardening cement, 676 kg/m³ (1140 lb/yd³) of sand, 1115 kg/m³ (1879 lb/yd³) of 25 mm (1 in.) nominal max. size coarse aggregate, a water to cement ratio of 0.46, a slump of 100 to 200 mm (4 to 8 in.), and a plasticizer and retarder. Initial set was at one hour (Pyle 2001).

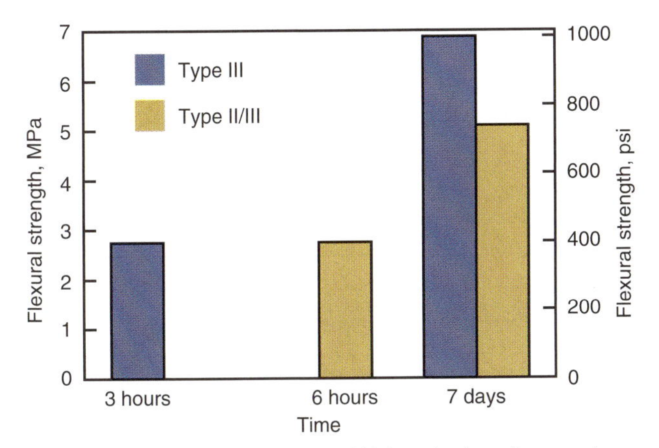

Figure 19-3. Strength development of high-early strength concrete mixtures made with 504 to 528 kg/m³ (850 to 890 lb/yd³) of Type III or Type II/III cement, a nominal maximum size coarse aggregate of 25 mm (1 in.), a water to cement ratio of 0.30, a plasticizer, a hydration control admixture, and an accelerator. Initial set was at one hour (Pyle 2001).

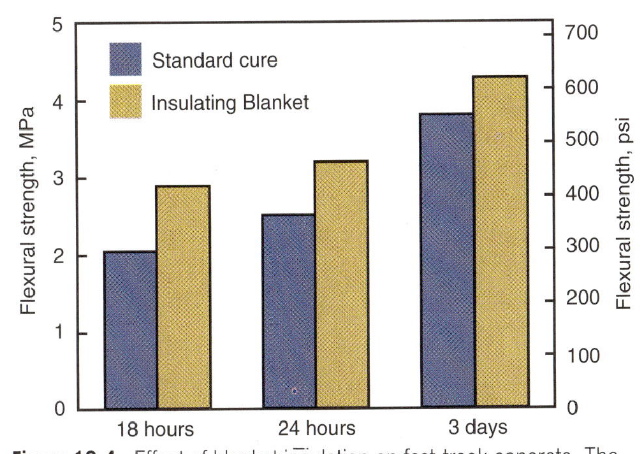

Figure 19-4. Effect of blanket insulation on fast-track concrete. The concrete had a Type I cement content of 421kg/m³ (710 lb/yd³) and a water to cement ratio of 0.30 (Grove 1989).

When designing early-strength mixtures, strength development is not the only criteria that should be evaluated; durability, early stiffening, camber of pretensioned members, autogenous shrinkage, drying shrinkage, temperature rise, and other properties also should be evaluated for compatibility with the project. Special curing procedures, such as fogging, may be needed to control plastic shrinkage cracking.

High-Strength Concrete

The definition of high strength changes over the years as concrete strength used in the field increases. This publication considers high-strength concrete (HSC) as a strength significantly beyond what is used in normal practice. About 90% of ready mixed concrete has a 28-day specified compressive strength ranging from 20 MPa (3000 psi) to 40 MPa (6000 psi), with most of it between 20 MPa (3000 psi) and 35 MPa (5000 psi).

Most high-strength concrete applications are designed for compressive strengths of 70 MPa (10,000 psi) or greater as shown in Tables 19-3 and 19-5. For bridges, the AASHTO LRFD Specifications (2010) state that the minimum allowable compressive strength for bridge decks and prestressed concrete members is 28 MPa (4000 psi). Therefore, HSC considered here has a minimum design strength of at least 55 MPa (8000 psi). For high strength concrete, stringent application of the best practices is required. Compliance with the guidelines and recommendations for preconstruction laboratory and field-testing procedures described in ACI 363.2 are essential. Concrete with a design strength of 131 MPa (19,000 psi) has been used in buildings (Figure 19-5).

Figure 19-5. The Two Union Square building in Seattle used concrete with a designed compressive strength of 131 MPa (19,000 psi) in its steel tube and concrete composite columns. High-strength concrete was used to meet a design criterion of 41 GPa (6 million psi) modulus of elasticity.

Table 19-5 (Metric). Mixture Proportions and Properties of Commercially Available High-Strength Concrete (Burg and Ost 1994)

Units	Mix number					
	1	2	3	4	5	6
Cement, Type I, kg/m3	564	475	487	564	475	327
Silica fume, kg/m3	—	24	47	89	74	27
Fly ash, kg/m3	—	59	—	—	104	87
Coarse aggregate SSD (12.5 mm crushed limestone), kg/m3	1068	1068	1068	1068	1068	1121
Fine aggregate SSD, kg/m3	647	659	676	593	593	742
HRWR Type F, liters/m3	11.6	11.6	11.22	20.11	16.44	6.3
HRWR Type G, liters/m3	—	—	—	—	—	3.24
Retarder, Type D, liters/m3	1.12	1.05	0.97	1.46	1.5	—
Water to cementing materials ratio	0.28	0.29	0.29	0.22	0.23	0.32
Fresh concrete properties						
Slump, mm	197	248	216	254	235	203
Density, kg/ m3	2451	2453	2433	2486	2459	2454
Air content, %	1.6	0.7	1.3	1.1	1.4	1.2
Concrete temp., °C	24	24	18	17	17	23
Compressive strength, 100 x 200-mm moist-cured cylinders						
3 days, MPa	57	54	55	72	53	43
7 days, MPa	67	71	71	92	77	63
28 days, MPa	79	92	90	117	100	85
56 days, MPa	84	94	95	122	116	—
91 days, MPa	88	105	96	124	120	92
182 days, MPa	97	105	97	128	120	—
426 days, MPa	103	118	100	133	119	—
1085 days, MPa	115	122	115	150	132	—
Modulus of elasticity in compression, 100 x 200-mm moist-cured cylinders						
91 days, GPa	50.6	49.9	50.1	56.5	53.4	47.9
Drying shrinkage, 75 by 75 x 285-mm prisms						
7 days, millionths	193	123	100	87	137	—
28 days, millionths	400	287	240	203	233	—
90 days, millionths	573	447	383	320	340	—
369 days, millionths	690	577	520	453	467	—
1075 days, millionths	753	677	603	527	523	—

Traditionally, the specified strength of concrete has been based on 28-day test results. However, in high-rise concrete structures, the process of construction is such that the structural elements in lower floors are not fully loaded for periods of a year or more. For this reason, compressive strengths based on 56- or 91-day test results are commonly specified in order to achieve significant economy in material costs. For bridges, the specified strength of concrete has also been based on 28-day test results. However, because of the use of fly ash and slag cement, which may hydrate slower than the cement, 56-day strengths have been specified on bridge projects.

When later ages are specified, supplementary cementing materials are usually incorporated into the concrete mixture. This produces additional benefits in the form of reduced heat generation during hydration.

With use of low-slump or no-slump mixtures, high compressive-strength concrete is produced routinely under careful control in precast and prestressed concrete plants. These stiff mixtures are placed in ruggedly-built forms and consolidated by prolonged vibration or shock methods. However, cast-in-place concrete uses more fragile forms that do not permit the same compaction procedures. Hence, more workable concretes are necessary to achieve the required compaction and to avoid segregation and honeycomb. Superplasticizing admixtures are invariably added to HPC mixtures to produce workable and often flowable mixtures.

Production of high-strength concrete may or may not require the purchase of special materials. The producer must know the factors affecting compressive strength and know how to vary those factors for best results. Each variable should be analyzed separately in developing a mix

Table 19-5 (Inch-Pound Units). Mixture Proportions and Properties of Commercially Available High-Strength Concrete (Burg and Ost 1994)

Units	Mix number					
	1	2	3	4	5	6
Cement, Type I, lb/yd^3	950	800	820	950	800	551
Silica fume, lb/yd^3	—	40	80	150	125	45
Fly ash, lb/yd^3	—	100	—	—	175	147
Coarse aggregate SSD (½ in. crushed limestone), lb/yd^3	1800	1800	1800	1800	1800	1890
Fine aggregate SSD, lb/yd^3	1090	1110	1140	1000	1000	1251
HRWR Type F, fl oz/yd^3	300	300	290	520	425	163
HRWR Type G, fl oz/yd^3	—	—	—	—	—	84
Retarder, Type D, fl oz/yd^3	29	27	25	38	39	—
Water to cementing materials ratio	0.28	0.29	0.29	0.22	0.23	0.32
Fresh concrete properties						
Slump, in.	7¾	9¾	8½	10	9¼	8
Density, kg/ lb/ft^3	153.0	153.1	151.9	155.2	153.5	153.2
Air content, %	1.6	0.7	1.3	1.1	1.4	1.2
Concrete temp., °F	75	75	65	63	62	74
Compressive strength, 4 x 8-in. moist-cured cylinders						
3 days, psi	8220	7900	7970	10,430	7630	6170
7 days, psi	9660	10,230	10,360	13,280	11,150	9170
28 days, psi	11,460	13,300	13,070	17,000	14,530	12,270
56 days, psi	12,230	13,660	13,840	17,630	16,760	—
91 days, psi	12,800	15,170	13,950	18,030	17,350	13,310
182 days, psi	14,110	15,160	14,140	18,590	17,400	—
426 days, psi	14,910	17,100	14,560	19,230	17,290	—
1085 days, psi	16,720	17,730	16,650	21,750	19,190	—
Modulus of elasticity in compression, 4 x 8-in. moist-cured cylinders						
91 days, million psi	7.34	7.24	7.27	8.20	7.75	6.95
Drying shrinkage, 3 by 3 x 11.5-in. prisms						
7 days, millionths	193	123	100	87	137	—
28 days, millionths	400	287	240	203	233	—
90 days, millionths	573	447	383	320	340	—
369 days, millionths	690	577	520	453	467	—
1075 days, millionths	753	677	603	527	523	—

design. When an optimum or near optimum is established for each variable, it should be incorporated as the remaining variables are studied. An optimum mix design is then developed keeping in mind the economic advantages of using locally available materials. Many of the materials considerations discussed below also apply to most high-performance concretes.

Cement

Selection of cement for high-strength concrete should not be based only on mortar-cube tests but should also include tests of comparative strengths of concrete at 28, 56, and 91 days. Cement that yields the highest concrete compressive strength at extended ages (91 days) is preferable. For high-strength concrete, the cement should produce a minimum 7-day mortar-cube strength of approximately 30 MPa (4350 psi).

Trial mixtures with cement contents between 400 kg/m^3 and 550 kg/m^3 (675 lb/yd^3 to 930 lb/yd^3) should be made for each cement being considered for the project. Amounts will vary depending on target strengths. Other than decreases in sand content as cement content increases, the trial mixtures should be as nearly identical as possible.

Supplementary Cementing Materials

Fly ash, silica fume, or slag cement are frequently used and are sometimes mandatory in the production of high-performance concrete. The strength gain obtained with these supplementary cementing materials cannot be attained by using additional cement alone. The addition of these supplementary cementitious materials greatly reduces permeability and improves durability. These supplementary cementing materials are usually added at dosage rates of 5% to 20% or higher by mass of cementing material. Some specifications only permit use of up to 10% silica fume, unless evidence is available indicating that concrete produced with a larger dosage rate will have satisfactory strength, durability, and volume stability. The water-to-cementing materials ratio should be adjusted so that equal workability becomes the basis of comparison between trial mixtures. For each set of materials, there will be an optimum cement-plus-supplementary cementing materials content at which strength does not continue to increase with greater amounts and the mixture becomes too sticky to handle properly. Blended cements containing fly ash, silica fume, slag, or calcined clay can be used to make high-strength concrete with or without the addition of supplementary cementing materials.

Aggregates

In high-strength concrete, careful attention must be given to aggregate size, shape, surface texture, mineralogy, and cleanness. For each source of aggregate and concrete strength level there is an optimum-size aggregate that will yield the most compressive strength per unit of cement. To find the optimum size, trial batches should be made with 19 mm (¾ in.) and smaller coarse aggregates and varying cement contents. Many studies have found that 9.5 mm to 12.5 mm (⅜ in. to ½ in.) nominal maximum-size aggregates give optimum strength. Combining single sizes of aggregate to produce the required grading is recommended for close control and reduced variability in the concrete.

In high-strength concretes, the strength of the aggregate itself and the bond or adhesion between the paste and aggregate become important factors. Tests have shown that crushed-stone aggregates produce higher compressive strength in concrete than gravel aggregate using the same size aggregate and the same cementing materials content. This is probably due to a superior aggregate-to-paste bond when using rough, angular, crushed material. For specified concrete strengths of 70 MPa (10,000 psi) or higher, the potential of the aggregates to meet design requirements must be established prior to use.

Coarse aggregates used in high-strength concrete should be free from detrimental coatings of dust and clay. Removing dust is important since it may affect the quantity of fines and consequently the water demand of a concrete mixture. Clay may affect the aggregate-paste bond. Washing of coarse aggregates may be necessary.

The quantity of coarse aggregate in high-strength concrete should be the maximum consistent with required workability. Because of the high percentage of cementitious material in high-strength concrete, an increase in coarse aggregate content beyond values recommended in standards for normal-strength mixtures is necessary and allowable.

In high-rise buildings and in bridges, the stiffness of the structure is an important structural concern. On certain projects a minimum static modulus of elasticity has been specified as a means of increasing the stiffness of a structure (Figure 19-5). The modulus of elasticity is not necessarily proportional to the compressive strength of a concrete. There are code formulas for normal-strength concrete and suggested formulas for high-strength concrete. The modulus achievable is affected significantly by the properties of the aggregate and also by the mixture proportions (Baalbaki and others 1991). If an aggregate has the ability to produce a high modulus, then the optimum modulus in concrete can be obtained by using as much of this aggregate as practical, while still meeting workability and cohesiveness requirements. If the coarse aggregate used is a crushed rock, and manufactured fine aggregate of good quality is available from the same source, then a combination of the two can be used to obtain the highest possible modulus.

Due to the high amount of cementitious material in high-strength concrete, the role of the fine aggregate (sand) in providing workability and good finishing characteristics is not as critical as in conventional strength mixtures. Sand with a fineness modulus (FM) of about 3.0 – considered a coarse sand—has been found to be satisfactory for producing good workability and high compressive strength. For specified strengths of 70 MPa (10,000 psi) or greater, FM should be between 2.8 and 3.2 and not vary by more than 0.10 from the FM selected for the duration of the project. Finer sand, say with a FM of between 2.5 and 2.7, may produce lower-strength, sticky mixtures.

High performance lightweight concrete has been used for bridges. This concrete typically uses normal-weight sand and lightweight coarse aggregate. Its lower mass also makes it an attractive option in seismic regions (Murugesh 2008 and Gilley 2008).

Admixtures

The use of chemical admixtures such as water reducers, retarders, high-range water reducers, or superplasticizers is necessary. They make more efficient use of the large amount of cementitious material in high-strength concrete and help to obtain the lowest practical water to cementing materials ratio. Chemical admixture efficiency must be evaluated by comparing strengths of trial batches. Also, compatibility between cement and supplementary cementing materials, as well as water-reducing and other admixtures, must be investigated by trial batches. From these trial batches, it will be possible to determine the workability, setting time, and amount of water reduction for given admixture dosage rates and times of addition.

The use of air-entraining admixtures where durability in a freeze-thaw environment is required is mandatory. However, air is not necessary or desirable in high-strength concrete protected from the weather, such as interior columns and shear walls of high-rise buildings. Because air entrainment decreases concrete strength of rich mixtures, testing to establish optimum air contents and spacing factors may be required. Certain high-strength concretes may not need as much air as normal-strength concrete for equivalent frost resistance. Pinto and Hover (2001) found that non-air-entrained, high-strength concretes had good frost and deicer-scaling resistance at a water to portland cement ratio of 0.25. Burg and Ost (1994) found good frost resistance with non-air-entrained concrete containing silica fume at a water to cementing materials ratio of 0.22 (Mix No. 4 in Table 19-5). However, this was not the case with other mixtures, including a portland-only mixture with a water to cement ratio of 0.28.

High-Performance Concrete Construction

Proportioning

The trial mixture approach is best for selecting proportions for high-performance concrete. To obtain high performance, it is necessary to use a low water to cementing materials ratio and, often, a high portland cement content. The unit strength obtained for each unit of cement used in a cubic meter (yard) of concrete can be plotted as strength efficiency to evaluate mixture designs.

The water requirement of concrete increases as the fine aggregate content is increased for any given size of coarse aggregate. Because of the high cementing materials content of these concretes, the fine aggregate content can be kept low. However, even with well-graded aggregates, a low water-cementing materials ratio may result in concrete that is not sufficiently workable for the job. If a superplasticizer is not used, the design should be revised. A slump of around 200 mm (8 in.) will provide adequate workability for most applications. ACI Committee 211 (2008), Farny and Panarese (1994), Nawy (2001), and Caldarone (2009) provide additional guidance on proportioning.

Mixing

High-performance concrete has been successfully mixed in transit mixers and central mixers. However, many of these concretes that have higher cementitious contents tend to be sticky and may cause build-up in these mixers, especially when silica fume is used. Where dry, uncompacted silica fume has been batched into a mixture, "balling" of the mixture has occurred and mixing has been incomplete. In these instances it has been necessary to experiment with the charging sequence, and the percentage of each material added at each step in the batching procedure. Batching and mixing sequences should be optimized during the trial mix phase. Where truck mixing is unavoidable, the best practice is to reduce loads to 90% of the rated capacity of the trucks.

Where there is no recent history of HPC mixtures that meet specified requirements, it is essential to first make laboratory trial mixtures to establish optimum proportions. At this stage, the properties of the mixture, such as workability, air content, density, strength, and modulus of elasticity can be determined. It is also important to determine how admixtures interact and their effects on concrete properties. Once laboratory mixture proportions have been determined, field trials using full loads of concrete are essential. They should be delivered to the site or to a mock-up to establish and confirm the suitability of the batching, mixing, transporting, and placing systems to be used.

For large projects or a mass concrete structure, a trial member may be required. One or more loads of the proposed mixture is cast into a trial member or mock-up. The fresh concrete is tested for slump, air content, temperature, and density. Casting the trial member or mock-up provides the opportunity to assess the suitability of the mixture for placing, compaction, and temperature gain. The trial member or mock-up can be instrumented to record temperatures and temperature gradients. It can also be cored and tested to provide correlation with standard cylinder test results. The cores can be tested to provide the designer with in-place strength and modulus values for reference during construction. The heat characteristics of the mixture can also be determined using a computer program, and the data used to determine how curing technology should be applied to the project.

Placing, Consolidation, Finishing, and Curing

Close liaison between the contractor and the concrete producer allows concrete to be discharged rapidly after arrival at the jobsite. Final adjustment of the concrete should be supervised by the concrete producer's technicians at the site, by a concrete laboratory, or by a consultant familiar with the performance and use of high-strength concrete.

Delays in delivery and placing must be eliminated. Sometimes it may be necessary to reduce batch sizes if placing procedures are slower than anticipated. Rigid surveillance must be exercised at the jobsite to prevent any addition of retempering water. Increases in workability should only be achieved by the addition of a superplasticizer. This should be done by the supplier's technician. The contractor must be prepared to receive the concrete and understand the consequences of exceeding the specified slump and water-cementitious materials ratio.

Consolidation is very important in achieving the potential of high-performance concrete. Concrete must be vibrated as quickly as possible after placement in the forms. High-frequency vibrators should be small enough to allow sufficient clearance between the vibrating head and reinforcing steel. Over-vibration of workable normal-strength concrete often results in segregation, loss of entrained air, or both. On the other hand, high-performance concrete without a superplasticizer will be relatively stiff and contain little air. Consequently, inspectors should be more concerned with under-vibration rather than over-vibration. Most high-strength concrete, particularly very high-strength-concrete, is placed at slumps of 180 mm to 220 mm (7 in. to 9 in.). Even at these slumps, some vibration is required to ensure compaction. The amount of compaction should be determined by onsite trials.

High-performance concrete can be difficult to finish. High cementitious materials contents, large dosages of admixtures, low water contents, and air entrainment all contribute to the concrete sticking to the trowels and other finishing equipment. When this occurs, finishing activities should be minimized. The finishing sequence should be modified to include the use of a fresno trowel in place of a bullfloat.

Curing of high-performance concrete is even more important than curing normal-performance concrete. Providing adequate moisture and favorable temperature conditions are recommended for a prolonged period, particularly when 56- or 91-day concrete strengths are specified.

Additional curing considerations apply with HPC. Where very low water-cement ratios are used in flatwork (slabs and overlays), and particularly where silica fume is used in the mixture, there will be little if any bleeding before or after finishing. In these situations it is imperative that fog curing or evaporation retarders be applied to the concrete immediately after the surface has been struck off. This is necessary to avoid plastic shrinkage cracking of horizontal surfaces and to minimize crusting. Fog curing, followed by 7 days of wet curing, has proven to be very effective.

It is inevitable that some vertical surfaces, such as columns, may be difficult to cure effectively. Where projects are fast-tracked, columns are often stripped at an early age to allow raising of self-climbing form systems. Concrete is thus exposed to early drying, sometimes within eleven hours after casting. Because of limited access, providing further curing is difficult and impractical.

Tests were conducted on column concrete to determine if such early exposure and lack of curing have any harmful effects. The tests showed that for a portland cement-slag-silica fume mixture with a specified strength of 70 MPa (10,000 psi), the matrix was sound and a very high degree of impermeability to water and chloride ions had been achieved (Bickley and others 1994). Nevertheless, the best curing possible is recommended for all HPC.

The temperature history of HPC is an integral part of its curing process. Advantage should also be taken of recent developments in curing technology. Temperature increases and gradients that will occur in a concrete placement can be predicted by procedures that provide data for this purpose. With this technique, measures to heat, cool, or insulate a concrete placement can be determined and applied to significantly reduce both micro- and macro-cracking of the structure and assure durability. The increasing use of these techniques will be required in most structures using HPC to assure that the cover concrete provides long term protection to the steel, to meet the intended service life of the structure.

Temperature Control

The quality, strength, and durability of HPC are highly dependent on its temperature history from the time of delivery to the completion of curing. In principle, favorable construction and placing methods will enable: (1) a low temperature at the time of delivery; (2) the smallest possible maximum temperature after placing; (3) minimum temperature gradients after placing; and (4) a gradual reduction to ambient temperature after maximum temperature is reached. Excessively high temperatures and gradients can cause excessively fast hydration and micro- and macro-cracking of the concrete. Keeping member temperature under 70°C (158°F) internal curing temperature helps prevent delayed ettringite formation (DEF) (see Chapter 11).

It has been a practice on major high-rise structures incorporating concrete with specified strengths of 70 MPa to 85 MPa (10,000 psi to 12,000 psi) to specify a maximum delivery temperature of 18°C (64°F) (Ryell and Bickley 1987). In summertime it is possible that this limit could only be met using liquid nitrogen to cool the concrete (see Chapter 16). Experience with very-high-strength concrete suggests that a delivery temperature of no more than 25°C (77°F), preferably 20°C (68°F), should be allowed. The specifier should state the required delivery temperature.

In HPC applications such as high-rise buildings, column sizes are large enough to be classified as mass concrete. Normally, excessive heat generation in mass concrete is controlled by using a low cement content. When high-cement-content HPC mixtures are used under these conditions, other methods of controlling maximum concrete temperature must be employed. Burg and Ost (1994) recorded temperature rise for 1220-mm (4-ft) concrete cubes using the mixtures in Table 19-5. A maximum temperature rise of 9.4°C to 11.7°C for every 100 kg of cement per cubic meter of concrete (10°F to 12.5°F for every 100 lb of cement per cubic yard of concrete) was measured. Burg and Fiorato (1999) monitored temperature rise in high-strength concrete caissons; they determined that in-place strength was not affected by temperature rise due to heat of hydration.

Quality Control

A comprehensive quality-control program is required at both the concrete plant and onsite to guarantee consistent production and placement of high-performance concrete. Inspection of concreting operations from stockpiling of aggregates through completion of curing is important. Closer production control than is normally obtained on most projects is necessary. Also, routine sampling and testing of all materials is particularly necessary to control uniformity of the concrete.

While tests on concrete should always be made in strict accordance with standard procedures, some additional requirements are recommended, especially where specified strengths are 70 MPa (10,000 psi) or higher. In testing high-strength concrete, some changes and more attention to detail are required. For example, cardboard cylinder molds, which can cause lower strength-test results, should be replaced with reusable steel or plastic molds. Capping of cylinders must be done with great care using appropriate capping compounds. Lapping (grinding) the cylinder ends is an alternative to capping. For specified strengths of 70 MPa (10,000 psi) or greater, end grinding to a flatness tolerance of 0.04 mm is recommended (Calderone and Burg 2009).

The physical characteristics of a testing machine can have a major impact on the result of a compression test. It is recommended that testing machines be extremely stiff, both longitudinally and laterally.

The quality control necessary for the production of high compressive strength concrete will, in most cases, lead to low variance in test results. Strict vigilance in all aspects of quality control on the part of the producer and quality testing on the part of the laboratory are necessary on high-strength concrete projects. For concretes with specified strengths of 70 MPa (10,000 psi) or greater, the coefficient of variation is the preferred measure of quality control.

Self-Consolidating Concrete

Self-consolidating concrete (SCC), also referred to as self-compacting concrete, is able to flow and consolidate under its own weight. At the same time it is cohesive enough to fill spaces of almost any size and shape without segregation or bleeding. This makes SCC particularly useful wherever placing is difficult, such as in heavily-reinforced concrete members or in complicated formwork.

This technology, developed in Japan in the 1980s, is based on increasing the amount of fine material, for example fly ash or limestone filler, without changing the water content. This changes the rheological behavior of the concrete. SCC must have a low yield value to ensure high flowability; a low water content ensures high viscosity, so the coarse aggregate can float in the mortar without segregating. To achieve a balance between deformability and stability, the total content of particles finer than the 150 µm (No. 100) sieve is typically high, usually about 520 kg/m³ to 560 kg/m³ (880 lb/yd³ to 950 lb/yd³). Generally, the higher the required flowability of the SCC, the higher the amount of fine material needed to produce a stable mixture. However, in some cases, a viscosity-modifying admixture (VMA) can be used instead of, or in combination with, an increased fine content to stabilize the concrete mixture. High-range water reducers based on polycarboxylate ethers are typically used to plasticize the mixture. Figure 19-6 shows an example of mixture proportions used in self-consolidating concrete as compared to a conventional concrete mixture.

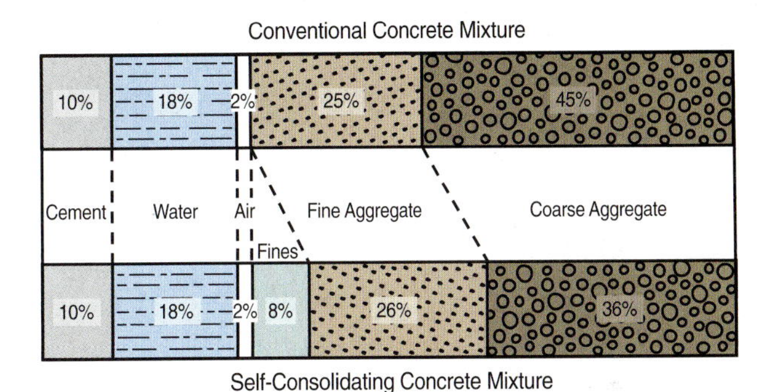

Figure 19-6. Examples of materials used in conventional concrete and self-consolidating concrete by absolute volume.

Since SCC is characterized by special fresh concrete properties, many new tests have been developed to measure flowability, viscosity, blocking tendency, self-leveling, and stability of the mixture (Skarendahl and Peterson 1999 and Ludwig and others 2001). The slump flow test (ASTM C1611, *Standard Test Method for Slump Flow of Self-Consolidating Concrete*) is performed to measure filling ability and stability. The test is performed similarly to the conventional slump test (ASTM C143). However, instead of measuring the slumping distance vertically, the mean diameter of the resulting concrete patty is measured horizontally. This number is recorded as the slump flow. The J-Ring test (ASTM C1621, *Standard Test Method for Passing Ability of Self-Consolidating Concrete by J-Ring*) measures passing ability. The J-Ring consists of a ring of reinforcing bar such that it will fit around the base of a standard slump cone (Figure 19-7). The slump flow with and without the J-Ring is measured, and the difference calculated.

ASTM C1610, *Standard Test Method for Static Segregation of Self-Consolidating Concrete Using Column Technique*, evaluates static stability of a concrete mixture by quantifying aggregate segregation. A column is filled with concrete and allowed to sit after placement. The column is then separated into three pieces. Each section is removed individually and the concrete from that section is washed

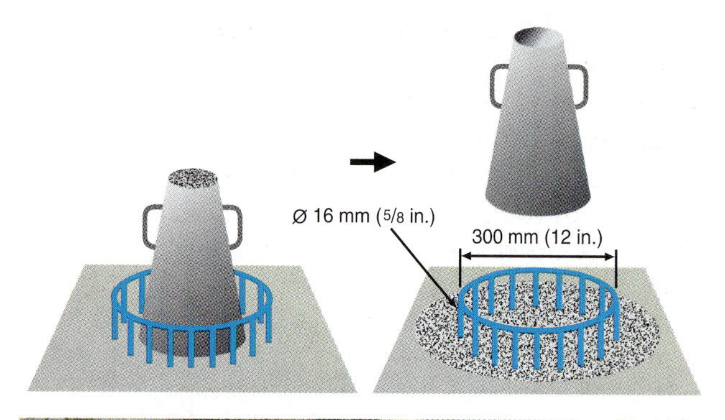

Figure 19-7. J-ring test. Photo courtesy of VDZ.

over a 4.75 mm (No. 4) sieve and the retained aggregate weighed. A non-segregating mixture will have a consistent aggregate mass distribution in each section. A segregating mixture will have higher concentrations of aggregate in the lower sections.

An earlier assessment of the segregation resistance is ASTM C1712, *Standard Test Method for Rapid Assessment of Static Segregation Resistance of Self-Consolidating Concrete Using Penetration Test*. A 45 g hollow cylinder device is placed on top of an inverted slump mold containing SCC. The distance the weight sinks in 30 seconds correlates to the static segregation resistance of the mixture. If the penetration depth is less than 10 mm, the mixture is considered segregation resistant. A penetration value above 25 mm signals a mixture that is probably prone to segregation.

Strength and durability of well-designed SCC are almost similar to conventional concrete. Without proper curing, SCC tends to have higher plastic shrinkage cracking than conventional concrete (Grube and Rickert 2001). Research indicates greater tensile creep for SCC, resulting in a reduced tendency to crack (Bickley and Mitchell 2001). The use of fly ash as a filler compared to limestone as a filler seems to be advantageous; it results in higher strength and higher chloride resistance (Bouzoubaa and Lachemi 2001 and Ludwig and others 2001).

The production of SCC is more expensive than regular concrete and it is difficult to keep SCC in the desired consistency over a long period. However, construction time is shorter and production of SCC is environmentally friendly (little noise, no vibration). Furthermore, SCC produces a good surface finish. These advantages make SCC particularly attractive for use in precasting plants.

SCC has been used successfully in tall buildings including the Trump International Hotel and Tower in Chicago, Illinois (Figure 19-8). Cast-in-place SCC has been used in the construction of inclined pylons for a cable-stayed pedestrian bridge, in Virginia (Lwin 2008). It has also been used for many drilled shaft foundations for bridges, including those of the new I-35W St. Anthony Falls Bridge in Minneapolis, Minnesota (Phipps 2008). SCC was also used to speed construction of precast prestressed bulb-tee girders in the replacement of the Biloxi Bay Bridge in Mississippi after the original bridge was destroyed by Hurricane Katrina (Carr 2008).

SCC has also been successfully used in a number of rehabilitation projects in Canada (Bickley and Mitchell 2001). Refer to ACI Committee 237, Khayat and Mitchell (2009), and Szecsy and Mohler (2009) for more information on SCC.

Figure 19-8. Completed in 2009, the 92 story Trump International Hotel and Tower, which was constructed with high-performance SCC, is the tallest building (at 1170 ft [1389 ft to the top of the spire]) built in North America since the completion of the Sears Tower in 1974.

Ultra-High Performance Concrete

Ultra-high performance concrete (UHPC) is also known as reactive powder concrete. Reactive-powder concrete was first patented by a French construction company in 1994. It is characterized by high strength and very low permeability, obtained by optimized particle packing and by a low water content.

The properties of UHPC are achieved by: (1) eliminating the coarse aggregates – only very fine powders are used (sand, crushed quartz, and silica fume), all with particle sizes between 0.02 and 300 µm; (2) optimizing the grain size distribution to densify the mixture; (3) using post-set heat-treatment to improve the microstructure; (4) addition of steel and synthetic fibers (about 2% by volume); and (5) use of superplasticizers to decrease the water to cement ratio – usually to less than 0.2 – while improving the rheology of the paste. See Figure 19-9 for a typical fresh UHPC.

Figure 19-9. Freshly-mixed ultra-high performance concrete.

The compressive strength of UHPC is typically around 200 MPa (29,000 psi), but can be produced with compressive strengths up to 810 MPa (118,000 psi) (Semioli 2001). However, the low comparative tensile strength requires prestressing reinforcement in severe structural service. Table 19-6 compares hardened concrete properties of RPC with those of an 80-MPa (11,600-psi) concrete.

The Federal Highway Administration (FHWA) studied multiple properties of UHPC, namely compressive and tensile strengths, creep and shrinkage, chloride ion penetration, and freeze-thaw durability. The 28-day compressive strengths ranged from 126 to 193 MPa (18,000 to 28,000 psi), depending upon whether or not a secondary heat treatment was used to further develop compressive strength. The tensile strength was approximately 6.2 MPa (900 psi) without secondary heat treatment and 9.0 MPa (1,300 psi) after secondary heat treatment. The UHPC showed excellent resistance to chloride ion penetration, exhibited good long-term creep and shrinkage behavior, and held up well in freeze-thaw testing (Graybeal 2006).

UHPC has been used in the beams and decks of U.S. bridges. The first bridge in the U.S. to use UHPC was the Mars Hill Bridge in Wapello County, Iowa. This 33.5-m (110-ft.) bridge used three 107-cm (42-in.) modified Iowa bulb-tee girders, and opened to traffic in 2006. The second bridge was the Cat Point Creek Bridge in Richmond County, Virginia. This ten-span bridge opened to traffic in 2008 and contained one span with five UHPC bulb-tee girders. The third bridge was the Jakway Park Bridge in

Figure 19-10. The Sherbrooke footbridge in Quebec, built in 1997, is North America's first reactive-powder ultra-high performance concrete structure.

Table 19-6. Typical Mechanical Properties of Reactive Powder Concrete (RPC) Compared to an 80-MPa Concrete (Perry 1998)

Property	Unit	80 MPa	RPC
Compressive strength	MPa (psi)	80 (11,600)	200 (29,000)
Flexural strength	MPa (psi)	7 (1000)	40 (5800)
Tensile strength	MPa (psi)		8 (1160)
Modulus of Elasticity	GPa (psi)	40 (5.8 x 10^6)	60 (8.7 x 10^6)
Fracture Toughness	10^3 J/m^2	<1	30
Freeze-thaw, ASTM C666	RDF	90	100
Carbonation depth: 36 days in CO_2	mm	2	0
Abrasion	10^{-12} m^2/s	275	1.2

Buchanan County, Iowa. This 15.7-m (51.5-ft.) long bridge used UHPC pi-girders. (Graybeal 2009), (Bierwagen 2009). Other uses for UHPC in bridges include waffle deck panels in Iowa and cast-in-place UHPC connections between full-depth deck panels in New York. See Li and Li (2010) for repair applications and Li (2010) for infrastructure applications. UHPC has also been used in pedestrian bridges (Figure 19-10) (Bickley and Mitchell 2001 and Semioli 2001). The low porosity of RPC also gives excellent durability and transport properties, which makes it a suitable material for the storage of nuclear waste (Matte and Moranville 1999). A low-heat type of reactive-powder concrete has been developed to meet needs for mass concrete pours for nuclear reactor foundation mats and underground containment of nuclear wastes (Gray and Shelton 1998).

References

AASHTO, LRFD *Bridge Design Specifications*, 5th Ed., American Association of State Highway and Transportation Officials, Washington, D.C., 2010.

ACI Committee 211, *Guide for Selecting Proportions for High-Strength Concrete Using Portland Cement and Other Cementitious Materials*, ACI 211.4-08, American Concrete Institute, Farmington Hills, Michigan, 2008, 25 pages.

ACI Committee 237, *Self-Consolidating Concrete*, ACI 237-07, American Concrete Institute, Farmington Hills, Michigan, 2007, 32 pages.

ACI Committee 363, *Guide to Quality Control and Testing of High-Strength Concrete*, 363.2R-98, American Concrete Institute, Farmington Hills, Michigan, 1998, 18 pages.

ASCE, *High-Performance Construction Materials and Systems*, Technical Report 93-5011, American Society of Civil Engineers, New York, April 1993.

Baalbaki, W.; Benmokrance, B.; Chaallal, O.; and Aïtcin, P.-C., "Influence of Coarse Aggregates on Elastic Properties of High Performance Concrete," *ACI Materials Journal*, Vol. 88, No. 5, 1991, pages 449 to 503.

Berra, M.; Ferrara, G.; and Tavano, S., "Behaviour of High-Erosion Resistant Silica Fume: Mortars for Repair of Hydraulic Structures," *Fly ash, silica fume, slag and natural pozzolans in concrete: Proceedings, Third International Conference, Trondheim, Norway*, (SP-114), Vol. 2, American Concrete Institute, Farmington Hills, Michigan, 1989, pages 827 to 847.

Bickley, John A., and Mitchell, Denis, *A State-of-the-Art Review of High Performance Concrete Structures Built in Canada: 1990—2000*, Cement Association of Canada, Ottawa, Ontario, May 2001, 122 pages.

Bickley, J.A.; Ryell, J.; Rogers, C.; and Hooton, R.D., "Some Characteristics of High-Strength Structural Concrete: Part 2," *Canadian Journal for Civil Engineering, National Research Council of Canada*, December 1994.

Bickley, J.A.; Sarkar, S.; and Langlois, M., "The CN Tower", *Concrete International*, American Concrete Institute, Farmington Hills, Michigan, August 1992, pages 51 to 55.

Bierwagen, D., "UHPC in Iowa", Presentation, Iowa Department of Transportation, Office of Bridges and Structures, September 2009.

Bouzoubaa, Nabil; Bilodeau, Alain; Sivasundaram, Vasanthy; Fournier, Benoit; and Golden, Dean M., "Development of Ternary Blends for High-Performance Concrete," *ACI Materials Journal*, Vol. 101, No. 1, January-February 2004, pages 19 to 29.

Bouzoubaa, N., and Lachemi, M., "Self-compacting concrete incorporating high volumes of class F fly ash. Preliminary results," Vol. 31, *Cement and Concrete Research*, Pergamon-Elsevier Science, Oxford, 2001, pages 413 to 420.

Burg, R.G., and Fiorato, A.E., *High-Strength Concrete in Massive Foundation Elements*, Research and Development Bulletin RD117, Portland Cement Association, 1999, 28 pages.

Burg, R.G., and Ost, B.W., *Engineering Properties of Commercially Available High-Strength Concretes (Including Three-Year Data)*, Research and Development Bulletin RD104, Portland Cement Association, 1994, 62 pages.

Caldarone, Michael A., *High-Strength Concrete: A Practical Guide*, 1st edition, Taylor and Francis, London, 2009, 272 pages.

Caldarone, M., and Burg, R.G., "Importance of End Surface Preparation when Testing High Strength Concrete Cylinders," *HPC Bridge Views*, Issue 57, September/October 2009.

Caldarone, Michael A.; Taylor, Peter C.; Detwiler, Rachel J.; and Bhide, Shrinivas B., *Guide Specification for High Performance Concrete for Bridges*, EB233, 1st edition, Portland Cement Association, Skokie, Illinois, 2005, 64 pages.

Carr, Mitchell K., "Self-Consolidating Concrete for Beams Speeds Biloxi Bridge Construction," *HPC Bridge Views*, Issue 50, July/August 2008.

Dunaszegi, Laszlo, "HPC for Durability of the Confederation Bridge," *HPC Bridge Views*, Federal Highway Administration and National Concrete Bridge Council, Portland Cement Association, September/October 1999, page 2.

Farny, James A., and Panarese, William C., *High-Strength Concrete*, EB114, Portland Cement Association, 1994, 60 pages.

Ferraris, Chiara F., and Lobo, Colin L., "Processing of HPC," *Concrete International*, American Concrete Institute, Farmington Hills, Michigan, April 1998, pages 61 to 64.

Gagne, R.; Chagnon, D.; and Parizeau, R., "L'Utilization du Beton a Haute Performance dans l'Industrie Agricole," *Proceedings of Seminar at Concrete Canada Annual Meeting*, Sherbrooke, Quebec, October 1994, pages 23 to 33.

Gagne, R.; Pigeon, M.; and Aïtcin, P.-C., "Durabilite au gel des betons de hautes performance mecaniques," *Materials and Structures*, Vol. 23, 1990, pages 103 to 109.

Gaines, Mark A., and Tragesser, Michelle L., "Hood Canal Bridge West-Half Retrofit and East-Half Replacement," *HPC Bridge Views*, Issue 48, March/April 2008.

Gilley, Rex, "Lightweight Concrete for the Route 33 Bridge over the Mattaponi River," *HPC Bridge Views*, Issue 49, May/June 2008.

Goodspeed, Charles H.; Vanikar, Suneel; and Cook, Raymond A., "High-Performance Concrete Defined for Highway Structures," *Concrete International*, American Concrete Institute, Farmington Hills, Michigan, February 1996, pages 62 to 67.

Gray, M.N., and Shelton, B.S., "Design and Development of Low-Heat, High-Performance Reactive Powder Concrete," *Proceedings, International Symposium on High-Performance and Reactive Powder Concretes*, Sherbrooke, Quebec, August 1998, pages 203 to 230.

Graybeal, Benjamin, *Material Property Characterization of Ultra-High Performance Concrete*, Federal Highway Administration, Report No. FHWA-HRT-06-103, August 2006.

Graybeal, Benjamin, "UHPC Making Strides," *Public Roads*, Vol. 72, Number 4, January/February 2009.

Grove, James D., "Blanket Curing to Promote Early Strength Concrete," *Concrete and Construction—New Developments and Management*, Transportation Research Record, 1234, Transportation Research Board, Washington, D.C., 1989.

Grube, Horst, and Rickert, Joerg, "Self compacting concrete—another stage in the development of the 5-component system of concrete," *Betontechnische Berichte* (*Concrete Technology Reports*), Verein Deutscher Zementwerke, Düsseldorf, 2001, pages 39 to 48.

Helland, S., "High Strength Concrete Used in Highway Pavements," *Proceedings of the Second International Symposium on High Strength Concrete*, SP-121, American Concrete Institute, Farmington Hills, Michigan, 1990, pages 757 to 766.

Higareda, Jose, "HPC for the Angeles Crest Bridge 1," *HPC Bridge Views*, Issue 62, July/August 2010.

Holland, T.C.; Krysa, Anton; Luther, Mark D.; and Liu, Tony C., "Use of Silica-Fume Concrete to Repair Abrasion-Erosion Damage in the Kinzua Dam Stilling Basin," *Proceedings of the Second International Conference on Fly Ash, Silica Fume, Slag and Natural Pozzolans in Concrete*, SP-91, American Concrete Institute, Farmington Hills, Michigan, 1986, pages 841 to 864.

Khayat, Kamal Henri, and Mitchell, Denis, "Self-Consolidating Concrete for Precast, Prestressed Concrete Bridge Elements," NCHRP Report 628, National Cooperative Highway Research Program, Transportation Research Board, Washington, D.C., 2009.

Knutson, Martin, and Riley, Randall, "Fast-Track Concrete Paving Opens Door to Industry Future," *Concrete Construction*, Addison, Illinois, January 1987, pages 4 to 13.

Lane, S.N., "HPC Lessons Learned and Future Directions," IABMAS 2010: *The Fifth International Conference on Bridge Maintenance, Safety and Management*, International Association on Bridge Maintenance and Safety, Philadelphia, Pennsylvania, July 11 to 15, 2010.

Li, M., and Li, V.C., "High-Early-Strength ECC for Rapid Durable Repair - Material Properties", *ACI Materials Journal*, American Concrete Institute, Farmington Hills, Michigan, June, 2010.

Li, V.C., "High-Ductility Concrete for Resilient Infrastructure", *Advanced Materials Journal*, July, 2010.

Ludwig, Horst-Michael; Weise, Frank; Hemrich, Wolfgang; and Ehrlich, Norbert, "Self compacting concrete—principles and practice," *Betonwerk- und Fertigteil-Technik (Concrete Plant+Precast Technology)*, Wiesbaden, Germany, June 2001, pages 58 to 67.

Lwin, M. Myint, "Self-Consolidating Concrete Bridge Applications are Expanding," *HPC Bridge Views*, Issue 50, July/August 2008.

Matte, V., and Moranville, M., "Durability of Reactive Powder Composites: Influence of Silica Fume on Leaching Properties of Very Low Water/Binder Pastes," *Cement and Concrete Composites 21*, 1999, pages 1 to 9.

Murugesh, Ganapathy, "Lightweight Concrete and the New Benicia-Martinez Bridge," *HPC Bridge Views*, Issue 49, May/June 2008.

Nawy, Edward G., *Fundamentals of High-Performance Concrete*, John Wiley and Sons, New York, 2001.

Perry, V., "Industrialization of Ultra-High Performance Ductile Concrete," *Symposium on High-Strength/High-Performance Concrete*, University of Calgary, Alberta, November 1998.

Phipps, Alan R., "HPC for 100-Year Life Span," *HPC Bridge Views*, Issue 52, November/December 2008.

Pinto, Roberto C.A., and Hover, Kenneth C., *Frost and Scaling Resistance of High-Strength Concrete*, Research and Development Bulletin RD122, Portland Cement Association, 2001, 75 pages.

Pyle, Tom, Caltrans, Personal communication on early strength concretes used in California, 2001.

Russell, Henry G., "ACI Defines High-Performance Concrete," *Concrete International*, American Concrete Institute, Farmington Hills, Michigan, February 1999, pages 56 to 57.

Russell, Henry G., and Ozyildirim, H. Celik, "Revising High-Performance Concrete Classifications," *Concrete International*, American Concrete Institute, Vol. 28, No. 28, August 2006, pages 43 to 49.

Russell, H.; Miller, R.; Ozyildirim, H.C.; and Tadros, M., *Compilation and Evaluation of Results from High Performance Concrete Bridge Projects*, Vol. 1: Final Report, Federal Highway Administration, Report No. FHWA-HRT-05-056, October 2006.

Ryell, J., and Bickley, J.A., "Scotia Plaza: High Strength Concrete for Tall Buildings," *Proceedings International Conference on Utilization of High Strength Concrete*, Stavanger, Norway, June 1987, pages 641 to 654.

Semioli, William J., "The New Concrete Technology," *Concrete International*, American Concrete Institute, Farmington Hills, Michigan, November 2001, pages 75 to 79.

Skarendahl, Å., and Peterson, Ö., *Self-Compacting Concrete —Proceedings of the first international RILEM-Symposium*, Stockholm, 1999, 786 pages.

Szecsy, Richard, and Mohler, Nathaniel, *Self-Consolidating Concrete*, IS546D, Portland Cement Association, Skokie, Illinois, 2009, 24 pages.

Tachitana, D.; Imai, M.; Yamazaki, N.; Kawai, T.; and Inada, Y., "High Strength Concrete Incorporating Several Admixtures," *Proceedings of the Second International Symposium on High Strength Concrete*, SP-121, American Concrete Institute, Farmington Hills, Michigan, 1990, pages 309 to 330.

Thomas, M.; Fournier, B.; and Folliard, K., *Report on Determining the Reactivity of Concrete Aggregates and Selecting Appropriate Measures for Preventing Deleterious Expansion in New Concrete Construction*, Federal Highway Administration, Report No. FHWA-HIF-09-001, April 2008.

CHAPTER 20
Special Types of Concrete

Special types of concrete are those with out-of-the-ordinary properties or those produced by unusual techniques. Concrete is by definition a composite material consisting essentially of a binding medium and aggregate particles, and it can take many forms. Table 20-1 lists many special types of concrete. In many cases the terminology of the listing describes the use, property, or condition of the concrete (brand names are not given). Some of the more common special types of concretes are discussed in this chapter.

Structural Lightweight Aggregate Concrete

Structural lightweight aggregate concrete is similar to normal-weight concrete except that it has a lower density. It is made with lightweight aggregates or with a combination of lightweight and normal-weight aggregates. The term "sand lightweight" refers to concrete made with coarse lightweight aggregate and natural sand.

According to ACI 213 (2003), structural lightweight concrete has an equilibrium density in the range of

Table 20-1. Some Special Types of Concrete

Special types of concrete made with portland cement		
Architectural concrete	High-early-strength concrete	Roller-compacted concrete
Autoclaved cellular concrete	High-performance concrete	Sawdust concrete
Centrifugally cast concrete	High-strength concrete	Self-compacting concrete
Colloidal concrete	Insulating concrete	Shielding concrete
Colored concrete	Latex-modified concrete	Shotcrete
Controlled-density fill	Low-density concrete	Shrinkage-compensating concrete
Cyclopean (rubble) concrete	Mass concrete	Silica-fume concrete
Dry-packed concrete	Moderate-strength lightweight concrete	Soil-cement
Epoxy-modified concrete	Nailable concrete	Stamped concrete
Exposed-aggregate concrete	No-slump concrete	Structural lightweight concrete
Ferrocement	Pervious (porous) concrete	Superplasticized concrete
Fiber concrete	Photocatalytic Concrete	Terrazzo
Fill concrete	Polymer-modified concrete	Tremie concrete
Flowable fill	Pozzolan concrete	Ultra High Performance Concrete
Flowing concrete	Precast concrete	Vacuum-treated concrete
Fly-ash concrete	Prepacked concrete	Vermiculite concrete
Gap-graded concrete	Preplaced aggregate concrete	White concrete
Geopolymer concrete	Reactive-powder concrete	Zero-slump concrete
Heavyweight concrete	Recycled concrete	
Special types of concrete not using portland cement		
Acrylic concrete	Gypsum concrete	Polymer concrete
Aluminum phosphate concrete	Latex concrete	Potassium silicate concrete
Calcium aluminate concrete	Magnesium phosphate concrete	Sodium silicate concrete
Epoxy concrete	Methyl methacrylate (MMA) concrete	Sulfur concrete
Furan concrete	Polyester concrete	

For more information on special types of concrete, please refer to http://www.concrete.org/Technical/CCT/ACI-Terminology.aspx

1120 kg/m^3 to 1920 kg/m^3 (70 lb/ft^3 to 120 lb/ft^3) and a minimum 28-day compressive strength of 17 MPa (2500 psi). For comparison, normal-weight concrete containing regular sand, gravel, or crushed stone has a density in the range of 2080 kg/m^3 to 2480 kg/m^3 (130 lb/ft^3 to 155 lb/ft^3). The density of structural lightweight aggregate concrete can be measured by following ASTM C567, *Standard Test Method for Determining Density of Structural Lightweight Concrete*. Structural lightweight aggregate concrete is used primarily to reduce the dead-load weight in concrete members, such as floors in high-rise buildings.

Structural Lightweight Aggregates

Structural lightweight aggregates are usually classified according to their production process because various processes produce aggregates with somewhat different properties. Processed structural lightweight aggregates should meet the requirements of ASTM C330, *Standard Specification for Lightweight Aggregates for Structural Concrete*, which includes:

- Rotary kiln expanded clays (Figure 20-1), shales, and slates
- Pelletized or extruded fly ash
- Expanded slags

Figure 20-1. Expanded clay.

Structural lightweight aggregates can also be produced by processing other types of material, such as naturally occurring pumice and scoria.

Structural lightweight aggregates have bulk densities significantly lower than normal-weight aggregates, ranging from 560 kg/m^3 to 1120 kg/m^3 (35 lb/ft^3 to 70 lb/ft^3), compared to 1200 kg/m^3 to 1760 kg/m^3 (75 lb/ft^3 to 110 lb/ft^3) for normal-weight aggregates. These aggregates may absorb 5% to 20% water by weight of dry material. To control the uniformity of structural lightweight concrete mixtures, the aggregates are prewetted (but not saturated) prior to batching.

Compressive Strength

The compressive strength of structural lightweight concrete can be related to the cementitious content at a given slump and air content, rather than to a water-to-cementitious materials ratio. This is due to the difficulty in determining how much of the total mix water is absorbed into the aggregate and thus not available for reaction with the cement. ACI 211.2 provides guidance on the relationship between compressive strength and cementitious content. Typical compressive strengths range from 20 MPa to 35 MPa (3000 psi to 5000 psi). High-strength concrete can also be made with structural lightweight aggregates.

In well-proportioned mixtures, the cementitious material content and strength relationship is fairly constant for a particular source of lightweight aggregate. However, the relationship will vary from one aggregate source or type to another. When information on this relationship is not available from the aggregate manufacturer, trial mixtures with varying cementitious materials contents are required to develop a range of compressive strengths, including the strength specified.

Figure 20-2 shows the relationshipbetween cement content and compressive strength. An example of a 28-MPa (4000-psi) structural lightweight concrete mixture with an equilibrium density of about 1800 kg/m^3 (112 lb/ft^3), a combination of natural sand and gravel, and a lightweight rotary kiln expanded clay coarse aggregate follows:

- 356 kg (600 lb) Type I portland cement
- 534 kg (900 lb) sand, ovendry
- 320 kg (540 lb) gravel (12.5 to 2.36 mm [½ in. to #8]), ovendry
- 356 kg (600 lb) lightweight aggregate (9.5 mm to 600 µm [⅜ in. to #30]), ovendry
- 172 kg (290 lb) mix water added
- 0.7 L (20 oz) water-reducing admixture
- 0.09 L (2.5 oz) air-entraining admixture
- 1 m^3 (1 yd^3) yield
- slump—75 mm (3 in.)
- air content—6%

Material proportions vary significantly for different materials and strength requirements.

Entrained Air

The use of entrained air can help provide lower density concrete mixtures. As with normal-weight concrete, entrained air in structural lightweight concrete ensures resistance to freezing and thawing and to deicer applications. It also improves workability, reduces bleeding and segregation, and may compensate for minor grading deficiencies in the aggregate.

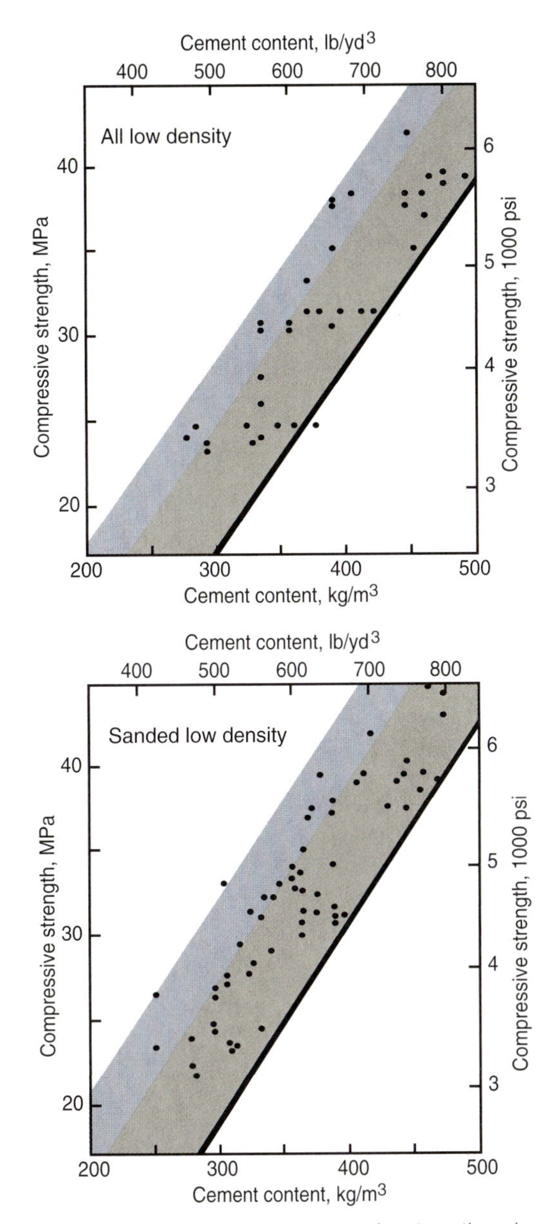

Figure 20-2. Relationship between compressive strength and cement content of field structural lightweight concrete using (top) lightweight fine aggregate and coarse aggregate or (bottom) lightweight coarse aggregate and normal-weight fine aggregate (data points represent actual project strength results using a number of cement and aggregate sources) (ACI 211.2).

The amount of entrained air should be sufficient to provide good workability to the plastic concrete and adequate freeze-thaw resistance to the hardened concrete. Air contents are generally between 5% and 8%, depending on the maximum size of coarse aggregate (and resulting paste content) used and the exposure conditions. Testing for air content should be performed by ASTM C173, *Standard Test Method for Air Content of Freshly Mixed Concrete by the Volumetric Method* (AASHTO T 196). The freeze-thaw durability is also significantly improved if structural lightweight concrete is allowed to dry before exposure to a freeze-thaw environment.

Specifications

Many suppliers of lightweight aggregates for use in structural lightweight aggregate concrete have information on suggested specifications and mixture proportions pertaining specifically to their manufactured product. Specifications for structural concrete typically state a minimum compressive strength, a maximum density, a maximum slump, and an acceptable range in air content.

Mixing

In general, mixing procedures for structural lightweight concrete are similar to those for normal-weight concrete; however, some of the more absorptive aggregates may require prewetting before use. Water added at the batching plant should be sufficient to produce the specified slump at the jobsite. Measured slump at the batch plant will generally be appreciably higher than the slump at the site. Pumping can especially aggravate slump loss if the aggregates are not saturated prior to pumping.

Workability and Finishability

Structural lightweight aggregate concrete mixtures can be proportioned to have the same workability, finishability, and general appearance as a properly proportioned normal-weight concrete mixture. Sufficient cement paste must be present to coat each particle, and coarse-aggregate particles should not separate from the mortar. Enough fine aggregate is needed to keep the freshly mixed concrete cohesive. If aggregate is deficient in minus 600 µm (No. 30) sieve material, finishability may be improved by: entraining air, using a portion of natural sand, increasing cement content, or using satisfactory mineral fines.

Slump

Due to lower aggregate density, structural lightweight aggregate concrete does not slump as much as normal-weight concrete with the same workability. A lightweight air-entrained mixture with a slump of 50 mm to 75 mm (2 in. to 3 in.) can be placed under conditions that would require a slump of 75 mm to 125 mm (3 in. to 5 in.) for normal-weight concrete. It is seldom necessary to exceed slumps of 125 mm (5 in.) for normal placement of structural lightweight aggregate concrete. With higher slumps, the large aggregate particles tend to float to the surface, making finishing difficult.

Vibration

As with normal-weight concrete, vibration can be used effectively to consolidate lightweight concrete. The same frequencies commonly used for normal-weight concrete are recommended. The length of time for proper consolidation varies, depending on mix characteristics. Excessive vibration causes segregation by forcing large aggregate particles to the surface.

Placing, Finishing, and Curing

Structural lightweight aggregate concrete is generally easier to handle and place than normal-weight concrete. A slump of 50 mm to 100 mm (2 in. to 4 in.) produces the best results for finishing. Greater slumps may cause segregation, delay finishing operations, and result in rough, uneven surfaces.

If pumped concrete is being considered, the specifier, suppliers, and contractor should all be consulted for field trials using the pump and mixture planned for the project. Adjustments to the mixture may be necessary; pumping pressure causes the aggregate to absorb more water, thus reducing the slump and increasing the density of the concrete. The relationship between the slump at the point of delivery (truck) and the point of placement (pump) should be correlated.

Finishing operations should be started earlier than for comparable normal-weight concrete, but finishing too early may cause surface defects. A minimum amount of floating and troweling should be done; magnesium finishing tools are preferred.

The same curing practices should be used for lightweight concrete as for normal-weight concrete. The two methods commonly used in the field are water curing (ponding, sprinkling, or using wet coverings) and preventing loss of moisture from the exposed surfaces (covering with waterproof paper, plastic sheets, or sealing with liquid membrane-forming compounds). Generally, 7 days of curing are adequate for ambient air temperatures above 10°C (50°F). Additionally, fully prewetted lightweight concrete aggregates provide a source of moisture for internal curing. See Chapter 15 and Bohan and Ries (2008) for more on internal curing with lightweight aggregate concrete. Lighweight aggregate concrete will have a longer period of drying time than normalweight concrete. If the concrete is to receive a moisture sensitive floor covering, this must be taken into consideration (see Kanare 2008).

Insulating and Moderate-Strength Lightweight Concretes

Insulating concrete is a lightweight concrete with an ovendry density of 800 kg/m^3 (50 lb/ft^3) or less. It is made with cementing materials, water, air, and with or without aggregate and chemical admixtures. The ovendry density ranges from 240 kg/m^3 to 800 kg/m^3 (15 lb/ft^3 to 50 lb/ft^3) and the 28-day compressive strength is generally between 0.7 MPa and 7 MPa (100 psi and 1000 psi). Cast-in-place insulating concrete is used primarily for thermal and sound insulation, roof decks, fill for slab-on-grade subbases, leveling courses for floors or roofs, firewalls, and underground thermal conduit linings.

Moderate-strength lightweight concrete has a density of 800 kg/m^3 to 1900 kg/m^3 (50 lb/ft^3 to 120 lb/ft^3) and has a compressive strength of approximately 7 MPa to 17 MPa (1000 psi to 2500 psi). It is made with cementing materials, water, air, and with or without aggregate and chemical admixtures. At lower densities, it is typically used as fill for thermal and sound insulation of floors, walls, and roofs. In these cases and is referred to as fill concrete. At higher densities it is used in cast-in-place walls, floors and roofs, and precast wall and floor panels.

For discussion purposes, insulating and moderate-strength lightweight concretes can be grouped as follows:

Group I is made with expanded aggregates such as perlite, vermiculite, or expanded polystyrene beads. Ovendry concrete densities using these aggregates generally range between 240 kg/m^3 and 800 kg/m^3 (15 lb/ft^3 and 50 lb/ft^3). This group is used primarily in insulating concrete. Some moderate-strength concretes can also be made from aggregates in this group.

Group II is made with aggregates manufactured by expanding, calcining, or sintering materials such as blast-furnace slag, clay, diatomite, fly ash, shale, or slate, or by processing natural materials such as pumice, scoria, or tuff. Ovendry concrete densities using these aggregates can range between 720 kg/m^3 and 1440 kg/m^3 (45 lb/ft^3 and 90 lb/ft^3). Aggregates in this group are used in moderate-strength lightweight concrete and some of these materials (expanded slag, clay, fly ash, shale, and slate) are also used in both moderate-strength and structural lightweight concrete (up to about 1900 kg/m^3 or 120 lb/ft^3 equilibrium density).

Group III concretes are made by incorporating into a cement paste or cement-sand mortar a uniform cellular structure of air voids that is obtained with preformed foam (ASTM C869, *Standard Specification for Foaming Agents Used in Making Preformed Foam for Cellular Concrete*), formed-in-place foam, special foaming agents, or polystyrene beads. Ovendry densities ranging between 240 kg/m^3 to 1900 kg/m^3 (15 lb/ft^3 to 120 lb/ft^3) are obtained by substitution of air voids for some or all of the aggregate particles. Air voids can consist of up to 80% of the volume. Cellular concrete can be made to meet the requirements of both insulating and moderate strength lightweight concrete.

Aggregates used in Groups I and II should meet the requirements of ASTM C332, *Standard Specification for Lightweight Aggregates for Insulating Concrete*. These aggregates have dry densities in the range of from 96 kg/m^3 to 1120 kg/m^3 (6 lb/ft^3 to 70 lb/ft^3) down to 16 kg/m^3 (1 lb/ft^3) for expanded polystyrene beads.

Table 20-2. Examples of Lightweight Insulating Concrete Mixtures

Type of concrete	Ratio: portland cement to aggregate by volume	Ovendry density, kg/m³ (pcf)	Type I portland cement, kg/m³ (lb/yd³)	Water-cement ratio, by mass	28-day compressive strength, MPa (psi), 150 x 300-mm (6 x 12-in.) cylinders
Perlite*	1:4	480 to 608 (30 to 38)	362 (610)	0.94	2.75 (400)
	1:5	416 to 576 (26 to 36)	306 (516)	1.12	2.24 (325)
	1:6	352 to 545 (22 to 34)	245 (414)	1.24	1.52 (220)
	1:8	320 to 512 (20 to 32)	234 (395)	1.72	1.38 (200)
Vermiculite*	1:4	496 to 593 (31 to 37)	380 (640)	0.98	2.07 (300)
	1:5	448 to 496 (28 to 31)	295 (498)	1.30	1.17 (170)
	1:6	368 to 464 (23 to 29)	245 (414)	1.60	0.90 (130)
	1:8	320 to 336 (20 to 21)	178 (300)	2.08	0.55 (80)
Polystyrene:** sand:					
0 kg/m³ (0 lb/yd³)	1:3.4	545 (34)‡	445 (750)	0.40	2.24 (325)
73 kg/m³ (124 lb/yd³)	1:3.1	625 (39)‡	445 (750)	0.40	2.76 (400)
154 kg/m³ (261 lb/yd³)	1:2.9	725 (44)‡	445 (750)	0.40	3.28 (475)
200 kg/m³ (338 lb/yd³)	1:2.5	769 (48)‡	474 (800)	0.40	3.79 (550)
Cellular* (neat cement)	—	625 (39)	524 (884)	0.57	2.41 (350)
	—	545 (34)	468 (790)	0.56	1.45 (210)
	—	448 (28)	396 (668)	0.57	0.90 (130)
	—	368 (23)	317 (535)	0.65	0.34 (50)
Cellular† (sanded)††	1:1	929 (58)	429 (724)	0.40	3.17 (460)
	1:2	1250 (78)	373 (630)	0.41	5.66 (820)
	1:3	1602 (100)	360 (602)	0.51	15.10 (2190)

* Reichard (1971).
** Source: Hanna (1978). The mix also included air entrainment and a water-reducing agent.
† Source: Gustaferro and others (1970).
†† Dry-rodded sand with a bulk density of 1600 kg/m³ (100 pcf).
‡ Air-dry density at 28 days, 50% relative humidity.

Mixture Proportions

Examples of mixture proportions for Group I and III concretes appear in Table 20-2. In Group I, air contents may be as high as 25% to 35%. The air-entraining agent can be prepackaged with the aggregate or added at the mixer. Because of the absorptive nature of the aggregate, the volumetric method (ASTM C173 or AASHTO T 196) should be used to measure air content.

Water requirements for insulating and fill concretes vary considerably, depending on aggregate characteristics, entrained air, and mixture proportions. An effort should be made to avoid excessive amounts of water in insulating concrete used in roof fills. Excessive water causes high drying shrinkage and cracks that may damage the waterproofing membrane. Accelerators containing calcium chloride should not be used where galvanized steel will remain in permanent contact with the concrete because of the potential for corrosion.

Mixture proportions for Group II concretes usually are based on volumes of dry, loose materials, even when aggregates are moist as batched. Satisfactory proportions can vary considerably for different aggregates or combinations of aggregates. Mixture proportions ranging from 0.24 m³ to 0.90 m³ of aggregate per 100 kg of cement (4 ft³ to 14 ft³ per 100 lb) can be used in lightweight concretes that are made with pumice, expanded shale, and expanded slag. Some mixtures, such as those for no-fines concretes, are made without fine aggregate but with total void contents of 20% to 35%. Cement contents for Group II concretes range between 120 kg/m³ to 360 kg/m³ (200 lb/yd³ to 600 lb/yd³) depending on air content, aggregate gradation, and mixture proportions.

No-fines concretes containing pumice, expanded slag, or expanded shale can be made with 150 kg/m³ to 170 kg/m³ (250 lb/yd³ to 290 lb/yd³) of water, total air voids of 20% to 35%, and a cement content of about 280 kg/m³ (470 lb/yd³).

Workability

Because of their high air content, lightweight concretes weighing less than 800 kg/m³ (50 lb/ft³) generally have excellent workability. Slumps of up to 250 mm (10 in.) usually are satisfactory for Group I and Group III concretes; appearance of the mix, however, may be a more reliable indication of consistency. Cellular concretes are handled as liquids; they are poured or pumped into place without further consolidation.

Mixing and Placing

All concrete should be mechanically mixed to produce a uniform distribution of materials of proper consistency and required density. In batch-mixing operations, various sequences can be used for introducing the ingredients. The preferred sequence is to first introduce the required amount of water into the mixer, then add the cement, air-entraining or foaming agent, aggregate, preformed foam, and any other ingredients.

Excessive mixing and handling should be avoided because they tend to break up aggregate particles, thereby changing density and consistency. Segregation is not usually a problem (though it could be for Group II) because of the relatively large amounts of entrained air in these mixtures.

Pumping is the most common method of placement, but other methods can be used. Finishing operations should be kept to a minimum; smoothing with a darby or bull-float is usually sufficient. Placement of insulating concretes should be done by workers experienced with these special concretes.

Periodic wet-density tests (ASTM C138, *Standard Test Method for Density (Unit Weight), Yield, and Air Content (Gravimetric) of Concrete* [AASHTO T 121]) at the jobsite can be performed to check the uniformity of the concrete. Variations in density generally should not exceed plus or minus 32 kg/m^3 (2 lb/ft^3). A close approximation of the ovendry density can be determined from the freshly mixed density.

Thermal Resistance

ASTM C177, *Test Method for Steady-State Heat Flux Measurements and Thermal Transmission Properties by Means of the Guarded-Hot-Plate Apparatus*, is used to determine values of thermal conductivity. This test method is the only appropriate method for determining the thermal conductivity of concrete; insulating to nomalweight. Other methods, such as ASTM C518, do not allow adequate time for the concrete to come to thermal equilibrium which is needed to ensure thermal mass effects are not included with the thermal conductivity. Figure 20-3 shows an approximate relationship between thermal resistance and density for a particular concrete mixture. The thermal conductivity of concrete increases with an increase in moisture content and density. See Brewer (1967) for additional density and conductivity relationships.

Strength

Strength requirements depend on the intended use of the concrete. For example, a compressive strength of 0.7 MPa (100 psi), or even less, may be satisfactory for insulation of underground steam lines. Roof-fill insulation requires sufficient early strength to withstand foot traffic. Compressive strengths of 0.7 MPa to 1.5 MPa (100 psi to 200 psi) are usually adequate for roof fills, but strengths up to 3.5 MPa (500 psi) are sometimes specified. In general, the strength of insulating concrete is of minor importance. Compressive strength of lightweight insulating concrete should be determined by the methods specified in ASTM C495, *Standard Test Method for Compressive Strength of Lightweight Insulating Concrete*, or ASTM C513, *Standard Test Method for Obtaining and Testing Specimens of Hardened Lightweight Insulating Concrete for Compressive Strength* (Withdrawn 2004).

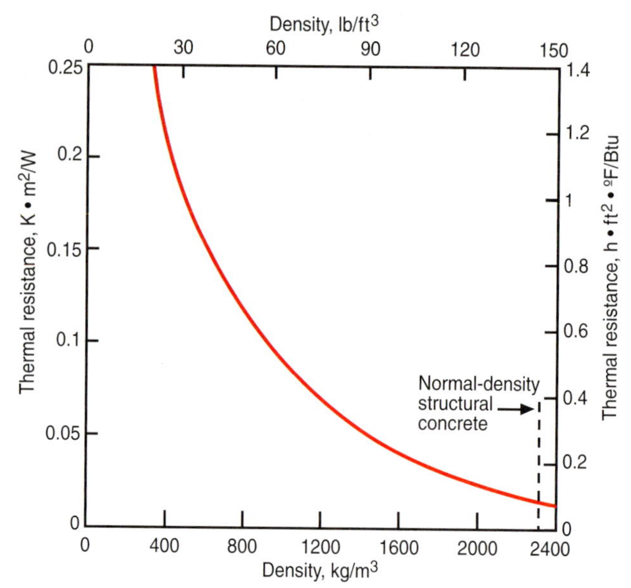

Figure 20-3. Thermal resistance of concrete versus density (PCA 1980).

Table 20-2 and Figure 20-4 give examples of the relationship between density and strength for lightweight insulating concretes. Figure 20-5 shows examples for cellular concrete containing sand. Mixtures with strengths outside the ranges shown can be made by varying the mixture proportions. Strengths comparable to those at 28 days would be obtained at 7 days with high-early-strength cement. The relationships shown do not apply to autoclaved products.

Resistance to Freezing and Thawing

Insulating and moderate-strength lightweight concretes normally are not required to withstand freeze-thaw exposure in a saturated condition. In service they are normally protected from the weather. Thus little research has been done on their resistance to freezing and thawing.

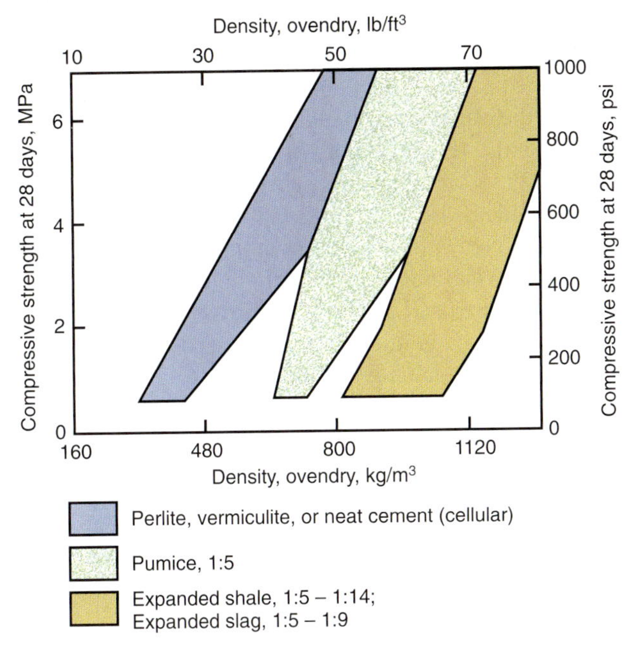

Figure 20-4. Approximate relationship between ovendry bulk density and compressive strength of 150 x 300-mm (6 x 12-in.) cylinders tested in an air-dry condition for some insulating and fill concretes. For the perlite and vermiculite concretes, mix proportions range from 1:3 to 1:10 by volume.

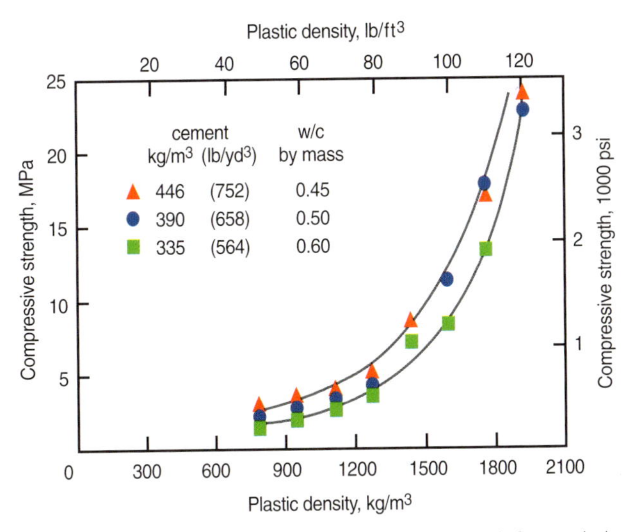

Figure 20-5. Plastic density versus compressive strength for sanded cellular concretes. Compressive strength was determined with 150 x 300-mm (6 x 12-in.) cylinders that were cured for 21 days in a 100% relative humidity moist room followed by 7 days in air at 50% RH (McCormick 1967 and ACI 523.3R).

Drying Shrinkage

The drying shrinkage of insulating or moderate-strength lightweight concrete is not usually critical when it is used for insulation or fill; however, excessive shrinkage can cause curling. In structural use, shrinkage should be considered. Moist-cured cellular concretes made without aggregates have high drying shrinkage. Moist-cured cellular concretes made with sand may shrink from 0.1% to 0.6%, depending on the amount of sand used. Autoclaved cellular concretes shrink very little on drying. Insulating concretes made with perlite or pumice aggregates may shrink 0.1% to 0.3% in six months of drying in air at 50% relative humidity; vermiculite concretes may shrink 0.2% to 0.45% during the same period. Drying shrinkage of insulating concretes made with expanded slag or expanded shale ranges from about 0.06% to 0.1% in six months.

Expansion Joints

Where insulating concrete is used on roof decks, a 25-mm (1-in.) expansion joint at the parapets and all roof projections is often specified. Its purpose is to accommodate expansion caused by the heat of the sun so that the insulating concrete can expand independently of the roof deck. Transverse expansion joints should be placed at a maximum of 30 m (100 ft) in any direction for a thermal expansion of 1 mm per meter (1 in. per 100 lin ft). A fiberglass material that will compress to one-half its thickness under a stress of 0.17 MPa (25 psi) is generally used to form these joints.

Autoclaved Cellular Concrete

Autoclaved cellular concrete (also called autoclaved aerated concrete, or AAC) is a special type of lightweight building material. It is typically manufactured from a mortar consisting of pulverized siliceous material (sand, slag, or fly ash), cement or lime, and water. A gas forming admixture, for example aluminum powder, is added to the mixture. The chemical reaction of aluminum with the alkaline water forms hydrogen bubbles. The hydrogen expands the mortar as macropores with a diameter of 0.5 mm to 1.5 mm (0.02 in. to 0.06 in.) form. The material is then pressure steam cured (autoclaved) over a period of 6 to 12 hours using a temperature of 190°C (374°F) and a pressure of 1.2 MPa (174 psi). This forms a hardened mortar matrix, which essentially consists of calcium silicate hydrates.

This porous mineral building material has densities between 300 kg/m³ and 1000 kg/m³ (19 lb/ft³ and 63 lb/ft³) and compressive strengths between 2.5 MPa and 10 MPa (300 psi and 1500 psi). Due to the high macropore content – up to 80% by volume – autoclaved cellular concrete has a thermal conductivity of only 0.15 W/(m•K) to 0.20 W/(m•K) (1 Btu•in./[h•ft²•°F] to 1.4 Btu•in./[h•ft²•°F]).

Autoclaved cellular concrete is produced in block or panel form for construction of residential or commercial buildings (Figure 20-6).

Additional information can be found in ACI 523.2R, *Guide for Precast Cellular Concrete Floor, Roof, and Wall Units.*

Figure 20-6. (top) Residential building constructed with autoclaved cellular concrete blocks. (bottom) Autoclaved cellular concrete block floating in water.

High-Density Concrete

High-density (heavyweight) concrete has a density of up to about 6400 kg/m^3 (400 lb/ft^3). Heavyweight concrete is used principally for radiation shielding but is also used for counterweights and other applications where high-density is important. As a shielding material, heavyweight concrete protects against the harmful effects of X-rays, gamma rays, and neutron radiation. Selection of concrete for radiation shielding is based on space requirements and on the type and intensity of radiation. Where space requirements are not important, normal-weight concrete will generally produce the most economical shield; where space is limited, heavyweight concrete will allow for reductions in shield thickness without sacrificing shielding effectiveness.

Type and intensity of radiation usually determine the requirements for density and water content of shielding concrete. Effectiveness of a concrete shield against gamma rays is approximately proportional to the density of the concrete; the heavier the concrete, the more effective the shield. On the other hand, an effective shield against neutron radiation requires both heavy and light elements. The hydrogen in water provides an effective light element in concrete shields. Some aggregates contain crystallized water, called fixed water, as part of their structure. For this reason, heavyweight aggregates with high fixed-water contents often are used if both gamma rays and neutron radiation are to be attenuated. Boron glass (boron frit) is also added to attenuate neutrons.

High-Density Aggregates

High-density aggregates such as barite, ferrophosphorus, goethite, hematite, ilmenite, limonite, magnetite, and degreased steel punchings and shot are used to produce high-density concrete. Where high fixed-water content is desirable, serpentine (which is slightly heavier than normal-weight aggregate) or bauxite can be used (see ASTM C637, *Standard Specification for Aggregates for Radiation-Shielding Concrete*, and ASTM C638, *Standard Descriptive Nomenclature of Constituents of Aggregates for Radiation-Shielding Concrete*).

Table 20-3 gives typical bulk density, relative density (specific gravity), and percentage of fixed water for some of these materials. The values are a compilation of data from a wide variety of tests or projects reported in the literature. Steel punchings and shot are used where concrete with a density of more than 4800 kg/m^3 (300 lb/ft^3) is required.

In general, selection of an aggregate is determined by physical properties, availability, and cost. Heavyweight aggregates should be reasonably free of fine material, oil, and foreign substances that may affect either the bond of paste to aggregate particle or the hydration of cement. For good workability, maximum density, and economy, aggregates should be roughly cubical in shape and free of excessive flat or elongated particles.

Table 20-3. Physical Properties of Typical High-Density Aggregates and Concrete

Type of aggregate	Fixed-water,* percent by weight	Aggregate relative density	Aggregate bulk density, kg/m^3 (pcf)	Concrete density, kg/m^3 (pcf)
Goethite	10–11	3.4–3.7	2080–2240 (130–140)	2880–3200 (180–200)
Limonite**	8–9	3.4–4.0	2080–2400 (130–150)	2880–3360 (180–210)
Barite	0	4.0–4.6	2320–2560 (145–160)	3360–3680 (210–230)
Ilmenite	†	4.3–4.8	2560–2700 (160–170)	3520–3850 (220–240)
Hematite	†	4.9–5.3	2880–3200 (180–200)	3850–4170 (240–260)
Magnetite	†	4.2–5.2	2400–3040 (150–190)	3360–4170 (210–260)
Ferrophosphorus	0	5.8–6.8	3200–4160 (200–260)	4080–5290 (255–330)
Steel punchings or shot	0	6.2–7.8	3860–4650 (230–290)	4650–6090 (290–380)

* Water retained or chemically bound in aggregates.
** Test data not available.
† Aggregates may be combined with limonite to produce fixed-water contents varying from about ½% to 5%.

Additions

Boron additions such as colemanite (hydrated calcium borate hydroxide) boron frits, and borocalcite are sometimes used to improve the neutron shielding properties of concrete. However, they may adversely affect setting and early strength of concrete; therefore, trial mixtures should be made with the addition under field conditions to determine suitability. Admixtures such as pressure-hydrated lime can be used with coarse-sand sizes to minimize any retarding effect.

Properties of High-Density Concrete

The properties of high-density concrete in both the freshly mixed and hardened states can be tailored to meet job conditions and shielding requirements by proper selection of materials and mixture proportions.

Except for density, the physical properties of heavyweight concrete are similar to those of normal-weight concrete. Strength is a function of water-cement ratio; thus, for any particular set of materials, strengths comparable to those of normal-weight concretes can be achieved. Typical densities of concretes made with some commonly used high-density aggregates are shown in Table 20-3. Because each radiation shield has special requirements, trial mixtures should be made with job materials and under job conditions to determine suitable mixture proportions.

Proportioning, Mixing, and Placing

The procedures for selecting mix proportions for heavyweight concrete are the same as those for normal-weight concrete. However, additional mixture information and sample calculations are given in ACI 211.1. Following are the most common methods of mixing and placing high-density concrete:

Conventional methods of mixing and placing often are used, but care must be taken to avoid overloading the mixer, especially with very heavy aggregates such as steel punchings. Batch sizes should be reduced to about 50% of the rated mixer capacity. Because some heavy aggregates are quite friable, excessive mixing should be avoided to prevent aggregate breakup and its resultant detrimental effects on workability and bleeding.

Preplaced aggregate methods can be used for placing normal and high-density concrete in confined areas and around embedded items. This practice will minimize segregation of coarse aggregate, especially steel punchings or shot. The method also reduces drying shrinkage and produces concrete of uniform density and composition. With this method, the coarse aggregate is preplaced in the forms and grout made of cement, sand, and water is then pumped through pipes to fill the voids in the aggregate.

Pumping of heavyweight concrete through pipelines may be advantageous in locations where space is limited. Heavyweight concretes cannot be pumped as far as normal-weight concretes because of their higher densities.

Puddling is a method whereby a 50-mm (2-in.) layer or more of mortar is placed in the forms and then covered with a layer of coarse aggregate that is rodded or internally vibrated into the mortar. Care must be taken to ensure uniform distribution of aggregate throughout the concrete.

Mass Concrete

Mass concrete is defined by ACI Committee 207 as "any volume of concrete in which a combination of dimensions of the member being cast, the boundary conditions, the characteristics of the concrete mixture, and the ambient conditions can lead to undesirable thermal stresses, cracking, deleterious chemical reactions, or reduction in the long-term strength as a result of elevated concrete temperature due to heat from hydration." Mass concrete includes not only low-cementitious-content concrete used in dams and other massive structures but also moderate- to high-cementitious-content concrete in structural members of bridges and buildings (Figure 20-7). Mass concrete placements require special considerations to reduce or control the heat of hydration and the resulting temperature rise to avoid damaging the concrete through excessive temperatures and temperature differences. These conditions can result in thermal damage and thermal cracking (Gajda 2007). Thermal shock cracking can also occur if these considerations are not employed for a sufficient period of time.

Figure 20-7. Large foundation placements as shown require mass-concrete precautions (Courtesy of Carson-Mitchell Construction Co.).

In mass concrete, temperature rise (Figure 20-8) results from the heat of hydration of cementitious materials. The temperature rise can be high, and concrete temperatures in excess of 70°C (158°F) can result in long-term damage of the concrete due to delayed ettringite formation (DEF) (see Chapter 11). In addition, as the interior concrete

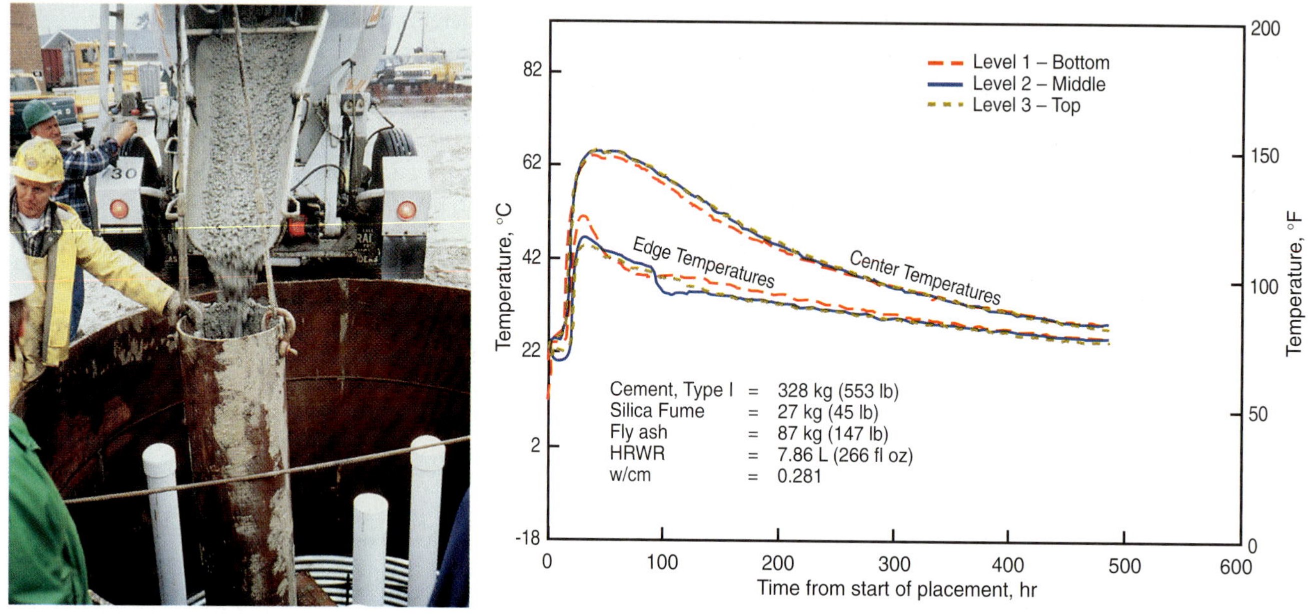

Figure 20-8. (left) A drilled pier (caisson), 3 m (10 ft) in diameter and 12.2 m (40 ft) in depth in which "low-heat" high-strength concrete is placed and (right) temperatures of this concrete measured at the center and edge and at three different levels in the caisson (Burg and Fiorato 1999).

increases in temperature and expands, the surface concrete may be cooling and contracting. This causes tensile stresses that may result in thermal cracks at the surface if the temperature differential between the surface and center is too great. The width and depth of cracks depends upon the temperature differential, physical properties of the concrete, and the reinforcing steel.

A definite member size beyond which a concrete structure should be classified as mass concrete is not readily available. Many large structural elements may be massive enough that heat generation should be considered. This is particularly critical when the minimum cross-sectional dimensions of a solid concrete member approach or exceed 1 meter (3 feet) or when cement contents exceed 355 kg/m³ (600 lb/yd³).

The temperature rise in a mass concrete placement is related to the initial concrete temperature (Figure 20-9), ambient temperature, size of the concrete element (its minimum dimension), and type and quantity of cementitious materials. Smaller concrete members less than 0.3 meters (1 ft) thick with moderate amounts of cementitious materials are typically of little concern as the generated heat is dissipated more rapidly than it is generated.

Internal concrete temperature gain can be controlled a number of ways: (1) a low cement content – 120 kg to 270 kg per cubic meter (200 lb to 450 lb per cubic yard); (2) large aggregate size – 75mm to 150 mm (3in. to 6 in.), well (uniformly) graded; (3) high coarse aggregate content – up to 80% of total aggregate; (4) low-heat-of-hydration cement; (5) low-heat pozzolans – where heat of hydration of a pozzolan is 40% to 90% that of cement; (6) reductions in the initial concrete temperature by cooling the concrete ingredients; (7) cooling the concrete through the use of embedded cooling pipes; (8) steel forms for rapid heat dissipation; (9) water curing; and (10) low lifts – 1.5 m (5 ft) or less during placement. Typically, several of the methods are used in combination.

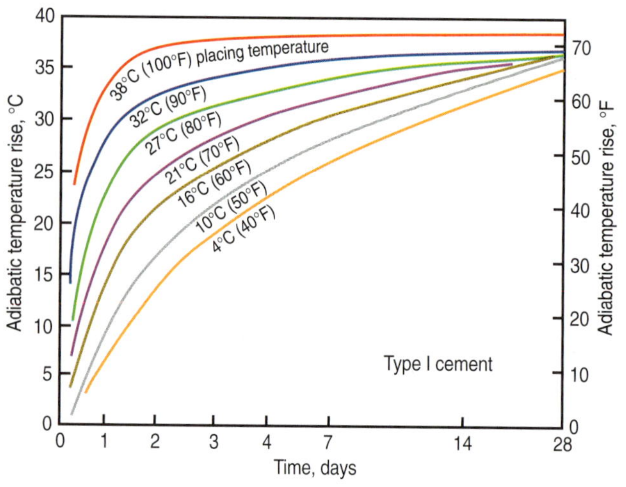

Figure 20-9. The effect of concrete-placing temperature on temperature rise in mass concrete with 223 kg/m³ (376 lb/yd³) of cement. Higher placing temperatures accelerate temperature rise (ACI 207.2R).

Massive structural reinforced concrete members with high cement contents (300 to 600 kg per cubic meter or 500 to 1000 lb per cubic yard) cannot use many of the placing techniques and controlling factors mentioned above to maintain the low temperatures needed to control cracking. For these concretes (often used in bridges, foundations, and power plants), a good technique is to (1) avoid external restraint from adjacent concrete elements, (2) reduce the size of the member by placing the concrete in multiple smaller pours with adequate time between pours to dissipate the temperature rise of the concrete, or (3) control internal differential thermal strains by preventing the concrete from experiencing an excessive temperature differential between the surface and the center. The latter is done by properly designing the concrete and either keeping the concrete surface warm through use of insulation, or reducing the internal concrete temperature by precooling the concrete or postcooling with internal cooling pipes. Insulation is generally always necessary, regardless of the season, in mass concrete applications to limit the temperature difference between the concrete mixture and ambient conditions.

Studies and experience have shown that by limiting the maximum temperature differential between the interior and exterior surface of the concrete to less than about 20°C (36°F), surface cracking can be minimized or avoided (FitzGibbon 1977 and Fintel and Ghosh 1978). Some sources indicate that the maximum temperature differential (MTD) for concrete containing granite or limestone (aggregates with a low coefficient of thermal expansion) should be 25°C and 31°C (45°F and 56°F), respectively (Bamforth 1981). The actual MTD for a particular mass concrete placement with a particular concrete mix design can be determined using equations in ACI 207 (2007), or by a stress/strain approach (Gajda 2007).

In general, an MTD of 20°C (36°F) should be assumed unless a demonstration or calculations based on physical properties of the actual concrete mixture and the geometry of the concrete member demonstrate that higher MTD values are allowable.

By limiting the temperature differential from the time the concrete is placed through the time the concrete cools to within 20°C (36°F) of the average air temperature, the concrete will cool slowly to ambient temperature with little or no surface cracking. However, this is true only if the member is not restrained by continuous reinforcement crossing the interface of adjacent or opposite sections of hardened concrete. Restrained concrete may crack due to eventual thermal contraction after the cool down. Unrestrained concrete should not crack if proper procedures are followed and if the temperature differential is monitored and controlled. If there is any concern over excess temperature differentials in a concrete member, the element should be considered as mass concrete and appropriate precautions taken.

Figure 20-10 illustrates the relationship between temperature rise, cooling, and temperature differentials for a section of mass concrete. As can be observed, if the forms (which are providing adequate insulation in this case) are removed too early, cracking will occur once the difference between interior and surface concrete temperatures exceeds the critical temperature differential of 20°C (36°F). If higher temperature differentials are permissible, the forms can be removed sooner. For large concrete placements, surface insulation may be required for several weeks or longer.

The maximum temperature rise can be estimated by approximation, if the concrete contains 300 to 600 kg of cement per cubic meter (500 to 1000 lb of Type I/II cement per cubic yard) and the least dimension of the member is 1.8 m (6 ft). This approximation (under normal, not adiabatic conditions) would be 15°C for every 100 kg of cement per cubic meter (16°F for every 100 lb of cement per cubic yard). For example, the maximum temperature of such an element made with concrete having 400 kg of Type I cement per cubic meter (674 lb of cement per cubic yard) and cast with an initial concrete temperature of 25°C (77°F) would be about

25°C + (15°C x 400/100) or 85°C

(77°F + [16°F x 674/100] or 185°F)

The effect of pozzolans on the temperature rise and maximum concrete temperature can also be estimated by a method given by Gajda (2007). Temperatures and temperature differences in mass concrete can also be calculated by a method in ACI 207.2R (2007).

The slow rate of heat exchange between concrete and its surroundings is due to the heat capacity and thermal diffusivity of concrete. Heat escapes from uninsulated concrete at a rate that is inversely proportional to the square of its least dimension. A 150-mm (6-in.) thick wall cooling from both sides will take approximately 1½ hours to dissipate 95% of its developed heat. A 1.5-m (5-ft) thick wall would take an entire week to dissipate the same amount of heat (ACI 207.2R-07). If internal cooling pipes were not used during the construction of the Hoover Dam, it would have taken more than 200 years to dissipate its internal heat.

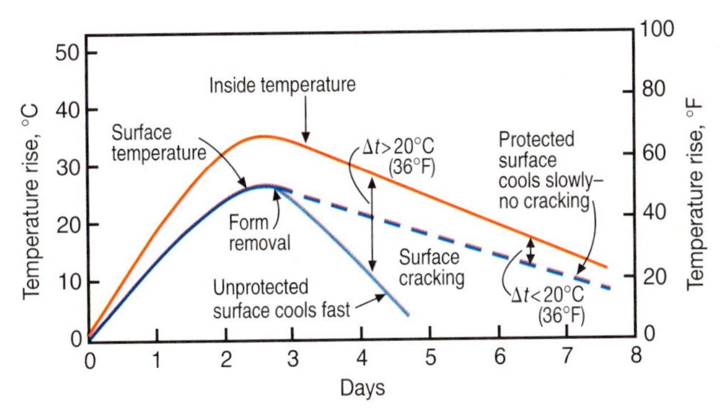

Figure 20-10. Potential for surface cracking after form removal, assuming a critical temperature differential, Δt, of 20°C (36°F). No cracking should occur if concrete is cooled slowly and Δt is less than 20°C (36°F) (Fintel and Ghosh 1978 and PCA 1987).

Inexpensive thermocouples or other types of temperature sensors can be used to monitor temperature and temperature differences in mass concrete placements. Specific locations for monitoring temperatures and temperature differences in typical mass concrete placements are described in Gajda (2007) and ACI 301-10.

Temperatures and temperature differences in mass concrete placements can be predicted by various methods, including the methods described above, the Schmidt Method in ACI 207.2R-07, commercially-available finite element software, and proprietary mass concrete software. A thermal control plan is now commonly submitted by the contractor to the owner/engineer prior to a mass concrete placement. The contents of a thermal control plan are described in Gajda (2007) and ACI 301-10.

Preplaced Aggregate Concrete

Preplaced aggregate concrete is produced by first placing coarse aggregate in a form and later injecting a cement-sand grout, usually with admixtures, to fill the voids. Properties of the resulting concrete are similar to those of comparable concrete placed by conventional methods. However, considerably less thermal and drying shrinkage can be expected because of the point-to-point contact of aggregate particles.

Coarse aggregates should meet requirements of ASTM C33, *Standard Specification for Concrete Aggregates* (AASHTO M 80). In addition, most specifications limit both the maximum and minimum sizes for example, 75-mm (3-in.) maximum and 12.5-mm (½ in.) minimum. Aggregates are generally graded to produce a void content of 35% to 40%. Fine aggregate used in the grout is generally graded to a fineness modulus of between 1.2 and 2.0, with nearly all of the material passing a 1.25 mm (No. 16) sieve.

The preplaced aggregate method has been used principally for restoration work and in the construction of reactor shields, bridge piers, and underwater structures. It has also been used in buildings to produce unusual architectural effects. Since the forms are completely filled with coarse aggregate prior to grouting, a dense and uniform exposed-aggregate facing is obtained when the surface is sandblasted, tooled, or retarded and wire-brushed at an early age.

Tests for preplaced aggregate concrete are given in ASTM C937, *Standard Specification for Grout Fluidifier for Preplaced-Aggregate Concrete* through ASTM C943, *Standard Practice for Making Test Cylinders and Prisms for Determining Strength and Density of Preplaced-Aggregate Concrete in the Laboratory*. Preplaced aggregate concrete is discussed in more detail in ACI 304R, *Guide for Measuring, Transporting, and Placing Concrete*, and ACI 304.1R, *Guide for the Use of Preplaced Aggregate Concrete for Structural and Mass Concrete Applications*.

No-Slump Concrete

No-slump concrete is defined as concrete with a consistency corresponding to a slump of 6 mm (¼ in.) or less. Such concrete, while very dry, must be sufficiently workable to be placed and consolidated with the equipment to be used on the job. The methods referred to here do not necessarily apply to mixtures for concrete masonry units or for compaction by spinning techniques.

Many of the basic laws governing the properties of higher-slump concretes are applicable to no-slump concrete. For example, the properties of hardened concrete depend principally on the ratio of water to cement, provided the mixture is properly consolidated.

Measurement of the consistency of no-slump concrete differs from that for higher-slump concrete. The slump cone is impractical for use with the drier consistencies. ACI 211.3, *Guide for Selecting Proportions for No-Slump Concrete*, describes three methods for measuring the consistency of no-slump concrete: (1) the Vebe apparatus; (2) the compacting-factor test; and (3) the Thaulow drop table. In the absence of the above test equipment, workability can be adequately judged by a trial mixture that is placed and compacted with the equipment and methods to be used on the job.

Intentionally entrained air is recommended for no-slump concrete where freeze-thaw durability is required. The amount of air-entraining admixture usually recommended for higher-slump concretes will produce lower air contents in no-slump concretes than in the higher-slump concretes. The lower volume of entrained air, however, generally provides adequate durability for no-slump concretes. While the volume of entrained air is not present, sufficient small air voids are present. This departure from the usual methods of designing and controlling entrained air is necessary for no-slump concretes.

For a discussion of water requirements and computation of trial mixtures, see ACI 211.3.

Roller-Compacted Concrete

Roller-compacted concrete (RCC) is a stiff, no-slump, nearly dry concrete that is compacted in place by vibratory roller or pneumatic tire rollers (Figure 20-11). RCC is a mixture of well-graded aggregates, cementitious materials, and water. Cementitious contents range from 60 to 360 kg per cubic meter (100 to 600 lb per cubic yard) depending on the application. Mixing is done with conventional batch mixers, continuous mixers, or in some instances tilting-drum truck mixers.

Figure 20-11. Vibratory rollers are used to compact roller-compacted concrete.

Applications for RCC fall into two distinct categories – water control structures (dams) and pavements. While the same term is used to describe both types of concrete use, the design and construction processes are different.

Water Control Structures

RCC can be used for the entire dam structure, or as an overtopping protection on the upper section and on the downstream face. The nominal maximum aggregate size can range from 50 mm (2 in.) up to 150 mm (6 in.). The zero slump mixture is produced in a high-capacity central-mixing plant near the site and delivered by truck or by conveyor belt. Cement content is usually lower than that used in a conventional concrete mixture, but similar to that of mass concrete. Compressive strengths ranging from 10 MPa to 24 MPa (1500 psi to 3500 psi) are typical for roller-compacted concrete dam and dam overtopping protection projects. The RCC mixture is transported by trucks or conveyor belts and spread by grader or bulldozer, followed by rolling with vibratory compactors. No forms are used. On most dam projects the upstream face is surfaced with higher strength conventional air-entrained concrete or precast concrete panels for improved durability and permeability.

An RCC dam has the advantage of a narrower foundation footprint compared to an earth or rock fill dam. In addition to the advantage of using less material, the dam is completed and placed in service much earlier and usually at a significant savings in overall cost as compared to an earth or rock fill structure.

Other water control RCC applications include use as an emergency spillway or overtopping protection for embankment dams, low permeable liner for settling ponds, bank protection, and grade control structure for channels and riverbeds. Over the past three decades RCC has been used for overtopping protection on more than 130 earthen dams (Abdo and Adaska 2007).

Pavements

The uses for RCC paving range from pavements such as mining operations, logging yards, intermodel and port facilities with thicknesses up to one meter (one yard) to city streets, shoulders, parking areas and auto manufacturing plants. See PCA (2003) and PCA (2002) for more information. The procedures for construction of an RCC pavement require tighter control than for dam construction (Arnold and Zamensky 2000). Cementitious content is typically lower than conventional concrete, ranging from 240 kg/m^3 to 360 kg/m^3 (400 lb/yd^3 to 600 lb/yd^3), while compressive strength is of the same order, 28 MPa to 40 MPa (4100 psi to 6000 psi). The nominal maximum aggregate size is limited to 19 mm (¾ in.) to provide a smooth, dense surface. For optional surface textures and improved ride quality, a 16 mm (⅝ in.) or 13 mm (½ in.) maximum size aggregate is recommended.

The zero slump mixture is usually produced in a continuous flow pugmill mixer at production rates as high as 450 metric tons (500 tons) per hour. It is possible to mix RCC in a central batch plant or in transit mixers. However, because of the extremely dry consistency of RCC, mixing times are extended and only 60 to 70 percent of the mixer capacity can be utilized. Specifications usually require that the mixture be transported, placed, and compacted within 60 minutes of the start of mixing. Ambient weather conditions may increase or decrease that time window.

RCC is typically placed in layers 100 mm to 250 mm (4 in. to 10 in.) in thickness using an asphalt-type paving machine. High-density paving equipment is preferred for layers thicker than 150 mm (6 in.) to reduce the need for subsequent compaction by rollers is reduced. Where a design calls for pavement thickness greater than 250 mm (10 in.), the RCC should be placed in multiple layers. In this type of construction, it is important that there be a minimum time delay in placing subsequent layers so that good bond is assured. Following placement by a paver, RCC can be compacted with a combination of vibratory steel-wheeled rollers and rubber-tired equipment.

Curing is vitally important in RCC pavement construction. The very low water content at the initial mixing stage means that an RCC mixture will dry out very quickly once it is in place. For most projects, a curing compound is the recommended method, although sprayed on asphaltic emulsion, plastic sheeting, and continuous water curing have been used in some cases.

High-performance RCC for areas subjected to high impact and abrasive loading were developed in the mid-1990s. These mixtures are based on obtaining the optimum packing of the various sizes of aggregate particles, and the addition of silica fume to the mixture (Marchand and others 1997 and Reid and Marchand 1998).

ACI addresses RCC in two guides – ACI 207.5, *Roller Compacted Mass Concrete* deals with RCC for water control structures and ACI 325.10 *Report on Roller Compacted Concrete Pavements*, covers new developments in RCC pavements. Holderbaum and Schweiger (2000) provide a guide for developing RCC specifications and commentary for spillways and embankments. Harrington and others (2010) provides a state-of-the-art guide to RCC pavements.

Soil-Cement

Soil-cement is a mixture of pulverized soil or granular material, cement, and water. Other terms applied to soil-cement include cement-treated base or subbase, cement stabilization, cement-modified soil, and cement-treated aggregate. The mixture is compacted to a high density, and as the cement hydrates the material becomes hard and durable. Soil-cement can be mixed in place using on-site matrials or mixed in a central plant using selected materials (often manufactured).

Soil-cement is primarily used as pavement base course for roads, streets, airports, and parking areas. A bituminous or portland cement concrete wearing course is usually placed over the soil-cement pavement layer. Soil-cement is also used as slope protection for earth dams and embankments, reservoir and ditch linings, deep-soil mixing, and foundation stabilization (PCA 2006) (Figure 20-12).

Figure 20-12. A dozer places a plant-mixed soil-cement on a slope for construction of a wastewater lagoon.

The soil material in soil-cement can be almost any combination of sand, silt, clay, and gravel or crushed stone. Local granular materials (such as slag, caliche, limerock, and scoria) plus a wide variety of waste materials (such as cinders, ash, and screenings from quarries and gravel pits) can be used to make soil-cement. Also, old granular-base roads, with or without their bituminous surfaces, can be recycled to make soil-cement.

Soil-cement should contain sufficient portland cement to resist deterioration from freeze-thaw and wet-dry cycling and sufficient moisture for maximum compaction.

Cement contents may range from as low as 2% to a high of 16% by dry weight of soil to be treated.

There are four basic steps in soil-cement construction: incorporating cement, mixing, compacting, and curing. The proper quantity of cement is added to the materials to be treated, and the necessary amount of water is mixed thoroughly using any of several types of mixing machines. Finally, the mixture is compacted with conventional road-building equipment to 96% to 100% of maximum dry density. See ASTM D558, *Standard Test Methods for Moisture-Density (Unit Weight) Relations of Soil-Cement Mixtures* (AASHTO T 134), and PCA 1992.

A light coat of bituminous material is commonly used to prevent moisture loss; it also helps bond and forms part of the bituminous surface. A common type of wearing surface for light traffic is a surface treatment of bituminous material and chips 13 mm to 19 mm (½ in. to ¾ in.) thick. For heavy-duty use and in severe climates a 38-mm (1½-in.) or greater asphalt mat is used.

Depending on the soil used, the 7-day compressive strengths of saturated specimens at the minimum cement content meeting soil-cement criteria are generally between 2 MPa to 5 MPa (300 psi and 800 psi). Like concrete, soil-cement continues to gain strength with age.

See ACI Committee 230 (2009) and PCA (1995) for detailed information on soil-cement construction.

Shotcrete

Shotcrete is mortar or small-aggregate concrete pneumatically projected onto a surface at high velocity (Figure 20-13). Also known as gunite and sprayed concrete, shotcrete was developed in 1911 and its concept is essentially unchanged even in today's use. The relatively dry mixture is consolidated by the impact force and can be placed on vertical or horizontal surfaces without sagging.

Figure 20-13. Shotcrete

Shotcrete is applied by a dry or wet process. In the dry process, a premixed blend of cement and damp aggregate is propelled by compressed air through a hose to a nozzle. Water is added to the cement and aggregate mixture at the nozzle and the intimately mixed ingredients are projected onto the surface. In the wet process, all the ingredients are premixed. The wet mixture is pumped through a hose to the nozzle, where compressed air is added to increase the velocity and propel the mixture onto the surface.

As the shotcrete mixture hits the surface, some coarser aggregates ricochet off the surface until sufficient paste builds up, providing a bed into which the aggregate can stick. To reduce overspray (mortar that attaches to nearby surfaces) and rebound (aggregate that ricochets off the receiving surface) the nozzle should be held at a 90° angle to the surface. The appropriate distance between nozzle and surface is usually between 0.5 m and 1.5 m (1.5 ft and 5 ft), depending on the material velocity.

Shotcrete is used for both new construction and repair work. It is especially suited for curved or thin concrete structures and shallow repairs, but can be used for thick members. The hardened properties of shotcrete are very operator dependent. Shotcrete has a density and compressive strength similar to normal- and high-strength concrete. Aggregate sizes up to 19 mm (3⁄4 in.) can be used, however most mixtures contain aggregates only up to 9.5 mm (3⁄8 in.). Aggregate amounts of 25% to 30% pea gravel are commonly used for wet mixtures (Austin and Robbins 1995).

Supplementary cementitious materials can also be used in shotcrete. They improve workability, chemical resistance, and durability. The use of accelerating admixtures allows build-up of thicker layers of shotcrete in a single pass. They also reduce the time of initial set. However, using rapid-set accelerators often increases drying shrinkage and reduces later-age strength (Gebler and others 1992).

Steel fibers are used in shotcrete to improve flexural strength, ductility, and toughness. They can be used as a replacement for wire mesh reinforcement in applications like rock slope stabilization and tunnel linings (ACI 506.1R). Steel fibers can be added up to 2% by volume of the total mix. Polypropylene fibers if used are normally added to shotcrete at a rate of 0.9 kg/m^3 to 2.7 kg/m^3 (1.5 lb/yd^3 to 4.5 lb/yd^3), but dosages up to 9 kg/m^3 (15 lb/yd^3) have also been used (The Aberdeen Group 1996).

Guidelines for the use of shotcrete are described in ACI 506R, ASCE (1995), Balck and others (2000) and www.shotcrete.org.

Shrinkage-Compensating Concrete

Shrinkage-compensating concrete is concrete containing expansive cement or an expansive admixture. The concrete produces expansion during hardening and thereby offsets the contraction occurring during drying (drying shrinkage). Shrinkage-compensating concrete is used in concrete slabs, pavements, structures, and repair work to minimize drying shrinkage cracks. Expansion of concrete made with shrinkage-compensating cement should be determined by the method specified in ASTM C878, *Standard Test Method for Restrained Expansion of Shrinkage-Compensating Concrete.*

The tensile capacity of the reinforcing steel in the structure restrains the concrete as the shrinkage compensating concrete expands. Upon shrinking due to drying contraction caused by moisture loss in hardened concrete, the tension in the steel is relieved; as long as the resulting tension in the concrete does not exceed the tensile strength of the concrete, no cracking will result. Shrinkage-compensating concrete can be proportioned, batched, placed, and cured similarly to normal concrete with some precautions. For example, it is necessary to assure the expected expansion by using additional curing. More information can be found in Chapter 3 and in ACI 223, *Standard Practice for the Use of Shrinkage-Compensating Concrete.*

Pervious Concrete

Pervious (porous or no-fines) concrete contains a narrowly graded coarse aggregate, little to no fine aggregate, and insufficient cement paste to fill voids in the coarse aggregate. This low water-cement ratio, low-slump concrete resembling popcorn is primarily held together by cement paste at the contact points of the coarse aggregate particles. This produces a concrete with a high volume of voids (15% to 35%) and a high permeability that allows water to flow through it at a rate of 81 L/min • m^2 to 730 L/min • m^2 (2 gal/min • ft^2 to 18 gal/min • ft^2) (ACI 522R and Tennis, Leming, and Akers 2004). According to ACI 522R, the compressive strength of pervious mixtures typically ranges from 2.8 MPa to 28 MPa (400 psi to 4000 psi).

Pervious concrete is used in hydraulic structures as drainage media, and in parking lots, pavements, and airport runways to reduce stormwater run off (Figure 20-14). It also recharges the local groundwater supply by allowing water to penetrate the concrete to the ground below. Pervious concretes have also been used in tennis courts and greenhouses. Leming, Malcom, and Tennis (2007) describe the fundamental hydrologic behavior of pervious concrete pavement systems.

Figure 20-14. Pervious concrete's key characteristic is the open-pore structure that allows high rates of water transmission.

As a paving material, pervious concrete is raked or slip-formed into place with conventional paving equipment and then roller compacted. Vibratory screeds or hand rollers can be used for smaller jobs. Joints are cut using bladed rollers or saw-cut like conventional concrete. In order to maintain its porous properties, the surfaces of pervious concrete are not floated or troweled, however it is recommended that the concrete be edged to mitigate raveling of the aggregates. As pervious concrete tends to be a low water content mixture with a very high surface area, immediate curing is required. Typically, the slab will be covered with plastic sheeting as soon as the concrete is compacted, jointed, and edged. ACI 522R recommends a maximum of 20 minutes from discharge to curing in the most favorable of conditions (low evaporation rates, see Chapter 16).

No-fines concrete is used in building construction (particularly walls) for its thermal insulating properties. For example, a 250-mm (10-in.) thick porous-concrete wall can have an R value of 0.9 (or 5 [inch-pound units]) compared to 0.125 (0.75) for normal concrete. No-fines concrete is also lightweight, 1600 kg/m^3 to 1900 kg/m^3 (100 lb/ft^3 to 120 lb/ft^3), and has low drying shrinkage properties (Malhotra 1976 and Concrete Construction 1983).

White and Colored Concrete

White Concrete

White portland cement is used to produce white concrete, a widely used architectural material (Figure 20-15). It is also used in mortar, plaster, stucco, terrazzo, and portland cement paint. White portland cement is manufactured from raw materials of low iron content; it conforms to ASTM C150, *Standard Specification for Portland Cement* (AASHTO M 85) even though these specifications do not specifically mention white portland cement.

Figure 20-15. Office building constructed with white cement concrete.

White concrete is made with aggregates and water that contain no materials that will discolor the concrete. White or light-colored aggregates can be used. Oil that could stain concrete should not be used on the forms. Care must be taken to avoid rust stains from tools and equipment. Curing materials that could cause stains must be avoided. Refer to Farny (2001) for more information.

Colored Concrete

Colored concrete can be produced using colored aggregates or by adding color pigments (ASTM C979, *Standard Specification for Pigments for Integrally Colored Concrete*) or both. When colored aggregates are used, they should be exposed at the surface of the concrete. This can be done several ways, for example, casting against a form that has been treated with a retarder. Unhydrated paste at the surface is later brushed or washed away. Other methods involve removing the surface mortar by sandblasting, waterblasting, bushhammering, grinding, or acid washing. If surfaces are to be washed with acid, a delay of approximately two weeks after casting is necessary. Colored aggregates may be natural rock such as quartz, marble, and granite, or they may be ceramic materials.

Pigments for coloring concrete should be pure mineral oxides ground finer than cement. They should be insoluble in water, free of soluble salts and acids, colorfast in sunlight, resistant to alkalies and weak acids, and virtually free of calcium sulfate. Mineral oxides occur in nature and are also produced synthetically. Synthetic pigments generally give more uniform results.

The amount of color pigments added to a concrete mixture should not be more than 10% of the mass of the cement. The amount required depends on the type of pigment and the color desired. For example, a dose of pigment equal to 1.5% by mass of cement may produce a pleasing pastel color, but 7% may be needed to produce a deep color. Use of white portland cement with a pigment will produce cleaner, brighter colors and is recommended in preference to gray cement, except for black or dark gray colors (Figure 20-16).

Figure 20-16. Colored concrete, used for a commuter rail platform.

To maintain uniform color, do not use calcium chloride, and batch all materials carefully by mass. To prevent streaking, a dry color pigment must be thoroughly blended with the cement. Mixing time should be longer than normal to ensure uniformity. Many color pigments are now sold in liquid form.

The addition of pigment in air-entrained concrete, may require an adjustment in the amount of air-entraining admixture to maintain the desired air content.

Dry-Shake Method. Slabs or precast panels that are cast horizontally can be colored by the dry-shake method. Prepackaged, dry coloring materials consisting of mineral oxide pigment, white portland cement, and specially graded silica sand or other fine aggregate are marketed ready for use by various manufacturers.

After the slab has been bullfloated once, two-thirds of the dry coloring material should be broadcast evenly by hand over the surface. The required amount of coloring material can usually be determined from previously cast sections. After the material has absorbed water from the fresh concrete, it should be floated into the surface. Then, the rest of the material should be applied immediately at right angles to the initial application, so that a uniform color is obtained. The slab should again be floated to work the remaining material into the surface.

Other finishing operations may follow depending on the type of finish desired. Curing should begin immediately after finishing; take precautions to prevent discoloring the surface. See Kosmatka and Collins (2004) for more information.

Photocatalytic Concrete

Recently introduced formulations of cement are able to neutralize pollution, in a process known as photocatalysis. Harmful smog can be turned into harmless compounds and washed away. Any concrete product is a potential application, because these cements are used in the same manner as regular portland cements. The technology (based on titanium dioxide additions) can be applied to white or gray cement and it performs like any other portland cement: it can be used in all varieties of concrete, including plaster. The only difference is that it is capable of breaking down smog or other pollution that has attached itself to the concrete substrate. As sunlight hits the surface, most organic and some inorganic pollutants are neutralized. They would otherwise eventually discolor the concrete surfaces. These products provide value through unique architectural and environmental performance capabilities.

The titanium-based catalyst is not consumed as it breaks down pollution, but continues to work. Typical products are oxygen, water, carbon dioxide, nitrate, and sulfate. Because rain washes away the pollution from the concrete surface, structures do not collect dirt and do not require chemical applications that are potentially harmful to the environment. Maintenance costs are reduced. This is true even for buildings in highly polluted locations – noted applications include the Air France headquarters at Roissy-Charles de Gaulle International Airport near Paris, a white concrete building that has remained white, and the Church of the Year 2000 in Rome.

Concrete products that are continually exposed to sunlight, like precast building panels, pavers, and roof tiles, are especially well-suited for manufacture with photocatalytic concrete. For instance, city streets made with special pavers are capable of reducing the pollution at its source – directly from the tailpipe. A study conducted in the Netherlands used photocatalytic concrete pavers on a section of a busy roadway and monitored the air quality 0.5 m to 1.5 m (19.5 in. to 58.5 in.) above the pavement at both a control area with normal pavers and of the test section. It was observed that the NO_X levels were reduced by 25% to 45% (Ballari, Yu, and Brouwers 2010). Certain state departments of transportation are also using these cements in roads and bridges to help mitigate pollution.

Polymer-Portland Cement Concrete

Polymer-portland cement concrete (PPCC), also called polymer-modified concrete, is essentially normal portland cement concrete to which a polymer or monomer has been added during mixing to improve durability and adhesion. Thermoplastic and elastomeric latexes are the most commonly used polymers in PPCC, but epoxies and other polymers are also used. In general, latex improves ductility, durability, adhesive properties, resistance to chloride-ion ingress, shear bond, and tensile and flexural strength of concrete and mortar. Latex-modified concretes (LMC) also have excellent freeze-thaw, abrasion, and impact resistance. Some LMC materials can also resist certain acids, alkalies, and organic solvents. Polymer-portland cement concrete is primarily used in concrete patching and overlays, especially bridge decks. See ACI 548.3R for more information on polymer-modified concrete and ACI 548.4 for LMC overlays.

Ferrocement

Ferrocement is a special type of reinforced concrete composed of closely spaced layers of continuous relatively thin metallic or nonmetallic mesh or wire embedded in mortar. It is constructed by hand plastering, shotcreting, laminating (forcing the mesh into fresh mortar), or a combination of these methods.

The mortar mixture generally has a sand-cement ratio of 1.5 to 2.5 and a water-cement ratio of 0.35 to 0.50. Reinforcement makes up about 5% to 6% of the ferrocement volume. Fibers and admixtures may also be used to modify the mortar properties. Polymers or cement-based coatings are often applied to the finished surface to reduce porosity.

Ferrocement is considered easy to produce in a variety of shapes and sizes; however, it is labor intensive. Ferrocement is used to construct thin shell roofs, swimming pools, tunnel linings, silos, tanks, prefabricated houses, barges, boats, sculptures, and thin panels or sections usually less than 25 mm (1 in.) thick (ACI 549R and ACI 549.1R).

References

ACI Committee 207, *Report on Thermal and Volume Change Effects on Cracking of Mass Concrete*, ACI 207.2R-07, ACI Committee 207 Report, American Concrete Institute, Farmington Hills, Michigan, 2007, 28 pages.

ACI Committee 207, *Guide to Mass Concrete*, ACI 207.1R-05, ACI Committee 207 Report, American Concrete Institute, Farmington Hills, Michigan, 2005, 30 pages.

ACI Committee 207, *Roller-Compacted Mass Concrete*, ACI 207.5R-99, ACI Committee 207 Report, American Concrete Institute, Farmington Hills, Michigan, 1999, 47 pages.

ACI Committee 211, *Guide for Selecting Proportions for No-Slump Concrete*, ACI 211.3R-03, ACI Committee 211 Report, American Concrete Institute, Farmington Hills, Michigan, 2003, 26 pages.

ACI Committee 211, *Standard Practice for Selecting Proportions for Structural Lightweight Concrete*, ACI 211.2-98, reapproved 2004, ACI Committee 211 Report, American Concrete Institute, Farmington Hills, Michigan, 1998, 20 pages.

ACI Committee 211, *Standard Practice for Selecting Proportions for Normal, Heavyweight, and Mass Concrete*, ACI 211.1-91, reapproved 2002, ACI Committee 211 Report, American Concrete Institute, Farmington Hills, Michigan, 1991, 38 pages.

ACI Committee 213, *Guide for Structural Lightweight Aggregate Concrete*, ACI 213R-03, ACI Committee 213 Report, American Concrete Institute, Farmington Hills, Michigan, 2003, 38 pages.

ACI Committee 223, *Standard Practice for the Use of Shrinkage-Compensating Concrete*, ACI 223-98, ACI Committee 223 Report, American Concrete Institute, Farmington Hills, Michigan, 1998 28 pages.

ACI Committee 230, *Report on Soil Cement*, ACI 230.1R-09, ACI Committee 230 Report, American Concrete Institute, Farmington Hills, Michigan, 2009, 28 pages.

ACI Committee 301, *Specifications for Structural Concrete*, ACI 301-10, ACI Committee 301 Report, American Concrete Institute, Farmington Hills, Michigan, 2010.

ACI Committee 304, *Guide for Measuring, Mixing, Transporting, and Placing Concrete*, ACI 304R-00, ACI Committee 304 Report, American Concrete Institute, Farmington Hills, Michigan, 2000, 41 pages.

ACI Committee 304, *Guide for Use of Preplaced Aggregate Concrete for Structural and Mass Concrete Applications*, ACI 304.1R-92, reapproved 2005, ACI Committee 304 Report, American Concrete Institute, Farmington Hills, Michigan, 1992, 19 pages.

ACI Committee 325, *Report on Roller-Compacted Concrete Pavements*, ACI 325.10R-95, reapproved 2001, ACI Committee 325 Report, American Concrete Institute, Farmington Hills, Michigan, 1995, 32 pages.

ACI Committee 506, *Committee Report on Fiber-Reinforced Shotcrete*, ACI 506.1R-08, ACI Committee 506 Report, American Concrete Institute, Farmington Hills, Michigan, 2008, 12 pages.

ACI Committee 506, *Guide to Shotcrete*, ACI 506R-05, ACI Committee 506 Report, American Concrete Institute, Farmington Hills, Michigan, 2005, 40 pages.

ACI Committee 522, *Report on Pervious Concrete*, ACI 522R-10, ACI Committee 522 Report, American Concrete Institute, Farmington Hills, Michigan, 2010, 38 pages.

ACI Committee 523, *Guide for Precast Cellular Concrete Floor, Roof, and Wall Units*, ACI 523.2R-96, ACI Committee 523 Report, American Concrete Institute, Farmington Hills, Michigan, 1996, 5 pages.

ACI Committee 523, *Guide for Cellular Concretes Above 50 pcf and for Aggregate Concretes Above 50 pcf with Compressive Strengths Less Than 2500 psi*, ACI 523.3R-93, ACI Committee 523 Report, American Concrete Institute, Farmington Hills, Michigan, 1993, 16 pages.

ACI Committee 548, *Report on Polymer-Modified Concrete*, ACI 548.3R-09, ACI Committee 548 Report, American Concrete Institute, Farmington Hills, Michigan, 2009, 39 pages.

ACI Committee 548, *Standard Specification for Latex-Modified Concrete (LMC) Overlays*, ACI 548.4-93, reapproved 1998, ACI Committee 548 Report, American Concrete Institute, Farmington Hills, Michigan, 1998, 6 pages.

ACI Committee 549, *Report on Ferrocement*, ACI 549R-97, reapproved 2009, ACI Committee 549 Report, American Concrete Institute, Farmington Hills, Michigan, 1997.

ACI Committee 549, *Guide for the Design, Construction, and Repair of Ferrocement*, ACI 549.1R-93, reapproved 2009, ACI Committee 549 Report, American Concrete Institute, Farmington Hills, Michigan, 1999, 30 pages.

Arnold, Terry, and Zamensky, Greg, *Roller-Compacted Concrete: Quality Control Manual*, EB215, Portland Cement Association, 2000, 58 pages.

ASCE, *Standard Practice for Shotcrete*, American Society of Civil Engineers, New York, 1995, 65 pages.

Austin, Simon, and Robbins, Peter, *Sprayed Concrete: Properties, Design and Application*, Whittles Publishing, U.K., 1995, 382 pages.

Balck, Lars; Gebler, Steven; Isaak, Merlyn; and Seabrook Philip, *Shotcrete for the Craftsman (CCS-4)*, American Concrete Institute, Farmington Hills, Michigan, 2000, 59 pages. Also available from PCA as LT242.

Ballari, M.; Yu, Qingliang; and Brouwers, H.J.H., "Experimental Study of the NO and NO_2 Degradation by Photo-Catalytically Active Concrete," *Catalysis Today*, available online, 2010.

Bamforth, P.B., "Large Pours," letter to the editor, *Concrete*, Cement and Concrete Association, Wexham Springs, Slough, England, February 1981.

Bohan, Richard P., and Ries, John, *Structural Lightweight Concrete*, IS032, Portland Cement Association, Skokie, Illinois, USA, 2008, 8 pages.

Brewer, Harold W., *General Relation of Heat Flow Factors to the Unit Weight of Concrete*, Development Department Bulletin DX114, Portland Cement Association, http://www.cement.org/pdf_files/DX114.pdf, 1967.

Burg, R.G., and Fiorato, A.E., *High-Strength Concrete in Massive Foundation Elements*, Research and Development Bulletin RD117, Portland Cement Association, 1999, 22 pages.

Concrete Construction, "Porous Concrete Slabs and Pavement Drain Water," *Concrete Construction*, Concrete Construction Publications, Inc., Addison, Illinois, September 1983, pages 685 and 687 to 688.

Farny, J.A., *White Cement Concrete*, EB217, Portland Cement Association, 2001.

Fintel, Mark, and Ghosh, S.K., "Mass Reinforced Concrete Without Construction Joints," presented at the Adrian Pauw Symposium on Designing for Creep and Shrinkage, Fall Convention of the American Concrete Institute, Houston, Texas, November 1978.

FitzGibbon, Michael E., "Large Pours for Reinforced Concrete Structures," Current Practice Sheets No. 28, 35, and 36, *Concrete*, Cement and Concrete Association, Wexham Springs, Slough, England, March and December 1976 and February 1977.

Gajda, John, *Mass Concrete for Buildings and Bridges*, EB547, Portland Cement Association, 2007, 44 pages.

Gebler, Steven H.; Litvin, Albert; McLean, William J.; and Schutz, Ray, "Durability of Dry-Mix Shotcrete Containing Rapid-Set Accelerators," *ACI Materials Journal*, May-June 1992, pages 259 to 262.

Gustaferro, A.H.; Abrams, M.S.; and Litvin, Albert, *Fire Resistance of Lightweight Insulating Concretes*, Research and Development Bulletin RD004, Portland Cement Association, http://www.cement.org/pdf_files/RD004.pdf, 1970.

Hanna, Amir N., *Properties of Expanded Polystyrene Concrete and Applications for Pavement Subbases*, Research and Development Bulletin RD055, Portland Cement Association, http://www.cement.org/pdf_files/RD055.pdf, 1978.

Hansen, Kenneth D., "A Pavement for Today and Tomorrow," *Concrete International*, American Concrete Institute, Farmington Hills, Michigan, February 1987, pages 15 to 17.

Harrington, Dale; Abdo, Fares; Adaska, Wayne; and Hazaree, Chetan, *Guide for Roller-Compacted Concrete Pavements*, National Concrete Pavement Technology Center, Ames, Iowa, 2010, 114 pages.

Holderbaum, Rodney E., and Schweiger, Paul G., *Guide for Developing RCC Specifications and Commentary*, EB214, Portland Cement Association, 2000, 80 pages.

Kanare, Howard M., *Concrete Floors and Moisture*, EB119, Portland Cement Association, Skokie, Illinois, and National Ready Mixed Concrete Association, Silver Spring, Maryland, USA, 2008, 176 pages.

Kosmatka, Steven H., and Collins, Terry C., *Finishing Concrete with Color and Texture*, PA124, Portland Cement Association, 2004, 72 pages.

Leming, M.L., Malcom, H.R., and Tennis, P.D., *Hydrologic Design of Pervious Concrete*, EB303, Portland Cement Association, 2007, 72 pages.

Malhotra, V.M., "No-Fines Concrete – Its Properties and Applications," *Journal of the American Concrete Institute*, American Concrete Institute, Farmington Hills, Michigan, November 1976.

Marchand, J.; Gagne, R.; Ouellet, E.; and Lepage, S., "Mixture Proportioning of Roller Compacted Concrete – A Review," *Proceedings of the Third CANMET/ACI International Conference*, Auckland, New Zealand, Available as: *Advances in Concrete Technology*, SP 171, American Concrete Institute, Farmington Hills, Michigan, 1997, pages 457 to 486.

McCormick, Fred C., "Rational Proportioning of Preformed Foam Cellular Concrete," *Journal of the American Concrete Institute*, American Concrete Institute, Farmington Hills, Michigan, February 1967, pages 104 to 110.

PCA, *Concrete Energy Conservation Guidelines*, EB083, Portland Cement Association, 1980.

PCA, *Concrete for Massive Structures*, IS128, Portland Cement Association, http://www.cement.org/pdf_files/IS128.pdf, 1987, 24 pages.

PCA, *Soil-Cement Construction Handbook*, EB003, Portland Cement Association, 1995, 40 pages.

PCA, *Soil-Cement Laboratory Handbook*, EB052, Portland Cement Association, 1992, 60 pages.

Reichard, T.W., "Mechanical Properties of Insulating Concretes," *Lightweight Concrete*, American Concrete Institute, Farmington Hills, Michigan, 1971, pages 253 to 316.

Reid, E., and Marchand, J., "High-Performance Roller Compacted Concrete Pavements: Applications and Recent Developments", *Proceedings of the Canadian Society for Civil Engineering 1998 Annual Conference*, Halifax, Nova Scotia, 1998.

Tennis, P.D.; Leming, M.L.; and Akers, D.J., *Pervious Concrete Pavements*, EB302, Portland Cement Association, 2004, 25 pages.

The Aberdeen Group, *Shotcreting—Equipment, Materials, and Applications*, Addison, Illinois, 1996, 46 pages.

Appendix

GLOSSARY

The intent of this Glossary is to clarify terminology used in concrete construction, with special emphasis on those terms used in *Design and Control of Concrete Mixtures*. Additional terminology that may not be in this book is included in the Glossary for the convenience of our readers. Other sources for terms include ACI Cement and Concrete Terminology (http://www.concrete.org/Technical/CCT/ACI-Terminology.aspx), and ASTM standards.

A

Absorption—see *Water absorption.*

Accelerating admixture—admixture that speeds the rate of hydration of hydraulic cement, shortens the normal time of setting, or increases the rate of hardening, of strength development, or both, of portland cement, concrete, mortar, grout, or plaster.

Addition—substance that is interground or blended in limited amounts into a hydraulic cement during manufacture (not at the jobsite) either as a "processing addition" to aid in manufacture and handling of the cement or as a "functional addition" to modify the useful properties of the cement.

Admixture—material, other than water, aggregate, hydraulic cement, and fibers used as an ingredient of concrete, mortar, grout, or plaster and added to the batch immediately before or during mixing.

Aggregate—granular mineral material such as natural sand, manufactured sand, gravel, crushed stone, aircooled blast-furnace slag, vermiculite, or perlite.

Air content—total volume of air voids, both entrained and entrapped, in cement paste, mortar, or concrete. Entrained air adds to the durability of hardened mortar or concrete and the workability of fresh mixtures.

Air entrainment—intentional introduction of air in the form of minute, disconnected bubbles (generally smaller than 1 mm) during mixing of portland cement concrete, mortar, grout, or plaster to improve desirable characteristics such as cohesion, workability, and durability.

Air-entraining admixture—admixture for concrete, mortar, or grout that will cause air to be incorporated into the mixture in the form of minute bubbles during mixing, usually to increase the material's workability and frost resistance.

Air-entraining portland cement—portland cement containing an air-entraining addition added during its manufacture.

Air void—entrapped air pocket or an entrained air bubble in concrete, mortar, or grout. Entrapped air voids usually are larger than 1 mm in diameter; entrained air voids are smaller. Most of the entrapped air voids should be removed with internal vibration, power screeding, or rodding.

Alkali-aggregate reactivity—production of expansive gel caused by a reaction between aggregates containing certain forms of silica or carbonates and alkali hydroxides in concrete.

Architectural concrete—concrete that will be permanently exposed to view and which therefore requires special care in selection of concrete ingredients, forming, placing, consolidating, and finishing to obtain the desired architectural appearance.

Autoclaved cellular concrete—concrete containing very high air content resulting in low density, and cured at high temperature and pressure in an autoclave.

B

Batching—process of weighing or volumetrically measuring and introducing into the mixer the ingredients for a batch of concrete, mortar, grout, or plaster.

Blast-furnace slag—nonmetallic byproduct of steel manufacturing, consisting essentially of silicates and aluminum silicates of calcium that are developed in a molten condition simultaneously with iron in a blast furnace.

Bleeding—flow of mixing water from a newly placed concrete mixture caused by the settlement of the solid materials in the mixture.

Blended hydraulic cement—cement containing combinations of portland cement, pozzolans, slag, and/or other hydraulic cement.

Bulking—increase in volume of a quantity of sand when in a moist condition compared to its volume when in a dry state.

C

Calcined clay—increase in volume of a quantity of sand when in a moist condition compared to its volume when in a dry state.

Calcined shale—shale heated to high temperature to alter its physical properties for use as a pozzolan or cementing material in concrete.

Carbonation—reaction between carbon dioxide and a hydroxide or oxide to form a carbonate.

Cellular concrete—high air content or high void ratio concrete resulting in low density.

Cement—see *Portland cement* and *Hydraulic cement*.

Cement paste—constituent of concrete, mortar, grout, and plaster consisting of cement and water.

Cementitious material (cementing material)—any material having cementing properties or contributing to the formation of hydrated calcium silicate compounds. When proportioning concrete, the following are considered cementitious materials: portland cement, blended hydraulic cement, fly ash, ground granulated blast-furnace slag, silica fume, calcined clay, metakaolin, calcined shale, and rice husk ash.

Chemical admixture—see *Admixture*.

Chemical bond—bond between materials resulting from cohesion and adhesion developed by chemical reaction.

Clinker—end product of a portland cement kiln; raw cementitious material prior to grinding

Chloride (attack)—chemical compounds containing chloride ions, which promote the corrosion of steel reinforcement. Chloride deicing chemicals are primary sources.

Coarse aggregate—natural gravel, crushed stone, or iron blast-furnace slag, usually larger than 5 mm (0.2 in.) and commonly ranging in size between 9.5 mm and 37.5 mm (⅜ in. to 1½ in.).

Cohesion—mutual attraction by which elements of a substance are held together.

Colored concrete—concrete containing white cement and/or mineral oxide pigments to produce colors other than the normal gray hue of traditional gray cement concrete.

Compaction—process of inducing a closer arrangement of the solid particles in freshly mixed and placed concrete, mortar, or grout by reduction of voids, usually by vibration, tamping, rodding, puddling, or a combination of these techniques. Also called consolidation.

Compressive strength—maximum resistance that a concrete, mortar, or grout specimen will sustain when loaded axially in compression in a testing machine at a specified rate; usually expressed as force per unit of cross sectional area, such as megapascals (MPa) or pounds per square inch (psi).

Concrete—mixture of binding materials and coarse and fine aggregates. Portland cement and water are commonly used as the binding medium for normal concrete mixtures, but may also contain pozzolans, slag, and/or chemical admixtures.

Consistency—relative mobility or ability of freshly mixed concrete, mortar, or grout to flow. (See also *Slump* and *Workability*.)

Construction joint—a stopping place in the process of construction. A true construction joint allows for bond between new concrete and existing concrete and permits no movement. In structural applications their location must be determined by the structural engineer. In slab on grade applications, construction joints are often located at contraction (control) joint locations and are constructed to allow movement and perform as contraction joints.

Contraction joint—weakened plane to control cracking due to volume change in a concrete structure. Joint may be grooved, sawed, or formed. Also known as a "Control joint."

Corrosion—deterioration of metal by chemical, electrochemical, or electrolytic reaction.

Creep—time-dependent deformation of concrete, or of any material, due to a sustained load.

Curing—process of maintaining freshly placed concrete mortar, grout, or plaster moist and at a favorable temperature for a suitable period of time during its early stages so that the desired properties of the material can develop. Curing assures satisfactory hydration and hardening of the cementitious materials.

D

Dampproofing—treatment of concrete, mortar, grout, or plaster to retard the passage or absorption of water, or water vapor.

Density—mass per unit volume; the weight per unit volume in air, expressed, for example, in kg/m^3 (lb/ft^3).

Durability—ability of portland cement concrete, mortar, grout, or plaster to resist weathering action and other conditions of service, such as chemical attack, freezing and thawing, and abrasion.

E

Early stiffening—rapidly developing rigidity in freshly mixed hydraulic cement paste, mortar, grout, plaster, or concrete.

Entrapped air—irregularly shaped, unintentional air voids in fresh or hardened concrete 1 mm or larger in size.

Entrained air—spherical microscopic air bubbles—usually 10 µm to 1000 µm in diameter—intentionally incorporated into concrete to provide freezing and thawing resistance and/or improve workability.

Epoxy resin—class of organic chemical bonding systems used in the preparation of special coatings or adhesives for concrete or masonry or as binders in epoxy-resin mortars and concretes.

Ettringite—needle like crystalline compound produced by the reaction of C_3A, gypsum, and water within a portland cement concrete.

Expansion joint—a separation provided between adjoining parts of a structure to allow movement.

F

Ferrocement—one or more layers of steel or wire reinforcement encased in portland cement mortar creating a thin-section composite material.

Fibers—thread or thread like material ranging from 0.05 to 4 mm (0.002 to 0.16 in.) in diameter and from 10 to 150 mm (0.5 to 6 in.) in length and made of steel, glass, synthetic (plastic), carbon, or natural materials.

Fiber concrete—concrete containing randomly oriented fibers in 2 or 3 dimensions through out the concrete matrix.

Fine aggregate—aggregate that passes the 9.5-mm (⅜-in.) sieve, almost entirely passes the 4.75-mm (No. 4) sieve, and is predominantly retained on the 75-µm (No. 200) sieve.

Fineness modulus (FM)—factor obtained by adding the cumulative percentages of material in a sample of aggregate retained on each of a specified series of sieves and dividing the sum by 100.

Finishing—mechanical operations like screeding, consolidating, floating, troweling, or texturing that establish the final appearance of any concrete surface.

Fire resistance—that property of a building material, element, or assembly to withstand fire or give protection from fire; it is characterized by the ability to confine a fire or to continue to perform a given structural function during a fire, or both.

Flexural strength—ability of solids to resist bending.

Fly ash—residue from coal combustion, which is carried in flue gases, and is used as a pozzolan or cementing material in concrete.

Forms—temporary supports for keeping fresh concrete in place until it has hardened to such a degree as to be self supporting (when the structure is able to support its dead load).

Freeze-thaw resistance—ability of concrete to withstand cycles of freezing and thawing. (See also *Air entrainment* and *Air-entraining admixture.*)

Fresh concrete—concrete that has been recently mixed and is still workable and plastic.

Functional resilience—longevity for continued use and adaptability to future use. Concepts of functional resilience include robustness, durability, enhanced disaster resistance, and longevity of which some components have been referred to as passive survivability.

G

Grading—size distribution of aggregate particles, determined by separation with standard screen sieves.

Grout—mixture of cementitious material with or without aggregate or admixtures to which sufficient water is added to produce a pouring or pumping consistency without segregation of the constituent materials.

H

Hardened concrete—concrete that is in a solid state and has developed a certain strength.

High-density concrete (heavyweight concrete)—concrete of very high density; normally designed by the use of heavyweight aggregates.

High-strength concrete—concrete with a design strength of at least 70 MPa (10,000 psi).

Honeycomb—term that describes the failure of mortar to completely surround coarse aggregates in concrete, leaving empty spaces (voids) between them.

Hydrated lime—dry powder obtained by treating quicklime with sufficient water to satisfy its chemical affinity for water; consists essentially of calcium hydroxide or a mixture of calcium hydroxide and magnesium oxide or magnesium hydroxide, or both.

Hydration—in concrete, mortar, grout, and plaster, the chemical reaction between hydraulic cement and water in which new compounds with strength-producing properties are formed.

Hydraulic cement—cement that sets and hardens by chemical reaction with water, and is capable of doing so under water. (See also *Portland cement.*)

I

Inch-pound units—units of length, area, volume, weight, and temperature commonly used in the United States during the 18th to 21st centuries. These include, but are not limited to: (1) length—inches, feet, yards, and miles; (2) area—square inches, square feet, square yards, and square miles; (3) volume—cubic inches, cubic feet, cubic yards, gallons, and ounces; (4) weight—pounds and ounces; and (5) temperature—degrees Fahrenheit.

Isolation Joint—separation that allows adjoining parts of a structure to move freely to one another, both horizontally and vertically.

J

Joint—see *Construction joint, Contraction joint, Isolation joint,* and *Expansion joint.*

K

Kiln—rotary furnace used in cement manufacture to heat and chemically combine raw inorganic materials, such as limestone, sand, and clay, into calcium silicate clinker.

L

Lightweight aggregate—low-density aggregate used to produce lightweight (low-density) concrete. Could be expanded or sintered clay, slate, diatomaceous shale, perlite, vermiculite, or slag; natural pumice, scoria, volcanic cinders, tuff, or diatomite; sintered fly ash or industrial cinders.

Lightweight concrete—low-density concrete compared to normal-density concrete.

Lime—general term that includes the various chemical and physical forms of quicklime, hydrated lime, and hydraulic lime. It may be high-calcium, magnesian, or dolomitic.

M

Masonry—concrete masonry units, clay brick, structural clay tile, stone, terra cotta, and the like, or combinations thereof, bonded with mortar, dry-stacked, or anchored with metal connectors to form walls, building elements, pavements, and other structures.

Masonry cement—hydraulic cement, primarily used in masonry and plastering construction, consisting of a mixture of portland or blended hydraulic cement and plasticizing materials (such as limestone, hydrated or hydraulic lime) together with other materials introduced to enhance one or more properties such as setting time, workability, water retention, and durability.

Mass concrete—any volume of structural concrete in which a combination of dimensions of the member being cast, the boundary conditions, the characteristics of the concrete mixture, and the ambient conditions can lead to undesirable thermal stresses, cracking, deleterious chemical reactions, or reduction in the long-term strength as a result of elevated concrete temperature due to heat from hydration.

Metakaolin—highly reactive pozzolan made from kaolin clays.

Metric units—also called System International (SI) Units. System of units adopted by most of the world by the 21st Century. These include but are not limited to: (1) length—millimeters, meters, and kilometers; (2) area—square millimeters and square meters; (3) volume—cubic meters and liters; (4) mass—milligrams, grams, kilograms, and megagrams; and (5) degrees Celsius.

Mineral admixtures—see *Supplementary cementitious materials.*

Modulus of elasticity—ratio of normal stress to corresponding strain for tensile or compressive stress below the proportional limit of the material; also referred to as elastic modulus, Young's modulus, and Young's modulus of elasticity; denoted by the symbol *E*.

Moist-air curing—curing with moist air (no less than 95% relative humidity) at atmospheric pressure and a temperature of about 23°C (73°F).

Mortar—mixture of cementitious materials, fine aggregate, and water, which may contain admixtures, and is usually used to bond masonry units.

Mortar cement—hydraulic cement, primarily used in masonry construction, consisting of a mixture of portland or blended hydraulic cement and plasticizing materials (such as limestone, hydrated or hydraulic lime) together with other materials introduced to enhance one or more properties such as setting time, workability, water retention, and durability. Mortar cement and masonry cement are similar in use and function. However, specifications for mortar cement usually require lower air contents and they include a flexural bond strength requirement.

N

Normal weight concrete—class of concrete made with normal density aggregates, usually crushed stone or gravel, having a density of approximately 2400 kg/m^3 (150 lb/ft^3). (See also *Lightweight concrete* and *High-density concrete.*)

No-slump concrete—concrete having a slump of less than 6 mm (¼ in.).

O

Overlay—layer of concrete or mortar placed on or bonded to the surface of an existing pavement or slab. Normally done to repair a worn or cracked surface. Overlays are seldom less than 25 mm (1 in.) thick.

P

Pavement (concrete)—highway, road, street, path, or parking lot surfaced with concrete. Although typically applied to surfaces that are used for travel, the term also applies to storage areas and playgrounds.

Permeability—property of allowing passage of fluids or gases.

Pervious concrete (no-fines or porous concrete)—concrete containing insufficient fines or no fines to fill the voids between aggregate particles in a concrete mixture. The coarse aggregate particles are coated with a cement and water paste to bond the particles at their contact points. The resulting concrete contains an interconnected pore system allowing storm water to drain through the concrete to the subbase below.

pH—chemical symbol for the logarithm of the reciprocal of hydrogen ion concentration in gram atoms per liter, used to express the acidity or alkalinity (base) of a solution on a scale of 0 to 14, where less than 7 represents acidity, and more than 7 alkalinity

Plastic cement—special hydraulic cement product manufactured for plaster and stucco application. One or more inorganic plasticizing agents are interground or blended with the cement to increase the workability and molding characteristics of the resultant mortar, plaster, or stucco.

Plasticity—that property of freshly mixed cement paste, concrete, mortar, grout, or plaster that determines its workability, resistance to deformation, or ease of molding.

Plasticizer—admixture that increases the plasticity of portland cement concrete, mortar, grout, or plaster.

Polymer-portland cement concrete—fresh portland cement concrete to which a polymer is added for improved durability and adhesion characteristics, often used in overlays for bridge decks; also referred to as polymer-modified concrete and latex-modified concrete.

Popout—shallow depression in a concrete surface resulting from the breaking away of pieces of concrete due to internal pressure.

Portland blast-furnace slag cement—hydraulic cement consisting of: (1) an intimately interground mixture of portland cement clinker and granulated blast-furnace slag; (2) an intimate and uniform blend of portland cement and fine granulated blast-furnace slag; or (3) finely ground blast-furnace slag with or without additions.

Portland cement—Calcium silicate hydraulic cement produced by pulverizing portland cement clinker, and usually containing calcium sulfate and other compounds. (See also *Hydraulic cement.*)

Portland cement plaster—a combination of portland cement-based cementitious material(s) and aggregate mixed with a suitable amount of water to form a plastic mass that will adhere to a surface and harden, preserving any form and texture imposed on it while plastic. See also *Stucco.*

Portland-pozzolan cement—hydraulic cement consisting of an intimate and uniform blend of portland cement or portland blast-furnace slag cement and fine pozzolan produced by intergrinding portland cement clinker and pozzolan, by blending portland cement or portland blast-furnace slag cement and finely divided pozzolan, or a combination of intergrinding and blending, in which the amount of the pozzolan constituent is within specified limits.

Pozzolan—siliceous or siliceous and aluminous materials, like fly ash or silica fume, which in itself possess little or no cementitious value but which will, in finely divided form and in the presence of moisture, chemically react with calcium hydroxide at ordinary temperatures to form compounds possessing cementitious properties.

Precast concrete—concrete cast in forms in a controlled environment and allowed to achieve a specified strength prior to placement on location.

Prestressed concrete—concrete in which compressive stresses are induced by high-strength steel tendons or bars in a concrete element before loads are applied to the element which will balance the tensile stresses imposed in the element during service. This may be accomplished by the following: Post-tensioning—a method of prestressing in which the tendons/bars are tensioned after the concrete has hardened; or Pre-tensioning—a method of prestressing in which the tendons are tensioned before the concrete is placed.

Q

Quality control—actions taken by a producer or contractor to provide control over what is being done and what is being provided so that applicable standards of good practice for the work are followed.

R

Reactive-powder concrete—high-strength, low-water and low-porosity concrete with high silica content and aggregate particle sizes of less than 0.3 mm.

Ready-mixed concrete—concrete manufactured for delivery to a location in a fresh state

Recycled concrete—hardened concrete that has been processed for reuse, usually as an aggregate.

Reinforced concrete—concrete to which tensile bearing materials such as steel rods or metal wires are added for tensile strength.

Relative density—a ratio relating the mass of a volume of material to that of water; also called specific gravity.

Relative humidity—The ratio of the quantity of water vapor actually present in the atmosphere to the amount of water vapor present in a saturated atmosphere at a given temperature, expressed as a percentage.

Retarder—an admixture that delays the setting and hardening of concrete.

Roller-compacted concrete (RCC)—a zero slump mixture of aggregates, cementitious materials and water that is consolidated by rolling with vibratory compactors; typically used in the construction of dams, industrial pavements, storage and composting areas, and as a component of composite pavements for highways and streets.

S

Scaling—disintegration and flaking of a hardened concrete surface, frequently due to repeated freeze-thaw cycles and application of deicing chemicals.

Segregation—separation of the components (aggregates and mortar) of fresh concrete, resulting in a nonuniform mixture.

Self-consolidating concrete—concrete of high workability that require little or no vibration or other mechanical means of consolidation, also called self-compacting concrete.

Set—the degree to which fresh concrete has lost its plasticity and hardened.

Silica fume—very fine noncrystalline silica which is a byproduct from the production of silicon and ferrosilicon alloys in an electric arc furnace; used as a pozzolan in concrete.

Shotcrete—mortar or small-aggregate concrete that is conveyed by compressed air through a hose and applied at high velocity to a surface. Also known as gunite and sprayed concrete.

Shrinkage—decrease in either length or volume of a material resulting from changes in moisture content, temperature, or chemical changes.

Shrinkage-compensating concrete—concrete containing expansive cement, or an admixture, which produces expansion during hardening and thereby offsets the contraction occurring later during drying (drying shrinkage).

Slag cement—hydraulic cement consisting mostly of an intimate and uniform blend of ground, granulated blast-furnace slag with or without portland cement or hydrated lime.

Slump—measure of the consistency of freshly mixed concrete, equal to the immediate subsidence of a specimen molded with a standard slump cone.

Slurry—thin mixture of an insoluble substance, such as portland cement, slag, or clay, with a liquid, such as water.

Soil cement—mixture of soil and measured amounts of portland cement and water compacted to a high density; primarily used as a base material under pavements; also called cement-stabilized soil.

Specific gravity—see *Relative density*.

Stucco—portland cement plaster and stucco are the same material. The term "stucco" is widely used to describe the cement plaster used for coating exterior surfaces of buildings. However, in some geographical areas, "stucco" refers only to the factory-prepared finish coat mixtures. (See also *Portland cement plaster.*)

Sulfate attack—most common form of chemical attack on concrete caused by sulfates in the groundwater or soil manifested by expansion and disintegration of the concrete.

Superplasticizer (plasticizer)—admixture that increases the flowability of a fresh concrete mixture.

Supplementary cementitious (cementing) materials—Cementitious material other than portland cement or blended cement. See *also Cementitious material.*

T

Tensile strength—stress up to which concrete is able to resist cracking under axial tensile loading.

U

Unit weight—density of fresh concrete or aggregate, normally determined by weighing a known volume of concrete or aggregate (bulk density of aggregates includes voids between particles).

V

Vibration—high-frequency agitation of freshly mixed concrete through mechanical devices, for the purpose of consolidation.

Volume change—Either an increase or a decrease in volume due to any cause, such as moisture changes, temperature changes, or chemical changes. (See also *Creep.*)

W

Water absorption—(1) The process by which a liquid (water) is drawn into and tends to fill permeable pores in a porous solid. (2) The amount of water absorbed by a material under specified test conditions, commonly expressed as a percentage by mass of the test specimen.

Water to cementing (cementitious) materials ratio—ratio of mass of water to mass of cementing materials in concrete, including portland cement, blended cement, hydraulic cement, slag, fly ash, silica fume, calcined clay, metakaolin, calcined shale, and rice husk ash.

Water to cement ratio (water-cement ratio and w/c)—ratio of mass of water to mass of cement in concrete.

Water reducer—admixture whose properties permit a reduction of water required to produce a concrete mix of a certain slump, reduce water-cement ratio, reduce cement content, or increase slump.

White portland cement—cement manufactured from raw materials of low iron content.

Workability—That property of freshly mixed concrete, mortar, grout, or plaster that determines its working characteristics, that is, the ease with which it can be mixed, placed, molded, and finished. (See also *Slump* and *Consistency.*)

X

Y

Yield—volume per batch of concrete expressed in cubic meters (cubic feet).

Z

Zero-slump concrete—concrete without measurable slump (see also *No-slump concrete*).

ASTM STANDARDS

ASTM International (ASTM) documents related to aggregates, cement, and concrete that are relevant to or referred to in the text are listed as follows and can be obtained at www.astm.org.

A416/A416M	Standard Specification for Steel Strand, Uncoated Seven-Wire for Prestressed Concrete
A421/A421M	Standard Specification for Uncoated Stress-Relieved Steel Wire for Prestressed Concrete
A615/A615M	Standard Specification for Deformed and Plain Carbon-Steel Bars for Concrete Reinforcement
A706/A706M	Standard Specification for Low-Alloy Steel Deformed and Plain Bars for Concrete Reinforcement
A722/A722M	Standard Specification for Uncoated High-Strength Steel Bars for Prestressing Concrete
A767/A767M	Standard Specification for Zinc-Coated (Galvanized) Steel Bars for Concrete Reinforcement
A775/A775M	Standard Specification for Epoxy-Coated Steel Reinforcing Bars
A820/A820M	Standard Specification for Steel Fibers for Fiber-Reinforced Concrete
A882/A882M	Standard Specification for Filled Epoxy-Coated Seven-Wire Prestressing Steel Strand
A884/A884M	Standard Specification for Epoxy-Coated Steel Wire and Welded Wire Reinforcement
A934/A934M	Standard Specification for Epoxy-Coated Prefabricated Steel Reinforcing Bars
A955/A955M	Standard Specification for Deformed and Plain Stainless-Steel Bars for Concrete Reinforcement
A1022/A1022M	Standard Specification for Deformed and Plain Stainless Steel Wire and Welded Wire for Concrete Reinforcement
A1035/A1035M	Standard Specification for Deformed and Plain, Low-carbon, Chromium, Steel Bars for Concrete Reinforcement
A1064/A1064M	Standard Specification for Steel Wire and Welded Wire Reinforcement, Plain and Deformed, for Concrete
C29/C29M	Standard Test Method for Bulk Density ("Unit Weight") and Voids in Aggregate
C31/C31M	Standard Practice for Making and Curing Concrete Test Specimens in the Field
C33/C33M	Standard Specification for Concrete Aggregates
C39/C39M	Standard Test Method for Compressive Strength of Cylindrical Concrete Specimens
C40	Standard Test Method for Organic Impurities in Fine Aggregates for Concrete
C42/C42M	Standard Test Method for Obtaining and Testing Drilled Cores and Sawed Beams of Concrete
C70	Standard Test Method for Surface Moisture in Fine Aggregate
C78/C78M	Standard Test Method for Flexural Strength of Concrete (Using Simple Beam with Third Point Loading)
C87	Standard Test Method for Effect of Organic Impurities in Fine Aggregate on Strength of Mortar
C88	Standard Test Method for Soundness of Aggregates by Use of Sodium Sulfate or Magnesium Sulfate
C91	Standard Specification for Masonry Cement
C94/C94M	Standard Specification for Ready-Mixed Concrete
C109/C109M	Standard Test Method for Compressive Strength of Hydraulic Cement Mortars (using 2 in. or [50 mm] Cube Specimens)
C114	Standard Test Methods for Chemical Analysis of Hydraulic Cement
C115	Standard Test Method for Fineness of Portland Cement by the Turbidimeter
C117	Standard Test Method for Materials Finer than 75-µm (No. 200) Sieve in Mineral Aggregates by Washing
C123	Standard Test Method for Lightweight Particles in Aggregate
C125	Standard Terminology Relating to Concrete and Concrete Aggregates
C127	Standard Test Method for Density, Relative Density (Specific Gravity), and Absorption of Coarse Aggregate
C128	Standard Test Method for Density, Relative Density (Specific Gravity), and Absorption of Fine Aggregate
C131	Standard Test Method for Resistance to Degradation of Small Size Coarse Aggregate by Abrasion and Impact in the Los Angeles Machine
C136	Standard Test Method for Sieve Analysis of Fine and Coarse Aggregates
C138/C138M	Standard Test Method for Density (Unit Weight), Yield, and Air Content (Gravimetric) of Concrete
C141/C141M	Standard Specification for Hydraulic Hydrated Lime for Structural Purposes
C142	Standard Test Method for Clay Lumps and Friable Particles in Aggregates
C143/C143M	Standard Test Method for Slump of Hydraulic Cement Concrete
C150/C150M	Standard Specification for Portland Cement
C151/C151M	Standard Test Method for Autoclave Expansion of Portland Cement
C156	Standard Test Method for Water Loss [from a Mortar Specimen] Through Liquid Membrane-Forming Curing Compounds for Concrete
C157/C157M	Standard Test Method for Length Change of Hardened Hydraulic-Cement, Mortar, and Concrete

C170/C170M	Standard Test Method for Compressive Strength of Dimension Stone
C171	Standard Specification for Sheet Materials for Curing Concrete
C172/C172M	Standard Practice for Sampling Freshly Mixed Concrete
C173/C173M	Standard Test Method for Air Content of Freshly Mixed Concrete by the Volumetric Method
C174/C174M	Standard Test Method for Measuring Thickness of Concrete Elements Using Drilled Concrete Cores
C177	Standard Test Method for Steady-State Heat Flux Measurements and Thermal Transmission Properties by Means of the Guarded-Hot-Plate Apparatus
C183	Standard Practice for Sampling and the Amount of Testing of Hydraulic Cement
C184	Standard Test Method for Fineness of Hydraulic Cement by the 150-µm (No. 100) and 75-µm (No. 200) Sieves (Withdrawn 2002)
C185	Standard Test Method for Air Content of Hydraulic Cement Mortar
C186	Standard Test Method for Heat of Hydration of Hydraulic Cement
C187	Standard Test Method for Normal Consistency of Hydraulic Cement
C188	Standard Test Method for Density of Hydraulic Cement
C191	Standard Test Method for Time of Setting of Hydraulic Cement by Vicat Needle
C192/C192M	Standard Practice for Making and Curing Concrete Test Specimens in the Laboratory
C204	Standard Test Method for Fineness of Hydraulic Cement by Air Permeability Apparatus
C215	Standard Test Method for Fundamental Transverse, Longitudinal, and Torsional Frequencies of Concrete Specimens
C219	Standard Terminology Relating to Hydraulic Cement
C226	Standard Specification for Air-Entraining Additions for Use in the Manufacture of Air-Entraining Hydraulic Cement
C227	Standard Test Method for Potential Alkali Reactivity of Cement-Aggregate Combinations (Mortar-Bar Method)
C230/C230M	Standard Specification for Flow Table for Use in Tests of Hydraulic Cement
C231/C231M	Standard Test Method for Air Content of Freshly Mixed Concrete by the Pressure Method
C232/C232M	Standard Test Methods for Bleeding of Concrete
C233	Standard Test Method for Air-Entraining Admixtures for Concrete
C235-68	Standard Method of Test for Scratch Hardness of Coarse Aggregate Particles (Withdrawn 1976)
C243	Standard Test Method for Bleeding of Cement Pastes and Mortars
C260	Standard Specification for Air-Entraining Admixtures for Concrete
C265	Standard Test Method for Calcium Sulfate in Hydrated Portland Cement Mortar
C266	Standard Test Method for Time of Setting of Hydraulic-Cement Paste by Gillmore Needles
C270	Standard Specification for Mortar for Unit Masonry
C289	Standard Test Method for Potential Alkali-Silica Reactivity of Aggregates (Chemical Method)
C293/C293M	Standard Test Method for Flexural Strength of Concrete (Using Simple Beam With Center-Point Loading)
C294	Standard Descriptive Nomenclature for Constituents of Concrete Aggregates
C295	Standard Guide for Petrographic Examination of Aggregates for Concrete
C305	Standard Practice for Mechanical Mixing of Hydraulic Cement Pastes and Mortars of Plastic Consistency
C309	Standard Specification for Liquid Membrane-Forming Compounds for Curing Concrete
C311	Standard Test Methods for Sampling and Testing Fly Ash or Natural Pozzolans for Use as a Mineral Admixture in Portland-Cement Concrete
C330/C330M	Standard Specification for Lightweight Aggregates for Structural Concrete
C332	Standard Specification for Lightweight Aggregates for Insulating Concrete
C341/C341M	Standard Practice for Length Change of Cast, Drilled, or Sawed Specimens of Hydraulic-Cement Mortar and Concrete
C342-97	Standard Test Method for Potential Volume Change of Cement-Aggregate Combinations (Withdrawn 2001)
C348	Standard Test Method for Flexural Strength of Hydraulic-Cement Mortars
C359	Standard Test Method for Early Stiffening of Portland Cement (Mortar Method)
C360-92	Test Method for Ball Penetration in Fresh Portland Cement Concrete (Withdrawn 1999)
C384	Standard Test Method for Impedance and Absorption of Acoustical Materials by the Impedance Tube Method
C387/C387M	Standard Specification for Packaged, Dry, Combined Materials for Mortar and Concrete
C403/C403M	Standard Test Method for Time of Setting of Concrete Mixtures by Penetration Resistance
C418	Standard Test Method for Abrasion Resistance of Concrete by Sandblasting

C430	Standard Test Method for Fineness of Hydraulic Cement by the 45-μm (No. 325) Sieve
C441	Standard Test Method for Effectiveness of Pozzolans or Ground Blast-Furnace Slag in Preventing Excessive Expansion of Concrete Due to the Alkali-Silica Reaction
C451	Standard Test Method for Early Stiffening of Hydraulic Cement (Paste Method)
C452/C452M	Standard Test Method for Potential Expansion of Portland-Cement Mortars Exposed to Sulfate
C457/C457M	Standard Test Method for Microscopical Determination of Parameters of the Air-Void System in Hardened Concrete
C465	Standard Specification for Processing Additions for Use in the Manufacture of Hydraulic Cements
C469/C469M	Standard Test Method for Static Modulus of Elasticity and Poisson's Ratio of Concrete in Compression
C470/C470M	Standard Specification for Molds for Forming Concrete Test Cylinders Vertically
C490/C490M	Standard Practice for Use of Apparatus for the Determination of Length Change of Hardened Cement Paste, Mortar, and Concrete
C494/C494M	Standard Specification for Chemical Admixtures for Concrete
C495	Standard Test Method for Compressive Strength of Lightweight Insulating Concrete
C496/C496M	Standard Test Method for Splitting Tensile Strength of Cylindrical Concrete Specimens
C511	Standard Specification for Mixing Rooms, Moist Cabinets, Moist Rooms, and Water Storage Tanks Used in the Testing of Hydraulic Cements and Concretes
C512	Standard Test Method for Creep of Concrete in Compression
C513-89 (1995)	Standard Test Method for Obtaining and Testing Specimens of Hardened Lightweight Insulating Concrete for Compressive Strength (Withdrawn 2004)
C535	Standard Test Method for Resistance to Degradation of Large-Size Coarse Aggregate by Abrasion and Impact in the Los Angeles Machine
C566	Standard Test Method for Total Evaporable Moisture Content of Aggregate by Drying
C567	Standard Test Method for Density of Structural Lightweight Concrete
C586	Standard Test Method for Potential Alkali Reactivity of Carbonate Rocks for Concrete Aggregates (Rock Cylinder Method)
C595/C595M	Standard Specification for Blended Hydraulic Cements
C597	Standard Test Method for Pulse Velocity Through Concrete
C617	Standard Practice for Capping Cylindrical Concrete Specimens
C618	Standard Specification for Coal Fly Ash and Raw or Calcined Natural Pozzolan for Use in Concrete
C637	Standard Specification for Aggregates for Radiation-Shielding Concrete
C638	Standard Descriptive Nomenclature of Constituents of Aggregates for Radiation-Shielding Concrete
C641	Standard Test Method for Staining Materials in Lightweight Concrete Aggregates
C642	Standard Test Method for Density, Absorption, and Voids in Hardened Concrete
C666/C666M	Standard Test Method for Resistance of Concrete to Rapid Freezing and Thawing
C671-94	Standard Test Method for Critical Dilation of Concrete Specimens Subjected to Freezing (Withdrawn 2003)
C672/C672M	Standard Test Method for Scaling Resistance of Concrete Surfaces Exposed to Deicing Chemicals
C682-94	Standard Practice for Evaluation of Frost Resistance of Coarse Aggregates in Air-Entrained Concrete by Critical Dilation Procedures (Withdrawn 2003)
C684	Standard Test Method for Making, Accelerated Curing, and Testing Concrete Compression Test Specimens
C685/C685M	Standard Specification for Concrete Made By Volumetric Batching and Continuous Mixing
C688	Standard Specification for Functional Additions for Use in Hydraulic Cements
C702	Standard Practice for Reducing Samples of Aggregate to Testing Size
C778	Standard Specification for Standard Sand
C779/C779M	Standard Test Method for Abrasion Resistance of Horizontal Concrete Surfaces
C786/C786M	Standard Test Method for Fineness of Hydraulic Cement and Raw Materials by the 300-μm (No. 50), 150-μm (No. 100), and 75-μm (No. 200) Sieves by Wet Methods
C796	Standard Test Method for Foaming Agents for Use in Producing Cellular Concrete Using Preformed Foam
C803/C803M	Standard Test Method for Penetration Resistance of Hardened Concrete
C805/C805M	Standard Test Method for Rebound Number of Hardened Concrete
C806	Standard Test Method for Restrained Expansion of Expansive Cement Mortar
C807	Standard Test Method for Time of Setting of Hydraulic Cement Mortar by Modified Vicat Needle
C823/C823M	Standard Practice for Examination and Sampling of Hardened Concrete in Constructions
C827/C827M	Standard Test Method for Change in Height at Early Ages of Cylindrical Specimens of Cementitious Mixtures

C845	Standard Specification for Expansive Hydraulic Cement
C856	Standard Practice for Petrographic Examination of Hardened Concrete
C869	Standard Specification for Foaming Agents Used in Making Preformed Foam for Cellular Concrete
C873/C873M	Standard Test Method for Compressive Strength of Concrete Cylinders Cast in Place in Cylindrical Molds
C876	Standard Test Method for Half-Cell Potentials of Uncoated Reinforcing Steel in Concrete
C878/C878M	Standard Test Method for Restrained Expansion of Shrinkage-Compensating Concrete
C881/C881M	Standard Specification for Epoxy-Resin-Base Bonding Systems for Concrete
C900	Standard Test Method for Pullout Strength of Hardened Concrete
C917	Standard Test Method for Evaluation of Cement Strength Uniformity From a Single Source
C918/C918M	Standard Test Method for Measuring Early-Age Compressive Strength and Projecting Later-Age Strength
C926	Standard Specification for Application of Portland Cement-Based Plaster
C928/C928M	Standard Specification for Packaged, Dry, Rapid-Hardening Cementitious Materials for Concrete Repairs
C937	Standard Specification for Grout Fluidifier for Preplaced-Aggregate Concrete
C938	Standard Practice for Proportioning Grout Mixtures for Preplaced-Aggregate Concrete
C939	Standard Test Method for Flow of Grout for Preplaced-Aggregate Concrete (Flow Cone Method)
C940	Standard Test Method for Expansion and Bleeding of Freshly Mixed Grouts for Preplaced-Aggregate Concrete in the Laboratory
C941	Standard Test Method for Water Retentivity of Grout Mixtures for Preplaced-Aggregate Concrete in the Laboratory
C942	Standard Test Method for Compressive Strength of Grouts for Preplaced-Aggregate Concrete in the Laboratory
C943	Standard Practice for Making Test Cylinders and Prisms for Determining Strength and Density of Preplaced-Aggregate Concrete in the Laboratory
C944/C944M	Standard Test Method for Abrasion Resistance of Concrete or Mortar Surfaces by the Rotating-Cutter Method
C953	Standard Test Method for Time of Setting of Grouts for Preplaced-Aggregate Concrete in the Laboratory
C979	Standard Specification for Pigments for Integrally Colored Concrete
C989	Standard Specification for Slag Cement for Use in Concrete and Mortars
C995-01	Standard Test Method for Time of Flow of Fiber-Reinforced Concrete Through Inverted Slump Cone (Withdrawn 2008)
C1012/C1012M	Standard Test Method for Length Change of Hydraulic-Cement Mortars Exposed to a Sulfate Solution
C1017/C1017M	Standard Specification for Chemical Admixtures for Use in Producing Flowing Concrete
C1018-97	Standard Test Method for Flexural Toughness and First-Crack Strength of Fiber-Reinforced Concrete (Using Beam With Third-Point Loading) (Withdrawn 2006)
C1038/C1038M	Standard Test Method for Expansion of Portland Cement Mortar Bars Stored in Water
C1040/C1040M	Standard Test Methods for In-Place Density of Unhardened and Hardened Concrete, Including Roller Compacted Concrete, By Nuclear Methods
C1059/C1059M	Standard Specification for Latex Agents for Bonding Fresh To Hardened Concrete
C1064/C1064M	Standard Test Method for Temperature of Freshly Mixed Portland Cement Concrete
C1073	Standard Test Method for Hydraulic Activity of Ground Slag by Reaction with Alkali
C1074	Standard Practice for Estimating Concrete Strength by the Maturity Method
C1077	Standard Practice for Laboratories Testing Concrete and Concrete Aggregates for Use in Construction and Criteria for Laboratory Evaluation
C1078-87 (1992)e1	Test Methods for Determining Cement Content of Freshly Mixed Concrete (Withdrawn 1998)
C1079-87 (1992)e1	Test Methods for Determining the Water Content of Freshly Mixed Concrete (Withdrawn 1998)
C1084	Standard Test Method for Portland-Cement Content of Hardened Hydraulic-Cement Concrete
C1105	Standard Test Method for Length Change of Concrete Due to Alkali-Carbonate Rock Reaction
C1116/C1116M	Standard Specification for Fiber-Reinforced Concrete
C1137	Standard Test Method for Degradation of Fine Aggregate Due to Attrition
C1138M	Standard Test Method for Abrasion Resistance of Concrete (Underwater Method)
C1150-96	Standard Test Method for The Break-Off Number of Concrete (Withdrawn 2002)
C1152/C1152M	Standard Test Method for Acid-Soluble Chloride in Mortar and Concrete
C1157/C1157M	Standard Performance Specification for Hydraulic Cement
C1170/C1170M	Standard Test Methods for Determining Consistency and Density of Roller-Compacted Concrete Using a Vibrating Table

C1202	Standard Test Method for Electrical Indication of Concrete's Ability to Resist Chloride Ion Penetration
C1218/C1218M	Standard Test Method for Water-Soluble Chloride in Mortar and Concrete
C1231/C1231M	Standard Practice for Use of Unbonded Caps in Determination of Compressive Strength of Hardened Concrete Cylinders
C1240	Standard Specification for Silica Fume Used in Cementitious Mixtures
C1252	Standard Test Methods for Uncompacted Void Content of Fine Aggregate (as Influenced by Particle Shape, Surface Texture, and Grading)
C1260	Standard Test Method for Potential Alkali Reactivity of Aggregates (Mortar-Bar Method)
C1263	Standard Test Method for Thermal Integrity of Flexible Water Vapor Retarders
C1293	Standard Test Method for Determination of Length Change of Concrete Due to Alkali-Silica Reaction
C1315	Standard Specification for Liquid Membrane-Forming Compounds Having Special Properties for Curing and Sealing Concrete
C1328	Standard Specification for Plastic (Stucco) Cement
C1329	Standard Specification for Mortar Cement
C1356	Standard Test Method for Quantitative Determination of Phases in Portland Cement Clinker by Microscopical Point-Count Procedure
C1362	Standard Test Method for Flow of Freshly Mixed Hydraulic Cement Concrete
C1365	Standard Test Method for Determination of the Proportion of Phases in Portland Cement and Portland-Cement Clinker Using X-Ray Powder Diffraction Analysis
C1383	Standard Test Method for Measuring the P-Wave Speed and the Thickness of Concrete Plates Using the Impact-Echo Method
C1435/C1435M	Standard Practice for Molding Roller-Compacted Concrete in Cylinder Molds Using a Vibrating Hammer
C1436	Standard Specification for Materials for Shotcrete
C1437	Standard Test Method for Flow of Hydraulic Cement Mortar
C1438	Standard Specification for Latex and Powder Polymer Modifiers for Hydraulic Cement Concrete and Mortar
C1439	Standard Test Methods for Evaluating Polymer Modifiers in Mortar and Concrete
C1451	Standard Practice for Determining Uniformity of Ingredients of Concrete From a Single Source
C1480/C1480M	Standard Specification for Packaged, Pre-Blended, Dry, Combined Materials for Use in Wet or Dry Shotcrete Application
C1524	Standard Test Method for Water-Extractable Chloride in Aggregate (Soxhlet Method)
C1542/C1542M	Standard Test Method for Measuring Length of Concrete Cores
C1543	Standard Test Method for Determining the Penetration of Chloride Ion into Concrete by Ponding
C1549	Standard Test Method for Determination of Solar Reflectance Near Ambient Temperature Using a Portable Solar Reflectometer
C1556	Standard Test Method for Determining the Apparent Chloride Diffusion Coefficient of Cementitious Mixtures by Bulk Diffusion
C1567	Standard Test Method for Determining the Potential Alkali-Silica Reactivity of Combinations of Cementitious Materials and Aggregate (Accelerated Mortar-Bar Method)
C1580	Standard Test Method for Water-Soluble Sulfate in Soil
C1581/C1581M	Standard Test Method for Determining Age at Cracking and Induced Tensile Stress Characteristics of Mortar and Concrete under Restrained Shrinkage
C1582/C1582M	Standard Specification for Admixtures to Inhibit Chloride-Induced Corrosion of Reinforcing Steel in Concrete
C1585	Standard Test Method for Measurement of Rate of Absorption of Water by Hydraulic-Cement Concretes
C1600/C1600M	Standard Specification for Rapid Hardening Hydraulic Cement
C1602/C1602M	Standard Specification for Mixing Water Used in the Production of Hydraulic Cement Concrete
C1603	Standard Test Method for Measurement of Solids in Water
C1604/C1604M	Standard Test Method for Obtaining and Testing Drilled Cores of Shotcrete
C1608	Standard Test Method for Chemical Shrinkage of Hydraulic Cement Paste
C1610/C1610M	Standard Test Method for Static Segregation of Self-Consolidating Concrete Using Column Technique
C1611/C1611M	Standard Test Method for Slump Flow of Self-Consolidating Concrete
C1621/C1621M	Standard Test Method for Passing Ability of Self-Consolidating Concrete by J-Ring
C1622/C1622M	Standard Specification for Cold-Weather Admixture Systems
C1646/C1646M	Standard Practice for Making and Curing Test Specimens for Evaluating Resistance of Coarse Aggregate to Freezing and Thawing in Air-Entrained Concrete
C1666/C1666M	Standard Specification for Alkali Resistant (AR) Glass Fiber for GFRC and Fiber-Reinforced Concrete and Cement
C1679	Standard Practice for Measuring Hydration Kinetics of Hydraulic Cementitious Mixtures Using Isothermal Calorimetry

C1697	Standard Specification for Blended Supplementary Cementitious Materials
C1698	Standard Test Method for Autogenous Strain of Cement Paste and Mortar
C1702	Standard Test Method for Measurement of Heat of Hydration of Hydraulic Cementitious Materials Using Isothermal Conduction Calorimetry
C1712	Standard Test Method for Rapid Assessment of Static Segregation Resistance of Self-Consolidating Concrete Using Penetration Test
D75/D75M	Standard Practice for Sampling Aggregates
D98	Standard Specification for Calcium Chloride
D345	Standard Test Method for Sampling and Testing Calcium Chloride for Roads and Structural Applications
D448	Standard Classification for Sizes of Aggregate for Road and Bridge Construction
D512	Standard Test Methods for Chloride Ion in Water
D516	Standard Test Method for Sulfate Ion in Water
D558	Standard Test Methods for Moisture-Density (Unit Weight) Relations of Soil-Cement Mixtures
D632-01	Standard Specification for Sodium Chloride (Withdrawn 2010)
D2240	Standard Test Method for Rubber Property-Durometer Hardness
D3042	Standard Test Method for Insoluble Residue in Carbonate Aggregates
D3398	Standard Test Method for Index of Aggregate Particle Shape and Texture
D3963/D3963M	Standard Specification for Fabrication and Jobsite Handling of Epoxy-Coated Steel Reinforcing Bars
D4263	Standard Test Method for Indicating Moisture in Concrete by the Plastic Sheet Method
D4580	Standard Practice for Measuring Delaminations in Concrete Bridge Decks by Sounding
D4791	Standard Test Method for Flat Particles, Elongated Particles, or Flat and Elongated Particles in Coarse Aggregate
D6910/D6910M	Standard Test Method for Marsh Funnel Viscosity of Clay Construction Slurries
D6942	Standard Test Method for Stability of Cellulose Fibers in Alkaline Environments
D7357	Standard Specification for Cellulose Fibers for Fiber-Reinforced Concrete
E11	Standard Specification for Wire Cloth and Sieves for Testing Purposes
E1155	Standard Test Method for Determining F_F Floor Flatness and F_L Floor Levelness Numbers
E1918	Standard Test Method for Measuring Solar Reflectance of Horizontal and Low-Sloped Surfaces in the Field
E1980-01	Standard Practice for Calculating Solar Reflectance Index of Horizontal and Low-Sloped Opaque Surfaces (Withdrawn 2010)
F1869	Standard Test Method for Measuring Moisture Vapor Emission Rate of Concrete Subfloor Using Anhydrous Calcium Chloride
F2170	Standard Test Method for Determining Relative Humidity in Concrete Floor Slabs Using in situ Probes
F2420	Standard Test Method for Determining Relative Humidity on the Surface of Concrete Floor Slabs Using Relative Humidity Probe Measurement and Insulated Hood
G59	Standard Test Method for Conducting Potentiodynamic Polarization Resistance Measurements
G109	Standard Test Method for Determining Effects of Chemical Admixtures on Corrosion of Embedded Steel Reinforcement in Concrete Exposed to Chloride Environments
IEEE/ASTM SI 10	Standard for Use of the International System of Units (SI): The Modern Metric System

AASHTO STANDARDS

American Association of State Highway and Transportation Officials (AASHTO) documents related to aggregates, cement, and concrete that are relevant to or referred to in the text are listed as follows and can be obtained at www.transportation.org.

M 6	Standard Specification for Fine Aggregate for Hydraulic Cement Concrete
M 43	Standard Specification for Sizes of Aggregate for Road and Bridge Construction
M 80	Standard Specification for Coarse Aggregate for Hydraulic Cement Concrete
M 85	Standard Specification for Portland Cement
M 92	Standard Specification for Wire-Cloth Sieves for Testing Purposes
M143	Standard Specification for Sodium Chloride
M 144	Standard Specification for Calcium Chloride
M 148	Standard Specification for Liquid Membrane-Forming Compounds for Curing Concrete
M 152	Standard Specification for Flow Table for Use in Tests of Hydraulic Cement
M 154	Standard Specification for Air-Entraining Admixtures for Concrete
M 157	Standard Specification for Ready-Mixed Concrete
M 171	Standard Specification for Sheet Materials for Curing Concrete
M 182	Standard Specification for Burlap Cloth Made from Jute or Kenaf and Cotton Mats
M 194	Standard Specification for Chemical Admixtures for Concrete
M 195	Standard Specification for Lightweight Aggregates for Structural Concrete
M 200	Standard Specification for Epoxy Protective Coatings
M 201	Standard Specification for Mixing Rooms, Moist Cabinets, Moist Rooms, and Water Storage Tanks Used in the Testing of Hydraulic Cements and Concretes
M 205	Standard Specification for Molds for Forming Concrete Test Cylinders Vertically
M 210	Standard Specification for Use of Apparatus for the Determination of Length Change of Hardened Cement Paste, Mortar, and Concrete
M 224	Standard Specification for Use of Protective Sealers for Portland Cement Concrete
M 231	Standard Specification for Weighing Devices Used in the Testing of Materials
M 233	Standard Specification for Boiled Linseed Oil Mixture for Treatment of Portland Cement Concrete
M 235	Standard Specification for Epoxy Resin Adhesives
M 240	Standard Specification for Blended Hydraulic Cement
M 241	Standard Specification for Concrete Made by Volumetric Batching and Continuous Mixing
M 284	Standard Specification for Epoxy-Coated Reinforcing Bars: Materials and Coating Requirements
M 295	Standard Specification for Coal Fly Ash and Raw or Calcined Natural Pozzolan for Use in Concrete
M 302	Standard Specification for Ground Granulated Blast-Furnace Slag for Use in Concrete and Mortars
M 307	Standard Specification for Silica Fume Used in Cementitious Mixtures
M 321	Standard Specification for High-Reactivity Pozzolans for Use in Hydraulic-Cement Concrete, Mortar, and Grout
PP 65	Standard Practice for Determining the Reactivity of Concrete Aggregates and Selecting Appropriate Measures for Preventing Deleterious Expansion in New Concrete Construction
R 39	Standard Practice for Making and Curing Concrete Test Specimens in the Laboratory
T 2	Standard Method of Test for Sampling of Aggregates
T 11	Standard Method of Test for Materials Finer Than 75-µm (No. 200) Sieve in Mineral Aggregates by Washing
T 19	Standard Method of Test for Bulk Density ("Unit Weight") and Voids in Aggregate
T 21	Standard Method of Test for Organic Impurities in Fine Aggregates for Concrete
T 22	Standard Method of Test for Compressive Strength of Cylindrical Concrete Specimens
T 23	Standard Method of Test for Making and Curing Concrete Test Specimens in the Field
T 24	Standard Method of Test for Obtaining and Testing Drilled Cores and Sawed Beams of Concrete
T 26	Standard Method of Test for Quality of Water to Be Used in Concrete
T 27	Standard Method of Test for Sieve Analysis of Fine and Coarse Aggregates
T 71	Standard Method of Test for Effect of Organic Impurities in Fine Aggregate on Strength of Mortar
T 84	Standard Method of Test for Specific Gravity and Absorption of Fine Aggregate
T 85	Standard Method of Test for Specific Gravity and Absorption of Coarse Aggregate

T 96	Standard Method of Test for Resistance to Degradation of Small-Size Coarse Aggregate by Abrasion and Impact in the Los Angeles Machine
T 97	Standard Method of Test for Flexural Strength of Concrete (Using Simple Beam with Third-Point Loading)
T 98	Standard Method of Test for Fineness of Portland Cement by the Turbidimeter
T 103	Standard Method of Test for Soundness of Aggregates by Freezing and Thawing
T 104	Standard Method of Test for Soundess of Aggregate by Use of Sodium Sulfate or Magnesium Sulfate
T 105	Standard Method of Test for Chemical Analysis of Hydraulic Cement
T 106	Standard Method of Test for Compressive Strength of Hydraulic Cement Mortar (Using 50-mm or 2-in. Cube Specimens)
T 107	Standard Method of Test for Autoclave Expansion of Hydraulic Cement
T 112	Standard Method of Test for Clay Lumps and Friable Particles in Aggregate
T 113	Standard Method of Test for Lightweight Pieces in Aggregate
T 119	Standard Method of Test for Slump of Hydraulic Cement Concrete
T 121	Standard Method of Test for Density (Unit Weight), Yield, and Air Content (Gravimetric) of Concrete
T 126	Making and Curing Concrete Test Specimens in the Laboratory
T 127	Standard Method of Test for Sampling and Amount of Testing of Hydraulic Cement
T 129	Standard Method of Test for Normal Consistency of Hydraulic Cement
T 131	Standard Method of Test for Time of Setting of Hydraulic Cement by Vicat Needle
T 133	Standard Method of Test for Density of Hydraulic Cement
T 134	Standard Method of Test for Moisture-Density Relations of Soil-Cement Mixtures
T 137	Standard Method of Test for Air Content of Hydraulic Cement Mortar
T 141	Standard Method of Test for Sampling Freshly Mixed Concrete
T 152	Standard Method of Test for Air Content of Freshly Mixed Concrete by the Pressure Method
T 153	Standard Method of Test for Fineness of Hydraulic Cement by Air Permeability Apparatus
T 154	Standard Method of Test for Time of Setting of Hydraulic Cement Paste by Gillmore Needles
T 155	Standard Method of Test for Water Retention by Liquid Membrane-Forming Curing Compunds for Concrete
T 157	Standard Method of Test for Air-Entraining Admixtures for Concrete
T 158	Standard Method of Test for Bleeding of Concrete
T 160	Standard Method of Test for Length Change of Hardened Hydraulic Cement Mortar and Concrete
T 161	Standard Method of Test for Resistance of Concrete to Rapid Freezing and Thawing
T 177	Standard Method of Test for Flexural Strength of Concrete (Using Simple Beam with Center-Point Loading)
T 178	Standard Method of Test for Portland-Cement Content of Hardened Hydraulic-Cement Concrete
T 185	Standard Method of Test for Early Stiffening of Hydraulic Cement (Mortar Method)
T 186	Standard Method of Test for Early Stiffening of Hydraulic Cement (Paste Method)
T 192	Standard Method of Test for Fineness of Hydraulic Cement by the 45-µm (No. 325) Sieve
T 196	Standard Method of Test for Air Content of Freshly Mixed Concrete by the Volumetric Method
T 197	Standard Method of Test for Time of Setting of Concrete Mixtures by Penetration Resistance
T 198	Standard Method of Test for Splitting Tensile Strength of Cylindrical Concrete Specimens
T 199	Standard Method of Test for Air Content of Freshly Mixed Concrete by the Chace Indicator
T 210	Standard Method of Test for Aggregate Durability Index
T 231	Standard Practice for Capping Cylindrical Concrete Specimens
T 248	Standard Method of Test for Reducing Samples of Aggregate to Testing Size
T 255	Standard Method of Test for Total Evaporable Moisture Content of Aggregate by Drying
T 259	Standard Method of Test for Resistance of Concrete to Chloride Ion Penetration
T 260	Standard Method of Test for Sampling and Testing for Chloride Ion in Concrete and Concrete Raw Materials
T 271	Standard Method of Test for Density of Plastic and Hardened Portland Cement Concrete In-Place by Nuclear Methods
T 276	Standard Method of Test for Measuring Early-Age Compression Strength and Projecting Later-Age Strength
T 277	Standard Method of Test for Electrical Indication of Concrete's Ability to Resist Chloride Ion Penetration
T 290	Standard Method of Test for Determining Water-Soluble Sulfate Ion Content in Soil

T 299	Standard Method of Test for Rapid Identification of Alkali-Silica Reaction Products in Concrete
T 303	Standard Method of Test for Accelerated Detection of Potentially Deleterious Expansion of Mortar Bars Due to Alkali-Silica Reaction
T 304	Standard Method of Test for Uncompacted Void Content of Fine Aggregate
T 309	Standard Method of Test for Temperature of Freshly Mixed Hydraulic Cement Concrete
T 318	Standard Method of Test for Water Content of Freshly Mixed Concrete Using Microwave Oven Drying
T 325	Standard Method of Test for Estimating the Strength of Concrete in Transportation Construction by Maturity Tests
T 336	Standard Method of Test for Coefficient of Thermal Expansion of Hydraulic Cement Concrete
TP 60	Coefficient of Thermal Expansion of Hydraulic Cement Concrete
TP 64	Standard Method of Test for Predicting Chloride Penetration of Hydraulic Cement Concrete by the Rapid Migration Procedure
TP 75	Standard Method of Test for Air-Void Characteristics of Freshly Mixed Concrete by Buoyancy Change

METRIC CONVERSION FACTORS

The following list provides the conversion relationship between U.S. customary units and SI (International System) units. The proper conversion procedure is to multiply the specified value on the left (primarily U.S. customary values) by the conversion factor exactly as given below and then round to the appropriate number of significant digits desired. For example, to convert 11.4 ft to meters: 11.4 x 0.3048 = 3.47472, which rounds to 3.47 meters. Do not round either value before performing the multiplication, as accuracy would be reduced. A complete guide to the SI system and its use can be found in IEEE/ASTM SI 10, Metric Practice.

To convert from	to	multiply by
Length		
inch (in.)	micrometer (µm)	25,400 E*
inch (in.)	millimeter (mm)	25.4 E
inch (in.)	meter (m)	0.0254 E
foot (ft)	meter (m)	0.3048 E
yard (yd)	meter (m)	0.9144
Area		
square foot (ft^2)	square meter (m^2)	0.09290304 E
square inch ($in.^2$)	square millimeter (mm^2)	645.2 E
square inch ($in.^2$)	square meter (m^2)	0.00064516 E
square yard (yd^2)	square meter (m^2)	0.8361274
Volume		
cubic inch ($in.^3$)	cubic millimeter (mm^3)	16.387064
cubic inch ($in.^3$)	cubic meter (m^3)	0.00001639
cubic foot (ft^3)	cubic meter (m^3)	0.02831685
cubic yard (yd^3)	cubic meter (m^3)	0.7645549
gallon (gal) U.S. liquid**	liter (L)	3.7854118
gallon (gal) U.S. liquid	cubic meter (m^3)	0.00378541
fluid ounce (fl oz)	milliliters (mL)	29.57353
fluid ounce (fl oz)	cubic meter (m^3)	0.00002957
Force		
kip (1000 lb)	kilogram (kg)	453.6
kip (1000 lb)	newton (N)	4,448.222
pound (lb) avoirdupois	kilogram (kg)	0.4535924
pound (lb)	newton (N)	4.448222
Pressure or stress		
pound per square foot (psf)	kilogram per square meter (kg/m^2)	4.8824
pound per square foot (psf)	pascal (Pa)†	47.88
pound per square inch (psi)	kilogram per square centimeter (kg/cm^2)	0.07031
pound per square inch (psi)	pascal (Pa)†	6,894.757
pound per square inch (psi)	megapascal (MPa)	0.00689476
Mass (weight)		
pound (lb) avoirdupois	kilogram (kg)	0.4535924
ton, 2000 lb	kilogram (kg)	907.1848
Mass (weight) per length		
kip per linear foot (klf)	kilogram per meter (kg/m)	0.001488
pound per linear foot (plf)	kilogram per meter (kg/m)	1.488
Mass per volume (density)		
pound per cubic foot (lb/ft^3)	kilogram per cubic meter (kg/m^3)	16.01846
pound per cubic yard (lb/yd^3)	kilogram per cubic meter (kg/m^3)	0.5933
Temperature		
degree Fahrenheit (°F)	degree Celsius (°C)	$t_C = (t_F - 32)/1.8$
degree Rankine (°R)	degree Kelvin (K)	$T_K = T_R/1.8$
degree Kelvin (K)	degree Celsius (°C)	$t_C = T_K - 273.15$
degree Rankine (°R)	degree Fahrenheit (°F)	$t_F = T_R - 450.67$
Energy and heat		
British thermal unit (Btu)	joule (J)	1055.056
calorie (cal)	joule (J)	4.1868 E
Btu/°F hr • ft^2	W/m^2 • °K	5.678263
kilowatt-hour (kwh)	joule (J)	3,600,000 E
British thermal unit per pound (Btu/lb)	calories per gram (cal/g)	0.55556
British thermal unit per hour (Btu/hr)	watt (W)	0.2930711
Permeability		
darcy	centimeter per second (cm/s)	0.000968
feet per day (ft/day)	centimeter per second (cm/s)	0.000352

* E indicates that the factor given is exact.
** One U.S. gallon equals 0.8327 Canadian gallon.
† A pascal equals 1.000 newton per square meter.

Note:
One U.S. gallon of water weighs 8.34 pounds (U.S.) at 60°F.
One cubic foot of water weighs 62.4 pounds (U.S.).
One milliliter of water has a mass of 1 gram and has a volume of one cubic centimeter.
One U.S. bag of cement weighs 94 lb.
The prefixes and symbols listed below are commonly used to form names and symbols of the decimal multiples and submultiples of the SI units.

Multiplication Factor	Prefix	Symbol
$1,000,000,000 = 10^9$	giga	G
$1,000,000 = 10^6$	mega	M
$1,000 = 10^3$	kilo	k
$1 = 1$	—	—
$0.01 = 10^{-2}$	centi	c
$0.001 = 10^{-3}$	milli	m
$0.000001 = 10^{-6}$	micro	µ
$0.000000001 = 10^{-9}$	nano	n

CEMENT AND CONCRETE RESOURCES

AASHTO	American Association of State Highway and Transportation Officials http://www.transportation.org
ACAA	American Coal Ash Association http://www.acaa-usa.org
ACBM	Center for Advanced Cement Based Materials (Northwestern University) http://acbm.northwestern.edu
ACI	American Concrete Institute http://www.concrete.org
ACPA	American Concrete Pavement Association http://www.pavement.com
ACPA	American Concrete Pipe Association http://www.concrete-pipe.org
ACPA	American Concrete Pumping Association http://www.concretepumpers.com/
ACPPA	American Concrete Pressure Pipe Association http://www.acppa.org
ACS	American Ceramic Society http://www.acers.org
AIT	International Ferrocement Society http://www.ferrocement.org
APA	Architectural Precast Association http://www.archprecast.org
API	American Petroleum Institute http://www.api.org
ANSI	American National Standards Institute http://www.ansi.org
ASCC	American Society of Concrete Contractors http://www.ascconline.org
ASCE	American Society of Civil Engineers http://www.asce.org
ASHRAE	American Society of Heating, Refrigerating, and Air-Conditioning Engineers, Inc. http://www.ashrae.org
ASI	American Shotcrete Association http://www.shotcrete.org
ASTM	American Society for Testing and Materials http://www.astm.org
BDZ	Bundesverband der Deutschen Zementindustrie e. V. http://www.bdzement.de
BFRL	Building and Fire Research Laboratory (NIST) http://www.bfrl.nist.gov
BRE	Building Research Establishment (UK) http://www.bre.co.uk
BSI	British Standards Institution http://www.bsi-global.com
CAC	Cement Association of Canada http://www.cement.ca
CCAA	Cement and Concrete Association of Australia http://www.concrete.net.au
CEMBUREAU	European Cement (Industry) Association http://www.cembureau.be
CEN	European Committee for Standardization http://www.cen.eu
CFA	Concrete Foundations Association http://www.cfana.org
CH	Concrete Homes http://www.concretehomes.com
CP Tech Center	National Concrete Pavement Technology Center http://www.cptechcenter.org
CRSI	Concrete Reinforcing Steel Institute http://www.crsi.org
CSA	Canadian Standards Association http://www.csa.ca
CSCE	Canadian Society for Civil Engineering http://www.csce.ca
CSDA	Concrete Sawing and Drilling Association http://www.csda.org
CSI	Construction Specifications Institute http://www.csinet.org
CSI	Cast Stone Institute http://www.caststone.org
DIN	Deutsches Institut für Normung e.V. (German Standards Institution) http://www.din.de
EPRI	Electric Power Research Institute http://www.epri.com
ERMCO	European Ready Mixed Concrete Association http://www.ermco.eu
ESCSI	Expanded Shale, Clay and Slate Institute http://www.escsi.org
FHWA	Federal Highway Administration http://www.fhwa.dot.gov/
FICEM	Federacion Interamericana del Cemento http://www.ficem.org
ICAR	International Center for Aggregates Research http://www.icar.utexas.edu
ICC	International Code Council http://www.iccsafe.org
ICFA	Insulating Concrete Forms Association http://www.forms.org/
ICMA	International Cement Microscopy Association http://www.cemmicro.org
ICPI	Interlocking Concrete Pavement Institute http://www.icpi.org
ICRI	International Concrete Repair Institute http://www.icri.org
IEEE	Institute of Electrical & Electronics Engineers http://www.ieee.org

IGGA	International Grinding & Grooving Association http://www.igga.net
IMCYC	Instituto Mexicano del Cemento y del Concreto A.C. http://www.imcyc.com
IPRF	Innovative Pavement Research Foundation http://www.iprf.org
ISO	International Standards Organization http://www.iso.org
JCI	Japan Concrete Institute http://www.jci-net.or.jp/
LTPP	Long Term Pavement Performance http://www.fhwa.dot.gov/pavement/ltpp/
MERL	U.S. Bureau of Reclamation Materials and Engineering Research Laboratory http://www.usbr.gov/pmts/materials_lab/
MRS	Materials Research Society http://www.mrs.org
NACE	National Association of Corrosion Engineers http://www.nace.org
NAHB	National Association of Home Builders http://www.nahb.com
NCMA	National Concrete Masonry Association http://www.ncma.org
NCPTC	National Concrete Pavement Technology Center http://www.cptechcenter.org/
NIBS	National Institute of Building Sciences http://www.nibs.org
NIST	National Institute of Standards and Technology http://www.nist.gov
NPCA	National Precast Concrete Association http://www.precast.org
NPCPA	National Pervious Concrete Pavement Association http://www.pervious.com
NRC	National Research Council http://sites.nationalacademies.org/NRC/
NRCC	National Research Council of Canada http://www.nrc.ca/
NRMCA	National Ready Mixed Concrete Association http://www.nrmca.org
NSA	National Slag Association http://www.nationalslag.org
NSSGA	National Stone, Sand & Gravel Association http://www.nssga.org
NTMA	National Terrazzo & Mosaic Association Inc. http://www.ntma.com
OP&CMIA	Operative Plasterers' and Cement Masons' International Association http://www.opcmia.org
OSHA	Occupational Safety and Health Administration http://www.osha.gov
PCA	Portland Cement Association http://www.cement.org
PCI	Precast/Prestressed Concrete Institute http://www.pci.org
PI	Perlite Institute http://www.perlite.org
PTI	Post-Tensioning Institute http://www.post-tensioning.org/
REMR	Repair, Evaluation, Maintenance, and Rehabilitation Research Program http://www.wes.army.mil/REMR
RILEM	International Union of Testing and Research Laboratories for Materials and Structures http://www.rilem.net
SCA	Slag Cement Association http://www.slagcement.org
SFA	Silica Fume Association http://www.silicafume.org
TAC	Transportation Association of Canada http://www.tac-atc.ca
TCA	Tilt-Up Concrete Association http://www.tilt-up.org
TRB	Transportation Research Board http://www.trb.org
TVA	The Vermiculite Association http://www.vermiculite.org/
USACE	U.S. Army Corps of Engineers http://www.usace.army.mil
VCCTL	Virtual Cement and Concrete Testing Laboratory http://ciks.cbt.nist.gov/vcctl
VDZ	Verein Deutscher Zementwerke (German Cement Works Association) http://www.vdz-online.de
WES	U.S. Army Engineer Waterways Experiment Station http://www.erdc.army.mil
WRI	Wire Reinforcement Institute http://www.wirereinforcementinstitute.org

Index

A

Abrams law, 7
Abrasion resistance, 173, 199–200, 380
  aggregate, 110
  cementitious materials, 79, 239
  curing, 303
  high-strength concrete, 375, 377, 380
  testing, 200, 364
Abrasive blasting, 292
Absolute volume,
  method of proportioning, 241, 243–244, 246-247, 256
  relative density, related to, 59, 107
  relative proportions by, 7, 389
Absorption, 79,
  aggregate, of, 97-99, 107-109, 111, 170, 264
  heat, of, 15, 307
  high-performance concrete, 377
  proportioning, use in, 244-249, 257
  sealers, 293, 311
  testing for, 197, 348, 366
Acid, 92, 291, 300, 410
  resistance to, 111, 222-223, 311, 381
  tannic, 93, 306
Acoustic performance, 18
Admixtures, 117–135, 157–164, 240, 321, 330, 386–387 7
  accelerating (set-accelerating), 128-129
  air-detrainers, 118, 121, 133, 158
  air-entraining, *see* Air-entraining admixtures
  alkali-aggregate reactivity inhibitors, 118, 131
  antiwashout, 118, 133, 280
  bonding, 118, 132
  bonding agents, 118, 132
  calcium chloride, *see* Calcium chloride
  coloring, 118, 131–132, 159, 410–411
  compatibility, 133–134
  corrosion-inhibiting, *see* Corrosion-inhibiting admixtures
  dampproofing, 118
  dispensing, 134–135
  fungicidal, 133
  gas-forming, 132–133
  germicidal, 133
  grouting, 132
  hydration-control, 129
  insecticidal, 133
  permeability-reducing, non-hydrostatic (PRAN), 130–131
  permeability-reducing, hydrostatic (PRAH), 130–131
  plasticizing, 122–124, 126–127
  polycarboxylate, 124–125, *see also* Admixtures, water-reducing, high-range
  pumping aids, 132
  retarding (set-retarding), 127
  shrinkage-reducing, 130
  storing, 134
  superplasticizing, 123–124, *see also* Admixtures, water-reducing, high-range
  viscosity-modifying, 133
  water-reducing, 122–127
  water-reducing, high-range, 122–127
  water-reducing, mid-range, 122–127
  workability-retaining, 129
Age-strength relationships, 298
Aggregates, 20, 22, 95–114, 235–236, 331–332, 346–349, 386, 395–399,
  abrasion and skid resistance, 110

absorption and surface moisture, 107

acid resistance, 111

alkali-aggregate reactivity, 113, 191, 208-214, 363–364, 381–382,

beneficiation, 113-114

bulk density (unit weight) and voids, 107

bulking, 108

characteristics, 99-111

classification, 96

coarse, *see* aggregate, coarse

D-cracking, 109

density, 107

fine, *see* aggregate, fine

fineness modulus, 106

fire resistance, 111

gap-graded, 105-106

geology, 95-96

grading, 100-105

gravel, *see* aggregate, coarse, gravel

harmful materials, 111-114

manufactured aggregate- 97

marine-dredged aggregate, 98

natural aggregate, 96-97

recycled concrete aggregate- 97-98

resistance to freezing and thawing, 108-110

sand, *see* aggregate, fine, sand

shrinkage, 110

specific gravity, 107

storage, 114

strength, 110

thermal properties, 111

wetting and drying properties, 110

Aggregate, coarse, 101-103

gravel, 96-97

Aggregate, fine, 101,

sand, 96-97

Air content, 118–122, 127, 157–164, 205–207, 236–238, 272, 317, 351-352, 381, 387, 396–397

air bubbles, 39, 120–122, 160–162, 196, 290, 352–354, 381

air-void clustering, 161-162,

air-void spacing, 157, 159, 162

control of, 121, 157-164

entrained air, 119–122, 127, 157, 159–160, 162–163, 202, 205, 236, 286, 289, 330–331

entrapped air, 120, 159, 237, 238, 282–284, 286, 360

freeze-thaw and scaling resistance, 206-207

impact on concrete properties, 121–122

mix design, 236

spacing factor, 157-164, 205–207, 352-353, 360, 381

strength, 122, 169

sulfate resistance, 122,

tests for, 351–353, 360, 416

Air-entraining admixtures, 75–76, 117–122, 133, 157–164, 236–237, 317, 352,

air content, control of, 121

air content, impact of on properties of concrete, 121–122

materials, 119–120

mechanism, 120–121

Air-entraining cement, 36, 39, 119

Air-void analysis, 352–353

Algae, 93, 160–161

Alite, 48

Alkali content, 80, 98, 158, 209, 212

Alkali-aggregate reaction 90, 98, 113, 191, 208–214, 363–364, 381–382

inhibitors, 118, 131

supplementary cementitious materials, effect of, 80–81

testing for, 362

See also Alkali-carbonate reaction *and* Alkali-silica reaction

Alkali-carbonate reaction, 113, 212, 214, 364

Alkali-silica reaction, 98, 113, 173, 208–214, 363-364, 381–382

cements, resistant, 41–42

inhibitors, 131

supplementary cementitious materials, effect of, 80–81

testing for, 363

Alkaline waters, 92, 160

Aluminum, 45, 131, 223, 272

embedments, 88, 91, 129, 216, 330

finishing tools, 287

gasformers, 118, 132, 401

oxides, 31, 34, 68

Architectural concrete, 40, 105, 113, 278, 279

Autoclave expansion, 56–57

Autoclaved cellular concrete, 395, 401–402

Autogenous shrinkage, 177–180, 362

B

Batching, 8, 90, 161, 263–267, 387

Belite, 32, 48

Belt conveyors, 163, 268, 272
Beneficiation of aggregates, 114
Blaine test, 56
Blankets, 309, 329, 338, 340
Blast resistance, 15, 380
Blast-furnace slag, 40–41, 69–70, 97
Bleeding, 74–75, 126, 155–157, 356
Blended hydraulic cements, 40–42
Blistering, 331
Bond, 118, 132, 280–282, 299
  bonded toppings, 280–282
  bonding admixtures, 118, 132
Bridges, 1–4, 13, 36, 273, 375–376, 383–384, 390, 403–404
Brooming, 289–290
Buckets, 268–270
Buggy, 269–270
Buildings, 1–4, 11–16, 36, 44, 375, 383, 390, 396, 401, 403, 411
Bulk density, 52, 69–71, 107, 366, 399
Bulk volume, 52, 236
Bulking of sand, 108
Bullfloating, 287
Burlap, 305–306
Bush hammering, 292

C

Calcined clay, 4, 72–84
Calcined shale, 72–84
Calcium aluminate, 43, 46, 186
Calcium aluminate cement, 43, 46, 186
Calcium chloride, 88, 91, 118, 128–129, 204
  calcium chloride test, 277–278, 364
Calcium silicate hydrate, 49, 72, 165, 197
Canadian cement, 65
Capillaries, 196, 201–202, 205
Carbon black, 118, 131
Carbonation, 79, 173, 190–191, 214–215, 339–340
Cast-in-place tests, 357
Cellular concrete, 395, 398, 400–402, 413–414
Cement, 1–9, 16–17, 20–25, 29–63
  availability, 46-47
  blended hydraulic, 40-41
  drinking water applications, 47
  hot, 63
  hydraulic, 29
  packaging, 61-62
  particle size distribution, 55-56
  portland, 29-39
  slag, 69-70
  special, 42-46
  specifying, 46-47
  storage, 62
  transportation, 61-62
  virtual testing, 61
  see also Portland cement, Blended hydraulic cement, Performance based hydraulic cement, Canadian and european cement specifications, Slag cement, and Special cements
Cement content, 40–41, 156–158, 239, 356, 361
Cementitious (cementing) materials, 4–8, 29–84, 158, 239, *see also* Supplementary cementitious materials
Central mixers, 161, 264, 387
Chemical admixtures, *see* Admixtures
Chemical phases (compounds) of cement, 48-51
Chemical resistance, 43, 81, 222, 409
Chemical shrinkage, 177–179, 362
Chemical stability, 132, 190
Chloride content, 91–94, 97–99, 128–129, 355, 361–362
Chloride resistance, 79, 390
Chloride tests, 355, 361-362
Chutes, 268–269, 272, 279, 321, 346
Cleaning, 12, 280–281, 293, 298–300, 302
Clear coatings, 292–293, 302
Clinker, 32–35, 48, 119, 165, 321
Coarse aggregate, *see* Aggregate, coarse
Coatings, 292–293, 302, 364, 386, 412
Coefficient of expansion, 185
Cold joints, 127, 272, 279–280, 286, 315
Cold weather concreting, 17, 39, 59, 327–342
  air-entrained concrete, 330-331
  concreting above ground, 335-336
  concreting on ground, 334-335
  duration of heating, 341
  effect of freezing on fresh concrete, 327-328
  enclosures, 336-338
  form removal and reshoring, 341
  heat of hydration, 328-329
  heaters, 338-341
  insulating materials, 338
  maturity concept, 341-342
  monitoring concrete temperature, 333-334
  special mixtures, 329-330

strength gain, 328
temperature of concrete, 331-333
Colored concrete, 118, 131–132, 174, 291–292, 410–411
Combined grading, 103, 105–106
Combustion emissions, 35-36
Compacting, 282, 350, 389, 392–395, 406–408
Compatibility, 133–134, 240, 321, 383, 386
Components of concrete, 8
Compounds, 32, 48–52, 165, 307–308, 311–312
Compression tests, 169, 187, 233, 359
accelerated, 355
Compressive strength, 233–235, 341–342, 353–355, 357–360, 382–386
Concrete strength, 77, 90–93, 265, 316, 328
Concrete temperature, 59, 166, 307, 315–336, 403–405
Consistency, 129, 153–155, 237–239, 349–350, 390
Consolidation, 153–155, 162–165, 282–286, 322, 387–390
Constituents, 6, 21–22, 157–158
Construction joints, 172, 280–281, 284, 287, 293–298
Construction practices, 12–13, 136, 262, 275
Contamination, 84, 97, 114–115, 134, 281
Continuous mixers, 265-267
Contraction joints, 172, 181, 288, 293–298
Control joints, 288, 294
Control tests, 345–346, 349
Conveying concrete, 268, 270
Cooling concrete, 315, 318–319, 324, 341, 404–405
Cores, 310, 357–360, 367, 372, 387
Coring, 357
Corrosion, 79, 90–93, 128–131, 140–141, 173, 215–218, 364
inhibiting admixtures, 117–118, 129–130, 135, 197, 217–218, 224, 245, 247
resistance, 79, 136, 140–141
Corrosive substances, 39, 111, 217
Cracking, 137, 167, 172–173, 179–191, 199
Crazing, 191, 287–289, 305–306, 321, 326
Creep, 79, 172, 189–191, 362, 390–391
Curing, 166, 184, 303–312, 325, 341
accelerated curing, 305, 309, 355
compounds, 184, 281, 304, 307–308, 311, 312, 325
field cured cylinders, 310
impervious paper, 304, 306–307,
match curing, 310
methods, 184, 208, 304, 306, 308–310, 313
period and temperature, 310-311
sealing compounds, 311- 312
Curling, 130, 185–187, 193, 340, 362, *see also* Warping
Cylinders, 169, 187, 189–190, 310, 353–360

D

D-cracking, 109–110, 173
Darbying, 286–288
Dedolomitization, 116, 212, 225
Defects, 286–290, 299–300, 369, 370–371, 398
Deflection, 144, 187–188, 331
Deicer resistance, 202-205
Deicing chemicals, 157, 198, 201–207
Delayed ettrignite formation, 222, 309
Deleterious substances, 97, 100, 291
Density, 12, 18, 118, 170–171
aggregates, 96, 99, 107–108, 235, 348
cement, 60
concrete, 170–171, 350–351, 360, 370, 396, 400, 402
fibers, 146
insulation, 337
subgrade, 276
See also Bulk density, Relative density, *and* Specific gravity
Deterioration, 172–173, 195, 199–224, 362
acid, 306
aggregate, 109–112
Differential scanning calorimetry, 62–63
Differential thermal analysis, 62, 211, 362
Diffusion, 196–197, 365–366, 377, 380–381
carbonation, 214
moisture, 168
sulfates, 220
Disaster resistance, 12, 14
Discoloration, 278, 293, 300, 306–307
calcium chloride, 128
corrosion, 215
Dropchutes, 269, 279
Dry-pack mortar, 299, 381
Drying rate of concrete, 167-168, 180-184,
Drying shrinkage, 172, 179–184, 189–190
admixtures, effect of, 126, 128
aggregates, effect of, 98-99
insulating lightweight concrete, 401
joints, 294, 298
mechanism, 167
shrinkage-compensating concrete, 409
shrinkage reducers, 118, 130

supplementary cementitious material, effect of, 79
tests, 362
Durability, 14, 153, 172–173, 195–224, 232–241, 380–381
abrasion, 79, 199–200
acid attack, 222–223
alkali-aggregate reactivity, 80, 113, 208–214
alkali-carbonate reaction, 113, 212, 214
alkali-silica reaction, 80, 113, 208–213
carbonation, 79, 214–215
corrosion, 79, 215–218
cracking, 199
deicers and anti-icers, 81–83, 202–208
delayed ettringite formation, 221–222
deterioration mechanisms, 199–224
diffusion, 196–198
erosion, 199–200
exposure categories, 198–199
factors affecting, 196–199
freezing and thawing, 81–83, 201–208
marine environment, 224
mitigation, 199–224
permeability, 79, 196–198
physical salt attack, 220–221
protective treatments, 199
salt crystallization, 220–221
seawater exposure, 223–224
sulfate attack, 38, 81, 218¬–220
Dusting, 167, 287–289, 310–311, 358
carbonation, 191, 339

E

Early stiffening, 55, 268, 355
Early strength, 77, 128–129, 382–383
curing, accelerated, 308–309
cements, 39, 41, 45
Earthquake Resistance, 14–15
Edging, 287–288
Efflorescence, 90, 99, 277, 300
Elastic deformation, 138, 187–190, 367
Enclosures, 309, 336–338
Energy performance, 11, 15–17, 35
Entrained air. *See* Air content
Entrapped air. *See* Air content
Epoxy, 132, 282, 297, 299
reinforcement, coated, 141–142, 144, 217–218
Ettringite, 49–52, 55. *See also* Delayed ettringite formation *and* Ettringite cements
Ettringite cements, 43, 46
European cement, 47
Evaporation rate, 324
Expansion joint, 401
Expansion of concrete, 177–180, 184–186, 191, 201–224, 362–364
Expansive cement, 43–46, 409
Exposed-aggregate, 105, 291–293
Exposure conditions, 172–173, 198–199, 232–234

F

False set, 55, 268
Ferrocement, 412
Fiber concrete, 145–150, 395
Fibers, 144–151
fiber-reinforced plastic reinforcement, 218
glass, 147–148
natural, 149–150
properties, 146
steel, 145–147
systems, 150
synthetic, 148–149
Fine aggregate, *see* Aggregate, fine
Finely-ground cements, 44
Fineness modulus, 101, 106, 236, 386
Fineness of portland cement, 55–56
Finishes, 17, 173–174, 290–293
Finishing, 9, 163, 286–300, 321–322
brooming, 289–290
bullfloating, 287
bushhammering, 292
cleaning concrete surfaces, 300
colored, 292
construction joints, 295–296
construction joints, horizontal, 296
contraction joints, 294–295
curing, 293
curing patches, 300
darbying, 287
defects, 299
edging, 287–288
exposed-aggregate, 291–292
filling floor joints, 297–298
floating, 288

formed surfaces, 290
grout cleandown, 290
highway straightedging, 287
isolation joints, 293–294
jointing, 287–288, 293–297
overlays, 299
paints and clear coatings, 292–293
patching and cleaning concrete, 298–300
patterns and textures, 290
premature, 288
protection, 293
removing forms, 298
sandblasting, 292
screeding, 286
slabs, 297
special surface finishes, 290–293
stains, 292–293
striking off, 286
troweling, 288–289
unjointed floors, 298
Fire resistance, 3, 14, 111, 168
Flash set, 55, 128, 333
Flexural strength, 170, 233, 353, 357–359
Floating, 288
Floor covering, 277, 307–308, 364, 398
Floors, 17, 186–187, 276–281, 287–289, 295–300
Flow test, 59, 389
Flowable fill, 395
Flowing concrete, 122–127
Fly ash, 34, 67–70, 72–84, 158–159, 356, 361–363
Fogging, 304–306, 325
Form removal, 39, 290, 298, 341, 405
Formed surfaces, 215, 282, 290, 298–299, 308
Free lime, 54, 58, 63
Freeze-thaw resistance (frost resistance), 108–110, 157–164, 173, 201–208, 363
Freezing fresh concrete, 327-328
Freshly-mixed concrete, 73, 263–273, 278–290, 349–356
Frost resistance, *see* Freeze-thaw resistance
Frozen subgrade, 276, 334
Functional resilience, 12, 14

G

Galvanized materials, 140-141, 142
Gamma radiography, 366
Gap-graded aggregates, 105–106, 347
Gas-forming, 117, 132–133
Geopolymer cements, 43, 45, 65, 395
Gillmore needle, 56
Glass fibers, 147, 149, 218
Grading of aggregates, 95–108, 158, 347–348, 386, 396
Gravel, *see* Aggregates, coarse, gravel
Grouting, 43–45, 117–118, 132, 280–281, 406
Grouting admixtures, 118, 132
Gypsum,
cement, 30–32, 40, 48–49, 51–52, 138, 165
aggregate, 96, 111,
soil and water, 218-221, 223,

H

Handling, 62, 84, 114, 134, 268–272, 286, 321-322
Harmful materials, 88, 92, 111–114, 198
Haul time, 162, 333
Heat induced delayed expansion, *see* Delayed ettrignite formation
Heat of hydration, *see* Hydration, heat of
Heaters, 338-341
Heavyweight aggregate concrete, 96, 170, 395, 402–403
High-density concrete, 402–403
High-durability concrete, 380-382
High early strength, 382-383
High-performance concrete, 375-392
blast resistance, 15
high-durability concrete, 380-382
high-early-strength concrete, 382-383
high-strength concrete, 383-387
proportioning, 387
self-consolidating concrete, 389-390
ultra-high performance concrete, 390-392
High-range water reducer, *see* Water Reducer
High-strength concrete, 383–387
Highway straightedging, 287
History of concrete, 4-6
Hot cement, 63
Hot weather, 315-326
chemical admixtures, 321
cooling concrete materials, 318
curing and protection, 325
effects of high concrete temperatures, 317-318
heat of hydration, 326
ice, 319-320
liquid nitrogen, 320

- plastic shrinkage cracking, 322-325
- precautions, 315-316
- supplementary cementing materials, 321
- transporting, placing, and finishing, 321-322

Hydration, 9, 73, 76-77, 165-166

Hydraulic cement, 29, 41

I

Ice, 319-320

Impact resistance, 79, 147, 375, 412

Indoor air quality, 17

Industrial wastewater, 92

Industry trends, 1-3

Inelastic deformation, 187

Infrared curing, 309–310

Inorganic salts, 92, 128

Insecticidal admixtures, 133

Insulating blankets, 309, 338

Internal curing, 308

Iron salts, 92

J

J-Ring test, 389–390

Joints, 172, 275, 287–288, 293-298

K

K-slump test, 350

Kelly-ball penetration test, 350

Kiln, 29, 31–32, 34–35

L

Laser screed, 285-286

Latex-modified concrete, 395

LEED®, 12-13

Life-Cycle Analysis, 22-25

Lightweight aggregates, 96–97, 308, 351, 395–398

Lightweight concrete, 96, 180, 308, 395–398, 401

Loss on ignition, 59, 72, 158

Low density, *see* Lightweight aggregate *and* Lightweight concrete,

M

Magnesium phosphate cement, 43, 46

Manufacture of portland cement, 29–34

Marine-dredged aggregate, 98–99

Masonry cement, 42–44, 47

Mass concrete, 403-405

Maturity, 304, 334, 336, 341–343, 355, 368

Maximum size aggregate, 103-104, 235

Membrane-forming compounds, 307–308

Metakaolin, 71-72, 73-84, 148, 158

Microwave adsorption test, 309-310, 348–349

Mid-range water reducing admixture, *see* Water reducer

Mineral admixtures, *see* Supplementary cementitious materials

Mix design, 231-240

Mixing concrete, 164, 264-267

Mixing water, 87-93
- effects of impurities on concrete properties, 90-92
- interaction with admixtures, 93
- organic impurities, 93
- sources, 88-90

Mixture proportioning, 241-258

Mobile batcher, 265, 267, 269

Moderate-strength lightweight concrete, 395, 398, 401

Modulus of elasticity, 170, 188

Modulus of rupture, 170, 359

Mortar cement, 42–43

Multiple fiber systems, 150

N

Natural fibers, 149–150

Natural pozzolans, 71-72

Nitrogen cooling, 316, 320,

No-fines concrete, *see* Pervious concrete

No-slump concrete, *see* Zero slump concrete

Nominal maximum size aggregate, 103

Non-potable water, 87

Nondestructive test, 310, 334, 360, 366, 371, 373

O

Oil-well cement, 43, 45

Ordering concrete, 263

Organic admixture content, 361

Organic impurities, 93, 111, 347

Osmotic pressure, 179, 202

Overlay, 280–282, 299, 369, 382

P

Painting concrete, 292–293

Patching concrete, 298-299

Patterned finishes, 290-292

Pavements, 3, 97, 271-272, 406

Performance based hydraulic cements, 41-42

Performance based specifications, 263

Permeability, 170-172, 196-197, 365-366
Permeable grid paver systems, 19
Pervious concrete, 18, 409-410
Petrographic examination, 210-214, 362
pH, 72-73, 190,214-216, 219, 223, 365
Photocatalytic concrete, 411
Pigments, 17, 131–132, 173-174, 292-293, 410–411
Placement of concrete, 268-273, 278-280, 282-286
Plastic cement, 43–44
Plastic concrete, 153, 284
Plastic sheet, 306, 364
Plastic shrinkage, 75, 145, 179, 315, 317, 322-325
Plasticity, *see* Plastic concrete
Plasticizer, *see* Water reducer *and* Superplasticizer
Poisson's ratio, 188, 362
Polymer-portland cement concrete, 412
Polypropylene fibers, 148, 150, 409
Popouts, 95, 109–113, 208–209, 347, 362
Porous concrete, *see* Pervious concrete
Portland cement, 1–3, 5–7, 29–39, 40-42, 48-63
- availability, 46-47
- air-entraining, 39
- alternative raw materials, use of, 34
- bulk density, 60
- chemical phases (compounds) of, 48-51
- combustion emissions from, 35
- compressive strength, 52-53
- consistency, 57
- density, 59
- differential scanning calorimetry, 61
- differential thermal analysis, 60-61
- energy and fuel, use of, 35
- fineness, 55-56
- heat of hydration, 57-59
- hot, 63
- land stewardship, 34
- loss on ignition, 59
- manufacture of, 29-32
- particle size, 55-56
- physical properties of, 52-60
- primary phases, 48-51
- relative density (specific gravity), 59
- selecting and specifying, 46
- setting time, 54-55
- solid waste reduction, 34-35
- soundness, 56-57
- storage, 62
- sustainability, role in, 34-36
- thermal analysis, 60-61
- thermogravametric analysis, 60
- transportation and packaging, 61-62
- types, 36-39
- use of tired derived fuel, 35
- virtual testing, 61
- water (evaporable and nonevaporable), 51-52
- white cement, 39-40

Pozzolans, 4, 37, 40, 47, 67, 71-73, *see also* Supplementary cementitious materials.
Preplaced aggregate concrete, 406
Prestressed concrete, 88, 143-144, 217, 240,
Proportioning, *see* Mixture proportioning
Pull-out test, 368,
Pull-off test 369
Pulse velocity test, 369
Pumping, 132, 162–163, 269, 271, 398
Pumping aids, 118, 132

Q

Quality control, 32, 341-342, 345, 389,

R

Radiation shielding, 170, 402
Rain protection, 293
Random cracking, 199, 295, 297
Rapid analysis machine, 356
Rapid hardening, 45, 165, 382–383
Rapid hardening cement, 45
Rate of hydration, 127–128, 165,
Reactive powder concrete, 390–392,
Reactivity of aggregates, 113, 210-211, 363–364
Ready mix concrete, 2, 13, 90, 263–267, 321
Rebound hammer test, 367
Recycled concrete, 20–21, 97–98, 395
Recycled-concrete aggregate, 97–98
Regulated-set cement, 45
Reinforcement, 103, 137–150, 217-218, 235, 278, 370
- bar marks, 138
- fibers, 144-150
- grades, 138
- prestressing steel, 143-144
- reinforcing bars, deformed, 140
- reinforcing bars, coated, 140

stress-strain curves, 138
welded wire reinforcement, 141-142
Relative density, *see* Specific gravity
Relative humidity, 166-168, 180-182, 184, 276-278, 304, 322-325, 364-365
Removal of forms, 298, 341, *see also* Reshoring
Repair, 195, 218, 299, 409
Reshoring, 298, 341
Resistivity, 216, 365-366,
Retarding admixtures, 118, 127-128, 134
Retempering, 121, 161–162, 267-268
RH testing, 364
Roller-compacted concrete, 395, 406–408,

S

Sack-rubbed finish, 290
Salamander heater, 340
Salt crystallization, 173, 191, 219, 220–221,
Sampling, 164, 346, 349, 389
Sand, *see* Aggregate, fine, sand
Sandblasting, 281, 292,
Saturated concrete, 208, 331
Sawed joints, 294
Scaling, 81-83, 173, 201-208, 311-312, 363, 381
Screeding, 285–286,
Screw spreader, 269, 271–272
Sealing compounds, 311–312
Seawater, 36–38, 90, 98, 198, 223–224,
Segregation, 74-75, 114, 133, 153-154, 268-269, 278-280, 283-286, 346, 376, 389-390, 400
Self-compacting concrete, *see* Self-consolidating concrete
Self-consolidating concrete, 389-390
Set-retarding, 127
Setting time, 54-55, 75, 127-129, 165-166, 317, 355,
Settlement, *see* Subsidence
Shielding concrete, 96, 395, 402-403
Shotcrete, 269, 271-272, 408-409
Shrink-mixed concrete, 265
Shrinkage, 79, 98, 110, 130, 172, 177-184
Shrinkage-compensating concrete, 44, 187, 395, 409,
Shrinkage-reducing admixtures, 130,
Sieve analysis, 100-103, 106, 347-348
Silica fume, 67, 70-71, 73-84
Site remediation, 20
Skid resistance, 110, *see also* Abrasion
Slag cement, 67, 69-70, 73-84
Slump, 122-123, 154-155, 237-238, 349
Sodium chloride, 90–91, 203
Sodium sulfate, 92, 110, 218-221, 363
Soil-cement, 395, 408
Solar reflectance, 16,
Soundness, 56-57, 108-110
Spacing factor, 157-164, 205–207, 352-353, 360, 381
Special cements, 42–46
applications, 43
calcium aluminate, 46
ettringite, 46
expansive, 44
finely ground (ultrafine), 44
geopolymer, 45
magnesium phosphate, 46
masonry and mortar, 42-43
natural, 46
oil-well, 45
plastic, 43
rapid hardening, 45
regulated-set, 45
sulfur, 46
water-repellent, 45
with functional additions, 45
Specific gravity,
aggregates, 99, 107, 402
calcined clay, 72
calcined shale, 72
cement, 59,
fly ash, 69,
metakaolin, 72
mixture proportions, 243
silica fume, 71
slag cement, 70
Sprayed concrete, *see* Shotcrete
Stability, 153-155, 389
Staining, 90, 92, 112-113, 307,
Stationary mixing, 264
Steam curing, 184, 190, 305, 308–309, 382,
Steel fibers, 145–147, 380, 409
Stockpiling, 97, 114
Stone, *see* Aggregate
Storage of cement, 62
Stormwater management, 18-19, 409-410
Strength, 77-78, 84, 91, 233, 382-386

Strength gain, 77-78, 166,169, 308-309, 328, 382
Strikeoff, 286
Structural lightweight aggregate concrete, 96, 395-398
Subbase, *see* Subgrade
Subgrade, 275-276, 321, 324, 334, 340-341
Subsidence, 178–179, 286
Sugar, 93
Sulfate resistance, 36-38, 41, 81, 173, 191, 218-224, 240, 363
Superplasticizers, 118-119, 123-127, 389, 391
Supplementary cementitious materials, 67-84
 availability, 84
 concrete mixture proportions, 84
 deicer-scaling resistance, 81
 fly ash, 68-69
 freshly mixed concrete, effects on, 73-77
 hardened concrete, effects on, 77-83
 hydraulic reactions, 73
 metakaolin, 71
 multi-cementitious systems, 84
 natural pozzolans, 71-72
 pozzolanic reactions, 72-73
 silica fume, 70-71
 slag cement, 69-70
 storage, 84
 strength, 77-79
Surface moisture,
 aggregate, 107-108, 245, 248, 258, 319, 348
 concrete, 322-325
Surface preparation, 275-276, 311-312,
Surface texture,
 aggregate, 99,106, 386
Sustainability, 6, 11-25, 34-36,
 acoustics, 17
 carbon dioxide sink, 21
 concrete ingredients, 21-22
 disaster resistance, 14
 durability, 14-15
 energy conservation and atmosphere, 11
 energy performance, 15
 functional resilience, 12
 indoor air quality, 17
 indoor environmental quality, 12
 LEED®, 12
 life cycle analysis, 22-25
 life cycle assessment and inventory, 22-25
 life cycle cost analysis, 22
 locally produced, 21
 Massachusetts Institute of Technology Concrete Sustainability Hub, 13, 24-25
 material quality and resources, 12
 rating systems, 12
 recycling, 20
 safety, 14
 stormwater management, 18-19
 site remediation, 20
 site selection and development, 12
 water management and resources, 11-12
Swelling, 179
Synthetic fibers, 148–149, 391

T

Temperature,
 calculations for fresh concrete, 319, 326
 curing, 310-311
 effects on air entrainment, 163-164
 effects on compressive strength, 317-318, 328
 effects on setting time, 317, 328
 effects on volume changes, 185-186, 294, 317, 331
 effects on water requirements,316-317
 effects on workability, 316-317
 mass concrete, 403-406
 testing, 350
Tensile strength, 137-138,170
Testing aggregates, 346-349
 grading, 347
 moisture content, 348-349
 objectionable fine material, 347
 organic impurities, 347
 sampling, 346
Testing freshly mixed concrete, 349-356
 accelerated compression tests to project later-age strength, 355
 air content, 351-352
 air-void analysis, 352-353
 bleeding, 356
 chloride content, 355
 consistency (slump), 349-350
 density and yield, 350-351
 portland cement content, 356
 sampling, 349
 slump, *see* consistency

strength specimens, 353-355
supplementary cementitious material content, 356
temperature measurement, 350
time of setting, 355
water-cement ratio, 356
water content, 356
Testing hardened concrete, 357-351
alkali-aggregate reactivity, 363-364
carbonation, 365
chloride content, 361-362
corrosion activity, 364
density, absorption, and voids, 360-361
freeze-thaw resistance, 363
moisture testing, 364-365
nondestructive test methods, 366-371
organic admixture content, 361
permeability, 365-366
petrographic analysis, 362
pH testing methods, 365
portland cement content, 361
strength tests, 357-360
supplementary cementitious material content, 361
tests for air content, 360
volume and length change, 362
Textured finishes, 290-291
Thaumasite, 219-220
Thermal analysis, 61–62, 362
Thermal expansion, 111, 184–186
Thermal resistance, 15, 336–338, 400
Thermal shrinkage, 185-186, 294, 317, 331
Thermogravimetric analysis, 60
Thermometers, 333, 350
Time of setting, 54-55, 165-166, 355
Tining, 9, 289
Tire derived fuel, 35
Tornado, hurricane and wind resistance, 14
Transporting concrete, 268-273, 321, 333, 387
equipment for, 268-271
work above ground level, 272
work at or below ground level, 272
Tremie placement, 280
Trial batch procedures, 245-246, 248-249,
Trial mixture design, 242-243
Troweling, 288-289,
Truck agitators, 265, 267, 269
Truck mixers, 264-265, 267-270
Truck-mixed concrete, 265, 267
Two-course floors, 281

U

Ultrafine cements, 44
Ultrasonic tests, 366, 369-370
Underwater concrete, 118, 133, 280
Uniformity, 153, 164, 366-367
Unjointed floors, 298
Urea, 203–204

V

Vapor barrier, 276–277
Vapor retarders, 276–277
Vermiculite, 96, 395, 398–399, 401
Vibration, 162-164, 282-286, 397, *see also* Consolidation
Vibratory roller, 275-276, 406-407
Vibratory screed, 285–286
Vicat apparatus, 57
Virtual cement testing, 61
Voids,
aggregate, 99, 103-107, 348,
concrete, 164, 171, 196-197, 282-286, 299,
testing, 360-361, 367, 369-370
Volcanic ash, 4, 71
Volume change, 8, 46, 168, 177–191, 362
alkali-aggregate reactivity, 191
autogenous shrinkage, 178-179
carbonation, 190-191
chemical changes, 177
chemical shrinkage, 177
compression strain, 187
creep, 189-190
curling, 186-187
deflection, 188
drying shrinkage, 180-184
early age, 177-179
elastic deformation, 187-189
high temperatures, 185-186
inelastic deformation, 187-189
low temperatures, 185
modulus of elasticity, 188
moisture changes, 180-182
plastic shrinkage, 179
poisson's ratio, 188

shear strain, 188
subsidence, 179
sulfate attack, 191
swelling, 179
temperature changes, 184-187
torsional strain, 188-189
warping, 186-187
Volumetric method,
proportioning, 241
batching, 264
air-test, 351

W

Walls, 268-269, 272, 279-281, 294-297
Warping, 130, 186–187, 362
Wash water, 89-90, 93, 160–161
Waste concrete, 20-21, 89, 97
Wastewater, 22, 89, 92
Water, *see* Mixing water
Water curing, 180, 308, 398, 404, 407
Water reducer, 117, 119, 122–128, 134, *see also* Admixtures
Water-cement ratio, 7-8, 196-198, 233-235
Water-cementitious (cementing) materials ratio, *see* water-cement ratio
Water-repellent, 43, 45, 118, 130-131, 293, 311-312
Watertightness, 170–171
Wear resistance, 99, 110, 239, 288–289
Welded wire reinforcement, 141-142
Wetting and drying, 8, 110, 180-181, 223–224, 325,
Wheelbarrows, 269-270
White cement, 17, 39, 40, 47, 51, 70, 82, 173, 410
White concrete, 17, 410
Windsor probe test, 367
Winter concreting, 327-342
Wood Fibers, 150
Workability, 74, 121-127, 129, 153-155, 237, 349-350, 399

X

X-ray diffraction, 50, 362
X-ray test, 366

Y

Yield,
concrete, 243, 350-351
reinforcement, 138-139, 141,
stress, fresh concrete, 155

Z

Zeolite cement, *see* Geopolymer cement
Zero-slump concrete, 395, 406-407